Atomistic Simulations of Glasses

# Atomistic Simulations of Glasses

## Fundamentals and Applications

*Edited by*

*Jincheng Du*
University of North Texas
Denton, TX, USA

*Alastair N. Cormack*
NYS College of Ceramics
Alfred University
Alfred, NY, USA

The American Ceramic Society

WILEY

A Joint Publication of the American Ceramics Society and John Wiley & Sons, Inc.

Published by John Wiley & Sons, Inc., Hoboken, New Jersey.
Published simultaneously in Canada.



*Registered Office*
John Wiley & Sons, Inc., 111 River Street, Hoboken, NJ 07030, USA

*Editorial Office*
John Wiley & Sons, Inc., 111 River Street, Hoboken, NJ 07030, USA

For details of our global editorial offices, customer services, and more information about Wiley products visit us at www.wiley.com.

Wiley also publishes its books in a variety of electronic formats and by print-on-demand. Some content that appears in standard print versions of this book may not be available in other formats.



*Library of Congress Cataloging-in-Publication Data*

Names: Du, Jincheng, editor. | Cormack, Alastair, N. editor.
Title: Atomistic simulations of glasses : fundamentals and applications / edited by Jincheng Du, University of North Texas, Denton, TX, USA, Alastair N. Cormack, NYS College of Ceramics, Alfred University, Alfed, NY, USA.
Description: First edition. | Hoboken, NJ, USA : Wiley-American Ceramic Society, 2022. | "Published simultaneously in Canada." | Includes bibliographical references.
Identifiers: LCCN 2022000615 (print) | LCCN 2022000616 (ebook) | ISBN 9781118939062 (cloth) | ISBN 9781118940235 (adobe pdf) | ISBN 9781118940242 (epub)
Subjects: LCSH: Glass–Analysis–Mathematics. | Glass–Mathematical models. | Chemical structure. | Molecules–Computer simulation. | Molecular dynamics–Data processing.
Classification: LCC QD139.G5 A86 2022 (print) | LCC QD139.G5 (ebook) | DDC 620.1/44–dc23/eng20220301
LC record available at https://lccn.loc.gov/2022000615
LC ebook record available at https://lccn.loc.gov/2022000616

Cover Design: Wiley
Cover Images: Courtesy of T.S. Mahadevan and Jincheng Du

Set in 8.5/11.5pt STIXTwoText by Straive, Chennai, India

MIX
Paper from responsible sources
FSC www.fsc.org FSC® C013604

10 9 8 7 6 5 4 3 2 1

# Contents

# List of Contributors

***Silvia Barbi***
Department of Science and Methods for Engineering
University of Modena and Reggio Emilia
Italy

***Mathieu Bauchy***
Physics of AmoRphous and Inorganic Solids Laboratory (PARISlab)
Civil and Environmental Engineering Department
University of California
Los Angeles, CA
USA

***Jamieson K. Christie***
Department of Materials
Loughborough University
Loughborough
UK

***Alastair N. Cormack***
Inamori School of Engineering
New York State College of Ceramics
Alfred University
Alfred, NY
USA

***Jean-Marc Delaye***
Département de recherche sur les technologies pour l'Enrichissement le Démantèlement et les Déchets
Commissariat à l'énergie atomique et aux énergies alternatives
Direction des Energies, Institut pour les Sciences et Technologies pour une Economie
Circulaire des Energies bas carbone
Bagnols-sur-Cèze CEDEX
France

***Lu Deng***
Department of Materials Science and Engineering
University of North Texas
Denton, TX
USA

***David A. Drabold***
Department of Physics and Astronomy
Nanoscale and Quantum Phenomena Institute
Ohio University
Athens, OH
USA

***Jincheng Du***
Department of Materials Science and Engineering
University of North Texas
Denton, TX
USA

***Lijie Guo***
National Centre for International Research on Green Metal Mining
BGRIMM Technology Group
Beijing
China

***Christian G. Hoover***
School of Sustainable Engineering and the Built Environment
Civil Engineering Department
Arizona State University
Tempe, AZ
USA

***Yushu Hu***
Physics of AmoRphous and Inorganic Solids Laboratory (PARISlab)
Civil and Environmental Engineering Department
University of California
Los Angeles, CA
USA

***Shinji Kohara***
Light/Quantum Beam Field
Research Center for Advanced Measurement and Characterization
National, Institute for Materials Science
Tsukuba, Ibaraki
Japan

***N. M. Anoop Krishnan***
Department of Civil Engineering
Indian Institute of Technology Delhi
New Delhi
India

*and*

Department of Materials Science and Engineering
Indian Institute of Technology Delhi
New Delhi
India

***Han Liu***
Physics of AmoRphous and Inorganic Solids Laboratory (PARISlab)
Civil and Environmental Engineering Department
University of California
Los Angeles, CA
USA

***Xiaonan Lu***
Pacific Northwest National Laboratory
USA

***Thiruvilla S. Mahadevan***
Department of Materials Science and Engineering
University of North Texas
Denton, TX
USA

***Matthieu Micoulaut***
Laboratoire de Physique Théorique de la Matière Condensée
Sorbonne Université
Paris Cedex 05
France

***Monia Montorsi***
Department of Science and Methods for Engineering
University of Modena and Reggio Emilia
Reggio Emilia
Italy

***Gavin Mountjoy***
School of Physical Sciences
University of Kent
Canterbury, Kent
UK

***Francesco Muniz-Miranda***
University of Modena and Reggio Emilia
Department of Chemical and Geological Sciences
Modena
Italy

***Alfonso Pedone***
University of Modena and Reggio Emilia
Department of Chemical and Geological Sciences
Modena
Italy

***Davide Presti***
University of Modena and Reggio Emilia
Department of Chemical and Geological Sciences
Modena
Italy

***László Pusztai***
Wigner Research Centre for Physics
Hungarian Academy of Sciences (WRCP HAS)
Budapest
Hungary

***Xusheng Qiao***
School of Materials Science and Engineering
Zhejiang University
Hangzhou
China

***Jessica M. Rimsza***
Geochemistry Department
Sandia National Laboratories
Albuquerque, NM
USA

***Morten M. Smedskjaer***
Department of Chemistry and Bioscience
Aalborg University
Aalborg
Denmark

***Francesco Tavanti***
University of Modena and Reggio Emilia
Department of Chemical and Geological Sciences
Modena
Italy

***Rajendra Thapa***
Department of Physics and Astronomy
Nanoscale and Quantum Phenomena Institute
Ohio University
Athens, OH
USA

***Xiuxia Xu***
School of Materials Science and Engineering
Zhejiang University
Hangzhou
China

***Kai Yang***
Physics of AmoRphous and Inorganic Solids Laboratory (PARISlab)
Civil and Environmental Engineering Department
University of California
Los Angeles, CA
USA

***Todd R. Zeitler***
Geotechnology & Engineering Department
Sandia National Laboratories
Albuquerque, NM
USA

***Junjie Zhao***
Department of Materials Science and Engineering
University of North Texas
Denton, TX
USA

*and*

School of Materials Science and Engineering
Zhejiang University
Hangzhou
China

***Qi Zhou***
Physics of AmoRphous and Inorganic Solids Laboratory (PARISlab)
Civil and Environmental Engineering Department
University of California
Los Angeles, CA
USA

# Preface

Computational materials science has evolved into an important branch of materials research, significantly contributing to all aspects of material design and discovery. This is especially true for inorganic glasses that find wide industrial and technological applications ranging from drinking vessels, window panes, and kitchenware in our everyday lives, to high technologies such as optical fiber communication, bioactive glasses for scaffolds, sealing glasses for fuel cells, and glass matrixes for nuclear waste disposal. Among various simulation methods used to study glass materials, atomistic simulations, both classical and quantum mechanical, play an important role in the understanding of complex structure of glasses and the structural origins of various glass properties.

Because of the lack of long-range order, molecular dynamics and Monte Carlo methods have been used to study the structures of glasses and their melts ever since computer simulation emerged as a scientific method in the 1960s. The field has evolved rapidly over the past several decades due to ever-increasing computational power and maturing of simulation algorithms, so the adoption of atomistic simulations of glasses has become more and more widespread in glass research, evidenced by its deployment in many university, industrial, and government laboratories.

With such a rapid growth in the applications of atomistic simulations, there has also been a concomitant increase in information accumulated in the literature. Although several textbooks are available on computational materials science or atomistic simulations in general, there lacks a reference and textbook that introduces the field of atomistic simulations of inorganic glasses: this is what initially motivated us to plan this book. We aim to provide a general introduction to the field of atomistic simulations of inorganic glass materials from simulation methodologies such as quantum mechanical and classical simulation methods and various approaches to analyzing the results, to a wide range of applications to different types of oxide glasses, e.g. silicates, phosphates, borosilicates, aluminosilicates, boroaluminosilicate, and nonoxide glass systems: halide and mixed anion glasses, as well as glass surfaces and glass–water interactions. With contributions from experts and active practitioners in the field, this book aims to provide a broad background of atomistic simulations of glass and amorphous materials to upper-level undergraduate students, graduate and postdoctoral researchers, and industrial practitioners. Students and researchers in related fields such as crystalline materials, nanostructured materials, and in general materials with complex structures will also find the book a helpful reference.

The book is divided into two parts. The first part covers various atomistic simulation and analysis methodologies and it contains five chapters. Classical simulation methods, i.e. those that rely on interatomic potentials to describe the energetics of a system, are introduced in Chapter 1, while those related to quantum mechanical simulation methods are introduced in Chapter 2. Principles of Reverse Monte Carlo-based glass simulations and examples of their applications to several glass and amorphous systems are presented in Chapter 3. Techniques available to analyze the results of atomistic simulations, such as structural characterization and the calculation of mechanical, dynamic, and other properties, are detailed in Chapter 4. Topological constraint theory and its coupling with atomistic simulations are discussed in Chapter 5.

The second part of the book describes the practical applications of atomistic simulations to various glass systems and it has 10 chapters. In most applications, both classical and first principles simulation methods are discussed.

Starting with a summary of the historical developments behind the rise of atomistic simulations to the study of glass materials in Chapter 6, simulations of silica and silicate glasses are covered in Chapter 7, borosilicate and boroaluminosilicate glasses in Chapter 8, in which the historical and latest developments of interatomic potentials for boron oxide are discussed. The most commonly used, including a few recently developed, borosilicate glasses potential parameters are compiled as Appendix in Chapter 8. Chapter 9 introduces simulations of multicomponent oxide glasses for nuclear disposal with simulations of related processes, such as radiation effects and glass alteration being presented. Chapter 10 covers simulations of phosphate glasses, ranging from phosphorous oxide, binary phosphate, aluminophosphate, and multicomponent glasses. Chapter 11 focuses on simulations of bioactive glasses, where traditional bioactive glasses such as 45S5, those modified with other oxides, and glass nanoparticles are covered. Chapter 12 introduces simulations of rare earth and transition metal-containing glasses, with a discussion of methods to characterize local environments and distributions of these cations, their effect on dynamic and electronic properties, and QSPR analysis. Three practical applications: rare earth-doped silicate and phosphate glasses for optical applications, transition metal-containing phosphate glasses as mixed electrical conductors, and borosilicate glasses for nuclear waste disposal are described. Chapter 13 covers simulations of halide and oxyhalide glasses, with topics covering early simulations of fluoride glasses, atomistic-level observation of phase separation, and its effect on nucleation and crystal growth. Chapter 14 covers simulations of glass surfaces and surface-environment interactions. Lastly, Chapter 15 introduces simulations of silicate glass–water reactions, where topics include quantum mechanical studies of reaction mechanisms and energetics of silicate glasses and the development of reactive potentials and their application to glass–water reactions along with a discussion of current challenges and potential solutions.

We are quite aware that the field is still rapidly evolving, with notable recent developments of artificial intelligence and machine learning methods applying to many fields of glass science and technology including atomistic simulations. They have already applied in areas such as machine learning based potentials and potential refinement, which are briefly discussed in the book for example in Chapter 7, and machine learning of glass properties based on atomistic simulation results. As these topics are still developing hence we did not include them in this book. It is also worth noting that the applications part of the book focuses on inorganic glasses, and particularly oxide glasses as they are the most studied and most common in practical applications. Non-oxide glasses such as halide and oxyhalide glasses are also covered (in Chapter 13) but other glass systems such as chalcogenide and bulk metallic glasses are not covered in the book due to space limitations.

This book would not be possible without the support of many, including those "behind the scene." We would like first to thank all the contributors. We are well aware that they each have busy schedules and we are very appreciative that they found the time to contribute to the book, which we hope will become an important reference in the field. We would also like to thank Mario Affatigato and others on the publication committee of the American Ceramic Society whose request initially motivated us to this book project. We would also thank the CTC of International Commission on Glass (ICG) with its support for a series of international workshops on challenges of MD simulations of glass and amorphous materials organized by its technical committee (TC27 atomistic simulations) that have facilitated discussions and development of the field.

We hope you find this book a useful reference in learning about and applying atomistic simulations in studying inorganic glass materials.

*Jincheng Du*
*Alastair N. Cormack*

# List of Abbreviations

| Acronym or Abbreviation | What it stands for |
|---|---|
| 3QMAS NMR | triple quantum magic angle spinning nuclear magnetic resonance |
| *a*-Si | amorphous silicon |
| AIMD | ab initio molecular dynamics |
| ASTM | American Society for Testing and Materials |
| AXS | anomalous X-ray scattering |
| BAD | bond angle distribution |
| BAFS | barium aluminum fluorosilicate glass |
| BB | bond-bending |
| BKS | van Beest–Kramer–van Santen |
| BMH | Born–Mayer–Huggins potential |
| BO | bridging oxygen |
| BOMD | Born–Oppenheimer molecular dynamics |
| BS | bond-stretching |
| CAS | calcium aluminosilicates |
| CMAS | calcium magnesium aluminosilicates |
| CN | coordination number of atoms |
| CNAS | calcium sodium aluminosilicates |
| COMB | charge optimized many-body potential |
| CPMD | Car–Parrinello molecular dynamics |
| CRN | continuous random network |
| CS | corner-sharing |
| DBO | double-bonded oxygen |
| DBX | Dell, Bray, and Xiao |
| DCR | diffuse charge reactive potentials |
| DFC | differential correlation function |
| DFT | density functional theory |
| DL_POLY | Daresbury Laboratory general purpose molecular dynamic simulation package |
| DOS | density of states |

*(Continued)*

| Acronym or Abbreviation | What it stands for |
|---|---|
| DT | Delaunay tetrahedron |
| ECMR | experimentally constrained molecular relaxation |
| EDFA | erbium-doped fiber amplifier |
| EDOS | electronic density of states |
| EFG | electric field gradient |
| EMD | equilibrium molecular dynamics |
| EPR | electron paramagnetic resonance |
| EPSR | empirical potential structure relaxation |
| ES | edge-sharing |
| EXAFS | extended X-ray absorption fine structure |
| FD | fluid dynamics |
| FEAR | force-enhanced atomic refinement |
| FEM | finite element analysis |
| FF | force field |
| FFT | fast Fourier transform |
| FFV | fractional free volume |
| FP | fluoride phosphate glass |
| FSDP | first sharp diffraction peak |
| FTIR | Fourier-transform infrared |
| FV | free volume |
| FWHM | full width at half maximum |
| g-$SiO_2$ | glassy silica |
| GAP | Gaussian approximation potential |
| GFA | glass-forming ability |
| GGA | generalized gradient approximation |
| GIPAW | gauge-including projector augmented wave |
| GST | $Ge_2Sb_2Te_5$ |
| GULP | General Utility Lattice Program |
| GW | Hedin's GW method for electronic structure properties |
| HF | Hartree–Fock |
| HOMO | highest occupied molecular orbital |
| HSE06 | Hyed–Scuseria–Ernzerhof hybrid functional |
| IPR | inverse participation ratio |
| IR | infrared |
| IRO | intermediate-range ordering |

| Acronym or Abbreviation | What it stands for |
|---|---|
| ISG | international simple glass |
| kMC, KMC | kinetic Monte Carlo |
| KNCMFATS | potassium, sodium, calcium, magnesium iron, and titanium oxide containing silicates |
| *l*-P | liquid phosphorus |
| *l*-Si | liquid silicon |
| *l*-$SiO_2$ | liquid silica |
| L–J, LJ | Lennard–Jones |
| LAMMPS | large-scale atomic/molecular massively parallel simulator |
| LD-DFT | ligand field density functional theory |
| LDA | local density approximation |
| LED | light emitting diode |
| LRO | Long-range order |
| LUMO | lowest unoccupied molecular orbital |
| MAE | mixed alkali effect |
| MAS | magic angle spinning |
| MAS–NMR | magic angle spinning nuclear magnetic resonance |
| MC | Monte Carlo |
| MD | molecular dynamics |
| MFA | mean-field approximation |
| ML | machine-learning |
| MM | molecular mechanics |
| MP2 | 2nd order Møller–Plesset perturbation theory |
| MQ | melt quench |
| MQ-MAS | multiple-quantum magic angle spinning |
| MRN | modified random network |
| msd, MSD | mean square displacement |
| NAPF | sodium alumino-phospho-fluorides |
| NAS | sodium aluminosilicates |
| NBO | non-bridging oxygen |
| NBOHC | non-bridging oxygen hole centers |
| NC | network connectivity |
| ND | neutron diffraction |
| NEMD | non-equilibrium molecular dynamics |
| NMR | nuclear magnetic resonance |

(*Continued*)

| Acronym or Abbreviation | What it stands for |
|---|---|
| NPT | constant number of atoms N, constant pressure P, constant temperature T ensemble |
| NVE | constant number of atoms N, constant volume V and constant energy E ensemble |
| NVT | constant number of atoms N, constant volume V and constant temperature T ensemble |
| OLCAO | orthogonalized linear combination of atomic orbitals |
| OLP | overlapping polarons tunneling |
| ONIOM | own N-layer integrated molecular orbital molecular mechanics |
| PAF | principal axis frame |
| PALS | positron annihilation lifetime spectroscopy |
| PAW | projected augmented wave |
| PBC | periodic boundary conditions |
| PBE | Perdew–Burke-Ernzerhof (exchange-correlation functional) |
| PBG | phosphate-based glasses |
| PD | persistence diagram |
| PES | potential energy surface |
| PME | particle mesh Ewald |
| PMMCS | Pedone-Malavasi-Menziani-Cormack-Segre potentials |
| POHC | phosphorous oxygen hole center |
| PP | pseudo potential or principal peak |
| QE | Quantum Espresso, an open-source electronic structure calculation code |
| QM | quantum mechanical |
| QPM | quasi-periodic models |
| QSPR | quantitative structure–property relationship |
| RDF | radial distribution function |
| RE | rare earth |
| REAPDOR | rotational echo, adiabatic passage, double resonance nmr |
| ReaxFF | Reactive Force Field |
| REDOR | rotational echo double resonance nmr |
| RF | resonance frequency |
| RI | rigid-ion |
| RMC | reverse Monte Carlo |
| RSL | Rahman–Stillinger–Lemberg |

| Acronym or Abbreviation | What it stands for |
|---|---|
| SANS | small angle neutron scattering |
| SAXS | small angle X-ray scattering |
| SM | shell model |
| SPC | space projected conductivity or simple point charge model for water |
| SPME | smooth particle-mesh Ewald |
| SRO | short-range order |
| STM | scanning tunneling microscopy |
| TBO | triply bonded oxygen or terminally bonded oxygen |
| TCAF | transverse-current autocorrelation-function |
| TCF | total correlation function |
| TCT | topological constraint theory |
| TD-DFT | time-dependent density functional theory |
| TDF | total distribution function |
| TEM | transmission electron microscopy |
| TM | transition metal |
| TMO | transition metal oxide |
| TMS | tetramethylsilane |
| TO | transverse optical frequency |
| TSSF | total scattering structure factor |
| UV | ultraviolet |
| VAC | velocity autocorrelation |
| VACF | velocity autocorrelation function |
| VASP | Vienna ab initio Simulation Package |
| VDOS | vibrational density of states |
| VIPR | vibrational inverse participation ratio |
| VPG | vanado-phosphate glasses |
| VSD | void size distribution |
| XANES | X-ray absorption near-edge structure |
| XPS | X-ray photoelectron spectroscopy |
| XRD | X-ray diffraction |
| Y&B | Yun and Bray model for borosilicate glass |
| YDB | Yun–Dell–Bray model for borosilicate glass |
| ZBL | Ziegler–Biersack–Littmark potentials |
| ZBLAN | $ZrF_4$-$BaF_2$-$LaF_3$-$AlF_3$-NaF fluoride glasses |
| ZPVE | zero-point vibrational energy |

# Part I

# Fundamentals of Atomistic Simulations

## 1

## Classical Simulation Methods

*Alastair N. Cormack*

*Inamori School of Engineering, New York State College of Ceramics, Alfred University, Alfred, NY, USA*

### 1.1 Introduction

Materials science is concerned with structure–property relations of (solid) materials with a view to engineering new materials with specific properties. Glasses are a subset of materials which are in widespread use, because of their properties, ranging from bottles and jars, and drinking vessels, on the one hand, to windows and electronic displays on the other. Glass fibers are used as thermal insulation in buildings, to strengthen wind turbine blades, and in telecommunications, with optical fibers replacing the previously ubiquitous copper cabling.

It is clearly desirable to have a thorough and complete understanding as possible of glass as a material. Perhaps surprisingly, given that it has been known as a man-made material for millennia, we know less about the materials science of glass than we do about some more modern materials. In large part, this is because of its disordered atomic structure which makes it difficult to probe experimentally in the same way that the more widespread crystalline materials can be investigated.

The principal, defining, characteristic of glasses is, as just noted, their structural disorder at the atomic scale. There is an ASTM definition of a glass [1], (*An inorganic product of fusion which has been cooled to a rigid condition without crystallizing*) basically positing that a glass is a material which has been cooled from a melt to a solid at a rate which prevents crystallization. Usually known as "melt-quench," this is the process which has been used since glasses were first made. whilst the melt is in thermal equilibrium, the solid is not; the disorder inherent in the dynamics of the melt is retained in the glass. The simple ASTM definition has been challenged recently as being inadequate; certainly, it has not changed much, if at all from its 1945 version. For example, it does not include glasses produced by other processing routes, such as the sol–gel process. In that process, the initial solution is made at room temperature (or close to it) and then dried to a gel, which, in turn, is heated to drive off residual water molecules. No melting is involved, so the temperatures are usually lower than those associated with the melt in the traditional melt-quench process.

*Atomistic Simulations of Glasses: Fundamentals and Applications*, First Edition.
Edited by Jincheng Du and Alastair N. Cormack.

In both processes, however, the final glass has a disordered atomic structure, introduced at the initial stages of processing. This disorder is associated with a lack of long-range translational symmetry, i.e. the glasses are noncrystalline, as well as being out of thermal equilibrium. Although, in this chapter, we are primarily concerned with inorganic glasses, there are glassy solid polymers and metallic glasses. The general discussion of the molecular dynamics (MD) technique is equally applicable to these other glasses, except, of course, that because the nature of bonding is different, the description of their interatomic forces will also be different. Most, if not all, of the applications noted earlier employ glasses with chemistries based on silica, $SiO_2$, with additional oxide components, such as alumina, boron oxide, alkali oxides, alkaline earth oxides among others. Indeed, one of the main attractions is the wide diversity of chemistries that can be made into glass. That diversity brings with it a wide range of properties that can be tuned by adjusting the compositions by fine degrees which are not possible with crystalline materials.

Figure 1.1 compares, in (a), the ordered, crystalline structure of cristobalite, a polymorph of silica, and in (b) the disordered structure of vitreous silica, from an MD simulation. The differences are clear.

It is a foundational premise of solid-state physics that, in order to be able to engineer materials with better, or specific, properties, it is necessary to know their atomic structure. With this, and an understanding of the interatomic forces, it is possible to calculate all the physical and thermodynamic properties of the solid.

The challenge for glass science is that because of their noncrystalline nature, it is not possible to determine their atomic structure completely from experimental methods, such as diffraction. Other approaches are therefore necessary, if the structure–property relationships of glasses are not to remain entirely empirical in nature. This is the role that has been fashioned for atomistic simulations.

The simulations provide a numerical way of probing the statistical mechanical phase space of the glassy materials. The resulting structures and their properties are, technically, ensemble averages, that is, they are averaged over the points in phase space. In the next

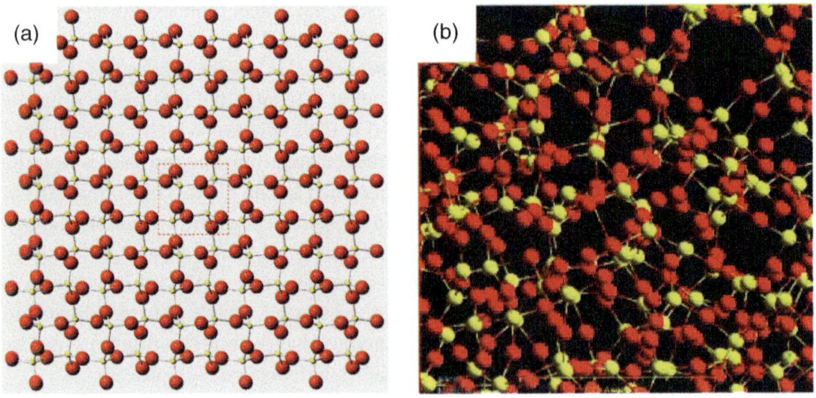


**Figure 1.1** Comparison of the ordered structure of cristobalite (a), and the disordered structure of vitreous silica (b).

section, the basics of the two types of atomistic simulation techniques, MD and Monte Carlo (MC) methods, which have been applied to glasses will be discussed.

## 1.2 Simulation Techniques

Simulations have sometimes been called numerical experiments, but that view has been challenged epistemologically on the grounds that they do not provide new knowledge of the physical world in the way that experiments do [2]. Philosophically, science is concerned with producing predictive explanations of nature and simulations are investigations, or tests, of these predictive explanations, or physical models, using computational methods.

The physical basis for the simulations lies in statistical mechanics and, more importantly, the models used to describe the interatomic forces in the glasses.

Although MC methods predate MD, it is the latter which has been more widely employed in simulating the glasses which are the subject of this volume, so we will discuss this technique first.

### 1.2.1 Molecular Dynamics (MD)

MD as a technique using computers was first described by Alder and Wainwright [3] for a collection of hard spheres. They noted the possibility of applying the method to solving "many problems in both equilibrium and non-equilibrium statistical mechanics." The basic question was how to handle the numerous interactions between particles in condensed systems, as opposed to those between molecules in dilute systems.

The idea behind MD is conceptually straightforward. Statistical mechanical phase space is sampled by calculating the positions and momenta of the atoms as a function of time. This time dependence marks a major difference to the MC method. Since the (Newtonian) equations of motion of the ensemble of particles cannot be solved simultaneously analytically with time as a continuous variable, an iterative approach is used. Each iteration represents a small increment of time, $\Delta t$, known as a timestep, over the previous iteration, and new positions and velocities are calculated in a process known as integrating the equations of motion.

#### 1.2.1.1 Integrating the Equations of Motion

At any point in time, the positions and velocities of the atoms are known. Hence, Newton's laws provide the way to determine by how much an atom will move during a small amount of time, $\Delta t$, known as a timestep, if the forces acting on the atom are known. The positions, and momenta, can therefore be generated; in this way, it is thus possible to sample statistical mechanical phase space and hence calculate thermodynamically related properties, such as structure and other physical properties, as discussed elsewhere in this volume.

It is obvious, therefore, that understanding, and being able to calculate, the interatomic forces is crucial to the success of the simulation. For classical simulations, this is the realm of the Born Model, which will be discussed in Section 1.3.

A Taylor expansion of $\mathbf{r}_i(t+\Delta t)$ allows the positions at $(t+\Delta t)$ to be obtained from the velocities and forces at time $t$:

$$\mathbf{r}_i(t+\Delta t) = \mathbf{r}_i(t) + \mathbf{v}_i\Delta t + \frac{1}{2}\boldsymbol{a}_i\Delta t^2 \tag{1.1}$$

Here, $\boldsymbol{a} = \mathbf{F}/m$, where $\mathbf{F}$ is the net force acting on an atom and is calculated from the interatomic potential model, and $\mathbf{v}$ is its velocity. Higher-order terms in the expansion are neglected, largely for computational convenience, but usually without much loss of accuracy. Adding a similar expansion for $\mathbf{r}_i(t-\Delta t) = \mathbf{r}_i(t) - \mathbf{v}_i\Delta t + \frac{1}{2}\boldsymbol{a}_i\Delta t^2$ yields

$$\mathbf{r}_i(t+\Delta t) = 2\mathbf{r}_i(t) + \mathbf{r}_i(t-\Delta t) + \boldsymbol{a}_i\Delta t^2 \tag{1.2}$$

Note that the velocities do not enter into this expression: the new positions can be calculated just from knowledge of the atom coordinates from the previous two timesteps, along with the interatomic forces, of course, and so they can be recorded as a function of time and used in ensemble averaging, to obtain the structure, among other properties. Although the velocities are not used to update the atomic positions, nevertheless they are needed for property calculations such as temperature and other momentum-based quantities like viscosity. In this method, the velocities $\mathbf{v}_i(t)$ are calculated from $\frac{\mathbf{r}_i(t+\Delta t)-\mathbf{r}_i(t-\Delta t)}{2\Delta t}$ and are only available after the $\mathbf{r}(t+\Delta t)$ have been determined, that is, one timestep later. This is somewhat of a drawback, as is the necessity of storing two sets of atomic coordinates, $\mathbf{r}_i(t)$ and $\mathbf{r}_i(t-\Delta t)$, although the latter is not such an issue now as it was in the early days of computer simulations.

This method of calculating the positions after each new timestep is originally due to Verlet [4], and is similar to the one used by Alder and Wainwright [3]. Variations of the original Verlet algorithm include the Leapfrog and Velocity Verlet algorithms.

In the Leapfrog Verlet algorithm, the velocities are calculated at $\left(t+\frac{1}{2}\Delta t\right)$ so we have

$$\mathbf{v}\left(t+\frac{1}{2}\Delta t\right) = \mathbf{v}\left(t-\frac{1}{2}\Delta t\right) + a(t)\Delta t \tag{1.3}$$

and

$$\mathbf{r}(t+\Delta t) = \mathbf{r}(t) + \mathbf{v}\left(t+\frac{\Delta t}{2}\right)\Delta t \tag{1.4}$$

The velocities are updated at half timesteps and are said to leap ahead of the positions, and then the positions subsequently leap over the velocities. The velocities, $\mathbf{v}(t)$, themselves are calculated from the half timestep velocities $\frac{\mathbf{v}\left(t+\frac{\Delta t}{2}\right)+\mathbf{v}\left(t-\frac{\Delta t}{2}\right)}{2}$ and are not known at the same time as the positions. This makes calculation of the total energy (potential and kinetic) at any particular point in time problematic.

The Velocity Verlet algorithm [5], on the other hand, gives the positions, velocities, and forces at the same time and is the one preferred by many MD packages, including LAMMPS and GROMACS. It is the default in DL_POLY.

The Velocity Verlet algorithm is similar to the original algorithm, except that the velocities are explicitly calculated, rather than being eliminated. In addition to Eq. (1.1) for the positions, the velocities are obtained from a knowledge of the forces at $t$ and $(t+\Delta t)$:

$$\mathbf{v}(t+\Delta t) = \mathbf{v}(t) + \Delta t\frac{\mathbf{f}(t+\Delta)t + \mathbf{f}(t)}{2m} \tag{1.5}$$

Initially, the positions, velocities, and forces are known at time $t$. The algorithm is usually implemented by first calculating new half-timestep velocities, at $\left(t+\frac{\Delta t}{2}\right)$ (by integrating the forces); the new positions at $(t+\Delta t)$ are then found from $\mathbf{r}(t)+\Delta t\mathbf{v}\left(t+\frac{\Delta t}{2}\right)$. Next, new forces are calculated, using the new $\mathbf{r}(t+\Delta t)$. Finally, the velocities at $(t+\Delta t)$ are obtained from $\mathbf{v}\left(t+\frac{\Delta t}{2}\right)+\frac{\Delta t}{2m}\mathbf{f}(t+\Delta t)$, to give Eq. (1.5). Note that the Velocity Verlet algorithm is, in fact, algebraically equivalent to the original Verlet approach without the elimination of the velocities.

Beeman's algorithm [6] is very similar to the original Verlet algorithm, but includes the forces at $(t-\Delta t)$ which provides for more accurate velocities and improved energy conservation, at the expense of added computation time. It has not been widely used in simulations of glasses, however.

Another class of integrators is known as predictor–corrector algorithms. As the name suggests, the positions, velocities, and acceleration (force) are predicted at the end of the next timestep, using Taylor expansions. Using the new positions, the forces at $(t+\Delta t)$ are then calculated. The differences between the actual and predicted forces are then used to correct the positions and velocities. These were occasionally used in early simulations but have largely fallen out of favor.

Finally, we mention stochastic integrators. These seem to be becoming more popular and are now found in a number of MD packages. In GULP, it is now the default integrator. As implemented, the method combines integrating the equations of motion with thermostat and barostat functions, to sample NPT ensembles [7].

The collected positions (and, if wanted, velocities and forces as well) as a function of time are known as an MD trajectory and provide the basis for all structural and diffusional details that can be calculated in subsequent analysis.

The temperature of the system typically obtained from the kinetic energies: $\frac{3}{2}kT=\frac{1}{2}mv^2$, which obviously requires the velocities that are calculated from the integration algorithms. Typically, target temperatures are maintained by requiring the velocities to conform to an appropriate (Maxwellian) distribution in an NVT or NPT ensemble using thermostats as noted below. For a system in equilibrium, the equipartition theorem provides a correlation between the kinetic and potential energies, and this is used, tacitly at least, to justify the use of temperature to define the kinetic energies. However, this is not the only way to calculate the temperature in statistical mechanics. Phase space is defined by both the positions and momenta of the particles, and all can be used, in various combinations, to define the temperature. It happens that using momenta, that is kinetic energy, is the simplest and most straightforward. It might be better to use the atomic coordinates to obtain the (effective) temperature of the structure, rather than rely on a correlation with the velocities of the atoms, but this is much more complicated and has not been widely explored. One might question whether a glass can actually be in equilibrium, since the undercooled liquid has, by definition, fallen out of equilibrium. However, experimentally, a fully annealed glass should obey the equipartition theorem, and this should be the case for simulations as well.

#### 1.2.1.2 Thermostats and Barostats

MD is at heart a statistical mechanics technique: the MD trajectory is an effort to sample the ensemble microstates, that is the different atomic configurations. Integrating the equations of motion produces by default the microcanonical, NVE, ensemble, in which the number of atoms, system volume, and energy are constant. This is because, in the absence of external forces, Newton's equations conserve energy and momentum.

However, in practice, one invariably wants to control the temperature, i.e. keep it constant, during an MD run, and, often, it is desirable to maintain a constant pressure, allowing the volume and thus density to vary. These conditions correspond to the canonical ensemble (NVT) and the isobaric-isothermal ensemble (NPT), respectively. Including thermal expansion is useful when considering explicit temperature effects, of course.

The temperature is controlled, theoretically, by connecting the system to a heat bath, with which energy is exchanged in order to keep the temperature constant. This is generally accomplished by constraining the kinetic energy distribution, through scaling the velocities. There are several ways to do this. The simplest is just to scale the velocities at each iteration as proposed by Nosé [8] and adapted by Hoover [9]. The equation of motion is augmented with a frictive term:

$$\frac{d^2r}{dt^2} = \frac{F}{m} - \frac{p_\xi}{Q}\frac{dr}{dt}$$

where the heat bath parameter $\xi$ has its own momentum $p_\xi$, with an equation of motion: $\frac{dp_\xi}{dt} = (T - T_0)$, with $T_0$ being the target temperature.

Another commonly used thermostat is a weak-coupling scheme suggested by Berendsen [10], the effect of which is to correct deviations of the temperature: $\frac{dT}{dt} = \frac{T_0 - T}{\tau}$. Here $\tau$ is a relaxation time which determines how fast deviations from the target temperature decay. The benefit of the Berendsen scheme is that the target temperature is reached quite quickly; the downside is that the ensemble obtained is not actually a proper canonical ensemble, although for practical purposes the differences may be small.

Similarly, pressure is usually controlled with a barostat, following, again, the suggestion of Berendsen [10]. The pressure is relaxed to the target pressure by $\frac{dP}{dt} = \frac{P^0 - P}{\tau_P}$. The coordinates, and simulation box parameters, are scaled every $n$ steps by $\mu_{ij} = \delta_{ij} - \frac{n\Delta t}{3\tau_p}\beta_{ij}\left[P^0_{ij} - P_{ij}(t)\right]$. Here $\beta$ is the isothermal compressibility of the system.

Barostats allow the system to reach a density that is a function of the potential model parameters being used, rather than having a density imposed on it.

It is to be noted that when comparing with experimental $T(r)$, it is often better to constrain the density to the experimental value, since some of the fine structure of the distribution functions, particularly bond angles, are sensitive to the density.

### 1.2.2 Monte Carlo (MC) Simulations

MD was not the first atomistic simulation technique; that honor goes to MC simulations [11, 12]. The basic MC method employs random sampling to approximate mathematical functions, such as integrals. It is particularly useful for probing complex systems where analytical solutions may not be available. As discussed in Chapter 6, the idea predates the advent of electronic computers. Random sampling was used in the late eighteenth century to estimate $\pi$. Currently, the MC method finds applications in fields as diverse as financial planning and weather forecasting.

Since we are focused here on the atomic scale structure of glasses, and related properties, the role that MC methods can play in statistical physics or statistical mechanics is of particular interest. Statistical mechanics provides a link between microscopic (i.e. atomic scale) and macroscopic (i.e. observable) properties. A macroscopic system, such as a piece of glass, obviously contains a very large number of atoms. At any instant in time, a complete description of the piece of glass (i.e. system configuration) requires knowing the spatial coordinates of the atoms, along with their momenta. If there are $N$ atoms, then the $6N$ coordinates – three position and three momentum coordinates for each atom – define a point in phase space. Any experimentally measured property is averaged over the various configurations adopted by the system during the time of the measurement. An equilibrium property, $\mathcal{A}$, would be obtained by averaging over all the points in phase space in a process known as ensemble averaging: integration over all configuration, or phase space.

In the simplest case, that of an isolated system in equilibrium, all of the configurations will have the same energy, and so be equally probable. Also, the number of atoms and volume will be fixed. Thus, the average of the macroscopic property $\mathcal{A} = \frac{1}{\hat{\Omega}}\sum_{C\in\hat{\Omega}} A(C)$, where $A(C)$ is the value of the property of configuration $C$ and $\hat{\Omega}$ is the total number of configurations. Since the number of configurations is very large, and very close together in phase space, formally, one could replace the sum with an integral over all phase space. That is, one could write $\mathcal{A} = \int_{\Omega} A(\Omega)d\Omega$. The main reason for doing this is to make the calculations more tractable, since an integral is often easier to evaluate than a sum. However, the integral cannot usually be evaluated analytically, so recourse to numerical methods is necessary, and this is where the MC method comes into play. As a numerical approach to calculating integrals, the MC method works in the following way. Suppose we have a function $f(x)$ to be integrated over the (continuous) range [0,1]: $I = \int_0^1 f(x)\,dx$. In the MC method, $f(x)$ is evaluated at a number of random points in the range [0,1], by generating a random number, $x_i$, between 0 and 1. $I$ is evaluated from $\frac{1}{N}\sum_{i=1}^{N} f(x_i)$. If $N$ is sufficiently large, then the sum is a reasonable approximation to $I$, with an error proportional to $1/N$. MC simulations were probably the first attempts to apply the power of computers to this aspect of statistical mechanics.

In the case of calculating $\mathcal{A}$, then, we are interested in obtaining a sufficiently large number of configurations, $C$, which should be chosen at random.

There are several ways to do this in practice. One is just to create configurations with sets of coordinates for all of the atoms in the system, using a random number generator and calculating the value of the property for each configuration. A variation on this approach would be to start with a set of coordinates and then create new configurations by adding randomly generated $\Delta x_i$ to each coordinate in the previous configuration. In any case, since each configuration is equally likely in this ensemble, the macroscopic property is found from averaging over all configurations created. In this case, where the next configuration to be created depends only on the one immediately preceding, the set of configurations are said to form a Markov chain [13].

Please note that we are not concerned here with the practical issues surrounding how the random numbers are obtained, or even whether the numbers so created are truly random. The point is to be able to generate a (large) number of configurations.

Notwithstanding, the ensemble created in the above fashion, keeping $N$, $V$, and $E$ constant, known as the microcanonical ensemble, is not the one that is most useful in simulations of glasses. Rather than constraining $E$, we want our simulated structures to be at a defined temperature. This is handled by assuming that the system is in contact with a heat bath. In this case, the ensemble average is given by $\mathcal{A} = \frac{1}{Z}\int_C A(x)d\mu(x)$, where $Z = \int_C \exp\left(\frac{-U(x)}{k_B T}\right) dx$ is the partition function. $U(x)$ is the potential energy of the system, and is a multidimensional function of all the positional coordinates. Not all configurations will contribute equally to the overall system energy, so the procedure outlined above will not readily produce a suitable approximation to the integral. Instead, some configurations – those with lower energies – will be more significant than others; this led Metropolis and colleagues to develop the idea of importance sampling [14]. As described above, the integral is replaced by a summation over the random sampling of configurations: $\langle A\rangle = \lim_{N\to\infty} \frac{\sum_{i=1}^{N} A(x_i)\mu(x_i)}{\sum_{i=1}^{N} \mu(x_i)}$.

To address this shortcoming, Metropolis and colleagues introduced an algorithm which biased the sampling to relevant volumes, in an approach called importance sampling. In this approach, new configurations are included in the sampling according to their Boltzmann weight, or importance. Potential configurations are evaluated against changes in the system energy, $U(x)$, in going from one configuration to the next. If the energy of the new configuration is lower than the previous one, the move is accepted, i.e. the new configuration is included. On the other hand, if the energy increased, then the new configuration would be accepted only if the change in energy, $\Delta H$, is associated with a Boltzmann factor, $\exp(-\Delta H/kT)$, that is larger than a randomly generated number. In this way, the Markov chain of configurations obtained conforms to the canonical distribution, ensuring that the average of $\mathcal{A}$ calculated is statistically mechanically sound. In classical MC, the system energy is generally calculated using Eq. (1.6) and a potential model of the kinds discussed in the Section 1.3.2.

Importance sampling via the Metropolis methodology was quickly adopted by the statistical mechanics community as it has been shown to speed up convergence. Notably, the technique has been widely adopted in numerical integration generally, particularly in financial analysis.

MC simulations, per se, have not been all that widely used to investigate inorganic glasses. Their principal merit is one of speed; with current computer power, new configurations can be created very quickly. However, time dependence is not included: the integration over phase space is by definition time-averaging. For this reason, MD has proved to be more popular than MC in simulations of glasses. Notwithstanding, the MC method has been used to simulate vitreous $B_2O_3$ [15], and to investigate dissolution of borosilicate glasses [16].

#### 1.2.2.1 Kinetic Monte Carlo

Notwithstanding, extensions aimed at overcoming this limitation have been made. Of particular interest here is the Kinetic Monte Carlo (KMC) technique, dating back to 1966 [17], which does introduce time dependence, by considering rates, which depend on energy barriers, rather than the energy differences between states. KMC is an application of transition state theory: it employs transition rates between configurations, i.e. states, instead of probabilities. In its earliest applications, it was sometimes known as Dynamical Monte Carlo [18]. It has been primarily employed in simulations of surface reactions.

In general, KMC simulations require prior knowledge of all possible processes that could occur, along with their rates. As an example, one might be interested in diffusion processes. If the diffusion constant is somewhat small, the atomic jump mechanisms involved might not be seen in standard MD simulations in the timespan of the simulated trajectory, usually limited to a few tens of nanoseconds. When atoms diffuse, that is, jump from one site to another in the structure, an energy barrier has to be overcome. The rate at which jumps occur depend on this energy barrier, according to $k_{SP} = vexp(-\Delta H_{SP}/k_B T)$, where $\Delta H_{SP}$ is the saddlepoint energy and $v$ represents the attempt frequency. The main advantage of KMC is that it can probe processes not accessible to MD simulations. The drawback is that knowledge of the rates of the processes have to be known in advance.

There have been no applications of KMC to inorganic glasses, as far as the author is aware.

#### 1.2.2.2 Reverse Monte Carlo

Technically, Reverse MC is not a type of simulation, but, rather, a data-fitting exercise. Notwithstanding, it is a useful tool for refining the structures of glasses, having been used for silica [19] and vitreous $B_2O_3$ [20], for example.

The approach is, like the original MC, straightforward. The positions of atoms, representing the glass structure, are changed at random, but this time the criterion for accepting the moves is whether the agreement is improved between the structure distribution functions calculated from the set of atoms (i.e. simulation box) and the experimental data. The experimental data is typically in the form of $S(q)$ or a real space counterpart such as $T(r)$, obtained from either X-ray or neutron scattering (or both).

Because of its data-fitting character, constraints are invariably included in the calculations. For example, the coordination number of silicon in a silicate glass will be constrained to be four, and bond lengths will be allowed to vary between physically sensible limits. Without the constraints, it will be possible to get excellent fits to the experimental data, from physically nonsensible structures.

This approach is discussed in more detail in Chapter 3.

## 1.3 The Born Model

In the classical, nonquantum, world, understanding the interatomic forces at play in materials come down to the nature of the bonding between the atoms in the solid. For oxide, and halide, glasses, the bonding is a mixture of ionic and covalent character, and their solid-state physics can be approached via the Born Model, which is discussed in this section.

As just noted, inorganic glass-forming materials generally exhibit iono-covalent bonding. For example, the silicon–oxygen bond is fairly rigid indicative of the electron-sharing nature of covalent bonds, and the tetrahedral geometry of the $SiO_4$ "building block" of silicates is also suggestive of predominantly covalent bonding. At the same time, however, the sharing of electrons between Si and O atoms is not complete, as it would be between atoms in chalcogenides, for example, so the atoms in silicates tend to have a residual net charge, and thus there is some ionicity to their bonding. Actually, there is quite a considerable amount of ionicity: quantum mechanical calculations of residual charges indicate a residual charge of ~0.6*e*. Furthermore, alkali and alkaline earth modifiers are very much cationic in nature, having given up, as typical, their outer valence electrons to form closed shell ions. The mixed ionic-covalent nature of bonding in inorganic solids allows classical atomistic simulations to be based on the Born Model of the solid.

This model, originally introduced by Born [21, 22] as the foundation of his lattice theory of ionic solids, specifically the alkali halides, treats atoms as point charges, with short-range forces acting between them, in addition to the electrostatic, or Coulombic, forces.

The energy of the structure, generically known as the lattice energy, can be expressed as:

$$E_{lattice} = \sum \frac{q_i q_j}{r_{ij}} + \sum V(r_{ij}) + \sum U(\theta_{ijk}) + \sum W(\phi_{ijkl}) + \dots \tag{1.6}$$

where the summations are over all the appropriate interactions in the solid. The first term represents the electrostatic interactions with the second being the interactions between pairs of atoms or ions. These are, in terms of the energy of interactions, the largest contributions to the overall lattice energy. However, bond angle interactions are often important in structures that are not 100% ionic because they help to define the local geometry, such as the tetrahedral character of the $SiO_4$ tetrahedron and the shape of molecular species such as carbonate or sulfate groups. Thus, the third and fourth terms represent bond-bending and torsional bond energy contributions. Whilst these do not always contribute much to the overall energy of the structure, their influence on the interatomic forces is significant.

Various analytical expressions have been proposed for the short-range potential dependencies on interatomic separation or bond angle, as discussed further in Section 1.3.

A key feature of the Born Model, in contrast, say, to models of rare gas solids or fluids, is that the atoms are charged, that is, chemically, they are ions having given up or received electrons from neighboring species. Summing up the electrostatic interactions is probably the most time-consuming step in evaluating the lattice energy because the relatively long-range $1/r$ dependency is slow to converge, mathematically. This means that it is not only the nearest neighbor interactions that need to be included in the sum.

### 1.3.1 Ewald Summation

Ewald realized that the mathematical problem of the slowly converging sum $\sum q/r$ could be solved by taking advantage of the properties of a periodic lattice [23]. His approach added to the array of point charges an array of normalized Gaussian charge distributions of opposite sign; this essentially canceled out the long-range contributions of the point charge array, leading to a rapid convergence. A second contribution consisted of an identical array of Gaussian charge distributions, but with opposite signs, so that the two Gaussian distributions canceled each other out (i.e. adding and subtracting the same charge distribution to the original point charge array). The second contribution is transformed into reciprocal space, in which it is rapidly convergent.

Accordingly, one can divide the charge density of point charges, $\rho(r) = \sum_i q_i \delta(r - r_i)$, into two terms $\rho = \rho_I + \rho_{II}$, where

$$\rho_I(r) = \sum_i q_i \left[\delta(r - r_i) - \alpha^3 \pi^{-\frac{3}{2}} \exp\left(-\alpha^2 |r - r_i|^2\right)\right] \tag{1.7}$$

and

$$\rho_{II}(r) = \sum_i q_i \alpha^3 \pi^{-\frac{3}{2}} \exp\left(-\alpha^2 |r - r_i|^2\right) \tag{1.8}$$

The second term in $\rho_I$ is the normalized Gaussian charge distribution, of half-width $\alpha^{-1}$, and opposite sign centered on each original point charge as just discussed. This is canceled by the identical Gaussian distribution but with the same charge as the point ion in $\rho_{II}$. (Note that this assumes that the system is electrostatically neutral, which is usually the case. However, the formalism can be adapted to cases in which there is a net charge, by adding a uniform charge distribution of opposite net charge.)

The potential field of $\rho$ can be split in the same way. Consequently, one can represent the Coulomb energy in two terms, $\Phi = \Phi_I + \Phi_{II}$:

$$\Phi_I = \frac{1}{2}\left\{ \sum_i q_i \left[ \sum_j{}' q_j \sum_{\mathbf{n}} \frac{\mathrm{erfc}(\alpha r_{ij,n})}{r_{ij,n}} - \frac{q_i \alpha}{\sqrt{\pi}} \right] \right\} \tag{1.9}$$

where $r_{ij,\,n} = |r_i - r_j + \mathbf{n}L|$ and

$$\Phi_{II} = \frac{1}{2\pi L} \sum_i \sum_j q_i q_j \sum_{\mathbf{G}}{}' \frac{\exp(-\pi^2 |\mathbf{G}|^2/\alpha^2 L^2)}{|\mathbf{G}|^2} \exp(i\mathbf{G}.\mathbf{r}_{ij}) \tag{1.10}$$

Here, $\mathbf{n}$ is a vector with integer components representing the unit cell in the lattice, and $\mathbf{G}$ is a reciprocal lattice vector. $\Phi_I$ is rapidly convergent in real space, essentially because the erfc term truncates the sum at large atomic separations. $\Phi_{II}$ is obtained by Fourier transforming the potential field of $\rho_{II}$ to get a sum which converges quickly in reciprocal space. An inverse FT is then applied to get $\Phi_{II}$. The $q\alpha/\sqrt{\pi}$ term in $\Phi_I$ is the self-electrostatic energy, which has to be subtracted from $\Phi$ since it is part of $\Phi_{II}$. It turns out that the value of the parameter $\alpha$ has opposite effects on the rate of convergence of the two sums, so an optimal value should be used to make the convergence rates about the same.

Use of Ewald summations speeds up the simulations considerably. Notwithstanding, it is still the most time-consuming part of computing interatomic forces, and it was also noted in early simulations that this computational cost could be minimized by restricting the interatomic separations, $r_{ij}$, to $L/2$, i.e. by truncating $\Phi_I$ at $|r_i - r_j| = L/2$. The practice of truncating $\Phi_I$ (and neglecting $\Phi_{II}$ completely by choosing a value for $\alpha$ judiciously) was common in the first simulations of glasses by Woodcock et al. [24], Soules [25], and Garofalini [26], for example. The errors introduced by this action could be countered by adjusting parameters in the short-range potentials.

However, with the advent of parallel computing architectures, the sizes of simulations which can be handled has increased substantially, and this creates an issue, since the computational cost of the reciprocal space term increases as $N^2$, where $N$ is the number of atoms in the simulation box. Several suggestions have been proposed to speed up this component of the simulations. The one used most commonly is the Particle Mesh Ewald (PME) approach [27], in which the charge density in the reciprocal space term is distributed over a three-dimensional grid of points with interpolations used between the grid points and atomic positions. A Fourier transform (using 3D FFT packages) is then used to obtain the reciprocal space energies. A further Fourier transform is employed to calculate the (real space) forces on the atoms. B-spline interpolations are typically used, in which case the method is called the Smooth Particle-Mesh Ewald, or SPME, method [28]. PME algorithms scale as $N \log(N)$ so are considerably quicker than the simple Ewald summation, and facilitate simulations of several thousand or tens of thousands, or even millions, of atoms. Contrast this with the 500 atoms discussed in Alder and Wainwright's original MD simulation in 1959!

Usually with atomistic simulations, periodic boundary conditions are applied; in the case of crystals, of course, these simply reflect the translational periodicity present, via the unit cell parameters and mean that instead of having to include every atom in the crystal, only a very small subset needs to be considered. But they also avoid the presence of surfaces and any artifacts associated with the termination of the structure at surfaces. This benefit is also adopted for simulations of glasses, since it is not feasible to calculate all the interactions between atoms in a sample of glass, which, like a crystal, contains of the order of Avogadro's Number of atoms.

For noncrystalline materials, cubic or orthogonal periodic boundary conditions are usually employed, but with the simulation box containing many more atoms than a typical crystalline unit cell.

Some comments about the size of the simulation box are appropriate. Early studies, such the pioneering work of Alder and Wainwright and the first simulations of silica, were severely limited in the number of atoms that could be handled by the computational resources available. This limitation, of course, still exists in principle, although the development of computer technology means that simulation boxes containing millions of atoms are now feasible. One might think that "the larger the better" argument applies to simulations as well as in other aspects of life. This is generally the case, because the use of periodic boundary conditions puts some artificial constraints on the relaxation of the atoms in the simulation box, particularly for structures which can be described in terms of (random) networks, such as silicates, aluminosilicates, and borate glasses. For example, the size of the largest (primitive) ring that can form in a periodic network is constrained by the length of the periodic repeat. Since the presence of larger rings allows for a greater relaxation of inter-tetrahedral bond angles, smaller simulations may not have an optimum T—O—T bond angle distribution. This is illustrated in Figure 1.2, which compares the tetrahedral ring size distribution for simulations containing 1500 and 12 000 atoms. The larger simulation was obtained from a $2 \times 2 \times 2$ "supercell" of the smaller one. Doubling the periodic repeat allows considerable relaxation which changes the topology of the network structure. On the other hand, larger simulations will require longer trajectories to thermalize their structure, so the amount of computer time needed increases. A reasonable trade-off would appear to be to have several thousand (say, 6000–10 000) atoms in the simulation box. One might note that this is still at least an order of magnitude or more than quantum mechanical methods can currently handle.

Wright [29] has argued that models of glasses derived using periodic boundary conditions are wrong. However, he cited in evidence simulations that used rather small (by contemporary standards) numbers of atoms, and therefore suffered from the inability to relax sufficiently, as discussed above. Whilst his preferred cluster model was able to be relaxed properly, it also included surface features related to the termination of the structure. The

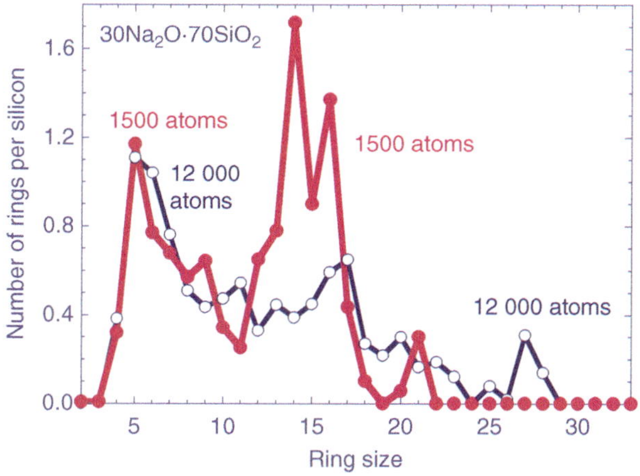


**Figure 1.2** The effect of simulation size is demonstrated by this comparison of the tetrahedral ring size distributions for simulations containing 1500 and 12 000 atoms.

ratio of surface atoms to bulk atoms was not negligible because of the small, finite size of the cluster. In the limit of an infinite simulation box size, of course, there would be no constraints on atomic relaxation, and Wright's argument would fail. Notwithstanding the "all models are wrong" position, there is obviously a great deal of usefulness to be derived from atomic scale simulations. The challenge is to find the most appropriate compromise between simulation size and computational resources to be consumed.

From the lattice energy, which is known as a function of atom positions and simulation box parameters, along with the interatomic potentials, all physical properties of the solid can be calculated. Both the first (i.e. interatomic forces) and the second derivatives of the lattice energy are used.

In the earliest models, the short-range forces were represented by force constants (as these could, in principle, be obtained from vibrational spectra); they are only defined for specific interatomic separations. Structural relaxation was not therefore possible and so force constants were soon superseded by interatomic potentials which could be expressed as a function of distance and thus allow atomic coordinates to vary.

The physical basis for the short-range forces lies in the repulsion between atoms as their electron densities overlap, i.e. Pauli repulsion, and, at longer interatomic separations, dipole–dipole and dipole–quadrupole attraction. These latter $r^{-6}$ and $r^{-8}$ potentials arise because at any given instant, electron densities are not completely spherical and thus form dipoles or quadrupoles, which themselves induce dipoles on neighboring atoms. Obviously, the magnitude of these terms will depend on the size of the atoms. As discussed in Chapter 6, dipole–quadrupole interactions were included in some of the early developments of potential models, by, e.g. Huggins and Mayer [30] among others. However, the magnitude of the quadrupole–dipole term is a fraction of the dipole–dipole term, and nowadays, its energy is subsumed into the term, by choosing suitable potential parameters.

### 1.3.2 Potentials

The choice, and parameterization, of interatomic potentials is key to successful simulations of glasses. The earliest work on potential model development was for crystalline materials, and this was informed by attempts to reproduce the vibrational spectra of molecules. The extension of models for molecules which could be described by pair interactions to condensed phases, which, in principle, adds many-body considerations was not initially thought straightforward. Eq. (1.6) is essentially an expansion of the (potential) energy hypersurface (i.e. a function of all of the atom coordinates in the structure) into 2-, 3-, ... $n$-body terms. Whilst the experience of modeling molecular interactions could be used for the two-body terms, little had been learnt about three-body, or bond-bending, terms or even higher-order terms.

One of the reasons for the introduction of three-body terms, in models of silica, for example is the high formal charges on the Si (and O) atoms, which promote linear bonds. Since, in the $SiO_4$ tetrahedral unit, the O—Si—O bond angle is 109.4°, additional constraints were necessary to prevent this angle from opening up under the influence of the strong electrostatic forces. Three-body interactions are harder to manage in MD than to include in static simulations, such as energy minimization. A simple way to reduce the effect of the electrostatic interactions is to reduce the magnitude of the charges on the atoms. From a mathematical point of view, the charges are just parameters in the expressions of potential energy. Whilst the formal oxidation states are an obvious choice, this is not a real physical constraint. As we have noted above, quantum mechanical calculations of charge density suggest a net charge on Si and O in silica of $\sim 2.4e$ and $-1.2e$, respectively. The use of these values was found to avoid the need for three-body terms, and so partial charge models are now the norm.

A number of different analytical representations of the short-range potentials have been used in the literature over the years. For an historical review, see Chapter 6. The earliest models assumed rigid elastic ("hard") spheres, but efforts such as fitting spectroscopic data, and reproducing the viscosity of rare gas fluids produced poor agreement with experiment. The step-function repulsion between atoms was supplanted with short-ranged repulsion, modeled with an inverse $n$th power law along with a longer (but still relatively short) ranged attraction, as championed by Lennard-Jones [31, 32] and Morse [33].

Born's original choice of an $n$th power law repulsion was replaced within a few years by an exponential expression, which better fit the compressibility data of the alkali halides. Combined expressions representing both the exponential repulsion and the dispersive attraction became known as Born–Mayer–Huggins (BMH) or Buckingham [34] potentials, after the authors who championed them. The analytical expressions of some common short-range potentials are gathered in Table 1.1. Note that the difference between the BMH and Buckingham forms is an explicit inclusion of an $r^{-8}$ dipole-quadrupole energy. Most MD codes can handle a large number of potentials types since they are used to model materials other than glasses, such as metals, biomolecules, or organic compounds. The manual for each package should be consulted to determine what exactly is implemented.

All of these expressions contain a number of variable parameters, which can be tuned to provide the optimum physical properties of a given composition. Note that, apart from the $r^{-8}$ term, the BMH and Buckingham expressions both contain the same number of variable

**Table 1.1** Common analytical expressions for the short-range potentials.

| | |
|---|---|
| Buckingham | $A_{ij}\exp(-r_{ij}/\rho_{ij}) - C_{ij}r_{ij}^{-6}$ |
| Born–Mayer–Huggins | $A_{ij}\exp[B_{ij}(\sigma_{ij} - r_{ij})] - C_{ij}r_{ij}^{-6} - D_{ij}r_{ij}^{-8}$ |
| Morse | $D_{ij}\left[[1 - \exp[-a_{ij}(r - \overline{r}_{ij})]]^2 - 1\right]$ |
| Lennard-Jones | $4\varepsilon_{ij}\left[\left(\frac{\sigma_{ij}}{r_{ij}}\right)^{12} - \left(\frac{\sigma_{ij}}{r_{ij}}\right)^{6}\right]$ |

parameters and are effectively mathematically equivalent. For crystalline materials, one could optimize these parameters to provide an equilibrium atomic structure (unit cell) from which mechanical properties, for example, could then be calculated. For glasses, this approach is not possible, since the structure cannot be uniquely determined experimentally. Indeed, one of the major applications of atomistic computer simulation techniques has been to elucidate the atomic scale structure of glasses, particularly those of multicomponent compositions.

For silica, a number of different potential models (i.e. sets of numerical values for variable parameters in the expressions for the short-range potentials in Table 1.1) have been developed by different authors at different times by different methods. The early parameters were derived empirically by fitting to known structural and physical properties. Later, quantum mechanical calculations were used to create energy surfaces from which the parameters were obtained. This was how the BKS model came about. Named after its authors, Van Beest et al. [35], the model used partial charges, permitting just a Buckingham pair potential description. Very recently, machine-learning techniques have been employed [36].

Of course, the major interest is in multicomponent compositions, not simply silica. For an alkali aluminosilicate glass, for example, one needs, in addition to parameters for the Si–O and O–O interactions, parameters for the Al–O, and alkali–oxygen interactions (assuming that cation–cation interactions are confined to being purely electrostatic). Whilst one could fit the additional parameters to data from specific crystalline compounds, it becomes tedious to have to do this for every new glass composition one wishes to model, and in any case glass compositions can vary widely and often do not correspond to crystalline phases.

This raises the question of the transferability of potentials, that is whether parameters derived for one composition, say lithium metasilicate, can be used for different compositions such as lithium disilicate or lithium trisilicate.

#### 1.3.2.1 Transferability of Potential Parameters: Self-consistent Sets

The need to derive, or otherwise obtain, suitable parameters for the additional interactions every time it is desired to investigate a new composition is rather tedious. If only one could refer to a database of potential parameters! Of course, the very early work on alkali halides attempted to do just that, albeit for a very limited composition space. In the crystalline arena, sets of potential for the alkaline earth fluorides were derived. A set of potential for a range of oxide compositions was developed by Lewis and Catlow [37], and Parker [38] found a set of parameters that reproduced the crystal structures of a number of pyroxenoid silicate minerals. All of these used the Buckingham form. None were suitable, however, for a generic silicate glass composition. The first effort to produce a self-consistent set of parameters for silicate compositions was that of David Teter. He used a number of crystalline silicate structures and elastic properties to fit, simultaneously, the pair potential parameters, also based on the Buckingham form. Partial charges were employed, to avoid the need for three-body terms, as discussed above. Although Teter's parameters have not been published in the literature as a full set, they have been used by the present author's group [39].

Although models using the Born–Mayer–Huggins or Buckingham analytical expressions are still widely used, Pedone et al. [40] derived a self-consistent set of parameters for silicate glasses, which was based on a Morse analytical form. Since its introduction, it has been widely adopted.

In efforts to obtain more accurate calculations of various mechanical properties for specific compositions, more complicated analytical expressions have been proposed. These invariably require larger numbers of potential parameters to provide the improved accuracy, and, unfortunately, do not usually lend themselves to application to other chemistries because the parameters are derived for the specific system under consideration.

All of the models mentioned, including BKS and its extensions, and the Teter and Pedone potentials assume that the ions are rigid. Classically, there are no electrons and so electronic polarizability is not addressed at all. Although (very) large simulation boxes can accommodate some of the flexibility imparted by allowing the ions to polarize, the absence of polarizability is quite a severe limitation, as discussed in the next section.

#### 1.3.2.2 Ion Polarizability

A major limitation of early potential models for inorganic solids, such as the halides and oxides, was the omission of ion polarizability, particularly for the anions. This drawback was manifested, inter alia, in the lack of LO–TO splitting of the optical branches of phonon spectra.

The first attempts to include polarizability were point polarizable models. These added a dipole moment to the point ions in the Born model, but these could lead to a polarization catastrophe. An improvement was proposed by Dick and Overhauser [41] who modeled ions as a core and a shell, represented by point charges, connected by an harmonic spring. The shell, with a charge $Y$, represented (loosely) the outer, valence electrons which are more deformable, and the core, with a charge $X$, (loosely) represented the inner electrons. The total charge, $X + Y$, equals the net charge on the ion, i.e. its formal oxidation state. A dipole is induced when the core and shell are displaced relative to each other, as illustrated in Figure 1.3.

The polarizability of an ion is given by $\alpha = Y^2/(k+R)$, where $Y$ is the shell charge, $k$ is the harmonic spring constant, and $R$ is a term reflecting the influence of interatomic forces acting on the ion. With the short-range potentials acting between shells, the polarization catastrophe was avoided.

The Shell Model became widely adopted in static simulations of oxides, particularly for point defect calculations in which charged defects will polarize their immediate surroundings. Although including shells effectively doubles the number of coordinates that must be included in the model structure, this is no longer a real difficulty for simulations which employ energy minimization.

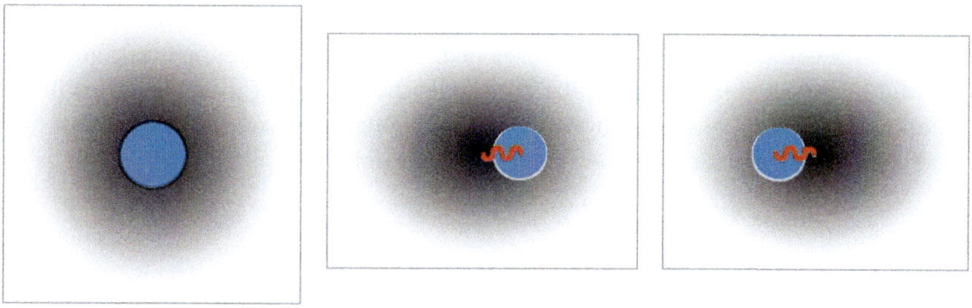

**Figure 1.3** Schematic of the Shell Model. On the left, an isolated atom in the absence of an electric field has a spherical electron density centered on the ion core. When two atoms approach each other, the electron densities may become displaced and lose their spherical symmetry, developing dipole moments. They are effectively polarized.

However, in the Shell Model, the shells have no mass, and this is a problem for MD, because the shells cannot then be treated as independent particles. To avoid them flying away from their cores, in each iteration of an MD run, the ions are first moved according to the equations of motion, and then the shells and cores are equilibrated. This additional step increases the cpu time of an iteration by about an order of magnitude, making the use of shell models for glass MD simulations practically unfeasible.

Inclusion of anion polarizability leads to improved inter-tetrahedral bond angle distributions, for a given size of simulation, so one way to counter its absence is to use larger numbers of atoms in the simulation. This comes, as we have noted, at the expense of significantly increased computational times.

#### 1.3.2.3 Potential Models for Borates

A consistent challenge for those modeling oxide glasses has been the ability, or otherwise, to handle borate glasses, for two main reasons. First, a major controversy in borate glass structural chemistry concerns the concentrations of boroxol rings in vitreous $B_2O_3$ [42]. From Raman spectra, it was deduced that around 70% of B should be in boroxol groups, but most, if not all, of the early studies reported boroxyl ring concentrations which were much lower than that. The controversy has not yet been completely resolved. Second, boron can change coordination from [3] to [4], with the addition of modifiers, but in complex compositions, such as the borosilicates, the change may not be analytically straightforward. Some groups avoided the problem by focusing on compositions for which it was known that all boron were in a single, say, tetrahedral coordination. In these, as in other more recent, studies, it was found necessary to vary some of the potential parameters as the composition changed.

In the last year or two, a renewed effort has led to a number of potential models for borate and borosilicate compositions. This is the subject of Chapters 8 and 9 in this volume.

#### 1.3.2.4 Modeling Reactivity: Electron Transfer

Another limitation of classical models is that, since electrons are not explicitly included, it is not possible to include electron transfer. Thus, chemical reactivity cannot be addressed. To counter this, models which allow a description of electron transfer by incorporating some degree of charge equilibration have been developed. That is, the charges on the (point) ions are allowed to change during the simulation. Currently, the best of these models is probably the ReaxFF model, developed by Van Duin and colleagues (see e.g. [43, 44]). In this approach, the overall system energy is composed of a number of contributions:

$$E_{system} = E_{bond} + E_{over} + E_{under} + E_{val} + E_{pen} + E_{tors} + E_{conj} + E_{vdWaals} + E_{Coulomb} \tag{1.11}$$

where $E_{bond}$ is a bond energy, with corrections for over- and undercoordinated atoms $E_{over}$ and $E_{under}$, $E_{val}$ is a valence angle energy, $E_{tors}$, a torsion angle energy, $E_{conj}$ is a conjugation energy, $E_{vdWaals}$ the van der Waals energy, and $E_{Coulomb}$ the Coulomb or electrostatic energy.

One drawback of the ReaxFF model is that the potential parameters are derived for specific composition spaces, so its transferability is severely restricted. Because of the complexity of the form of the interatomic potentials, it is by no means trivial to build ReaxFF models for different chemistries.

## 1.4 Calculation of Observables

Typically, the information desired from an MD simulation of glasses is (i) the structure and related dynamic properties such as vibrational spectra, and (ii) diffusion properties, i.e. details of long-range ion transport in the structure.

### 1.4.1 Atomic Structure

Since all of the atomic coordinates are known, it is straightforward to show the structure graphically, as in Figure 1.1, for example, using any available graphics package. For comparison with experimental data, though, a more quantitative approach is necessary.

Because of the lack of translational periodicity, the structure of a glass cannot be described in terms of a unit cell and its contents. Instead, distribution functions are required. These basically describe the range of interatomic distances and are based on the radial density function, g($r$). The principle is sketched in Figure 1.4. The radial density $\rho(r)$ represents the number density of atoms at a distance $r$ from other atoms in the structure. g($r$) can count

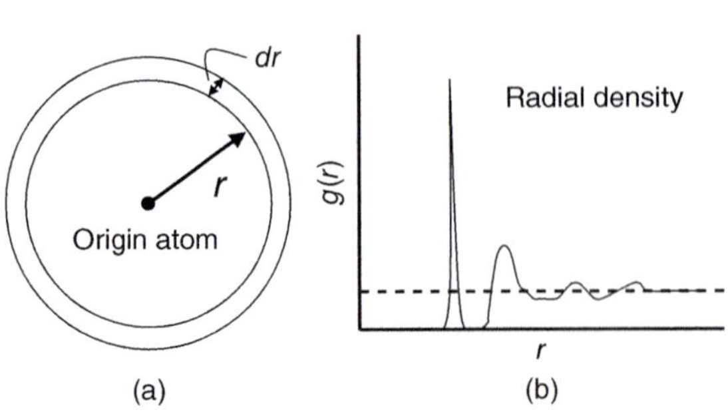


**Figure 1.4** (a) Principle of radial distribution function: $g(r)$ counts the number of atoms in a shell of thickness $dr$ at a distance $r$ from an atom. (b) $g(r)$ is plotted for a non-crystalline material.

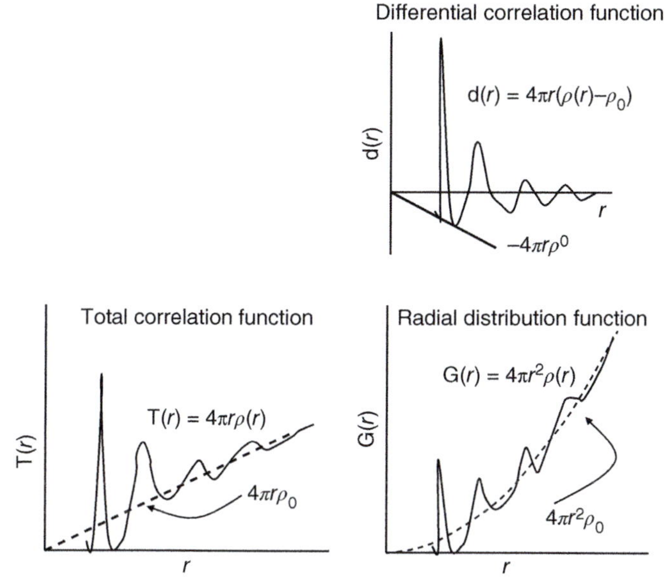


**Figure 1.5** Different radial distribution functions derived from the radial density distribution. Neutron scattering studies of silicates have generally produced $T(r)$.

all types of atom from the atom at the origin, or only those of one type. The radial density is then turned into a radial distribution function: $g(r) = \rho(r)/\rho_0$, where $\rho_0$ is the overall density of atoms.

The radial density is calculated from the number of atoms, $N(r + dr)$, in a shell of thickness $dr$ at a distance $r$ from the atom of interest, at the origin, that is $\rho(r) = \langle N(r + dr)\rangle/4\pi r^2 \mathrm{d}r$, where the angular brackets mean that an ensemble average is taken over all the timesteps in the MD trajectory.

The radial density represents the number density of atoms at a distance $r$ from other atoms in the structure. $g(r)$ can count all types of atom in the structure from the atom at the origin, or only those of one type. The radial distribution function is then turned into a specific distribution function that can be measured. There are several variations on the basic theme, shown in Figure 1.5; each has its own attractions. For glasses, the most common is the total distribution function, $T(r)$, since it is the distribution function most readily obtained from neutron scattering data. When it comes to visualizing the atomic structure, this is obviously done in real space, so real space distributions functions are generally used in conjunction with the reciprocal space scattering data. Figure 1.6 shows $T(r)$ obtained from a $60SiO_2{\cdot}20Al_2O_3{\cdot}20Er_2O_3$ simulated glass, and Figure 1.7 shows some pair ($T(r)$) distribution functions from the same simulation. The data were averaged over 500 timesteps. The differences between the Si—O, Al—O, and Er—O bond lengths are clear.

However, it is as well to remember that scattering data are collected in reciprocal space. MD simulations are also conducted in real space, so comparing MD structures with experimental data involves a transformation from either reciprocal space (i.e. experimental versus MD $T(r)$) or from real space (i.e. experimental versus MD $S(q)$). Mathematically, such transformations are simply Fourier Transforms. However, the transformations are not exact, because, on the one hand, scattering experiments only sample a fraction of

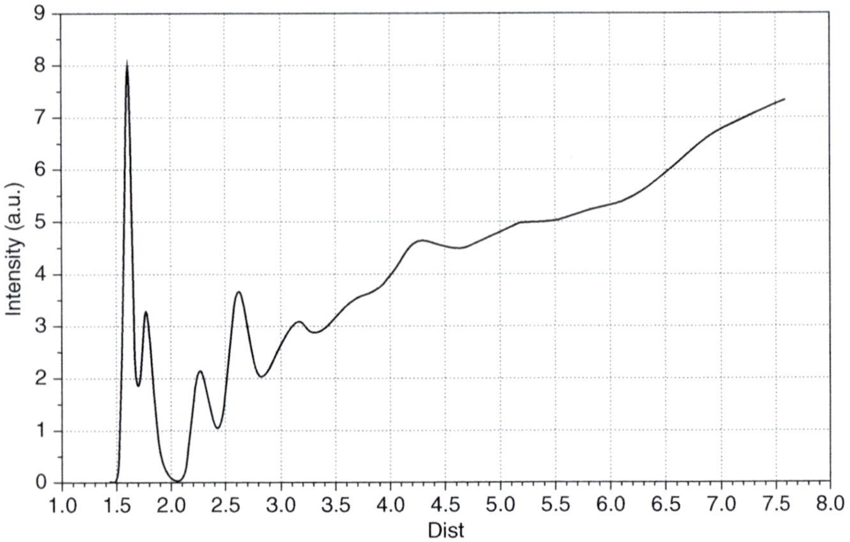


**Figure 1.6** $T(r)$ calculated from 500 timesteps, at 300 K, from a simulation of a $60SiO_2 \cdot 20Al_2O_3 \cdot 20Er_2O_3$ glass.

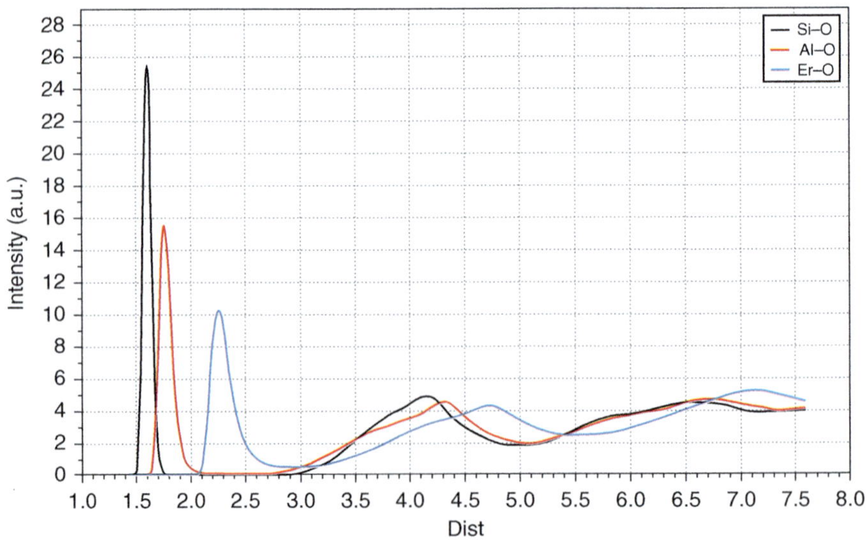


**Figure 1.7** Pair $T$(r) distribution functions for a $60SiO_2 \cdot 20Al_2O_3 \cdot 20Er_2O_3$ simulated glass.

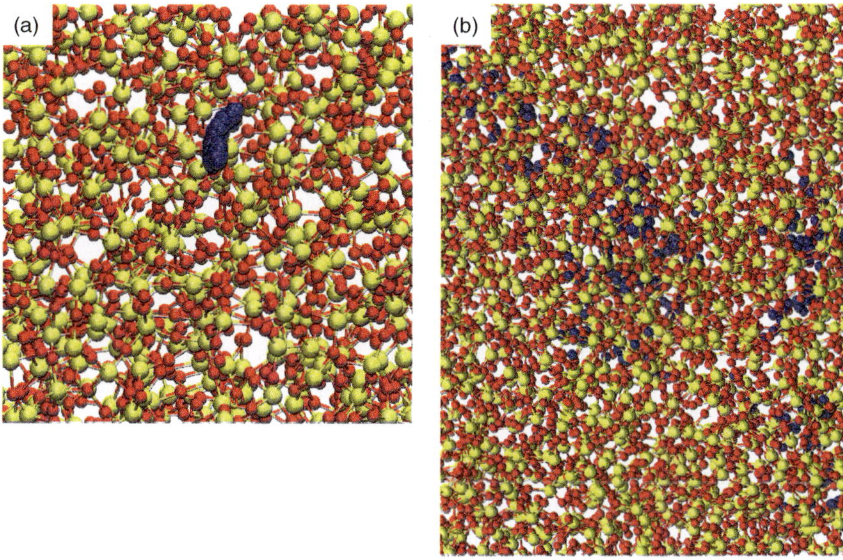


**Figure 1.8** Trajectories of a single Li ion (blue spheres) in a $15Li_2O{\cdot}85SiO_2$ glass at (a) 300 K and (b) 1100 K over 501 timesteps. The Si (yellow) and O (red) positions are those of the last timestep. Note the difference in scale between (a) and (b).

reciprocal space and, on the other hand, the range of interatomic distances sampled in MD simulations is limited by the size of the simulation box.

MD simulations are an excellent way to examine the atomic jump mechanisms underlying diffusional processes. Because diffusion is an activated process, diffusing ions move more easily at higher temperatures. This is demonstrated in Figure 1.8, which shows the positions of a single Li over 501 timesteps, at 300 and 1100 K, in a $15Li_2O{\cdot}85SiO_2$ glass. The Si and O positions are those at the last timestep. At 300 K, the Li ion remains quite localized, with predominantly Debye–Waller behavior (i.e. thermal motion). At 1100 K, though, the Li ion is able to explore a wide range of sites in the structure.

Calculating diffusion coefficients is a natural outcome from MD simulations, since they can be obtained from the trajectories – the recording of atomic positions as a function of time, according to the random walk theory:

$$D = \frac{1}{6t}\langle r^2(t)\rangle \tag{1.12}$$

where

$$\langle r^2(t)\rangle = \frac{1}{N_O N_I}\sum_{j=1}^{N_O}\sum_{i=1}^{N_I}[\mathbf{r}_i(t+t_j) - \mathbf{r}_i(t_j)]^2 \tag{1.13}$$

is summed over all $N_I$ species of the atom whose diffusion coefficient is being sought. Better accuracy is obtained when summing over a number of different time origins, $N_O$. In principle, each time step can be used as a new time origin.

A plot of the mean square displacement (msd), $\langle r^2(t)\rangle$, against time will thus yield the diffusion coefficient. Furthermore, plotting $D$ calculated at different temperatures will yield the activation energy for migration.

An important consideration, which is often overlooked, is that the length of the MD trajectory has to be long enough to enable long-range migration to take place. A typical alkali jump distance, from one alkali site to a neighboring site is ~3 Å, so a minimum $r^2$ would be 9 or 10 Å$^2$. Anything less than this is just Debye–Waller thermal motion. The way to test if long-range diffusion has occurred is to consider the derivative of the msd as a function of time. In the diffusive regime, it should be a constant. This illustrated in Figure 1.9, which shows $\langle r^2(t)\rangle$ at long times (Figure 1.9b), and at short times (Figure 1.9a) for Li in a $15Li_2O{\cdot}85SiO_2$ glass. At long times, the plot is linear, but at short times, it is not. (Note the units of time in plots (a) and (b).) Figure 1.9c shows the derivative of the msd, plotted for the first 1000 ps. The derivative decreases rapidly and levels off to a constant value, which is related, of course, to the diffusion coefficient. Although it may look as though it falls to zero, this is only because of the magnitude of the derivative/diffusion coefficient in the relevant units. (Li diffusion coefficients in lithium silicates are ~$10^{-9}$ or $10^{-10}$ m$^2$/s at 1100 K.) From Figure 1.9a, it looks as though one could fit a straight line to the first part of the curve, giving the impression of a constant diffusion coefficient. This would be misleading (and give an erroneous value for $D$), as can be seen from the whole of the 1.5 μs time span. Many of the very early simulations were, necessarily, of very short

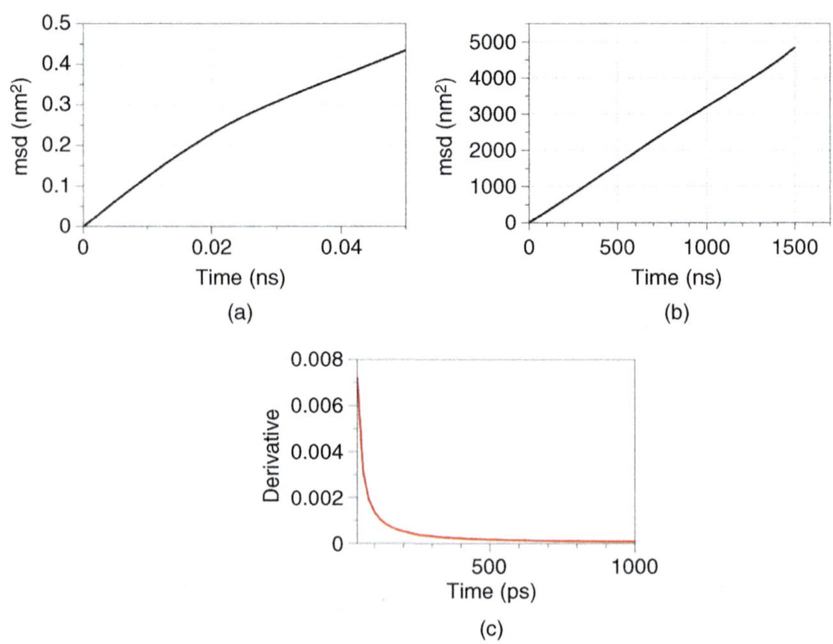


**Figure 1.9** Mean square displacements (msd) for Li in a $15Li_2O{\cdot}85SiO_2$ glass at 1200 K. In (a) the first few ps of the trajectory are shown. In (b) the msd over 1500 ns is plotted. In (c) the derivative of the msd over the first 1000 ps is given.

time duration; the diffusion coefficients obtained from them actually correspond to the very short time behavior seen in Figure 1.9a.

The diffusion coefficient obtained from Eq. (1.12) is, technically, the self-diffusion coefficient since it arises from random displacements, and is to be compared to the diffusion coefficient measured in tracer diffusion experiments. Frequently, though, diffusion coefficients are derived from ionic conductivity measurements – because they are much easier – via the Nernst–Einstein relation. Because of the electric field applied in ionic conductivity experiments, $D_\sigma$ is not the same as $D^*$. The ratio $D_\sigma / D^*$ is known as the Haven ratio and is indicative of correlations between successive atom jumps. It should be noted that the Nernst–Einstein relation strictly only holds if the atom jump mechanisms are the same in both the ionic conductivity and tracer diffusion.

For a more detailed discussion of the use of atomistic simulations to calculate the properties of glasses, please see Chapter 4 and the contributions in Part II of this volume.

### 1.4.2 Hyperdynamics and Peridynamics

Major limitations of atomistic MD are the restricted length and time scales imposed by the dimensions of the simulation box on the one hand and the computational costs of integrating the equations of motion on the other.

MD trajectories are practically limited to several nanoseconds, which means that, in probing diffusive processes, some mechanisms are unlikely be seen. Thermodynamically, this is because the relevant activation energies are too large in relation to the simulated temperature. Sometimes this limitation can be overcome through transition state theory, but this requires knowledge of the initial and final states, which is not generally available in glasses because of the complexity of the energy landscape.

To address the problem, Voter [45] derived a general method, which he called hyperdynamics, for accelerating MD simulations to cope with infrequent events. The basic idea is to add a bias potential, which has the effect of reducing the magnitude of the energy barrier by raising the energy of the initial energy basin, or potential well of the diffusing species, but retaining its local shape.

As well as limited trajectories, atomistic simulations are also practically limited to the nanoscale. But many problems of interest are larger, meso- or macro-scale phenomena, yet which still have an underlying atomic scale origin, and it would be nice to be able to address these multiscale processes, using MD. This is the purpose of peridynamics [46–48]. The general idea is that, notwithstanding that peridynamics is a continuum theory, the forces involved can be discretized, making them similar to those used in traditional MD. It has been implemented in LAMMPS.

## 1.5 Glass Formation

The standard approach to creating glasses in silico is superficially similar to the experimental melt-quench process. An initial structure is melted at a high temperature, with a sufficiently long MD run-time to ensure that the melt structure is in thermal equilibrium. There are several ways to generate the initial bulk structure, which are then frequently used to develop models of glass surfaces and fibers.

### 1.5.1 Bulk Structure

The first method is to melt a crystal structure by heating it up until it is completely disordered; this, though, requires an appropriate crystal structure and this is not usually available for many compositions.

The more common method is to populate the simulation box with atoms whose positions are generated from random numbers. The main pitfall to avoid is inserting atoms too close to previously placed atoms, but with care, this is not a problem.

The initial structure is then equilibrated at a high temperature to ensure that any artifacts that might have been introduced into the initial configuration have been removed, and then the resulting melt is then cooled down to room temperature (or other desired temperature) by reducing the temperature and re-equilibrating the structure at the new temperature. Typical decrements in temperature range from 10 to 100 K, although larger $\Delta T$ have been used. We have found 100 K to be most practicable. As important, however, as the $\Delta T$ is the length of the MD equilibration time at each temperature since that defines the nominal cooling rate. Runs of a few hundred ps to several ns are typical. At melt temperatures, these are adequate for equilibration, but as the temperature is decreased to the glass transition range, shorter run-times will result in structures falling out of equilibrium more quickly. This is also similar to the experimental practice and experience.

It is worth emphasizing that the physical mechanisms for removing thermal energy in a simulated glass are not the same as those active in an experimental melt-quench process. This means that it is not always appropriate to compare experimental and simulated cooling rates, and to impart to simulated glass structures characteristics, such as fictive temperature, inferred from experimental conditions.

Whilst it is the case that the internal temperatures of simulated glasses are generally higher than the nominal MD temperature, this has as much to do with the small numbers of atoms in the simulation box and the related absence of ion polarizability restricting structural relaxation. Structures that have not been able to relax fully at a given temperature will, of course, have characteristics of higher temperature structures, i.e. a fictive temperature. (Note that the fictive temperature of a completely annealed glass is the temperature at which it is annealed.)

### 1.5.2 Surfaces and Fibers

The reactivity of glasses takes place almost exclusively at their surfaces. MD simulations are also able to investigate surface structures and model glass fibers, by suitably adjusting the dimensions of periodic boundary repeat distances, as shown in Figure 1.10.

The first models of surfaces [49] simply removed the periodic repeat perpendicular to the surface. This was possible because of the nature of the potential model which used a truncated Coulomb sum, and not an Ewald sum to obtain the electrostatic energy. Whilst

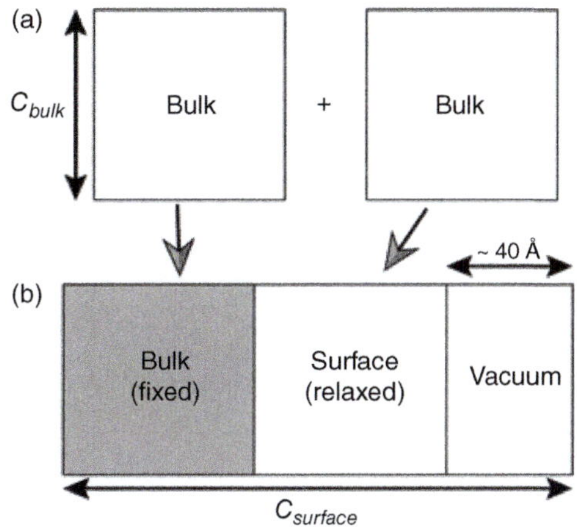


**Figure 1.10** Simulation box setup for (a) surface and (b) fiber simulations. In (b) the periodic repeat is expanded in two dimensions, as illustrated by the green box.

2-D Ewald sum methodologies do exist [50, 51], they are not as simple to implement and are rarely used.

As indicated in Figure 1.10, model surfaces can be created by introducing a vacuum gap which is accomplished by lengthening the periodic repeat perpendicular to the surface, thus retaining the 3-D periodicity and Ewald sums. In effect, an array of repeating slabs is generated. The vacuum gap has to be large enough to avoid interactions between one slab and its adjacent neighbors. It is important to note that the bottom portion of the slab should be frozen in order to prevent the slab from "floating" in the simulation box. Freezing a portion of the atoms, which is handled by requiring the velocities (or momenta) of the atoms to be zero, in the configuration of the bulk structure, also provides a constraint on the relaxation of the surface atoms. In effect, an interface between the bulk structure and surface region is created.

In the scheme shown in Figure 1.10, the frozen section is simply the original bulk simulation box, to which has been added an image box, thus maintaining the structural integrity between bulk and surface regions.

Fibers are created in a similar fashion, except that two of the periodic repeats are increased; the structure is thus continuous only along the axis of the fiber, as shown in Figure 1.11.

## 1.6 Geometry Optimization and Property Calculations

Static lattice calculations, while common in simulations of crystalline materials, are not so common in simulations of glasses, although they have been used to calculate physical properties, notably mechanical properties and elastic constants. The latter are related to the second derivatives of the lattice energy with respect to strain: $C_{ij} = \frac{1}{V}\left(\frac{\partial^2 E}{\partial \varepsilon_i \partial \varepsilon_j}\right)$.

Similarly, other properties such as piezoelectric constants can be derived.

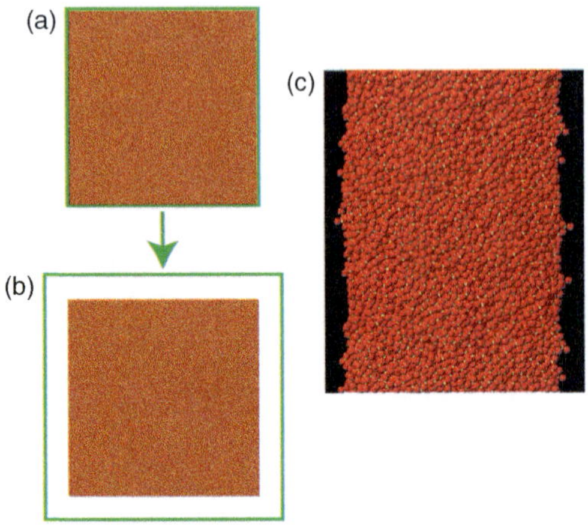


**Figure 1.11** Formation of a fiber. The green outlines in the left panel represent the simulation box. In the lower left panel, the periodic repeat has been lengthened relative to the bulk in two dimensions introducing a vacuum gap to create a fiber (seen end on). The right panel depicts the resulting fiber along its axis.

For glasses, which are noncrystalline, having no long-range translational periodicity, the $C_{ij}$ are, in principle, not defined. Instead, the basic mechanical properties are described by elastic moduli, specifically the bulk and shear moduli, and the Poisson ratio. For crystalline materials, the elastic moduli can be derived from the elastic constants, and this approach is often used for glasses, since the periodic boundary conditions impart a quasicrystalline character.

It is also possible to obtain the elastic moduli directly from the MD trajectory, without recourse to calculating $C_{ij}$. In addition, the elastic moduli can be calculated from simulations in which the structure is strained, or subject to a tensile stress [52].

Energy minimization is a way to find the ground state energy, albeit at a temperature of nominally absolute zero because temperature is not included in static lattice calculations. The standard way to accomplish this is to use a Newton–Raphson technique [53]. This requires the inverse matrix of second derivatives of the energy; the calculation of this matrix is a time-consuming exercise, particularly for large numbers of atoms in the simulation box. Instead of calculating and inverting this matrix in every iteration in the Newton–Raphson method, Fletcher and Powell [54] developed an algorithm for updating the matrix each iteration, without inversion. Whilst these algorithms only approximate the new matrix of second derivatives, they are accurate enough so that complete inversion is only required at every $n$th iteration, where $n$ is typically taken to be 10. It is to be noted that the Newton–Raphson approach is not guaranteed to find a global minimum, but only a local minimum. Given the structure of energy landscapes for glasses, this could be a problem. Other approaches to finding global minima include simulated annealing [55] and genetic algorithms [56, 57].

## References

**1** ASTM (2015). *ASTM C162-05(2015), Standard Terminology of Glass and Glass Products.* West Conshohocken, PA: ASTM International.

**2** Beisbart, C. and Norton, J.D. (2013). Why Monte Carlo simulations are inferences and not experiments. *Int. Stud. Philos. Sci.* 26: 403–422.

3 Alder, B.J. and Wainwright, T.E. (1959). Studies in molecular dynamics I. General method. *J. Chem. Phys.* 31: 459–467.

4 Verlet, L. (1967). Computer "experiments" on classical fluids. I. Thermodynamical properties of Lennard-Jones molecules. *Phys. Rev.* 159: 98–103.

5 Swope, W.C., Andersen, H.C., Berens, P.H., and Wilson, K.R. (1982). A computer simulation method for the calculation of equilibrium constants for the formation of physical clusters of molecules: application to small water clusters. *J. Chem. Phys.* 76: 637–649.

6 Beeman, D. (1976). Some multistep methods for use in molecular dynamics calculations. *J. Comput. Phys.* 20: 130–139.

7 Bussi, G., Zykova-Timan, T., and Parrinello, M. (2009). Isothermanl-isobaric molecular dynamics using stochastic velocity rescaling. *J. Chem. Phys.* 130: 074101.

8 Nosé, S. (1984). A molecular dynamics method for simulations in the canonical ensemble. *Mol. Phys.* 81: 255–268.

9 Hoover, W.G. (1985). Canonical dynamics: equilibrium phase-space distributions. *Phys. Rev. A* 31: 1695–1697.

10 Berendsen, H.J.C., Postma, J.P.M., van Gunsteren, W. et al. (1984). Molecular dynamics with coupling to an external bath. *J. Chem. Phys.* 81: 3684.

11 R.D. Richtmyer, J.v. Neumann (1947) Statistical methods in neutron diffraction, in: Los Alamos Scientific Laboratory Report LAMS-551.

12 Metropolis, N. and Ulam, S. (1949). The Monte Carlo method. *J. Am. Stat. Assoc.* 44: 335–341.

13 Gagniuc, P.A. (2017). *Markov Chains: From Theory to Implementation and Experimentation*. Hoboken, NJ: Wiley.

14 Metropolis, N., Rosenbluth, A.W., Rosenbluth, M.N., and Teller, A.H. (1953). Equation of state calculations by fast computing machines. *J. Chem. Phys.* 21: 1087–1092.

15 Teter, M.P. (1997). Ab initio based studies of borate glasses. In: *Borate Glasses, Crystals and Melts* (ed. A.C. Wright, S.A. Feller and A.C. Hannon), 407–416. Sheffield: Society of Glass Technology.

16 Kerisit, S. and Du, J. (2019). Monte Carlo simulation of borosilicate glass dissolution using molecular dynamics-generated glass structures. *J. Non-Cryst. Solids* 522: 119601.

17 Young, W.M. and Elcock, E.W. (1966). Monte Carlo studies of vacancy migration in binary ordered alloys: I. *Proc. Phys. Soc.* 89: 735–746.

18 Meng, B. and Weinberg, W.H. (1966). Dynamical Monte Carlo studies of molecular beam epitaxial growth models: interfacial scaling and morphology. *Surf. Sci.* 364: 151–163.

19 Keen, D.A. and McGreevy, R.L. (1990). Structural modelling of glasses using reverse Monte Carlo simulation. *Nature* 344: 423–425.

20 Swenson, J. and Börjesson, L. (1997). Fraction of boroxol rings in vitreous boron trioxide. *Phys. Rev. B* 55: 11138–11143.

21 Born, M. and Landé, A. (1918). Uber die Berechnung der Kompressibilität regulärer Kristalle aus der Gittertheorie. *Verh. Dtsch. Phys. Ges.* 20: 210–216.

22 Born, M. and Huang, K. (1954). *Dynamical Theory of Crystal Lattices*. Oxford University Press.

23 Ewald, P.P. (1921). Die Berechnung optischer und elektrostatischer Gitterpotentiale. *Ann. Phys.* 369: 253–287.

24 Woodcock, L.V., Angell, C.A., and Cheeseman, P. (1976). Molecular dynamics studies of the vitreous state: simple ionic systems and silica. *J. Chem. Phys.* 65: 1565.

**25** Soules, T.F. (1979). A molecular dynamic calculation of the structure of sodium silicate glasses. *J. Chem. Phys.* 71: 4570–4578.

**26** Garofalini, S.H. (1982). Molecular dynamics simulation of the frequency spectrum of amorphous silica. *J. Chem. Phys.* 76: 3189–3192.

**27** Darden, T., York, D., and Pedersen, L. (1993). Particle mesh Ewald: an $N$ log($N$) method for Ewald sums in large systems. *J. Chem. Phys.* 98: 10089.

**28** Essmann, U., Perera, L., Berkowitz, M.L. et al. (1995). A smooth particle mesh Ewald method. *J. Chem. Phys.* 103: 8577.

**29** Wright, A.C. (2017). Can a periodic boundary model reproduce the longer-range density fluctuations in a real amorphous material? *J. Non-Cryst. Solids* 461: 113–128.

**30** Huggins, M.L. and Mayer, J.E. (1933). Interatomic distances in crystals of the alkali halides. *J. Chem. Phys.* 1: 643–646.

**31** Lennard-Jones, J.E. and Chapman, S. (1925). On the forces between atoms and ions. *Proc. R. Soc. London, Ser. A* 109: 584–597.

**32** Lennard-Jones, J.E. and Dent, B.M. (1926). The forces between atoms and ions. II. *Proc. R. Soc. London, Ser. A* 112: 230–234.

**33** Morse, P.M. (1929). Diatomic molecules according to the wave mechanics. II. Vibrational levels. *Phys. Rev.* 34: 57.

**34** Buckingham, R.A. (1937). The quantum theory of atomic polarization II – the van der Waals energy of two atoms. *Proc. R. Soc. London, Ser. A* 160: 113–126.

**35** van Beest, B.W., Kramer, G.J., and van Santen, R.A. (1990). Force fields for silicas and aluminophosphates based on ab initio calculations. *Phys. Rev. Lett.* 64: 1955–1958.

**36** Liu, H., Li, Y., Fu, Z. et al. (2020). Exploring the landscape of Buckingham potentials for silica by machine learning: soft vs hard interatomic potentials. *J. Chem. Phys.* 152: 051101.

**37** Lewis, G.V. and Catlow, C.R.A. (1985). Potential models for ionic oxides. *J. Phys. C: Solid State Phys.* 18: 1149–1162.

**38** Parker, S.C., Catlow, C.R.A., and Cormack, A.N. (1984). Structure prediction of silicate minerals using energy-minimization techniques. *Acta Crystallogr.* B40: 200–208.

**39** Du, J. and Cormack, A.N. (2004). The medium range structure of sodium silicate glasses: a molecular dynamics simulation. *J. Non-Cryst. Solids* 349: 66–79.

**40** Pedone, A., Malavasi, G., Menziani, M.C. et al. (2006). A new self-consistent empirical interatomic potential model for oxides, silicates, and silicas-based glasses. *J. Phys. Chem. B* 110: 11780–11795.

**41** Dick, B.G. and Overhauser, A.W. (1958). Theory of the dielectric constants of alkali halide crystals. *Phys. Rev.* 112: 90–103.

**42** Cormack, A.N. and Park, B. (2000). Molecular dynamics simulations of borate glasses. *Phys. Chem. Glasses* 41: 272–277.

**43** van Duin, A.C.T., Dasgupta, S., Lorant, F., and Goddard, W.A. III, (2001). ReaxFF: a reactive force field for hydrocarbons. *J. Phys. Chem. A* 105: 936–9409.

**44** Fogarty, J.C., Aktulga, H.M., Grama, A.Y. et al. (2010). A reactive molecular dynamics simulation of the silica-water interface. *J. Chem. Phys.* 132: 174704.

**45** Voter, A.F. (1997). Hyperdynamics: accelerated molecular dynamics of infrequent events. *Phys. Rev. Lett.* 78: 3908–3911.

**46** Seleson, P., Parks, M.L., Gunzburger, M., and Lehoucq, R.B. (2009). Peridynamics as an upscaling of molecular dynamics. *Multiscale Model. Simul.* 8: 204–227.

**47** Parks, M.L., Lehoucq, R.B., Plimpton, S.J., and Silling, S.A. (2008). Implementing peridynamics within a molecular dynamics code. *Comput. Phys. Commun.* 179: 777–783.

**48** Silling, S.A. (2000). Reformulation of elasticity theory for discontinuities and long-range forces. *J. Mech. Phys. Solids* 48: 175–209.

**49** Garofalini, S.H. (1983). A molecular dynamics simulation of the vitreous silica surface. *J. Chem. Phys.* 78: 2069–2072.

**50** Parry, D.E. (1975). The electrostatic potential in the surface region of an ionic crystal. *Surf. Sci.* 49: 433–440.

**51** Parry, D.E. (1976). Errata: the electrostatic potential in the surface region of an ionic crystal. *Surf. Sci.* 54: 195–195.

**52** Pedone, A., Malavasi, G., Cristina Menziani, M. et al. (2008). Molecular dynamics studies of stress-strain behavior of silica glass under a tensile load. *Chem. Mater.* 20: 4356–4366.

**53** Householder, A.S. (1953). *Principles of Numerical Analysis*. New York, NY: McGraw-Hill.

**54** Fletcher, R. and Powell, M.J.D. (1963). A rapidly convergent descent method for function minimisation. *Comput. J.* 6: 163–184.

**55** van Laarhoven, P.J.M. and Aarts, E.H.L. (1987). *Simulated Annealing: Theory and Applications*. Dordrecht: Springer Netherlands.

**56** Holland, J.H. (1992). *Adaption in Natural and Artificial Systems: An Introductory Analysis with Applications to Biology, Control and Artificial Intelligence*. Cambridge, MA: MIT Press.

**57** Bush, T.S., Catlow, C.R.A., and Battle, P.D. (1995). Evolutionary programming techniques for predicting inorganic crystal structures. *J. Mater. Chem.* 5: 1269–1272.

# 2

# Ab Initio Simulation of Amorphous Materials

*Rajendra Thapa and David A. Drabold*

*Department of Physics and Astronomy, Nanoscale and Quantum Phenomena Institute, Ohio University, Athens, OH, USA*

## 2.1 Introduction

### 2.1.1 Big Picture

Experiments on materials determine our state of knowledge about the physical world. Such experiments come in all shapes and sizes. Diffraction of X-rays, neutrons, or electrons convey essential information about local ordering of atoms, extended X-ray fine absorption structure (EXAFS) gives specific information about local structural order derived from the electronic structure of the system. The electronic density of states (DoS) can be indirectly measured by photoemission experiments or inferred from optical absorption. The electronic tail states are built from valence and conduction states near the Fermi level and are the key to doping, transport, and optics. The vibrations of the network are essential to understanding thermal and mechanical properties.

These experiments never uniquely identify the structure of matter in an amorphous state (in the sense of providing atomic coordinates), and many such measurements require intricate computations to interpret their meaning. Experimentally validated computer models can convey reliable atomistic information about the material.

We argue then that to understand disordered phases of condensed matter, a full set of experiments and realistic computer models are required. Naturally, obtaining convergence between experiments and theory is necessary and can sometimes be an iterative process. As a point of logic, the more experiments we have, the more information we possess about the material. The information intrinsic to a neutron diffraction experiment is very different than a measurement of optical absorption or vibrations. In principle as each new experiment is performed, we should create models anew to jointly satisfy *all* the experiments. And it is not uncommon to find that one experiment is relatively "easy" to represent and others might be quite difficult. We show that usually highly accurate "ab initio" interatomic interactions are required for these complex systems.

*Atomistic Simulations of Glasses: Fundamentals and Applications*, First Edition.
Edited by Jincheng Du and Alastair N. Cormack.

## 2.1.2 The Limits of Experiment

### 2.1.2.1 The Scourge of Averaging

Most measurements on materials involve macroscopic numbers of atoms, and thus are an average over myriad local environments, hence the smoothing (compared to crystals) of measured observables. Local probes of materials are important and are especially relevant to amorphous materials if the spatial resolution of the experiment is fine enough to detect variation in local environments. Perhaps the ultimate example of a local probe is atomic resolution scanning tunneling microscopy (STM). Techniques like micro Raman can probe materials on a c. 1 μm length scale. Obviously enough, all "spectroscopic" techniques face related challenges: to get a measurable spin-resonance, Raman, IR...signal, many atoms are needed to enable the measurement, and the signal is necessarily averaged over these. Electronic and transport measurements suffer analogously. So for disordered systems, experiments do not usually offer direct information about atomic scale disorder, but instead provide "sum rules" (structural, electronic, and otherwise) that a good computer model must satisfy.

### 2.1.2.2 Diffraction

For crystals, it is usually possible to "invert" the structure from diffraction measurements. The intuitive reason is that the information contained in a palisade of Bragg peaks [1] is vast – mathematically speaking it is approximately a collection of Dirac $\delta$ functions with intensity and energy precisely defined. By contrast, the diffraction structure factor $S(q)$ (or the pair-correlation function $g(r)$) is smooth, and the information is commensurately reduced. For sufficiently large distances in space, $g(r)$ asymptotically approaches unity, implying no further structural information. Protein crystallography is one of the great triumphs of modern science, with more than 100 000 entries in the Protein Database and 20 Nobel Prizes [2]. Yet this glittering success depends entirely on the incredible possibility of crystallizing the proteins, and thus cannot help us with our noncrystalline systems.

Diffraction data provides only pair correlations, and it is also blind to chemical order: only the total structure factor is usually measured, so that in a material with (say) two atomic species, we get information about all pair distances with no chemical specificity. The method of isotopic substitution, successfully demonstrated by Salmon and coworkers [3], exploits the fact that neutron scattering lengths can depend strongly on the particular isotope present and thus enables the extraction of partial pair correlations. The method is an important advance, but is not always readily applied to arbitrary materials (since it depends on the availability of possibly rare isotopes to fabricate samples large enough to produce high-quality diffraction data). Satisfying diffraction data is necessary but certainly *not* sufficient to guarantee that a computer model is realistic.

### 2.1.2.3 Spectroscopic Information

A wide variety of spectroscopic experiments are available and provide valuable information about materials. Meaningful comparisons with models require accurate computations of the electronic, optical, and possibly vibrational properties of the computer models. And, of course, spectroscopic methods are subject to the scourge of averaging!

### 2.1.3 Synergy Between Experiment and Modeling

Disordered materials are a classic arena for the interplay between experiment, theory, and simulation. While we have criticized the limitations of information carried by experiment, there are serious limitations in modeling too (systems so small that they are not obviously physically relevant or adequately sampling the local bonding environments – the opposite of the "averaging" charge lodged above against experiments), inaccuracies in atomic interactions, basic challenges to computing physical observables reliably (as for example accurately computing an optical spectrum), simulation times that are many orders of magnitude removed from all but the fastest spectroscopies. We face a difficult limit: we need accurate interactions, large systems, and often extended (say nanosecond) simulation time or a thorough exploration of configuration space that will sample representative local structures. Every "computational knob" should be on maximum for these calculations.

Thus, the respective warts of the experiment and modeling paradigms are limiting in opposite ways, and if we can bridge the gap between them, then we can gain new and reliable insight into these materials and learn about the limitations of the methods themselves [4].

### 2.1.4 History of Simulations and the Need for Ab Initio Methods

The first recognizable molecular dynamics (MD) simulation was carried out by a Swedish astronomer, Erik Holmberg, during World War II [5]. He took advantage of the identical radial $r^{-2}$ dependence of the light intensity with distance from a point source, and the gravitational force to enable simulations based upon an "optical analogy" of the gravitational interaction. Holmberg constructed a clever apparatus in which two groups of 37 lamps each (each lamp representing a star, each group representing a galaxy) interacted optically, and the light intensity at each lamp ("star") was measured with selenium photocells. Since the light intensity was proportional to the interstellar forces, the time evolution of a galactic collision could then be simulated with a "mechanical Euler's method," and Holmberg was able to see spiral arms and draw inferences about galactic scattering and capture probability.

Early simulations of liquids and materials used *ad hoc* potentials and moved atoms according to classical mechanics (Newton's second law). The primitive computers (and smart scientists) of the time could directly simulate the dynamics of systems with hundreds or thousands of atoms or molecules [6]. While the potentials were crude by modern standards, an impressive amount was learned, and the experience gained with empirical potentials with integration methods, size artifacts, alternate "ensembles" (meaning simulations with different fixed thermodynamic variables), and many other aspects of simulation transferred immediately to the ab initio simulations we emphasize here. A classic work in this area is the book of Allen and Tildesley [7].

The first MD simulations used empirical potentials: functions made up to mimic some known behavior about a material. Such a potential might be designed to reproduce the right ground state crystal structure, melting points, etc. The trick for such calculations is applying the potential to a structure with different local bonding – as for example carrying out a simulation for an amorphous solid or say a small molecule for a potential that was designed to fit a crystal. Such simulations are hit or miss, and often fail qualitatively. A potential that works in all salient bonding environments is called "transferable." Some potentials with directional covalent bonding have limited transferability in different bonding environment.

Why is it so difficult to make a generally reliable interatomic potential? The answer is short: *electrons*. Consider the ground state conformations for a series of small Si molecules: $Si_2$, $Si_3$, $Si_4$ ... Start with a pair of Si atoms. Pair interactions are obviously adequate, and the ground state separation between the atoms is at the minimum in the pair potential energy function $V_2(\mathbf{x}_1, \mathbf{x}_2)$ for the dimer. Now add a third atom in the vicinity of the pair. To minimize the energy of the trimer, covalent bonds form between all three atoms (and it turns out the minimum energy configuration is an isosceles triangle with an apex angle of about 80° [8]). Note that the presence of atom 3 changes the *pair* interactions – a different $V_2$ function is required depending on the position of atom 3 ($\mathbf{x}_3$). From elementary chemistry, new linear combinations of the s and p orbitals on all three atoms emerge depending on the positions of all three atoms, so the pair potential picture fails, and it is obvious that at least triplet interactions are required $V_3(\mathbf{x}_1, \mathbf{x}_2, \mathbf{x}_3)$. This argument can be extended *mutatis mutandis* by adding more and more atoms. In practice for accurate computations, approach fails for practical cluster sizes.

The epiphany needed to break out of this scheme was to admit and treat the origin of the complexity: the electronic structure of the system. The first generation of such methods used single-particle Linear Combination of Atomic Orbitals "tight-binding" recipes for the electronic structure.[1] The earliest such calculations used simplified forms for total energies derived from recursion [13] or moment expansions [14, 15] of local electronic densities of states. Soon thereafter, full diagonalization methods were used to compute total energies, for which the electronic component of the total energy came from summing up the electronic eigenvalues of the tight-binding Hamiltonian up to the Fermi level, and a balancing repulsive interaction between nuclei was added to obtain a total energy (the potential energy of the configuration). If we think of this "tight-binding total energy functional," it is clear that it is capable of representing much more complex and spatially nonlocal behavior than an empirical potential. Sankey and Allen [16] were the first to implement this in 1985, now what we call "tight-binding MD." So even though the tight-binding Hamiltonians of the time were imperfect, the work was a great advance since many of the "electronic structure" aspects of the computation were correctly handled.

Ab initio of course means "from the beginning." This is basically ill-defined, since the *beginning* is subjective. Our purist friends down the hall might want us to start with quarks. What it means in practice is a parameter-free approach to computing the electronic ground state energy and associated quantities like forces. Parameter free means no experiments are fit to, and the prescription for the electronic structure and computation of total energy derives from a general theory of an inhomogeneous electron gas as in a solid or molecule (or in principle, but rarely in practice, many-body wavefunctions).

At about the same time as Sankey and Allen, Car and Parrinello [17] did the first calculations adopting the fundamentally appealing approach of using density functional theory to compute total energies and forces. This was the first truly ab initio simulation, and the

1 The tight-binding models themselves were "empirical," and required extensive fitting to a variety of experiments that limit their transferability. One should never underestimate the insights possible for thoughtful tight-binding calculations, see the classic work of Harrison [12].

ab initio schemes in wide use today are remote descendants of this fundamental work, and where the local orbital methods are concerned also the work of Allen and Sankey. All of these methods assume the Born–Oppenheimer [18] approximation (separating the ionic and electronic degrees of freedom) and the additional approximation of classical mechanics for ionic motion.

We should also comment that there is a wider world than just MD simulations. Direct MD is sorely limited by the tiny time step (of order 1 fs ($10^{-15}$ s)) to correctly integrate the classical equations of motion. A very important problem is the effective exploration of configuration space for a complex system. Various "energy landscape" methods are now in common use including some with ab initio interactions [4].

### 2.1.5 The Difference Between Ab Initio and Classical MD

Choosing to carry out an ab initio simulation requires some understanding of the electronic structure of the system under study. The details of which (of many) ab initio codes to employ, technical questions of Brillouin zone integration (to compute total energies, forces, densities of states, etc.), selection of pseudopotentials,[2] basis sets (plane wave cutoffs or local orbitals employed), density functional used, and so on, all must be considered. We think that the best attitude here for a beginner facing a specific new problem is to first search the literature for ab initio simulations on similar systems, and invoke the same approximations for the new problem. The wonderful database AFLOW "An automatic framework for high-throughput materials discovery" (aflow.org) offers a vast amount of information on structures, band structures, and many other physical quantities as well as high-specific information about settings for ab initio codes required for a given problem. Frankly, the quickest way to learn ab initio methods is probably to begin by collaborating with one of the many groups using these methods.

Conceptually, there is no difference between classical MD with an empirical potential and ab initio MD. The distinction is the origin and form of the interatomic potential. For an ab initio simulation, the potential and forces derive directly from the electronic structure of the system as the result of a typically computationally demanding computation. Also, "typical" ab initio simulations have poor scaling with system size, with computational cost scaling as at least the cube of the number of atoms. Empirical potentials can usually be formulated to scale linearly with system size, so from a practical point of view, one is constrained to moderate-sized systems with ab initio calculations. If one is interested in high precision calculation of local structure, defects, etc., it is likely that an ab initio method must be used. If instead, one is interested in the low-q tail of the X-ray structure factor (or large distance pair correlations), or say the distribution of nanovoids in a glass, a large model will be required, and empirical potentials will probably be the only game in town.

There are many codes available. Some are free for academic use (such as SIESTA [19], FIREBALL [20], CPMD [17], Quantum Espresso [21], and ABINIT [22]). Others such as VASP require a license. All of these codes are highly impressive resulting from many years

2 Psuedopotentials are almost universally used to avoid explicitly treating atomic core electrons. For a full discussion, see the treatment by Martin [18].

of work by excellent researchers and are "mature." SIESTA and FIREBALL both use a local orbital basis set constructed from optimized pseudoatomic orbitals, the other codes use a plane-wave basis set. By no means is this list of codes exhaustive. Local orbital codes offer considerable flexibility with respect to the basis set (ranging from minimal "single-zeta" basis sets to complicated multizeta basis with polarization functions, etc.), and this can be especially helpful for approximate (in the sense of limited basis set) calculations on large systems. Plane wave codes are usually the easiest to use for small systems (up to a few hundred atoms) and come with a well-tested library of pseudopotentials. Highly useful work can be readily carried out on modest workstations, and in some cases (notably VASP [23, 24]), graphics processing unit (GPU) acceleration is very helpful. Of course, supercomputers at national centers devote many cycles to ab initio MD!

From a practical point of view, no ab initio code is entirely a black box. The Achille's heel of ab initio simulations is the fact that the Kohn–Sham equations are nonlinear. This means that a self-consistent field (SCF) solution must be found, and this requires some kind of iteration scheme. In certain cases (highly problem and basis set dependent), the iteration may not converge to the solution and may oscillate between two states. In our personal experience, the code VASP does well on SCF iterations and rarely fails to find the solution. If one is interested in metals, far more care is needed with Brillouin zone integrations. Local orbital codes have to be used with care to ensure that the basis set is adequate to represent interatomic interactions, but again, there is a wealth of published work that can help to guide the researcher and the payoff in efficiency relative to a plane wave code may be substantial.

Besides providing the same basic information as a classical MD simulation (say, coordinates as a function of time), there are "fringe benefits": an approximate electronic structure for the model at any time step, local atomic charges, charge densities, etc. This allows the researcher to link to the important class of experiments associated with the electronic structure.

### 2.1.6 Ingredients of DFT

We considered producing a short review of DFT. One of us (DAD) has written at least two such summaries [25, 26]. We will largely eschew this task in this chapter, since there are many masterful treatments available. The Nobel lecture of Kohn [27] is easily readable and historically significant. Richard M. Martin [18] has written the gold standard treatment of DFT and *many* important developments that have arisen in the last 20–30 years. The best recent history from another acknowledged master is the recent paper in Reviews of Modern Physics by R.O. Jones [28], who among other things pointed out Dirac's clear and prescient grasp of the concept of DFT prior to 1930.

Still, to make this chapter somewhat self-contained, we will tersely summarize the basic ideas of DFT and leave the equations to Martin and Jones. Hohenberg and Kohn [29] showed that the electronic density was a suitable function from which to exactly compute the ground state energy $E = E(\rho(\mathbf{r}))$ of *any* inhomogeneous electron gas, that there was a *universal* functional for the energy of the gas, and provided the variational principle to determine $E(\rho)$ on the road toward practical methods. The primacy of this one scalar field $\rho(\mathbf{r})$, rather than an unimaginably complex many-electron wave function[3] was profound both philosophically (e.g. to give a better way to "think" about such problems), and also practically, because it led to methods that are readily handled by modern computers. The method was made practical by Kohn and Sham [30], who converted the Hohenberg–Kohn theorem into an effective single-particle problem with a complicated mean or "self-consistent" field in the spirit of the Hartree [31] or Hartree–Fock approximation [32]. The form of the "Kohn–Sham Hamiltonian" is only approximately known and relies on accurate quantum Monte Carlo calculations on the homogeneous electron gas [33]. The complicated single-particle problem is then solved on some basis set, most commonly plane waves or local orbitals. With the Kohn–Sham eigenvalues and orbitals available, it is easy to compute the single-particle density matrix, total energies, forces, and indeed many other properties. Such computations typically scale at best with the cube of the number of electrons, so it becomes computationally expensive fast. Attempts to improve the scaling have been somewhat successful, always based ultimately upon the spatial locality of interatomic interactions: at some point, moving a sufficiently remote atom at $\mathbf{R}'$ will make no significant difference on at atom at $\mathbf{R}$ if $|\mathbf{R} - \mathbf{R}'|$ is sufficiently large [9]. As the electronic structure determines the interatomic interactions, it must also determine this range. The quantity providing this information turns out to be the decay of the single-particle density matrix. Depending on the accuracy needed, the range of the density matrix is at least several Å in semiconductors and farther in metals [10, 11]. A real-space localized representation can be exploited for this purpose too, the "maximally localized" Wannier functions [34], variants of which may be computed efficiently by projection into the occupied electronic subspace [35, 36]. These ranges are fundamental properties of materials and give direct insight into the nonlocality of the interatomic interactions.

### 2.1.7 What DFT Can Provide

Standard DFT codes produce a cornucopia of data: total energies, forces, structure, dynamics (these can usually be quite well represented), and atomic charges. The electronic structure is a trickier undertaking if one considers excited states, since DFT is a *ground state* method. To even estimate a hypothetical optical gap, additional methods beyond DFT are required. Of course, excited states are required for optical and transport calculations. The band gap problem has been practically addressed by computations with "hybrid functionals," especially the popular Heyd et al. [37] ("HSE06") approximation. For correlated electrons, computationally demanding methods are possible such as the GW approximation, "dynamical mean field theory" and other topics treated by Martin et al. [38].

### 2.1.8 The Emerging Solution for Large Systems and Long Times: Machine Learning

While our palantir is probably murkier than most, we can make one prediction: machine-learning (ML) methods will transform this field. The revolution is already underway. Of all

3 Kohn argues that it does not make sense to even try to compute many-electron wavefunctions if the number of electrons exceeds about 100 [27].

the myriad potential uses of ML, we are thinking of a specific one: ML-based interatomic potentials. There have been several attempts here. We mention the one that we know the most about, the Gaussian approximation potential (GAP) potential [39], owing to our good fortune to collaborate with the pioneers of the method, and direct the reader to the burgeoning literature on the subject.

A naive but helpful way to think about ML approaches to interatomic potentials is to interpret it as a sophisticated interpolation scheme. An ML potential is of course only as good as the information that is "fed" to it. The information takes the form of a large number of atomic configurations, and for each of these, the total energy and forces on the atoms are computed from an accurate (probably DFT) code. The idea is that if enough of these configurations is available, then one could undertake a new simulation only using information in the database – if "new" configurations were close enough to existing conformations in the database to enable sufficiently accurate "interpolation." Of course, the devil is in the details: what does the metric on configuration space "close enough" mean and how does one really "interpolate." After a lot of very ingenious work [40], there are quantitative answers to these questions, and furthermore, what is not *a priori* obvious is that the method is quite practical.

The GAP potentials are truly of DFT quality, especially for silicon, and *ML has completely leapfrogged DFT in one critical way* – it is rigorously "order-N" – with computational cost scaling linearly with the number of atoms. So ML opens the door to truly enormous simulations and with essentially no loss in accuracy [41]. This linear scaling is a profound feature of the method that guarantees that it will eventually supplant DFT for production runs on large systems. DFT will become the "teacher" of the ML algorithms, and DFT will still have to be used for *terra incognito* – for which no database to learn from exists.

ML can, of course, provide information beside forces and energies. For example, Cerlotti and coworkers [42] have shown that the electronic DoS can be "learned" and applied this to 100 000-atom simulations of exotic phase changes in disordered phases of silicon under high pressure [41].

We excise an illustration from Ref. [41]: we show a snapshot from a 100 000 atom model of disordered silicon at a pressure of 12 GPa in Figure 2.1. At this particular pressure,

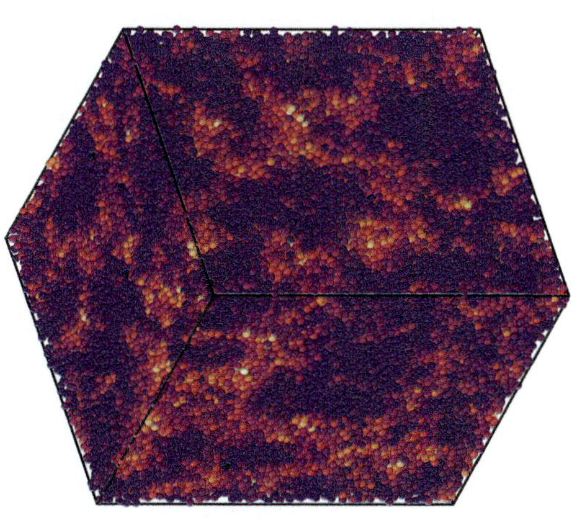

**Figure 2.1** Snapshot of a 100 000 atom model of disordered silicon under 12 GPa pressure created with OVITO [43], as in Reference [41]. The atoms are color coded according to their coordination: light and dark atoms having high and low coordination, respectively, revealing polyamorphism at this pressure.

polyamorphism is revealed with low- and high-density amorphous regions coexisting. The colors indicate coordination (darker, lower coordination). For a full treatment of complex pressure-induced transitions, we refer the reader to the original paper.

### 2.1.9 A Practical Aid: Databases

In the United States, the so-called "Materials Genome Initiative" was introduced more than a decade ago to develop computational methods for materials design to bring advanced technological materials to market sooner. This stimulated efforts in many sectors of academe and produced useful new methods exploiting high-performance computing, such as data mining. Impressive databases such as AFLOW now hold information on millions of materials and include relatively scarce calculations like band structures and thermal properties. It is an ideal site to pick up the threads of an earlier calculation that is a helpful start for something new.

## 2.2 Methods to Produce Models

How does one make a computer model of a complex material? In this section, we try to answer these basic questions and explain how the computer simulations are done at present.

In crystals where atoms are arranged periodically, determining the positions of atoms is simple (unless the unit cell itself is complex), and it is easy to imagine filling space with such a unit cell using the lattice vectors. In amorphous systems, devoid of long-range order between the atoms, determining the structure can be daunting. The first point to note is that the arrangement of atoms in amorphous systems is *not* random: it lacks long-range order but almost always has well defined short-range order. The tendency to order decays with distance: second neighbor distances exhibit more variation than the first neighbor, and so on. To illustrate this, the atomic arrangement in a $A_2O_3$ type glass proposed by Zachariasen [44] has been shown in Figure 2.2. Such ordering (or lack thereof) on various length scales is well discussed by Elliott [45].

Once a model of an amorphous material is obtained, the first step in the analysis is to compare the experiments. The decay of spatial pair-correlations with distance is a first test that the model must pass since this is an observable inferred from diffraction experiments [4]. The lack of crystal-like long-range order obviously makes model generation in amorphous materials challenging, but on the brighter side, this inherent disorder in atomic positions leads to emergent phenomena unknown in crystals and render amorphous materials to a wide range of applicability. As we stated above, favorable agreement of a model with diffraction data is necessary but is not sufficient. As we discuss below with "Reverse Monte Carlo," it is relatively easy to make computer models that agree nicely with a structure factor or pair correlation function, but the atomic structure generated from such a fitting has high density of unphysical structures, e.g. chemically unrealistic bond, atypical values of bond lengths, bond angles. This emphasizes the importance of comparing models with experiments measuring diverse physical properties.

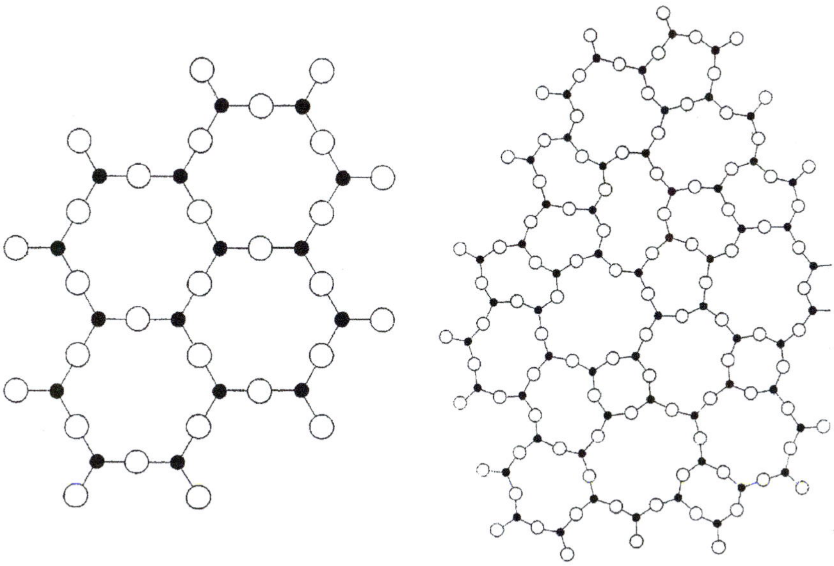


**Figure 2.2** The classic illustration of the continuous random network model: two-dimensional representation of atomic arrangements in crystalline (a) and glassy $A_2O_3$ [44] (b).

The rest of this chapter is organized as follows: First, we discuss the simulation methods used to produce models of amorphous materials, comparing one with other. Second, we discuss how the structural, electronic, and vibrational properties are analyzed from the models. Appropriate comparison with its crystalline counterpart is made whenever possible to create a better understanding of the subject being discussed. We illustrate all this with specific amorphous materials that we hope are both illustrative and interesting in their own right.

### 2.2.1 Simulation Paradigm: Melt Quench

What we call the "simulation paradigm" is the idea of directly simulating the transition of the material from the liquid phase to an amorphous solid. The melt-quench method, as the name suggests, involves creating a high-temperature liquid melt of the material and then rapidly quenching it, resulting in a structure with disorder "frozen in" [4]. A starting random configuration of atoms is chosen (with atoms being disallowed to be unphysically close). Periodic boundary conditions (PBC) are always employed so that what we are really modeling is a crystal with a large unit cell. This system is heated to a temperature well above its melting point to produce a liquid melt and to make sure that there is diffusive ion dynamics. The melt is cooled to room temperature in subsequent cycles of cooling, using some form of dissipative dynamics, and then equilibration. The structure is ultimately subjected to a force minimization to reduce forces to near zero. The hope is that the final model consistent with experimental observables.

There are many details to be selected for the melt quenching, especially the choice of cooling rate. There is no real *a priori* theory for selecting these rates, and the usual way to proceed is to try several cooling rates and verify that the results are acceptably converged (e.g. the topology does not change significantly with slower quenches). The cooling rates used in simulations are always drastically faster than the physical cooling rates, and it is interesting that melt-quench simulations work well for many materials, despite this difference. For a highly detailed study of the effect of cooling rates on structure, see the ML simulations of Bernstein et al. for amorphous silicon [46].

As an example, a melt-quench simulation of Cu-doped $Ta_2O_5$ [47], made with a cooling rate of 100 K/ps, is presented here. This material is a promising candidate for conducting bridge random access memory (CBRAM) memory devices. The Cu atoms tend to cluster, and their connectivity can be modified electrochemically, creating insulating and conducting phases. Models produced with melt quenching at different cooling rates were able to capture the process of the segregation of the Cu atoms from the forming tantala network as the melt was cooled. The presence of the interconnected Cu atoms in the network is evident in the Cu–Cu correlation in Figure 2.3.

The melt-quench method always begs difficult questions. (i) Is the model big enough? (ii) Are there artifacts from the use of periodic boundary conditions (PBC)? (iii) Are the cooling rates used in the simulation acceptably "slow"? and (iv) Is it sufficient to compute energies and forces using only the Γ point of the Brillouin zone associated with the periodic boundary conditions?

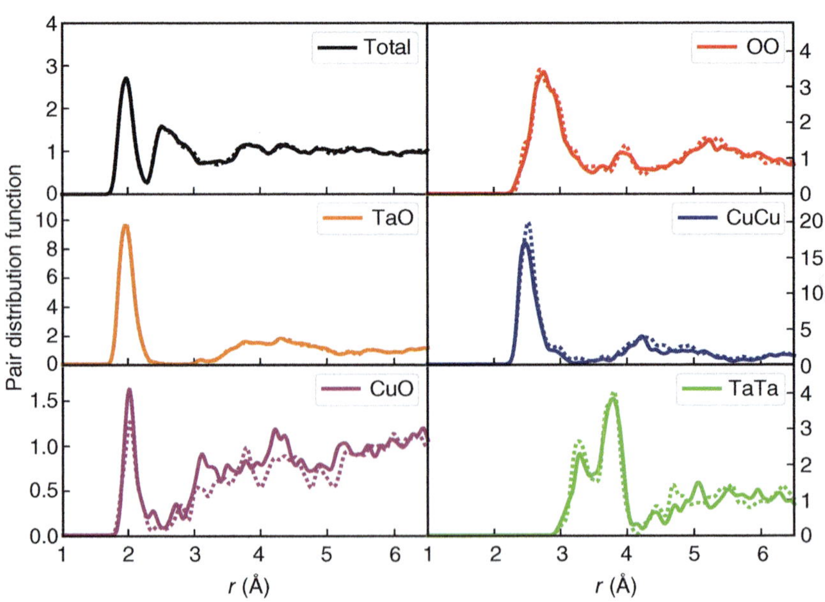


**Figure 2.3** Pair distribution function for Model I (solid line) and Model II (dotted line) of amorphous Cu-doped tantala [47] with same number of atoms and differing only in their cooling rates.

First, the term "big enough" can be subjective and the size of the model that can be generated depends significantly on the nature of the interatomic interactions used during the melt-quench process. For example, performing DFT-based calculation on VASP limits the system size to a few hundreds atoms. On the other hand, calculation with empirical potentials can have millions of atoms in the system. As usual, there is an intelligent balancing required among size, time scales, and accuracy. The choices depend also on the questions the researcher is attempting to answer.

The use of PBC allows us to perform MD of the bulk material by removing surface effects and rendering the system as a crystal with a large unit cell. One should be careful with the system size when using the PBC: the size of the supercell must be greater than the cutoff distance of the interatomic potential to avoid self-interaction among atoms and their images in other cells. The PBC allows meaningful information on a length-scale half the size of the simulation box [25]. There are analogous artifacts for computations of lattice vibrations [25]. Also, size and PBC artifacts are strongly **k** and **r**-dependent in diffraction data: $S(\mathbf{k})$ and $g(\boldsymbol{r})$, respectively. To check if the system is free of artifacts from PBC, one should perform the simulations with different-sized simulation box and compare the results.

Cooling rate is a critical parameter that determines the credibility of the generated amorphous models. Overly short cooling rates can produce several artifacts, like remnants of the liquid melt, because the system will not have enough time to thoroughly explore the configuration space. Typical cooling rates used for melt quench (MQ) methods using VASP is of the order of 100–200 K/ps. Also note that classical molecular dynamics code could use a much slower cooling rate but is still far from comparable to the actual process in lab.

Calculations of energy and forces require summation over the Brillouin zone and the accuracy depends solely on the way the zone is sampled. However, for amorphous systems that lack translational invariance are usually represented by large supercells containing hundreds of atoms. Such large supercells have a commensurately small Brillouin zone [48] and thus the calculation using a single $k$-point ($\Gamma$-point) may be acceptable in amorphous systems. This point is explored in detail in Ref. [49].

### 2.2.2 Information Paradigm

The idea of this method is to produce models that agree with experimental information. It differs from the melt-quench method as the idea is to construct models directly using implied experimental information instead of trying to simulate the glass formation process. The idea was first adopted by Kaplow et al. [50] to make vitreous Se models consistent with X-ray radial distribution function (RDF). The idea was later developed by McGreevy and Pusztai [51] who named it Reverse Monte Carlo (RMC). A more recent idea is the empirical potential structure relaxation (EPSR) by Soper that starts with an empirical potential which is modified recursively by the difference between the model and the experimental data [52].

This idea of this "inverse modeling" scheme is to find a set of atomic coordinates that minimizes the discrepancy between the model and the experiment. The quantity measuring this discrepancy, for a set of atomic coordinates $\vec{R}$, is defined as follows:

$$\chi^2(\mathbf{R}) = \sum_i \frac{[F_E(Q_i) - F_m(\mathbf{R}, Q_i)]^2}{\sigma_i^2} \tag{2.1}$$

where $F_E(Q_i)$ is the experimental value of the quantity at point $Q_i$ and $F_m(\vec{R}, Q_i)$ is the value calculated from the model. $\sigma_i$ is a nonnegative quantity associated with the experimental error. In simple words, $\chi^2$ measures the goodness of fit and the inverse modeling is merely a problem of finding a set of atomic coordinates that render a minimum $\chi^2$ (e.g. small enough to fit within experimental error bars). Note that in practice there are a vast number of such minima, and it begs the question: "which minimum do we accept"? One can get out of a trapped configuration with a Monte Carlo minimization [53] accepting an atomic move that increases $\chi^2$ with a positive nonzero probability.

An attractive feature of this method is that it does not require interatomic potentials for the minimization and thus presents particular promise in systems where the energy calculations are prohibitively expensive. The approach honors experimental data by construction, the weakness however is that none of these algorithms took a course in freshman chemistry – so a model of amorphous $SiO_2$ made in this way would have many O–O and Si–Si bonds, which are essentially physically nonexistent.[4]

### 2.2.3 Teaching Chemistry to RMC: FEAR

How do we include all the information available during the process of model construction, viz. experimental information and accurate interatomic interactions? This might be achieved through unification of simulation and information paradigm thereby changing the problem to the minimization of the quantity

$$\phi = E(\mathbf{R}) + \chi^2\Lambda \tag{2.2}$$

where $\Lambda(> 0)$ determines the relative weighting of the experiment and potential [54]. We do not follow this program because determining $\Lambda$ is ambiguous, and we find better results with alternative techniques. The first alternative, called experimentally constrained molecular relaxation (ECMR) [55], employs a self-consistent iteration scheme that involved completely converged RMC minimization and energy minimization. Often, the configuration gets trapped in a local minimum and requires a new start.

A successful technique for minimizing $\phi$ in Eq. (2.2) is called *force-enhanced atomic refinement* (FEAR) [56–58]. In contrast to the completely converged minimization in ECMR, FEAR involves separate incremental minimization $E$ and $\chi^2$. The fact that the models generated agree reasonably well with experiment and also sit in an energy minimum of a suitable interatomic interactions greatly increases the credibility of the models. FEAR modeling shows particular promise in reducing the computational effort needed to produce models compared to the melt-quench process.

A schematic diagram of the FEAR technique has been given in Figure 2.4. It has been applied to a wide range of amorphous materials ranging from *a*-Si [56] and *a*-C [59] to bulk metallic glasses [60] and sodium-silicates [61]. Recently FEAR has been employed in a high precision study of annealing induced changes in the intermediate range ordering

4 *Ad hoc* constraints can be included, introducing a "penalty" for homopolar bonds, for example. This is somewhat unsatisfying however, since it introduces bias from the modeler.

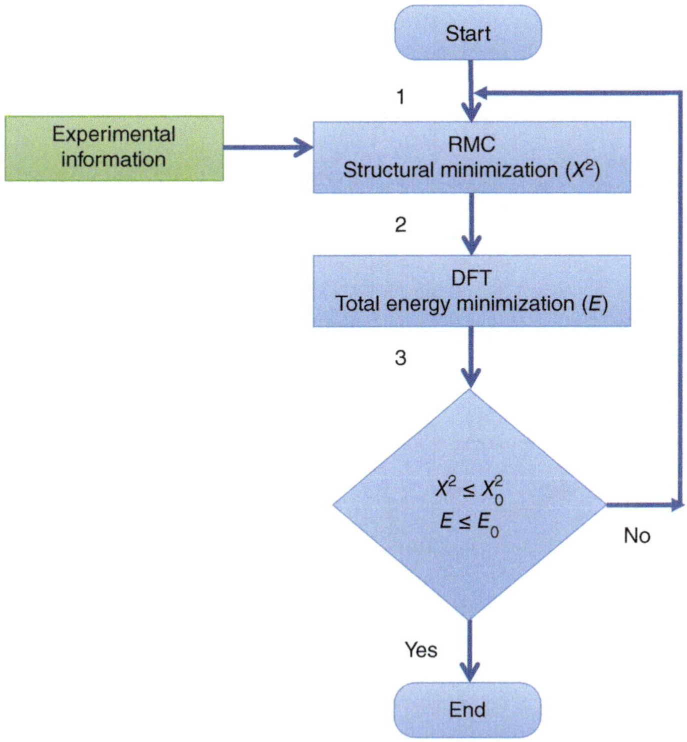


**Figure 2.4** Flowchart of FEAR method.

of zirconia-doped tantala, a possible candidate for laser interferometer gravitational-wave observatory (LIGO) mirror coatings [62].

It is reasonable to describe FEAR as a way to "invert" the scattering data to obtain a realistic computer model. Of course, this is true only in a statistical sense – there is of course no hope of *uniquely* inverting pair correlations into coordinates in space, but with the inclusion of chemical information incorporated through the partial energy minimizations, a highly representative result reflective both of the experiment and chemistry can be obtained.

A representative example of FEAR modeling in $Pd_{40}Ni_{40}P_{20}$ bulk metallic glass is presented here. Two models were generated for this work, one with 200 atoms using melt-quench technique and the other with 300 atoms using FEAR. Figure 2.5 suggests that the models generated have successfully reproduced the structural features implied by the experiment as confirmed by the agreement of the RDF of the models with the experiment. Since this multicomponent system has several (unknown) partial pair-correlation functions (thereby making the total correlation less informative than for elemental system), we also compute EXAFS to gain valuable first shell information. The experimental data [64] is shown by red dots connecting line, FEAR, and MQ models are represented by the blue and

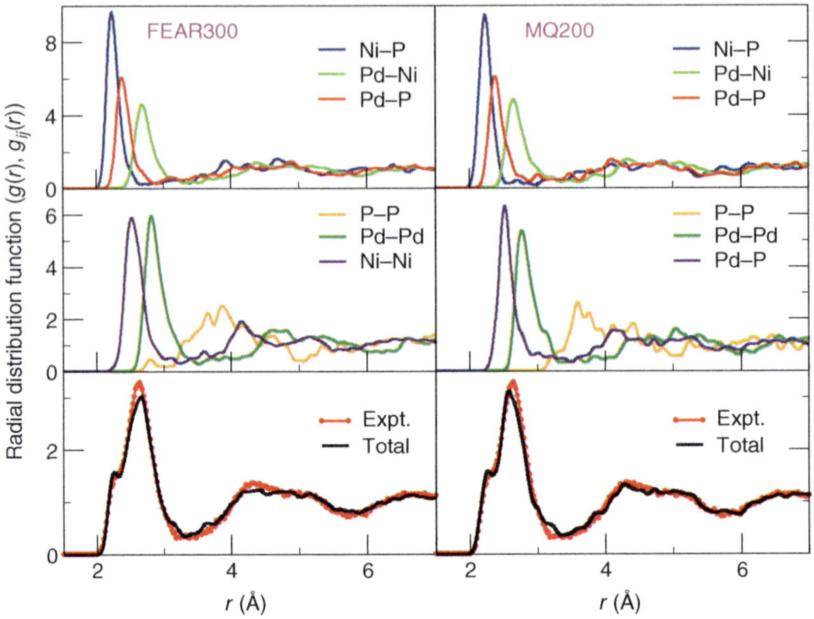


**Figure 2.5** We show comparison of our $Pd_{40}Ni_{40}P_{20}$ models, made with melt-quench (MQ200) and FEAR (FEAR300) [60] method, with the experiment [63]. The radial distribution function ($g(r)$) is in good agreement with the experiment and the partial RDF are also consistent with previous literature.

black lines, respectively. The inset in each figure shows the EXAFS spectrum, i.e. $\left(k^2\chi(k)\right)$. The FEAR300 model shows a better agreement with the experimental EXAFS spectra. The corresponding Fourier transformations of FEAR300 and MQ200 is plotted alongside experimental values [64], and it is observed that FEAR300 model is qualitatively in better agreement with the experiment. Electronic properties and a comparison between EXAFS experiment and theory is provided in (Figure 2.6).

The FEAR model, despite being 100 atoms larger than the melt quench, used only 16% of the total computer time used by the melt-quench model. This computational efficiency could be utilized to produce larger computer models of materials with a identical computational resource.

### 2.2.4 Gap Sculpting

A common drawback of the melt-quench MD simulation is that it produces structural models with greater disorder than the real material. The excess disorder can range from acute short-range defects like coordination defects to more subtle defects in medium range like

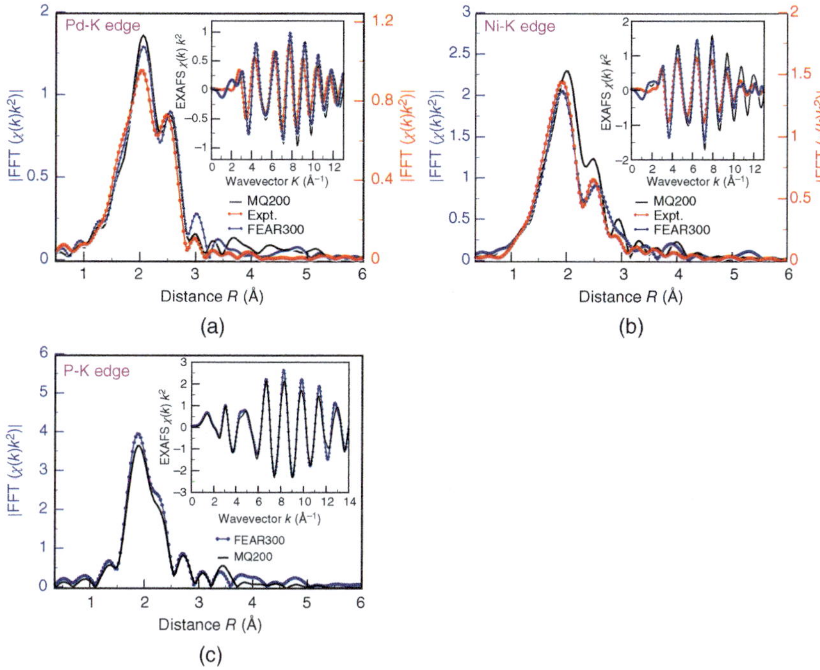


**Figure 2.6** Comparison of EXAFS spectra: (a) Pd-K-edge, (b) Ni-K-edge, and (c) P-K-edge EXAFS spectra of $Pd_{40}Ni_{40}P_{20}$.

presence of irregular-sized rings. While the extent of the discrepancy depends on the systems as well as modeling approach, the issue is universal. Often, one possesses experimental information about optical gaps and defects. As such, it may also be desirable to explore the possibility of obtaining a desired (usually defect free) band gap. To enable this, we discuss "Gap Sculpting."

Many approaches have been proposed to remedy the above drawback – most focus either on better potential/pseudopotential or on better sampling of the potential energy landscape. Gap sculpting, proposed in 2016 and applicable to all electronic structure-based MD simulation, takes a completely different approach.

To understand the method of gap sculpting, it is necessary to start by appreciating the correlation between structural defects and electronic defects. It is common knowledge that the short-range structural defects like dangling and floating bonds, anomalous bond angles, etc., often give rise to defect states in the electronic band gap. The Urbach tails, observed as the exponential tapering of band edges on various crystalline and amorphous materials, are attributed to disorders in the network, e.g. long- and short-bond lengths [65], disorder at the medium range [66]. It has been shown by many researchers that atomic models with improved structural orders are also the ones with few electronic defects [67]. Gap sculpting is based on the converse of this observation – a model with fewer electronic defects should also be the one with fewer structural defects. In that regard, gap sculpting modifies the atomic forces in MD simulation so that the electronic defects in the band gap are explicitly disfavored. In practice, this takes the form of a biased dynamics where the total forces on atoms are given by [68, 69]

$$\mathbf{F}_{\alpha}^{BS} = -\sum_{i}^{occ} \left\langle \Psi_i(r) \left| \frac{\partial H}{\partial R_{\alpha}} \right| \Psi_i(r) \right\rangle = -\sum_{i}^{occ} \frac{\partial \lambda_i}{\partial R_{\alpha}}. \tag{2.3}$$

Here, $\lambda_i$ is the eigenvalue of the Hamiltonian $H$ corresponding to $i$th eigenstate $\Psi_i$, and $R_{\alpha}$ are the $3N$ positional degrees of freedom. The gradient $\frac{\partial \lambda_i}{\partial R_{\alpha}}$ provides the direction in configuration space for maximum increase of $\lambda_i$ – so gives information about how to move atoms to maximally affect $\lambda_i$. Eigenstates with eigenvalues in an *a priori* known energy range can be forced out of that range using a biasing force given by [69]:

$$\mathbf{F}_{\alpha}^{bias} = -\sum_{i:E_{min}<\lambda_i<E_{max}} \gamma g(\lambda_i) \left\langle \Psi_i(r) \left| \frac{\partial H}{\partial R_{\alpha}} \right| \Psi_i(r) \right\rangle. \tag{2.4}$$

This biasing force pushes the states below and above the Fermi level in opposite directions. The quantity $\gamma$ is merely a knob to control the magnitude of the biasing forces. The method also is ideal and frankly easier to use to study metallization-closing gap rather than opening. In this case, one shepherds electronic level toward the Fermi level. It is important to note here that the biasing forces derived from the standard Hellmann–Feynman forces which are already computed in any electronic-structure-based MD simulation, and thus the method adds negligible computational overhead compared to the standard MD simulation.

As an example of the application of the metallization technique, we described an insulator to metal transition achieved in amorphous $(GeSe_3)_{1-x}Ag_x$ [70] glasses using suitable biasing forces. The melt-quench models made using VASP [71] with PBE [72] exchange-correlation functional and had 240 atoms with $x = 0.15$ and 0.25. Conducting structures were obtained by annealing the starting configuration using biased forces at 700 K for 18 ps. The electronic states in the energy range $[\epsilon_F - 0.4\,\text{eV}, \epsilon_F + 0.4\,\text{eV}]$ were included in the computation of the biasing forces and a $\gamma = 3.0$ was used. The biasing potential moves the electronic states at the band edges into the band-gap region. This causes the force biased structure to have states in the band gap (obtained with MD without biasing forces) thereby inducing metallic characters. After obtaining the "biased models," we relaxed and annealed the models with "unbiased" forces. The DoS and the inverse participation ratio (IPR) (measuring the localization of the states) are compared between the model with no biased forces (insulating model) and the model with biased forces (metallized model) are shown in Figure 2.7. The presence of extended states in the metallized models clearly suggests conducting behavior. The final metallic models were found to be almost energetically identical to the insulating phase, an interesting "electronic polyamorphism."

## 2.3 Analyzing the Models

Once a model is created, it is subject to a different analysis depending upon researchers' interest. The analysis performed test the validity and credibility of the model by allowing

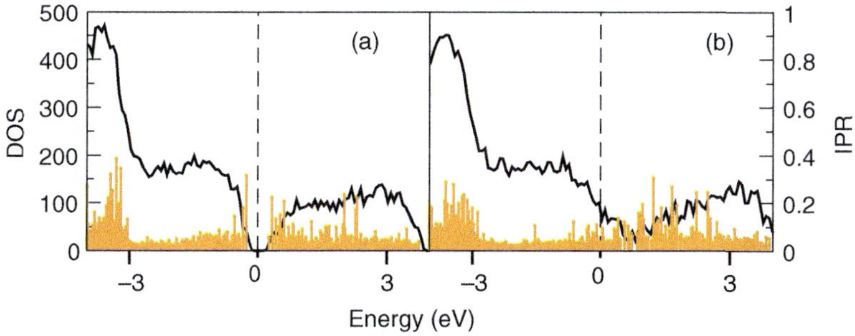


**Figure 2.7** The (black curve) electronic density of states (DoS) and (orange drop lines) inverse participation ratio (IPR) of the insulating model (a) and the metallized model (b). The energy axis for all data sets is shifted to have Fermi level at 0 eV (highlighted by the broken vertical line).

comparison with the experiment. The results obtained from the analysis often guide future theoretical as well as experimental studies.

### 2.3.1 Structure

Structural analysis of the models is the first pivotal test amid the lack of a well-defined positions of the atoms like in crystals. Since amorphous materials inherently lack long-range order, characteristic of crystalline materials, the order usually extends up to the second nearest set of neighbors, sometimes called the medium-range order. The first test is to check whether or not the arrangement of atoms represent an amorphous systems. Other test include, but are not limited to, bonding environment calculation, coordination number, Voronoi polyhedra analysis, and voids.

#### 2.3.1.1 Radial Distribution Function

The RDF is defined as:

$$g(\mathrm{r}) = \frac{N}{4\pi \mathrm{r}^2 V} \frac{dn(\mathrm{r})}{d\mathrm{r}} \tag{2.5}$$

where $N$ and $V$ denote the total number of atoms and the volume of the simulation box, respectively. $dn(r)$ denotes the number of atoms located at distances lying between $r$ and $r + dr$ [73]. The RDF $g(r)$ gives the probability of finding an atom in a shell of thickness $d\,r$ at a distance $r$ from a reference atom. For systems having more than one atom type, the partial RDF is defined as follows:

$$g_{\alpha\beta}(\mathrm{r}) = \frac{Nc_\alpha}{4\pi \mathrm{r}^2 V} \frac{dn_{\alpha\beta}(\mathrm{r})}{d\mathrm{r}} \tag{2.6}$$

where $c_\alpha$ is the concentration of $\alpha$ atomic species. The functions $g_{\alpha\beta}(r)$ denotes the normalized probability of finding an atom of type $\beta$ at a distance $r$ from the reference $\alpha$ atom.

The qualitative information portrayed by the RDF can be easily understood from Figure 2.8 comparing the RDF in amorphous and crystalline form of Carbon at nearly

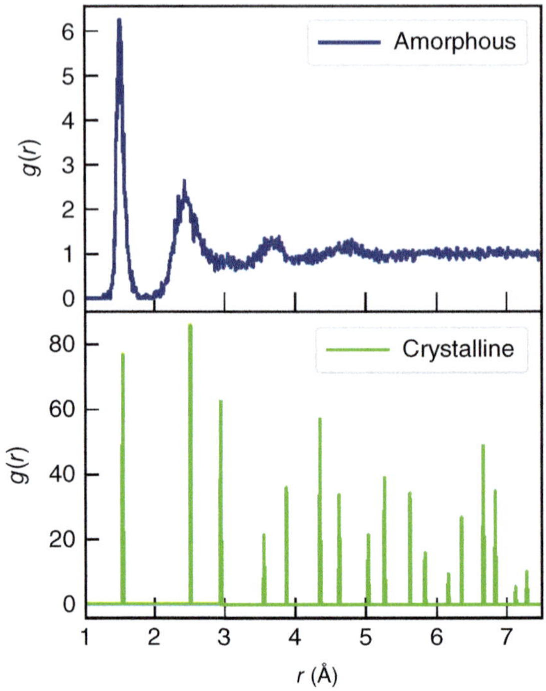


**Figure 2.8** Comparison of RDF of crystalline (diamond) and amorphous Carbon [59] at 3.50 g/cc.

same density. In crystalline materials, if one leaves out positional uncertainty arising from thermal vibrations, which is not a crude approximation at room temperature, then the probability of finding an atom at certain distance from another is precisely known because of the inherent long-range order/periodicity. This certainty is marked by sharp peaks in the RDF at specific distances and vanishing elsewhere as seen in Figure 2.8. However, when a system is amorphous, the sharp peaks are broadened, and there is little order after 4.0 Å in amorphous Carbon as seen in Figure 2.8. A basic point from Figure 2.8 is the identical location of the first peak in both amorphous and crystalline carbon suggesting the presence of a well-defined short-range order, though with modest bond-length and bond-angle distortions [74], in amorphous material and thereby invalidating preconception of a random distribution of atoms in amorphous materials that a first time reader may have in mind.

It is a common practice in the amorphous research community to take the position of the minimum after the first peak as defining coordination in amorphous systems and the area under the first peak gives an estimate of the coordination number of an atom in the sample. The second peak gives information on a longer length scale: in transition metal oxides like a-$Ta_2O_5$, the first peak gives idea of the metal–oxygen bonds, while the second peak provides information about the connection among the metal–oxygen polyhedra [62].

A lot of different definitions of the same or similar symbols and their additive and/or multiplicative constants within the correlation functions often cause confusion. To clear any confusion with the definitions and to compare several different definitions used by different research groups, we suggest the reader to go through the work of David Keen comparing these functions [75]. Also, for readers interested in the reciprocal space counterpart of the RDF, the structure factor $S(Q)$, we suggest going through Refs. [51, 73, 75].

#### 2.3.1.2 Voronoi Analysis

Computer models represent a set of particles (atoms, molecules) in space and hence their properties are governed by the ordering of such particles. Providing a compact but full characterization of the structure of materials is still a challenge. Voronoi analysis is a step forward in that direction. It determines the local structure of a particle considering the arrangement of the neighbors around that central particle [76]. For a fixed set of particles, a Voronoi cell of the central particle is the region in space closer to that particle than any other. This 3D geometry is characterized by a set of indices, called Voronoi indices, written in the form $\langle n_3\ n_4\ n_5\ n_6 \rangle$, where $n_i$ denotes the number of faces with $i$ edges present in the Voronoi cell. A perfect icoshedra has $\langle 0\ 0\ 12\ 0 \rangle$ Voronoi indices. The analysis of Voronoi indices often provide useful information on the Voronoi cell dynamics which can be used to study glass-forming ability and glass transition in glasses.

Voronoi analysis is applicable to a wide range of materials ranging from defected crystals and deformed polycrystals to glasses and liquids [76]. It can be used to study how the crystalline structure is compromised at higher temperatures [76]. In a bulk metallic glass ($Cu_{46}\ Zr_{46}\ Al_8$), a class of amorphous materials, it has been used to study the evolution of certain Voronoi indices near the glass transition temperature which has slow dynamics believed to be responsible for the excellent glass-forming ability of the material [77].

### 2.3.2 Electronic Structure

Amorphous materials have widespread electronic applications ranging from solar cells to light sensors and optical memory devices to CBRAM. Electronic structure not only serves to validate (or invalidate) the model but also can be used as *a priori* information to produce materials with desired electronic properties [68–70] as discussed in Section 2.2.4. The change in electronic structure with doping is important, as it renders materials with potentially important applications.

#### 2.3.2.1 Electronic Density of States

The lack of periodicity and long-range order in amorphous materials rules out the possibility of describing electrons as Bloch states as in crystals. However, the DoS is a equally valid description of the electron states in both crystals and amorphous materials and is defined as follows:

$$g(E) = \frac{1}{V}\sum_{n}\delta\left(E - E_n\right) \tag{2.7}$$

where $g(E)$ is the DoS per unit volume per unit energy interval and $V$ is the volume of the system [78]. The electronic DoS in amorphous materials is qualitatively similar to that of its crystal counterpart except that that sharp edges expected in crystals are smoothed out due to disorder into "band tails" [65, 74].

In analogy to crystals, amorphous materials can be classified as insulators, semiconductors, and conductors based on the nature of the DoS and the spatial extent of the wavefunctions. Amorphous semiconductors like a-Si possess conduction and valence band and a clear band gap, whereas for bulk metallic glass like $Pd_{40}Ni_{40}P_{20}$ [60], which is conducting, there is no band gap. Alloying can sometimes transform an insulating host conducting by creating states in the band gap of the insulating host. For example, Cu-doping in $Ta_2O_5$ to be conducting by generating states in the band gap arising mostly from the Cu-atoms as shown in Figure 2.9. The partial DoS show contribution of individual atomic species to the total DoS. It is only the states near the Fermi level that play a significant role in determining conduction mechanisms.

Apart from the information about whether the material is conducting or not, DoS also gives valuable information about the presence of defects and impurities. For example, the localized states in the band gap of a-Si is a consequence of the intrinsic threefold and fivefold defects, often known as dangling and floating bonds, present in the model [67, 79].

Electronic band gap evaluated with local density approximation (LDA), generalized gradient approximation (GGA), and PBE functionals never agree closely to the experimental value since DFT is a ground state theory. One can overcome this shortcoming by using hybrid functional like Hyed–Scuseria–Ernzerhof (HSE06) [37, 80] that produce band gaps close to the experimental values [81].

### 2.3.2.2 Inverse Participation Ratio

Another quantity of importance is the IPR [82] which indicates whether the electron states are extended (distributed throughout the sample) or concentrated/localized at some atoms. Due to the inherent disorder present in amorphous materials, electronic states can be localized in a certain region or a certain site. If one has the contributions to the electronic states (or eigenvectors) from different atomic orbitals, the IPR for a state ($\psi_n$) can be calculated as follows:

$$\mathcal{I}(\psi_n) = \frac{\sum_{orb} |a_n^{orb}|^4}{\left(\sum_{orb} |a_n^{orb}|^2\right)^2} \tag{2.8}$$

where $a_n^{orb}$ is the contribution to $\psi_n$ from a particular local orbital and the sum runs over the atomic orbitals of all the atoms present. The mathematical form implies that localized states have high IPR value. For example, if $\psi_n$ has a contribution arising only from a particular orbital of a particular atom, the IPR would be unity. On the other hand, a completely extended state produces a value of $(1/N)$, i.e. evenly distributed over $N$ atoms. Localized states (near the Fermi level) yield zero d.c. conductivity, while extended states provide a finite conductivity at 0 K. In other words, the only way for conduction between localized states is by means of thermally assisted "hopping" [78]. IPR is important but qualitative and depends, for example on the basis set employed. An example of extended states in Cu-doped tantala [47] is shown in Figure 2.9.

### 2.3.2.3 Space Projected Conductivity

Conduction in amorphous materials depends mainly on two quantities: the DoS providing information about the species contributing near the Fermi-level and the localization of

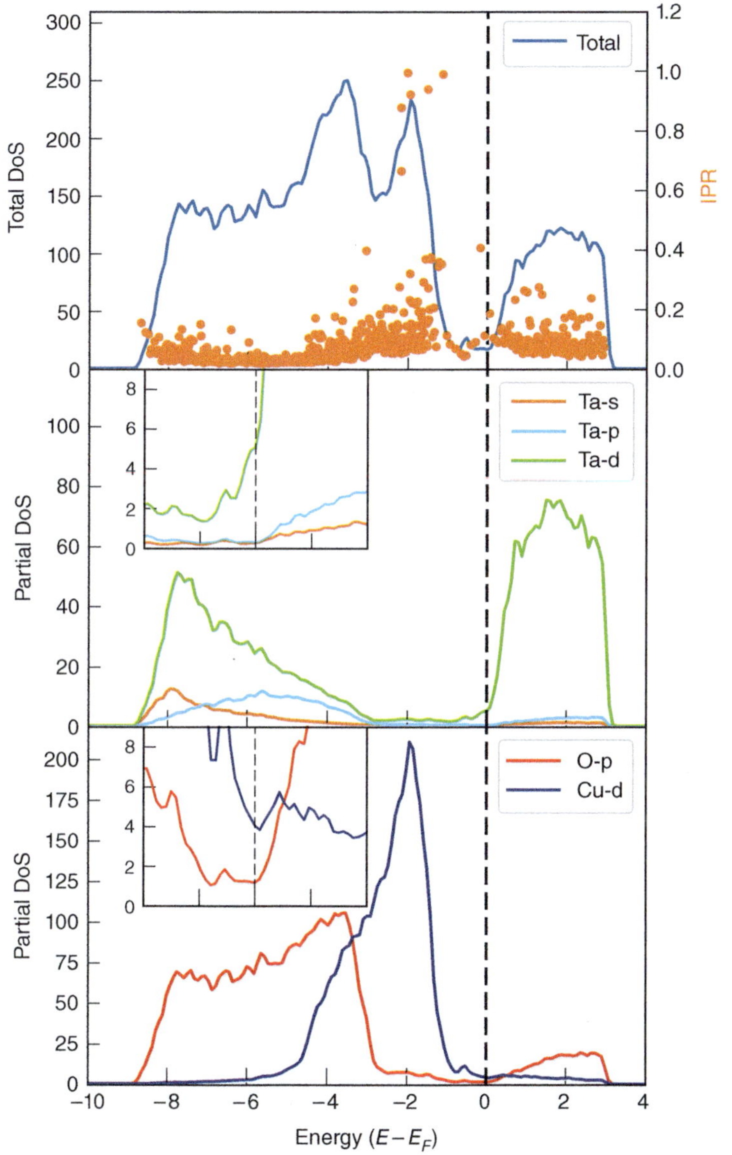


**Figure 2.9** Electronic DoS and inverse participation ratio (IPR) of Cu-doped $Ta_2O_5$ indicating conducting behavior due to the states created in the gap by the added Cu atoms [47].

the electronic states and momentum matrix elements between Kohn–Sham states near the Fermi level. Recently, a spatial decomposition of the Kubo–Greenwood [83, 84] developed by our group provides information about conducting paths in real space by projecting the conductivity onto real space grids. By introducing a discrete grid in space, we show that the quantity:

$$\zeta(\mathbf{x}) = \left| \sum_{\mathbf{x}'} \Gamma(\mathbf{x}, \mathbf{x}') \right| \tag{2.9}$$

provides such information at the spatial grid point **x** and for which

$$\Gamma(\mathbf{x}, \mathbf{x}') = \sum_{ij\alpha} g_{ij} \xi_{ij}^{\alpha}(\mathbf{x}) (\xi_{ij}^{\alpha}(\mathbf{x}'))^{*}. \tag{2.10}$$

Here, $g_{ij}$ is defined in Prasai et al. [85] and $\xi_{ij}^{\alpha}(\mathbf{x}) \equiv \psi_i^{*}(\mathbf{x}) p^{\alpha} \psi_j(\mathbf{x})$ is a complex-valued function, $\psi_i(\mathbf{x})$ is the $i$th Kohn–Sham eigenfunction and $p^{\alpha} = \frac{\hbar}{i} \frac{\partial}{\partial x_{\alpha}}, (\alpha = x, y, z)$. We have used this approach to describe transport in a solid electrolyte material [85], Cu-doped *a-alumina* [86], and most recently on other materials [87]. For a more detailed discussion on space projected conductivity (SPC), we suggest the reader to go through Ref. [87].

As an example of this technique, we have shown the SPC calculations of Cu-doped tantala models discussed in Section 2.2.1 in Figure 2.10. The grayscale shows that interconnected Cu-atoms are form the most active conduction part in the otherwise insulating tantala host.

### 2.3.3 Vibrational Properties

Vibrations of atoms in solids, both crystalline and amorphous, form an important part of analysis because it is directly related to transport and thermal properties. The vibrational behavior of amorphous solids are similar to their crystalline counterpart except for the

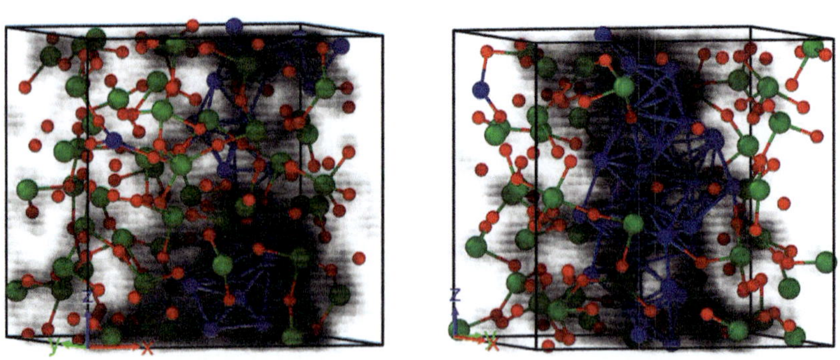

(a) (b)

**Figure 2.10** Space projected conductivity scalar field for Model I (a) and Model II (b) of Cu-doped tantala models [47] shown in grayscale. Those parts of the network containing the interlinked Cu-atoms are clearly more conducting in both models. The electrical conductivity of Model II is about five times that of Model I because of the more robust Cu filament of Model II.

fact that the lack of periodicity smears out the sharp features (Van-Hove singularities) in amorphous materials [88].

Vibrational density of states (VDoS) is an observable that can be measured experimentally most directly using inelastic neutron scattering. Most often, the vibrational properties are calculated using the harmonic approximation and small finite displacements [88]. VDoS can also be calculated from the velocity auto correlation function that uses the trajectory given by a thermal MD simulation [25]. Details of the theory behind calculation of the normal modes can be found elsewhere [88, 89]. Standard MD codes such as VASP and LAMMPS [90] offer such computations. The quantity of importance is the VDoS defined as follows:

$$g(\omega) = \frac{1}{3N}\sum_{i=1}^{3N}\delta\left(\omega - \omega_i\right) \tag{2.11}$$

where $N$ is the number of atoms and $\omega_i$ is the $3N$ eigenfrequencies. Thus, evaluating the VDoS reduces to a problem finding the $\omega_i$'s for the system. The calculation of these eigenfrequencies are a routine task in most simulation packages including VASP [71].[5] In physical terms, the quantity $g(\omega)d\omega$ represents the number of modes within the interval $\left[\omega, \omega + d\omega\right]$. The delta function on the right-hand side of Eq. (2.11) is approximated by a Gaussian with a suitable width.[6]

In multiatomic systems, the contribution to the VDoS from each atomic species, called partial VDoS, can be calculated as follows:

$$g_\alpha(\omega) = \frac{1}{3N}\sum_{i=1}^{N_\alpha}\sum_{n}|e_i^n|^2\delta\left(\omega - \omega_n\right) \tag{2.12}$$

where, $e_i^n$ are the eigenvectors of the normal modes, $N_\alpha$ is the number of atom of species $\alpha$. The partial VDoS must satisfy $g(\omega) = \sum_\alpha g_\alpha(\omega)$, a relation that can be used to check if the partial VDoS is correctly calculated. The general trend in the partial VDoS is that lighter atoms contribute toward higher-frequency vibrations while heavier atoms contribute toward lower-frequency vibrations.

Presence of disorder in the arrangement of atoms in amorphous materials can often lead to localization of vibrational modes with some analogy to electron states. While the VDoS is an experimental observable, the localization of the vibrational modes are not easily observable in experiments. The extent of localization of vibrational modes in computer models can be measured with a quantity called vibrational inverse participation ratio (VIPR) defined as follows:

$$v\left(\omega_n\right) = \frac{\sum_{i=1}^{N}|\mathbf{u}_n^i|^4}{\left(\sum_{i=1}^{N}|\mathbf{u}_n^i|^2\right)^2} \tag{2.13}$$

5 Note for VASP users: Relax the model with zero pressure relaxation before the vibration calculations. Often times, problems arise when there are more than three imaginary eigenfrequencies in VASP output after the vibrational calculation. To get rid of this problem, I recommend relaxing the system to a better energy and force convergence (lower EDIFF and EDIFFG) and rerunning the vibrational calculation.

6 $\sigma \approx 1$–$2\%$ of the maximum eigenfrequency

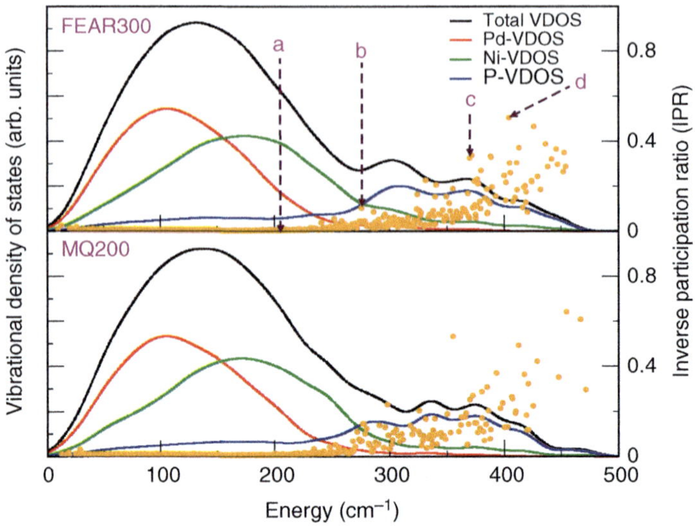


**Figure 2.11** Vibrational DoS and VIPR of two different models of $Pd_{40}Ni_{40}P_{20}$ bulk metallic glass [60].

where $(\mathbf{u}_n^i)$ is the displacement vector of the $i$th atom in the normal mode frequency $\omega_n$. By definition, if a particular normal mode is completely localized on a single atom, the VIPR would be unity, while on the other hand vibrational modes that are extended over large number of atoms have smaller VIPR value. Often times in literature, researchers use the term "participation ratio" defined as the inverse of VIPR.

A representative example for the VDoS and VIPR for a bulk metallic glass $(Pd_{40}Ni_{40}P_{20})$ has been shown in Figure 2.11. The figure shows localization of high-frequency normal modes that are localized on the lightest atoms, i.e. P-atoms.

## 2.4 Conclusion

We have outlined some of the aspects of ab initio simulations of disordered systems. We explained the similarities and differences of ab initio and classical methods. We intentionally avoided the mathematical intricacy of a first principles approach as this is well treated elsewhere. We described methods for creating computer models of materials and emphasized the differences between direct "melt quench" and information-based approaches. We discussed FEAR in some detail. We also pointed out the possibility of exploring models with desired optical/electronic properties in the Gap Sculpting section. We then discussed some common methods for characterizing models and/or comparing to experiments including structural, electronic, and vibrational signatures of models. In the spirit of analyzing models, in systems for which electrical conductivity is relevant, we described a recently developed method to project conductivity onto real space grids to determine those parts of a material that may provide conduction.

## Acknowledgments

We thank the National Science Foundation (NSF) for support under DMR award 1507670. We also acknowledge Extreme Science and Engineering Discovery Environment (XSEDE), supported by NSF grant number ACI-1548562, for providing computational resources under DMR-190008P. We offer special thanks to Dr. Kiran Prasai for his help with the manuscript. We would also like to thank Dr. Volker Deringer for providing us the figure for the disordered silicon under pressure, Figure 2.1.

## References

1 Bragg, W.H. and Bragg, W.L. (1913). The reflection of X-rays by crystals. *Proc. Phys. Soc. A* 88: 428–438.

2 Wlodawer, A., Minor, W., Dauter, Z., and Jaskolski, M. (2013). Protein crystallography for aspiring crystallographers or how to avoid pitfalls and traps in macromolecular structure determination. *FEBS J.* 280 (22): 5705–5736.

3 Rowlands, R., Zeidler, A., Fischer, H.E., and Salmon, P.S. (2019). Structure of the intermediate phase glasses $GeSe_3$ and $GeSe_4$: the deployment of neutron diffraction with isotope substitution. *Front. Mater.* 6: 1–13.

4 Drabold, D.A. (2009). Topics in the theory of amorphous materials. *Eur. Phys. J. B* 68 (1): 1–21.

5 Holmberg, E. (1941). On the clustering tendencies among the nebulae. II. A study of encounters between laboratory models of stellar systems by a new integration procedure. *Astrophys. J.* 94: 385. Provided by the SAO/NASA Astrophysics Data System.

6 Rahman, A. (1964). Correlations in the motion of atoms in liquid argon. *Phys. Rev.* 136: A405–A411.

7 Allen, M.P. and Tildesley, D.J. (2017). 2e. Oxford University Press. Clarendon Press, Oxford, UK. 640.

8 Sankey, O., Niklewski, D., Drabold, D., and Dow, J. (1990). Molecular-dynamics determination of electronic and vibrational spectra, and equilibrium structures of small Si clusters. *Phys. Rev. B: Condens. Matter* 41: 12750–12759.

9 Prodan, E. and Kohn, W. (2005). Nearsightedness of electronic matter. *Proc. Natl. Acad. Sci. U.S.A.* 102 (33): 11635–11638.

10 Prasai, K., Biswas, P., and Drabold, D.A. (2016). Electrons and phonons in amorphous semiconductors. *Semicond. Sci. Technol.* 31 (7): 073002.

11 Zhang, X. and Drabold, D.A. (2001). Properties of the density matrix from realistic calculations. *Phys. Rev. B* 63: 233109.

12 Harrison, W.A. (1980). *Electronic Structure and the Properties of Solids: The Physics of the Chemical Bond*. San Francisco, CA: Freeman.

13 Matthew, W., Foulkes, C., and Haydock, R. (1989). Tight-binding models and density-functional theory. *Phys. Rev. B* 39: 12520–12536.

14 Sutton, A.P., Finnis, M.W., Pettifor, D.G., and Ohta, Y. (1988). The tight-binding bond model. *J. Phys. C: Solid State Phys.* 21 (1): 35–66.

**15** Carlsson, A.E. (1990). Beyond pair potentials in elemental transition metals and semiconductors. In: *Solid State Physics*, (eds. H. Ehrenreich and D. Turnbull) vol. 43, 1–91. Academic Press.

**16** Sankey, O.F. and Allen, R.E. (1986). Atomic forces from electronic energies via the Hellmann-Feynman theorem, with application to semiconductor (110) surface relaxation. *Phys. Rev. B* 33: 7164–7171.

**17** Car, R. and Parrinello, M. (1985). Unified approach for molecular dynamics and density-functional theory. *Phys. Rev. Lett.* 55: 2471-2474.

**18** Martin, R.M. (2020). *Electronic Structure: Basic Theory and Practical Methods*, 2e. Cambridge University Press.

**19** Soler, J.M., Artacho, E., Gale, J.D. et al. (2002). The SIESTA method for ab initio order-N materials simulation. *J. Phys. Condens. Matter* 14 (11): 2745–2779.

**20** Lewis, J.P., Jelínek, P., Ortega, J. et al. (2011). Advances and applications in the FIREBALL ab initio tight-binding molecular-dynamics formalism. *Phys. Status Solidi B* 248 (9): 1989–2007.

**21** Giannozzi, P., Baroni, S., Bonini, N. et al. (2009). Quantum Espresso: a modular and open-source software project for quantum simulations of materials. *J. Phys. Condens. Matter* 21 (39): 395502. (19pp).

**22** Gonze, X., Amadon, B., Anglade, P.-M. et al. (2009). ABINIT: First-principles approach to material and nanosystem properties. *Comput. Phys. Commun.* 180: 2582–2615.

**23** Hacene, M., Anciaux-Sedrakian, A., Rozanska, X. et al. (2012). Accelerating VASP electronic structure calculations using graphic processing units. *J. Comput. Chem.* 33 (32): 2581–2589.

**24** Hutchinson, M. and Widom, M. (2012). VASP on a GPU: Application to exact-exchange calculations of the stability of elemental boron. *Comput. Phys. Commun.* 183 (7): 1422–1426.

**25** Drabold, D.A. (2000). Molecular dynamics simulation of network glasses. In: *Insulating and Semiconducting Glasses* (ed. P. Boolchand), 607–652. Singapore: World Scientific.

**26** Drabold, D.A. (2004). Recent developments in modeling of amorphous materials. In: *Non-Crystalline Materials for Optoelectronics* (ed. M. Mitkova, M.N. Kozicki, G. Lucovsky, and M. Popescu). Burachest: INOE, 441–466.

**27** Kohn, W. (1999). Nobel lecture: electronic structure of matter—wave functions and density functionals. *Rev. Mod. Phys.* 71: 1253–1266.

**28** Jones, R.O. (2015). Density functional theory: its origins, rise to prominence, and future. *Rev. Mod. Phys.* 87: 897–923.

**29** Hohenberg, P. and Kohn, W. (1964). Inhomogeneous electron gas. *Phys. Rev.* 136: B864–B871.

**30** Kohn, W. and Sham, L.J. (1965). Self-consistent equations including exchange and correlation effects. *Phys. Rev.* 140: A1133–A1138.

**31** Hartree, D.R. (1928). The wave mechanics of an atom with a non-coulomb central field. Part I. Theory and methods. *Proc. Cambridge Philos. Soc.* 24 (1): 89. Provided by the SAO/NASA Astrophysics Data System.

**32** Fock, V. (1930). Approximation method for the solution of quantum mechanical multibody problems. *Z. Phys.* 61 (1–2): 126–148.

**33** Ceperley, D.M. and Alder, B.J. (1980). Ground state of the electron gas by a stochastic method. *Phys. Rev. Lett.* 45: 566–569.

34 Marzari, N. and Vanderbilt, D. (1997). Maximally localized generalized Wannier functions for composite energy bands. *Phys. Rev. B* 56: 12847–12865.

35 Stephan, U. and Drabold, D.A. (1998). Order-*n* projection method for first-principles computations of electronic quantities and Wannier functions. *Phys. Rev. B* 57: 6391–6407.

36 Stephan, U., Martin, R.M., and Drabold, D.A. (2000). Extended-range computation of Wannier-like functions in amorphous semiconductors. *Phys. Rev. B* 62: 6885–6888.

37 Heyd, J., Scuseria, G.E., and Ernzerhof, M. (2003). Hybrid functionals based on a screened Coulomb potential. *J. Chem. Phys.* 118 (18): 8207.

38 Martin, R.M., Reining, L., and Ceperley, D.M. (2016). *Interacting Electrons: Theory and Computational Approaches*. Cambridge University Press.

39 Deringer, V.L., Caro, M.A., and Csányi, G. (2019). Machine learning interatomic potentials as emerging tools for materials science. *Adv. Mater.* 31 (46): 1902765.

40 Butler, K.T., Davies, D.W., Cartwright, H. et al. (2018). Machine learning for molecular and materials science. *Nature* 559 (7715): 547–555.

41 Deringer, V.L., Bernstein, N., Csányi, G. et al. (2021). Origins of structural and electronic transitions in disordered silicon. *Nature* 589: 59–64.

42 Mahmoud, C.B., Anelli, A., Csányi, G., and Ceriotti, M. (2020). Learning the electronic density of states in condensed matter. *Phys. Rev. B* 102: 235130.

43 Stukowski, A. (2009). Visualization and analysis of atomistic simulation data with OVITO–the open visualization tool. *Modell. Simul. Mater. Sci. Eng.* 18 (1): 015012.

44 Zachariasen, W.H. (1932). The atomic arrangement in glass. *J. Am. Chem. Soc.* 54: 3841–3851.

45 Elliott, S.R. (1984). Strucutre. In: *Physics of Amorphous Materials*, (ed. P. Paufler) 53–134. New York: Longman Scientific and Technical.

46 Bernstein, N., Bhattarai, B., Csányi, G. et al. (2019). Quantifying chemical structure and machine-learned atomic energies in amorphous and liquid silicon. *Angew. Chem. Int. Ed.* 58 (21): 7057–7061.

47 Thapa, R., Bhattarai, B., Kozicki, M.N. et al. (2020). Structure and charge transport of amorphous cu doped $Ta_2O_5$: an ab initio study. *Phys. Rev. Mater.* 4: 064603.

48 Kratzer, P. and Neugebauer, J. (2019). The basics of electronic structure theory for periodic systems. *Front. Chem.* 7: 106.

49 Drabold, D.A., Dow, J.D., Fedders, P.A. et al. (1990). Convergence of force calculations for noncrystalline Si. *Phys. Rev. B* 42: 5345–5348.

50 Kaplow, R., Rowe, T.A., and Averbach, B.L. (1968). Atomic arrangement in vitreous selenium. *Phys. Rev.* 168: 1068–1079.

51 McGreevy, R.L. and Pusztai, L. (1988). Reverse Monte Carlo simulation: a new technique for the determination of disordered structures. *Mol. Simul.* 1 (6): 359–367.

52 Soper, A.K. (2005). Partial structure factors from disordered materials diffraction data: an approach using empirical potential structure refinement. *Phys. Rev. B* 72: 104204.

53 Metropolis, N., Rosenbluth, A.W., Rosenbluth, M.N. et al. (1953). Equation of state calculations by fast computing machines. *J. Chem. Phys.* 21 (6): 1087–1092.

54 Habeck, M., Rieping, W., and Nilges, M. (2006). Weighting of experimental evidence in macromolecular structure determination. *Proc. Natl. Acad. Sci. U.S.A.* 103 (6): 1756–1761.

55 Biswas, P., Tafen, D., and Drabold, D. (2005). Experimentally constrained molecular relaxation: the case of glassy $GeSe_2$. *Phys. Rev. B* 71: 02.

**56** Pandey, A., Biswas, P., and Drabold, D.A. (2016). Inversion of diffraction data for amorphous materials. *Sci. Rep.* 6: 33731.

**57** Pandey, A., Biswas, P., and Drabold, D.A. (2015). Force-enhanced atomic refinement: structural modeling with interatomic forces in a reverse Mont Carlo approach applied to amorphous Si and $SiO_2$. *Phys. Rev. B* 92: 155205.

**58** Pandey, A., Biswas, P., Bhattarai, B., and Drabold, D.A. (2016). Realistic inversion of diffraction data for an amorphous solid: the case of amorphous silicon. *Phys. Rev. B* 94: 235208.

**59** Bhattarai, B., Pandey, A., and Drabold, D.A. (2018). Evolution of amorphous carbon across densities: an inferential study. *Carbon* 131: 168–174.

**60** Bhattarai, B., Thapa, R., and Drabold, D.A. (2019). Ab initio inversion of structure and the lattice dynamics of a metallic glass: the case of $PD_40$ $Ni_40$ $P_20$. *Modell. Simul. Mater. Sci. Eng.* 27 (7): 075002.

**61** Zhou, Q., Du, T., Guo, L. et al. (2020). New insights into the structure of sodium silicate glasses by force-enhanced atomic refinement. *J. Non-Cryst. Solids* 536: 10.

**62** Prasai, K., Jiang, J., Mishkin, A. et al. (2019). High precision detection of change in intermediate range order of amorphous zirconia-doped tantala thin films due to annealing. *Phys. Rev. Lett.* 123: 045501.

**63** Hruszkewycz, S.O. (2009). The effect of fluctuation microscopy constraints on reverse Monte Carlo models of metallic glass structure. The Johns Hopkins University. PhD thesis.

**64** Kumar, V., Fujita, T., Konno, K. et al. (2011). Atomic and electronic structure of $Pd_{40}$ $Ni_{40}$ $P_{20}$ bulk metallic glass from ab initio simulations. *Phys. Rev. B* 84: 134204.

**65** Pan, Y., Inam, F., Zhang, M., and Drabold, D.A. (2008). Atomistic origin of urbach tails in amorphous silicon. *Phys. Rev. Lett.* 100: 206403.

**66** Amato, A., Terreni, S., Granata, M. et al. (2020). Observation of a correlation between internal friction and urbach energy in amorphous oxides thin films. *Sci. Rep.* 10: 9.

**67** Igram, D., Bhattarai, B., Biswas, P., and Drabold, D.A. (2018). Large and realistic models of amorphous silicon. *J. Non-Cryst. Solids* 492: 27–32.

**68** Prasai, K., Biswas, P., and Drabold, D. (2015). Sculpting the band gap: a computational approach. *Sci. Rep.* 5: 06.

**69** Prasai, K., Biswas, P., and Drabold, D.A. (2016). Electronically designed amorphous carbon and silicon. *Phys. Status Solidi A* 213 (7): 1653–1660.

**70** Prasai, K., Chen, G., and Drabold, D. (2017). Amorphous to amorphous insulator-metal transition in $GeSe_3$:Ag glasses. *Phys. Rev. Mater.* 1: 03.

**71** Kresse, G. and Furthmuller, J. (1996). Efficient iterative schemes for ab initio total-energy calculations using a plane-wave basis set. *Phys. Rev. B* 54: 11169–11186.

**72** Perdew, J.P., Burke, K., and Ernzerhof, M. (1996). Generalized gradient approximation made simple. *Phys. Rev. Lett.* 77: 3865–3868.

**73** Roux, S.L. and Petkov, V. (2010). *ISAACS* – interactive structure analysis of amorphous and crystalline systems. *J. Appl. Crystallogr.* 43 (1): 181–185.

**74** Drabold, D.A. and Abtew, T. (2006) Defects in amorphous semiconductors: amorphous silicon. In: *Theory of Defect in Semiconductors, Topics in Applied Physics* (ed. D.A. Drabold and S.K. Estreicher), 245–268. Berlin Heidelberg, New York: Springer.

**75** Keen, D.A. (2001). A comparison of various commonly used correlation functions for describing total scattering. *J. Appl. Crystallogr.* 34 (2): 172–177.

**76** Lazar, E.A., Han, J., and Srolovitz, D.J. (2015). Topological framework for local structure analysis in condensed matter. *Proc. Natl. Acad. Sci. U.S.A.* 112 (43): E5769–E5776.

**77** Thapa, R., Subedi, K.N., Bhattarai, B., and Drabold, D.A. (2021) Forced enhanced atomic refinement modeling of the metallic glass $Cu_46$ $Zr_46$ $Al_8$. *Phys. Status Solidi B* 258 2000415.

**78** Elliott, S.R. (1984). Electrons. In: *Physics of Amorphous Materials* (ed. S.R. Elliott), 173–246. New York: Longman Scientific and Technical.

**79** Pantelides, S.T. (1987). Gap states in amorphous silicon-dangling and floating bonds. *J. Non-Cryst. Solids* 97–98: 79–82. Amorphous and liquid semiconductors.

**80** Heyd, J., Scuseria, G.E., and Ernzerhof, M. (2006). Erratum: "hybrid functionals based on a screened coulomb potential" [J. Chem. Phys. 118, 8207 (2003)]. *J. Chem. Phys.* 124 (21): 219906.

**81** Chimata, R. (2010). Optical properties of materials calculated from first principles theory. Uppsala University, Dissertation.

**82** Ziman, J.M. (1979). *Models of Disorder*. Cambridge University Press.

**83** Kubo, R. (1957). Statistical mechanical theory of irreversible processes. I. General theory and simple applications to magnetic and conduction problems. *J. Phys. Soc. Jpn.* 12: 570–586.

**84** Greenwood, D.A. (1958). The Boltzmann equation in the theory of electrical conduction in metals. *Proc. Phys. Soc.* 71: 585–596.

**85** Prasai, K., Subedi, K.N., Ferris, K. et al. (2018). Spatial projection of electronic conductivity: the example of conducting bridge memory materials. *Phys. Status Solidi RRL* 12 (9): 1800238.

**86** Subedi, K.N., Prasai, K., Kozicki, M.N., and Drabold, D.A. (2019). Structural origins of electronic conduction in amorphous copper-doped alumina. *Phys. Rev. Mater.* 3: 065605.

**87** Subedi, K.N., Prasai, K., and Drabold, D.A. (2021) Space-projected conductivity and spectral properties of the conduction matrix. *Phys. Status Solidi B* 258: 2000438.

**88** Elliott, S.R. (1984). Vibrations. In: *Physics of Amorphous Materials*, (ed. P. Paufler) 135, 172. New York: Longman Scientific and Technical.

**89** Bhattarai, B. and Drabold, D.A. (2016). Vibrations in amorphous silica. *J. Non-Cryst. Solids* 439: 6–14.

**90** Plimpton, S. (1995). Fast parallel algorithms for short-range molecular dynamics. *J. Comput. Phys.* 117 (1): 1–19. lammps.sandia.gov.

# 3

# Reverse Monte Carlo Simulations of Noncrystalline Solids

*Shinji Kohara[1] and László Pusztai[2]*

[1] *Light/Quantum Beam Field, Research Center for Advanced Measurement and Characterization, National Institute for Materials Science, Tsukuba, Ibaraki, Japan*
[2] *Wigner Research Centre for Physics Hungarian Academy of Sciences (WRCP HAS), Budapest, Hungary*

## 3.1 Introduction – Why RMC Is Needed?

The absence of translational symmetry and the very much present complexity in the structure of glassy, liquid, and amorphous materials make glass science a challenging field. Diffraction and X-ray absorption experimental techniques are available for providing direct information on the two-body (or "pairwise") correlations between atomic positions. Direct experimental information on higher-order correlations cannot be gained. For this reason, one must turn to computational methods of structural modeling, with the aim of producing large sets of atomic coordinates ("particle configurations") that are consistent with measured diffraction signals within errors. Our preferred method for this purpose is the Reverse Monte Carlo (RMC) method [1] of structural modeling.

From particle configurations, in turn, it is possible to calculate structural quantities of arbitrary detail – with the notion that any such quantity would *not* be a piece of experimental result (but "only" a piece that is consistent with experimental results) and that it would be unique only at the level of two-body correlations. In other words, any detailed description of the local structure related to experimental (diffraction and X-ray absorption) data (via appropriate structural modeling methods) would only be one of the many possibilities that fulfill the condition of being consistent with measured two-particle properties.

It is because of this nonuniqueness *of the measured data* that correlations appearing due to the presence of diffraction data should be clearly identified. This identification can be carried out by generating *reference systems*, also containing many particles, which would correspond only to the definition of the glassy – covalent – topology and to the presence of hard-sphere like interactions. Such reference systems would contain correlations resulting from purely topological (covalent glasses) and/or excluded volume (metallic glasses) effects.

*Atomistic Simulations of Glasses: Fundamentals and Applications*, First Edition.
Edited by Jincheng Du and Alastair N. Cormack.

Only thorough comparison between higher-order correlations characteristic to the reference and diffraction-data-related particle configurations can features be revealed that are genuine attributes of the real system.

In this chapter, we describe a protocol for extracting higher-order correlations between atoms from diffraction data, with an emphasis on the various routes of calculating relevant correlation functions. The process consists of the following stages:

***(a) Experimental (X-ray and/or neutron diffraction, or EXAFS) data***

↓

***(b) 3D configurations of thousands of particles***

*(fully consistent with the above data)*

↓

***(c) Detailed characterization of the atomic level structure***

*(partial radial distribution functions, partial coordination numbers, bond angle distributions, etc.)*

The main part of this chapter is concerned with the route from stage (a) to stage (b) (Section 3.2) and from stage (b) to stage (c) (Section 3.3). Purely experimental issues would exceed the framework of the chapter and therefore, they are not discussed here in any detail. Readers interested in experimental aspects are kindly directed to refs. [2–6].

Before going into further details, it is important to remember that this contribution focuses on the determination of structural features that are related to atomic positions in *disordered* solids (covalent and metallic glasses); for ordered crystalline phases, well-known methods of crystallography are available (see, e.g. refs. [7–10]).

One more thing to note before delving into the subject matter: very fast, we had to realize that, within the framework of the present text, it would be impossible to provide a full coverage of all RMC modeling work on noncrystalline solids. After some hesitation, we decided to focus mostly on the method itself, and on actual materials that have been touched upon during our own carriers. Also, we have tried to pick examples where the amount of input (and "external") information (c.f. Section 2.2), was maximized (i.e. where diffraction data AND information from other kinds of experiments, e.g. EXAFS, NMR, and/or computer simulations, have been considered SIMULTANEOUSLY by RMC modeling). We duly apologize to researchers whose related works have been left out this time.

## 3.2 Reverse Monte Carlo Modeling

RMC was conceived as a simple computational tool for constructing large, 3D structural models that are consistent (within the estimated level of their errors) with the *total scattering structure factors* (*tssf*-s) obtained from diffraction experiments (for a thorough definition of these – and related – functions, see ref. [11]). Via random movements of particles, the difference (calculated similarly to the $\chi^2$-statistics) between experimental and model total scattering structure factors is minimized. From the particle configurations, *partial radial distribution functions* (*prdf*), as well as other structural characteristics (neighbor distributions, cosine distribution of bond angles, see e.g. ref. [12]) can be calculated.

### 3.2.1 The Basic RMC Algorithm

The RMC algorithm may be given as follows (refs. [1, 13–16]):

1. Start with an initial configuration with periodic boundary conditions. The positions of the $N$ (>5000) atoms should be consistent with the topology of the covalent network already at the start. Assume some tolerance for the (covalently) bonded distances. Naturally, the density of the simulated systems should be equal to the experimental density. The excluded volume of atoms is also taken into account, in the form of closest approach ("cutoff") distances.
2. Calculate the partial radial distribution functions for this initial configuration as

$$g_{ij}^{C,o}(r) = \frac{n_{ij}^{C,o}(r)}{4\pi\rho_j r^2 dr} \tag{3.1}$$

   where $n_{ij}^{C,o}$ is the number of atoms of type $j$ at a distance between $r$ and $r+d\,r$ from a central atom of type $i$, averaged over all atoms of type $i$ as centers. The superscripts $C$ and $o$ refer to "calculated" and "old," respectively. $\rho_j$ is the number density of particles type $j$.
3. Compose the total radial distribution function(s) from the partials, weighted according to the concentrations and scattering powers of atoms:

$$G^{C,o}(r) = \sum_i \sum_j c_i c_j \bar{b}_i \bar{b}_j \left( g_{ij}^{C,o}(r) - 1 \right) \tag{3.2}$$

   where $c$'s are molar fractions and $b$'s are the (neutron) scattering lengths of the components. (Note that for X-rays, a more complicated way has to be followed since the X-ray scattering power of atoms depends on the scattering vector; for the sake of simplicity, throughout this contribution the formalism of neutron scattering [11] will be applied.) Transform to the total scattering structure factor (*tssf*) as

$$F^{C,o}(Q) - 1 = \frac{4\pi\rho}{Q} \int_0^\infty rG^{C,o}(r) \sin Qr dr \tag{3.3}$$

   ($\rho$ is the bulk atomic number density). It is worth remembering that in the RMC procedure, transforms from $r$ to $Q$ space occur only, which makes it easier to handle truncation errors: the extent of the available $r$-range can be increased by using larger models.
4. Calculate the difference between the experimental total scattering structure factor, $F^E(Q)$, and that calculated from the configuration, $F^C(Q)$ as

$$\chi_o^2 = \sum_k \frac{\{F^{C,o}(Q_k) - F^E(Q_k)\}^2}{\sigma^2} \tag{3.4}$$

   where the sum is over all the experimental points and $\sigma$ is the – estimated – experimental error, which functions as a "control parameter" for the simulation. By prescribing how close a fit to experimental data is required, $\sigma$ controls the ratio of the numbers of accepted/attempted moves [13, 14].
5. Move one atom at random.
6. Calculate the new radial distribution function, $G^{C,n}(r)$ and total scattering structure factor, $F^{C,n}(Q)$, and the new difference between model and experiment, $\chi^2_n$.

7. If ${\chi^2}_n < {\chi^2}_o$ the move is accepted and the new configuration becomes the old configuration. If ${\chi^2}_n > {\chi^2}_o$ then the move is accepted with a probability of $\exp\left\{-\left(\chi_n^2 - \chi_o^2\right)/2\right\}$.
8. Repeat from Step 5.

In case of reliable data and meaningful input parameters, $\chi^2$ will decrease until it reaches its minimum value that is linked to the given value of $\sigma$. After reaching this minimum, $\chi^2$ will oscillate so that the ratio of the numbers of accepted and attempted moves will not decrease. (For a demonstration of what an acceptable agreement between experimental and model structure factors is, see Figure 3.1 [17, 18].)

During its extensive use, RMC modeling has helped us to clarify a number of issues concerning structural disorder in general. Some of these points may be worth repeating here:

(a) There are always more than one structural model that would fit a given (set of) data – and this has nothing to do with RMC. RMC has just happened to be the technique which started to produce different models for a given set of diffraction data (see, e.g. refs. [19, 20]), but it should be recognized that it is the diffraction data that allow for such a diversity of structures. Therefore, one must explore, as fully as possible, the "configuration space" available for models connected to a (set of) measurement(s).
(b) In practice, there is always more than one (set of partial) radial distribution function(s) corresponding to a given (set of) *tssf*(-s), even when the number of independent experiments equals the number of partials (i.e. when there are three data sets for a two component ($A_xB_y$), or six data sets for a three component ($A_xB_yC_z$) system, etc.). This fact, demonstrated, for instance, in refs. [21–23], is even less appreciated than the one under (a). This is just the consequence of the fact that there are no "perfect data": no data would ever span an infinite scattering vector range (which would be necessary for a perfect Fourier transform) and no data would ever be error-free, either (which would be a prerequisite for a unique separation of the partial *rdf*-s). That is, one has to consider more than one solution in terms of the radial distribution functions, as well. The severely information-deficient case when the number of independent measurements is

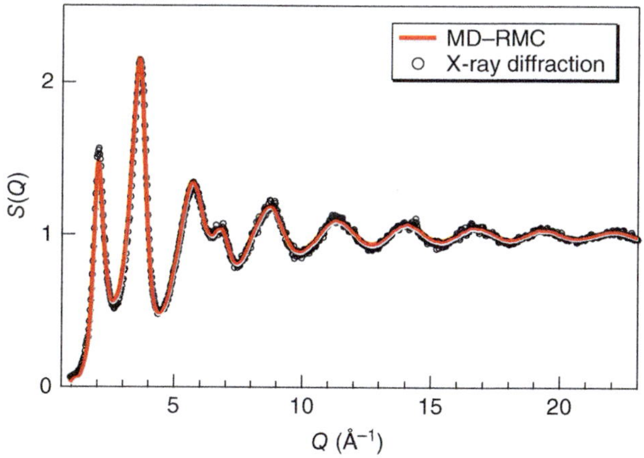


**Figure 3.1** Structure factor, *S*(*Q*), for amorphous silicon obtained by X-ray diffraction measurement (open circle) [17] and MD–RMC modeling (red curve) [18].

less (many times, MUCH less) than the number of *prdf*-s will be dealt with in a separate subsection (see below).

(c) One way of attempting to meet the above requirements is the introduction of geometrical constraints (i.e. ones that can be formulated on the basis of the particle coordinates). These constraints can, for instance, represent specific ideas about the microscopic structure of the material in question. In this way, one will be able to tell if a given idea is consistent with existing measurements or not. That is, RMC can help us to select ideas/models which are (or are not) acceptable, on a very strict basis. This concept for molecular liquids may be elaborated by using, for instance, the RMC++ software [15].

(d) Modeling the total scattering structure factor by RMC will help to spot if there are problems (systematic errors) with the *tssf* in question (a clear demonstration of this feature can be seen in, for instance, ref. [23]). Experience shows that if sensibly executed RMC runs do not lead to a satisfactory agreement with *tssf*-s, then it is quite likely that it is the experimental results that should be reconsidered (or at least, reanalyzed; a more theoretical approach to this problem can be found in ref. [24]).

(e) As (at least, partly) a consequence of (b), it should always be the total scattering structure factor and not the (partial) radial distribution function(s) that are applied as input data for RMC. First of all, *tssf*-s constitute the primary experimental information, whereas (p)*rdf*-s will always contain errors related to the direct Fourier transform from reciprocal to real space. Moreover, a given (set of partial) *rdf*(-s) has to be considered as only one (out of the many possible) interpretation of the *tssf*(-s). Thus, modeling only one (set of partial) *rdf*(-s) will certainly be insufficient for a proper coverage of possible structures that are consistent with the experimental data in question.

### 3.2.2 Information Deficiency

Although points (a) and (b) above have already touched upon the issue, this short sub-section has received a separate heading due to its extraordinary importance for glasses, particularly for covalent glasses.

Materials that are important in practice contain almost always more than two kinds of atoms – many times, more than three. Under such circumstances, it is impossible in practice to perform as many scattering and EXAFS experiments that would equal the number of partial radial distribution functions in a system (6 for a three-component, 10 for a four-component, 15 for a five-component glass, etc.).

It is therefore a *must* that at least the largest possible number of experiments for a given multicomponent material should be performed so that the information deficiency is minimized. In addition, every single piece of available external information should be included in the process of RMC structural modeling, including high-level (classical and quantum) computer simulations. The RMC algorithm does allow the use of external information, in the form of constraints/restraints (see, e.g. [14–16]).

For the above reason, in Section 3.4, only examples where the problem of information deficiency is appropriately handled will be described (and even mentioned) – even though quite a few RMC studies have appeared over the past three decades where this has not been the case. For instance, there are a number of investigations (by unnamed authors…) where one single structure factor was the basis for showing all *prdf*-s of a many-component covalent glass; this practice should be halted immediately.

### 3.2.3 Preparation of Reference Structures: Hard Sphere Monte Carlo

Hard sphere Monte Carlo (MC) calculations, with the same parameters (density; bond-length and coordination constraints; cutoffs; etc.) as the RMC runs, are very useful additions to a RMC-based study. The easiest way for performing them is running RMC without fitting experimental data. These calculations provide reference structures which possess all features that originate to excluded volume (pure size) effects. For covalent glasses, the desired topology may be prepared via hard-sphere MC runs with coordination and angular constraints. Differences between hard sphere and RMC structures are characteristic to the nature of intermolecular interactions. From a practical point, hard-sphere MC calculations with all sensible constraints can provide reasonable initial configurations for RMC.

This – rather simple – way of interpreting diffraction results and corresponding RMC models has proved to be remarkably useful for covalent [20, 25, 26], as well as metallic [27–30] glasses. For molecular liquids, hard-sphere MC simulations are absolutely essential for a thorough analysis of orientational correlations, as exemplified by, e.g. refs. [31–33].

### 3.2.4 Other Methods for Preparing Suitable Structural Models

Empirical potential structure refinement (EPSR) [34–37] is, similarly to RMC modeling, an inverse method for producing structures that are consistent with experimental data. In EPSR, an initial (set of) pairwise interatomic potential(s) is(/are) iterated by a scheme that involves standard (Metropolis) MC [38] simulation. The EPSR process is successful if after the final MC stage, the structural quantity of choice (in many cases, some form of partial structure factor, see, e.g. ref. [39], instead of the *tssf*) agrees with the corresponding experimental results. The quality of EPSR fit is frequently comparable to that of RMC fits (see, e.g. ref. [40]) and in such cases, the EPSR structure represents also reasonable energetic relations; this is not guaranteed for RMC structures. However, in quite a few instances (see, e.g. refs. [41, 42]), there are sizable deviations between EPSR and diffraction results. Apart from the personal involvement in developing RMC, this – somewhat unpredictable – behavior of the EPSR method is the reason why the simpler RMC technique is discussed more in detail in this chapter.

Standard techniques of computer simulation, like MC and molecular dynamics (MD) may also be appropriate ways of connecting diffraction results and detailed structural information (for a detailed account on computer simulations, see ref. [38]). The main requirement is that total scattering structure factors calculated from MC or MD particle configurations should be consistent with experimental *tssf*-s at the same level as RMC based functions. This condition cannot frequently be met, although in some cases, it may be demonstrated that MD and RMC configurations are in semiquantitative agreement [43, 44]. In such cases, MD configurations may be exploited for calculating characteristics of the local structure. (Note that diffraction experiments still have to be performed, as consistency with them can only be checked this way.)

## 3.3 Topological Analyses

In this section, we present three important novel topological analysis tools for glassy materials: ring statistics, cavity, and persistent homology analyses. Naturally, well-established simpler methods, like calculating distributions of bond angles, may still be widely applied (for proper usage of bond angle distributions, see, e.g. [20–31]).

### 3.3.1 Ring Statistics

The history of ring statistics calculations is relatively long. The most widely used computer code was developed by Roux and Jund [45, 46]. There is a variety of criteria for a ring: King [47], shortest path [48, 49], primitive [50, 51] (or irreducible [52]), and strong rings [50, 51]. Details are described in refs. [45, 46]. Figure 3.2a–d show primitive ring statics in $SiO_2$ crystals, α-cristobalite ($d$ = 2.327 g/cm$^3$), $\alpha$-quartz ($d$ = 2.655 g/cm$^3$), and coesite ($d$ = 2.905 g/cm$^3$), together with that of glassy (g)-$SiO_2$ ($d$ = 2.2 g/cm$^3$) obtained from RMC–MD modeling [18]. α-cristobalite shows only sixfold rings, consisting of six $SiO_4$ tetrahedra, while α-quartz has a large fraction of eightfold rings in addition to sixfold rings. On the other hand, both coesite and g-$SiO_2$ exhibit a distribution of different size rings, which is a sign of topological disorder according to Gupta and Cooper [53].

### 3.3.2 Cavity Analysis

The distribution of cavities in glassy materials is an important property for understanding physicochemical properties. Although positron annihilation lifetime spectroscopy (PALS) has been employed to perform a direct observation of the behavior of cavities [54], calculation of the distribution of cavities on the basis of atomic configurations obtained by computer modeling becomes important. Heimbach et al. have recently developed a dedicated code for cavity analysis, in which it is possible to calculate cavity distributions [55]. Figure 3.3 compares the modification of cavity distributions in alkali silicate glasses [56]. It is to be noted that the absolute value of cavity volume depends on the definition of minimum size of a cavity, which makes it difficult to compare with results from PALS.

### 3.3.3 Persistent Homology Analysis

Recently, topological data analysis has rapidly progressed and has provided several tools for analyzing multiscale data in physical and biological fields [57]. Following the landmark study of Hirata et al. [58], Hiraoka et al. applied persistent homology for disordered materials to understand the homology of rings, which feature cannot be detected by conventional ring statistics analysis [57]. This mathematical tool, based on the persistence diagram (PD), was developed to capture shapes of multiscale data. The input to the PDs is given by atomic configurations and the output is expressed as 2D histograms. Then, specific distributions such as curves and islands in the PDs identify meaningful shape characteristics of the given atomic configuration.

The homology of atomic configurations can be investigated by the PDs obtained by using the HomCloud software package [59]. Given a set of points in space, persistent homology captures its topological multiscale structures, and the structures identified are compactly expressed in the PD. The construction of the PD follows the process schematically depicted in Figure 3.4a. We first replace each point by a sphere and increase the radius from zero to a sufficiently large value, which corresponds to the changing resolution of input $x$, $y$, $z$ coordinates of atoms. Then, we record the pair of radii ($b$, $d$) at which a ring in a specific

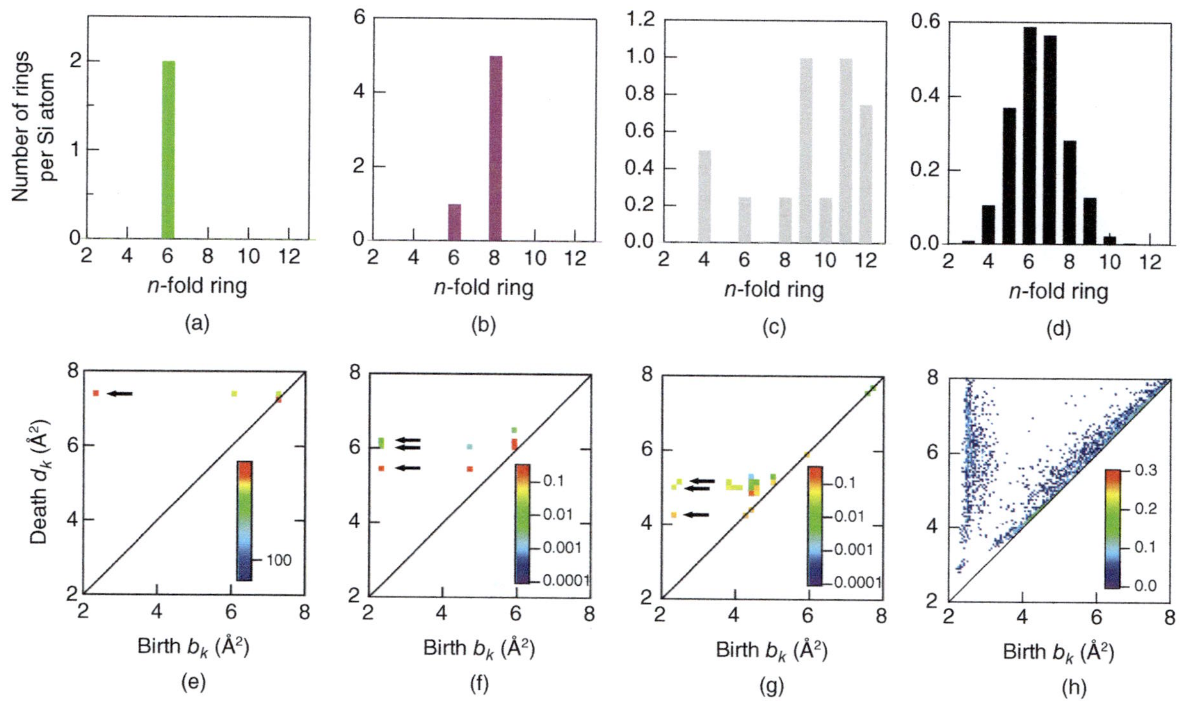


**Figure 3.2** Primitive ring statistics for (a) α-cristobalite, (b) α-quartz, (c) coesite, and (d) *g*-$SiO_2$ [18]. The Si-centric PDs for (e) α-cristobalite, (f) α-quartz, (g) coesite, and (h) *g*-$SiO_2$ [18].

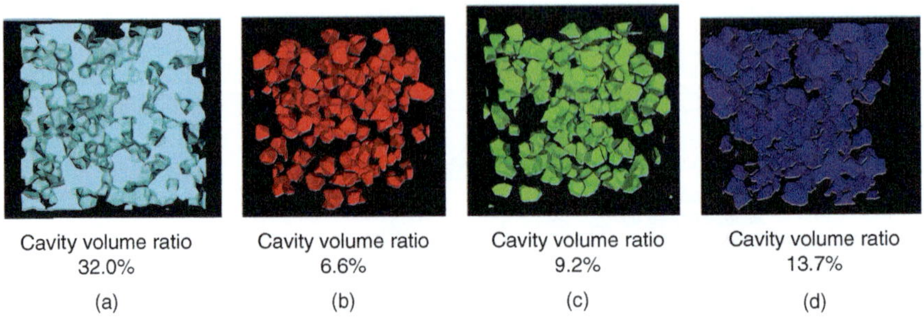


**Figure 3.3** Visualization of surface cavities. (a) $SiO_2$ glass, (b) Na100 glass, (c) Na50K50 glass, (d) K100 glass. Source: Onodera et al. [56].

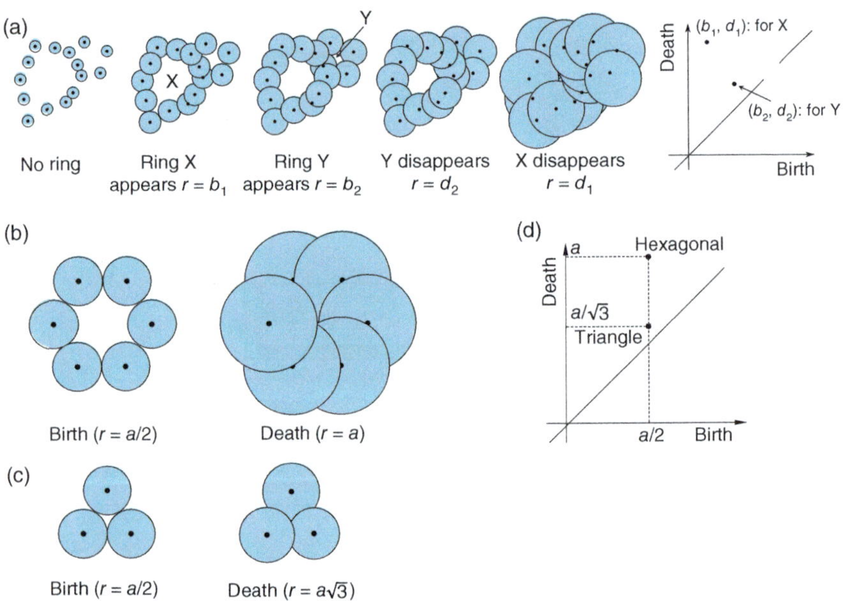


**Figure 3.4** Persistent homology and PD [18]. (a) The increasing sequence of spheres for input data (left). The PD (right) is obtained as a histogram counting the number of rings on the birth-death plane. (b and c) The appearance and disappearance of a ring for a regular hexagon/triangle. (d) The pairs of birth and death radii for hexagon and triangle in the one-dimensional PD.

location appears (birth) and disappears (death), respectively. The PD is a histogram of the birth/death plane with counts of rings at the coordinate $(b, d)$. This construction enables one not only to count the number of rings but also to characterize their shapes at multiscale, by observing characteristic shapes of the diagrams themselves.

Typical examples of birth/death pairs for typical regular structures are shown in Figure 3.4b–d. For a regular hexagonal arrangement of points, in which the distance between points is $a$, the ring appears at radius $a/2$ and disappears at radius $a$, as shown in Figures 3.4b. For a regular triangular configuration, the ring appears at $a/2$ and disappears at $\sqrt{1/3}a \approx 0.577a$, as shown in Figure 3.4c. The one-dimensional PD for regular hexagonal/triangular points is shown in Figure 3.4d. In this chapter, PDs are being used for investigating rings and polyhedral formations in atomic configurations. We also note that the detected rings are recorded during the computation of the diagrams, and hence we can explicitly identify their geometric shapes.

Figure 3.2e–h show the Si-centric PDs calculated from the crystal structures for α-cristobalite, α-quartz, and coesite, together with that of g-$SiO_2$ obtained from RMC–MD modeling [18]. A systematic change in the Si-centric PDs with density is observed for the crystalline phases in Figure 3.2e–g. In contrast to the crystalline phases, the PD for g-$SiO_2$ has a vertical profile along the death axis at $b_k \sim 2.2$ Å$^2$, which is thought to be a signature of the formation of a –Si–O–Si–O– glass network [57]. The comparison between Si-centric PDs and primitive ring size distributions shown in Figure 3.2 provides us with comprehensive topological information, because ring size distribution analysis is sensitive to the ring size, whereas PDs make it possible to reveal the shape of rings. Since all forms of silica have corner-sharing $SiO_4$ tetrahedral motifs, a comparison of α-cristobalite, α-quartz, and coesite with g-$SiO_2$ in the Si-centric PDs suggests that the glass has not only the homology with a crystalline phase with comparable density (α-cristobalite) but also the homology with higher-density crystalline phases (α-quartz and coesite). It is known that g-$SiO_2$ has a distribution of ring sizes (topological disorder, [53] see Figure 3.2d) [60, 61], but Figure 3.2a–c indicate that crystalline phases exhibit topological disorder with increasing density, because ring size distributions become broad with increasing density. Moreover, both α-quartz and coesite have large rings in spite of small death values. It is concluded from a combination of PD analysis and conventional ring statistical analysis that the vertical profile along the death axis observed in the Si-centric PD for g-$SiO_2$ is the result of disorder, because the small death value in the glass implies that the arrangement of $SiO_4$ tetrahedra is locally more densely packed in the glass than in α-cristobalite, whose density is comparable to g-$SiO_2$.

## 3.4 Applications

### 3.4.1 Single Component Liquid and Amorphous Materials: Amorphous (A)- and Liquid (L-) Silicon (Si), and L-Phosphorous (P) Under High Pressure and High Temperature

#### 3.4.1.1 *l*-Si and *a*-Si

Amorphous silicon (*a*-Si) was one of the first materials that has been extensively studied by the RMC method (see, e.g. [20, 25, 30]). Some characteristics of the – assumed and expected – tetrahedral network structure, such as the large number of tetrahedral bond angles, have been found even in "unconstrained" RMC models [20]. However, perhaps the most spectacular early finding was that it had been possible to reproduce experimental diffraction data by a structure that did not contain one single fourfold coordinated Si atom (although tetrahedral angles have appeared in large numbers) [20]. Accordingly, it could be shown convincingly that, in some cases (such as of *a*-Si), even very different structural models are consistent with diffraction data [20, 25], which was an important conclusion from these early investigations.

Here, we intend to show some findings from a rather recent RMC study of amorphous silicon. Figure 3.1 shows the structure factor, $S(Q)$, of *a*-Si obtained by X-ray diffraction [17], together with results of MD–RMC modeling [18]. During the modeling, fourfold silicon and the width of Si—Si—Si bond angles were constrained to the original MD data, which provide good agreement with the measurement. It is well known that *a*-Si comprises a full tetrahedral network. The comparison of X-ray $S(Q)$ between *a*-Si and *l*-Si is shown in Figure 3.5a. The two prominent peaks of *a*-Si, observed at $Q = 2.0\ \text{Å}^{-1}$ and $Q = 3.6\ \text{Å}^{-1}$, merge into a single broad peak in the liquid. This change may be associated with the increase of the coordination number to 5.7, which results in a higher density in the liquid. This behavior is well understood in the Si-centric PDs shown in Figure 3.5b,c. The PD for *a*-Si (Figure 3.5b) exhibits a vertical profile along the death axis, which is the signature of a tetrahedral network. On the other hand, the PD profile for *l*-Si, obtained from RMC modeling, is broad with a short death value that suggests a densely packed structure [18].

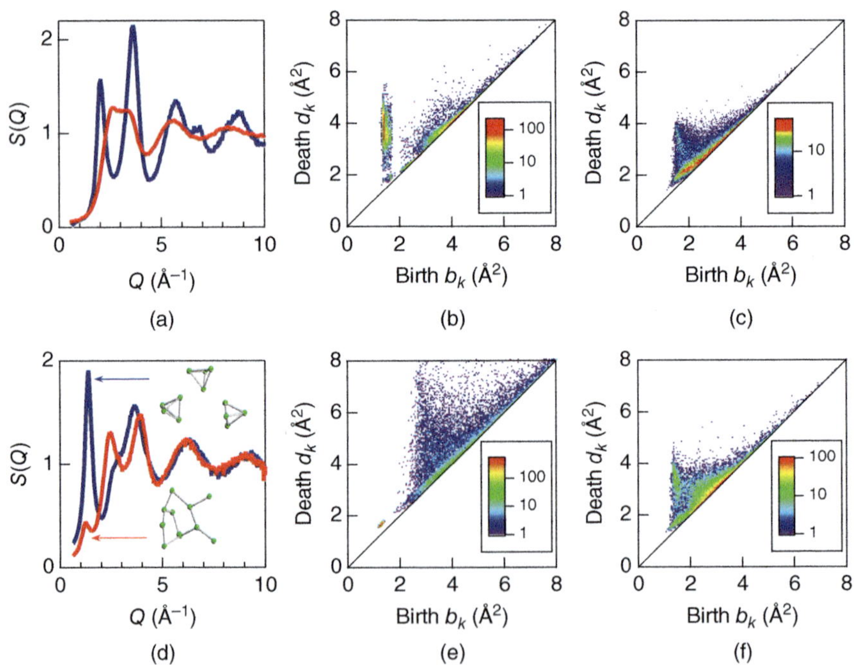


**Figure 3.5** (a) X-ray $S(Q)$ for *a*-Si (blue curve) [17] and *l*-Si (red curve, 1770 K) [18], Si-centric PDs for (b) *a*-Si [17] and (c) *l*-Si (1770 K) [18]. (d) X-ray $S(Q)$ for *l*-P at 0.96 GPa/1328 K (blue curve, low-density molecular liquid) and 1.01 GPa/1323 K (red curve, high-density network liquid) [62]. P-centric PDs for *l*-P at (e) 0.96 GPa/1328 K (low-density molecular liquid) and at (f) 1.01 GPa/1323 K (high-density network liquid) [18].

#### 3.4.1.2 *l*-P Under High Pressure and High Temperature

The X-ray $S(Q)$ for *l*-P under high pressures and high temperatures [62] are shown in Figure 3.5d. A first-order phase transition is observed between a low-density molecular liquid (0.96 GPa/1328 K) and a high-density network liquid (1.01 GPa/1323 K). The low-density molecular liquid data (blue curve) exhibits a first sharp diffraction peak (FSDP) at $Q = 1.35\,\text{Å}^{-1}$ in the X-ray $S(Q)$ that diminishes in the high-density network data (red curve). On the contrary, a very sharp principal peak (PP) is observed at $Q = 2.4\,\text{Å}^{-1}$, reflecting the increased packing fraction of phosphorous atoms [63] in the high-pressure form. This behavior is obviously reflected in the P-centric PDs (Figure 3.5a, b) [18] in which the profile of the low-pressure form is indistinct, while that of the high-pressure form reflects a densely packed structure associated with the diminishing FSDP. Moreover, we can see an excellent contrast in PDs between network formed by tetrahedra in *a*-Si (Figure 3.5b) and isolated tetrahedra in *l*-P (Figure 3.5e).

### 3.4.2 Oxide Glasses

#### 3.4.2.1 $SiO_2$ Glass

Silica ($SiO_2$) can be classified as glass-former according to Zachariasen [64] and Sun [65], and hence g-$SiO_2$ is one of the most important glassy materials in both fundamental and engineering sciences. The structure of g-$SiO_2$ has been studied widely by both experimental and theoretical approaches [66].

Figure 3.6 shows neutron-weighted, $S^N(Q)$ [67], and X-ray-weighted, $S^X(Q)$ [18], total structure factors of g-$SiO_2$ together with the total and partial structure factors, $S_{ij}(Q)$ obtained by MD–RMC modeling [18]. A three-peak ($Q_1$ [FSDP], $Q_2$ [PP], and $Q_3$) structure [68] is observed in the $S^N(Q)$, but a PP cannot be observed in the $S^X(Q)$ due to oxygen–oxygen correlation's low weighting factor for X-rays.

It is well known that the short-range structure in g-$SiO_2$ is a $SiO_4$ tetrahedron, and the interconnections of tetrahedra form a network with shared oxygen atoms at the corner. This polyhedral motif is manifested by the FSDP and the PP in the $S(Q)$ [69]. The FSDP was first discussed in 1976 [70], although it appears that the name "FSDP" was first used by Phillips in 1981 [71]. An interpretation of diffraction peaks, including the FSDP, was attempted in the 1980s [71, 72] and details are discussed in several papers [73–77]. It is known that the FSDP of silica glass is related to the formation of a random network, suggested by Zachariasen [64], and the model was extended to silicate glasses as illustrated in Figure 3.7 of Ref. [77] by Mei et al. It was confirmed that intermediate-range ordering (IRO) arises from the periodicity of boundaries between successive small cages in the network, formed by connected, regular $SiO_4$ tetrahedra with shared oxygen atoms at the corners. The IRO is thus associated with the formation of a ring structure and cavities [78]. The second maximum, the PP, reflects the size of the local-network-forming motif, whereas the FSDP indicates the arrangement of these motifs on an intermediate range, according to Zeidler and Salmon [68]. Another interpretation of the FSDP was recently proposed by Shi and Tanaka, who discussed local tetrahedral ordering in covalent liquids and glasses [79]. Since the origin of the FSDP has been discussed for a long time as mentioned above, it is possible to estimate the periodicity and correlation length given by $2\pi/Q_{FSDP}$ of ~4.15 Å and $2\pi/\Delta Q_{FSDP}$ of ~9.9 Å for g-$SiO_2$.

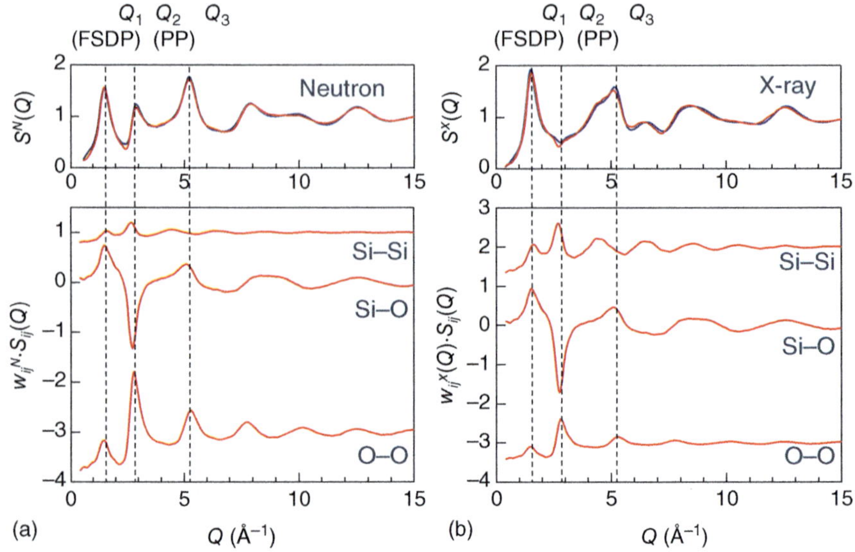


**Figure 3.6** Neutron- (a) [67] and X-ray (b)-weighted total structure factors [18], $S(Q)$, for $g$-$SiO_2$ together with the total and partial structure factors, $S_{ij}(Q)$ obtained from the MD–RMC modeling [18].

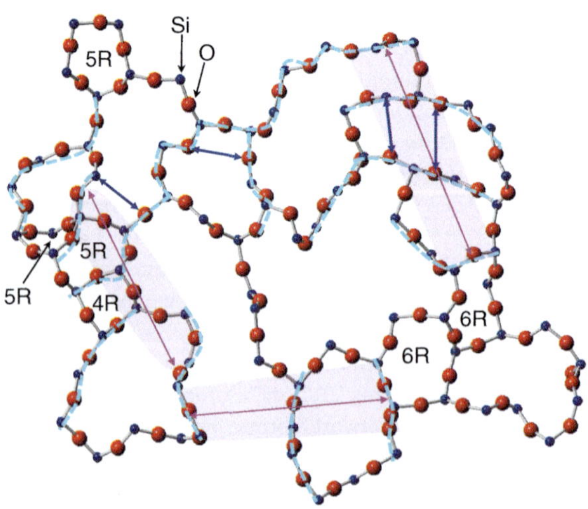


**Figure 3.7** RMC–MD-generated atomic configuration for $g$-$SiO_2$ [18]. The thickness of the cell is approximately 9 Å and only the atoms belonging to the network are shown.

These periodicity and correlation length are visible in the atomic arrangement obtained by MD–RMC modeling (see Figure 3.7) [18].

The experimental $S^X(Q)$ (upper) and MD-RMC-generated Faber–Ziman partial structure factors, $S_{ij}(Q)$ (lower), of g-$SiO_2$ (blue curves), and $l$-$SiO_2$ (red curves) [80, 81] are shown in Figure 3.8a. Although the peaks are broader, it appears that the FSDP is still dominant in

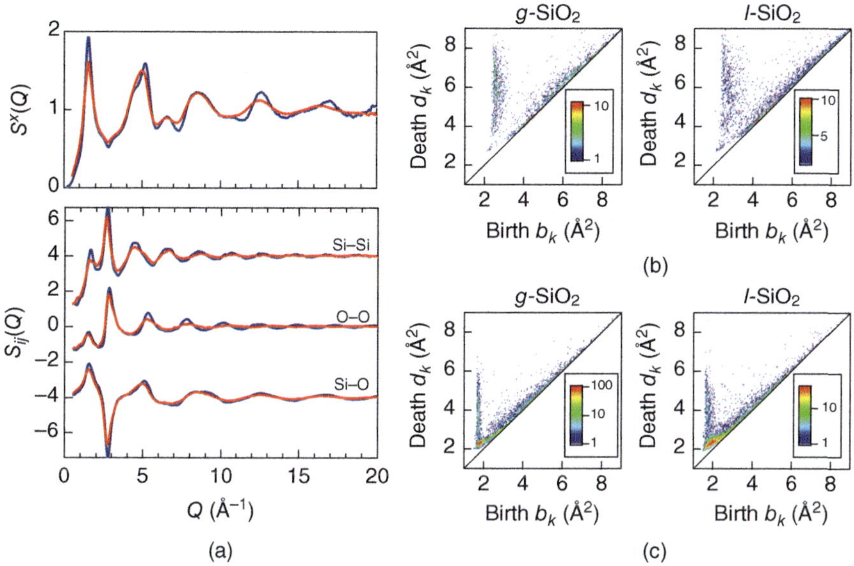


**Figure 3.8** (a) X-ray total structure factors (upper), $S^X(Q)$, for $g$-$SiO_2$ [18] and $l$-$SiO_2$ (2323 K) [81] together with the MD–RMC-generated partial structure factors (lower), $S_{ij}(Q)$, for $g$-$SiO_2$ (blue curves) and $l$-$SiO_2$ (red curves, 2323 K) [18]. Si-centric PD (b) for $g$-$SiO_2$ and $l$-$SiO_2$. O-centric PD (c) for $g$-$SiO_2$ and $l$-$SiO_2$ [18].

the liquid. Both the Si-and O-centric PDs, shown in Figure 3.8b,c, and the X-ray $S(Q)$ (top) and the partial structure factors, $S_{ij}(Q)$ (bottom), in Figure 3.8a show only small differences between the glassy and liquid phases, suggesting that the $SiO_4$ network with corner-sharing is very massive even in the liquid phase [81]. In addition, the density difference is very small between liquid ($d = 2.1\,g/cm^3$) and glass ($d = 2.2\,g/cm^3$). This feature can be reasonably understood from the very high viscosity of $l$-$SiO_2$ [82] and hence it is consistent with the concept of a "strong liquid," as proposed by Angell [83]. Moreover, it is worth mentioning that the behavior in $SiO_2$ is very different from that of Si, in which Si is mostly octahedrally coordinated in the liquid phase as mentioned above.

#### 3.4.2.2 $R_2O$–$SiO_2$ Glass (R = Na, K)

It is well known that alkali mixing causes several unusual phenomena; that is, when an alkali ion is gradually replaced by another one, some physical properties vary in an extremely nonlinear trend. For instance, a composition containing equally mixed alkali ions has a pronounced maximum electrical resistivity and a substantially lower viscosity as a concave function of the alkali content, although the variation in the molar volume is almost linear with the alkali content [84].

The origin of the mixed alkali effect has been studied widely by many experimental and theoretical techniques [85]. A landmark study using RMC modeling based on neutron diffraction and NMR data, reported by Wicks et al. in 1997 [86], concluded that the Modified

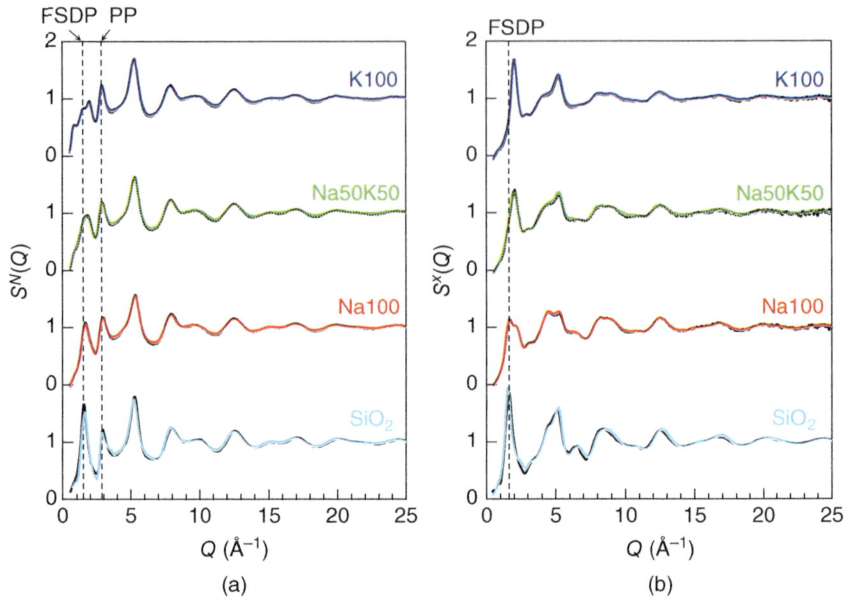


**Figure 3.9** (a) Neutron total structure factor $S^N(Q)$ and (b) X-ray total structure factor $S^X(Q)$ for a series of $K_2O$–$SiO_2$ glasses [56]. Black solid curve, experimental data; colored curve, RMC–MD model. Average experimental $S^{N,X}(Q)$ for the Na50K50 glass calculated using the Na100 glass and K100 glass data are plotted as a dashed black curve. The RMC–MD and experimental data are practically indistinguishable. The $S^{N,X}(Q)$ for the Na50K50 glass is roughly identical to the average of those for the Na100 and K100 glasses. It is worth mentioning that both the FSDP and PP are visible in the neutron structure factors, $S(Q)$, while X-ray $S(Q)$ exhibits only the FSDP, because PP is largely taken by the oxygen-related correlation, which can be easily detected by neutrons.

Random Network (MRN) theory suggested by Greaves [87] is consistent with the data set used in Ref. [86]. Very recently Onodera et al. revisited this topic, with the aim of revealing the origin of the mixed alkali effect on the basis of topology that is hidden in the pairwise correlations [56].

Figure 3.9 compares the neutron and X-ray structure factors, $S^{N,X}(Q)$, between the 22.7$R_2O$–77.3$SiO_2$ glasses (R = Na, K) and $SiO_2$ glass. A sharp FSDP is observed at $Q \sim 1.5$ Å$^{-1}$ in the $S^{N,X}(Q)$ for $SiO_2$ glass, whereas the FSDP becomes broader when alkali ions are added. The FSDP also appears in the $S^N(Q)$ for the alkali silicate glasses, but the shape of the peak is very complicated. Indeed, an extra peak is observed at $Q \sim 0.8$ Å$^{-1}$ in the case of the K100 glass. On the other hand, the FSDP is not visible in the $S^X(Q)$, for either the Na50K50 or K100 glasses, due to the large weighting factors of potassium for X-rays. The PP is found only in the $S^N(Q)$ because the PP in silicate glasses is thought to reflect oxygen–oxygen correlation that are easily detected by neutrons. It is worth noting that the average of the diffraction profiles of the Na100 and K100 glasses (black broken curve) is almost identical to the experimental data for the Na50K50 glass (green curve) for

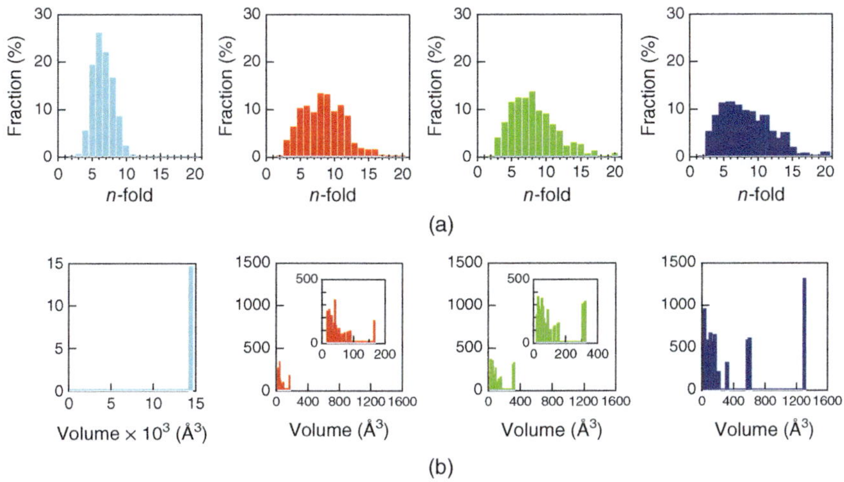


**Figure 3.10** (a) Primitive ring statistics, (b) weighted surface cavity histograms for a series of $K_2O-SiO_2$ glasses [56]. Cyan; $SiO_2$ glass, red; Na100 glass, green; Na50K50 glass, blue; K100 glass. The insets highlight the distributions of small cavities.

both $S^N(Q)$ and $S^X(Q)$, implying that the mixed alkali effect cannot be detected easily in the diffraction data.

Therefore, it is necessary to analyze the RMC–MD-generated atomistic configurations to extract structural information. Figure 3.10a depicts the primitive $(Si–O)_n$ ring statistics. It is well known that glassy $SiO_2$ has a broad ring size distribution [61], although six-atom rings, the only possible ring structure in the crystalline phase with comparable density (cristobalite), are dominant. This broad distribution is the result of topological disorder according to Cooper and Gupta [53], and hence, this is a very important characteristic of the glass. In the case of alkali silicate glasses, we can see a broader distribution due to the introduction of alkali ions that break down the Si–O network, which is consistent with findings of previous studies [86, 88]. The cavity surfaces are visualized in Figure 3.3. $SiO_2$ glass, which yields the largest cavity volume ratio of 32%, exhibits a string-like cavity. The cavity volume ratio in the alkali silicate glasses increases monotonically with increasing $K_2O$ amount. This ratio cannot provide any evidence for the mixed alkali effect in the intermediate-range structure, similarly to NMR data [89].

In order to shed light on the mixed alkali effect, the cavity volume distribution was calculated (Figure 3.10b). The volume of the largest cavity in $SiO_2$ glass is 14 206 Å$^3$, and the maximum cavity volumes for the Na100, Na50K50, and K100 glasses are 166, 323, and 1310 Å$^3$, respectively, representing the nonlinear variation of the structure with increasing $K_2O$ content. Indeed, Figure 3.10b demonstrates that the maximum cavity size in the Na50K50 glass is obviously smaller than the average value of the Na100 and K100 glasses. Assuming that the size difference between sodium and potassium ions is 0.4 Å, the smaller size cavities should prevent the migration of potassium ions in the Na50K50 glass, which provides an insight to discuss mixed alkali effect.

In order to obtain more detailed information and reconcile the mixed alkali effect, we applied a novel topological approach, the persistent homology analysis. The alkali-centric PDs of Na50K50 glass for Na-centric, Na/K-centric, and K-centric data, shown in Figure 3.11, reveal striking differences. None of the Na-centric (Figure 3.11a) or K-centric data (Figure 3.11c) has any characteristic profiles, while the Na/K-centric PD (Figure 3.11b) has two profiles at $d_k \sim 5$ and $\sim 10\,\text{Å}^2$. These profiles demonstrate that Na and K are highly correlated with edge-sharing oxygens in the Na50K50 glass, similar to a Na–O–K pattern. Indeed, a connectivity analysis [90] confirmed that 81% of Na, K, and O atoms form a network in the Na50K50 glass.

The profiles shown in Figure 3.11d–f are compared with the partial radial distribution functions, $g_{ij}(r)$, for the series of alkali silicate glasses in Figure 3.11g. As seen in Figure 3.11g (left), we can hardly observe differences between $g_{\text{NaNa}}(r)$, $g_{\text{NaK}}(r)$, and $g_{\text{KK}}(r)$ because these functions contain only one-dimensional information. On the other hand, three-body correlations as a function of distance in Figure 3.11g (right) show the formation of specific correlations between Na and K ions, as mentioned above. Thus, the PD enables us to extract the characteristic topology in these glasses.

In order to obtain more detailed structural information associated with the breakdown of the silicate network, alkali-oxygen polyhedral, with the formation of edge-sharing NBO atoms in alkali silicate glasses, are visualized in Figure 3.12a–c. We believe that the visualization should provide us with crucial information for uncovering the origin of the mixed alkali effect. Contrary to similar schematic drawings in the previous studies [87, 91, 92], we depicted atomistic configurations from the structural models that are consistent with both NMR and diffraction data. A comparison between the Na100 and K100 glasses suggests that the shape of the Na–O polyhedra is planar, whereas that of the K–O polyhedra is truly three dimensional.

To understand the nature of alkali-oxygen polyhedra, we calculated the numbers of the bridging oxygen and nonbridging oxygen (BO and NBO) atoms around the alkali ions. The alkali-NBO coordination number is ~2 in both the Na100 and K100 glasses, and the alkali-BO coordination number is 2.1 in the Na100 glass. On the other hand, the K–BO coordination number in the K100 glass is 2.4, and this trend is enhanced in the Na50K50 glass, in which the K–BO coordination number is almost 3. The K–O polyhedra that are highly coordinated to the BO and Na–O polyhedra are highlighted in black in Figure 3.12b, where the potassium atom is trapped by a higher number of BO atoms (see Figure 3.12d). It is likely that these specific configurations involving NBO atoms are a bottleneck for ionic conduction because the trapped potassium ions disturb ionic migration.

Based on MD simulations of lithium potassium metasilicate glasses, it has been reported that each kind of alkali ion has an independent conduction pathway [93]. Accordingly, Habasaki et al. found that the transfer of alkali ions through different ionic sites has a low probability and proposed an interpretation with restricted ion jumping between sites occupied by different types of ions. This seemed to be necessary due to a mismatch between occupation energies of unlike alkali ions [94]. This intriguing finding that the local environment of potassium ions in highly coordinated K–O polyhedra is significantly different from that of sodium ions in the four-coordinated Na–O polyhedra, successfully proves their hypothesis. In addition, specific correlations between the different ions that are found in

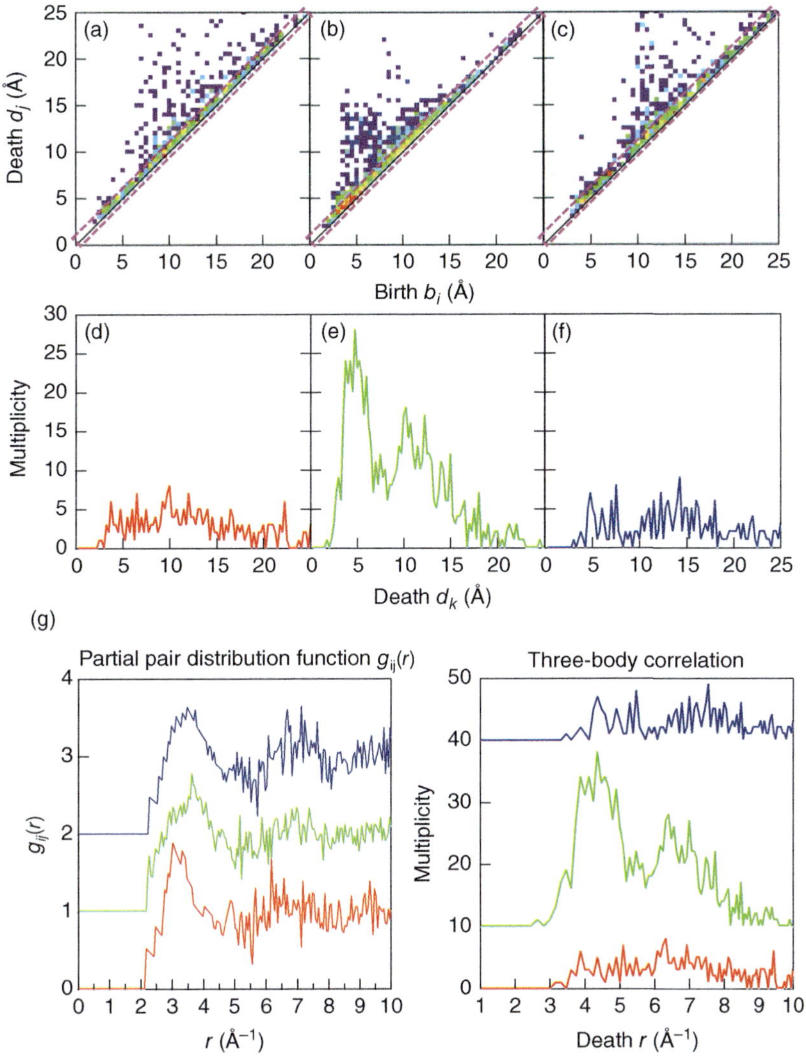


**Figure 3.11** (a) Na-centric, (b) Na/K-centric, and (c) K-centric PDs for a series of $K_2O-SiO_2$ glasses [56]. The boxed regions follow the diagonal, and the associated profiles are plotted for (d) Na-centric, (e) Na/K-centric, and (f) K-centric [56]. (g) Comparison between partial pair distribution functions, $g_{ij}(r)$ and three-body correlations extracted from the persistent diagram along with the diagonal lines for Na-centric (red), Na/K-centric (green), and K-centric (blue) [56].

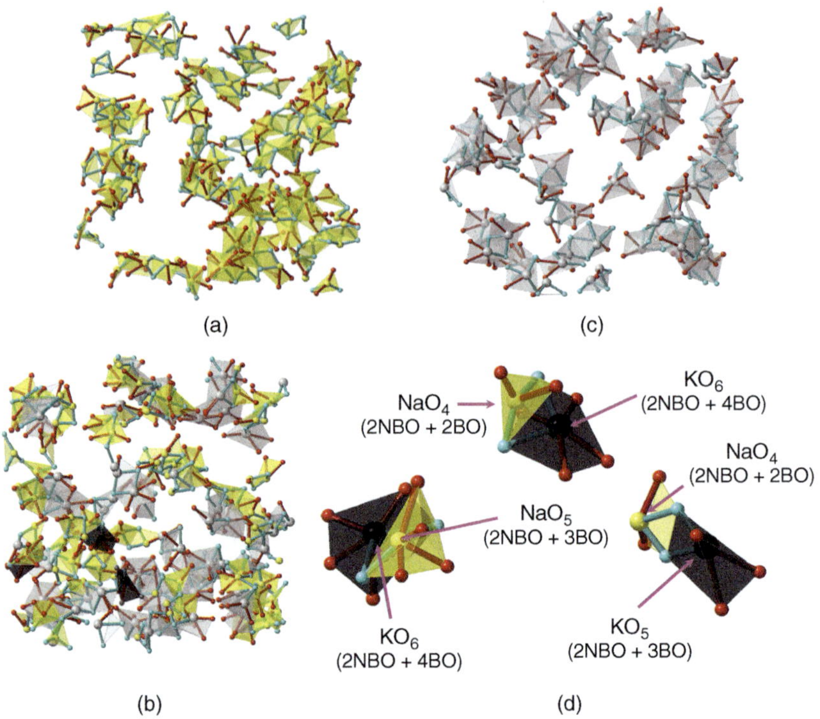


**Figure 3.12** Visualization of alkali-oxygen polyhedra around nonbridging oxygen atoms in the alkali silicate glasses [56]. (a) Na100 glass, (b) Na50K50 glass, (c) K100 glass, and (d) typical bottle neck structures consist of Na–O and K–O polyhedra. Red; bridging oxygen (BO), cyan; NBO, yellow; Na, gray; K. Alkali-oxygen polyhedra with the formation of edge-sharing of NBO atoms are visualized. Typical K–O polyhedra with increased BO coordination are highlighted in black.

the Na50K50 glass by persistent homology (Figure 3.11g [right]) and connectivity analyses, profoundly inhibited ionic migration in the Na50K50 glass. Hence, the local structure associated with alkali connectivity is the intrinsic origin of the mixed alkali effect, which supports the MRN model proposed by Greaves with a bottleneck structure [87].

#### 3.4.2.3 $CaO–Al_2O_3$ Glass

Glass formation in the $CaO–Al_2O_3$ system is an important phenomenon because the material does not contain a typical network former. RMC modeling was combined with density functional theory (DFT)-based simulations to investigate the relationship between the atomistic/electronic structures and the glass-forming ability (GFA) in the $50CaO–50Al_2O_3$ (50CaO) and the $64CaO–36Al_2O_3$ (64CaO) glasses. These compositions were chosen because the GFA of the 64CaO glass is much higher than that of the 50CaO glass, although the composition difference is small.

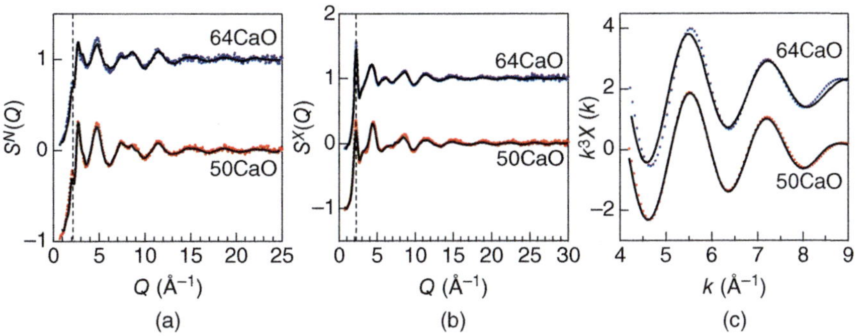


**Figure 3.13** Neutron (a) and X-ray (b) total structure factors $S(Q)$ and EXAFS $k^3\chi(k)$ (c) data for 50CaO and 64CaO glasses. The EXAFS $k^3\chi(k)$ data were obtained by back Fourier transformation of the first correlation peak in real space. Circles, experimental data; curve, DFT–RMC model [95, 96].

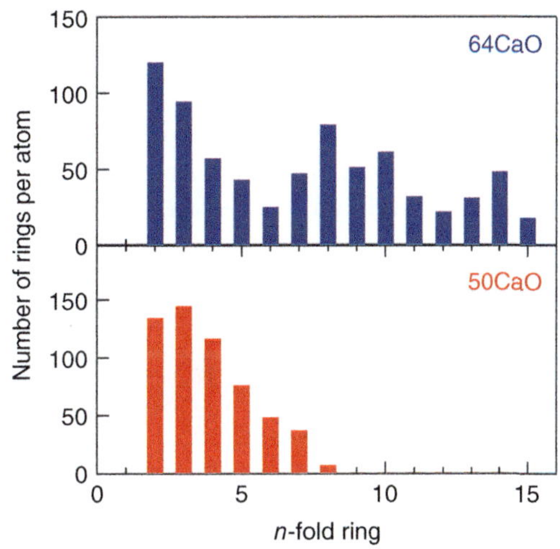


**Figure 3.14** Distribution of -Al(Ca)–O–Al(Ca)–O–Al(Ca)– rings in 50CaO and 64CaO glasses obtained by RMC–DFT simulation [95, 96].

Figures 3.13a–3.14c show neutron and X-ray total structure factors $S^{N,X}(Q)$ [95, 96] and EXAFS $k^3\chi(k)$ data, respectively, of 50CaO and 64CaO glasses as colored dots, together with results of RMC–DFT simulations (black curves). Differences in terms of diffraction and EXAFS data are small between the two compositions, implying that the short-range structure is similar in the two materials. Both the neutron and X-ray total structure factors, $S(Q)$, show an FSDP at $Q \sim 2.2$ Å$^{-1}$, but the X-ray data show a sharper peak, perhaps because X-rays are more sensitive to cation–cation correlation than neutrons. The prepeaks provide information on the ordering of the centers of Al–O and Ca–O polyhedra. In addition, the FSDP for the 64CaO glass is higher than that for the 50CaO glass in the X-ray data. The partial pair distribution functions, $g_{ij}(r)$, of the above CaO–$Al_2O_3$ glasses obtained by RMC–DFT simulation show very small differences between the 50CaO glass and 64CaO glass. The Al–O coordination number, $N_{Al-O}$, calculated up to 2.5 Å is 4.26 for the 50CaO glass and 4.14 for the 64CaO glass, respectively. The Ca–O coordination number $N_{Ca-O}$ calculated up to 2.8 Å is 5.02 for the 50CaO glass and 4.92 for the 64CaO glass, respectively.

Figure 3.14 compares the distributions of –Al(Ca)–O–Al(Ca)–O–Al(Ca)– rings in CaO–$Al_2O_3$ glasses. The 64CaO glass (high GFA) has a wide ring-size distribution, while the 50CaO glass (low GFA) has a very narrow ring size distribution. This trend can be understood on the basis of the concept of topological order–disorder according to Gupta and Cooper [53]. This characteristic ring-size distribution, and particularly the formation of large rings, indicates the existence of a cage structure [97] at the eutectic composition of 64CaO glass. Moreover, it is suggested that the topological disorder in the 64CaO glass is a signature of a stronger liquid [83] than that of the 50CaO composition. This scenario can reasonably explain the high GFA of the 64CaO glass without the need for the presence of a traditional network former, as proposed in Zachariasen's original theory [64].

The "C12A7" (63.2CaO–36.8$Al_2O_3$): $e^-$ electrode glass has been synthesized from a strongly reduced high-temperature melt by Kim et al. [97]. It is assumed that the removal of excess O atoms from $AlO_5$ and $AlO_6$ units in the C12A7 melt results in the formation of an extended cage structures in the electride version of the glass, which can host solvated electrons. To shed light on the early stage of the formation of the electride glass, the highest occupied molecular orbital (HOMO) and lowest unoccupied molecular orbital (LUMO) single-particle states of electrons were computed for the 64CaO glass by DFT calculations, as illustrated in Figure 3.15a,b. The characteristics of the HOMO and LUMO states are considerably different: while the former is located across atoms and bonds (Figure 3.15a), the latter is associated with a cavity (cage), forming a spin-paired state analogous to an F-center

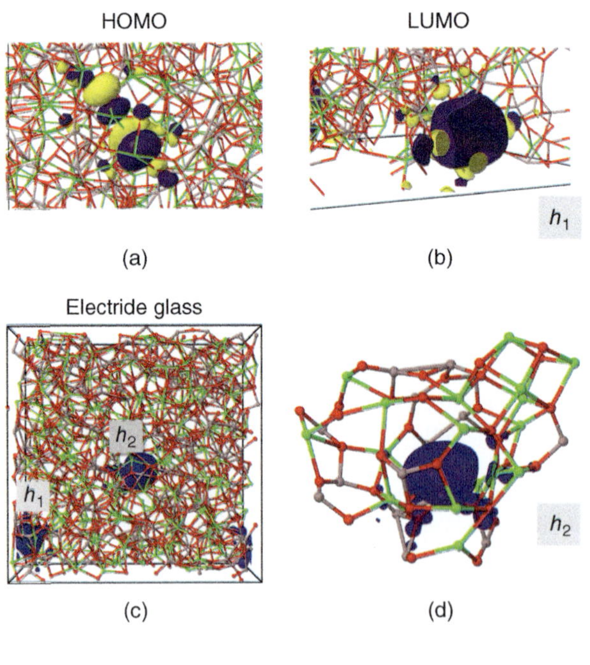


**Figure 3.15** Close-up visualizations of (a) the HOMO and (b) LUMO four single-particle electron states in the 64CaO glass. $h_1$ indicates the cavity (cage) occupied by the LUMO. Yellow and magenta represent the different signs of the wave-function nodes. (c) Simulation box and the electron spin density of the 64CaO glass with one oxygen removed at $h_2$, i.e. with two additional electrons. (d) Cage structure around the spin density of one electron corresponding to the $h_2$ cavity (close-up of (c)). Al, gray; Ca, green; O, red [95, 96].

in a vacancy site of crystalline MgO ($h_1$ in Figure 3.15b). The computed HOMO–LUMO gap is 1.71 eV.

Since it is well known that DFT calculations at this level underestimate band gaps by approximately a factor of two, this value is consistent with the typical insulating electronic behavior and transparent properties of the glass. Furthermore, our RMC–DFT model suggests that the first three cage-trapped LUMO states appear as impurity states below the onset of the conduction band. Following the idea of bipolaron states and a conducting electride glass, we removed one oxygen atom from the $h_2$ site in Figure 3.15b, thus releasing two additional electrons from the Al/Ca cations while keeping the total charge neutral, and then the structure was optimized for several spin configurations by DFT. In the spin-degenerate case, where there is no distinction concerning the "spin" of electrons, these electrons occupy a cavity vacated by the removed oxygen (marked as $h_2$), yielding a HOMO state similar to that of the LUMO state of the parent system. On the other hand, the removal of the spin degeneracy (triplet spin configuration) leads to a 0.97 eV more energetically favorable electronic configuration, where the two additional electrons have the same spin and are located in well-separated cavities (Figure 3.15c, $h_1$ and $h_2$). This procedure was repeated for two, three, and four removed oxygen atoms (four, six, and eight additional electrons), and in all cases, the separated (solvated) electrons in individual cages were energetically more favorable than the F-center-like states (two electrons in one cavity). An example of the cage structure around a single electron (spin density) is presented in Figure 3.15d. These cases confirm that by removing oxygen from the standard stoichiometry, one can achieve local spin states in the 64CaO glass. Moreover, the gradual removal of O increases the number of impurity states within the electronic band gap, which leads to changes in conductivity due to a narrower band gap. Although this is a simulation of an early stage in the formation of electride glass, it is demonstrated that the combination of RMC modeling and DFT calculation is a powerful tool for understanding the glassy structure at both the atomistic and electronic levels.

### 3.4.3 Chalcogenide Glasses (with a Focus on the $Ge_2Sb_2Te_5$ Alloy)

Although data storage and distribution via the Internet is dominant nowadays, DVD/Blu-ray (BD) systems are still indispensable media in our daily lives. Information is stored on a DVD/BD in the form of microscopic bits in a thin layer of a polycrystalline $Ge_2Sb_2Te_5$ (GST) alloy. The bits can have a disordered, amorphous structure (recorded) or an ordered, crystalline structure (erased). The phase change between the two phases lasts only a few tens of nanoseconds and may be triggered by a laser pulse [98]. The combination of synchrotron radiation high-energy X-ray diffraction, element-specific anomalous X-ray scattering (AXS) measurements, and DFT-MD/RMC simulations on GST has made it possible to determine the structures of both phases and allowed the establishment of a model to explain the rapid phase change and its high durability, and in particular, the stability of the amorphous phase.

Figure 3.16a shows a comparison of the X-ray total structure factors $S(Q)$ for crystalline ($c$-) GST and amorphous ($a$)-GST [99]. The $S(Q)$ of $c$-GST exhibits sharp Bragg peaks with a diffuse scattering pattern, while only the latter appears for $a$-GST. It was confirmed that the $S(Q)$ from the RMC model (solid curve) is in good agreement with experimental data (open

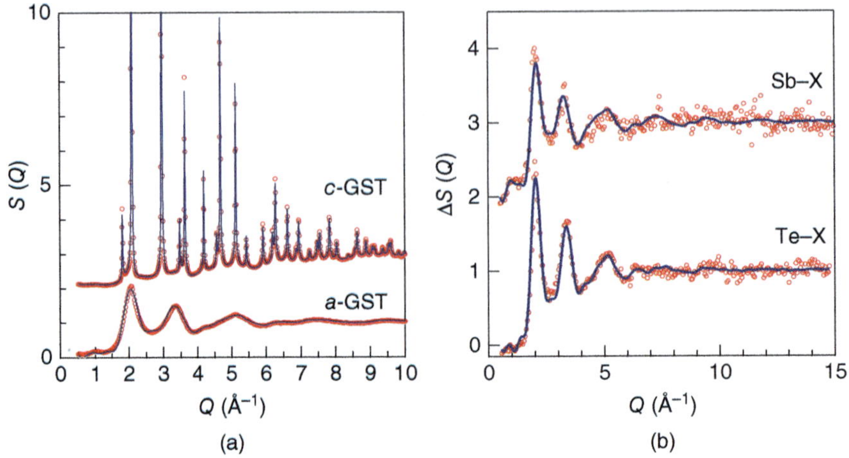


**Figure 3.16** Comparison between the experimental data (open circles) and the results of RMC modeling (solid curve). (a) X-ray total structure factors $S(Q)$ for *c*-GST and *a*-GST. (b) $\Delta S(Q)$ for *a*-GST measured at Sb and Te K absorption edges [99]. The $S(Q)$ for *c*-GST and $\Delta S(Q)$ for Sb are shifted upward by two units for clarity. X: Ge, Sb, Te.

circles). The differential structure factors $\Delta S(Q)$ of *a*-GST at the appropriate Sb and Te edges, obtained from AXS measurements, are shown in Figure 3.16b. There are distinct differences between the Sb- and Te-related data at $Q < 6\,\text{Å}^{-1}$, suggesting that AXS can distinguish Sb (atomic number of 51)-related and Te (atomic number of 52)-related correlations beyond the nearest-neighbor distance. The differential structure factor of Sb, $\Delta S_{\text{Sb}}(Q)$ exhibits an FSDP at $Q \sim 1\,\text{Å}^{-1}$. This peak is also observed in the X-ray $S(Q)$ data for both the amorphous and crystalline phases (owing to diffuse scattering present in *c*-GST), as shown in Figure 3.16a, although the FSDP is not prominent in comparison with that of a typical glass-forming material such as $SiO_2$ glass.

In order to investigate the atomic ordering formed by Ge—Te and Sb—Te bonds in detail, connectivities between atoms were calculated for various maximum distances ($r_{max}$) within which atomic pairs were considered to be connected. As can be seen in Figure 3.17a, about 60% of the Ge—Te bonds form a continuous network when $r_{max}$ is 3.2 Å, while Sb—Te bonds do not. This distance corresponds to the bond length determined by DFT–MD simulation [100]. Atomic configurations of Ge—Te and Sb—Te with bonds considered up to a distance of 3.2 Å are shown in Figure 3.17b,c, respectively. It is observed that Ge—Te bonds form a network, which stabilizes the amorphous phase, whereas Sb—Te bonds do not. Furthermore, Ge—Te bonds form large fractions of fourfold (highlighted in blue) and sixfold (highlighted in light blue) rings. On the other hand, about 70% of the Sb–Te pairs form a so-called "pseudonetwork," as can be seen in Figure 3.17a,d, when $r_{max}$ is 3.5 Å (in this case, the coordination numbers of Te around Ge ($N_{\text{Ge–Te}}$) and Sb ($N_{\text{Sb–Te}}$) are 4.1 and 3.1, respectively). In other words, the Sb–Te pseudonetwork becomes visible when the correlation distance is increased to 3.5 Å, whereas such a feature cannot be observed for Ge–Te connectivity.

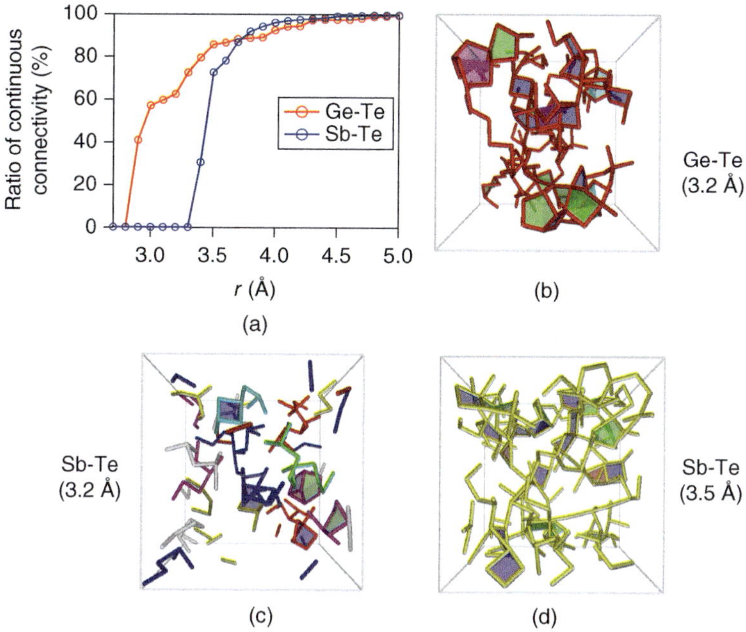


**Figure 3.17** RMC-generated atomic configurations and connectivity of Ge−Te and Sb−Te in *a*-GST [99]. (a) Connectivity of Ge−Te and Sb−Te for different values of $r_{max}$. (b) Atomic configuration of Ge−Te (connectivity of Ge−Te for $r_{max}$ = 3.2 Å). (c) Atomic configuration of Sb−Te (connectivity of Sb−Te for $r_{max}$ = 3.2 Å). (d) Atomic configuration of Sb−Te (connectivity of Sb−Te for $r_{max}$ = 3.5 Å). The threefold, fourfold, fivefold, and sixfold rings are highlighted by red, blue, green, and light blue colors, respectively.

Figure 3.18 illustrates the phase-change process based on the structural features in *a*-GST. As highlighted in red in the figure, Ge—Te bonds substantially contribute to the network comprised by fourfold rings. Therefore, Ge and Te can be recognized as network-forming elements that ensure the long-term stability of the amorphous phase at room temperature. The Ge–Te network may remain in the crystalline phase, which explains the rapid crystallization. As highlighted in yellow in Figure 3.18, the Sb–Te correlations beyond the nearest coordination distance can form a Sb–Te pseudonetwork. This unusual atomic ordering in terms of Sb–Te correlations is ascribed to the combination of two positively charged atoms (Ge, 0.22 electrons; Sb, 0.32 electrons; Te, 0.22 electrons [100]) that allows critical nuclei to form in the amorphous phase, via the formation of Sb—Te bonds by small atomic displacements of antimony and tellurium atoms.

The structure of the GST alloy has received much interest so that very different RMC modeling studies by other groups have also been conducted. The approach of Jóvári et al. [101, 102] differs from the one described above in that instead of emphasizing AXS, a large amount of EXAFS results (centered on each element) have been combined with X-ray and neutron diffraction data. This way, five independent data sets were available for the six

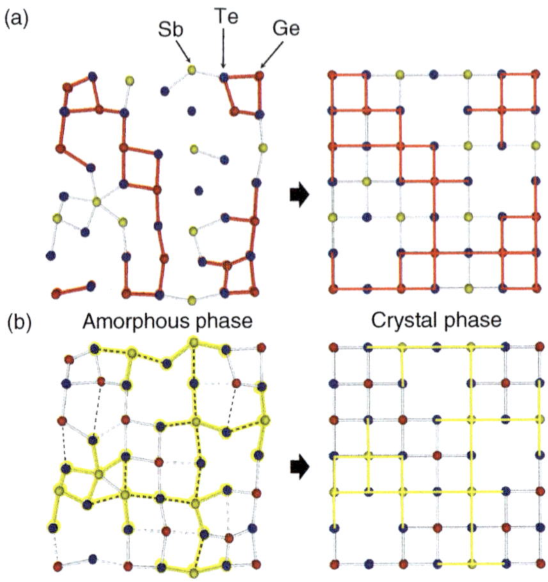


**Figure 3.18** Schematic drawing of phase-change process in *a*-GST. (a) Highlights the Ge–Te network and (b) highlights the Sb–Te pseudonetwork. The sticks represent Ge–Te (red) and Sb–Te (yellow) bonds up to 3.2 Å and the dotted yellow lines show Sb–Te correlations of 3.2–3.5 Å in the amorphous phase. Source: Based on Ohara et al. [99].

partial radial distribution functions present in the system, which is, to our best of knowledge, the most complete set of experimental data used for structural investigations of GST. Perhaps the most significant finding from these EXAFS-based studies was that "wrong" homopolar bonds appeared, most intriguingly for Ge–Ge pairs.

Although we do not wish to go into any further details here, studying chalcogenide glasses by RMC is by no means restricted to GST: very recent examples for the EXAFS-based [103–107] and AXS-based [108] approaches are provided in references [103–108].

### 3.4.4 Metallic Glasses

Even though this chapter was intended to highlight the application of RMC modeling for covalent glasses, a short summary on metallic glasses may be appropriate.

Metallic glasses (or amorphous metallic alloys) were among the earliest example materials that have been considered by RMC. This is partly because early "unconstrained" RMC calculations were well suited to the nature of metallic glasses: unlike covalent glasses, they have a very high packing fraction, with no "empty space" between atoms so that excluded volume effects were thought to be dominant.

Following a very early (and probably premature) investigation of one of the present-authors [109], during the mid-1990s, quite a few RMC studies have appeared (see, e.g. [27–29, 110]), dominantly on Ni-containing amorphous alloys. The reason for choosing this particular group of materials was that the very different neutron scattering lengths of Ni-isotopes made it possible to perform isotopic substitution neutron diffraction experiments. This meant partial structure factors and radial distribution functions for two-component alloys (containing Ni and B, P, and Nb, for instance) were available via a purely experimental route (three partials, three independent neutron diffraction measurements). A very nice consistency between partial radial distribution functions obtained from the traditional (purely experimental) and RMC-based routes was found for the $Ni_{62}Nb_{38}$ metallic glass [111]; unfortunately, sufficient amount of diffraction data of similar quality cannot be gathered routinely.

Very recently, EXAFS-based [112, 113] and AXS-based [114] approaches have both been present, and they are expected to remain in use in the future.

## 3.5 Conclusion

The RMC method of structural modeling [1, 14–16, 115, 116] has been introduced. Novel ways of characterizing glassy topology (ring statistics, cavity analysis, and persistent homology analysis, see Section 3.3) have been described in some detail. Examples of RMC-based structural studies have been provided, mostly for covalent glasses, where these novel tools have been made use of.

Arguably, the major problem concerning structure determination of functional disordered materials is *information deficiency*. Here we have demonstrated what must be done if one wishes to tackle this difficulty with even just a moderate success: as many pieces of direct experimental evidence (X-ray and neutron diffraction, AXS, EXAFS) as it is physically possible must be gathered. Direct experimental data should be augmented by available external information (e.g. from spectroscopic methods and/or high-level computer simulations) and then the RMC method can be used to find structures that are consistent with all the input information. To the best of our knowledge, such RMC structures are the most reliable representations of real amorphous structures.

## Acknowledgments

The authors are grateful to the National Research, Development, and Innovation Office (NKFIH) of Hungary for supporting this work via Grant No. KH 130425, JSPS KAKENHI Grant Numbers 20H05878, 20H05881, JST PRESTO, Japan Grant Number JPMPR15N4, and the "Materials Research by Information Integration" Initiative ($MI^2I$) project of the Support Program for Starting Up Innovation Hub from JST.

## References

**1** McGreevy, R.L. and Pusztai, L. (1988). *Mol. Simul.* 1: 359.
**2** Squires, G.L. (1978). *Introduction to the Theory of Thermal Neutron Scattering*. Cambridge: Cambridge University Press.
**3** Sköld, K. and Price, D.L. (1986). *Neutron Scattering*. Orlando, FL: Academic Press.
**4** Elliott, S.R. (1990). *Physics of Amorphous Materials*, 2e. Harlow: Longman Scientific & Technical.

5 Egelstaff, P.A. (1994). *An Introduction to the Liquid State*, 2e. New York, USA: Oxford University Press.
6 Fisher, H.E., Barnes, A.C., and Salmon, P.S. (2006). *Rep. Prog. Phys.* 69: 233.
7 Ladd, M.F.C. and Palmer, R.A. (1977). *Structure Determination by X-Ray Crystallography*. New York: Plenum Press.
8 Giacovazzo, C., Monaco, H.L., Artioli, G. et al. (1992). *Fundamentals of Crystallography*. Oxford: Oxford University Press.
9 Mak, T.C.W. (1997). *Crystallography in Modern Chemistry: A Resource Book of Crystal Structures*. Chichester: Wiley.
10 Nield, V.M. and Keen, D.A. (2001). *Diffuse Neutron Scattering from Crystalline Materials*. Oxford: Clarendon Press.
11 Keen, D.A. (2001). *J. Appl. Crystallogr.* 34: 172.
12 McGreevy, R.L. and Pusztai, L. (1990). *Proc. R. Soc. London, Ser. A* 430: 221.
13 McGreevy, R.L. (2001). *J. Phys. Condens. Matter* 13: R877.
14 Evrard, G. and Pusztai, L. (2005). *J. Phys. Condens. Matter* 17: S1.
15 Gereben, O., Jóvári, P., Temleitner, L., and Pusztai, L. (2007). *J. Optoel. Adv. Mater.* 9: 3021.
16 Gereben, O. and Pusztai, L. (2012). *J. Comput. Chem.* 33: 2285.
17 Laaziri, K., Kycia, S., Roorda, S. et al. (1999). *Phys. Rev. Lett.* 82: 3460.
18 Onodera, Y., Kohara, S., Tahara, S. et al. (2019). *J. Ceram. Soc. Jpn.* 127: 853.
19 Howe, M.A. and McGreevy, R.L. (1991). *Phys. Chem. Liq.* 24: 1.
20 Gereben, O. and Pusztai, L. (1994). *Phys. Rev. B* 50: 14136.
21 Pusztai, L. and McGreevy, R.L. (1998). *J. Phys. Condens. Matter* 10: 525.
22 Pusztai, L. (1999). *Phys. Rev. B* 60: 11851.
23 Pusztai, L. (2000). *Phys. Rev. B* 61: 28.
24 Gereben, O., Pusztai, L., and Baranyai, A. (1995). *Phys. Scr.* T57: 69.
25 Pusztai, L. and Gereben, O. (1995). *J. Non-Cryst. Solids* 192, 193: 640.
26 Jóvári, P., Delaplane, R.G., and Pusztai, L. (2003). *Phys. Rev. B* 67: 172201.
27 Pusztai, L. and Sváb, E. (1993). *J. Phys. Condens. Matter* 5: 8815.
28 Duine, P.A., Sietsma, J., Thijsse, B.J., and Pusztai, L. (1994). *Phys. Rev. B* 50: 13240.
29 Pusztai, L. and Gereben, O. (1994). *Mat. Sci. Eng.* A179/180: 433.
30 Gereben, O. and Pusztai, L. (1996). *Phys. Chem. Liq.* 31: 159.
31 Pusztai, L. and McGreevy, R.L. (1997). *Mol. Phys.* 90: 533.
32 Pothoczki, S., Temleitner, L., Jóvári, P. et al. (2009). *J. Chem. Phys.* 130: 064503.
33 Pothoczki, S., Temleitner, L., and Pusztai, L. (2015). *Chem. Rev.* 115: 13308.
34 Soper, A.K. (1996). *Chem. Phys.* 202: 295.
35 Soper, A.K. (1998). *J. Mol. Liq.* 78: 179.
36 Soper, A.K. (2005). *Phys. Rev. B* 72: 104204.
37 Youngs, T. (2019). *Mol. Phys.* 117: 3464.
38 Allen, M.P. and Tildesley, D.J. (1987). *Computer Simulation of Liquids*. Oxford: Clarendon Press.
39 Imberti, S., Bott, A., Bruni, F. et al. (2005). *J. Chem. Phys.* 122: 194509.
40 McLain, S., Soper, A.K., and Luzar, A. (2006). *J. Chem. Phys.* 124: 074502.
41 Hardacre, C., Holbrey, J.D., McMath, S.E.J. et al. (2003). *J. Chem. Phys.* 118: 273.
42 Botti, A., Bruni, F., Mancinelli, R. et al. (2008). *J. Chem. Phys.* 128: 164504.
43 Mile, V., Pusztai, L., Dominguez, H., and Pizio, O. (2009). *J. Phys. Chem. B* 113: 10760.

44 Mile, V., Gereben, O., Kohara, S., and Pusztai, L. (2010). *J. Mol. Liq.* 157: 36.
45 Roux, S. and Jund, P. (2010). *Comput. Mater. Sci.* 49: 70.
46 Roux, S. and Jund, P. (2011). *Comput. Mater. Sci.* 50: 1217.
47 King, S.V. (1967). *Nature* 213: 1112.
48 Guttman, L. (1990). *J. Non-Cryst. Solids* 116: 145.
49 Franzblau, D.S. (1991). *Phys. Rev. B* 44: 4925.
50 Goetzke, K. and Klein, H.J. (1991). *J. Non-Cryst. Solids* 127: 215.
51 Yuan, X. and Cormack, A.N. (2002). *Comput. Mater.* 24: 343.
52 Wooten, F. (2002). *Acta Cryst. A* 58: 346.
53 Gupta, P.K. and Cooper, A.R. (1990). *J. Non-Cryst. Solids* 123: 14.
54 Ono, M., Hara, K., Fujinami, M., and Ito, S. (2012). *Appl. Phys. Lett.* 101: 164103.
55 Heimbach, I., Rhiem, F., Beule, F. et al. (2017). *J. Comput. Chem.* 38: 389.
56 Onodera, Y., Takimoto, Y., Hijiya, H. et al. (2019). *NPG Asia Mater.* 11: 75.
57 Hiraoka, Y., Nakamura, T., Hirata, A. et al. (2016). *Proc. Natl. Acad. Sci. U.S.A.* 113: 7035.
58 Hirata, A., Kang, L.J., Fujita, T. et al. (2013). *Science* 341: 376.
59 Obayashi, I. (2018). HomCloud. (Hiraoka Laboratory, Advanced Institute for Materials Research, Tohoku University, Japan.) https://www.wpi-aimr.tohoku.ac.jp/hiraoka_labo/homcloud/index.en.html.
60 Rino, J.P., Ebbsjö, I., Kalia, R.K. et al. (1993). *Phys. Rev. B* 47: 3053.
61 Kohara, S., Akola, J., Morita, H. et al. (2011). *Proc. Natl. Acad. Sci. U.S.A.* 108: 14780.
62 Katayama, Y., Mizutani, T., Utsumi, W. et al. (2000). *Nature* 403: 170.
63 Salmon, P.S. (2018). Densification mechanism of oxide glasses and melts. In: *Magma under Pressure: Advances in High-Pressure Experiments on Structure and Properties of Melts* (ed. Y. Kono and C. Sanloup), 347–348. Amsterdam: Elsevier.
64 Zachariasen, W.H. (1932). *J. Am. Chem. Soc.* 54: 3841.
65 Sun, K.-H. (1947). *J. Am. Ceram. Soc.* 30: 277.
66 Greaves, G.N. and Sen, S. (2007). *Adv. Phys.* 56: 1.
67 ISIS Disordered Materials Database. http://www.alexhannon.co.uk.
68 Zeidler, A. and Salmon, P.S. (2016). *Phys. Rev. B* 93: 214204.
69 Salmon, P.S., Martin, R.A., Mason, P.E., and Cuello, G.J. (2005). *Nature* 435: 75.
70 Wright, A.C. and Leadbetter, A.J. (1976). *Phys. Chem. Glasses* 17: 122.
71 Phillips, J.C. (1981). *J. Non-Cryst. Solids* 43: 37.
72 Price, D.L., Moss, S.C., Reijers, R. et al. (1988). *J. Phys. C: Solid State Phys.* 21: L1069.
73 Elliott, S.R. (1991). *Nature* 354: 445.
74 Wright, A.C. (1994). *J. Non-Cryst. Solids* 179: 84.
75 Salmon, P.S. (1994). *Proc. R. Soc. London Ser. A* 445: 351.
76 Gaskell, P.H. and Wallis, D.J. (1996). *Phys. Rev. Lett.* 76: 66.
77 Mei, Q., Benmore, C.J., Sen, S. et al. (2008). *Phys. Rev. B* 78: 144204.
78 Kohara, S., Akola, J., Patrikeev, L. et al. (2014). *Nat. Commun.* 5: 5892.
79 Shi, R. and Tanaka, H. (2019). *Sci. Adv.* 5: 3194.
80 Mei, Q., Benmore, C.J., and Weber, J.K.R. (2007). *Phys. Rev. Lett.* 4: 057802.
81 Takada, A., Richet, P., Catlow, C.R.A., and Price, G.D. (2004). *J. Non-Cryst. Solids* 345, 346: 224.
82 Urbain, G., Bottinga, Y., and Richet, P. (1982). *Geochim. Cosmochim. Acta* 46: 1061.
83 Angell, C.A. (1995). *Science* 267: 1924.

84 Isard, J.O. (1969). *J. Non-Cryst. Solids* 1: 235.
85 Habasaki, J., Leon, C., and Ngai, K.L. (2017). *Dynamics of Glassy, Crystalline and Liquid Ionic Conductors*. Switzerland: Springer International Publishing.
86 Wicks, J.D., McGreevy, R.L., and Börjesson, L. (1997). *Phase Transitions* 61: 195.
87 Greaves, G.N. (1985). *J. Non-Cryst. Solids* 71: 203.
88 Du, J. and Corrales, L.R. (2006). *J. Non-Cryst. Solids* 352: 3255.
89 Maekawa, H., Maekawa, T., Kawamura, K., and Yokokawa, T. (1991). *J. Non-Cryst. Solids* 127: 53.
90 Kohara, S., Ohno, H., Takata, M. et al. (2010). *Phys. Rev. B* 82: 134209.
91 Greaves, G.N. and Ngai, K.L. (1995). *Phys. Rev. B* 52: 6358.
92 Park, B. and Cormack, A.N. (1999). *J. Non-Cryst. Solids* 255: 112.
93 Habasaki, J., Okada, I., and Hiwatari, Y. (1995). *J. Non-Cryst. Solids* 183: 12.
94 Habasaki, J., Okada, I., and Hiwatari, Y. (1996). *J. Non-Cryst. Solids* 208: 181.
95 Akola, J., Kohara, S., Ohara, K. et al. (2013). *Proc. Natl. Acad. Sci. U.S.A.* 110: 10129.
96 Kohara, S. (2017). *J. Ceram. Soc. Jpn.* 125: 799.
97 Kim, S.W., Shimoyama, T., and Hosono, H. (2011). *Science* 333: 71.
98 Wuttig, M. and Yamada, N. (2007). *Nat. Mater.* 6: 824.
99 Ohara, K., Temleitner, L., Sugimoto, K. et al. (2012). *Adv. Funct. Mater.* 22: 2251.
100 Akola, J. and Jones, R.O. (2008). *J. Phys. Condens. Matter* 20: 465103.
101 Jóvári, P., Kaban, I., Steiner, J. et al. (2007). *J. Phys. Condens. Matter* 19: 335212.
102 Jóvári, P., Kaban, I., Steiner, J. et al. (2008). *Phys. Rev. B* 77: 035202.
103 Pethes, I., Chahal, R., Nazabal, V. et al. (2018). *J. Non-Cryst. Solids* 484: 49.
104 Pethes, I., Nazabal, V., Chahal, R. et al. (2019). *J. Non-Cryst. Solids* 505: 340.
105 Pethes, I., Nazabal, V., Ari, J. et al. (2019). *J. Alloys Compd.* 774: 1009.
106 Jóvári, P., Piarristeguy, A., Pradel, A. et al. (2019). *J. Alloys Compd.* 771: 268.
107 Pethes, I., Piarristeguy, A., Pradel, A. et al. (2020). *J. Alloys Compd.* 834: 155097.
108 Stellhorn, J.R., Paulus, B., Hosokawa, S. et al. (2020). *Phys. Rev. B* 101: 214110.
109 Pusztai, L. and Naturf, Z. (1991). *Zeitschrift für Naturforschung A* 46a: 69.
110 Lamparter, P. (1995). *Phys. Scr.* T57: 72.
111 Pusztai, L. and Sváb, E. (1993). *J. Non-Cryst. Solids* 156–158: 973.
112 Kaban, I., Jóvári, P., Kokotin, V. et al. (2013). *Acta Mater.* 61: 2509.
113 Kaban, I., Jóvári, P., Escher, B. et al. (2015). *Acta Mater.* 100: 369.
114 Hosokawa, S., Bérar, J.-F., Boudet, N. et al. (2019). *Phys. Rev. B* 100: 054204.
115 Tucker, M.G., Keen, D.A., Dove, M.T. et al. (2007). *J. Phys. Condens. Matter* 19: 335218.
116 Playford, H.Y., Owen, L.R., Levin, I., and Tucker, M.G. (2014). *Annu. Rev. Mater. Res.* 44: 429.

# 4

# Structure Analysis and Properties Calculations

*Davide Presti, Francesco Muniz-Miranda, Francesco Tavanti, and Alfonso Pedone*

*University of Modena and Reggio Emilia, Department of Chemical and Geological Sciences, Modena, Italy*

## 4.1 Introduction

Glasses are of great importance in both the field of technological industries and geoscience. Irrespective of their application, the major challenge of researchers engaged in glass science is that of determining the relationships among composition, structure, and the final properties.

In this respect, computational techniques are becoming very powerful tools for glass manufactures since they permit to predict the structure and physicochemical properties of glasses thus allowing the development of new materials, reducing costs and the time to market.

In Chapters 1–3, several computational techniques employed to generate glass structural models and all the information (position, velocities, and acceleration of the atoms) necessary to predict accurately the properties of glasses have been described.

In this chapter, we shall discuss the salient features of glass structures, the statistical distribution of parameters used to analyze their structure, and the computational procedures and statistical formulas used to compute the most important properties of glasses.

Among these, we will first focus our attention on the calculation of X-ray and neutron diffraction, and vibrational and nuclear magnetic resonance (NMR) spectra because the ability to compute such spectra has become fundamental for validating the computational structural models through a direct comparison with the experimental counterparts.

If the structural model is good enough, the thermodynamical, transport, and mechanical properties that are so important for the applicability of glasses in different fields can be predicted. Therefore, in the second part of this chapter, we will describe the way in which the aforementioned properties can be computed, and we will show the accuracy reachable by exploiting case studies already reported in the recent literature.

*Atomistic Simulations of Glasses: Fundamentals and Applications*, First Edition.
Edited by Jincheng Du and Alastair N. Cormack.

## 4.2 Structure Analysis

### 4.2.1 Salient Features of Glass Structures

The structure of glasses and other noncrystalline solids is characterized by the absence of periodicity, symmetry, and long-range order, which instead define crystalline solids. This implies that unlike crystalline materials where the structure can be detailed specified by a single unit cell (the lattice parameters and fractional coordinates of unique atoms as defined by the space group) to define the structure of a glass the coordinates for all the atoms present should be specified. This is clearly impossible for any real sample and the maximum that can be extracted from diffraction experiments is a one-dimensional correlation function. Albeit computer simulations allow us to know the position of atoms in relatively large boxes, the structure of a modeled glass must be quantitatively analyzed in terms of structural parameter distributions and one-dimensional correlation functions. This is due to the fact that the visualization of the models by means of graphical user interfaces does not provide any quantity directly comparable with experimental evidences.

Therefore, in this section, we will describe the structure of glasses and the more important structural parameters used to define them.

Before the 1930s, glasses were meant to be an ensemble of nanocrystals (~20 Å) disposed with no reciprocal order, but in 1932, Zachariasen [1] proposed a valid model for the structure of oxide-based glasses of composition $T_xO_y$, where T is a network former element (Si, P, B, Ge, etc., …). The basic idea was derived from the observation that mechanical properties of glasses are similar to those of crystals of the same composition. Hence, Zachariasen concluded that atoms in a glass are linked together by the same forces as in crystals.

This assumption led to the elaboration of the continuous random network [1], the rules of which are (i) oxygen atoms are linked to no more than two former atoms, T; (ii) the oxygen coordination around T is small (i.e. three or four); (iii) oxygen polyhedra do share corners but not edges or faces; and (iv) for a 3D network at least three corners are shared.

On the bases of this theory, cation–oxygen polyhedra like $SiO_4$ in vitreous silica or $BO_3$ in vitreous boron oxide are regular as in crystals and corner-linked through "bridging" oxygens (BOs). However, these structural units are not connected in an orderly fashion because of the statistical distribution of unconstrained dihedral T–BO–T angles between the coordination polyhedra compared to those in crystals [2].

When a second network former is introduced into the structure of a single component glass (for example, the addition of $B_2O_3$ or $P_2O_5$ in $SiO_2$ glass), a mixed continuous network of two structural units will form with a cation intermixing depending on the miscibility of the two oxides. Instead, the addition of alkaline and alkaline-earth ions results in the disruption of the T—O—T bonds and the formation of negatively charged nonbridging oxygens is connected only to one network former cation following the reaction $\text{T–O–T} + M_2O \geq \text{2T–NBO–M}$. This leads to a modification (depolymerization) of the network, and thus such cations are called network modifiers. Traditionally, such cations were thought to occupy holes in the continuous random network, but extended X-ray absorption fine structure (EXAFS) and diffraction measurements have revealed that they tend to assume a local environment similar to that in crystals. These observations showed

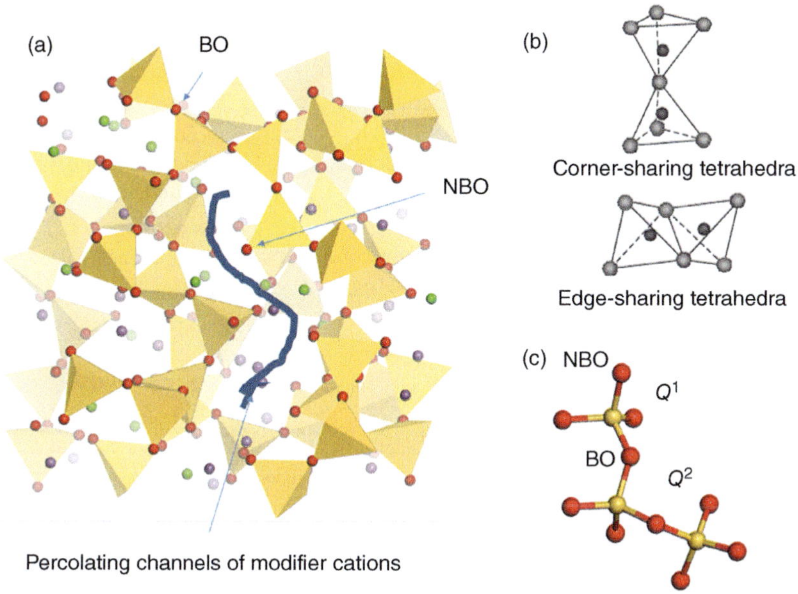


**Figure 4.1** (a) The structure of a soda-lime silica glass obtained by MD simulations showing the $SiO_4$ tetrahedra (yellow) BO, NBO, and percolating channels formed by modifier cations (Na = violet spheres and Ca = green spheres). (b) Corner-sharing and edge-sharing tetrahedral. (c) $Q^1$ and $Q^2$ species for silicon.

that continuous random network theory is no longer valid, and the modified random network theory was proposed [3].

According to this theory, the structure of alkaline silicate glasses can be envisaged as a combination of partly depolymerized network and ionically packed modifying oxides, and at high concentrations, the modifiers percolate through the bulk of the glass and form primary pathways, or channels, which play a key role in important properties of glasses. The structure of a soda-lime silicate glass obtained by molecular dynamic simulations showing these channels as well as the non-bridging oxygens (NBO) and BO is shown in Figure 4.1.

Beyond the former and modifier oxides, *intermediate oxides* ($R_mO_n$) provide cations, R, which can behave as network former or extra-framework cation depending on the glass's overall composition.

For example, when $Al_2O_3$ is introduced into a modified silicate glass composition, Al atoms predominantly form negatively charged $(AlO_{4/2})^{1-}$ units that are charge-compensated by the modifying cations, and the concentration of nonbridging oxygens is reduced in proportion. Geometrically, a compensated continuous random network [4] should be returned for aluminosilicate glasses with a tectosilicate composition that is for an equimolar ratio of alkali or alkaline-earth oxide and $Al_2O_3$. When a network charge compensation is required and no modifier cations (M) are available, intermediate cations (R) assume higher coordination and exit the glass network.

**Table 4.1** Classification of the range order in amorphous solids of $T_nO_m$ stoichiometry.

| Range | Order | Characteristic parameters |
|---|---|---|
| Short range | Structural units or coordination polyhedron ($TO_n$) | Internal coordination number<br>T—O bond lengths and<br>O—T—O bond angles |
| Intermediate range | Interconnection/relative orientation of adjacent units<br>Network topology | T–T distances<br>T—O—T bond angles<br>Network connectivity<br>Short path ring size |
| Long range | Longer range fluctuations | Composition<br>Density<br>Morphology |

The theories mentioned above were formulated to describe the conventional oxide glasses quenched from the melt in a metastable thermodynamic equilibrium at relatively low cooling rates and thus these are not necessarily applicable to other materials (amorphous semiconductors and chalcogenide glasses) prepared by using techniques involving high nonequilibrium processes such as vapor deposition [5].

For systems with no well-defined directional bonding such as metallic and ionic glasses, the structure is much more properly described by a random packing of spheres. In analogy with the crystalline close packing of sphere used to describe the structure of metallic and ionic crystal structures, the important concepts are those of radius ratio, packing density, and hole filling.

## 4.2.2 Classification of the Range Order in Amorphous Solids

The random network and random sphere packing theories describe qualitatively the structure of amorphous solids. To describe quantitatively the structure of glasses, it is necessary to deal with statistically averaged distribution of structural parameters, which are summarized in Table 4.1.

The structural order is usually classified in short, intermediate, and long ranges.

### 4.2.2.1 Short-Range Order

The most important structural property to be determined for any glass is the identity of any structural unit that is present in the system. For example, in vitreous silica, the structural unit is the $SiO_4$ tetrahedron, whereas in more complex multicomponent glasses, other polyhedra must be defined.

The complete characterization of a given structural unit or coordination polyhedron $TO_n$ involves the definition of internal parameters: the coordination number, the distribution of the T—O bond distances, and O—T—O bond angles arising from both static and thermal disorder.

In some cases, the overall structure can be rationalized by the interconnection of superstructural units like the $B_3O_6$ boroxol groups in borate glasses and the specification of other internal bond torsional angles that may be necessary.

In random packed systems where coordination polyhedral are not clearly identifiable, the first coordination sphere is usually analyzed in terms of Voronoi polyhedra [6, 7].

#### 4.2.2.2 Intermediate-Range Order

The intermediate-range order in glasses describes the interconnection, the relative orientation of adjacent units, and the network topology and order beyond adjacent units. The investigation of the second coordination sphere around a structural unit $TO_n$ that can be surrounded by a second $TO_n$ unit and/or a $MO_m$ unit usually starts with the determination of the T–T, T–M distances, and coordination numbers.

In order to describe quantitatively the amount of intermixing among structural units and/or coordination polyhedra or the tendency of a structural unit/coordination polyhedra to be surrounded by a specific cation, a few descriptors have been proposed in the past [8–13].

For example, suppose one wants to determine in a soda-calcia-phosphosilicate glass if Ca ions have a different affinity to Si or P and if the former cation is in large excess with respect to the second. If this is the case, the *CN*(Ca–Si) is obviously greater than the *CN*(Ca–P) because it is statistically most probable to find a silicon ion around a calcium ion rather than phosphorous one. Therefore, the best way to determine the Ca affinity toward Si and P is to compute the normalized *CN*(Ca–Si)/*CN*(Ca–P) ratio with respect to the total number of Si and P atoms in the simulation box. When the ratio $R_{Ca}^{Si/P} = \frac{CN(Ca-Si)}{CN(Ca-P)} \times \frac{n°P}{n°Si}$ (where $n$(P) and $n$(Si) represent the number of P and Si atom in the models, respectively) is >1, Ca has a greater affinity to Si, whereas when the ratio is <1 Ca prefers P coordination.

Another structural parameter that gives an idea of the interconnection between adjacent structural units or polyhedra is the connection mode, which is the number of atoms ($C$) in common between the adjacent units. In simple covalent systems, the most frequent arrangement is corner sharing with ($C = 1$), although edge-sharing ($C = 2$) arrangements do exist in chalcogenide glasses.

In oxide glasses, the number $n$ of bridging oxygens associated with the $[TO_4]$ tetrahedron is an important index in the description of the intermediate glass structure. $Q^n$ is defined as a quaternary (tetrahedral) species with $n$ BO associated with it. For example, a $Q^2$ species of silicon is a Si with 2 BO and 2 NBO in its tetrahedron. Therefore, the $Q^n$ distribution represents the percentage of tetrahedra with $n$ BO. Experimentally, the $Q^n$ distribution is determined by NMR and Raman techniques [14, 15].

Once the $Q^n$ distribution is obtained, the overall degree of polymerization can be summarized by the so-called "network connectivity" $NC = \sum_{n=1}^{4} Q^n \cdot n$.

Moreover, computer simulations give the possibility to determine network topology by specifying the connectivity matrix as shown in Figure 4.2. However, as with the atomic coordinates, it is impossible to specify the connectivity matrix for any real material and thus it is necessary to work with simpler statistical averages.

For example, in oxide glasses, it is possible to detail the composition of $Q^n$ distributions by computing the quantities $Q^{n,ijkm}$, where the upper suffixes ($ijkm$) indicate the type of network formers bonded to the central species as well as the partitioning of modifier ions

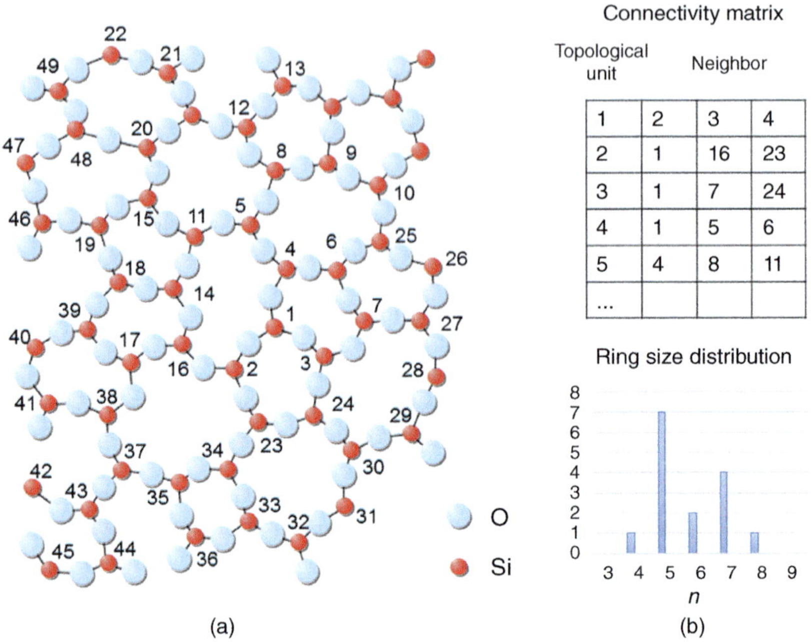


**Figure 4.2** (a) 2D silica glass network with the connectivity matrix defining its topology and (b) ring size distribution of this network.

around different $Q^n$ species useful, for example, for explaining solubility data and chemical durability [16].

Another quantity usually employed to define the network topology is the ring statistics obtained by the shortest path analysis [17]. In silicate glasses, an $n$-membered ring is usually defined as the number of tetrahedra involved in the smallest closed chains formed by BO and network formers in the glass network. Therefore, rings that can be decomposed into two or more smaller rings are not counted in the distribution. The ring size distribution of the network reported in Figure 4.2a is shown in Figure 4.2b.

The ring size distribution is not experimentally measurable even though some information on the presence of small rings can be gained by vibrational and NMR spectroscopy. Despite this, most of the structural models in the literature are frequently discussed in terms of it [18, 19]. Another quantity that can give information on the intermediate range order in glasses is the void size distribution (VSD), that is the distribution of the occurrence of voids (defined as holes) of radius $r$. This is also important for many diffusion-related properties since it can be indirectly derived from gas diffusion measurements [20].

Some years ago, we developed an algorithm based on the geometric Voronoi analysis to compute the VSD of disordered materials [21]. This analysis requires that atoms are considered as spheres.

The Voronoi analysis consists of the partitioning of the space in polyhedra, each containing the volume nearest to a given atom. The vertices of these polyhedra are equidistant to at least four atoms in the network so that it is possible to define spheres centered at the vertices that correspond to the interstitial voids.

The free volume (FV) of different glasses can be analyzed in terms of the Delaunay description of the void space into a disordered system [22].

In this description, a set of four atoms contiguous to each other forms a Delaunay tetrahedron (DT) and the whole set of these DT fill the space without gaps and overlaps. The FV is computed by summing all the void spaces inside the DTs that cover the entire box of the simulation. Finally, the fractional free volume (FFV) of the system is defined as the fraction of FV over the total volume of the box.

#### 4.2.2.3 Long-Range Order

It is well known that amorphous solids lack in the long-range order understood as the periodicity and translation symmetry found in crystals. Therefore, when one speaks about the long-range order in glasses, it usually refers to fluctuation in density and/or composition arising from clustering of group of atoms and phase separation. Fluctuation in glass composition and density at the nanoscale has been recently observed by classical molecular dynamics simulations in phosphosilicate glasses [12, 23–26].

Figure 4.3 reports the structures of two phosphosilicate glasses with different amounts of $P_2O_5$ and shows how the substitution of $SiO_2$ by $P_2O_5$ leads to nanoscale heterogeneities formed by Ca–Na phosphate-rich domains and silicate-rich domains. It is important to highlight that very large simulation boxes must be used if one is interested in investigating these fluctuations.

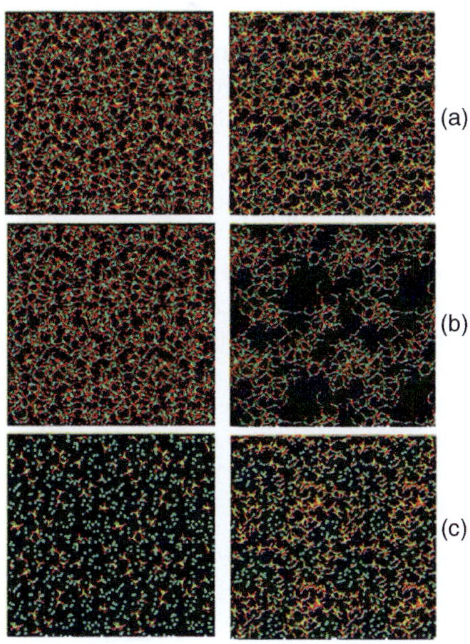


**Figure 4.3** Structure of glasses with composition $(Na_2O)_{0.24}(CaO)_{0.27}[(SiO_2)_{0.94}(P_2O_5)_{0.06}]_{0.49}$ (left panels) and $(Na_2O)_{0.24}(CaO)_{0.27}[(SiO_2)_{0.74}(P_2O_5)_{0.26}]_{0.49}$ (right panels): (a) only Si (turquoise), P (yellow), and O (red) atoms are displayed; (b) only the silicate network is displayed; (c) only the phosphate groups are displayed, with the Si atoms represented as spheres. Source: Reprinted with permission from Tilocca et al. [12]. Copyright 2007 American Chemical Society.

### 4.2.3 Real Space Correlation Functions

Generally, the analysis of the structure of glasses starts with the calculation of one-dimensional real space correlation functions, which allow not only to determine bond distances and coordination numbers in the first and second neighbors but also to validate the structural model by direct comparison with those extracted from X-ray and neutron diffraction experiments.

The pair distribution function (PDF, $\rho_{ij}(r)$) describes on average the number density of the atomic species $j$ at a distance $r$ from the central atomic species $i$ relative to the total number density of the system.

The PDF can be expressed in the following form:

$$\rho_{ij}(r) = \frac{\langle N_j(r,\Delta r)\rangle}{\frac{1}{2}N_i\rho_j^0 V(r,\Delta r)} = \frac{\sum_{k=1}^{M} N_j^k(r,\Delta r)}{M1/2N_i\rho_j^0 V(r,\Delta r)} \tag{4.1}$$

In the second formulation, $N_j^k(r, \Delta r)$ is the number of atoms of the $j$-th species found at the instant $t_k$ in a spherical shell of radius $r$ and thickness $\Delta r$, with the shell centered on an atom of the $i$-th species; $V(r,\Delta r)$ is the volume of the shell; $N_i$ is the number of atoms of the $i$-th species; $\rho_j^0 = \frac{N_j}{V} = c_j\rho_0$ is the density of atoms of the $j$-th species. $c_j$ is the fraction of atoms $j$ in the system, whereas $\rho_0 = N/V$ is the average number density of the material (in atoms Å$^{-3}$).

The PDF describes the distribution of distances between pairs of atoms contained in the simulation box and its analysis provides the most probable distance (location of the peak maximum) between two atoms and the average distance, by averaging the distances $r_1$ and $r_2$ at the half width, half maximum of the first peak as shown in Figure 4.4. Usually, in oxide glasses, the PDFs of network former – oxygens pairs – present a first peak very sharp that goes to zero soon after the first maximum, whereas broader functions are observed for modifier-oxygen pairs. In the first case, the most probable distance usually coincides with the average distance, while in the second case, the two distances are different.

The number of neighboring species $j$ of species $i$ between $r_A$ and $r_B$ is given by the integration:

$$CN_{ij}(r_A, r_B) = \rho_j^0 \int_{r_A}^{r_B} g_{ij}(r)dr = 4\pi\rho_j^0 \int_{r_A}^{r_B} \rho_{ij}(r)r^2 dr = 4\pi c_j\rho_0 \int_{r_A}^{r_B} \rho_{ij}(r)r^2 dr \tag{4.2}$$

where $g_{ij}(r) = 4\pi r^2\rho_{ij}(r)$ is the radial PDF. If one is interested in calculating the number of oxygen ions in the first coordination shell of a cation X, the integration runs between $r_A = 0$ and $r_B = r_{cut}$; the latter is the distance at the minimum after the first peak of the PDF (see Figure 4.4).

Very important correlation functions that can be extracted from X-ray and Neutron diffraction experiments are the radial distribution function (RDF, $G(r)$), the total correlation function (TCF, $T(r)$), and the differential correlation function (DFC, $D(r)$) which are defined as follows:

$$G(r) = \sum_{i,j} c_i c_j g_{ij}(r) = 4\pi r^2 \sum_{i,j} c_i c_j \rho_{ij}(r) \tag{4.3}$$

$$T(r) = \sum_{i,j} c_i t_{ij}(r) = 4\pi r \sum_{i,j} c_i c_j \rho_0 \rho_{ij}(r)$$

$$t_{ij}(r) = 4\pi r\rho_j^0\rho_{ij}(r) \tag{4.4}$$

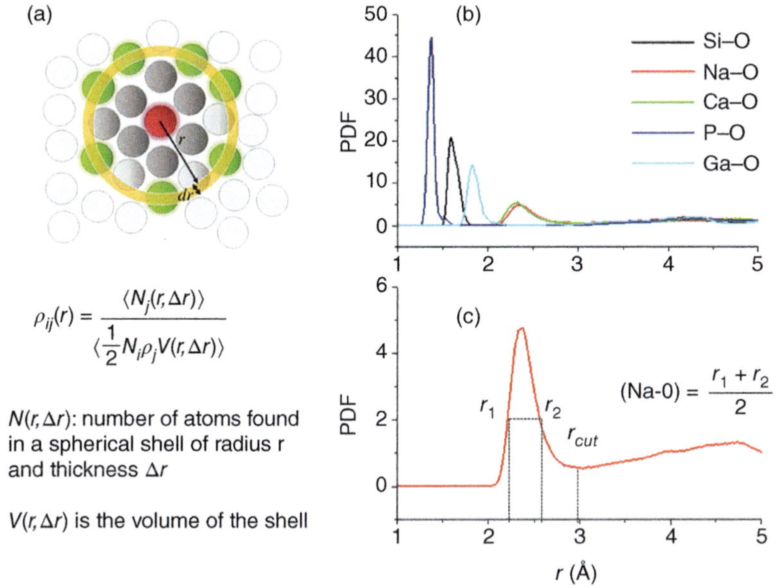


**Figure 4.4** (a) Distribution of atoms and definitions of the pair distribution function. (b) PDF for the Si–O, Na–O, Ca–O, P–O, and Ga–O pairs in the Ga-containing 45S5 bioglass simulated in Ref. [10]. (c) Na–O PDF showing the cutoff used to compute the coordination number of the firs oxygen shell of Na and the way to compute the average Na–O bond distance.

$$D(r) = \sum_{i,j} c_i d_{ij}(r) = \sum_{i,j} c_i[t_{ij}(r) - t_j^0(r)] = 4\pi r \sum_{i,j} \rho_j^0(\rho_{ij}(r) - 1)$$
$$d_{ij}(r) = 4\pi r \rho_j^0[\rho_{ij}(r) - 1] \tag{4.5}$$

where $t_j^0(r) = 4\pi r \rho_j^0$ is the average density contribution and the summation runs on the number of species in the system.

Among these functions, the $T(r)$ is usually extracted from X-ray and Neutron experiments since it is in this function that experimental broadening is both symmetric and $r$-independent.

Therefore, an efficient way to summarize and validate the results of the MD simulations of disordered glasses consists of comparing the computed TCF $T^{sim}(r)$ with appropriate diffraction data, $T^{exp}(r)$, the accuracy of the fit being measured by the $R_x$ factor proposed by Wright [27]:

$$R_x = \left[ \frac{\sum_{i=1}^{N} \left(T^{exp}(r_i) - T^{sim}(r_i)\right)^2}{\sum_{i=1}^{N} (T^{exp}(r_i))^2} \right] \tag{4.6}$$

However, in order to compare the simulated TCF with the experimental one, the former must be broadened to account for the limit of the momentum transfer ($Q_{max}$) in diffraction experiments [28].

To do so, the partial correlation function $t_{ij}(r)$ is first broadened by convoluting with a component peak function $P_{ij}(r)$ defining the experimental resolution:

$$t'_{ij}(r) = \int_0^\infty t_{ij}(r')[P_{ij}(r - r') - P_{ij}(r + r')]dr' \tag{4.7}$$

For neutron experiments, the peak component function $P_{ij}^N(r)$ si defined as follows:

$$P_{ij}^N(r) = \frac{\overline{b_i}\,\overline{b_j}}{\pi} \int_0^{Q_{max}} M(Q)\cos(rQ)dQ \tag{4.8}$$

where $\overline{b_i}$ is the average neutron scattering length of atom $i$, $Q_{max}$ is the maximum value of the experimental scattering vector, and $M(Q)$ is a modification function.

A Lorch function is usually employed as modification function since it provides a gradual cutoff:

$$M(Q) = \begin{cases} \dfrac{\sin \Delta r}{\Delta r Q} & Q \le Q_{max} \\ 0 & Q > Q_{max} \end{cases} \tag{4.9}$$

Instead, in the case of X-ray diffraction experiments, the component peak function assumes the form:

$$P_{ij}^X(r) = \frac{1}{\pi} \int_0^{Q_{max}} \frac{f_i(Q) f_j(Q)}{f_e^2(Q)} M(Q)\cos(rQ)dQ \tag{4.10}$$

$$f_e(Q) = \frac{\sum_i |f_i(Q)|}{\sum_i Z_i}$$

where $f_i(Q)$ is the X-ray form factor of atom $i$, whereas $f_e(Q)$ is a sharpening function used to transform electron–electron correlation function usually obtained by X-ray diffraction to pseudo-atomic correlation functions [28].

Figure 4.5 shows the comparison of experimental and simulated TCF $T(r)$ for the vitreous silica and the $30Na_2O{\cdot}70SiO_2$ glass [31]. The $R_x$ values for these systems are 5.5% and 5.1%, respectively, when the Pedone-Malavasi-Menziani-Cormack-Segre (PMMCS) force field is used in classical MD simulations [32, 33].

## 4.3 Spectroscopic Properties: Validating the Structural Models

### 4.3.1 X-ray and Neutron Diffraction Spectra

The total scattering structure factor that defines the scattering of a beam of neutrons from an ensemble of disordered atoms for a given scattering vector Q is given by the formula:

$$F^N(Q) = \sum_{i,j=1}^{n} c_i c_j \overline{b_i}\,\overline{b_j}[A_{ij}(Q) - 1] \tag{4.11}$$

where $A_{ij}(Q)$ are the Faber–Ziman partial structure factors defined as follows:

$$A_{ij}(Q) = 1 + 4\pi\rho_0 \int_0^\infty r^2[\rho_{ij}(r) - 1]\frac{\sin Qr}{Qr}dr \tag{4.12}$$

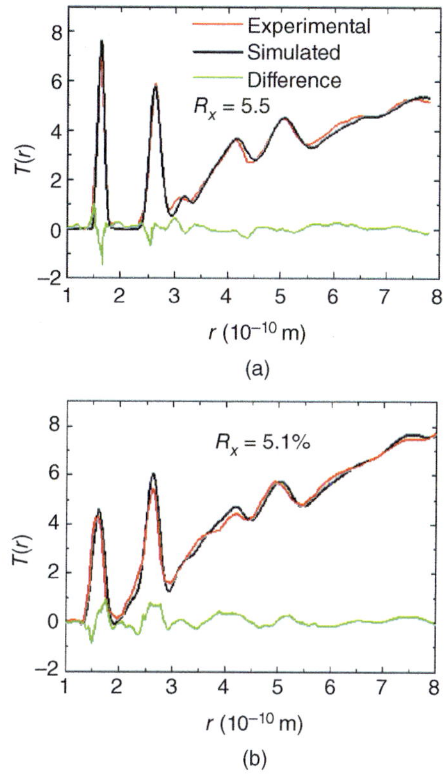


**Figure 4.5** (a) Computed total distribution function of $SiO_2$ glass, broadened using the experimental $Q_{max} = 45.2$ Å$^{-1}$, compared with the neutron diffraction data digitalized from Grimley et al. [29]. (b) Computed total distribution function of the $30Na_2O70SiO_2$ glass, broadened using the experimental $Q_{max} = 22.88$ Å$^{-1}$, compared with the neutron diffraction data digitalized from Wright et al. [30]. Source: Simulated data were taken from Pedone et al. [31] and Pedone et al. [32].

However, because of the finite size of simulation cells, the integration is usually performed to a maximum value $R_{max}$ that equals half the smallest side of the simulation box. To reduce the ripples at low $Q$ values arising because of the size effect, the Lorch-type windows function is employed and Eq. (4.12) becomes

$$A_{ij}(Q) = 1 + 4\pi\rho_0 \int_0^{R_{max}} r^2[\rho_{ij}(r) - 1]\frac{\sin Qr}{Qr}\frac{\sin(\pi r/R)}{\pi r/R}dr \tag{4.13}$$

Sometime, in neutron experiments, the normalized total-scattering structure factor, $S^N(Q)$, is collected instead of $F^N(Q)$. The former is defined as

$$S^N(Q) - 1 = \frac{F^N(Q)}{\left(\sum_{i=1}^{n} c_i\overline{b}_i\right)^2} \tag{4.14}$$

which, when rearranged, leads to

$$S^N(Q) = \left(\sum_{i=1}^{n} c_i\overline{b}_i\right)^{-2} \sum_{i,j=1}^{n} c_i c_j \overline{b}_i \overline{b}_j A_{ij}(Q) \tag{4.15}$$

The normalized X-ray total scattering structure factor, $S^X(Q)$, can be derived by exchanging the coherent bound neutron scattering lengths, $\overline{b}_j$ with Q-dependent X-ray scattering

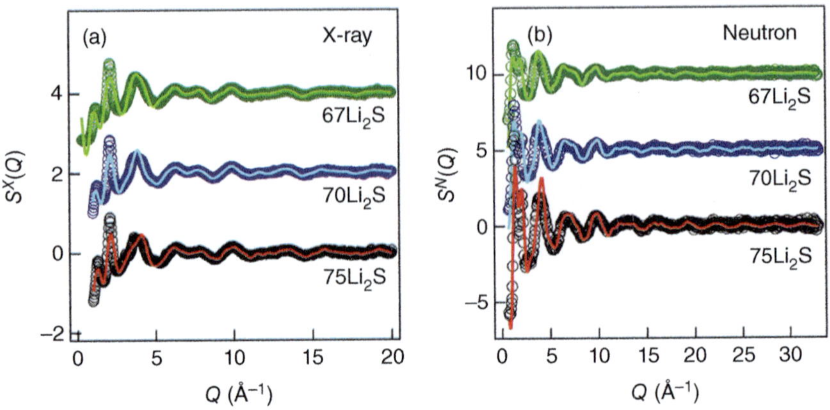


**Figure 4.6** Total structure factors $S(Q)$ at room temperature for $Li_2S-P_2S_5$ glasses derived from (a) X-ray and (b) neutron diffraction. Circles, experimental data; lines, DFT/RMC model. Source: Reprinted with permission from Ohara et al. [34]. Copyright © 2016, Rights Managed by Nature Publishing Group.

factors, $f_i(Q)$ in Eq. (4.15) to obtain

$$S^X(Q) = \sum_{i,j=1}^{n} c_i c_j f_{ij}(Q) A_{ij}(Q) \tag{4.16}$$

where

$$f_{ij}(Q) = \frac{f_i(Q) f_j(Q)}{\left[\sum_{i=1}^{n} c_i f_i(Q)\right]^2} \tag{4.17}$$

Figure 4.6 shows the comparison between experimental and simulated (a) X-ray and (b) neutron diffraction total structure factors $S(Q)$ at room temperature for $Li_2S-P_2S_5$ glasses. The simulated spectra are based on density functional theory (DFT)/reverse Monte Carlo (RMC) models generated by Ohara et al. [34].

### 4.3.2 Vibrational Spectra

To compute vibrational spectra, two methods can be applied, the so-called "static method" in which the mass-weighted Hessian matrix is computed on the optimized geometry of the system, and the dynamical method in which vibrational properties are computed exploiting molecular dynamics simulations and time-correlation functions. In the next two paragraphs, we briefly describe the theoretical background behind these two methods.

#### 4.3.2.1 The "Static" Method

The so-called "static" method to compute vibrational (IR, Raman) properties of amorphous (and any other kind of) systems relies on the harmonic approximation.

In this framework, the frequencies of the normal modes together with the displacements of the individual atoms are calculated at both quantum mechanics (QM) and molecular mechanics (MM) levels using the Hessian matrix of the second derivatives:

$$H_{i,j} = \left( \frac{\partial^2 V(\mathbf{r})}{\partial r_i \partial r_j} \right)_0 \tag{4.18}$$

where $\boldsymbol{r}$ is the collective $(x,y,z)$ Cartesian displacement coordinates of atoms $i$ and $j$, and the second derivatives are carried out on local minima of the potential energy surface.

The Hessian is first converted to the equivalent force-constant matrix (**F**) in mass-weighted coordinates ($u_i = m_i^{1/2} r_i$), as follows:

$$\mathbf{F} = \mathbf{M}^{-1/2} \cdot \mathbf{H} \cdot \mathbf{M}^{\frac{1}{2}} \tag{4.19}$$

where **M** is a diagonal matrix of dimension $3N \times 3N$, containing the atomic masses of the atoms in the systems:

$$\mathbf{M} = \begin{bmatrix} m_1 & \cdots & 0 \\ \vdots & \ddots & \vdots \\ 0 & \cdots & m_{3N} \end{bmatrix} \tag{4.20}$$

Subsequently, the matrix **F** is diagonalized through a similitude transformation

$$\mathbf{L}^{-1} \cdot \mathbf{F} \cdot \mathbf{L} = \begin{bmatrix} \lambda_1 & \dots & 0 \\ 0 & \ddots & 0 \\ 0 & \dots & \lambda_{3N} \end{bmatrix} \equiv \boldsymbol{\Lambda} \tag{4.21}$$

$$\mathbf{L} = \begin{bmatrix} l_{11} & \cdots & l_{1,3N} \\ \vdots & \ddots & \vdots \\ l_{3N,1} & \cdots & l_{3N,3N} \end{bmatrix}$$

where $\boldsymbol{\Lambda}$ is the matrix containing the eigenvalues $\lambda_k$ ($k = 1, \dots, 3N$), whereas **L** is a matrix containing the $k$-th column and the eigenvectors corresponding to the eigenvalue $\lambda_k$.

It is important to know that if the Hessian is defined in terms of Cartesian coordinates, then six of these eigenvalues correspond to translational and rotational motion of the entire system and will be equal to zero.

The frequency of each normal mode is computed from the relationship:

$$\nu_k = \frac{\sqrt{\lambda_k}}{2\pi} \tag{4.22}$$

whereas the $k$-th normal mode ($Q_k$) is given by the linear combination:

$$Q_k = \sum_{j=1}^{3N} l_{kj} u_j \tag{4.23}$$

Therefore, in the quantum picture, the energy levels of the $k$-th normal mode and the total energy of a system with $3N - 6$ vibrational degrees of freedom are given by

$$E_{\nu_k} = \left( \nu_k + \frac{1}{2} \right) \hbar\omega_k \rightarrow E_{vib} = \sum_{k=1}^{3N-6} \left( \nu_k + \frac{1}{2} \right) \hbar\omega_k \tag{4.24}$$

with $v_k = 0, 1, 2\ldots$; notably, when the system is in its fundamental state ($v_k = 0$), the vibrational energy is not null, and it is called *zero-point vibrational energy* (*ZPVE*):

$$E_{ZPVE} = \frac{1}{2}\sum_{k=1}^{3N-6} \hbar\omega_k \tag{4.25}$$

Such a quantity is relevant for small-to-large molecular systems, but becomes negligible when considering extended systems (thousands of atoms) as, for example, a simulation box for calculations on glassy/amorphous materials, or a finite temperature – where thermal energy contributions dominate over ZPVE.

Vibrational IR intensities can be derived considering the *transition dipole moment* between an initial and a final (vibrational) state (the wave functions $\psi_i$ and $\psi_f$, respectively) which is, in bra-ket notation, $\langle\psi_i^k|\boldsymbol{\mu}|\psi_f^k\rangle$. The dipole strength $T_{i\to f}^k$ for a single excitation is thus $T_{i\to f}^k = |\langle\psi_i^k|\boldsymbol{\mu}|\psi_f^k\rangle|^2$.

$T_{i\to f}^k$ enters the expression of the *molar attenuation coefficient* of infrared radiation:

$$\varepsilon(v) \propto N_A \sum_k p_i^k\, T_{i\to f}^k \tag{4.26}$$

where $N_A$ is the Avogadro's number and $p_i^k$ is the fractional population for the state $i$. This provides a physical quantity for the intensity of IR bands, which is determined experimentally through the famous macroscopic law of Lambert–Beer: $A(v) = \varepsilon(v)M\ell$, where $M$ and $\ell$ stand for the molar concentration of the sample in solution and the path length of radiation through the sampling cell.

For ordered and disordered periodic systems (insulators only), thus defined in the context of a unit cell – or a simulation box – with periodic boundary conditions, an effective numerical method used to calculate the dipole strength is the Berry phase approach [35–37].

It consists of computing the derivative of the dipole moment with respect to the Cartesian atomic displacements (the *atomic effective Born charge tensor* $\mathbf{Z}_{k,i}$, for a set of atomic coordinates $i = (x, y, z)$), as polarization differences between the equilibrium and deformed geometry.

The oscillator strength for the $k$-th mode $T_k$ between particles $A$ and $B$, thus becomes

$$T_k = \frac{4\pi}{V}\frac{\mathbf{Z}_{k,A}\mathbf{Z}_{k,B}}{v_k^2} \tag{4.27}$$

where $V$ is the unit cell volume and $v_k^2$ the transverse (TO) frequency.

In Figure 4.7, the IR spectra of the 45S5 Bioglass® [38] computed with the aforementioned method at the B3LYP level is compared with the experimental one. Figure 4.7 shows that if a good structural model is available, periodic DFT calculations provide IR spectra in very good agreement with experimental ones and allows the assignments of the major and more important bands [39, 40].

#### 4.3.2.2 The "Dynamical" Method

Vibrational properties such as IR spectra and vibrational density of states can be computed also in a time-dependent approach by exploiting molecular dynamics simulations and time-correlation functions.

For example, IR spectrum of a system can be computed via the Fourier transformation of the dipole autocorrelation function:

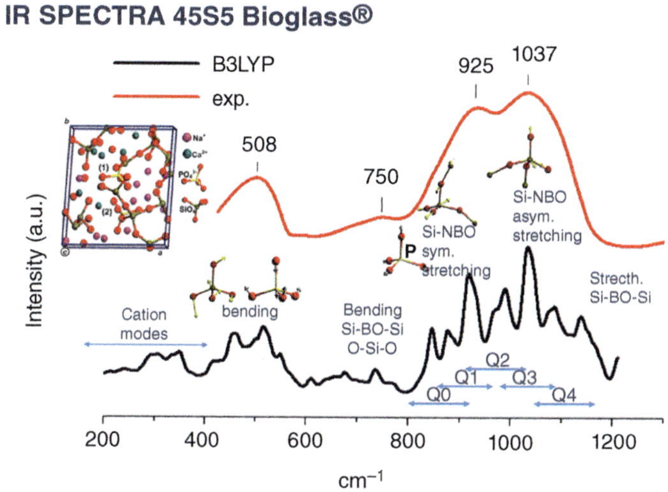


**Figure 4.7** Experimental (red line) and B3LYP (black line) IR spectra of the 45S5 Bioglass®.

$$IR(\omega) = \int_{-\infty}^{\infty} c_{dip}(t)e^{-i\omega t}dt \tag{4.28}$$

with

$$c_{dip}(t) = \frac{\langle \boldsymbol{\mu}_{TOT}(t) \cdot \boldsymbol{\mu}_{TOT}(0)\rangle}{\langle \boldsymbol{\mu}_{TOT}(0) \cdot \boldsymbol{\mu}_{TOT}(0)\rangle} \tag{4.29}$$

$$\boldsymbol{\mu}_{TOT}(t) = \sum_{i=1}^{N} \boldsymbol{\mu}_i(t)$$

where $\boldsymbol{\mu}_i(t)$ is the dipole molecule of particle $i$ at time $t$.

Instead, the phonon spectra (also called vibrational density of states, VDOS) of a system can be simulated by Fourier transforming the velocity autocorrelation function [41]:

$$VDOS(\omega) \propto 2\int_{0}^{\infty} C_{vv}(t)\,\cos(\omega t)\,dt \tag{4.30}$$

where

$$C_{vv}(t) = \frac{1}{N}\sum_{i}^{N} \frac{\langle \mathbf{v}_i(t) \cdot \mathbf{v}_i(0)\rangle}{\langle \mathbf{v}_i(0) \cdot \mathbf{v}_i(0)\rangle} \tag{4.31}$$

where the ensemble average on velocity with respect to time, $\langle \mathbf{v}_i(t) \cdot \mathbf{v}_i(0)\rangle$, is indicated by brackets.

Figure 4.8 reports the comparison between the experimental and computed (at the QM and MM level) VDOS of vitreous-$SiO_2$ with this method.

In this particular case, it is possible to note that the VDOS spectrum calculated with a molecular mechanics force field (FFSiOH), parametrized to model properly silica-based glasses, furnishes a very good estimate of the band shapes, with respect to a quantum calculation at DFT level adopting the local density approximation.

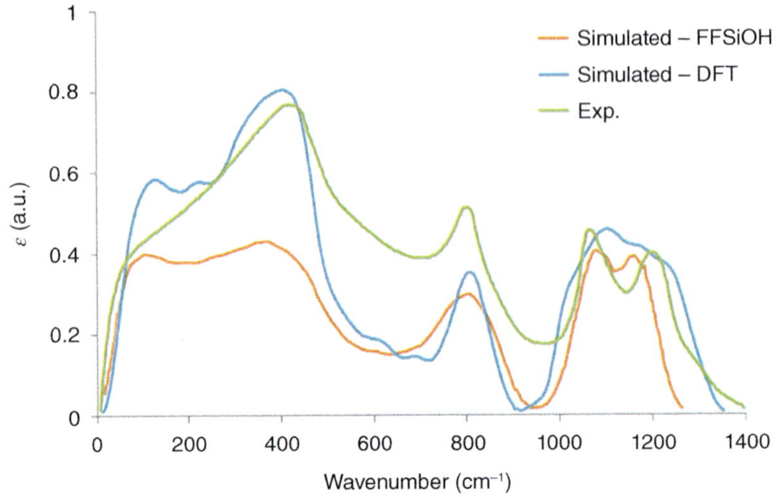


**Figure 4.8** Comparison between the simulated VDOS from density functional theory [42] (blue line), and from MM calculations with the FFSiOH force-field (orange line) [43] with experimental VDOS determined through neutron inelastic scattering experiments, Refs. [44, 45].

### 4.3.3 NMR Spectra

As for the computational simulation of solid-state NMR, the first step is to compute the *chemical shielding* (CS) tensor of the active NMR nuclei constituting the system under investigation and, in case quadrupolar species are present, also the *electric field gradient* (EFG) tensor's eigenvalues and eigenvectors. These properties can be computed after geometry optimization at the ab initio quantum mechanical level, by using the gauge including projector augmented wave (GIPAW) approach [46].

Once these properties are computed, the NMR spectra are simulated by using software [47, 48] based on *spin-effective Hamiltonians*, that allow to compute the resonance frequency (RF) of a single $m$-to-$n$ spin-transition as follows:

$$\nu_{m,n} = \langle m|H_{eff}|m\rangle - \langle n|H_{eff}|n\rangle; \quad H_{eff} = H_{Zeeman} + H^1_{quad} + H^2_{quad} + H_{CS} \tag{4.32}$$

where the effective Hamiltonian of the equation above contains only the most relevant magnetic interaction contributions, i.e. Zeeman effect, first-, and second-order quadrupolar and chemical shift. The powder line shape for a single NMR-active nucleus $I_i$ can be thus computed by employing the following equation:

$$I_i(\nu) = \int_0^{\pi} \int_0^{2\pi} G\left(\nu - \nu_{m,n}(\phi, \theta)\right) \sin\theta d\theta d\phi \tag{4.33}$$

where the Euler angles $\phi, \theta$ relate the CS and quadrupolar tensors of single nuclei to the principal axis frame (PAF) of the rotor. $G$ is a Gaussian or Lorentzian-type convolution function on frequencies of single $m, n$ spin transitions, that finally gives the simulated NMR spectrum.

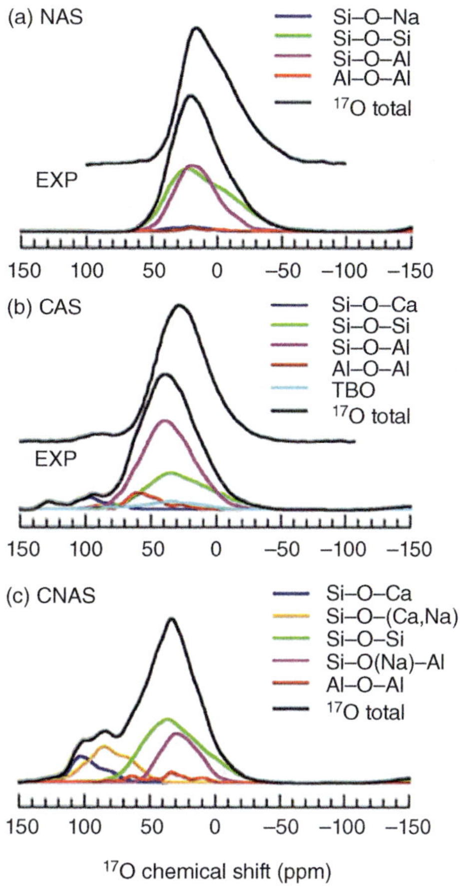


**Figure 4.9** Theoretical $^{17}$O MAS NMR spectra at 14.1 T for (a) Na-Al-Silica (NAS) (b) Ca-Al-Silica (CAS), and (c) Ca-Na-Al-Silica (CNAS) glasses. The contributions of different O sites are highlighted in color. Source: Experimental MAS-NMR spectra for (a) and (b) were taken from Les and Stebbins [49] (NAS) and Ref. [50] (CAS). Reprinted with permission from Pedone et al. [51]. Copyright 2012 American Chemical Society.

Finally, for 2D-NMR spectra, the direct summation method, where $I(\nu_1, \nu_2) = I(\nu_1) \times I(\nu_2)$ can be efficiently employed for each crystal orientation.

Figures 4.9 and 4.10 show the $^{17}$O MAS and 3QMAS NMR spectra of sodium and calcium silicate and aluminosilicate [51, 52].

A good reproduction of the experimental NMR spectra allowed the validation of the structural models generated and helped the experimentalists in the interpretation of the spectra, thus opening the possibility to obtain quantitative structural information of highly disordered systems from NMR experiments [11, 53, 54].

## 4.4 Transport Properties

Transport properties are related to the flow of material, energy, or momentum from one region to another of nonequilibrium systems in which a gradient of concentration, energy,

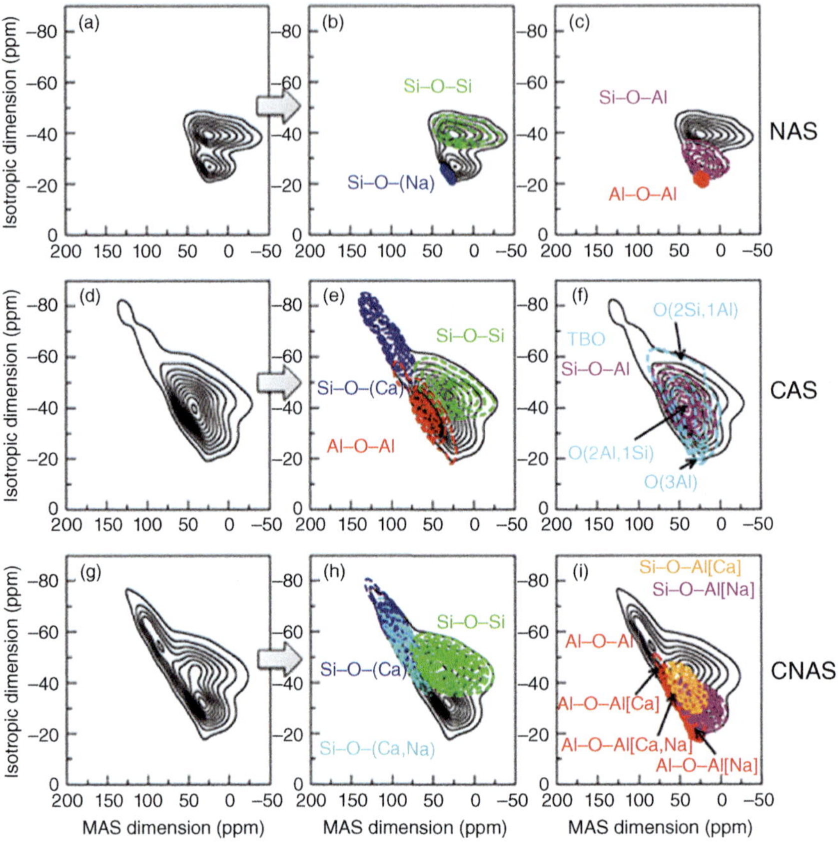


**Figure 4.10** Theoretical $^{17}O$ 3QMAS NMR for NAS (a), CAS (d) and CNAS (g) glasses. The contributions of different O sites are highlighted in colour in panels (b)–(c) for NAS, (e)–(f) for CAS and (h)–(i) for CNAS glasses. Source: Reprinted with permission from Pedone et al. [51]. Copyright 2012 American Chemical Society.

or momentum is present. This means that the transport properties relate the response of a system to a perturbation. For example, the diffusion constant relates the flux of mass of a certain species with its concentration gradient in the systems, and the thermal conductivity relates the heat flux with the temperature gradient, whereas the viscosity of a material arises from a momentum gradient applied to the system.

These nonequilibrium properties can be obtained by equilibrium molecular dynamics (EMD) simulations exploiting the microscopic local fluctuations occurring even in equilibrium systems or by, in a straight way, nonequilibrium molecular dynamics (NEMD) simulations.

In the framework of EMD simulations, the general transport coefficient, $\gamma$, can be computed by the Green–Kubo relation [55, 56] as the infinite time integral of an equilibrium correlation function of the variable $A$ related to the perturbation term:

$$\gamma = \int_0^\infty dt\langle \dot{A}(t)\dot{A}(0)\rangle \tag{4.34}$$

Associated with the expressions of this type, there is also an Einstein's relation of the form

$$2t\gamma = \langle (A(t) - A(0))^2 \rangle \tag{4.35}$$

This expression holds for large $t$ if compared to the correlation time of $A$.

### 4.4.1 Diffusion Coefficient and Diffusion Activation Energy

The Green–Kubo relation for the diffusion coefficient is

$$D = \frac{1}{3}\int_0^\infty dt\langle \mathbf{v}_i(t) \cdot \mathbf{v}_i(0)\rangle \tag{4.36}$$

where $\mathbf{v}_i$ is the velocity of the centre of mass of molecule $i$. The Einstein relation valid for long times is

$$2tD = \frac{1}{3}MSD(t) \tag{4.37}$$

where MSD is the Mean Square Displacement which is given by

$$MSD(t) = \left\langle (r_i(t) - r_i(0))^2 \right\rangle \tag{4.38}$$

where $r_i(t)$ and $r_i(0)$ are the positions of the $i$-th atom at time $(t + t_0)$ and at the reference time $t_0$, respectively, and the brackets represent the ensemble average over the total number of atoms.

Plotting the MSD versus the simulation time, two regions can be found: the first region is called the *ballistic transport region*, where the transport takes place when not enough collisions occurred and the random walk regime is still not reached. In this region, the MSD is not linearly dependent on the simulation time, but usually the MSD goes as the time squared ($MSD \approx t^2$) (see Figure 4.11). The second region is characteristic of longer simulation times, and it is called the *random walk diffusion*. Here the MSD is proportional to the simulation time through the diffusion coefficient, which is at last the slope of the MSD at long times, see Figure 4.11.

The same calculation may be done for a subset of $n$ atoms (with $n \leq N$) of the total system in order to take into account for different atom species, such as for only Li ions in a $LiAlSi_2O_6$ glass [58].

The diffusion process depends not only on the atomic species involved but also on the system temperature and on the number of vacancies available.

The diffusion activation energy, $Q_d$, is the barrier that the atom has to overcome to break interatomic bonds and to jump into a vacancy, as schematically described in Figure 4.11. An atom overpassing an activation energy barrier requires much more energy, $Q_d \sim 1$ eV/atom, than the average thermal energy ($k_B T = 0.026$ eV at room temperature). This process takes place when a large fluctuation in energy is condensed in a small volume and the atom has enough energy to "jump" into a vacancy. Considering a frequency attempt, $R_0$, which is

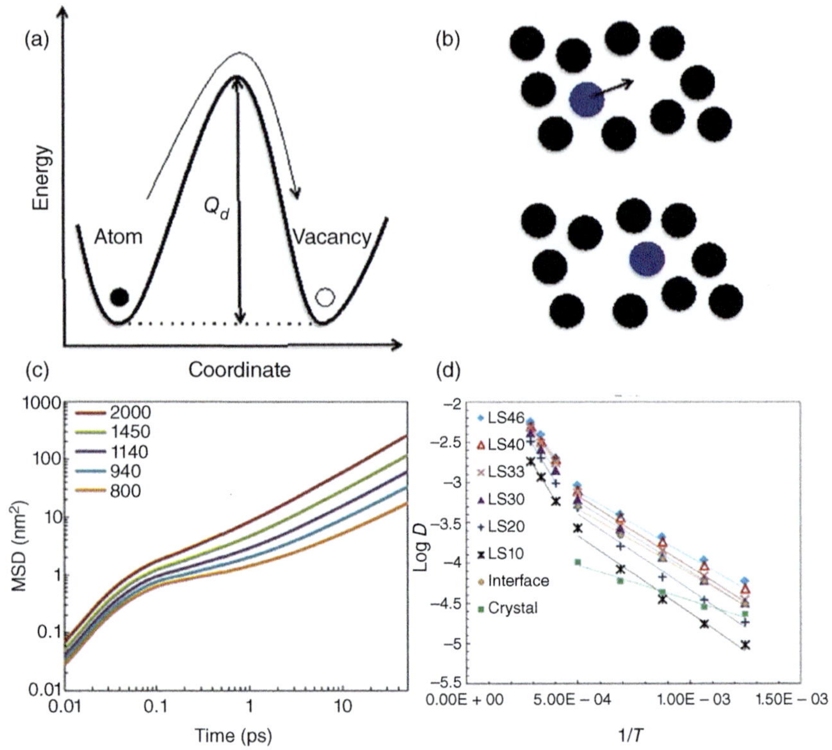


**Figure 4.11** In panel (a) the schematic representation of the diffusion activation energy and in panel (b) the motion of an atom, in blue, from its reference position to the vacancy. In (c) the MSD versus simulation time for Lithium in $Li_2Si_2O_5$ glass at various temperatures. The ballistic transport regime can be observed up to 0.1 ps. In (d) the logarithmic plot of diffusion coefficient versus 1/T of Li ions in lithium silicate glasses with different amount of $Li_2O$. Source: Figures (c) and (d) were obtained from Ren and Du [57]. Copyright © 2021 by Elsevier. Reprinted by permission of Elsevier.

proportional to the frequency of atomic vibrations, the probability of such fluctuation, $R_j$, is given by the Arrhenius equation:

$$R_j(T) = R_0 e^{-\frac{Q_d}{k_B T}} \tag{4.39}$$

From Eq. (4.39), the diffusion coefficient temperature-dependent, $D(T)$, can be derived

$$D(T) = D_0 e^{-\frac{Q_d}{RT}} \tag{4.40}$$

where $D_0$ is the diffusion coefficient temperature-independent, $R$ is the gas constant, and $T$ is the absolute temperature expressed in Kelvin [59]. Taking the logarithm to the base 10 (or the natural logarithm), of Eq. (4.40), we obtain that

$$\log D = \log D_0 - \frac{Q_d}{2.3R}\left(\frac{1}{T}\right) \tag{4.41}$$

Therefore, plotting $\log D$ versus $1/T$, a straight line with slope of $-Q_d/2.3R$ and intercept $\log D_0$ is obtained and both $D_0$ and $Q_d$ can be estimated.

Several works have been focused on the determination of the energy barriers for diffusion of first and second group cations such as Li, Na, K, Ca in oxide glasses [58, 60–63].

### 4.4.2 Viscosity

Shear viscosity can be calculated using the Green–Kubo relation in EMD or NEMD simulations or the Transverse-current autocorrelation-function (TCAF). The latter method will not be described here and thus the interested reader is referred to the paper by Palmer [64].

The Green–Kubo shear viscosity calculation is based on the relation

$$\eta = \frac{V}{k_B T} \int_0^\infty dt \langle P_{\alpha\beta}(t) \cdot P_{\alpha\beta}(0) \rangle \tag{4.42}$$

where $\eta$ is the shear viscosity, $V$ is the volume, $T$ is the temperature, $k_B$ is the Boltzmann's constant, and $P_{\alpha\beta}$ are the off-diagonal components of the pressure tensor $P$. Considering a $N$-particle system, the microscopic stress tensor is given by

$$P_{\alpha\beta} = \frac{1}{V}\left( \sum_{i=1}^{N} \frac{p_{i\alpha} p_{i\beta}}{m_i} + \sum_{i=1}^{N} \sum_{j>i}^{N} r_{ij\alpha} f_{ij\beta} \right) \tag{4.43}$$

where $m_i$ is the mass of the $i$-th atom, $p_{i\alpha}$ is the $\alpha$-component of the momentum of the $i$-th atom, and $r_{ij\alpha}$ is the $\alpha$-component of the distance between atoms $i$ and $j$, whereas $f_{ij\beta}$ is the $\beta$-component of the force between atom $i$ and $j$.

Since the **P** tensor is a property of the system as a whole no additional averaging over the $N$ particles is possible and as a consequence, the viscosity is subjected to a greater statistical error than the diffusion coefficient. Some improvements can be reached by averaging over different components or by computing the bulk viscosity which is defined as follows:

$$\eta_V = \frac{V}{9k_B T} \sum_{\alpha\beta} \int_0^\infty dt \langle \delta P_{\alpha\alpha}(t) \delta P_{\beta\beta}(0) \rangle \tag{4.44}$$

Another possible method to compute the shear viscosity is by using NEMD simulations. These simulations are performed applying an external force toward a direction on the system in order to create a shear flow. The viscosity is then computed as the response of the system to the external force obtaining a very fast convergence respect to EMD simulations, but resulting in the loss of supplementary information and in the need to run simulations with different directions of forces to compute the viscosity along other axis.

### 4.4.3 Thermal Conductivity

The thermal conductivity, $\lambda$, describes the relation between the heat current, $J_\alpha$, and the temperature gradient, $\frac{\partial T}{\partial x_\beta}$ through the Fourier's law

$$J_\alpha = -\sum_\beta \lambda_{\alpha\beta} \frac{\partial T}{\partial x_\beta} \tag{4.45}$$

The measure of $\lambda$ by experiments is typically done imposing a heat current to the system and measuring the resulting temperature gradient, while in computer simulations, two main approaches are possible: the "direct method" based on NEMD simulations and the Green–Kubo method based on EMD simulations.

The "direct method" is similar to the experimental procedure and it relies in the application of a temperature gradient across the simulation box which is divided into three main regions as shown in Figure 4.12. The same amount of energy is added ($+\Delta\varepsilon$) and removed ($-\Delta\varepsilon$) on two different areas creating a heat current across the region in the middle of the box [65]. This method produces a heat current along a single crystal lattice dimension obtaining a value for $\lambda$ for that direction. To obtain $\lambda$ for the others directions, new simulations with different simulation cells must be carried out. In the direct method, it is important to note that very large temperature gradients, in the order of $10^9$ K/m, are used that are over the experimental range. This leads to the inability to apply the Fourier's law due to nonlinear response effects of the system. Another important aspect is that the calculation must be performed after the system has reached a steady-state current flow meaning that the temperature is constant during time. It is worthy to note that the system size could affect the thermal conductivity measurement when the phonon mean-free path is comparable to the box length, as reported by Schelling et al. [66].

Figure 4.12b shows the dependence of thermal conductivity of silica glass computed by Jund et al. [65] and reveals that their data are in very good agreement with experimental ones when a finite size correction is applied to the simulations.

In the framework of EMD simulations, the thermal conductivity is computed by using the Green–Kubo relation:

$$\lambda_{\alpha\beta} = \frac{1}{Vk_BT^2}\int_0^{t_{max}} dt\langle J_\alpha(t)J_\beta(0)\rangle \tag{4.46}$$

where $\alpha$ and $\beta$ are the Cartesian components $x$, $y$, $z$, and the brackets denote an ensemble average, or, in the case of MD simulations, a time average.

In a practical MD simulation with a timestep $\Delta t$, the formula usually employed is

$$\lambda_{\alpha\beta}(t_{max}) = \frac{\Delta t}{Vk_BT^2}\sum_{m=1}^{M}(N-m)^{-1}\sum_{n=1}^{N-m} J_\alpha(m+n)J_\beta(n) \tag{4.47}$$

where $t_{max} = M\Delta t$ and $J_\alpha(m+n)$ are the $\alpha$-component of the heat current at the MD step $m+n$. $M$ is the number of steps of integration that must be less than the total number of steps $N$.

The heat current is written as

$$\boldsymbol{J}(t) = \frac{d}{dt}\left(\sum_i \boldsymbol{r}_i(t)\varepsilon_i(t)\right) \tag{4.48}$$

where $\varepsilon_i$ is the site energy, and $\boldsymbol{r}_i$ is the vector position at time $t$ of atom $i$. In the case of a pairwise potential, $U_2(r_{ij})$, between atoms $i$ and $j$, the site energy is given by

$$\varepsilon_i = \frac{1}{2}m_i\boldsymbol{v}_i^2 + \frac{1}{2}\sum_j U_2(r_{ij}) \tag{4.49}$$

For this definition, the thermal current can be written as

$$\boldsymbol{J}(t) = \sum_i \boldsymbol{v}_j\varepsilon_i + \frac{1}{2}\sum_i\sum_{j\neq i}\boldsymbol{r}_{ij}(\boldsymbol{F}_{ij}\cdot\boldsymbol{v}_i) \tag{4.50}$$

where $\boldsymbol{F}_{ij}$ is the force on atom $i$ from atom $j$ given by the pair potential. $J_i$ is also called the Irving–Kirkwood heat flux [67].

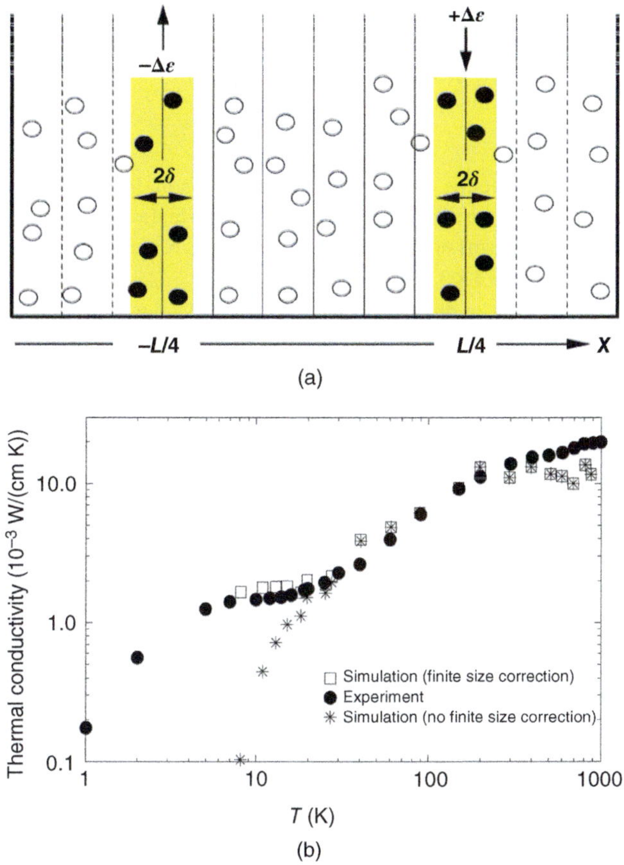


**Figure 4.12** In (a) schematic representation of the NEMD simulation box with length $L$, while regions into which energy is added/removed have length $2\delta$. In (b) the temperature dependence of thermal conductivity of silica glasses [65]. Source: Reprinted figure with permission from Jund and Jullien [65]. Copyright (1999) by the American Physical Society.

Shelling et al. used both the direct method and the Green–Kubo method for computing the thermal conductivity of silicon and found that the two different methods yield essentially the same results.

However, the advantage of the second method with respect to the former is that it allows calculating the heat current for all directions using a single simulation.

## 4.5 Mechanical Properties

Most of the potential applications of glasses are hindered by their poor mechanical properties. In fact, since glasses are usually regarded as "brittle" materials, they are seldom employed in applications requiring high levels of tensile stress [68]. Thus, in order to either predict or corroborate experimental data on these systems and provide useful insights for new applications, accurate ways to compute mechanical properties are needed. Here we shall focus mainly on calculating the elastic constant of these materials and simulating the stress–strain diagrams and the fracturing mechanism.

### 4.5.1 Elastic Constants

The fundamental quantity needed to define elastic properties is the strain, i.e. the relative change in the position of points within a body that has undergone deformation due to some external force. It can be expressed in terms of the strain tensor, which is represented microscopically in Cartesian coordinates as a $3 \times 3$ matrix:

$$\varepsilon = \begin{pmatrix} \dfrac{\partial u_x}{\partial x} & \dfrac{1}{2}\left(\dfrac{\partial u_z}{\partial y} + \dfrac{\partial u_y}{\partial x}\right) & \dfrac{1}{2}\left(\dfrac{\partial u_x}{\partial z} + \dfrac{\partial u_z}{\partial x}\right) \\ \dfrac{1}{2}\left(\dfrac{\partial u_y}{\partial x} + \dfrac{\partial u_z}{\partial y}\right) & \dfrac{\partial u_y}{\partial y} & \dfrac{1}{2}\left(\dfrac{\partial u_y}{\partial z} + \dfrac{\partial u_z}{\partial y}\right) \\ \dfrac{1}{2}\left(\dfrac{\partial u_z}{\partial x} + \dfrac{\partial u_x}{\partial z}\right) & \dfrac{1}{2}\left(\dfrac{\partial u_z}{\partial y} + \dfrac{\partial u_y}{\partial z}\right) & \dfrac{\partial u_z}{\partial z} \end{pmatrix} \tag{4.51}$$

with $\boldsymbol{u} = (u_x, u_y, u_z)$ being the infinitesimal displacement vector along the three principal Cartesian direction $x$, $y$, and $z$. This tensor is clearly symmetric, since $\varepsilon_{ij} = \varepsilon_{ji}$, thus only six components are needed to define it. Due to this, it is often reported and treated not as a full-fledged $3 \times 3$ matrix but as a six-components vector, as

$$\varepsilon = (\varepsilon_{xx}, \varepsilon_{yy}, \varepsilon_{zz}, \varepsilon_{xy}, \varepsilon_{xz}, \varepsilon_{yz}) = (\varepsilon_1, \varepsilon_2, \varepsilon_3, \varepsilon_4, \varepsilon_5, \varepsilon_6) \tag{4.52}$$

Hereafter, we shall refer to it in the latter vector form to adhere to common practice.

In order to compute mechanical or elastic properties from static calculations, three basic methods are available: second derivative, constant stress minimization, and constant strain minimization. All these techniques can be adopted to calculate the stiffness matrix ($\boldsymbol{C}$) and the compliance matrix ($\boldsymbol{S}$). With the first method, the structures of the glasses are optimized via energy minimization and the elastic properties are calculated from the curvature of the energy surface at the minimum, thus performing a second derivative calculation to obtain the stiffness matrix elements as

$$C_{ij} = \frac{1}{V}\left(\frac{\partial E}{\partial \varepsilon_i \partial \varepsilon_j}\right) \tag{4.53}$$

with $V$ and $E$ representing the total volume and the energy of the system, respectively, and $\varepsilon_i$ being the $i$-component of the strain vector as defined before. For isotropic materials, the stiffness has only three types of nonvanishing components (in turn, depending on only two elements, $a$ and $b$), since it can be represented in following $6 \times 6$ matrix form:

$$\boldsymbol{C} = \begin{pmatrix} a & b & b & 0 & 0 & 0 \\ b & a & b & 0 & 0 & 0 \\ b & b & a & 0 & 0 & 0 \\ 0 & 0 & 0 & c & 0 & 0 \\ 0 & 0 & 0 & 0 & c & 0 \\ 0 & 0 & 0 & 0 & 0 & c \end{pmatrix} \tag{4.54}$$

with elements $a = C_{11}$, $b = C_{12}$, and $c = \frac{1}{2}(a - b)$.

From this 6 × 6 stiffness matrix, the compliance matrix can then be computed with simple matrix inversion

$$\boldsymbol{S} = \boldsymbol{C}^{-1} \tag{4.55}$$

while, in turn, the Young's modulus components can be calculated from the compliance matrix as

$$E_i^{-1} = \sum_{j=1}^{3} S_{ij}\delta_{ji} \tag{4.56}$$

with $i$ and $j$ being two of the three principal components and $\delta_{ij}$ being the Kronecker's delta, such that $\delta_{ij} = 1$ if $i = j$, and 0 otherwise. The Young's modulus describes the response of a rigid system to a strain, and as a consequence basically yields the force (per unit area) that is needed to stretch (or compress) a rigid system; in the case of validity of Hook's harmonic law, the Young's modulus can be shown being proportional to the force constant $k = \sqrt{\omega/m}$ of the harmonic spring. It has also to be noticed that for isotropic materials, the three components of $E$ should be equal.

One can also define other scalar moduli, such as the bulk modulus $B$ yielding the pressure change due to a volume change. $B$ in Reuss notation is constructed from the upper-left quadrant of the compliance matrix $\boldsymbol{S}$ as

$$B^{-1} = \sum_{i,j=1}^{3} S_{ij} \tag{4.57}$$

while in Voight notation, it can be expressed as combination of matrix elements $C_{ij}$ from the upper-left quadrant of the stiffness matrix as

$$B = \frac{1}{9}\sum_{i,j=1}^{3} C_{ij} \tag{4.58}$$

Also the Shear modulus $G$ can be built, using the following equivalent formulas in Reuss and Voight notations, respectively,

$$G = \frac{15}{4(S_{11} + S_{22} + S_{33} - S_{12} - S_{13} - S_{23}) + 3(S_{44} + S_{55} + S_{66})}$$
$$G = \frac{1}{15}[C_{11} + C_{22} + C_{33} + 3(C_{44} + C_{55} + C_{66}) - C_{12} - C_{13} - C_{23}] \tag{4.59}$$

Poisson's ratio matrix elements are computed for six components from the compliance matrix as the ratio

$$\upsilon_{xy} = \frac{S_{21}}{S_{11}}, \upsilon_{xz} = \frac{S_{31}}{S_{11}}, \ldots, \upsilon_{zy} = \frac{S_{32}}{S_{22}} \tag{4.60}$$

Both $G$ and $\boldsymbol{\upsilon}$ describe the reaction of the system to stress, albeit along different directions. The isotropic Poisson's ratio is then obtained by averaging these six components. For isotropic materials, following the simplified expression for the **C** matrix, one can find that the following relations between Young's modulus, Bulk modulus, Shear modulus, and Poisson's ratios are satisfied:

$$B = \frac{E}{3(1 - 2\upsilon)}, G = \frac{E}{2(\upsilon + 1)} \tag{4.61}$$

**Table 4.2** Elastic properties of three alkali-doped silicate glasses (25% $M_2O$, M = Li, Na, K) and one aluminosilicate glass.

| | *E* (GPa) | | *G* (GPa) | | υ | | *B* (GPa) | |
|---|---|---|---|---|---|---|---|---|
| | calc | exp | calc | exp | calc | exp | calc | exp |
| Li | 78.52 | 78.44 | 31.87 | 31.74 | 0.232 | 0.236 | 48.84 | 48.93 |
| Na | 58.13 | 59.77 | 23.26 | 24.26 | 0.250 | 0.232 | 38.52 | 37.16 |
| K | 46.59 | 46.45 | 18.49 | 18.43 | 0.263 | 0.260 | 32.22 | 32.31 |
| Aluminosilicate | — | — | 95.60 | 93.30 | 0.280 | 0.300 | — | — |

Source: Data are taken from Pedone et al. [31], Pedone et al. [62], Bauchy [70], and Bansal and Doremus [71].

Some molecular dynamics codes, such as for example general utility lattice program (GULP) [69], come including routines to easily compute all these quantities. One can obtain very high levels of confidence with values close to the experimental ones, as shown in the following Table 4.2, which reports the elastic constant for alkali-silicate and aluminosilicate glasses previously investigated. As can be appreciated, the value of the Young's, Bulk, and Shear moduli as well as that of the Poisson's ratio are predicted within an error of 0.1–3% with respect to the experimental values.

### 4.5.2 Stress–Strain Diagrams and Fracture Mechanism

As hinted before, a major impediment for further advancement in the production of silicate-based glasses with specific properties suitable to new industrial applications is the fact that glasses undergo a catastrophic fracture (also called "brittle fracture") when subject to significant stress. As a consequence, a deeper understanding of the fracture mechanisms and intrinsic strength of silicate glasses is of crucial importance for the design of new ultrastrong materials for new applications in energy, medicine, building, and communication systems.

To describe the tensions inside a solid, it is thus relevant computing a quantity called stress tensor. The latter is defined as the following matrix:

$$\sigma_{ij} = -\frac{1}{V}\sum_{a}\left(\frac{p_{ia}p_{ja}}{m_a} + r_{ia}f_{ja}\right) \tag{4.62}$$

where $V$ is the total volume, $m_a$ is the mass of atom $a$, and $p_{ia}$, $r_{ia}$, and $f_{ia}$ are the $i$ components of the linear momentum, position, and force acting on particle $a$, respectively. Usually, it is often reported as a so-called "nominal" or "engineering" stress tensor calculated as

$$\sigma_{nom} = \sigma\frac{S(t)}{S_0} \tag{4.63}$$

with $S(t)$ and $S_0$ being the instantaneous and initial cross-sectional area perpendicular to the direction of the applied external force, respectively.

Within a molecular dynamics simulation, it is possible to apply a systematic strain to the system and follow how stress change with time. This, in turn, yields information about the breaking mechanism.

However, as will be shown, some care has to be exerted. In fact, the simulated fracturing process is tremendously dependent on the sample size (i.e. the cell dimensions) and the computational protocol applied. This process is graphically illustrated in Figure 4.13 for a silicate glass. As can be appreciated (Panel b), the stress changes continuously with the increase of the strain, and so does the Si—O bond length (Panel c). Voids start appearing (Panel a) after 1.8 ns of simulation, at $\varepsilon = 0.2$. After this value of the strain is reached, moreover, the stress starts decreasing, albeit with many oscillations. The fracturing process continues gradually until $\varepsilon > 0.35$, when the complete breaking eventually occurs and the stress becomes equal to that at $\varepsilon = 0$ and does not change further with the strain increase.

This description of the breaking event is rather far from what is experimentally expected and observed and presents a spurious "nanoductility." With the latter term, we refer to plastic properties that become apparent only at the nanoscale: while recently nanoductility has been in fact detected experimentally [73], the molecular dynamics simulations yield a nanoductility which is order of magnitudes greater than the experimental one [72]. The computational scheme can be improved so as to have a fracturing dynamics closer to the observed one following two ways.

The first way is to increase the sample size. As shown in Ref. [74], the increasing of the sample size of a silicate glass from 13 000 simulated atoms to 30 000 atoms (red line, Panel b) completely alters the stress–strain diagram, reported in Figure 4.14. In fact, with the 30 000 atoms sample, a sudden drop in the stress occurring during the fracture of the glass is observed, as expected. As a consequence, the fracture occurs at a strain increase of 1%, whereas in the previous case (Figure 4.13) an increase of 15% was needed to achieve the same goal.

Anyway, there is no a priori way to know how large the sample has to be in order to yield the correct breaking dynamics: for example, in Ref. [74], it is shown that for silica glasses, 30 000 atoms seems to be enough, but for soda-silicate glasses, 60 000 atoms should be adopted to reproduce the brittle fracture, at least. In any case, increasing the sample size allows improving the reproduction of the correct fracture dynamics. As an alternative, since increasing the number of simulated atoms can be computationally unfeasible, a constrained dynamics can be attempted: within this scheme, when the elongating force is applied along an axis (say, $z$), the $x$ and $y$-side length are frozen at their starting values. In this way, the breaking occurs abruptly as expected, because the formation and coalescence of voids inside the glass is greatly sped up.

However, it has to be remarked that this procedure requires an ad hoc and, thus, spurious, enforcing of constraints: the most scientifically sound way still remains enlarging the sample size.

In any case, these analyses allow studying a seldom-investigated subject: the microscopic origin of the nanoductility. While it is known that glasses are subject to brittle fracture, at the nanoscale, there is some evidence of a more ductile behavior, that is more affine to that of metals [73]. As discussed so far, unconstrained molecular dynamic simulations on small samples tend to display this behavior, which, albeit spurious and due to the limited

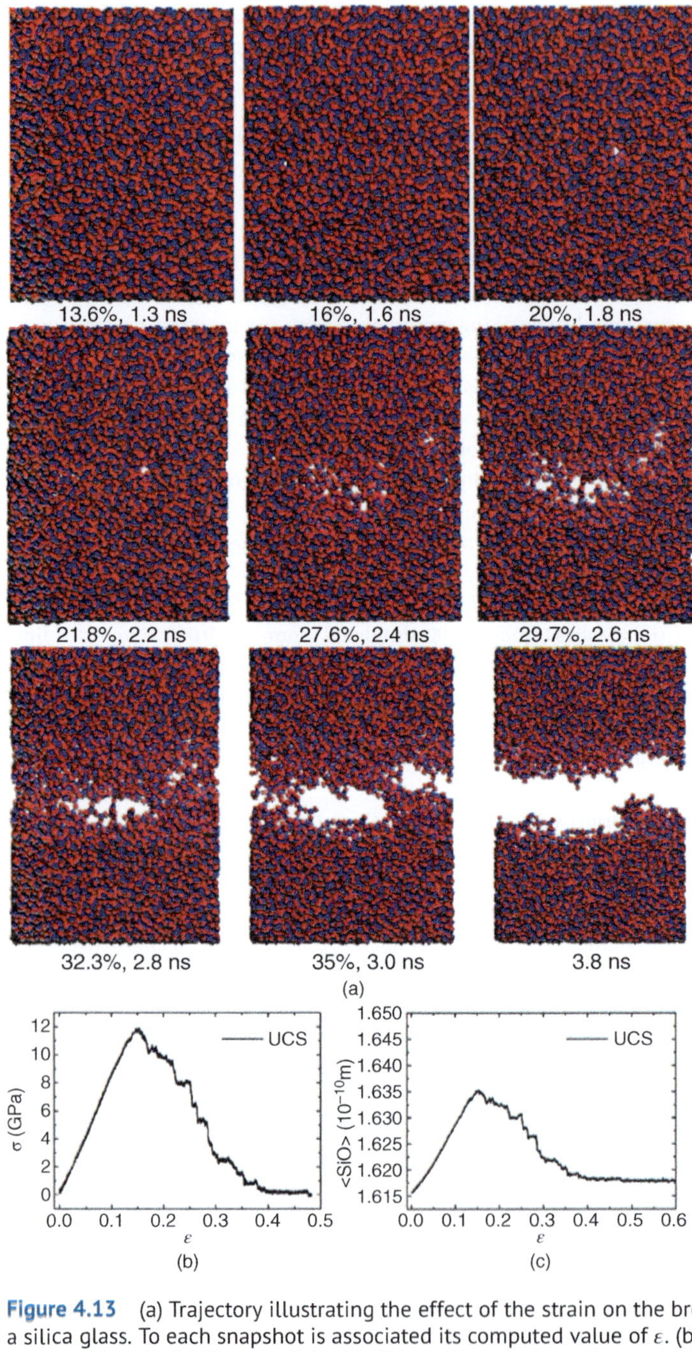


**Figure 4.13** (a) Trajectory illustrating the effect of the strain on the breaking process of a silica glass. To each snapshot is associated its computed value of $\varepsilon$. (b) Variation of stress and strain. (c) Change of the Si—O bond length along the vertical axis as a function of the strain. Source: Reprinted with permission from Pedone et al. [72]. Copyright 2008 American Chemical Society.

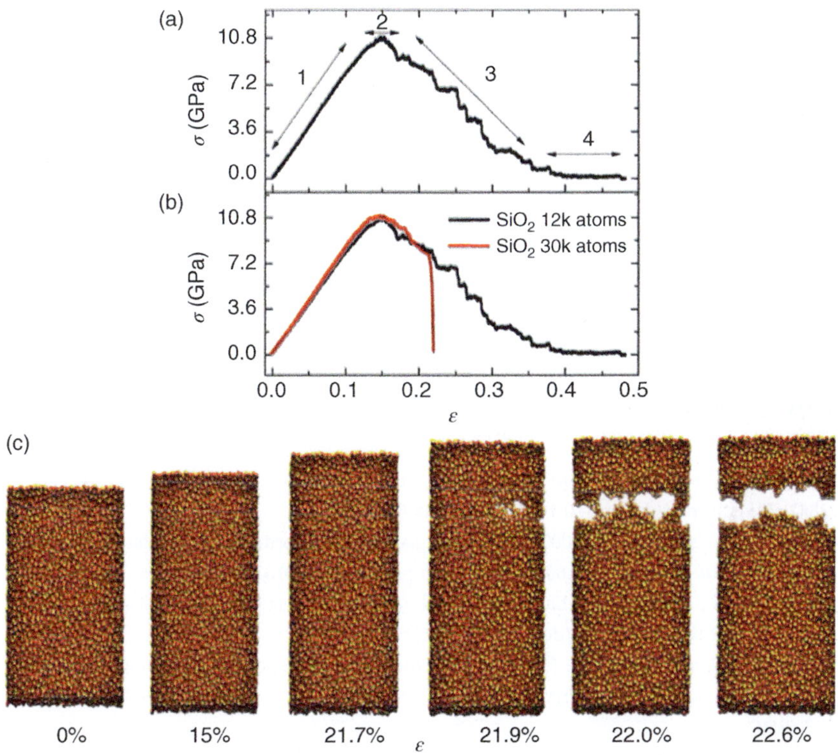


**Figure 4.14** (a) Stress–strain plot for a simulation of 13 000 atoms of silicate glass. Various regions are highlighted: in 1, the systems display elastic behavior, 2 is the so-called "yielding region," 3 is the unstable region, and 4 is the zero-stress region. (b) Comparison of stress–strain plots for a simulation of 13 000 atoms of silicate glass and another one with 30 000 comprising atoms. (c) Snapshots taken from the 30 000-atoms simulation and corresponding strain values. Source: Reprinted with permission from Pedone et al. [74] Copyright 2015 American Chemical Society.

sample size, can nevertheless give insights on nanoductility. In fact, as reported in Ref. [74], nanoductility is related to the presence and stability of "nanovoids," which, in turn, are mainly due to Si atoms with a coordination number other than the expected 4. These under- or overcoordinated Si atoms "are able to sustain plastic deformation by converting them from high-energy point defects to more stable structures" [74]. These computational findings are in agreement with later molecular dynamics studies carried out by Wang et al. [75], where it is stated that "due to the heterogeneity of the local rigidity, plastic events occur in different regions of the glass, and ultimately merge to form the crack, resulting in a nanoductile fracture," even if "the heterogeneity of topological constraints remains limited to a nanometric scale."

## 4.6 Concluding Remarks

In this chapter, we have reviewed the methodologies and computational procedures and approaches used in silico study of the structural, dynamical, thermodynamical, transport, and mechanical properties of inorganic glasses.

For each property that was treated, we have furnished examples taken from the literature in order to give an idea to the reader of the accuracy and reliability that can be easily reached up to now.

We hope this contribution will be useful to students and newcomers in glass science community interested in the simulation of glass properties.

## References


1 Zachariasen, W.H. (1932). The atomic arrangement in glass. *J. Am. Chem. Soc.* 54: 3841–3851. https://doi.org/10.1021/ja01349a006.

2 Greaves, G.N. and Sen, S. (2007). Inorganic glasses, glass-forming liquids and amorphizing solids. *Adv. Phys.* 56: 1–166. https://doi.org/10.1080/00018730601147426.

3 Greaves, G.N., Fontaine, A., Lagarde, P. et al. (1981). Local structure of silicate glasses. *Nature* 293: 611–616. https://doi.org/10.1038/293611a0.

4 Greaves, G.N. and Ngai, K.L. (1995). Reconciling ionic-transport properties with atomic structure in oxide glasses. *Phys. Rev. B* 52: 6358–6380. https://doi.org/10.1103/PhysRevB.52.6358.

5 Singh, S., Ediger, M.D., and de Pablo, J.J. (2013). Ultrastable glasses from in silico vapour deposition. *Nat. Mater.* 12: 139–144. https://doi.org/10.1038/nmat3521.

6 Fukunaga, T., Itoh, K., Otomo, T. et al. (2006). Voronoi analysis of the structure of Cu–Zr and Ni–Zr metallic glasses. *Intermetallics* 14: 893–897. https://doi.org/10.1016/j.intermet.2006.01.006.

7 Gedeon, O. and Liška, M. (2002). Voronoi polyhedra analysis of MD simulated silicate glasses. *J. Non-Cryst. Solids* 303: 246–252. https://doi.org/10.1016/S0022-3093(02)00951-1.

8 Lusvardi, G., Malavasi, G., Tarsitano, F. et al. (2009). Quantitative structure–property relationships of potentially bioactive fluoro phospho-silicate glasses. *J. Phys. Chem. B* 113: 10331–10338. https://doi.org/10.1021/jp809805z.

9 Pedone, A., Malavasi, G., and Menziani, M.C. (2009). Computational insight into the effect of CaO/MgO substitution on the structural properties of phospho-silicate bioactive glasses. *J. Phys. Chem. C* 113: 15723–15730. https://doi.org/10.1021/jp904131t.

10 Malavasi, G., Pedone, A., and Menziani, M.C. (2013). Study of the structural role of gallium and aluminum in 45S5 bioactive glasses by molecular dynamics simulations. *J. Phys. Chem. B* 117: 4142–4150. https://doi.org/10.1021/jp400721g.

11 Gambuzzi, E., Pedone, A., Menziani, M.C. et al. (2014). Probing silicon and aluminium chemical environments in silicate and aluminosilicate glasses by solid state NMR spectroscopy and accurate first-principles calculations. *Geochim. Cosmochim. Acta* 125: 170–185. https://doi.org/10.1016/j.gca.2013.10.025.

**12** Tilocca, A. and Cormack, A.N. (2007). Structural effects of phosphorus inclusion in bioactive silicate glasses. *J. Phys. Chem. B* 111: 14256–14264. https://doi.org/10.1021/jp075677o.

**13** Christie, J.K., Malik, J., and Tilocca, A. (2011). Bioactive glasses as potential radioisotope vectors for in situ cancer therapy: investigating the structural effects of yttrium. *Phys. Chem. Chem. Phys.* 13: 17749–17755. https://doi.org/10.1039/C1CP21764J.

**14** Schneider, J., Mastelaro, V.R., Panepucci, H., and Zanotto, E.D. (2000). $^{29}$Si MAS–NMR studies of $Q^n$ structural units in metasilicate glasses and their nucleating ability. *J. Non-Cryst. Solids* 273: 8–18. https://doi.org/10.1016/S0022-3093(00)00139-3.

**15** Le Saoût, G., Simon, P., Fayon, F. et al. (2002). Raman and infrared study of $(PbO)_x(P_2O_5)_{(1-x)}$ glasses. *J. Raman Spectrosc.* 33: 740–746. https://doi.org/10.1002/jrs.911.

**16** Tilocca, A., Cormack, A.N., and de Leeuw, N.H. (2007). The structure of bioactive silicate glasses: new insight from molecular dynamics simulations. *Chem. Mater.* 19: 95–103. https://doi.org/10.1021/cm061631g.

**17** Yuan, X. and Cormack, A.N. (2002). Efficient algorithm for primitive ring statistics in topological networks. *Comput. Mater. Sci.* 24: 343–360. https://doi.org/10.1016/S0927-0256(01)00256-7.

**18** Du, J. and Cormack, A.N. (2004). The medium range structure of sodium silicate glasses: a molecular dynamics simulation. *J. Non-Cryst. Solids* 349: 66–79. https://doi.org/10.1016/j.jnoncrysol.2004.08.264.

**19** Yuan, X. and Cormack, A.N. (2001). Local structures of MD-modeled vitreous silica and sodium silicate glasses. *J. Non-Cryst. Solids* 283: 69–87. https://doi.org/10.1016/S0022-3093(01)00363-5.

**20** Shackelford, J.F. (1999). Gas solubility in glasses – principles and structural implications. *J. Non-Cryst. Solids* 253: 231–241. https://doi.org/10.1016/S0022-3093(99)00355-5.

**21** Cleaver, F.L. (1965). On a theorem of Voronoï. *Trans. Am. Math. Soc.* 120: 390–400. https://doi.org/10.1090/S0002-9947-1965-0184909-6.

**22** Malavasi, G., Menziani, M.C., Pedone, A., and Segre, U. (2006). Void size distribution in MD-modelled silica glass structures. *J. Non-Cryst. Solids* 352: 285–296. https://doi.org/10.1016/j.jnoncrysol.2005.11.022.

**23** Fayon, F., Duée, C., Poumeyrol, T. et al. (2013). Evidence of nanometric-sized phosphate clusters in bioactive glasses as revealed by solid-state $^{31}$P NMR. *J. Phys. Chem. C* 117: 2283–2288. https://doi.org/10.1021/jp312263j.

**24** Lusvardi, G., Malavasi, G., Cortada, M. et al. (2008). Elucidation of the structural role of fluorine in potentially bioactive glasses by experimental and computational investigation. *J. Phys. Chem. B* 112: 12730–12739. https://doi.org/10.1021/jp803031z.

**25** Lusvardi, G., Malavasi, G., Menabue, L. et al. (2005). A computational tool for the prediction of crystalline phases obtained from controlled crystallization of glasses. *J. Phys. Chem. B* 109: 21586–21592. https://doi.org/10.1021/jp0546857.

**26** Tilocca, A., Cormack, A.N., and de Leeuw, N.H. (2007). The formation of nanoscale structures in soluble phosphosilicate glasses for biomedical applications: MD simulations. *Faraday Discuss.* 136: 45–55. https://doi.org/10.1039/B617540F.

**27** Wright, A.C. (1993). The comparison of molecular dynamics simulations with diffraction experiments. *J. Non-Cryst. Solids* 159: 264–268. https://doi.org/10.1016/0022-3093(93)90232-M.

**28** Wright, A.C. (1988). Neutron and X-ray amorphography. *J. Non-Cryst. Solids* 106: 1–16. https://doi.org/10.1016/0022-3093(88)90218-9.

**29** Grimley, D.I., Wright, A.C., and Sinclair, R.N. (1990). Neutron scattering from vitreous silica IV. Time-of-flight diffraction. *J. Non-Cryst. Solids* 119: 49–64. https://doi.org/10.1016/0022-3093(90)90240-M.

**30** Wright, A.C., Shakhmatkin, B.A., and Vedishcheva, N.M. (2001). The chemical structure of oxide glasses: a concept consistent with neutron scattering studies? *Glass Phys. Chem.* 27: 97–113. https://doi.org/10.1023/A:1011366508857.

**31** Pedone, A., Malavasi, G., Cormack, A.N. et al. (2007). Insight into elastic properties of binary alkali silicate glasses; prediction and interpretation through atomistic simulation techniques. *Chem. Mater.* 19: 3144–3154. https://doi.org/10.1021/cm062619r.

**32** Pedone, A., Malavasi, G., Menziani, M.C. et al. (2006). A new self-consistent empirical interatomic potential model for oxides, silicates, and silica-based glasses. *J. Phys. Chem. B* 110: 11780–11795. https://doi.org/10.1021/jp0611018.

**33** Pedone, A. (2009). Properties calculations of silica-based glasses by atomistic simulations techniques: a review. *J. Phys. Chem. C* 113: 20773–20784. https://doi.org/10.1021/jp9071263.

**34** Ohara, K., Mitsui, A., Mori, M. et al. (2016). Structural and electronic features of binary $Li_2S$–$P_2S_5$ glasses. *Sci. Rep.* 6: 21302. https://doi.org/10.1038/srep21302.

**35** Dall'Olio, S., Dovesi, R., and Resta, R. (1997). Spontaneous polarization as a Berry phase of the Hartree–Fock wave function: the case of $KNbO_3$. *Phys. Rev. B* 56: 10105–10114. https://doi.org/10.1103/PhysRevB.56.10105.

**36** Baranek, P., Zicovich-Wilson, C.M., Roetti, C. et al. (2001). Well localized crystalline orbitals obtained from Bloch functions: the case of $KNbO_3$. *Phys. Rev. B* 64: 125102. https://doi.org/10.1103/PhysRevB.64.125102.

**37** Noel, Y., Zicovich-Wilson, C.M., Civalleri, B. et al. (2001). Polarization properties of ZnO and BeO: An *ab initio* study through the Berry phase and Wannier functions approaches. *Phys. Rev. B* 65: 014111. https://doi.org/10.1103/PhysRevB.65.014111.

**38** Hench, L.L., Splinter, R.J., Allen, W.C., and Greenlee, T.K. (1971). Bonding mechanisms at the interface of ceramic prosthetic materials. *J. Biomed. Mater. Res.* 5: 117–141. https://doi.org/10.1002/jbm.820050611.

**39** Corno, M. and Pedone, A. (2009). Vibrational features of phospho-silicate glasses: periodic B3LYP simulations. *Chem. Phys. Lett.* 476: 218–222. https://doi.org/10.1016/j.cplett.2009.06.039.

**40** Pedone, A., Charpentier, T., Malavasi, G., and Menziani, M.C. (2010). New insights into the atomic structure of 45S5 bioglass by means of solid-state NMR spectroscopy and accurate first-principles simulations. *Chem. Mater.* 22: 5644–5652. https://doi.org/10.1021/cm102089c.

**41** Allen, M.P. and Tildesley, D.J. (1989). *Computer Simulation of Liquids*. Clarendon Press.

**42** Giacomazzi, L. and Pasquarello, A. (2007). Vibrational spectra of vitreous $SiO_2$ and vitreous $GeO_2$ from first principles. *J. Phys. Condens. Matter* 19: 415112. https://doi.org/10.1088/0953-8984/19/41/415112.

**43** Pedone, A., Malavasi, G., Menziani, M.C. et al. (2008). FFSiOH: a new force field for silica polymorphs and their hydroxylated surfaces based on periodic B3LYP calculations. *Chem. Mater.* 20: 2522–2531. https://doi.org/10.1021/cm703437y.

**44** Buchenau, U., Prager, M., Nücker, N. et al. (1986). Low-frequency modes in vitreous silica. *Phys. Rev. B* 34: 5665–5673. https://doi.org/10.1103/PhysRevB.34.5665.

**45** Carpenter, J.M. and Price, D.L. (1985). Correlated motions in glasses studied by coherent inelastic neutron scattering. *Phys. Rev. Lett.* 54: 441–443. https://doi.org/10.1103/PhysRevLett.54.441.

**46** Pickard, C.J. and Mauri, F. (2001). All-electron magnetic response with pseudopotentials: NMR chemical shifts. *Phys. Rev. B* 63: 245101. https://doi.org/10.1103/PhysRevB.63.245101.

**47** Charpentier, T., Kroll, P., and Mauri, F. (2009). First-principles nuclear magnetic resonance structural analysis of vitreous silica. *J. Phys. Chem. C* 113: 7917–7929. https://doi.org/10.1021/jp900297r.

**48** Pedone, A., Charpentier, T., and Menziani, M.C. (2010). Multinuclear NMR of $CaSiO_3$ glass: simulation from first-principles. *Phys. Chem. Chem. Phys.* 12: 6054–6066. https://doi.org/10.1039/b924489a.

**49** Lee, S.K. and Stebbins, J.F. (2000). The structure of aluminosilicate glasses: high-resolution $^{17}O$ and $^{27}Al$ MAS and 3QMAS NMR study. *J. Phys. Chem. B* 104: 4091–4100. https://doi.org/10.1021/jp994273w.

**50** Stebbins, J.F. and Xu, Z. (1997). NMR evidence for excess non-bridging oxygen in an aluminosilicate glass. *Nature* 390: 60–62. https://doi.org/10.1038/36312.

**51** Pedone, A., Gambuzzi, E., and Menziani, M.C. (2012). Unambiguous description of the oxygen environment in multicomponent aluminosilicate glasses from $^{17}O$ solid state NMR computational spectroscopy. *J. Phys. Chem. C* 116: 14599–14609. https://doi.org/10.1021/jp304802y.

**52** Pedone, A. (2016). Recent advances in solid-state NMR computational spectroscopy: the case of alumino-silicate glasses. *Int. J. Quantum Chem.* https://doi.org/10.1002/qua.25134.

**53** Gambuzzi, E., Pedone, A., Menziani, M.C. et al. (2012). Calcium environment in silicate and aluminosilicate glasses probed by $^{43}Ca$ MQMAS NMR experiments and MD-GIPAW calculations. *Solid State Nucl. Magn. Reson.* https://doi.org/10.1016/j.ssnmr.2015.04.003.

**54** Gambuzzi, E., Charpentier, T., Menziani, M.C., and Pedone, A. (2014). Computational interpretation of $^{23}Na$ MQMAS NMR spectra: a comprehensive investigation of the Na environment in silicate glasses. *Chem. Phys. Lett.* 612: 56–61. https://doi.org/10.1016/j.cplett.2014.08.004.

**55** Green, M.S. (1954). Markoff random processes and the statistical mechanics of time-dependent phenomena. II. Irreversible processes in fluids. *J. Chem. Phys.* 22: 398–413. https://doi.org/10.1063/1.1740082.

**56** Kubo, R. (1957). Statistical-mechanical theory of irreversible processes. I. General theory and simple applications to magnetic and conduction problems. *J. Phys. Soc. Jpn.* 12: 570–586. https://doi.org/10.1143/JPSJ.12.570.

**57** Du, J. and Chen, C.-H. (2012). Structure and lithium ion diffusion in lithium silicate glasses and at their interfaces with lithium lanthanum titanate crystals. *J. Non-Cryst. Solids* 358: 3531–3538. https://doi.org/10.1016/j.jnoncrysol.2012.05.015.

**58** Ren, M. and Du, J. (2016). Structural origin of the thermal and diffusion behaviors of lithium aluminosilicate crystal polymorphs and glasses. *J. Am. Ceram. Soc.* https://doi.org/10.1111/jace.14292.

**59** Callister, W.D. and Rethwisch, D.G. (2010). *Materials Science and Engineering: An Introduction*, 8e. Wiley.

**60** Du, J. and Xiang, Y. (2012). Effect of strontium substitution on the structure, ionic diffusion and dynamic properties of 45S5 bioactive glasses. *J. Non-Cryst. Solids* 358: 1059–1071. https://doi.org/10.1016/j.jnoncrysol.2011.12.114.

**61** Malavasi, G., Lusvardi, G., Pedone, A. et al. (2007). Crystallization kinetics of bioactive glasses in the $ZnO–Na_2O–CaO–SiO_2$ system. *J. Phys. Chem. A* 111: 8401–8408. https://doi.org/10.1021/jp071528u.

**62** Pedone, A., Malavasi, G., Cormack, A.N. et al. (2008). Elastic and dynamical properties of alkali-silicate glasses from computer simulations techniques. *Theor. Chem. Acc.* 120: 557–564. https://doi.org/10.1007/s00214-008-0434-7.

**63** Pedone, A., Malavasi, G., Menziani, M.C. et al. (2008). Role of magnesium in soda-lime glasses: insight into structural, transport, and mechanical properties through computer simulations. *J. Phys. Chem. C* 112: 11034–11041. https://doi.org/10.1021/jp8016776.

**64** Palmer, B.J. (1994). Transverse-current autocorrelation-function calculations of the shear viscosity for molecular liquids. *Phys. Rev. E* 49: 359–366. https://doi.org/10.1103/PhysRevE.49.359.

**65** Jund, P. and Jullien, R. (1999). Molecular-dynamics calculation of the thermal conductivity of vitreous silica. *Phys. Rev. B* 59: 13707–13711. https://doi.org/10.1103/PhysRevB.59.13707.

**66** Schelling, P.K., Phillpot, S.R., and Keblinski, P. (2002). Comparison of atomic-level simulation methods for computing thermal conductivity. *Phys. Rev. B* 65: 144306. https://doi.org/10.1103/PhysRevB.65.144306.

**67** Irving, J.H. and Kirkwood, J.G. (1950). The statistical mechanical theory of transport processes. IV. The equations of hydrodynamics. *J. Chem. Phys.* 18: 817–829. https://doi.org/10.1063/1.1747782.

**68** Wondraczek, L., Mauro, J.C., Eckert, J. et al. (2011). Towards ultrastrong glasses. *Adv. Mater.* 23: 4578–4586. https://doi.org/10.1002/adma.201102795.

**69** Gale, J.D. (1997). GULP: a computer program for the symmetry-adapted simulation of solids. *J. Chem. Soc. Faraday Trans.* 93: 629–637. https://doi.org/10.1039/a606455h.

**70** Bauchy, M. (2014). Structural, vibrational, and elastic properties of a calcium aluminosilicate glass from molecular dynamics simulations: the role of the potential. *J. Chem. Phys.* 141: 024507.

**71** Bansal, N. and Doremus, R. ISBN 9780120781409(1986). *Handbook of Glass Properties*, 1e. http://store.elsevier.com/Handbook-of-Glass-Properties/Narottam-Bansal/isbn-9780120781409/ (accessed 21 June 2016).

**72** Pedone, A., Malavasi, G., Menziani, M.C. et al. (2008). Molecular dynamics studies of stress–strain behavior of silica glass under a tensile load. *Chem. Mater.* 20: 4356–4366. https://doi.org/10.1021/cm800413v.

**73** Célarié, F., Prades, S., Bonamy, D. et al. (2003). Glass breaks like metal, but at the nanometer scale. *Phys. Rev. Lett.* 90: 075504. https://doi.org/10.1103/PhysRevLett.90.075504.

**74** Pedone, A., Menziani, M.C., and Cormack, A.N. (2015). Dynamics of fracture in silica and soda-silicate glasses: from bulk materials to nanowires. *J. Phys. Chem. C* 119: 25499–25507. https://doi.org/10.1021/acs.jpcc.5b08657.

**75** Wang, B., Yu, Y., Lee, Y.J., and Bauchy, M. (2015). Intrinsic nano-ductility of glasses: the critical role of composition. *Glass Sci.* 2: 11. https://doi.org/10.3389/fmats.2015.00011.

# 5

# Topological Constraint Theory of Glass: Counting Constraints by Molecular Dynamics Simulations

*Yushu Hu[1], Han Liu[1], Kai Yang[1], Qi Zhou[1], Christian G. Hoover[2], N. M. Anoop Krishnan[3,4], Morten M. Smedskjaer[5], Matthieu Micoulaut[6], Lijie Guo[7], and Mathieu Bauchy[1]*

[1] *Physics of AmoRphous and Inorganic Solids Laboratory (PARISlab), Civil and Environmental Engineering Department, University of California, Los Angeles, CA, USA*
[2] *School of Sustainable Engineering and the Built Environment, Civil Engineering Department, Arizona State University, Tempe, AZ, USA*
[3] *Department of Civil Engineering, Indian Institute of Technology Delhi, New Delhi, India*
[4] *Department of Materials Science and Engineering, Indian Institute of Technology Delhi, New Delhi, India*
[5] *Department of Chemistry and Bioscience, Aalborg University, Aalborg, Denmark*
[6] *Laboratoire de Physique Théorique de la Matière Condensée, Sorbonne Université, Paris Cedex 05, France*
[7] *National Centre for International Research on Green Metal Mining, BGRIMM Technology Group, Beijing, China*

## 5.1 Introduction

Topological constraint theory (TCT, or rigidity theory) [1, 2] has been a key enabler in glass science and engineering over the past decades – both from a fundamental perspective and a practical standpoint. TCT builds on the idea that structural glasses should be considered as networks rather than just a collection of atoms. This point can be illustrated by considering a subway map. If a passenger wants to travel from station A to station B, the detail of the position of each station in the entire subway network is of little relevance. Rather, it is more convenient to treat the subway map as a network, wherein some nodes (the subway stations) are connected to each other (by some subway lines). In that sense, the knowledge of the connectivity among stations is key for the passenger to identify the optimal itinerary between stations A and B – or to identify alternative itineraries should any of the subway lines be closed by assessing the existence of closed-path rings in the network.

A similar case can be made for glassy atomic structures. The atomic structure of a glass is encoded in the knowledge of (i) the type (element) and (ii) Cartesian position $(x, y, z)$ of each atom – which is accessible as an output of molecular dynamics (MD) simulations [3, 4]. This information offers a full, thorough description of the glass structure. However, this information is usually unmanageably large (i.e. the position of $N$ atoms is associated with $3N$ real numbers). More importantly, this information does not separate the structural features that eventually affect glasses' macroscopic properties from those that do not [2]. Clearly, certain

*Atomistic Simulations of Glasses: Fundamentals and Applications*, First Edition.
Edited by Jincheng Du and Alastair N. Cormack.

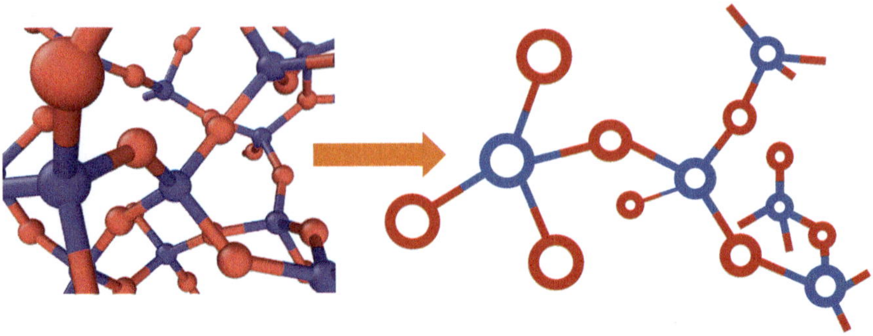

**Figure 5.1** Topological constraint theory simplifies complex disordered atomic networks into simpler mechanical trusses made of nodes (atoms) connected to each other by constraints (chemical bonds).

structural features (e.g. bond distances, bond angles, and coordination numbers) have a more direct relationship with macroscopic properties than others. The main assumption behind TCT is that – like subway maps – glasses should be described as networks (i.e. based on the connectivity between atoms) rather than as a collection of atoms with specific positions (Figure 5.1). In this picture, atoms are solely considered as nodes, which are connected to each other by some topological constraints (i.e. the chemical bonds) [2]. The success of TCT largely relies on the fact that, in many cases, this theory manages to "separate the wheat from the chaff," that is, to isolate the key topological features that affect macroscopic properties to the first order, while filtering out less-relevant structural details that, ultimately, only have a second-order effect on the properties [4].

The level of simplification offered by TCT (i.e. describing glasses only based on the connectivity of the atoms rather than on the exact position of each atom) has enabled the development of a series of analytical models aiming to predict glass properties as a function of their composition or to pinpoint optimal compositions exhibiting unusual properties [2, 5–8]. TCT has also been extended to a broader range of materials, including concrete [9–12], fly ash [13, 14], minerals [15–17], phase-change materials [18], amorphous semiconductors [19, 20], proteins [21–23], etc. Nevertheless, the most iconic success of TCT is probably the design of Corning® Gorilla® Glass 3 (a scratch- and damage-resistant glass used in billions of smartphones and tablets), which was discovered in silico through the optimization of its atomic network topology before anything was actually melted in a laboratory [24–26].

Despite the success of TCT, its potential is often limited by our ability to accurately describe the topology of the atomic network of glasses. This primarily arises not only from a lack of knowledge regarding composition–structure relationships in glasses but also from some degree of uncertainty on how to properly interpret the structure so as to build a meaningful topological model for a given family of glasses [27]. In addition, although experiments can offer some useful inputs (e.g. atomic coordination numbers), they often do not provide a full access to the glass structure [28] and, hence, sometimes do not yield all the information that is required to build a topological model. In that regard, atomistic simulations are an ideal complement to TCT since they offer a direct access to the glass structure (i.e. positions $x$, $y$, $z$ of each atom) [3]. The main objective of this chapter is to review a series of methods aiming to describe the rigidity of glasses by means of atomistic simulations, that is, to accurately and robustly enumerate the density of topological constraints in a given glass network based on the analysis of atomic configurations or trajectories.

## 5.2 Background on Topological Constraint Theory

### 5.2.1 Rigidity of Mechanical Networks

The topological description of glass networks builds on the concept of Maxwell–Lagrange mechanical stability [29]. In short, mechanical trusses can be described as flexible (underconstrained), stressed-rigid (overconstrained), or isostatic (ideally constrained) – the three states of rigidity of a network (Figure 5.2). The state of rigidity of a given truss can be determined by comparing the number of mechanical constraints $N_c$ (i.e. the number of red sticks in Figure 5.2) with the initial number of degrees of freedom of the truss nodes. In three-dimensions, the initial number of degrees of freedom is given by $3N$ (i.e. three translation directions per node), where $N$ is the number of nodes. The number of remaining internal modes of deformation (or floppy modes) present within the truss $F$ after the application of the constraints is then given by

$$F = 3N - N_c - 6 \tag{5.1}$$

This Eq. (5.1) captures the fact that, starting from the situation wherein each node can freely move along three directions, each constraint $N_c$ effectively removes one internal degree of freedom. Note that the term "6" corresponds to the six macroscopic degrees of freedom of a rigid structure in three-dimension (i.e. three rigid-body translations and rotations, respectively). The state of rigidity of a mechanical truss is then determined by the value of $F$, namely, the network is: (i) flexible if $F > 0$, (ii) stressed-rigid if $F < 0$, and (iii) isostatic (or statically determinate) if $F = 0$. According to this classification, flexible networks exhibit a positive number of internal modes of deformation and, as such, can be deformed without any external energy. In contrast, isostatic and stressed-rigid trusses are fully rigid. In addition, in stressed-rigid trusses, all the constraints cannot be satisfied at the same time – just like the angles of a triangle cannot be varied to arbitrary values if the dimensions of the three edges are already fixed. In this case, some internal eigenstress will form, since some of the constraints will be under tension, whereas others will be under compression. Such internal eigenstress does not result in any macroscopic stress as the constraints under tension and compression mutually compensate each other – so that the

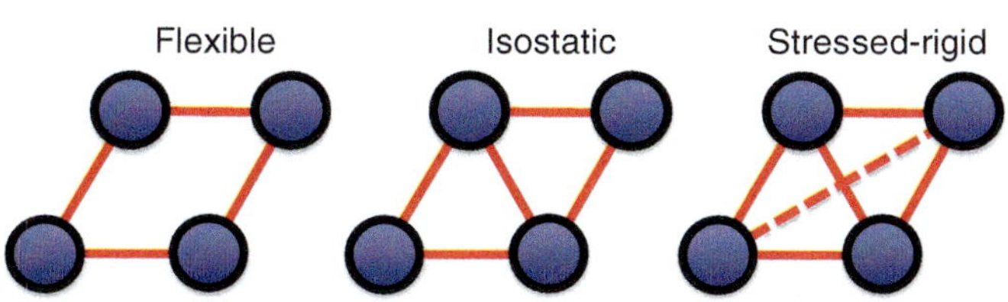


Figure 5.2 The three states of rigidity of a mechanical network. The dashed red line denotes a redundant constraint that is here under tension.

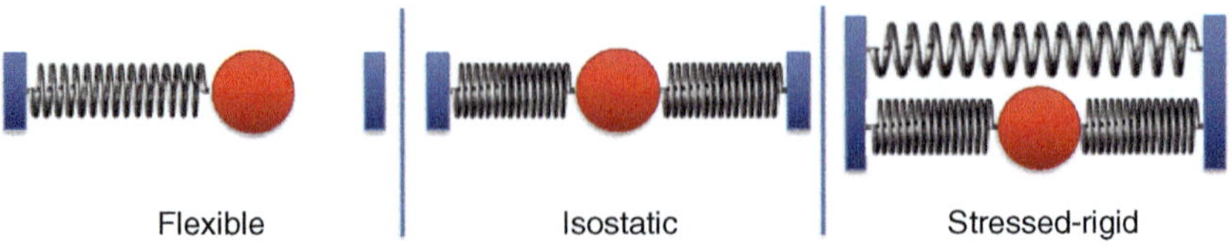


**Figure 5.3** Illustration of the origin of the internal eigenstress that is present in stressed-rigid structures.

overall structure is at zero macroscopic stress (Figure 5.3). In between the flexible and stressed-rigid regimes, the isostatic state corresponds to a network that is rigid (i.e. with no internal mode of deformation) but unstressed (i.e. with no eigenstress).

### 5.2.2 Application to Atomic Networks

In 1979 [1], Phillips adopted the Maxwell–Lagrange stability framework to disordered atomic networks – the birth of the TCT of glass. The mathematical foundation of Phillips' TCT was later refined by Thorpe in 1983 [30–32]. This analogy between mechanical truss and atomic network relies on the idea that atoms can be considered as nodes that are connected to each other by some mechanical constraints – wherein each constraint removes some internal degrees of freedom by preventing the relative motion between atoms. Phillips' TCT considers two kinds of constraints, namely (i) the radial two-body bond-stretching (BS) constraints that maintain interatomic distances fixed around their average values and (ii) the angular three-body bond-bending (BB) constraints that keep angles fixed (Figure 5.4). BS and BB constraints are essentially equivalent to harmonic springs that fix the values of the radial and angular interatomic bonds in the network.

### 5.2.3 Constraint Enumeration Under Mean-Field Approximation

In covalent networks (e.g. chalcogenide glasses) at low temperature, all the BS and BB constraints are intact and, as such, their numbers solely depend on the coordination number $r$ of each atom [33]. In detail, for a given atom, the number of BS constraints is given by $r/2$ (note that each radial constraint is necessarily shared by two atoms), while the number of BB constraints is given by $2r - 3$ for $r \geqslant 2$ (since a triplet of atoms corresponds to one BB constraint and two new BB constraints that are then needed to fix the direction of each additional neighbor). Onefold coordinated atoms are not associated with any BB constraint. Assuming that all the atoms have a coordination number larger or equal to 2, the total number of

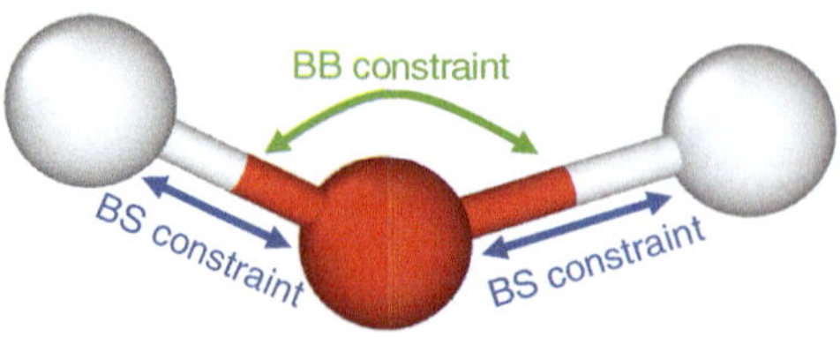


**Figure 5.4** Schematic illustrating the role of the radial bond-stretching (BS) and angular bond-bending (BB) constraints in fixing the values of interatomic distances and angles, respectively.

constraints $N_c$ is given by

$$N_c = \sum_i \left[N_i\left(\frac{r_i}{2} + 2r_i - 3\right)\right] = \sum_i \left[N_i\frac{5r_i}{2}\right] - 3N \tag{5.2}$$

where $N$ is the total number of atoms and $N_i$ is the number of atoms having a coordination number $r_i$. The equation makes it possible to calculate the total number of internal degrees of freedom $F$ (from Eq. (5.1)) and, therefore, to determine the state of rigidity of a glass.

However, due to the large number of atoms present in a given glass network, it is convenient to adopt a mean-field approximation and to solely rely on the average number of internal modes of deformation per atom, $f = F/N$, to determine the state of rigidity of a glass. This metric can be expressed as follows:

$$f = 3 - n_c - \frac{6}{N} = 3 - n_c \tag{5.3}$$

where $n_c = N_c/N$ is the average number of constraints per atom. Note that the term $6/N$ becomes infinitely small for a large number of atoms and, hence, can be ignored. The $n_c$ metric then solely depends on the fraction atom $x_i = N_i/N$ having a coordination $r_i$:

$$n_c = \sum_i \left[x_i\frac{5r_i}{2}\right] - 3 \tag{5.4}$$

Based on Maxwell's stability criterion, glass atomic networks are classified as (i) flexible if $f > 0$ (i.e. $n_c < 3$), (ii) stressed-rigid if $f < 0$ (i.e. $n_c > 3$), and (iii) isostatic if $f = 0$ (i.e. $n_c = 3$). Due to the excess of degrees of freedom, flexible glasses exhibit some internal floppy modes of deformation (wherein the number of floppy modes per atom is given by $f = 3 - n_c$) [34–37]. In contrast, due to the excess of constraints, stressed-rigid glasses present some internal stress arising from the fact that some constraints mutually depend on each other (wherein the number of redundant constraints per atom is given by $n_c - 3$). In turn, isostatic glasses are free of both floppy modes and internal stress [38, 39].

A simplified criterion capturing the state of rigidity of glasses can be expressed in terms of their average coordination number $\langle r\rangle$:

$$\langle r\rangle = \sum_i x_i r_i \tag{5.5}$$

which yields the following expression for the number of constraints per atom:

$$n_c = \frac{5\langle r\rangle}{2} - 3 \tag{5.6}$$

Assuming that all BS and BB constraints are active and that the network does not comprise of any onefold coordinated atom, an isostatic network ($n_c = 3$) is then achieved for the average coordination number $\langle r\rangle = 2.4$. In the flexible domain, the number of floppy modes per atom is given by

$$f = 3 - n_c = \frac{5}{2}(2.4 - \langle r\rangle) \tag{5.7}$$

### 5.2.4 Polytope-Based Description of Glass Rigidity

Rather than describing a glass network in terms of connected atoms, Gupta and Cooper proposed (nearly at the same time as Phillips) an alternative viewpoint [40]. This framework is based on a topological description of glass networks in terms of interconnected polytopes [41], wherein a disordered network is flexible if it exhibits a nonnegative number of degrees of freedom per vertex $f$ and is rigid if the number of degrees of freedom is zero (or lower). In the following, we focus on the Phillips formulation of TCT, since Phillips' constraint-based formulation and Gupta's polytope-based formulation are mathematically equivalent [4].

### 5.2.5 Impact of Temperature

The constraint enumeration captured in Eq. (5.6) is only valid in the absence of any one-fold coordinated atoms in covalent networks (i.e. featuring directional bonds) and at low temperature. In turn, since each constraint is associated with a given bonding energy, it can become thermally broken upon increasing temperature. The effect of temperature on topological constraints was rationalized in 2009 by Mauro and coworkers who introduced the framework of temperature-dependent constraints [42, 43]. In detail, each constraint is associated with a given free energy and, therefore, can be active or thermally broken depending on the temperature. Namely, all constraints are active at low temperature and thermally broken at very high temperature. In between these two regimes, the transition from active to thermally broken occurs at a given onset temperature $T_c$ [44]. Note that different constraints exhibit different free energy values and, thereby, break at different temperatures. For instance, angular BB constraints are typically weaker than radial BS constraints and, as such, are associated to lower onset temperatures [45, 46].

### 5.2.6 Need for Molecular Dynamics Simulations

In select glasses, the number of constraints can be analytically calculated as a function of composition (see Ref. [2]). However, in the case of most glasses, a direct analytical enumeration is made extremely challenging (if not impossible) due to the following complications:

(1) The coordination number of each atom – a key input for the constraint enumeration (see Eq. (5.6)) – is not always known or cannot be analytically predicted without relying on systematic experiments. For instance, in the case of borosilicate glasses, B atoms can be threefold or fourfold coordinated depending on the composition [47, 48]. In the case of modified silicate glasses, Ca atoms typically exhibit a coordination number that is between 4 and 8, thereby not following the Octet rule [49, 50].

(2) In addition, it has been shown that the effective number of BS constraints created by an atom is not always equal to its geometrical coordination number (that is as calculated from the number of neighbors inside its first coordination shell) [46, 50, 51].

(3) Isolated atoms or molecules (e.g. water molecules in the case of partially hydrated glasses [12, 52–55]) are not part of the atomic network and, therefore, do not contribute to its macroscopic rigidity. Therefore, they should be excluded from the constraint enumeration.

(4) Topological constraints are not always homogeneously distributed throughout the glass network, which may limit the use of analytical meanfield approaches [56–60].

(5) Atoms located near the glass surface may exhibit a connectivity that differs from atoms in the bulk and, hence, may be associated with a different number of constraints [61, 62].

Due to their localized nature, the experimental characterization of the topology of surface atoms is extremely challenging.

(6) As mentioned earlier, each constraint is associated with a given free energy and, hence, can be intact or broken depending on temperature, i.e. the amount of available thermal energy [42]. Based on the competition between constraint energy and thermal energy, weaker angular constraints (like the Si—O—Na bond in sodium silicate) can be broken even at 300 K [46, 63]. Determining the onset temperature at which each type of constraint break is a complex task [46].

(7) More generally, besides the effect of temperature, the constraint enumeration can be affected by external stimuli applied on glasses, e.g. external pressure [64–67], irradiation [15, 68], or ion exchange [69]. All of these effects must be accounted for to establish a reliable topological model of glasses subjected to extreme conditions.

For all these reasons, one cannot just rely on unproven guesses to enumerate the number of constraints [70]. In the following, we review how the analysis of MD simulations can overcome these limitations and offer a robust basis to conduct a constraint enumeration.

## 5.3 Counting Constraints from Molecular Dynamics Simulations

### 5.3.1 Constraint Enumeration Based on the Relative Motion Between Atoms

Although a direct computation of the bonding energy between pairs and triplets of atoms (i.e. the "cause" of topological constraints) would be desirable, it is often more convenient to track the relative motion between atoms (i.e. the "consequence" of topological constraints) [18, 46, 71]. Indeed, since the nature of the topological constraints acting on the atoms is not always known, the analysis of the interatomic motion can be used to infer the existence of constraints, which are preventing such interatomic motion. This "reverse" analysis is based on the very similar principle that allows one to infer the magnitude of the gravity force by analyzing the parabolic trajectory of a projectile. The main idea of this method is that active BS or BB constraints tend to maintain interatomic bond distances or bond angles fixed around their average values, respectively (Figure 5.5). In turn, the existence of a pronounced interatomic motion is suggestive of the absence of any underlying constraints. This method has been widely applied to chalcogenide and oxide glasses [46, 64, 71–75] and, more recently, to atomic-scale models of cement hydrates [12, 39, 76].

#### 5.3.1.1 Enumeration of Bond-Stretching Constraints

We first focus on the identification of the radial BS constraints. To obtain the number of BS constraints acting on a central atom 0, we first determine their $N$ nearest neighbors (e.g. $N = 6$) and rank them in terms of increasing average distance from the central atom 0 (where atom 1 is the nearest neighbor of atom 0). The value of $N$ is arbitrary but should be large enough to include all the potential nearest neighbors of each atom. For each neighbor, we then compute their distance from atom 0 as a function of time, as well as (i) the average

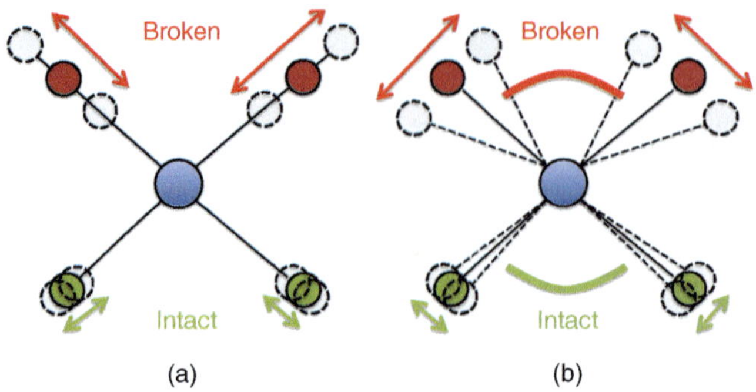


**Figure 5.5** Illustration of the use of molecular dynamics simulations to compute the number of (a) radial bond-stretching and (b) angular bond-bending constraints per atom.

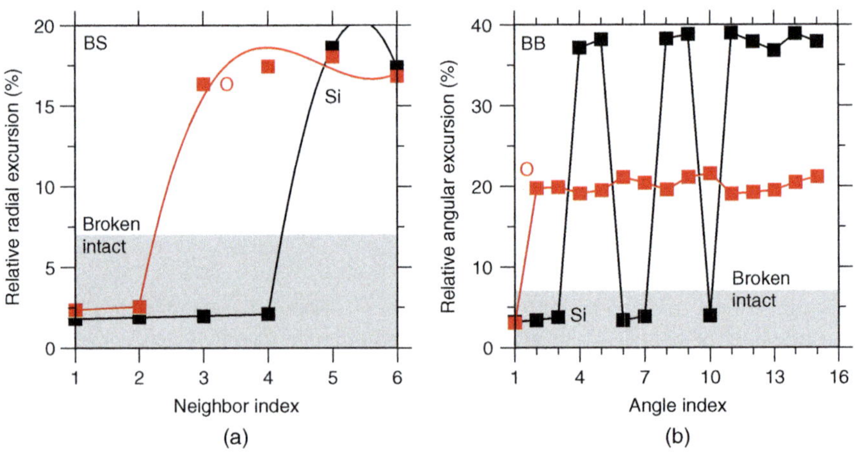


**Figure 5.6** (a) Relative radial excursion of the neighbors around central Si and bridging O atoms, respectively. (b) Relative angular excursion of each angle formed around central Si and bridging O atoms, respectively.

distance $d$ and (ii) the standard deviation of this distance $\sigma_r$. The value of $\sigma_r$ corresponds to the radial excursion (e.g. in Å) of each neighbor around the central atom. Based on these values, we then calculate the relative radial excursion of each neighbor as $\sigma_r/d$ (in %).

As an illustration of this method, Figure 5.6a shows the relative radial excursions obtained in a sodium silicate glass [46]. We observe a clear distinction between pairs of atoms exhibiting low (around 2%) and large (around 17%) relative radial excursions. These two cases are associated with intact and broken radial constraints, respectively. In detail, the fact that a given neighbor exhibits a low relative radial excursion suggests that there exists an underlying BS constraint that maintains the distance between these two atoms fixed. In contrast, the fact that a given neighbor exhibits a large relative radial excursion suggests the absence of any underlying constraint (which would otherwise prevent such a large interatomic motion). In the present case, this analysis yields expected results, that is,

(i) each Si atom creates four BS constraints with its four nearest O neighbors while (ii) each bridging oxygen (BO) atom creates two BS constraints with its two nearest Si neighbors.

Due to the clear gap between the relative radial excursion of intact and broken constraints (see Figure 5.6a), this analysis does not significantly depend on the specific choice of the threshold between intact and broken constraints. Nevertheless, a relative radial excursion threshold around 7% typically yields good results [12, 46], which is fairly close to the Lindemann criterion [77] – although the exact value of this threshold may depend on the forcefield.

#### 5.3.1.2 Enumeration of Bond-Bending Constraints

We then focus on the enumeration of the BB constraints. The method used to enumerate the BB constraints is largely similar to that used to compute the BS constraints [71]. The $N$ nearest neighbors form $N(N-1)/2$ independent angles around the central atom 0 (i.e. angles 102, 103, 104, 203, etc.). We then track the value of each angle over time so as to compute the (i) average $\theta$ and (ii) standard deviation $\sigma_\theta$ of each angle. The value of $\sigma_\theta$ corresponds to the angular excursion (in °) of each angle formed around the central atom. Based on these values, we then calculate the relative angular excursion of each angle as $\sigma_\theta/\theta$ (in %).

As an illustration of this method, Figure 5.6b shows the relative angular excursions obtained in a sodium silicate glass [46]. Like in the case of the BS constraints, we observe a clear distinction between angles exhibiting low (around 5%) and large (20%–40%) relative angular excursions. These two cases are associated with intact and broken angular constraints, respectively. Namely, the fact that a given triplet of atoms exhibits a low relative angle excursion suggests that there exists an underlying BB constraint that maintains the angle between these three atoms fixed. In contrast, the fact that a given triplet of atoms exhibits a large relative angular excursion suggests the absence of any underlying constraint (which would otherwise prevent such a large variation in their interatomic angle). In the present case, this analysis yields again expected results, that is, (i) each Si atom creates five independent BB constraints with its four nearest O neighbors (i.e. angles 102, 103, 104, 203, 204, and 304 – but note that one of these angles is fully redundant with the others), while (ii) each BO atom creates only a single BB constraint with its two nearest Si neighbors (i.e. angle 102).

As observed in the case of BS constraints, there exists a clear gap between the relative angular excursions of intact and broken BB constraints (see Figure 5.6b) so that this analysis does not significantly depend on the specific choice of the threshold between intact and broken constraints. Nevertheless, a relative radial excursion threshold around 7% typically again yields good results [12, 46].

#### 5.3.1.3 Effect of Temperature on the Topological Constraints

We now focus on assessing the effect of temperature on the number of topological constraints. As mentioned in Section 5.2.5, each constraint is associated with a given free energy and, hence, eventually breaks as temperature increases. Building a temperature-dependent topological model for a given glass family requires the knowledge of the onset temperature at which each type of constraint breaks [42, 43, 78] or, more accurately, the fraction of intact and broken constraints as a function of temperature [46].

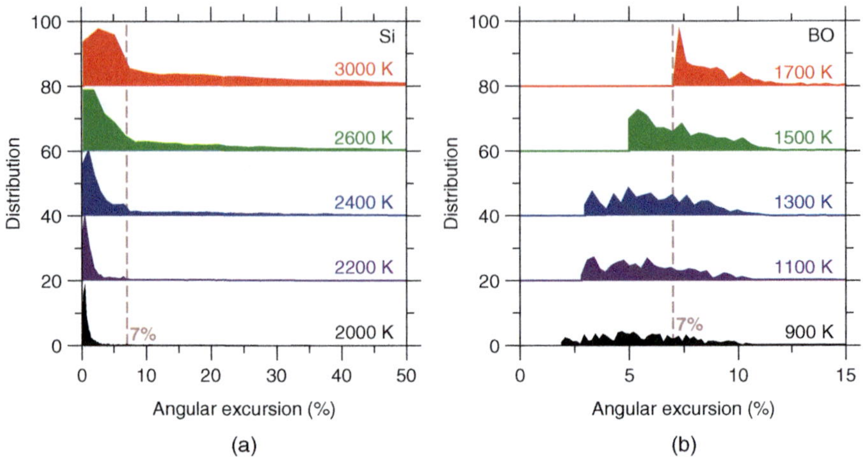


**Figure 5.7** Distribution of relative angular excursions of the angles forming around (a) Si atoms (i.e. four O–Si–O angles) and (b) bridging O (BO) atoms (i.e. one Si–BO–Si angle) in a $(CaO)_{50}(SiO_2)_{50}$ glass [44].

The motion-based enumeration method presented above can be extended so as to track the effect of temperature on topological constraints. In detail, we compute the relative radial and angular excursions associated with each BS and BB constraint as a function of temperature. This yields a distribution of excursions for each type of atom and for each temperature. As an illustration of this method, Figure 5.7 shows the distributions of relative angular excursions associated with Si and BO atoms (i.e. four O–Si–O angles around each Si atom and one Si–BO–Si angle around each BO) in a $(CaO)_{50}(SiO_2)_{50}$ glass. In general, we find that the relative angular excursions around Si atoms are lower than those around BO atoms – in agreement with the fact that intrapolytope (inside the $SiO_4$ polytopes) BB constraints are expected to be stronger than interpolytope (in between the $SiO_4$ polytopes) BB constraints [42, 46, 79]. We then note that both of these distributions gradually shift toward larger relative angular excursions as temperature increases. This signals the fact that, upon increasing temperature, BB constraints become less active since their propensity to maintain the angles fixed around their average value decreases.

In detail, we observe some notable differences between the behaviors of Si and BO angular constraints upon increasing temperature. First, in the case of the O–Si–O BB constraints, we observe that the maximum position of the distribution only shows a moderate shift toward larger relative angular excursion upon increasing temperature. However, a long tail expanding toward very large angular excursions gradually form upon increasing temperature. This signals that, as temperature increases, a larger number of angular BB constraints remain fairly unaffected, while, in turn, a fraction of the BB constraints break in a dramatic fashion (as evidenced by a very large angular excursion). Indeed, it has previously been shown that, around each Si atom, the BB constraints involving the fourth O neighbor (i.e. angles 104, 204, and 304) are the first to break, while the other BB constraints (i.e. angles 102, 103, and 203) remain active even until very high temperature [46]. In contrast, Si–BO–Si BB constraints

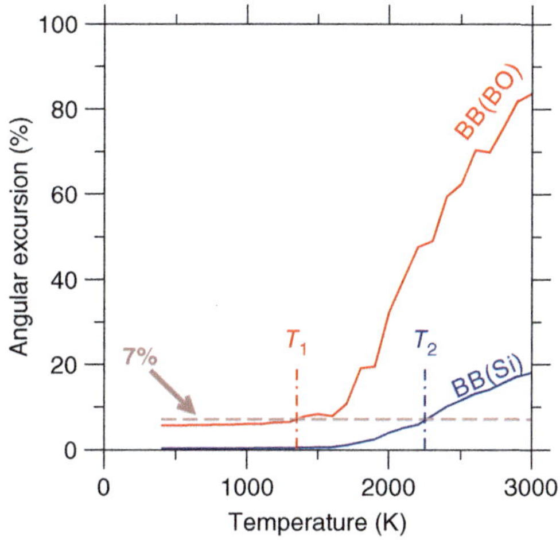


**Figure 5.8** Average relative angular excursion associated with the bond-bending (BB) constraints created around Si and bridging O (BO) atoms (i.e. O–Si–O and Si–BO–Si angles, respectively) as a function of temperature. The horizontal line indicates the threshold relative angular excursion value (7%) separating active (<7%) from thermally broken (>7%) constraints [44].

exhibit a different behavior upon increasing temperature – as the relative angular excursion distributions gradually shift "in-block" toward larger values upon increasing temperature. This suggests that these constraints collectively break as temperature increases.

Altogether, the angular excursion distributions presented in Figure 5.7 make it possible to determine the fraction of active/inactive constraints as a function of temperature [44]. The fraction of active/inactive constraints can be readily obtained by comparing their angular excursion to the threshold value given above, i.e. 7% – that is the critical relative radial/angular excursion discriminating intact from broken constraints.

The onset temperature associated with each constraint can then be estimated by plotting the average relative angular excursion associated with each type of constraint as a function of temperature (Figure 5.8). Overall, we note that the average relative angular excursions of Si and BO atoms exhibit a fairly bilinear behavior, that is they are fairly constant at low temperature, and then tend to increase linearly with increasing temperature. The onset temperature can be estimated as the temperature at which the average relative angular excursion crosses the threshold value, namely, 7%. This analysis yields a notably different onset temperature value for Si–BO–Si and O–Si–O BB constraints, namely, $T_1 = 1350$ K and $T_2 = 2250$ K. This echoes the fact that O–Si–O BB constraints are stronger than Si–BO–Si constraints (namely, they are associated with a higher free energy) and, hence, can sustain higher temperatures before breaking.

#### 5.3.1.4 Limitations

Altogether, this method offers a direct access to the number of BS and BB constraints created by each atom as a function of temperature. Note that this analysis must be conducted on each atom individually, especially in situations where a given element can exhibit varying coordination numbers (and, hence, varying number of constraints). One should also keep in mind that the number of BS constraints for a given atom offered by this analysis does not always match with its geometric coordination number – that is the number of atoms present in its first coordination shell. For instance, Na atoms usually show a high geometrical coordination number (around 6), but typically feature a single BS constraint [46]. This suggests that a simple proximity-based analysis solely relying on a distance cutoff does not always properly capture the topological environment of a given atom.

Furthermore, it is worth noting that the present method does not always properly account for mutually redundant constraints (as commonly observed in crystals). For instance, consider a triplet of atoms {1, 2, 3}. If there exist 2 BS constraints between atoms 1–2 and 2–3 and 1 BB constraint for the angle 123, then the relative radial motion between atoms 1 and 3 may be low – despite the fact that there exists no BS constraint acting between atoms 1 and 3. This arises from the fact that the triangle formed by atoms {1, 2, 3} is already fully defined by the knowledge of the distance between the pairs of atoms 1–2 and 2–3 and the value of the 123 angle – so that the 1–3 distance is here fully redundant. Nevertheless, the fairly large angular flexibility observed in glasses typically mitigates such difficulties [63, 80]. As an additional difficulty, due to the high cooling rate used in MD simulations, simulated glasses often exhibit higher fictive temperatures than those obtained experimentally [81, 82] – so that this analysis may overestimate the onset temperature associated with each type of constraint (which may need to be rescaled). Performing several MD simulations while using different (decreasing) cooling rates (e.g. from 100 to 0.001 K/ps) and extrapolating predictions toward experimental cooling rates (e.g. around 1 K/s) can overcome these limitations [83, 84].

## 5.3.2 Computation of the Internal Stress

### 5.3.2.1 Stress Per Atom Formalism

As an alternative route to a direct constraint enumeration, MD simulations can be used to track some signatures of the flexible to stressed-rigid transition upon increasing rigidity (i.e. increasing connectivity) – which can help to identify isostatic glass compositions that feature an optimally connected network. In the following, we detail the computation of two useful signatures of rigidity transitions, namely, internal stress (a signature of stressed-rigid network) and floppy modes (a signature of flexible networks).

We first focus on the computation of the internal stress. To this end, we rely on the concept of "stress per atom" introduced by Egami [85, 86]. Note that, strictly speaking, stress is ill-defined for individual atoms and can only be meaningfully calculated for an ensemble of atoms. Nevertheless, an atomic-level stress tensor $\sigma_i^{\alpha\beta}$ can be defined for each atom $i$ by expressing the contribution of each atom to the virial of the system:

$$\sigma_i^{\alpha\beta} = \frac{1}{V_i}\sum_j r_{ij}^{\alpha} \cdot F_{ij}^{\beta} \tag{5.8}$$

where $V_i$ is the volume of atom $i$, $r_{ij}$ is the distance between atoms $i$ and $j$, $F_{ij}$ is the interatomic force applied by atom $j$ on atom $i$, and the indexes $\alpha$ and $\beta$ refer to the projections of these vectors along the Cartesian directions $x$, $y$, or $z$. Note that the definition of the atomic volume is somehow arbitrary – the atomic volume is here solely introduced so as to obtain a quantity that has the unit of a stress, i.e. N/m$^2$ or eV/Å$^3$. It is common to define $V_i$ as the Voronoi volume of each atom. Altogether, this atomic stress can be readily computed from MD based on the knowledge of the interatomic distances and forces [87]. Note that the existence of some atomic stress does not imply that the system is experiencing a state of macroscopic stress. Indeed, even if a glass is macroscopically at zero pressure, some bonds may locally be under compression while others may locally be under tension – so that bonds under tension and compression may eventually mutually compensate each other at the macroscale. Although the decomposition of the virial in terms of individual atomic contributions does not have a clear physical interpretation, it provides a convenient means to capture the local state of stress (i.e. compression or tension) experienced by each atom. As such, several previous studies have utilized this approach to quantify the extent of local instabilities within the atomic network [11, 15, 39, 88–94].

A local hydrostatic stress undergone by each atom can be defined from the trace of the stress tensor:

$$\sigma_i = \left(\sigma_i^{xx} + \sigma_i^{yy} + \sigma_i^{zz}\right)/3 \tag{5.9}$$

By convention, the hydrostatic stress is usually taken as positive if the atom is experiencing a local state of tension and negative if the atom is experiencing a local state of compression. A local atomic shear stress can also be defined based on the von Mises definition of shear stress as follows:

$$\tau_i = \sqrt{\left(\sigma_i^{xy}\right)^2 + \left(\sigma_i^{xz}\right)^2 + \left(\sigma_i^{yz}\right)^2 / 3} \tag{5.10}$$

#### 5.3.2.2 Stress in Isolated Clusters

When computing the internal stress present in a stressed-rigid glass, one must distinguish the "intrinsic atomic stress" (i.e. the localized stress that would be undergone by an atom belonging to an isolated cluster, that is which is not part of a fully connected atomic network) from the "network stress" (i.e. the internal stress arising from the existence of mutually dependent constraints in the network). This distinction arises from the observation that even an isolated $SiO_4$ unit would present some internal stress due to a competition between Si–O attraction and O–O repulsion [95, 96]. As such, the topological network stress acting on each atom can be isolated by calculating the difference between the total stress (computed in a given bulk glass) and the intrinsic stress that would still be present in an isolated cluster [95].

The computation of the intrinsic stress requires one to simulate a series of isolated clusters that capture the range of local environments found in the glass of interest. For instance, in modified silicate glasses, one must simulate all the possible $Q^n$ units, wherein a $Q^n$ unit represents an Si atom that is surrounded by $n$ BO atoms and $4 - n$ nonbridging O (NBO) atoms (Figure 5.9). Note that some charge-compensating atoms (e.g. Na) are required to ensure that the isolated clusters are charge-neutral. The initial configuration of each cluster can be created by manually placing the atoms at their designated position, that is based on the average values of the Si—O and Na—O bond lengths, as well as those of the O–Si–O, Si–BO–Si, and Si–NBO–Na interatomic angles [66]. The atoms can then be placed at the center of a large cubic box with fixed volume (e.g. a side of 50 Å) with no periodic boundary conditions. The isolated clusters should be properly relaxed (e.g. 300 K run in the constant number (N), volume (V) and temperature (T), microcanonical ensemble [*NVT*] ensemble for 100 ps, subsequently cooling from 300 to 0 K with a cooling rate of 1 K/ps while fixing to zero the linear and angular momentum of the cluster to avoid any drift of the system, and final energy minimization). To ensure consistency, it is important to use the same forcefield and simulation parameters for the simulations of the isolated clusters and bulk glass structure.

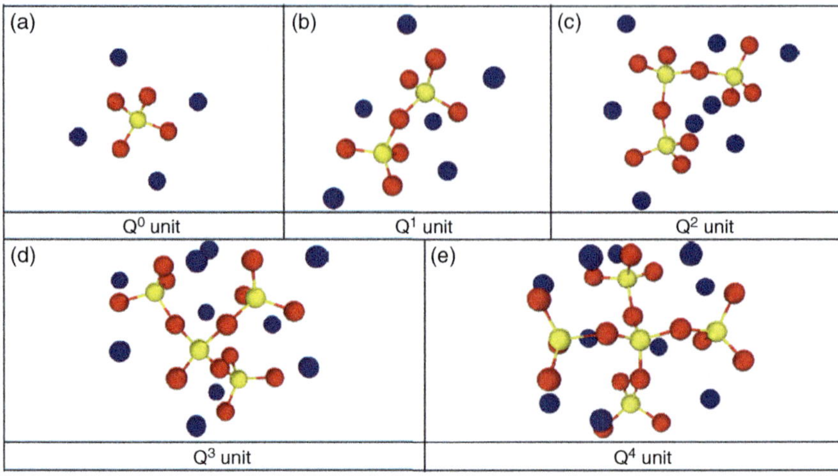


**Figure 5.9** Snapshots of isolated (a) $Q^0$, (b) $Q^1$, (c) $Q^2$, (d) $Q^3$, and (e) $Q^4$ unit clusters relevant to silicate glasses. Si, O, and Na atoms are indicated in yellow, red, and blue, respectively.

#### 5.3.2.3 Network Stress as a Signature of Stressed-Rigid Networks

We now illustrate the atomic stress computation by taking the example of simulated sodium silicate glasses with varying composition [66, 97, 98]. Binary sodium silicate glasses, $(Na_2O)_x(SiO_2)_{1-x}$, offer an ideal archetypical system to investigate glass rigidity as their connectivity can be tuned by changing their composition $x$ [99]. In this system, the number of constraints per atom is given by (see Ref. [2] for details):

$$n_c = (11 - 10x)/3 \tag{5.11}$$

This enumeration captures the fact that the overall degree of polymerization of sodium silicate glasses decreases upon increasing sodium concentration ($x$), since each sodium atom results in the formation of an NBO. This system is found to be isostatic ($n_c = 3$) for $x = 20\%$ [2, 27], in agreement with experimental observations [100, 101].

Figure 5.10a shows the hydrostatic stress per BO as a function of the glass composition, as obtained in isolated clusters (reference intrinsic stress) and in bulk glasses. As mentioned above, the atomic stress is found to be nonzero, even in isolated clusters [95]. The difference between these two stresses yields the internal topological network stress, which arises from the existence of mutually dependent constraints in the atomic network (shown in Figure 5.10b). We observe that, overall, the magnitude of internal stress monotonically decreases with increasing sodium concentration. However, the break in slope at 20% of $Na_2O$ signals the occurrence of the unstressed-to-stressed transition (i.e. unstressed and stressed at high and low $Na_2O$ content, respectively). This analysis confirms that the computation of the internal stress acting in glasses makes it possible to identify isostatic glass compositions (i.e. $x = 20\%$ in the present case). Note that the network stress does not fully drop to zero in the flexible (unstressed) domain due to the existence of some residual entropic stress in the

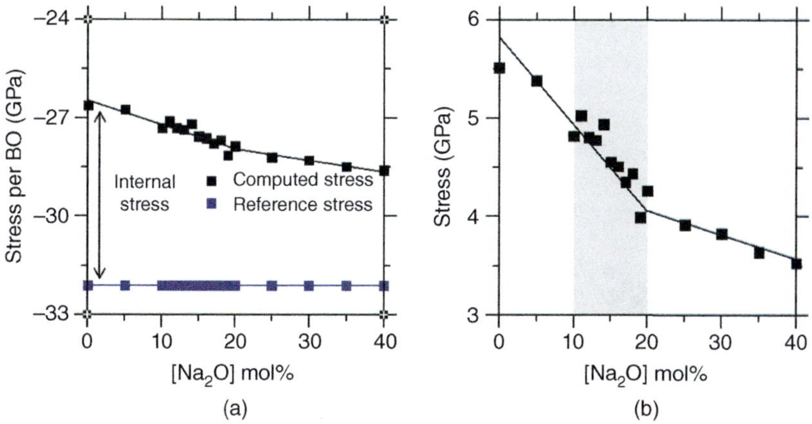


**Figure 5.10** (a) Stress per bridging O (BO) computed in bulk sodium silicate glasses ("computed stress") and in isolated clusters ("reference stress") as a function of composition. The difference between these two stress values yields the topological "network stress," which is plotted in panel (b). The gray area indicates an approximate range of "intermediate phase" compositions wherein sodium silicate glasses were found to exhibit an isostatic character [93, 100].

glass arising from the high cooling rate used in MD simulations [95]. The contribution of this entropic stress can be formally captured by repeating this analysis for varying cooling rates – since this entropic stress decreases upon decreasing cooling rate as the glass reaches a lower fictive temperature [95].

### 5.3.3 Computation of the Floppy Modes

We now focus on methods aiming at assessing the existence of floppy modes (a manifestation of atomic mobility) within glasses [31]. The level of internal flexibility of a given network can be determined by computing the extent of atomic mobility following an external stimulus [15, 39, 102]. In detail, we first ensure that all the atoms reach their local minimum in the energy landscape by subjecting the glass to an energy minimization (which is equivalent to relaxing the glass at 0 K) so that the glass reaches its inherent configuration. Starting from this configuration, we then apply an instantaneous energy bump by manually increasing the average kinetic energy of the atoms. Due to the equipartition of the energy, half of the energy is used to increase the temperature, while the other half is used to increase the potential energy of the system through some atomic displacements. The atomic motion in the constant number (N), volume (V) and energy (E), canonical ensemble (*NVE*) ensemble resulting from this energy bump is then tracked to assess the level of atomic mobility within the network. The level of atomic motion can be estimated by computing the final mean square displacement shown by the atoms after the energy bump [103].

This method is based on the following physical picture. Starting from their local minimum of energy, atoms are able to move to a new position if their kinetic energy is sufficient to overcome some local energy barriers within the energy landscape [104, 105]. Namely, an

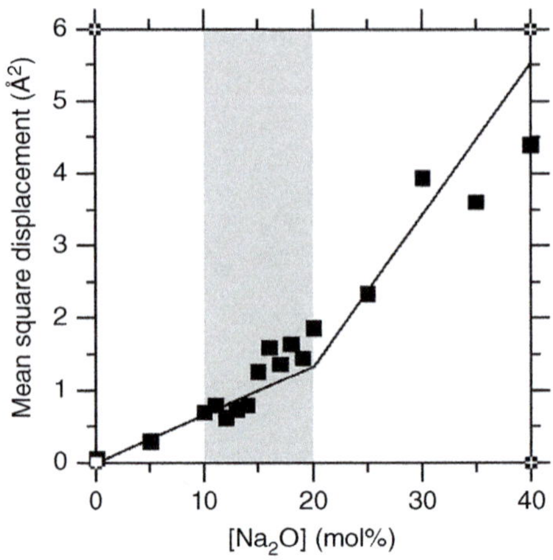


**Figure 5.11** Mean square displacement of the atoms in sodium silicate glasses following an "energy bump" (a proxy for the atomic mobility within the network) as a function of the composition of the glass. The gray area indicates an approximate range of "intermediate phase" compositions wherein sodium silicate glasses were found to exhibit an isostatic character [93, 100].

atom will exhibit some motion if it has access to a low-energy barrier, whereas it will only vibrate around its average position if it is only surrounded by large energy barriers. As such, this method can be used to indirectly sample the local roughness of the energy landscape [102]. In particular, the existence of low-energy barriers is indicative of some floppy internal modes of freedom within the network – so that the level of atomic mobility following an energy bump is a direct indication of the density of floppy modes. Note that the magnitude of the energy bump should be chosen based on the targeted height of the energy barriers to be investigated. Namely, a very small energy bump would not result in any noticeable atomic motion since none of the atoms would be able to escape their local minimum of energy, while, in turn, a very large energy bump would permanently displace all the atoms in the network by breaking topological constraints (i.e. the glass would melt). Hence, the level of the energy bump should be chosen so as to be larger than the typical energy barrier associated with a floppy mode, but lower than the energy of a topological constraint. In the case of sodium silicate glasses, an energy bump of 0.2 eV/atom can be used to track the density of floppy modes without breaking any interatomic constraints.

Figure 5.11 illustrates the outcome of this method by taking the example of simulated sodium silicate glasses. We observe that the postbump mean square displacement of the atoms increases monotonically upon increasing $Na_2O$ content. However, the break in slope at 20% of $Na_2O$ signals the occurrence of the flexible-to-rigid transition (i.e. flexible and rigid at high and low $Na_2O$ content, respectively). Once again, this analysis confirms that the computation of the internal atomic mobility within glass networks makes it possible to identify isostatic glass compositions (i.e. $x = 20\%$ in the present case).

### 5.3.4 Dynamical Matrix Analysis

Finally, we review the constraint enumeration method introduced by Mauro and Varshneya [106, 107], which relies on the computation of the dynamical matrix (or Hessian matrix). This method is based on the idea that constraints and floppy modes are encoded in the topography of the energy landscape and, specifically, in the local curvature of the energy landscape for each atom around its average position [108]. In brief, topological constraints prevent the motion of atoms and, hence, result in the formation of barriers within the landscape, while floppy modes manifest themselves as channels (i.e. low-energy barriers) between different configurations [109–111]. As such, for a given atom, the vicinity of a channel should manifest itself as low local curvature in the landscape, whereas the local landscape around a fully constrained atom should exhibit a high curvature.

Starting from a properly equilibrated glass configuration wherein the atoms are located at their local minimum of energy (i.e. inherent configuration at 0 K), the dynamical matrix is computed as the derivative of the total potential energy $U$ with respect to the Cartesian coordinated of each atom [106, 107]:

$$H = \begin{pmatrix} \frac{\partial^2 U}{\partial x_1^2} & \frac{\partial^2 U}{\partial x_1 \partial x_2} & \cdots & \frac{\partial^2 U}{\partial x_1 \partial x_{3N}} \\ \frac{\partial^2 U}{\partial x_2 \partial x_1} & \frac{\partial^2 U}{\partial x_2^2} & \cdots & \frac{\partial^2 U}{\partial x_2 \partial x_{3N}} \\ \vdots & \vdots & \ddots & \vdots \\ \frac{\partial^2 U}{\partial x_{3N} \partial x_1} & \frac{\partial^2 U}{\partial x_{3N} \partial x_2} & \cdots & \frac{\partial^2 U}{\partial x_{3N}^2} \end{pmatrix}_{x_i = x_i^0} \tag{5.12}$$

where $x_i$ represents the $i$-th coordinate of the ensemble of the $3N$ coordinates $\{x_i\}$ of the $N$ atoms and $x_i^0$ refers to the equilibrium coordinates (i.e. in the inherent configuration). Hence, the dynamical matrix is essentially a real symmetric $3N{\cdot}3N$ matrix representing the curvature of the potential energy landscape along each of $3N$ dimensions. The energy curvatures associated with independent modes of deformation can be calculated by diagonalizing the dynamical matrix, i.e. by computing the eigenvalues $b$ and eigenvectors $\vec{V}$ satisfying:

$$H\vec{V} = b\vec{V} \tag{5.13}$$

where the eigenvalue $b_i$ corresponds to the curvature of the potential energy landscape along the direction of its associated eigenvector $\vec{V}_i$.

Since constraints and floppy modes are associated with high and low curvatures, respectively, the fractions of constraints and floppy modes within the $3N$ atomic degrees of freedom can be estimated by ranking the eigenvalues $b_i$ in terms of increasing value (i.e. increasing curvature). In chalcogenide glasses, the critical curvature value separating floppy modes from constraints was estimated to be at $b_0 = 20\ \mathrm{J/m^2}$ (which corresponds to a vibrational frequency of 2.0 THz in the harmonic approximation [106]). The six macroscopic degrees of freedom should also be removed from this enumeration. Altogether, the total number of floppy modes can be estimated as the number of eigenvalues satisfying $b_i < b_0$.

As an illustration of this method, Figure 5.12 shows the computed fraction of floppy (soft) modes as a function of the average coordination number $<r>$ in simulated Ge–Se and As–Se glasses [106, 107]. As expected, the fraction of floppy modes shows a minimum plateau in the stressed-rigid domain ($<r> > 2.4$) and increases linearly with decreasing $<r>$ in the flexible regime ($<r> < 2.4$) [112–119]. This signals that once the glass becomes flexible upon decreasing connectivity, the network starts to comprise of some internal modes of deformation associated with low energy curvature. The linear dependence between the fraction of floppy modes and $2.4 - <r>$ is a natural consequence of Eq. (5.7). Overall, this method offers a complementary pathway to track flexible-to-rigid transitions in glasses.

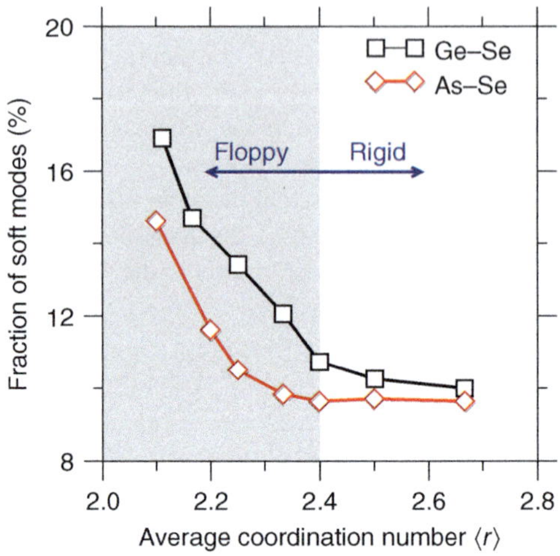


**Figure 5.12** Computed fraction of floppy modes in simulated Ge–Se and As–Se glasses as a function of the average coordination number $\langle r \rangle$. The gray window (i.e. $\langle r \rangle < 2.4$) indicates the flexible domain [106, 107].

## 5.4 Conclusion

The MD-based methods presented in this chapter offer a complete set of complementary approaches enabling an accurate description of the topology and rigidity of glasses. As such, when developing a topological model for a given family of glasses using atomistic simulations, it is extremely valuable to ensure that all these viewpoints on glass rigidity (e.g. direct constraints calculation, internal stress and floppy modes computation, dynamical matrix approach) mutually support each other. The harmony shown by these alternative, yet complementary approaches is key to ensure that topological models of glasses do not rely on any unproven guesses and, rather, are based on a sound structural and topological description of glasses.

Finally, it should be noted that, even though MD simulations are pivotal to obtain accurate inputs for topological models, these inputs can often serve to develop fully analytical models aiming to predict glass properties as a function of composition or temperature. In that regard, the combination of TCT and MD simulations can ultimately allow one to avoid having to use MD when exploring the properties of glasses within a given compositional envelope. In the future, we envision that MD simulations and network topology should inform or be informed by machine learning [120–122]. Clearly, a synergetic combination of simulations, topological approaches, artificial intelligence algorithms, and data-driven knowledge shows great promises to significantly accelerate the discovery of novel glasses with enhanced properties or unusual functionalities.

## Acknowledgments

The authors acknowledge some financial support for this research provided by the National Science Foundation under Grants No. 1762292, 1826420, and 1928538, as well as the International Cooperation on Scientific and Technological Innovation Programs of BGRIMM under Grant No. 2017YFE0107000.

## References


1 Phillips, J.C. (1979). Topology of covalent non-crystalline solids. 1. Short-range order in chalcogenide alloys. *J. Non-Cryst. Solids* 34: 153–181. https://doi.org/10.1016/0022-3093(79)90033-4.

2 Bauchy, M. (2019). Topological constraint theory and rigidity of glasses. In: *21st Century Nanoscience – A Handbook* (ed. K.D. Sattler), 13-1–13-20. CRC Press https://doi.org/10.1201/9780367333003-13.

3 Massobrio, C., Du, J., Bernasconi, M., and Salmon, P.S. (ed.) (2015). *Molecular Dynamics Simulations of Disordered Materials*. Cham: Springer International Publishing https://doi.org/10.1007/978-3-319-15675-0.

4 Mauro, J.C. (2011). Topological constraint theory of glass. *Am. Ceram. Soc. Bull.* 90: 31–37.

5 Smedskjaer, M.M., Mauro, J.C., and Yue, Y. (2010). Prediction of glass hardness using temperature-dependent constraint theory. *Phys. Rev. Lett.* 105: 115503. https://doi.org/10.1103/PhysRevLett.105.115503.

6 Mauro, J.C., Tandia, A., Vargheese, K.D. et al. (2016). Accelerating the design of functional glasses through modeling. *Chem. Mater.* 28: 4267–4277. https://doi.org/10.1021/acs.chemmater.6b01054.

7 Mauro, J.C. (2018). Decoding the glass genome. *Curr. Opin. Solid State Mater. Sci.* 22: 58–64. https://doi.org/10.1016/j.cossms.2017.09.001.

8 Smedskjaer, M.M. and Bauchy, M. (2015). Sub-critical crack growth in silicate glasses: role of network topology. *Appl. Phys. Lett.* 107: 141901. https://doi.org/10.1063/1.4932377.

9 Bauchy, M. (2017). Nanoengineering of concrete via topological constraint theory. *MRS Bull.* 42: 50–54. https://doi.org/10.1557/mrs.2016.295.

10 Liu, H., Du, T., Krishnan, N.M.A. et al. (2019). Topological optimization of cementitious binders: advances and challenges. *Cem. Concr. Compos.* 101: 5–14. https://doi.org/10.1016/j.cemconcomp.2018.08.002.

11 Zhao, C., Zhou, W., Zhou, Q. et al. (2021). Topological origin of phase separation in hydrated gels. *J. Colloid Interface Sci.* 590: 199–209. https://doi.org/10.1016/j.jcis.2021.01.068.

12 Bauchy, M., Abdolhosseini Qomi, M.J., Bichara, C. et al. (2014). Nanoscale structure of cement: viewpoint of rigidity theory. *J. Phys. Chem. C* 118: 12485–12493. https://doi.org/10.1021/jp502550z.

13 Song, Y., Yang, K., Chen, J. et al. (2021). Machine learning enables rapid screening of reactive fly ashes based on their network topology. *ACS Sustainable Chem. Eng.* 9: 2639–2650. https://doi.org/10.1021/acssuschemeng.0c06978.

14 Oey, T., Plante, E.C.L., Falzone, G. et al. (2020). Topological controls on aluminosilicate glass dissolution: complexities induced in hyperalkaline aqueous environments. *J. Am. Ceram. Soc.* 103: 6198–6207. https://doi.org/10.1111/jace.17357.

**15** Wang, B., Krishnan, N.M.A., Yu, Y. et al. (2017). Irradiation-induced topological transition in $SiO_2$: structural signature of networks' rigidity. *J. Non-Cryst. Solids* 463: 25–30. https://doi.org/10.1016/j.jnoncrysol.2017.02.017.

**16** Hsiao, Y.-H., La Plante, E.C., Krishnan, N.M.A. et al. (2017). Effects of irradiation on Albite's chemical durability. *J. Phys. Chem. A* 121: 7835–7845. https://doi.org/10.1021/acs.jpca.7b05098.

**17** Pignatelli, I., Kumar, A., Field, K.G. et al. (2016). Direct experimental evidence for differing reactivity alterations of minerals following irradiation: the case of calcite and quartz. *Sci. Rep.* 6: 20155. https://doi.org/10.1038/srep20155.

**18** Micoulaut, M., Raty, J.-Y., Otjacques, C., and Bichara, C. (2010). Understanding amorphous phase-change materials from the viewpoint of Maxwell rigidity. *Phys. Rev. B* 81: 174206. https://doi.org/10.1103/PhysRevB.81.174206.

**19** King, S.W., Bielefeld, J., Xu, G. et al. (2013). Influence of network bond percolation on the thermal, mechanical, electrical and optical properties of high and low-k a-SiC:H thin films. *J. Non-Cryst. Solids* 379: 67–79. https://doi.org/10.1016/j.jnoncrysol.2013.07.028.

**20** King, S.W., Ross, L., and Lanford, W.A. (2018). Narrowing of the Boolchand intermediate phase window for amorphous hydrogenated silicon carbide. *J. Non-Cryst. Solids* 499: 252–256. https://doi.org/10.1016/j.jnoncrysol.2018.07.045.

**21** Phillips, J.C. (2004). Constraint theory and hierarchical protein dynamics. *J. Phys. Condens. Matter.* 16: S5065–S5072. https://doi.org/10.1088/0953‐8984/16/44/004.

**22** Phillips, J.C. (2009). Scaling and self-organized criticality in proteins: Lysozyme C. *Phys. Rev. E* 80: https://doi.org/10.1103/PhysRevE.80.051916.

**23** Rader, A.J., Hespenheide, B.M., Kuhn, L.A., and Thorpe, M.F. (2002). Protein unfolding: rigidity lost. *Proc. Natl. Acad. Sci. U.S.A.* 99: 3540–3545. https://doi.org/10.1073/pnas.062492699.

**24** Ball, P. (2015). Material witness: concrete mixing for gorillas. *Nat. Mater.* 14: 472–472. https://doi.org/10.1038/nmat4279.

**25** Wray, P. (2013). Gorilla Glass 3 explained (and it is a modeling first for Corning!). *Ceramic Tech Today* http://ceramics.org/ceramictechtoday/2013/01/07/gorilla-glass-3-explained-and-it-is-a-modeling-first-for-corning.

**26** Onbaşlı, M.C., Tandia, A., and Mauro, J.C. (2018). Mechanical and compositional design of high-strength corning Gorilla® glass. In: *Handbook of Materials Modeling Applications: Current and Emerging Materials* (ed. W. Andreoni and S. Yip), 1–23. Cham: Springer International Publishing https://doi.org/10.1007/978-3-319-50257-1_100-1.

**27** Bauchy, M. (2019). Deciphering the atomic genome of glasses by topological constraint theory and molecular dynamics: a review. *Comput. Mater. Sci.* 159: 95–102. https://doi.org/10.1016/j.commatsci.2018.12.004.

**28** Pandey, A., Biswas, P., and Drabold, D.A. (2016). Inversion of diffraction data for amorphous materials. *Sci. Rep.* 6: 33731. https://doi.org/10.1038/srep33731.

**29** Maxwell, J.C. (1864). L. On the calculation of the equilibrium and stiffness of frames. *Philos. Mag. Ser. 4* 27: 294–299. https://doi.org/10.1080/14786446408643668.

**30** Thorpe, M.F. (1983). Continuous deformations in random networks. *J. Non-Cryst. Solids* 57: 355–370. https://doi.org/10.1016/0022-3093(83)90424-6.

**31** Thorpe, M.F. and Duxbury, P.M. (ed.) (1999). *Rigidity Theory and Applications*, 1e. Springer.

**32** Thorpe, M.F., Jacobs, D.J., Chubynsky, M.V., and Phillips, J.C. (2000). Self-organization in network glasses. *J. Non-Cryst. Solids* 266: 859–866. https://doi.org/10.1016/S0022-3093(99)00856-X.

**33** Chakravarty, S., Almutairi, B.S., Chbeir, R. et al. (2020). Progress, challenges, and rewards in probing melt dynamics, configurational entropy change, and topological phases of group V- and group IV-based multicomponent sulfide glasses. *Phys. Status Solidi B* 257: 2000116. https://doi.org/10.1002/pssb.202000116.

**34** Boolchand, P., Bresser, W., Zhang, M. et al. (1995). Lamb–Mossbauer factors as a local probe of floppy modes in network glasses. *J. Non-Cryst. Solids* 182: 143–154. https://doi.org/10.1016/0022-3093(94)00540-0.

**35** Chen, P., Holbrook, C., Boolchand, P. et al. (2008). Intermediate phase, network demixing, boson and floppy modes, and compositional trends in glass transition temperatures of binary As–S system. *Phys. Rev. B* 78: 224208. https://doi.org/10.1103/PhysRevB.78.224208.

**36** Micoulaut, M., Malki, M., Simon, P., and Canizares, A. (2005). On the rigid to floppy transitions in calcium silicate glasses from Raman scattering and cluster constraint analysis. *Philos. Mag.* 85: 3357–3378. https://doi.org/10.1080/14786430500157029.

**37** Novita, D.I., Boolchand, P., Malki, M., and Micoulaut, M. (2009). Elastic flexibility, fast-ion conduction, boson and floppy modes in $AgPO_3$–AgI glasses. *J. Phys. Condens. Matter.* 21: 205106. https://doi.org/10.1088/0953-8984/21/20/205106.

**38** Wang, F., Mamedov, S., Boolchand, P. et al. (2005). Pressure Raman effects and internal stress in network glasses. *Phys. Rev. B* 71: 174201. https://doi.org/10.1103/PhysRevB.71.174201.

**39** Zhou, Q., Wang, M., Guo, L. et al. (2019). Intermediate phase in calcium–silicate–hydrates: mechanical, structural, rigidity, and stress signatures. *Front. Mater.* 6: https://doi.org/10.3389/fmats.2019.00157.

**40** Gupta, P.K. and Cooper, A.R. (1990). Topologically disordered networks of rigids polytopes. *J. Non-Cryst. Solids* 123: 14–21. https://doi.org/10.1016/0022-3093(90)90768-H.

**41** Sen, S. and Mason, J.K. (2019). Topological constraint theory for network glasses and glass-forming liquids: a rigid polytope approach. *Front. Mater.* 6: https://doi.org/10.3389/fmats.2019.00213.

**42** Gupta, P.K. and Mauro, J.C. (2009). Composition dependence of glass transition temperature and fragility. I. A topological model incorporating temperature-dependent constraints. *J. Chem. Phys.* 130: 094503. https://doi.org/doi:10.1063/1.3077168.

**43** Mauro, J.C., Gupta, P.K., and Loucks, R.J. (2009). Composition dependence of glass transition temperature and fragility. II. A topological model of alkali borate liquids. *J. Chem. Phys.* 130: 234503. https://doi.org/10.1063/1.3152432.

**44** Hu, Y., Liu, Z., Yang, K. et al. (2021). Rigidity theory of glass: determining the onset temperature of topological constraints by molecular dynamics. *J. Non-Cryst. Solids* 554: 120614. https://doi.org/10.1016/j.jnoncrysol.2020.120614.

**45** Bauchy, M., Qomi, M.J.A., Bichara, C. et al. (2015). Rigidity transition in materials: hardness is driven by weak atomic constraints. *Phys. Rev. Lett.* 114: 125502. https://doi.org/10.1103/PhysRevLett.114.125502.

**46** Bauchy, M. and Micoulaut, M. (2011). Atomic scale foundation of temperature-dependent bonding constraints in network glasses and liquids. *J. Non-Cryst. Solids* 357: 2530–2537. https://doi.org/10.1016/j.jnoncrysol.2011.03.017.

**47** Smedskjaer, M.M., Mauro, J.C., Youngman, R.E. et al. (2011). Topological principles of borosilicate glass chemistry. *J. Phys. Chem. B* 115: 12930–12946. https://doi.org/10.1021/jp208796b.

**48** Wang, M., Anoop Krishnan, N.M., Wang, B. et al. (2018). A new transferable interatomic potential for molecular dynamics simulations of borosilicate glasses. *J. Non-Cryst. Solids* 498: 294–304. https://doi.org/10.1016/j.jnoncrysol.2018.04.063.

**49** Bauchy, M. (2014). Structural, vibrational, and elastic properties of a calcium aluminosilicate glass from molecular dynamics simulations: the role of the potential. *J. Chem. Phys.* 141: 024507. https://doi.org/10.1063/1.4886421.

**50** Yang, K., Hu, Y., Li, Z. et al. (2021). Analytical model of the network topology and rigidity of calcium aluminosilicate glasses. *J. Am. Ceram. Soc.* 104: 3947–3962.

**51** Gin, S., Wang, M., Bisbrouck, N. et al. (2020). Can a simple topological-constraints-based model predict the initial dissolution rate of borosilicate and aluminosilicate glasses? *npj Mater. Degrad.* 4: 1–10. https://doi.org/10.1038/s41529-020-0111-4.

**52** Abdolhosseini Qomi, M.J., Bauchy, M., Ulm, F.-J., and Pellenq, R.J.-M. (2014). Anomalous composition-dependent dynamics of nanoconfined water in the interlayer of disordered calcium-silicates. *J. Chem. Phys.* 140: 054515. https://doi.org/10.1063/1.4864118.

**53** Krishnan, N.M.A., Wang, B., Falzone, G. et al. (2016). Confined water in layered silicates: the origin of anomalous thermal expansion behavior in calcium-silicate-hydrates. *ACS Appl. Mater. Interfaces* 8: 35621–35627. https://doi.org/10.1021/acsami.6b11587.

**54** Abdolhosseini Qomi, M.J., Krakowiak, K.J., Bauchy, M. et al. (2014). Combinatorial molecular optimization of cement hydrates. *Nat. Commun.* 5: 4960. https://doi.org/10.1038/ncomms5960.

**55** Bauchy, M., Abdolhosseini Qomi, M.J., Ulm, F.-J., and Pellenq, R.J.-M. (2014). Order and disorder in calcium–silicate–hydrate. *J. Chem. Phys.* 140: 214503. https://doi.org/10.1063/1.4878656.

**56** Bauchy, M. and Micoulaut, M. (2013). Percolative heterogeneous topological constraints and fragility in glass-forming liquids. *Europhys. Lett.* 104: 56002. https://doi.org/10.1209/0295‐5075/104/56002.

**57** Wang, B., Yu, Y., Wang, M. et al. (2016). Nanoductility in silicate glasses is driven by topological heterogeneity. *Phys. Rev. B* 93: 064202. https://doi.org/10.1103/PhysRevB.93.064202.

**58** Micoulaut, M. and Bauchy, M. (2017). Evidence for anomalous dynamic heterogeneities in isostatic supercooled liquids. *Phys. Rev. Lett.* 118: 145502. https://doi.org/10.1103/PhysRevLett.118.145502.

**59** Thorpe, M.F., Chubynsky, M.V., Jacobs, D.J., and Phillips, J.C. (2001). Non-randomness in network glasses and rigidity. *Glass Phys. Chem* 27: 160–166. https://doi.org/10.1023/A:1011336511583.

**60** Jacobs, D.J. and Thorpe, M.F. (1995). Generic rigidity percolation – the pebble game. *Phys. Rev. Lett.* 75: 4051–4054. https://doi.org/10.1103/PhysRevLett.75.4051.

61 Yu, Y., Krishnan, N.M.A., Smedskjaer, M.M. et al. (2018). The hydrophilic-to-hydrophobic transition in glassy silica is driven by the atomic topology of its surface. *J. Chem. Phys.* 148: 074503. https://doi.org/10.1063/1.5010934.

62 Wang, Z., Du, T., Anoop Krishnan, N.M. et al. (2020). On the equivalence of vapor-deposited and melt-quenched glasses. *J. Chem. Phys.* 152: 164504. https://doi.org/10.1063/5.0006590.

63 Wang, M., Wang, B., Bechgaard, T.K. et al. (2016). Crucial effect of angular flexibility on the fracture toughness and nano-ductility of aluminosilicate glasses. *J. Non-Cryst. Solids* 454: 46–51. https://doi.org/10.1016/j.jnoncrysol.2016.10.020.

64 Bauchy, M. and Micoulaut, M. (2013). Transport anomalies and adaptative pressure-dependent topological constraints in tetrahedral liquids: evidence for a reversibility window analogue. *Phys. Rev. Lett.* 110: 095501. https://doi.org/10.1103/PhysRevLett.110.095501.

65 Smedskjaer, M.M., Bauchy, M., Mauro, J.C. et al. (2015). Unique effects of thermal and pressure histories on glass hardness: structural and topological origin. *J. Chem. Phys.* 143: 164505. https://doi.org/10.1063/1.4934540.

66 Bauchy, M. (2012). Structural, vibrational and thermal properties of densified silicates: insights from molecular dynamics. *J. Chem. Phys.* 137: 044510. https://doi.org/10.1063/1.4738501.

67 Mantisi, B., Bauchy, M., and Micoulaut, M. (2015). Cycling through the glass transition: evidence for reversibility windows and dynamic anomalies. *Phys. Rev. B* 92: 134201. https://doi.org/10.1103/PhysRevB.92.134201.

68 Krishnan, N.M.A., Wang, B., Sant, G. et al. (2017). Revealing the effect of irradiation on cement hydrates: evidence of a topological self-organization. *ACS Appl. Mater. Interfaces* 9: 32377–32385. https://doi.org/10.1021/acsami.7b09405.

69 Wang, M., Smedskjaer, M.M., Mauro, J.C. et al. (2017). Topological origin of the network dilation anomaly in ion-exchanged glasses. *Phys. Rev. Appl.* 8: 054040. https://doi.org/10.1103/PhysRevApplied.8.054040.

70 Micoulaut, M., Bauchy, M., and Flores-Ruiz, H. (2015). Topological constraints, rigidity transitions, and anomalies in molecular networks. *Mol. Dyn. Simul. Disord. Mater.* 275–311.

71 Bauchy, M., Micoulaut, M., Celino, M. et al. (2011). Angular rigidity in tetrahedral network glasses with changing composition. *Phys. Rev. B* 84: 054201.

72 Bauchy, M. (2012). Topological constraints and rigidity of network glasses from molecular dynamics simulations. *Am. Ceram. Soc. Bull.* 91: 34–38A.

73 Bauchy, M., Micoulaut, M., Boero, M., and Massobrio, C. (2013). Compositional thresholds and anomalies in connection with stiffness transitions in network glasses. *Phys. Rev. Lett.* 110: 165501. https://doi.org/10.1103/PhysRevLett.110.165501.

74 Micoulaut, M. and Bauchy, M. (2013). Anomalies of the first sharp diffraction peak in network glasses: evidence for correlations with dynamic and rigidity properties. *Phys. Status Solidi B* 250: 976–982. https://doi.org/10.1002/pssb.201248512.

75 Bauchy, M. and Micoulaut, M. (2015). Densified network glasses and liquids with thermodynamically reversible and structurally adaptive behaviour. *Nat. Commun.* 6: 6398. https://doi.org/10.1038/ncomms7398.

**76** Bauchy, M., Wang, B., Wang, M. et al. (2016). Fracture toughness anomalies: viewpoint of topological constraint theory. *Acta Mater.* 121: 234–239. https://doi.org/10.1016/j.actamat.2016.09.004.

**77** Lindemann, F.A. (1984). The calculation of molecular eigen-frequencies. http://ntrs.nasa.gov/search.jsp?R=19840027015 (accessed 24 June 2016).

**78** Smedskjaer, M.M., Mauro, J.C., Sen, S., and Yue, Y. (2010). Quantitative design of glassy materials using temperature-dependent constraint theory. *Chem. Mater.* 22: 5358–5365. https://doi.org/10.1021/cm1016799.

**79** Shi, Y., Tandia, A., Deng, B. et al. (2020). Revisiting the Makishima–Mackenzie model for predicting the Young's modulus of oxide glasses. *Acta Mater.* 195: 252–262. https://doi.org/10.1016/j.actamat.2020.05.047.

**80** Zhang, M. and Boolchand, P. (1994). The central role of broken bond-bending constraints in promoting glass formation in the oxides. *Science* 266: 1355–1357. https://doi.org/10.1126/science.266.5189.1355.

**81** Du, J. (2015). Challenges in molecular dynamics simulations of multicomponent oxide glasses. In: *Molecular Dynamics Simulations of Disordered Materials* (ed. C. Massobrio, J. Du, M. Bernasconi and P.S. Salmon), 157–180. Springer International Publishing.

**82** Huang, L. and Kieffer, J. (2015). Challenges in modeling mixed ionic-covalent glass formers. In: *Molecular Dynamics Simulations of Disordered Materials* (ed. C. Massobrio, J. Du, M. Bernasconi and P.S. Salmon), 87–112. Springer International Publishing https://doi.org/10.1007/978-3-319-15675-0_4.

**83** Li, X., Song, W., Yang, K. et al. (2017). Cooling rate effects in sodium silicate glasses: bridging the gap between molecular dynamics simulations and experiments. *J. Chem. Phys.* 147: 074501. https://doi.org/10.1063/1.4998611.

**84** Bhaskar, P., Kumar, R., Maurya, Y. et al. (2020). Cooling rate effects on the structure of 45S5 Bioglass: insights from experiments and simulations. *J. Non-Cryst. Solids* 534: 119952. https://doi.org/10.1016/j.jnoncrysol.2020.119952.

**85** Vitek, V. and Egami, T. (1987). Atomic level stresses in solids and liquids. *Phys. Status Solidi B* 144: 145–156. https://doi.org/10.1002/pssb.2221440113.

**86** Egami, T. (2011). Atomic level stresses. *Prog. Mater Sci.* 56: 637–653. https://doi.org/10.1016/j.pmatsci.2011.01.004.

**87** Thompson, A.P., Plimpton, S.J., and Mattson, W. (2009). General formulation of pressure and stress tensor for arbitrary many-body interaction potentials under periodic boundary conditions. *J. Chem. Phys.* 131: 154107. https://doi.org/10.1063/1.3245303.

**88** Yu, Y., Wang, M., Smedskjaer, M.M. et al. (2017). Thermometer effect: origin of the mixed alkali effect in glass relaxation. *Phys. Rev. Lett.* 119: 095501. https://doi.org/10.1103/PhysRevLett.119.095501.

**89** Yu, Y., Mauro, J.C., and Bauchy, M. (2017). Stretched exponential relaxation of glasses: origin of the mixed-alkali effect. *Am. Ceram. Soc. Bull.* 96: 34–36.

**90** Yu, Y., Wang, M., Anoop Krishnan, N.M. et al. (2018). Hardness of silicate glasses: atomic-scale origin of the mixed modifier effect. *J. Non-Cryst. Solids* 489: 16–21. https://doi.org/10.1016/j.jnoncrysol.2018.03.015.

**91** Liu, H., Dong, S., Tang, L. et al. (2019). Effects of polydispersity and disorder on the mechanical properties of hydrated silicate gels. *J. Mech. Phys. Solids* 122: 555–565. https://doi.org/10.1016/j.jmps.2018.10.003.

**92** Zhao, C., Zhou, W., Zhou, Q. et al. (2020). Precipitation of calcium–alumino–silicate–hydrate gels: the role of the internal stress. *J. Chem. Phys.* 153: 014501. https://doi.org/10.1063/5.0010476.

**93** Song, W., Li, X., Wang, M. et al. (2020). Dynamic and stress signatures of the rigid intermediate phase in glass-forming liquids. *J. Chem. Phys.* 152: 221101. https://doi.org/10.1063/5.0007093.

**94** Zhou, Q., Du, T., Guo, L. et al. (2020). Role of internal stress in the early-stage nucleation of amorphous calcium carbonate gels. *Appl. Sci.* 10: 4359. https://doi.org/10.3390/app10124359.

**95** Li, X., Song, W., Smedskjaer, M.M. et al. (2019). Quantifying the internal stress in over-constrained glasses by molecular dynamics simulations. *J. Non-Cryst. Solids* 1: 100013. https://doi.org/10.1016/j.nocx.2019.100013.

**96** Liu, H., Li, Y., Fu, Z. et al. (2020). Exploring the landscape of Buckingham potentials for silica by machine learning: soft vs hard interatomic forcefields. *J. Chem. Phys.* 152: 051101. https://doi.org/10.1063/1.5136041.

**97** Bauchy, M. and Micoulaut, M. (2011). From pockets to channels: density-controlled diffusion in sodium silicates. *Phys. Rev. B* 83: 184118. https://doi.org/10.1103/PhysRevB.83.184118.

**98** Du, J. and Cormack, A. (2004). The medium range structure of sodium silicate glasses: a molecular dynamics simulation. *J. Non-Cryst. Solids* 349: 66–79. https://doi.org/10.1016/j.jnoncrysol.2004.08.264.

**99** Zhou, Q., Du, T., Guo, L. et al. (2020). New insights into the structure of sodium silicate glasses by force-enhanced atomic refinement. *J. Non-Cryst. Solids* 536: 120006. https://doi.org/10.1016/j.jnoncrysol.2020.120006.

**100** Vaills, Y., Qu, T., Micoulaut, M. et al. (2005). Direct evidence of rigidity loss and self-organization in silicate glasses. *J. Phys. Condens. Matter.* 17: 4889–4896. https://doi.org/10.1088/0953‐8984/17/32/003.

**101** Micoulaut, M. (2008). Constrained interactions, rigidity, adaptative networks, and their role for the description of silicates. *Am. Mineral.* 93: 1732–1748. https://doi.org/10.2138/am.2008.2903.

**102** Krishnan, N.M.A., Wang, B., Yu, Y. et al. (2017). Enthalpy landscape dictates the irradiation-induced disordering of quartz. *Phys. Rev. X* 7: 031019. https://doi.org/10.1103/PhysRevX.7.031019.

**103** Bauchy, M., Guillot, B., Micoulaut, M., and Sator, N. (2013). Viscosity and viscosity anomalies of model silicates and magmas: a numerical investigation. *Chem. Geol.* 346: 47–56. https://doi.org/10.1016/j.chemgeo.2012.08.035.

**104** Tang, L., Liu, H., Ma, G. et al. (2021). The energy landscape governs ductility in disordered materials. *Mater. Horiz.* https://doi.org/10.1039/D0MH00980F.

**105** Tang, L., Ma, G., Liu, H. et al. (2020). Bulk metallic glasses' response to oscillatory stress is governed by the topography of the energy landscape. *J. Phys. Chem. B* 124: 11294–11298. https://doi.org/10.1021/acs.jpcb.0c08794.

**106** Mauro, J.C. and Varshneya, A.K. (2007). Modeling of rigidity percolation and incipient plasticity in germanium–selenium glasses. *J. Am. Ceram. Soc.* 90: 192–198. https://doi.org/10.1111/j.1551‐2916.2006.01374.x.

**107** Mauro, J.C. and Varshneya, A.K. (2007). Multiscale modeling of arsenic selenide glass. *J. Non-Cryst. Solids* 353: 1226–1231. https://doi.org/10.1016/j.jnoncrysol.2006.09.067.

**108** Naumis, G.G. (2005). Energy landscape and rigidity. *Phys. Rev. E* 71: 026114. https://doi.org/10.1103/PhysRevE.71.026114.

**109** Pignatelli, I., Kumar, A., Alizadeh, R. et al. (2016). A dissolution-precipitation mechanism is at the origin of concrete creep in moist environments. *J. Chem. Phys.* 145: 054701. https://doi.org/10.1063/1.4955429.

**110** Bauchy, M., Wang, M., Yu, Y. et al. (2017). Topological control on the structural relaxation of atomic networks under stress. *Phys. Rev. Lett.* 119: 035502. https://doi.org/10.1103/PhysRevLett.119.035502.

**111** Pignatelli, I., Kumar, A., Bauchy, M., and Sant, G. (2016). Topological control on silicates' dissolution kinetics. *Langmuir* 32: 4434–4439. https://doi.org/10.1021/acs.langmuir.6b00359.

**112** Micoulaut, M., Kachmar, A., Bauchy, M. et al. (2013). Structure, topology, rings, and vibrational and electronic properties of GeSe glasses across the rigidity transition: a numerical study. *Phys. Rev. B* 88: 054203. https://doi.org/10.1103/PhysRevB.88.054203.

**113** Bauchy, M., Kachmar, A., and Micoulaut, M. (2014). Structural, dynamic, electronic, and vibrational properties of flexible, intermediate, and stressed rigid As–Se glasses and liquids from first principles molecular dynamics. *J. Chem. Phys.* 141: 194506. https://doi.org/10.1063/1.4901515.

**114** Boolchand, P., Bauchy, M., Micoulaut, M., and Yildirim, C. (2018). Topological phases of chalcogenide glasses encoded in the melt dynamics. *Phys. Status Solidi B* 255: 1800027. https://doi.org/10.1002/pssb.201800027.

**115** Chbeir, R., Bauchy, M., Micoulaut, M., and Boolchand, P. (2019). Evidence for a correlation of melt fragility index with topological phases of multicomponent glasses. *Front. Mater.* 6: https://doi.org/10.3389/fmats.2019.00173.

**116** Shatnawi, M.T.M., Farrow, C.L., Chen, P. et al. (2008). Search for a structural response to the intermediate phase in $Ge_xSe_{1-x}$ glasses. *Phys. Rev. B* 77: 094134. https://doi.org/10.1103/PhysRevB.77.094134.

**117** Sreeram, A.N., Varshneya, A.K., and Swiler, D.R. (1991). Microhardness and indentation toughness versus average coordination number in isostructural chalcogenide glass systems. *J. Non-Cryst. Solids* 130: 225–235. https://doi.org/10.1016/0022-3093(91)90358-D.

**118** Varshneya, A.K. and Mauro, D.J. (2007). Microhardness, indentation toughness, elasticity, plasticity, and brittleness of Ge–Sb–Se chalcogenide glasses. *J. Non-Cryst. Solids* 353: 1291–1297. https://doi.org/10.1016/j.jnoncrysol.2006.10.072.

**119** Swiler, D., Varshneya, A.K., and Callahan, R. (1990). Microhardness, surface toughness and average coordination number in chalcogenide glasses. *J. Non-Cryst. Solids* 125: 250–257.

**120** Liu, H., Fu, Z., Yang, K. et al. (2019). Machine learning for glass science and engineering: a review. *J. Non-Cryst. Solids* 4: 100036. https://doi.org/10.1016/j.nocx.2019.100036.

**121** Liu, H., Zhang, T., Krishnan, N.M.A. et al. (2019). Predicting the dissolution kinetics of silicate glasses by topology-informed machine learning. *npj Mater. Degrad.* 3: 1–12. https://doi.org/10.1038/s41529-019-0094-1.

**122** Yang, K., Xu, X., Yang, B. et al. (2019). Predicting the Young's modulus of silicate glasses using high-throughput molecular dynamics simulations and machine learning. *Sci. Rep.* 9: 8739. https://doi.org/10.1038/s41598-019-45344-3.

# Part II

# Applications of Atomistic Simulations in Glass Research

# 6

# History of Atomistic Simulations of Glasses

*Alastair N. Cormack*

*Inamori School of Engineering, New York State College of Ceramics, Alfred University, Alfred, NY, USA*

## 6.1 Introduction

The objective of this chapter is to discuss the origins of the applications of atomistic computer simulations to inorganic glasses and to chart the development of the techniques over the years to the present time. The overall current state of affairs is described in the other chapters of this book, so no attempt is made here to be comprehensive; indeed, space limitations also render that approach impracticable.

As a starting point, one might reasonably ask: what are simulations? Simply put, simulations are just sets of computations; a more elaborate description would be that a simulation is the investigation of a physical model using numerical, or computational, methods. Atomistic simulations, then, are investigations at the atomic scale of models of condensed matter. In the current case, these are noncrystalline solids, i.e. glasses, and, frequently, their corresponding melts.

Simulations have, in fact, been around for a long time: Buffon's Needle dates from the eighteenth century and is generally considered the progenitor of Monte Carlo techniques. In the Buffon's Needle simulation, a rod, or a needle, is dropped onto an array of lines whose separation is the same as the length of the rod. It can be shown analytically that the probability that the rod intersects a line is $2/\pi$; repeatedly dropping the rod onto the array of lines allows an increasing more accurate estimation of $2/\pi$, and hence, $\pi$. The model here is the analysis that leads to the probability of $2/\pi$. Because the (numerical) experiment includes a random element, it also shares some similarities with Monte Carlo techniques.

The point of Buffon's Needle is that it is a way of obtaining a numerical value for $\pi$. Reliable models not only describe accurately things that have been experimentally measured but also permit predictions about the outcome of future experiments. The usefulness of simulations is that they provide a means of probing the model's predictions in circumstances that are not readily accessible. This was recognized by those who developed the Monte Carlo technique in the 1940s [1, 2].

*Atomistic Simulations of Glasses: Fundamentals and Applications*, First Edition.
Edited by Jincheng Du and Alastair N. Cormack.

It may be reasonably argued that the field of atomistic simulations as we know it today largely owes its birth to the activities of the nuclear fuel industry which came on the heels of the Manhattan project. The overall driving force was the need to be able to determine the properties of materials such as nuclear fuels under conditions not accessible in the laboratory either because of the level of radioactivity or because of operating temperatures. The most common fuels are uranium-235 and Pu-239, usually used in their oxide forms. The physical model underlying the simulations was the crystalline nature of these solids, and, more specifically, the details of the interatomic forces providing their cohesion. A key factor here was the dominant contribution of electrostatic interactions in inorganic solids generally. Such a model for crystals containing positively and negatively charged atoms was first proposed by Max Born [3]. The Born Model, as it is widely known, is rooted in classical physics, not in the least because quantum theory was still mostly in embryonic form at that time.

At the same time as models of inorganic solids were being developed, there was growing interest in understanding the properties of ionic liquids or melts [4]. It was appreciated that the nature of the interatomic forces in these materials would be similar to those of the crystalline solid counterparts. The physical properties of these materials also naturally invited the application of the new computer-based atomistic simulation techniques.

Most simulations of inorganic glasses are still based on the Born Model, which is grounded in classical physics, solid-state theory, along with computational capability has improved to the point where it is possible to use quantum mechanical descriptions of the interatomic forces to simulate glasses. This topic is discussed in detail in Chapter 2 of this volume.

The structure of glass has long been a subject of discussion, even before Zacharaisen's seminal 1932 paper [5]. Among the first atomistic simulations (but not computer-based), one might point to the bent wire spring model of Evans and King [6] or the hand-built ball-and-stick model of Bell and Dean [7]. In both cases, the distribution functions obtained from measuring the bond lengths and bond angles favorably compared with the experimental X-ray scattering data of Warren et al. [8–10].

The first application of computer-based atomistic simulations to glass-forming systems was that of molecular dynamics (MD) to $BeF_2$ [11], a halide analogue to silica. Both $BeF_2$ and $SiO_2$ have structures which are based on tetrahedral coordination polyhedra linked to other tetrahedra through corner-sharing, to form a polymerized network. $BeF_2$ was chosen because it could reasonably be considered to have ionic bonding and so could build on the tremendous amount of work that had been done on the closely related alkali halides (see below).

The Born Model, as discussed in Chapter 1, treats the atoms as point charges, with short-range potentials, both repulsive and attractive, acting between them. The dominant contribution to the cohesive energy is the Coulomb, or electrostatic, energy, with additional attractive dipole–dipole and dipole–quadrupole interactions, which depend on $r^{-6}$ and $r^{-8}$, respectively. These short-range attractive potentials were first given the name van der Waals by Lennard-Jones [12].

The nature, and most appropriate analytical form, for the repulsive potential has been the subject of intense investigation, as will be discussed below.

## 6.2 Simulation Techniques

A typical crystal – or comparable piece of glass – will contain a very large number of atoms, of the order of Avogadro's Number, $N_A$, give or take a factor of a 1000 or so. An ensemble of $N_A$ atoms will have $6N_A$ microstates or configurations. Considering the behavior of such a large number of atoms simultaneously represents a significant challenge. Statistical mechanics tells us that the observed property of an ensemble of atoms is related to the average of the (macroscopic) values of the property of its various microstates or configurations:

$$\langle O \rangle = \frac{1}{\Omega} \sum_C O_C \tag{6.1}$$

where $\Omega$ is the total number of states, and $O_C$ is the macroscopic value of configuration $C$. This trivial case of an isolated system is known as the microcanonical ensemble. Since it is not practical to calculate (or measure, for that matter) the value of every single $O_C$, statistical sampling is necessary.

Atomistic simulations were developed to provide such sampling.

### 6.2.1 Monte Carlo Techniques

Although molecular dynamics is by far the most widely used atomistic simulation technique employed today in the modeling of inorganic glasses, it was not the first simulation technique to make use of electronic digital computers. This honor went to Monte Carlo simulations, which is founded on the mathematics of statistical sampling, using random numbers.

The basic approach is pick random configurations to include in the summation in the equation. A sufficiently large number of configurations (or sample size), approximating $\Omega$ in Eq. (6.1), will give a value for $\langle O \rangle$ that is within a required error. In the trivial, isolated system, all configurations would be equally likely. However, most of the time one is interested in temperature effects, and the canonical ensemble is appropriate. In this case, each configuration has its own energy, which will be different from that of other configurations, and not all configurations will contribute equally to the sum in the equation. It doesn't make sense to choose configurations completely at random; instead, one wants to pick those that contribute most. This is known as importance sampling, first introduced by Metropolis et al. [13]. For a more detailed description, please refer to Chapter 1.

Perhaps surprisingly, Monte Carlo simulations have rarely been applied to glass-forming systems – a notable exception being vitreous $B_2O_3$ [14] – so our primary focus will be on the history of molecular dynamics and its application to inorganic glass structures and their properties. (Very recently, Monte Carlo techniques have been used to model dissolution behavior in borosilicate glasses, but with atomic structures produced using molecular dynamics [15].)

### 6.2.2 Molecular Dynamics

Molecular dynamics is another approach to generating microstates or configurations. The ideas behind molecular dynamics were first developed by Alder and Wainwright [16]. They recognized that the solving, simultaneously, of the equations of motion of a set of particles was ideally suited to attack by computer, since analytical methods were not available. It had already been demonstrated that computers could be successfully used for Monte Carlo simulations.

The main difference between the two simulation techniques is that, in molecular dynamics, the next configuration is obtained from the previous one by calculating the new positions and velocities from the equations of motion, that is, it is time-dependent or dynamic, whereas Monte Carlo is basically a static method.

In both the techniques, the critical piece of physics is the description of the interatomic forces which govern the energy of the system, and also, in the case of molecular dynamics, the motion of the atoms.

As computers became faster, and the availability of increasingly large amounts of data storage, the computer programs use to run the MD simulations became more sophisticated. Additional algorithms for integrating the equations of motion were introduced, and a variety of thermostats and barostats were developed. Choosing which ones to use became a matter of setting parameters in an input control file.

A discussion of the developments in computer hardware and software is included in Section 6.5.

## 6.3 Classical Simulations: Interatomic Potentials

As we have noted, atomistic simulations rely on accurate representations of the forces acting between atoms. This is true whether that representation is embodied in classical or quantum physics.

Historically, the development of atomistic solid-state theory was spurred by the discovery and application of X-ray diffraction to the determination of crystal structures (alkali halides in particular). Born [3] was the first to argue that the lattice energy of ionic crystals was composed of two contributions, namely the electrostatic Coulomb term, which is attractive, and a repulsive term. This he expressed in the form $b/r^n$, largely on the basis of multipole effects.

Madelung [17], in 1918, had shown that the electrostatic, or Coulomb, energy could be written as $U = \alpha q_c q_a / R$, where $\alpha$ is a constant, now known as the Madelung constant, $q_c$ and $q_a$ are the charges on the cation and anion, respectively, and $R$ is the nearest neighbor cation–anion interionic separation. The numerical value of $\alpha$ is a function of the crystal structure.

The potential between two ions could then be written as

$$\Phi(r) = -\frac{\alpha e^2}{r} + \frac{b}{r^n} \tag{6.2}$$

where $r$ was the lattice spacing (in the alkali halide structure). The value of $n$ was an unknown. In 1918, Born and Landé [3], having first suggested that $n = 5$ might be a reasonable approximation (perhaps because that is what Maxwell used in his treatment of kinetic theory), found that fitting to the experimental compressibilities of the alkali halides required a value of $n$ closer to nine. No real theoretical basis, however, had been suggested for the inverse power law description of the repulsive energy and the most appropriate mathematical form for the repulsive interactions was the subject of much discussion at the time.

In fact, as the kinetic theory of gases was being developed, three types of model interactions between molecules were suggested. In one, termed Maxwellian molecules, the molecules were supposed to repel each other according to an inverse fifth power law, the second assumed that the molecules were rigid elastic spheres (the hard sphere model of today), and the third behaved as rigid elastic spheres with (weak) attractive force between them [18]. For the most part, comparison with the experimental data becoming available ruled out the Maxwellian and the rigid elastic sphere models. Thus, the idea that there were both attractive and repulsive interactions between atoms was, in some way, confirmed. However, the nature, if not origin, of the nonelectrostatic attractive interactions did not receive as much attention as the repulsive forces, until later on.

In the 1920s, in a series of papers [18–20], Lennard-Jones argued that, for the rare gases, an inverse $n$th power law for the repulsive forces produced better fits to the data such as viscosity than the hard sphere approach which had hitherto been deemed adequate. But he also included attractive forces in the form of a third inverse power law, for no other reason than it made tractable the integrals he encountered [18]. He concluded [12] that a repulsive inverse power law, which allowed for compressible atoms, was superior to the hard sphere model. The success of that endeavor led him to suppose that the nature of the repulsive potentials in alkali halide crystals could also be expressed in terms of an inverse $n$th power law, on the grounds that their electron configurations were the same [21]. Interestingly, Lennard-Jones found that different crystals required different values of $n$ to get conformity with experimental data, depending on the electronic structure (of the cation). He argued that the attractive van der Waals terms could be neglected because they were so much weaker than the electrostatic forces.

In 1927, Pauling [22] pointed out that "the wave mechanics" gave for the potential energy of two ions (H and Cl, specifically) at a distance $R$ apart as

$$\Phi_{\mathrm{HCl}} = -\frac{e^2}{R} + \frac{e^2}{18R}e^{-\xi}(\xi^5 + 2\xi^4 + 12\xi^3 + 48\xi^2 + 120\xi + 144) \tag{6.3}$$

in which $\xi = (2/3)[(Z - S_S)/a_0]R$. The first term is obviously the Coulomb potential; the second comes from "the penetration of the hydrogen ion in the chloride ion, and the resultant repulsion between two similarly charged nuclei." Noting that the potential energy of an ionic crystal could be written as $\Phi = -\alpha(e^2z^2)/R + \phi(R)$, where $\phi(R)$ is the potential "of the repulsive forces," he reasonably observed that Eq. (6.2) suggested a simple form for $\phi(R)$, namely

$$\phi(R) = \beta_m R^m e^{-\gamma R} \tag{6.4}$$

Notwithstanding, he actually opted instead for an inverse power law, noting that that was "customary," being "widely applied by Born and coworkers." Thus, he wrote the potential energy between two atoms, $A$ and $B$, as follows:

$$\Phi = \frac{z_A z_B e^2}{r_{AB}} + \beta_{AB}\frac{(r_A + r_B)^n}{r_{AB}^n}B_0 \tag{6.5}$$

in which $r_A$ and $r_B$ are "standard radii" characteristic of the ions, $B_0$ is a universal constant and $\beta_{AB}$ had the form

$$\left(1 + \frac{z_i}{N_i} + \frac{z_j}{N_j}\right) \tag{6.6}$$

For the alkali halides, the $\beta_{AB}$ had the values of 1 for $M^+ - X^-$, 0.75 for $X^- - X^-$, and 1.25 for $M^+ - M^+$. A point of interest here is that at that time the concept of ionic radius was still to be developed. A key goal of applying Born's model was, in fact, to determine the absolute sizes of ions in the solid, not just their relative sizes. Pauling argued [23] that the ratio of the radii was a significant factor in explaining the properties of the alkali halides. (Earlier, he had shown that the ratio of the radii was central in determining the coordination number of the cation.) The underlying point here is that, at the scale of interatomic distances in solids, the differences between exponential and inverse power law descriptions are not so great that the available experimental data was adequate to distinguish between them. This was to change shortly.

In 1930, London [24] showed that dipole–dipole resonance potentials exist even between nominally spherically symmetric atoms (such as rare gas atoms, for example) and that this was the main contribution to the nonelectrostatic attractive potential between a pair of atoms.

In 1932 Born and Mayer [25] reported that an exponential form, $B\exp(r/\rho)$, for the repulsive energy provided better fit to the data for alkali halides. They noted that both the inverse power law and the exponential form had two unknowns that had to be determined. After some manipulation, they arrived at an expression which included parameters which they described as lengths characteristic of the two ions, $r_1$ and $r_2$: $C_{12}b\exp((r_1 + r_2 - r)/\rho)$. The preexponential factor $C_{12}$ was that derived by Pauling (Eq. (6.6)) and $b$ a universal constant (cf. the $B_0$ of Pauling in Eq. (6.2)). They also added into their expression for the lattice energy the London $C/r^6$ term, although it was suggested that London's formula gave values that were too small.

Margenau [26] went further than London and argued that the dipole–quadrupole term should also be included (but that higher terms in the multipole expansion, such as quadrupole–quadrupole interactions, could be neglected). This was followed up by Mayer [27] who calculated, from ultraviolet absorption data the constants $C$ and $D$ in the $C/r^6$ and $D/r^8$ expressions for the dipole–dipole and dipole–quadrupole terms in the alkali halides.

These two contributions to the attractive potential were included in an analysis by Huggins and Mayer [28] who wrote the potential on an ion in a crystal as

$$\Phi(r) = A(r) + B(r) \tag{6.7}$$

with

$$A(r) = -\alpha e^2/r - C/r^6 - D/r^8 \tag{6.8}$$

for the attractive components. The values of the van der Waals constants $C$ and $D$ were those calculated by Mayer [27]. They assumed, for the repulsive potential, the formula

$$B(r) = c_{ij}bb_ib_je^{-r_{ij}/\rho} = c_{ij}be^{\frac{(r_i + r_j - r_{ij})}{\rho}} \tag{6.9}$$

for which they calculated refined values for radii of the ions and the repulsion parameters $b_i$ and $b_j$ in the expression, for alkali halide crystals. In Eq. (6.9), $\rho$ is an empirically determined constant, $c_{ij}$ are the Pauling coefficients containing the dependence of the repulsion on the charges of the ions, $r_i$ and $r_j$ are what they termed ionic radii; $b$ was arbitrarily chosen to have the value of $10^{-12}$ ergs, an old-fashioned energy unit. $B(r)$ was written in terms of both nearest neighbor and next-nearest neighbor interactions: $B(r) = B_{+-}(r) + [B_{--}(r) + B_{++}(r)]/2$. It will be appreciated that restricting the range of interatomic distances to just nearest (cation–anion), and next-nearest (cation–cation and anion–anion) neighbors, separations effectively introduces a cutoff beyond which the potential is assumed to be zero. Huggins and Mayer assumed a single value for $\rho$ of $0.345 \times 10^{-8}$ cm for all alkali halides, and then derived values for $r_i$ and $r_j$. The Huggins–Mayer form for the attractive potential, $A(r)$, was taken up by later workers.

Shortly afterward, Buckingham [29] used quantum mechanics to calculate the van der Waals parameters. He pointed out that the magnitude of the dipole–quadripole energy was only about 20–25% of the dipole–dipole energy. (Mayer had noted that in the alkali halides, the $D/r^8$ potential was only about 10–20% that of the $C/r^6$ potential.) Buckingham further observed that van der Waals energies deduced from experiments were usually expressed in terms of $C/r^6$ anyway, and thus invariably contained contributions from quadripole–dipole and higher terms. The practical significance of this appears later, as empirically fitting potential parameters became common.

In 1938, Buckingham [30] had observed that little was known about "the strong repulsion between atoms at close range due to interpenetration of electron shells." He noted that the repulsive potential could be written as $P(r)e^{-r/\rho}$, where $P(r)$ was a polynomial, commenting that quantum theoretical calculations on closed shell ions (helium and neon) suggested that treating $P(r)$ as a constant would be adequate, as evidenced by the work of Born and Mayer and Huggins. Previously, Buckingham [29] had used the quantum theory of polarization to calculate the van der Waals parameters for a series of alkali–alkali interactions, following Mayer, and Margenau.

The next stage in the development of interatomic potentials was the work of Tosi and Fumi in 1964 [31, 32]. They argued that a universal value for $\rho$, assumed by Huggins and Mayer, limited the fit between calculated and experimental compressibilities for individual salts. A consequence of allowing $\rho$ to vary for each compound was that $b$ was also no longer an arbitrary constant, but became a determinable parameter, albeit still a universal parameter.

Somewhat later, in 1972, Busing [33] used an expression for the repulsive energy similar to that of Huggins and Mayer,

$$b(r) = \left(1 + \frac{z_i}{N_i} + \frac{z_j}{N_j}\right) be^{\frac{\left(r_i^0 + r_j^0 - r_{ij}\right)}{\rho}} \tag{6.10}$$

to "interpret the crystal structure of $Li_2BeF_4$." He took the values of $r^0$, $\rho$, and $b$ from Tosi and Fumi. Note the explicit Pauling factor. He did two calculations, one which included a $c_i c_j r_{ij}^{-6}$ van der Waals term, with values for $c_{Li}$ and $c_F$ calculated by Mayer (and a theoretical value for $c_{Be}$). The other model excluded the van der Waals term. Busing did not include the quadrupole–dipole $D/r^8$ term. No explanation was given, but in otherwise unrelated simulations on organic molecules, only the dipole–dipole term was included in the potential model.

Busing noted that the model including the $r^{-6}$ van der Waals term gave excellent agreement, even though the $c_F$ value he took from Mayer was nominally three times too large, concluding that it "tends to compensate for ionic polarization or covalent bonding effects" which were omitted from the model, an observation similar to that of Buckingham earlier (and, probably, another reason for avoiding the additional computational complexity of including the $r^{-8}$ term).

In a companion paper [11], Rahman et al. used Busing's parameters, but without the van der Waals term, in molecular dynamics simulations of molten LiF, $BeF_2$, and $Li_2BeF_4$. The omission of the attractive dispersion term was not explained, but may be assumed that it was to reduce the computation time. Woodcock [4] had noted that including the dispersion times increased computation times by 10–15%, but only contributed ~5% to the total potential energy. The simulations of $BeF_2$ were the first of a tetrahedral network structure.

The evolution of analytical forms for the repulsive potential continued as attention expanded from the alkali halides to oxides.

### 6.3.1 Potential Models for Silica

In the first application of molecular dynamics (MD) to an oxide glass, namely vitreous silica, Woodcock, Cheeseman, and Angell [34] in 1976 followed the approach of Rahman et al. [11], in using Busing's form for the repulsive potential (Eq. (6.10)), taking the same value of $\rho$ as for $BeF_2$, but without the van der Waals term and used a truncated electrostatic sum "with some loss of accuracy" to save computation times.

They had to derive their own values for the silicon and oxygen $r^0$ parameters. These were adjusted over three trials to improve the agreement between the calculated and experimental average interatomic distances, that is, the maxima in Si–O, O–O, and Si–Si pair distribution functions. The final Si–O distance was well reproduced as they noted, but the model O–O distance was somewhat overestimated and the Si–Si separation slightly underestimated. The coordination number of Si was also underestimated. Nevertheless, the overall agreement was very encouraging. With the benefit of hindsight, of course, one might consider the agreement quite fortuitous. We now know that using full oxidation state charges of +4 and −2 without bond-bending terms leads to rather distorted geometries, but the absence of three-body potentials was probably compensated by the small size of the simulation box (containing just 54 cations and 108 anions) and the truncated Coulomb sum. Of course, that size was dictated by the limits of available computational resources.

They were also just as interested in modeling the liquid, that is, the silica melt, as the solid, glassy state. However, from a basic structural viewpoint, vitreous silica is not that interesting: it has long been known to consist of a collection of corner-sharing $SiO_4$ tetrahedra, forming a completely polymerized network.

Notwithstanding, this pioneering work can be said to have started the field of atomistic simulations of silicate and other, related, network oxide glasses, although subsequent work focused more on the structure of the glass and its properties, and in time expanded to more complex chemistries. Since Woodcock et al.'s paper [34], there have been more than 20 different classical potential models for silica proposed. Not all have been used in molecular dynamics simulations, however. Here we will focus only on those which have had some impact on the field.

In 1979, Soules [35] reported simulations of a series of sodium silicates, including silica. He wrote the repulsive term as $A_{ij}\exp(-r/\rho)$, with $A_{ij} = (1 + z_i/n_i + z_j/n_j)b\exp(r_i + r_j/\rho)$, that is, the Huggins–Mayer form (Eq. (6.6)), with the Pauling prefactor (Eq. (6.3)), using the same parameters for Si and O as Woodcock et al., except that value of $r_{Si}$ was adjusted to give a better fit to the experimental Si—O bond distance of 1.62 Å. The actual value for $r_{Si}$ was not recorded, but rather included in the overall (revised) value for $A_{ij}$.

Unlike Woodcock et al., Soules gave an explicit expression for the truncated Coulomb sum:

$$z_i z_j e^2 \frac{[1 - erf(r/\eta L)]}{r} - \frac{\left(z_i^2 + z_j^2\right) e^2}{\sqrt{\pi}\eta L} \tag{6.11}$$

In Eq. (6.8), $L$ is the length of the simulation box side, and an optimal value of $\eta$ was determined to be 0.35. One might note that in subsequent reports, the second term in Eq. (6.8) was dropped.

This was followed in 1981 by Mitra et al. [36], who used an explicit expression for the interionic forces, rather than a potential model from which the forces would be calculated. The forces were described by the Eq. (6.12):

$$F(r_{ij}) = \frac{q_i q_j}{r_{ij}^2}\left[1 + sign(q_i q_j)\left(\frac{\sigma_i + \sigma_j}{r_{ij}}\right)^n\right] \tag{6.12}$$

where $q$ is the ionic charge, $\sigma$ the ionic radius, and $n$ represents the hardness of the repulsive part of the pair interaction, which took the value 10. This is also a two-body model.

In 1982, Garofalini [37] reported calculations of the frequency spectrum of vitreous silica. Using molecular dynamics simulations, he obtained the velocity–velocity auto-correlation function which was Fourier transformed to obtain the frequency spectrum. His repulsive potential and initial parameters were the ones used by Soules, i.e. the same as Woodcock et al., but for the modified $r_{Si}$, although he investigated the effects of varying the pre-exponential factors $A_{Si\text{-}Si}$ and $A_{O\text{-}O}$. He also used a truncated sum for the Coulomb interactions: $\frac{z_i z_j e^2}{r_{ij}} erfc\left(\frac{r_{ij}}{\eta L}\right)$, but with a value for $\eta$ of 0.175 instead of 0.35 adopted by Soules. (For the nonmathematically astute, the complementary error function, erfc, is simply (1 − erf).)

Also in 1982, Ching [38] created (static) models for amorphous $SiO_2$ by taking existing structural models for *a*-Si, inserting oxygens in the Si—Si bonds, and relaxing the structure using a Keating type potential for both Si and O:

$$V = \sum_l V_l^{Si} + \sum_n V_n^{O} \tag{6.13}$$

$$V_l^{Si} = \frac{3}{16}\frac{\alpha}{d^2}\sum_{i=1}^{4}(\mathbf{r}_{l,i}\cdot\mathbf{r}_{l,i} - d^2)^2 + \frac{3}{8}\frac{\beta_1}{d^2}\sum_{(i,i')}^{6}\left(\mathbf{r}_{l,i}\cdot\mathbf{r}_{l,i'} - d^2\cos^{-1}\phi\right)^2 \tag{6.14}$$

$$V_n^{O} = \frac{3}{16}\frac{\alpha}{d^2}\sum_{j=1}^{2}(\mathbf{r}_{n,j}\cdot\mathbf{r}_{n,j} - d^2)^2 + \frac{3}{8}\frac{\beta_2}{d^2}\sum_{(j,j')}^{6}\left(\mathbf{r}_{n,j}\cdot\mathbf{r}_{n,j'} - d^2\cos^{-1}\theta\right)^2 \tag{6.15}$$

where $\mathbf{r}_{l,i}$ is the vector from atom $l$ to its nearest neighbor atom $i$, $\varphi$ is the ideal O–Si–O, O–Si–Si, or Si–Si–Si angle, $\theta$ is the average S–O–Si angle, and $d$ is the ideal Si—O or Si—Si bond length.

He described his models as Quasi-Periodic Models (QPM), commenting that "Although construction of structural models for $a$-$SiO_2$ started more than 15 years ago, the present results are the first time that finite models with periodic boundary conditions (PBC) have been constructed for $a$-$SiO_2$" [38]. This was not strictly correct, of course, because the earlier molecular dynamics simulations of Woodcock et al. [34] and Soules [35] also used periodic boundary conditions; their simulation boxes are the equivalent of Ching's finite models. The earlier structural models to which Chin referred would be those of Evans and King [6] and Bell and Dean [7].

Ching then calculated the electronic structures of these three structural models using a direct-space orthogonalized linear combination of atomic orbitals (OLCAO) method [39].

As a bit of a digression, it is to be noted that potential models for crystalline oxides and silicate minerals were being developed at the same time. Although the standard Born model, with Born–Mayer repulsive and van der Waals attractive potentials worked reasonably well for many oxides and silicates, for silica itself, it was soon recognized that the use of full oxidation state charges meant that, when including silicon, the Coulomb sum dominated the potential and the resulting interatomic forces caused wider bond angles. To counter this effect, three body potentials were introduced, as shown by Sanders et al. in 1983 [40], Feuston and Garofalini in 1988 [41], and subsequently by Vashishta et al. [42].

In 1988, Tsuneyuki et al. derived potential parameters for silica from Hartree–Fock self-consistent field (SCF) calculations on model molecular clusters [43]. These calculations suggested partial charges of $q = 0.6e$, and they used the following functional form for the short-range potentials, as suggested by Ida [44], following Gilbert [45]:

$$U(r) = f_0(b_i + b_j)\exp(a_i + a_j - r/(b_i + b_j)) - c_i c_j / r^6 \tag{6.16}$$

Here $a_i$ are effective (ionic) radii and $b_i$ softness parameters; $f_0$ is "the standard force" = 1 kcal /Å.mol. This was the first model to use partial charges.

A similar approach, of fitting to quantum mechanical energy surfaces, was adopted by van Beest, Kramers, and van Santen [46], who assumed a pairwise expression for the interatomic parameters, of the form

$$V(r_{ij}) = \frac{q_i q_j}{r_{ij}} + A_{ij}\exp(-b_{ij} r_{ij}) - C_{ij} r_{ij}^{-6} \tag{6.17}$$

They also introduced a generic short-range potential cutoff, which was not limited to specific nearest, or next-nearest, neighbor distances. The parameters in what has become known as the BKS potential were derived not just by fitting to energy surfaces of small molecules obtained from quantum mechanical calculations but by also optimizing the values of bulk parameters through comparison of the calculated elastic constants and unit-cell dimensions with the experimental values for quartz. They argued that the fact that the short-range potentials extended beyond nearest–neighbor interactions meant that it was essential to incorporate bulk information.

Critically, they did not attempt to express the $A_{ij}$ parameters in terms of other parameters as had Born and Mayer or Huggins and Mayer (see Eq. (6.6)); it was treated as a variable parameter in the fitting process, along with the $b_{ij}$ and $C_{Si-O}$. Initially, the $C_{O-O}$ was set to zero, but it was subsequently found necessary to let it vary, in order to be able to reproduce the properties of bulk quartz. It might be noted that the values for both the $C$ parameters were really quite large: $C_{Si-O}$ was 133.5381 and $C_{O-O}$ was 175.0000. Consequently, the results are rather sensitive to the magnitude of the short-range cutoff, although no value for the cutoff was quoted in either Ref. [46] or a more complete discussion of their parameter derivation in Ref. [47]. Subsequently, Vollmayer Kob and Binder [48] and Horbach and Kob [49] used a cutoff of 5.5 Å when they employed the BKS potential in simulations of silica melts.

The main attraction of the BKS model was that it did not need three-body terms. Instead, it used partial charges in place of formal oxidation state charges. In this way, the dominance of the electrostatic forces, which promoted distortion of the $SiO_4$ tetrahedral geometry, was sufficiently reduced to avoid the need for the bond-bending terms. At the same time, they fit parameters for aluminum phosphate analogues of silica. However, the partial charges for Al (1.4) and P (3.4) were not consistent with those obtained for silica (2.4 and −1.2 for Si and O, respectively), so it was not possible to use these parameters for aluminosilicates.

A comprehensive study on silica, by Horbach and Kob [49] using the BKS potential, from high temperature melt to the vitreous state showed that the $T_g$ predicted from the BKS potential was somewhat higher than the experimental value. This was ascribed to the nominal cooling rate. Carré et al. [50] subsequently (in 2008) tried to improve on the BKS model by refitting the parameters.

Horbach, Kob, and Binder [51] used the BKS potential to calculate an effective density of states $g(\nu)$ from the velocity autocorrelation function within the harmonic approximation. They found a reasonable approach at temperatures below 300 K but less so at higher temperatures. They argued that for a reliable description of the low-frequency part of $g(\nu)$, i.e. where the boson peak is observed, it was essential to use large systems for the simulations and small cooling rates.

One difference between simulating crystalline and glassy systems was that for crystals, the experimental unit cell parameters (i.e. density) were usually used, whereas for glasses, the experimental density is a function of cooling rate. Simulations of glasses therefore used constant pressure conditions, rather than constant volume conditions. It turned out that both the Leslie et al. and BKS potentials did not reproduce very accurately experimental densities of vitreous silica. This limitation was overcome by adjusting the magnitude of the $C$ parameter in the attractive part of the Buckingham potential.

The potential model introduced by Vashishta et al. [42], mentioned above, was more complicated than that of Sanders et al. [40]. Their two-body potential term, $V_2$, consisted of three terms. One represented steric repulsion due to ion size; a second involved Coulomb interactions designed to take into account charge transfer, and the third, charge–dipole interactions, intended to include the effects of electronic polarizabilities: $V_2 = \frac{H_{ij}}{r^{\eta_{ij}}} + \frac{Z_i Z_j}{r} - \frac{\frac{1}{2}\left(\alpha_i Z_j^2 + \alpha_j Z_i^2\right)}{r^4} e^{-r/r_{4s}}$, where $H_{ij}$ and $\eta_{ij}$ are the strengths and exponents of steric repulsion, $Z_i$ and $\alpha_i$ are the effective charge and electronic polarizability of ion $i$ and $r_{4s}$ is a decay length, taken to be a few Ångstroms.

Their expression for the three-body interaction was $V_3 = B_{ijk} f(r_{ij}, r_{ik}) p(\theta_{ijk}, \overline{\theta}_{ijk})$. $B_{ijk}$ is the strength of the three-body interaction, the functions $f(r_{ij}, r_{ik})$ represent bond-stretching and

$p(\theta_{ijk}, \overline{\theta}_{ijk})$, the effects of bond-bending. Overall, 18 parameters had to be determined, many more than needed for the BKS model and other models with only two-body potentials. One of the reasons for the complexity of the Vashishta et al. model was their attempt to include explicitly effects of electronic polarizability, especially for the $O^{2-}$ ion.

Their two MD simulations on vitreous silica used simulation boxes containing 648 or 5184 atoms. The latter was the largest of its kind at the time.

In 1993, Wilson and Madden [52] showed how the inclusion of a realistic description of polarization effects in essentially ionic models (with full formal charges) of several network-forming materials corrected the M—O—M bond-angle distribution. They argued that it was the absence of polarization, and not neglect of covalency effects, that underlay the deficiencies of the Born model.

For example, when they added polarizability to the potential model of Woodcock et al., the peak in the Si—O—Si bond angle distribution shifted to lower angles, bringing it more into line with the experimental value.

Other efforts aimed at including polarizability include those of Tangney and Scandolo [53], and the fluctuating-charge Morse stretched potential suggested by Demiralp, Cagin, and Goddard [54].

Although Wilson and Madden [55] went on to show how their polarizable ionic model could be incorporated into molecular dynamics simulations, using a scheme analogous to the quantum mechanical Car–Parrinello approach, its relative complexity meant that it was not widely adopted.

The absence of polarizability is accommodated to some extent by increased contributions from the attractive dispersive terms, as noted by Lennard-Jones and, later, Buckingham. Larger simulation sizes allow greater structural relaxation and will also mitigate the shortcomings of not including polarizability.

#### 6.3.1.1 Silica: Quantum Mechanical Simulations

We have seen how Tsuneyuki et al. [43] and van Beest et al. [46] employed energy surfaces obtained from quantum mechanical calculations to derive parameters for classical potential models. These calculations were based on molecular clusters, such as $H_4SiO_4$ or $H_6Si_2O_7$, using Hartree–Fock theory, and followed Gibbs and colleagues [56–58] who used similar calculations on small clusters to investigate bond length and bond angle variations in a range of silicate minerals.

The first complete molecular dynamics simulation of silica using quantum mechanics to calculate the interatomic forces was that of Sarnthein et al. [59] who used the approach of Car and Parrinello [60] with density functional theory to model a silica melt and amorphous silica in 1995. They found good agreement with experimental values for their calculated vibrational density of states. That work was followed in 1998 with calculations of the dynamic structure factor, again finding very good agreement with experiment [61].

Simulations of an amorphous silica surface were presented in 2000 by Ceresoli et al. [62] and Mischler et al. in 2002 [63].

For further discussions of the development and application of quantum mechanical simulations, please refer to other contributions in this volume, particularly Chapters 2 and 14.

### 6.3.2 Modified Silicates and Aluminosilicates

The proof-of-concept work of Woodcock et al. naturally spurred the application of the MD technique to other compositions.

In 1979, Soules followed Woodcock et al.'s paper, with a report on the simulation of a series of sodium silicates, a soda-lime silicate, and a sodium borosilicate glass [35]. He also modeled a mixed Na–K silicate glass. The simulation boxes contained around 200 atoms. He used a version of the Born–Mayer–Huggins potential (without the attractive dispersion terms) and a modified Coulomb sum, as discussed above, Eq. (6.11). The short-range repulsive potential parameters for silica were those of Woodcock et al. but with the magnitude of the $A_{Si\text{-}O}$ parameter adjusted to give better agreement with the experimental Si—O bond distance of 1.62 Å; the parameters for the modifier cations were taken from those derived by Fumi and Tosi [32] for the alkali halides.

He claimed good agreement with available X-ray diffraction data and with Woodcock et al.'s results on silica. However, his calculated coordination numbers for the modifiers were somewhat low, around 4.5 for sodium, for example. Si was slightly overcoordinated and the existence of free oxygens was noted. There were also non-negligible (~2%) concentrations of three-coordinated oxygens in the sodium silicates; even silica had 0.7%.

All the boron in the $9B_2O_3{\cdot}5Na_2O{\cdot}48SiO_2$ glass were tetrahedrally coordinated.

In 1981, Soules and Varshneya [64] published a more extensive study of borosilicate glasses, again with simulations of ~200 atoms. They found boron in both three [3]- and four [4]-fold coordination and correlated the relative concentrations with the sodium content.

Mitra and Hockney (1983) employed the interionic force model used earlier (Eq. (6.12)) for silica for sodium silicates [65]. The additional parameters needed for the sodium were apparently taken from their simulations on the melting of NaCl.

In 1987, Tesar and Varshneya [66] ran MD simulations on some alkali silicates, using the potential parameters from the work of Soules [35], and Soules and Varshneya [64], with some adjustments to the preexponential factor $A_{ij}$ to "correct first neighbor distances." Notably, their simulations contained ~800 ions, considerably larger than the earlier work. Their idea was to identify compositional fluctuations that might act as precursors to phase separation. Perhaps not surprisingly, given the small size of the simulation boxes used, as the authors acknowledged, no such fluctuations were observed.

Also in 1987, Ching's group [67] reported structural models for sodium silicate glasses. These were obtained by replacing units of $SiO_2$ with $Na_2O$ in their structure of amorphous silica derived, in turn, from a model of *a*-Si as noted above. These models were then used [68] to calculate their electronic structure with a direct-space OLCAO method that had been used earlier for amorphous silica.

In 1990, Zirl and Garofalini [69] reported simulations of sodium aluminosilicate glasses. They included three-body terms as introduced by Feuston and Garofalini [41] along with the truncated Coulomb sum. Although the number of atoms in their simulation box was relatively small, ranging from 327 to 370, their trajectories were quite long for the time, with runs of 40 ps at some temperatures during the cooling cycle. They found [3]-coordinated oxygen with no change in Al coordination at $R = Al/Na > 1$, as well as Al—O—Al bonds, in contravention to Lowenstein's rule observed for crystalline aluminosilicates.

Habasaki and colleagues presented a series of simulations of a lithium silicate, $Li_2SiO_3$ glass, beginning in 1992, in order to investigate the Li ion dynamics. A Gilbert-Ida form of the short-range potential was employed (Eq. (6.9)). Partial ionic charges were used. These were 2.60 for Si, 0.80 for Li, and −1.40 for O; these are not self-consistent, and so could not be used for other lithium silicate compositions.

The author's group started simulations of silicate glasses in the early 1990s, beginning with alkali silicates [70–72].

These early simulations used a rigid ion two- and three-body formal oxidation state potential model of the form proposed by Vessal et al. [73]. The two-body terms used a four-range Buckingham form and a somewhat complicated form for the three-body term, as follows:

$$E_{ijk} = \frac{1}{4} A_{ijk} B_{ijk}^2 \exp\left(\frac{-r_{ij}}{\rho_1}\right) \exp\left(\frac{-r_{jk}}{\rho_2}\right) \tag{6.18}$$

for $r_{ij}$ and $r_{jk} < t$, and where $A_{ijk} = \frac{k_{ijk}}{2}(\phi_0 - \pi)^2$, $B_{ijk} = (\phi_0 - \pi)^2 - (\phi - \pi)^2$, with $\phi$ the calculated bond angle, $\phi_0$ an equilibrium bond angle, $k_{ijk}$ a three-body spring constant and $t$ the three-body spline point. If either $r_{ij}$ or $r_{jk} > t$, but less than $t_C$, then the relevant exponential would be replaced with $P(x) = G_1 + H_1 x + I_1 x^2 + J_1 x^3 + K_1 x^4 + L_1 x^5$ with $x = r^2 - t^2$. Among other things, the complexity was introduced to avoid discontinuities in potential and forces if bond lengths were to exceed the cutoffs specified for the designation of the two Si—O bonds comprising the Si—O—Si bond angle in the three-body potential.

The parameters for Si–O and O–O were obtained by fitting to the structure and properties of α-quartz. For the alkali–oxygen parameters a simple exponential repulsive form was used with the parameters obtained by fitting to their respective crystal structures. The parameters were used unchanged in the subsequent MD simulations.

The simulation boxes contained from 691 to 927 atoms, depending on the alkali content, with each composition including 216 silicon atoms. The complexity of the interatomic potential model meant that trajectories were restricted to 10 000 timesteps at each temperature.

These simulations were the first to report a systematic analysis of the compositional dependence of the glass structure, particularly with respect to the inhomogeneous distribution of the alkali and non-bridging oxygen (NBO) species, which was found to follow the Modified Random Network Model of Greaves [74], rather than an extension of the Zachariasen model proposed by Warren [75, 76].

In addition, Huang performed simulations of mixed (Na, K) silicates [77]. He introduced the idea of using separate thermostats for the alkalis and the Si and O in the tetrahedral network. This allowed the alkali ions to migrate more quickly through a network structure at room temperature, in an early (and crude) attempt at hyperdynamics. Figure 6.1 shows in (a), the mean square displacement (MSD) of alkali ions at a nominal 3000 K, along with those of Si and O, at 300 K, and in (b), the derivatives of the MSD, indicating that constant long-range migration (i.e. constant diffusion coefficient) had been reached.

The idea had some experimental support from NMR studies [78, 79] which indicated that the motion of the alkalis was indeed largely decoupled from those of silicon and oxygen in the glass as it was heated through the glass transition.

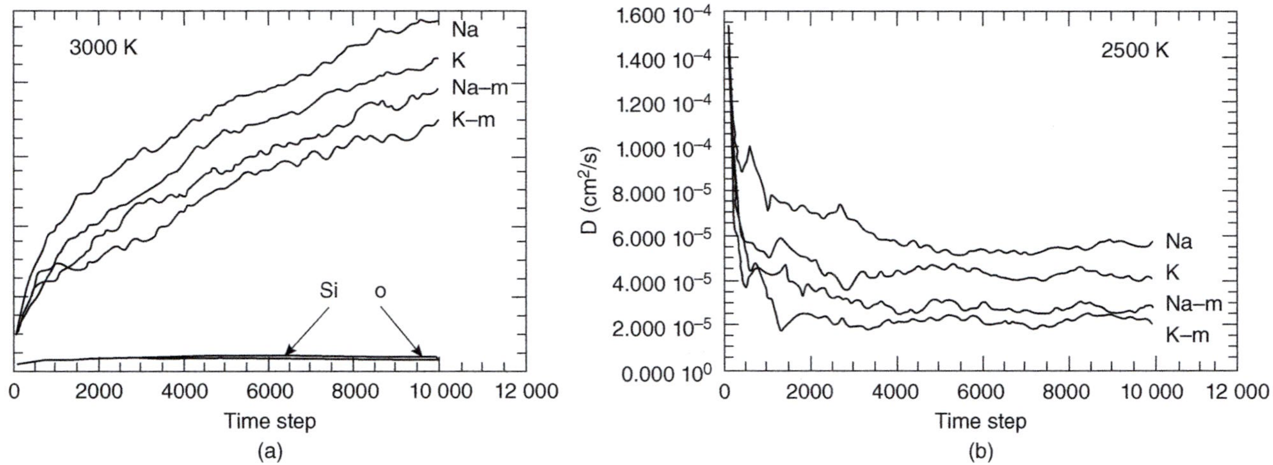


**Figure 6.1** (a) Mean square displacements in Na, K silicates. (b) The derivatives of the MSD in (a) for the alkali ions. The nominal temperature for the alkali ions is given in the figures; the temperatures of the Si and O atoms in (a) was kept at 300 K. Source: Huang and Cormack [77].

Cao [80] extended the group's simulations to compositions including calcium and aluminum using the silica potential model of Huang. The parameters for aluminum and calcium were taken from the compilation of Lewis and Catlow [81].

Since using three-body terms was computationally expensive, the idea of adopting and adapting the two-body BKS potential was attractive. However, although van Beest et al. had presented parameters for Al and Na, one quirk of their model was that the charges on Na and Al were not scaled to those of Si and O, and so could not be used for compositions other than those for which they were derived (i.e. silica, $AlPO_4$, and two Na containing zeolites [47]). Thus, to use the BKS silica potentials for a range of modified silicates, modifier–oxygen interactions were necessary, and such parameters were derived by Yuan for R–O (where R = Li, Na, K) potentials using an energy minimization technique to fit the R–O interaction parameters to the structures of several crystalline alkali silicates with the BKS Si–O and O–O parameters.

For example, the crystal structures of α- and β-$Na_2Si_2O_5$, and the $Na_2SiO_3$ were used to obtain the Na–O parameters. Yuan [82] presented a very detailed analysis of the local structure in silica and sodium silicates, including the Si—O—Si bond angle distributions, comparing the BKS and Vessal potential models.

Cao found, consistent with Zirl and Garofalini, a violation of Lowenstein's rule in his aluminosilicate simulations [80]. He also performed a series of simulations of mixed alkali (Li, Na) aluminosilicates, which showed that the different coordination environments of Li and Na were largely maintained in the mixed alkali structure. Additionally, a more detailed examination revealed that the Li—O bond length distribution was slightly narrower in the mixed alkali structure, i.e. the packing of the oxygens around the Li was more efficient than in the single alkali structure, resulting in a lower site energy, as had also been observed by Huang in the mixed Na, K silicates (using a different potential model).

The development of self-consistent sets of potential parameters which could be used on a range of compositions without modification was a major step forward, since it enabled the systematic investigation of structure as a function of composition and calculation of physical properties.

The introduction of the two-body partial charge BKS potential for silica reduced the computation time per timestep significantly. However, and just discussed, it was not possible to employ the BKS potential for broader compositions without additionally deriving potential parameters for the modifier interactions. Although this had been done for a few cations, it was not only an ad hoc effort but also somewhat time-consuming.

Fortunately, the need to undertake potential parameter derivations was made redundant by the work of David Teter [83], who had obtained a self-consistent set of parameters for oxides (and fluorides) by fitting to a wide range of crystal structures and properties. He used a simple Buckingham form with partial charges of 0.6*e*. The reduction from formal oxidation state charges meant that the dominance of the electrostatic interactions was reduced sufficiently to avoid needing three-body bond-bending terms.

There is a drawback to this approach and that is a limitation to the accuracy of the simulations. For any given composition, it is no doubt possible to refine the parameters from the self-consistent set and get better agreement with (known) properties. However, this limitation is usually a price worth paying in order to be able to scan composition space.

Electronic polarizability is generally excluded from molecular dynamics simulations. The reasons for this are twofold. First, its inclusion adds to the computational complexity, no matter in what form it is included. Shell model potentials increase the computation time by an order of magnitude, making long simulations even longer. Second, and probably more significantly, polarizabilities are composition and structure-dependent, and are thus model parameters that are not generally transferable from one composition/structure to another.

Teter's potentials were rapidly adopted by the author's group, as well as others. The availability of such potentials spurred the simulation of a wide range of compositions, of increasing chemical complexity, as the power of computers expanded rapidly. For example, using the Teter potential parameters, the authors' group produced a model of Bioglass®; see Chapter 11 for further details.

In 1994, Matsui [84] developed a set of potential parameters for the CaO–MgO–$Al_2O_3$–$SiO_2$ (CMAS) system by fitting to a series of calcium and magnesium oxide, silicate and aluminosilicate crystal structures. To preserve transferability, the (partial) charges were constrained, and a value of $q = 0.945e$ for Ca and Mg was found. These were later applied to MD simulations of silicate melts [85].

In 2006, Pedone et al. [86] presented a new self-consistent set of potential parameters, again produced by fitting to a wide range of oxide and silicate crystal structures and properties. As with the Teter set, partial charges ($q = 0.6e$) were used, but for the repulsive potentials, a Morse analytical form was employed:

$$R(r) = D_{ij}\left[\left[1 - \exp\left[-a_{ij}\left(r - \bar{r}_{ij}\right)\right]\right]^2 - 1\right] \tag{6.19}$$

An additional $\frac{C_{ij}}{r^{12}}$ term was included to prevent atoms getting too close to each other under high temperature and pressure conditions.

Pedone et al. used this to calculate the mechanical properties of a range of silicate glasses and went on to the model of the fracture mechanisms in silica and sodium silicates [87]. The mechanical properties of silica had been the subject of a study by Huang and Kieffer who developed a rather sophisticated model specifically for the purpose [88].

In 2007, Guillot and Sator [89] fitted a set of potential parameters ($z_i$, $B_{ij}$, $q_{ij}$, and $C_{ij}$) for aluminosilicates containing potassium, sodium, calcium, magnesium iron, and titanium (KNCMFATS system), by constraining MD simulations to reproduce at best the density at zero pressure and the structure of a selected set of 11 natural silicate melts covering a large range of compositions. They found that several sets of potential parameters were able to reproduce satisfactorily (at least for one given state point) the structure and the density of most of the investigated silicate melts. Their method did not produce a unique value for the effective charge of oxygen but rather a range of preferential values ($-1.2e < z_O < -0.8e$). For compatibility with the parameters of Matsui [84] for the calcium magnesium alumino-silicate (CMAS) system, they used the same effective charge for oxygen ($z_O = -0.945e$) and optimized the other potential parameters to describe their extended KNCMFATS system. Their simulation box contained ~1000 atoms, but they found no significant size effects in runs containing ~8000 atoms.

In the simulated silicate melts ($T = 2273$ K), Al was found clearly to be a network former, with ~75% in tetrahedral coordination, 4–12% of [3]Al, and 12–22% of [5]Al with negligible numbers of [6]Al. They noted that experimental spectroscopic data indicated a substantial fraction of [5]Al which the simulations reproduced, but did not comment on the concentrations of [3]Al which they found. Ti was present in [4], [5], and [6] coordination. They commented that although Ti is not regarded as a network former, the relatively strong Ti—O bonds exerted a stabilizing influence on the structure.

Further, their calculated diffusivities were somewhat larger than the experimental values of silicate melts, which observation they used to argue that the viscosities of the simulated melts were somewhat underestimated, a point also made by Micoulaut et al. [90] for simulations of silica and germania glasses close to $T_g$.

Potential development continues to be an active area of research. As noted earlier, one of the limitations of the self-consistent sets of potential parameters is the accuracy with which any specific composition can be modeled. It is always possible to fine-tune the parameters for such specific compositions to produce more accurate simulations.

Because of their significance, lead silicate compositions were an early subject of molecular dynamics simulations. In 1990, Damodaran, Rao, and Rao simulated a $PbO{\cdot}SiO_2$ glass and melt using a Born–Mayer–Huggins type repulsive potential, along with the Ewald sum for the Coulomb energy. Although some of the parameters were taken from Soules [91], those for lead came from work on alkali halides [92] and $PbF_2$ [93].

In 1996, Cormier et al. [94] adapted the approach of Mitra et al., who had described the forces (rather than the potentials) acting between two atoms as noted above, to undoped, and $Yb^{3+}$-doped $PbSiO_3$ glass. While the parameters for Si and O were taken from Mitra et al.'s work [36], those of Pb and P came from their earlier work on $Pb(PO_3)_2$ [95]. Their ionic radii were determined from the criterion that the short-range structure (average inter-atomic distances and coordination number of all possible atomic pairs) of the simulated glass mimicked those of the parent crystalline compound.

The simulation boxes contained about 5000 atoms. This model had been used previously for Eu-doped silica and Eu-doped $Na_2Si_2O_5$ glass [96].

Peres et al. [97], in 1997, modeled a $PbO{\cdot}SiO_2$ glass (undoped and doped with $Er^{3+}$) with the addition of a three-body potential, of the form:

$$V_3(r_{ij}, r_{ik}, \theta_{ijk}) = \left[\lambda_{jik} \exp\left(\frac{\gamma_{ij}}{r_{ij} - r_i^c} + \frac{\gamma_{ik}}{r_{ik} - r_i^c}\right)\left(\cos\theta_{jik} + \frac{1}{3}\right)^2\right] \tag{6.20}$$

where $i$ is the central atom bonded to atoms $j$ and $k$, $\lambda_{jik}$, $\gamma_{ij}$, $\gamma_{ik}$, and $r_i^c$ are constants with $\theta_{jik}$ the angle between the bonds $r_{ij}$ and $r_{jk}$.

The parameters for Si and O were taken from Feuston and Garofalini [41], with those for Pb and Er being adjusted to reproduce the rdfs of crystalline PbO and $Er_2O_3$. These simulations boxes had ~2500 atoms.

A 2001 study by Rybicki et al. [98] modeled a series of lead silicates, $x\text{PbO}\cdot(1-x)\text{SiO}_2$, with $x$ ranging from 0.1 to 0.9. They adopted the potential parameters used by Damodaran et al. and compared their simulated structures with their EXAFS results, noting good agreement.

In 2000, Nanba et al. [99] modeled the structure of sodium germanate glasses, having developed a three-body potential, incorporating a multibody potential function with two energy minima in the angular term for a given three-atom unit which allowed a transformation between different coordination states, such as four- and sixfold coordinated Ge sites.

### 6.3.3 Borate Glasses

One important glass-forming system, namely $B_2O_3$, has proved to be difficult to model correctly. Two key structural aspects of borate glasses are the existence of boroxol rings, and the conversion of [3]-coordinated boron cations to [4]-coordination when modifiers are introduced. The earliest potential models did not provide for the [3] to [4] coordination changes as alkalis were introduced, and all models undercounted the fraction of boron in boroxol rings. This latter point proved to be quite controversial. Estimates from a series of spectroscopic studies suggested that the fraction of boron in boroxol rings was very high – about 70–75%, but no simulations have ever come close to that number.

The first simulation of vitreous $B_2O_3$ was by Soules in 1980 [100], as part of the immediate aftermath of the Woodcock et al. paper. He used a simple pair potential model with "ionic size parameters determined empirically," that is, the one he had used for silica glasses as discussed above; the Coulomb sum was modified by an error function: $(z_i z_j e^2/r)\,\mathrm{erfc}(r/0.35L)$, where $L$ is the size of the simulation box, 13.73 Å (Eq. (6.11)), without the second term). Unfortunately, the structure produced did not compare well with the experimental data of Mozzi and Warren [101], and also did not include any boroxol rings. Soules suggested that boroxol rings were low temperature structural features and would not be present in the higher (fictive) temperature simulations, a view that was not embraced by experimentalists. However, the potential model and (very) small size of the simulation really mitigated against the formation of boroxol rings. The boroxol ring diameter is about 5.5 Å. This is only slightly smaller than half of the simulation box length indicating that geometric frustration would limit boroxol ring formation.

This was followed by Amini, Mitra, and Hockney (1981) [102], who used an interionic force law similar to the one used for silica (Eq. (6.12) in their MD simulation. Two system sizes were studies; one had just 64 molecules (320 atoms), with the other having 640 molecules of $B_2O_3$. After cooling from high temperature, they found only 0.6% of the atoms were "oddly coordinated." But they also found no boroxol groups, or three-membered $BO_3$ rings.

In 1985, Hirao and Soga [103] reported an MD study of $Eu^{3+}$-doped sodium borate glasses and their fluorescence spectra. They included a boron–boron interaction of the form $-A\exp[-C(r-0.239)^2]$ in addition to the modified Born–Mayer–Huggins repulsive potential. The electrostatic energies (the Ewald sum) were tabulated in advance, and the Coulomb force was computed by linear interpolation. The MD run was for 7000 timesteps with simulation boxes containing 60 oxygen atoms and the corresponding number of boron, sodium, and europium cations depending on the glass composition. The electronic energy levels of $Eu^{3+}$ were calculated using a point charge model for the crystal field, with only nearest neighbor oxygens being included. They concluded that a more refined model of the local field that included covalency was required in order to interpret the spectroscopic properties of their simulated glasses.

Inoue, Aoki, and Yasui [104] used a simple $B\exp(-r/\rho)$ form for the repulsive potential, along with three-body bond-bending terms to include covalent effects, to model vitreous $B_2O_3$ and $Na_2O{\cdot}B_2O_3$. The shape of the potential was calculated by the INDO method. Their simulation box contained about 100 atoms. They observed some boroxol rings in $B_2O_3$, suggesting that a slower rate of quenching would increase the concentration, and their structure of $Na_2O{\cdot}B_2O_3$ included a triborate ring, a di-triborate ring, and a diborate group.

In 1988, Xu, Kawamura, and Yokokawa [105] used the Gilbert–Ida form (Eq. (6.16), which they described as the Busing approximation of the Born–Mayer–Huggins form) for the repulsive interactions to model $B_2O_3$ and a series of sodium borates, with simulation box sizes of ~715 to 750 atoms. Although they claimed good agreement with the results measured by NMR and their calculated composition dependence of the fraction of the boron ions having tetrahedral coordination, their simulations underestimated the number of $N_4$ at lower sodium concentrations and overestimated them at higher Na concentrations. They also reported a complete absence of boroxol rings (whether planar or not). Speculating that their structures should have contained a few, on statistical grounds, they suggested that "a further identification is necessary for the peaks in the Raman spectra of both glasses and crystals," given that no boroxol rings are observed in crystalline $B_2O_3$.

In 1988, Abramo, Carini, and Pizzimenti [106] reported MD simulations of silver borate glasses. They observed the conversion of boron coordination from [3] to [4] as the Ag concentration increased (up to ~25 mol%) and the Ag seemed to be homogeneously distributed throughout the structure.

Soppe et al. [107] used the Soules-type BMH potential:

$$V_{ij}(r) = A_{ij}\exp(-r/\rho) + \frac{z_i z_j e^2}{r} erfc(r/\eta L) \tag{6.21}$$

with $A_{ij} = 20.3545\ \mathrm{kJ/mol} \times \left(1 + \frac{z_i}{n_i} + \frac{z_j}{n_j}\right) \times \exp((r_i + r_j)/\rho)$. A range of cooling rates were studied. Slower quench rates resulted in greater numbers of BOB angles approaching 120°. But with simulation boxes containing just 206 atoms under NVE conditions, with a box edge of 13.73 Å, no boroxol rings were observed. In a related study, they constrained half the B–O–B and half the O–B–O angles to 120°. Again, no boroxol rings were observed. They noted that adjacent $BO_3$ triangles preferred to be noncoplanar.

In their report [108] of studies of lithium borates and Li borates with chlorine $[(B_2O_3)_{0.6}(Li_2O)_{0.3}(LiCl)_{0.1}]$, the numerical values of the $A_{ij}$ were simply recorded, without reference to the equation above.

Verhoef and den Hartog [109, 110] used a number of different potential models in their MD simulations of vitreous $B_2O_3$. They concluded that the structural properties were not very sensitive to the interaction potentials, and that three-body interactions appeared to be necessary for a satisfactory simulation of all vibrational modes in the glass network. They also found no boroxol rings.

Takada, Catlow, and Price [111, 112] proposed yet another new potential, which reproduced the crystal structures and properties of both $B_2O_3$ polymorphs and applied it to MD simulations of the glassy phase. Their model, with a simulation box containing 480 atoms, employed Buckingham-type repulsive potentials, but with a Morse form for the B–O interaction, and they included a simple three-body, bond-bending term in order to express more covalence effects. However, that model produced some fourfold coordinated boron, which is not found in vitreous $B_2O_3$. To correct that problem, they developed a coordination-dependent model based on the fact that B—O bond lengths depend quite

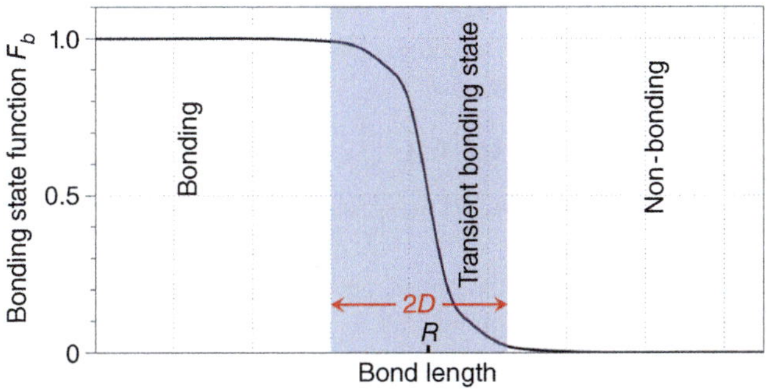


**Figure 6.2** Schematic bonding state function variation with bond length in a coordination-dependent potential model.

strongly on coordination number. They defined a bonding state function:

$$F_b = \begin{cases} 1 & r < (r - D) \\ \{1 - \sin \pi(r - R)/2D\} & (R - D) \leq r \leq (R + D) \\ 0 & (R + D) < r \end{cases} \tag{6.22}$$

where the range $R - D$ to $R+D$ represents the distance over which bond formation and breaking occurs. This is depicted in Figure 6.2.

A coordination function was then defined: $F_c = \sum_i F_{b,i}$, where the sum is over the number of B—O bonds of the atom concerned. Although this approach is in principle applicable to both cations and anions, Takada et al. found that considering just the oxygen coordination environment was sufficient. Thus, the actual potential for the B–O interaction is a weighted sum of those of the two possible oxygen coordination states: $V(r) = (1-\alpha)V_2(r) + \alpha V_3(r)$, where $\alpha = F_c - 2$.

The complexity of this potential limited the size of the simulations largely to 270 atoms although some with 480 atoms were run. This potential no longer produced [4] boron, but at the expense of a density that was lower than the experimental one.

They did find significant numbers of boroxol rings, but with a maximum of around 50% still considerably lower than experiments suggest. They argued that the discrepancy could be ascribed to three main reasons. First, the (small) size of the simulation boxes limits the ability of the structure to disorder completely; second, the limited length of the MD trajectories restricts the ability to model all of the processes that may be occurring; and, third, the nature of the interatomic potentials themselves puts a constraint to the accuracy of the interatomic forces and hence the simulations. They suggested that ab initio simulations would offer a way to circumvent these limitations. However, this turned out not to be the case, as noted below.

Instead, it was found that structures, based on the cesium enneaborate ($Cs_2O{\cdot}9B_2O_3$) crystal structure which consists of two interpenetrating networks, could be constructed that had both a high fraction of boroxol rings ($f \sim 0.8$) and the correct density of vitreous $B_2O_3$ [113, 114].

In 1996, Fernández-Perea, Bermejo, and Senent [115] derived potentials from Hartree–Fock calculations on model $B_3O_3(OH)_3$ structures. They also included a four-body term to stabilize the planar boroxol ring of the form:

$$V_{\text{O-B-O-B}}(\tau) = A_1[1 - \cos(2\tau)] + A_2[1 - \cos(8\tau)] \tag{6.23}$$

where $\tau$ is the dihedral angle subtended by the O–B–O and B–O–B planes within the boroxol ring.

They found just two boroxol rings in their simulation box (which contained only 420 atoms, though), but they saw an improved agreement with the static structure factor.

In 1997 Swenson and Börjesson [116] used Reverse Monte Carlo to fit a structure with 2160 atoms to experimental scattering data. They concluded that it was not possible to reproduce the neutron scattering factor with a structure containing more than ~30% of boron in boroxol rings.

The next attempt was by Michael Teter, in 1997, using a potential model including polarization, derived from a series of quantum mechanical simulations [14]. He did not perform a molecular dynamics simulation, but, instead, used the Monte Carlo technique, with 1000 atoms. He found only 12% of boron in boroxol rings.

In 2000, Cormack and Park [117] adapted the potential model of Takada et al. [111] for use with alkali borates and borosilicates. In defining the coordination number of the oxygen atoms, alkali ions were also included, in addition to the boron (and silicon in borosilicates). They used a larger value for the transition bonding region than Takada et al. The complexity of the potential tested limits of the then-available computer resources, restricting the size of the simulation box usually to 405 atoms, although on occasion a box containing 1010 atoms was used.

While this model reproduced the concentration of [4]B with increasing alkali content, the fraction of B in boroxol rings was again low, at around 22%. Notwithstanding, several superstructural units were observed in MD structures, suggesting that if larger size simulations had been possible, more interesting results might have been obtained.

In order to model borosilicate glasses, they found it necessary to develop new parameters for the Si–O interactions, which were consistent with the O–O potential parameters of the boron oxide model.

The outcome common to all of these simulations of borate glasses, except the later ones of Takada [113, 114], was a value for the fraction of borons in boroxol rings that was very much lower than experimental estimates, in many cases being close to zero. As the melts were cooled in the simulations, boroxol rings did not form at the expected rates. This is almost certainly because the sizes of the simulations were much too small to accommodate the formation of boroxol rings, although Takada also suggested that nominal cooling rates may also be a factor.

The only way to get close to the experimental values of $f$ and a reasonable density appears to be to configure starting structures with those values, which is possible if the structure of an alkali enneaborate is used, as mentioned above.

Huang and Kieffer (2006) [118] developed a coordination-dependent variable charge model for $B_2O_3$, based on that they had developed for silica. They were able to produce melts containing various amounts of boroxol rings, and found that the experimental $S(Q)$

of vitreous $B_2O_3$ could be fit more or less equally well with structures in which the fraction of boroxol rings varied from 10% to 75%.

The lack of a universally applicable model for the B–O interaction potential meant that researchers interested in borate and borosilicate compositions tended to focus on more restricted compositional spaces. Particularly for borosilicates, these tended to be those in which the boron coordination number was always four so that models did not have to cope with variable coordination states. Such models are described in Chapters 8 and 9 of this volume.

Very recently, there has been a renewed effort in developing potential models for borate glasses, driven by the need to understand the variation of structure across a range of borosilicate compositions (see, for example Tuheen et al. [119]). Although based on the Buckingham form of the repulsive potential, they tend to be more complicated, involving both variable charges and potential parameters, as the composition changes, in a manner similar to the earlier models, but encompassing a much broader range of compositional space. These are also considered in Chapter 8.

#### 6.3.3.1 Borates: Quantum Mechanical Simulations

The failure of all classical potential models to reproduce the experimental fraction of borons in boroxol rings led Takada et al. [112] to suggest that quantum mechanical simulations themselves would be necessary to explain the experimental data. Although QM calculated energy surfaces had been used to derive classical potentials, these also produced but low numbers of boroxol rings. The first quantum mechanical simulation of vitreous $B_2O_3$ was that of Umari and Pasquarello [120]. Using a Car–Parrinello approach, an initial configuration of 160 atoms, based on the crystalline $B_2O_3$-II structure, was melted and cooled to room temperature. The resulting structure had two unconnected boroxol rings. Although they admitted that this was probably not a true representation of f, the fraction of boron in boroxol rings, they argued from their analysis of Raman and $^{11}B$ NMR spectra for a fraction $f$ of ~0 : 75.

Then, in 2008, Ferlat et al. [121] used the SIESTA code [122] to model $B_2O_3$. They generated a random network of $BO_{3/2}$ triangles in a cubic box at the glass density (1.84 g/cm$^3$). This structure, contained 100 atoms and 4 boroxol rings ($f \sim 30\%$), but in the final glass structure, only 3 remained. A second structural model used previously with classical potentials and with $f \sim 75\%$ was also simulated with SIESTA. MD simulations were carried out at 300 K on systems containing 80 and 320 atoms for durations of 20 and 7 ps, respectively They concluded that, at the glass density, boroxol-rich structures were more stable than boroxol-poor ones and noted that the presence of boroxols favored low-density structures which explained the higher energetic stability of boroxol-rich structures at the glass densities. They argued that, as the liquid was quenched, boroxols tended to form as compensation for the thermal densification that would otherwise occur.

## 6.4 Simulations of Surfaces

The surface structures of glasses play a key role in determining the chemical behavior of glasses. Unfortunately, probing surface structures at the atomic scale is experimentally very difficult, and it was recognized early on that atomistic simulations could be an important tool in characterizing surface structures of glasses.

The first simulations of glass surfaces were reported by Garofalini in 1983 [123]. The surface was created by cooling a periodic system to 1500 K, and then removing the periodic repeat along the $z$-direction. Since a truncated Coulomb interaction was used, Ewald sums were not necessary. The bottom 100 atoms of the 288 atoms were frozen as the system was cooled further. He reported that the outermost atoms at the surface were predominantly oxygen of which a number were nonbridging.

This was followed in 1990 by simulations of water surface interactions [124]. The potential model previously developed for silica was supplemented by additional pair interactions for Si–H, O–H, and H–H include terms of the same form used by Rahman, Stillinger, and Lemberg (RSL) in their pair potential for water,

$$V(r_{ij}) = a_{ij}/\{1 + \exp[b_{ij}/(r_{ij} - c_{ij})]\} \tag{6.24}$$

where $a$, $b$, and $c$ are adjustable parameters. Three-body terms for Si--O–H and H–O--H configurations were also added. By this time, larger simulation boxes of 648 atoms were possible.

Zirl and Garofalini extended their surface simulations to include aluminosilicates [125].

Garofalini's group [126] also modeled adsorbate interactions at silica and fluorinated silica surfaces. Fluorine-doped surfaces were created by replacing an NBO with an F ion. Alkali ions were then introduced. A key observation was the reduced reactivity of the alkali cations on the fluorinated surfaces because of a reduction of preferential bonding sites.

Yuan [127] used the BKS potential, along with his alkali–oxygen parameters, to model the surface structure of silica and alkali silicates. Because this model used the whole of the Ewald summation of the Coulombic interactions, simply removing the periodicity along one direction was not practical (because in 2-D the conditional convergence of the Ewald sums is much harder to get around), surfaces were created by increasing the periodic repeat distance, effectively introducing a vacuum gap); the general approach is depicted in Figure 6.3. (This approach was later modified by putting two cubic bulk simulation boxes together and freezing the lower one.) Yuan also extended his simulations to include a series of soda-lime silicate glass surfaces [128].

The key observations were that the resultant surface structures after relaxation were rich in alkali, which had migrated to the surface, in addition to concentrations of NBO and [3]Si not seen in bulk structures.

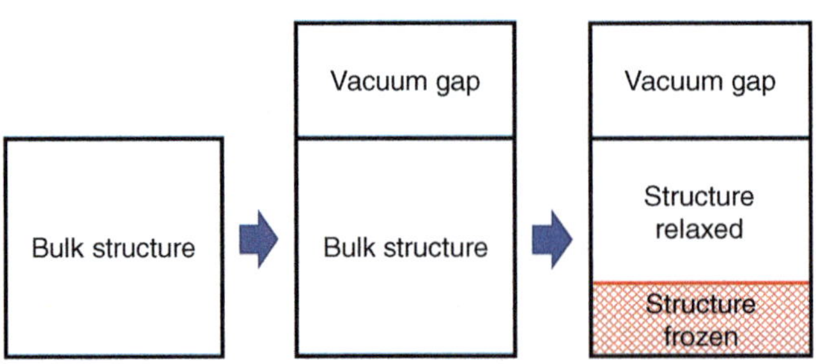


**Figure 6.3** Schematic setup for modeling glass surfaces.

An attempt to create surfaces by fracturing a bulk glass has been considered [127, 129]. In this approach, the simulated box is doubled along the *z*-direction and then the magnitude of the periodic repeat along that direction was increased incrementally, and the structure equilibrated after each change. Only atom positions in the middle of the doubled simulation box were allowed to relax; the others were frozen in their bulk positions. At some point, the bonds could not be strained further, and they broke, creating surfaces in which the bonds were no longer strained because of the elongation of the simulation box, but only by inadequate structural reconfigurations at the newly created surfaces.

Mischler et al. (2002) reported [63] a study of amorphous silica surfaces using both classical MD, with the BKS potential, and quantum mechanical Car–Parrinello MD. The surfaces were created with an approach similar to that of Yuan. However, instead of simply increasing the periodic repeat along *z*, they used an oblong box, which was then cut in such a way that only oxygens were on the surface; these were then saturated with H atoms and a 4.5 Å region was frozen at $T = 0$ K. A vacuum gap was then introduced next to the free surface. The sizes of the simulation boxes were necessarily small, since the same structures were used for both the classical and quantum mechanical simulations.

The surface structure of a complex $SiO_2 \cdot B_2O_3 \cdot Al_2O_3 \cdot ZrO_2 \cdot Na_2O \cdot CaO$ glass was modeled by Abbas et al. (2003) [130]. They found that just like Yuan [127, 128] for the simpler alkali silicates and soda-lime silicates, alkali cations tended to migrate toward the surface, lowering the coordination number of the trivalent elements in the subsurface layer and the outer layer was also enriched in oxygen.

Leed and Pantano (2003) modeled [131] the adsorption of water on silica and sodium silicate glass surfaces. The surfaces were created by removing the periodic repeat along the *z*-direction (as did Garofalini) and continuing the simulation for a while to allow the surface to reconstruct, thus removing many dangling bonds created on initial surface formation. The simulations did not allow water dissociation to occur but was modeled by putting a proton on NBOs and OH molecules on [3] silicon atoms.

Du [132] used the Teter potentials to simulate silica and alkali silicate surfaces and their interaction with water. Because this model did not allow water molecular dissociation, surfaces were hydrolyzed "manually," that is by adding protons to surface NBO and OH molecules to undercoordinated Si. Two- and three-rings were also broken manually with H and OH being added to the resulting [3]Si and NBO. Appropriate potential parameters were obtained by fitting to the relevant hydroxide crystal structures.

The idea of increasing the magnitude of a periodic repeat to produce a surface was extended by Adkins [133] to create models of glass fibers, which were then strained in tension to breaking point.

Simulations of fibers have also been reported by Yuan and Huang [134].

## 6.5 Computer Science and Engineering

The development and widespread deployment of atomistic simulations would not have been possible, of course, without the concomitant evolution of computers. It is therefore worth reviewing these developments.

The earliest computers utilized for simulations were large machines, generically known as mainframes, produced by companies such as IBM, and shortly thereafter, DEC, Amdahl, CDC, Cray, among others, and installed in central facilities. These were specialized machines which required dedicated personnel to operate them. They required software, both to run them, i.e. their operating systems, and to provide the instructions as to what to calculate, or compute, and each manufacturer had their own (proprietary) operating system and associated software.

### 6.5.1 Software

The "brains" of the computer – central processing units (cpus) – received instructions in the form of machine code, or assembly language, which told the computers what to do: add numbers together, for example, or write out the results of the computations. Since machine code is not very user friendly, so-called "high-level" languages were introduced. These allowed instructions to be written in a manner that could be (readily) understood in contrast to the binary ones and zeros of machine code. The conversion of user instructions, embodied in the high-level language, into machine code, is a process known as compiling. The high-level languages had their own grammar and syntax which also evolved over time. Their advantage lies in the fact that a programmer could write a single program which could then be compiled to run on different computers. The original high-level programming languages were ALGOL, first introduced in 1960 and which introduced the concept of structured programming. Hard on the heels of ALGOL came FORTRAN which was specifically designed for scientific applications. It has been able to evolve and adapt with developments in computer science. ALGOL and other languages have faded away.

Initially, operating systems were provided by the manufacturers of computers; each manufacturer had their own version. At that time, it was generally thought that an operating system had to be written in an assembly language (specific to the manufacturer, naturally) so that it could function effectively and gain access to the hardware.

Originally developed as a multitasking system for minicomputers and mainframes in the mid-1970s, UNIX is now one of the most popular operating systems in the world. It was initially written in machine code, for DEC PDP minicomputers, but in 1973 was rewritten in the higher-level language C, contrary to the idea that an operating system's complexity and sophistication required it to be written in assembly language. This made it portable and changed the history of operating systems.

One of the biggest reasons for using Unix is networking capability. With other operating systems, additional software had to be purchased for networking. With Unix, networking capability is simply part of the operating system; this feature was an enabler of the linking together of groups of computers into clusters ("Beowulf" clusters) that were the forerunners of massively parallel computers that are found in most contemporary super-computer centers.

As part of a drive to make software freely available, rather than being constrained by the foibles of commercial producers, Linux was created. Although it was not originally meant to be a UNIX-derivative, being written from scratch, many of the commands that are found in Linux are also found in UNIX. Today, Linux is essentially a UNIX lookalike and is used on most computers and computer clusters used for scientific computing, including atomistic simulations.

The other key language to have lasted as long as FORTRAN is C and its offspring, including C++. As we have seen, the relationship between UNIX/Linux and C is a symbiotic one. UNIX operating systems are written in C, and most of Linux is also in C.

C instructions are compiled to assembly code, therefore, depending on the complexity of the code and on the compiler optimization capabilities, C code may run as fast as assembly.

Other programming languages, like Python and Perl, use also compilers or interpreters that are written in C.

The second relevant advance in computer science came in the form of computer programs which contained the various algorithms and instructions on how to run the simulations. Molecular dynamics programs can run to thousands of lines of code because they are designed to have a number of capabilities. Efficient programming is a skilled activity, so it is understandable that only a relatively small number of programs have survived the evolutionary process.

As the scale of computing resources has expanded, the range of simulations and analysis that can be done has also expanded. Atomistic computer simulation codes have thus grown in size to accommodate these different capabilities. Fortran is still the high-level language of choice and it itself has undergone substantial evolution. A programmer versed only in Fortran 66 (dating from ~1966) would hardly recognize code written in Fortran 95 (dating from ~1995) or more recent implementations.

As the codes became more complex, their maintenance became more complicated and time-consuming. Users found it less convenient to maintain their own in-house programs, and through a process of evolution, only a few programs survived, and these tended to have the resources of large institutions behind them: Sandia National Laboratory in the US for LAMMPS [135]; the UK Science and Technology Facilities Council's Daresbury Laboratory in the case of DL_POLY [136]. Another package often used for classical simulations is GROMACS [137], which was created and developed with financial support from a number national agencies. These three programs are freely available for download, under open-source licensing.

The most widely used program for quantum mechanical simulations is VASP [138], which is not free but licenses may be purchased for a one-time payment. Other programs that have been used recently for quantum mechanical simulations of glasses include Quantum Espresso [139], SIESTA [122], and cp2k [140], which are open-source. CASTEP [141] is now free for academics, but commercial and industrial users are still required to purchase a license. CRYSTAL [142] is still licensed software.

Finally, the role that computer graphics has played in improving our understanding of glass structure by being able to visualize the output from molecular dynamics simulations must be acknowledged. The development of computer graphics was largely driven by the entertainment industry – arcade video games, initially and then computer-based gaming applications. But science was quick to recognize the advantages, with companies like Biosym (now part of Biovia Materials Studio, via Accelerys) and Molecular Simulations

Inc., producing visualization software such as Insight II and Cerius[2]. Coupled with specialized hardware such as produced by Silicon Graphics, visualizing atomic structures became relatively straightforward. Two early examples are shown in Figure 6.4.

### 6.5.2 Hardware

At the time of Woodcock et al.'s paper, computers were large machines, housed in special facilities and expensive, both to run and to use. From the viewpoint of scientific computing, the rate-limiting step, apart from the cost, was the very limited amount of memory that was available in which to store data needed for the calculations. For example, in the early 1980s, when the author needed 1 MB of random access memory (RAM) for a calculation, the job could only be run at the weekend, because the operating system had to be removed from one of the two coupled IBM mainframes in order to free up the 1 MB.

For MD simulations, this limit of available RAM severely restricted the sizes of the simulation boxes which could be used. At the same time, the cost of cpu cycles resulted in short – by contemporary standards – trajectories. For their silica simulations, Woodcock et al. considered the 2400 timestep run at 300 K to be "long." At higher temperatures, they ran for just 1400 timesteps. Soules' 2500 timestep runs, with 162 atoms, took five hours of cpu time on a Honeywell 625 computer. Contrast this with the 25.5 hours it takes for the author to run 100 ns trajectories (50 million timesteps), with 6000 atoms, on a 10-year-old workstation with two 10-core Intel Xeon cpus.

As semiconductor technology advanced, according to the predictions of Moore's law, cpu speeds increased and their size shrank, enabling a generation of smaller computers, as well as faster ones. This made it possible, by the mid-1980s, for individual departments to have their own computer. Indeed, the introduction of reduced instruction set computer (RISC) architecture produced desktop workstations, such as the DEC Alphastation, and the rise of companies like Silicon Graphics, which specialized in graphical capabilities. In fact, the desire for better computer graphics (in, e.g. pinball machines and other arcade games) was one of the principle drivers in this respect.

Probably, the single most important outcome of the evolutions in semiconductor technology was the introduction of the personal computer (pc), which was made possible by the dramatic reductions in cost which accompanied Moore's Law expansion. Large mainframes and supercomputers became once again speciality machines. The shift toward distributed parallel architectures with which we are now familiar, had begun.

The introduction of pcs began a movement of scientific computational capabilities away from central computing centers to individual laboratories and desktops. Among other things, personal computers were made possible by the ever-decreasing price of RAM and of storage devices, i.e. hard disc drives.

Originally developed for DEC computers, the Unix operating system came to include networking as a standard component, so the idea emerged of linking a collection of computers together in the form of what were called Beowulf clusters and distributed, or parallel, processing was born. Such a design requires the ability of passing information between the nodes (i.e. cpus on separate mother boards) in a process known as message passing. Although the Unix operating system could handle this, additional hardware was still

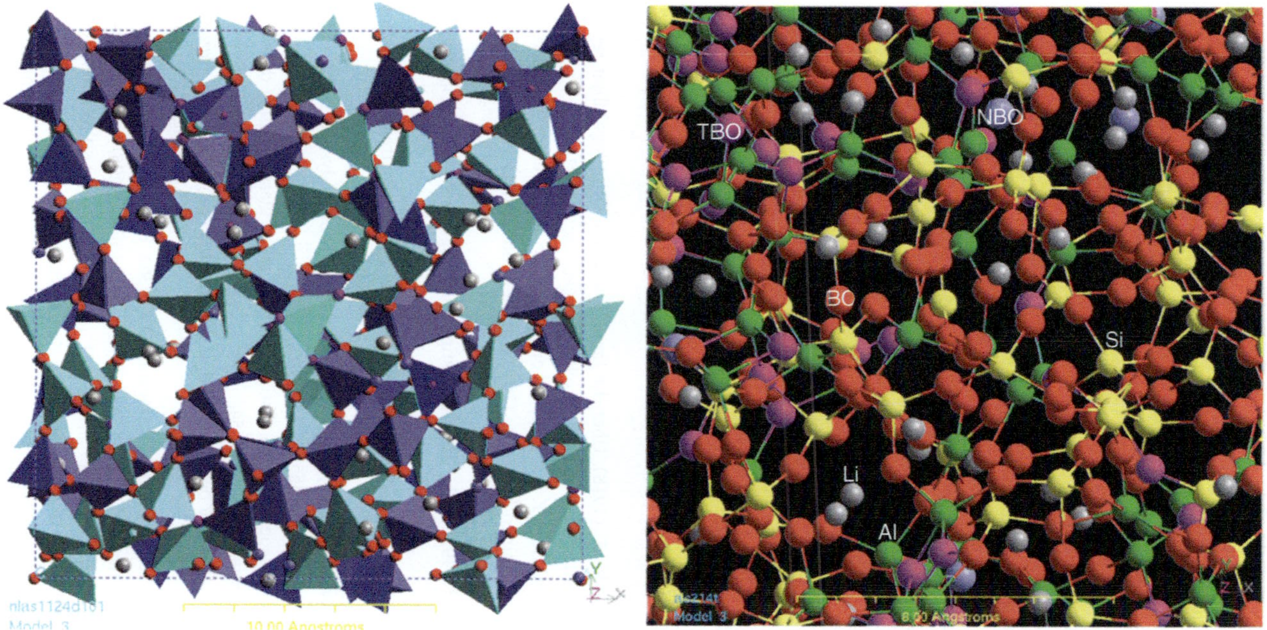


**Figure 6.4** Two different representations of glass structures produced by Cerius[2] software in the late 1990s. In the days before multimedia projectors, thermal printers enabled color transparencies.

required; here, again, the advances in semiconductor technology made possible dedicated processors designed to move data from cpu to cpu. The next stage saw new motherboard designs, for example, to include two cpu units (although multiple processor motherboards had been implemented earlier, by, e.g. Silicon Graphics). Cpu speeds continued to increase, although nowadays, it is the number of cpus in a cluster which is the rate-limiting step, rather than the cpu speed.

Currently, all major central supercomputing facilities house massively parallel computer clusters. To take but one example, the Frontera supercomputer at the Texas Advance Computing Center (TACC) has 448 448 processing cores [140, 143].

The end result of these hardware developments is that very cost-effective computational facilities are available that enable researchers to perform atomistic simulations within their own laboratories. The option to access national supercomputer facilities is still there, of course, for simulations of a scale too large for laboratory-based clusters, such as million atom simulations, for example.

## References


1 Metropolis, N. (1987). The beginning of the Monte Carlo method. *Los Alamos Science Special Issue* 125–130.

2 Metropolis, N. and Ulam, S. (1949). The Monte Carlo method. *J. Am. Stat. Assoc.* 44: 335–341.

3 Born, M. and Landé, A. (1918). Uber die Berechnung der Kompressibilität regulärer Kristalle aus der Gittertheorie. *Verh. Dtsch. Phys. Ges.* 20: 210–216.

4 Woodcock, L.V. and Singer, K. (1971). Thermodynamic and structural properties of liquid ionic salts obtained by Monte Carlo computation. Part 1.—Potassium chloride. *Trans. Faraday Soc.* 67: 12–30.

5 Zachariasen, W.H. (1932). The atomic arrangement in glass. *J. Am. Chem. Soc.* 54: 3841–3851.

6 Evans, D.L. and King, S.V. (1966). Random network model of vitreous silica. *Nature* 212: 1353–1354.

7 Bell, R.J. and Dean, P. (1966). Properties of vitreous silica: analysis of random network models. *Nature* 212: 1354–1356.

8 Warren, B.E. and Biscoe, J. (1938). The structure of silica glass by X-ray diffraction studies. *J. Am. Ceram. Soc.* 21: 49–54.

9 Warren, B.E. (1934). X-ray determination of the structure of glass. *J. Am. Ceram. Soc.* 17: 249–254.

10 Warren, B.E., Krutter, H., and Morningstar, O. (1936). Fourier analysis of the X-ray patterns of vitreous silica and $B_2O_3$. *J. Am. Ceram. Soc.* 19: 202–206.

11 Rahman, A., Fowler, R.H., and Narten, A.H. (1972). Structure and motion in liquid $BeF_2$, $LiBeF_3$, and LiF from molecular dynamics calculations. *J. Chem. Phys.* 57: 3010–3011.

12 Lennard-Jones, J.E. and Dent, B.M. (1926). The forces between atoms and ions.—II. *Proc. R. Soc. Lond. A, Math. Phys. Charac.* 112: 230–234.

**13** Metropolis, N., Rosenbluth, A.W., Rosenbluth, M.N. et al. (1953). Equation of state calculations by fast computing machines. *J. Chem. Phys.* 21: 1087–1092.

**14** Teter, M.P. (1997). Ab initio based studies of borate glasses. In: *Borate Glasses, Crystals and Melts* (ed. A.C. Wright, S.A. Feller and A.C. Hannon), 407–416. Sheffield, UK: Society of Glass Technology.

**15** Kerisit, S. and Du, J. (2019). Monte Carlo simulation of borosilicate glass dissolution using molecular dynamics-generated glass structures. *J. Non-Cryst. Solids* 522: 119601.

**16** Alder, B.J. and Wainwright, T.E. (1959). Studies in molecular dynamics I. General method. *J. Chem. Phys.* 31: 459–467.

**17** Madelung, E. (1918). Das elektrische Feld in Systemen von regelmäßig angeordneten Punktladungen. *Phys. Z.* 19: 524–532.

**18** Jones, J.E. (1924). On the determination of molecular fields. – I. From the variation of the viscosity of a gas with temperature. *Proc. R. Soc. A Math. Phys. Eng. Sci.* 106: 441–462.

**19** Jones, J.E. (1924). On the determination of molecular fields.—II. From the equation of state of a gas. *Proc. R. Soc. A Math. Phys. Eng. Sci.* 106: 463–477.

**20** Jones, J.E. (1924). On the determination of molecular fields. III.—From crystal measurements and kinetic theory data. *Proc. R. Soc. A Math. Phys. Eng. Sci.* 106: 709–718.

**21** Lennard-Jones, J.E. and Taylor, P.A. (1925). Some theoretical calculations of the physical properties of certain crystals. *Proc. R. Soc., Ser. A* 109: 476–508.

**22** Pauling, L. (1927). The sizes of ions and the structure of ionic crystals. *J. Am. Chem. Soc.* 49: 765–790.

**23** Pauling, L. (1928). The influence of relative ionic sizes on the properties of ionic compounds. *J. Am. Chem. Soc.* 50: 1036–1045.

**24** London, F. (1930). Zur Theorie und Systematik der Molekularkräfte. *Zeits. f. Physik* 63: 245–279.

**25** Born, M. and Mayer, J.E. (1932). Zur Gittertheorie der Ionenkristalle. *Z. Angew. Phys.* 75: 1–18.

**26** Margenau, H. (1931). The role of quadrupole forces in van der Waals attractions. *Phys. Rev.* 38: 747–755.

**27** Mayer, J.E. (1933). Dispersion and polarizability and the van der Waals potential in the alkali halides. *J. Chem. Phys.* 1: 270–279.

**28** Huggins, M.L. and Mayer, J.E. (1933). Interatomic distances in crystals of the alkali halides. *J. Chem. Phys.* 1: 643–646.

**29** Buckingham, R.A. (1937). The quantum theory of atomic polarization II – the van der Waals energy of two atoms. *Proc. R. Soc. Lond. A Math. Phys. Sci.* 160: 113–126.

**30** Buckingham, R.A. (1938). The classical equation of state of gaseous helium, neon and argon. *Proc. R. Soc. Lond. Ser. A Math. Phys. Sci.* 168: 264–283.

**31** Fumi, F.G. and Tosi, M.P. (1964). Ionic sizes and Born repulsive parameters in the NaCl-type alkali halides – I: the Huggins–Mayer and Pauling forms. *J. Phys. Chem. Solids* 25: 31–43.

**32** Tosi, M.P. and Fumi, F.G. (1964). Ionic sizes and Born repulsive parameters in the NaCl-type alkali halides – II: the generalized Huggins–Mayer form. *J. Phys. Chem. Solids* 25: 45–52.

**33** Busing, W.R. (1972). Interpretation of the crystal structure of $Li_2BeF_4$ in terms of the Born–Mayer–Huggins model. *J. Chem. Phys.* 57: 3008–2010.

34 Woodcock, L.V., Angell, C.A., and Cheeseman, P. (1976). Molecular dynamics studies of the vitreous state: simple ionic systems and silica. *J. Chem. Phys.* 65: 1565.

35 Soules, T.F. (1979). A molecular dynamic calculation of the structure of sodium silicate glasses. *J. Chem. Phys.* 71: 4570–4578.

36 Mitra, S.K., Amini, M., Fincham, D., and Hockney, R.W. (1981). Molecular dynamics simulation of silicon dioxide glass. *Philos. Mag. B* 43: 365–372.

37 Garofalini, S.H. (1982). Molecular dynamics simulation of the frequency spectrum of amorphous silica. *J. Chem. Phys.* 76: 3189–3192.

38 Ching, W.Y. (1982). Theory of amorphous $SiO_2$ and $SiO_x$. I. Atomic structural models. *Phys. Rev. B* 26: 6610–6621.

39 Ching, W.Y. (1982). Theory of amorphous $SiO_2$ and $SiO_x$. II. Electron states in an intrinsic glass. *Phys. Rev. B* 26: 6622–6632.

40 Sanders, M.J., Leslie, M., and Catlow, C.R.A. (1984). Interatomic potentials for $SiO_2$. *J. Chem. Soc., Chem. Commun.* 1271–1273.

41 Feuston, B.P. and Garofalini, S.H. (1988). Empirical three-body potential for vitreous silica. *J. Chem. Phys.* 89: 5818–5824.

42 Vashishta, P., Kalia, R.K., Rino, J.P., and Ebbsjö, I. (1990). Interaction potential for $SiO_2$: a molecular-dynamics study of structural correlations. *Phys. Rev. B* 41: 12 197–12 209.

43 Tsuneyuki, S., Tsukada, M., Aoki, H., and Matsui, Y. (1988). First-principles interatomic potential of silica applied to molecular dynamics. *Phys. Rev. Lett.* 61: 869–872.

44 Ida, Y. (1976). Interionic repulsive force and compressibility of ions. *Phys. Earth Planet. Inter.* 13: 97–104.

45 Gilbert, T.L. (1968). Soft-sphere model for closed-shell atoms and ions. *J. Chem. Phys.* 49: 2640–2642.

46 van Beest, B.W., Kramer, G.J., and van Santen, R.A. (1990). Force fields for silicas and aluminophosphates based on ab initio calculations. *Phys. Rev. Lett.* 64: 1955–1958.

47 Kramer, G.J., Farragher, N.P., van Beest, B.W., and van Santen, R.A. (1991). Interatomic force fields for silicas, aluminophosphates, and zeolites: derivation based on ab initio calculations. *Phys. Rev. B: Condens. Matter* 43: 5068–5080.

48 Vollmayr, K., Kob, W., and Binder, K. (1996). Cooling-rate effects in amorphous silica: a computer simulation study. *Phys. Rev. B* 54: 15808–15827.

49 Horbach, J. and Kob, W. (1999). Static and dynamic properties of a viscous silica melt. *Phys. Rev. B* 60: 3169–3181.

50 Carré, A., Horbach, J., Ispas, S., and Kob, W. (2008). New fitting scheme to obtain effective potential from Car–Parrinello molecular-dynamics simulations: application to silica. *Europhys. Lett.* 82: 17001–17006.

51 Horbach, J., Kob, W., and Binder, K. (1999). Specific heat of amorphous silica within the harmonic approximation. *J. Phys. Chem. B* 103: 4104–4108.

52 Wilson, M. and Madden, P.A. (1993). Polarization effects in ionic systems from first principles. *J. Phys. Condens. Matter* 5: 2687–2706.

53 Tangney, P. and Scandolo, S. (2002). An ab initio parametrized interatomic force field for silica. *J. Chem. Phys.* 117: 8898.

54 Demiralp, E., Çağin, T., and Goddard III, W.A.G. (1999). Morse stretch potential charge equilibrium force field for ceramics: application to the quartz-stishovite phase transition and to silica glass. *Phys. Rev. Lett.* 82: 1708–1711.

55 Wilson, M. and Madden, P.A. (1993). Short- and intermediate-range order in $MCl_2$ melts: the importance of anionic polarization. *J. Phys. Condens. Matter* 5: 6833–6844.
56 Tossel, J.A. and Gibbs, G.V. (1978). The use of molecular-orbital calculations on model systems for the prediction of bridging-bond-angle variations in siloxanes, silicates, silicon nitrides and silicon sulphides. *Acta Crystallogr.* A34: 463–472.
57 Newton, M.D. and Gibbs, G.V. (1979). Ab initio calculated geometries and charge distributions for $H_4SiO_4$ and $H_6Si_2O_7$ compared with experimental values for silicates and siloxanes. *Phys. Chem. Miner.* 6: 221–246.
58 Gibbs, G.V. (1982). Molecules as models for bonding in silicates. *Am. Mineral.* 67: 421–450.
59 Sarnthein, J., Pasquarello, A., and Car, R. (1995). Structural and electronic properties of liquid and amorphous $SiO_2$: an ab initio molecular dynamics study. *Phys. Rev. Lett.* 74: 4682–4685.
60 Car, R. and Parrinello, M. (1985). Unified approach for molecular dynamics and density-functional theory. *Phys. Rev. Lett.* 55: 2471–2474.
61 Pasquarello, A., Sarnthein, J., and Car, R. (1998). Dynamic structure factor of vitreous silica from first principles: comparison to neutron-inelastic-scattering experiments. *Phys. Rev. B* 57: 14133.
62 Ceresoli, D., Bernasconi, M., Iarlori, S. et al. (2000). Two-membered silicon rings on the dehydroxylated surface of silica. *Phys. Rev. Lett.* 84: 3887–3890.
63 Mischler, C., Kobb, W., and Binder, K. (2002). Classical and ab-initio molecular dynamic simulation of an amorphous silica surface. *Comput. Phys. Commun.* 147: 222–225.
64 Soules, T.F. and Varshneya, A.K. (1981). Molecular dynamic calculations of a sodium borosilicate glass structure. *J. Am. Ceram. Soc.* 64: 145–150.
65 Mitra, S.K. and Hockney, R.W. (1983). Molecular dynamics simulation of the structure of soda silica. *Philos. Mag. B* 48: 151–167.
66 Tesar, A.A. and Varshneya, A.K. (1987). Molecular dynamics simulation of alkali-silicate glass structures. *J. Chem. Phys.* 87: 2986–2989.
67 Murray, R.A., Song, L.W., and Ching, W.Y. (1987). Structural models for $(Na_2O)_x(SiO_2)_{1-x}$ glasses with periodic boundaries. *J. Non-Cryst. Solids* 94: 133–143.
68 Murray, R.A. and Ching, W.Y. (1987). Electronic structure of sodium silicate glasses. *J. Non-Cryst. Solids* 94: 144–159.
69 Zirl, D.M. and Garofalini, S.H. (1990). Structure of sodium aluminosilicate glasses. *J. Am. Ceram. Soc.* 73: 2848–2856.
70 Huang, C. and Cormack, A.N. (1990). The structure of sodium silicate glass. *J. Chem. Phys.* 93: 8180.
71 Huang, C. and Cormack, A.N. (1991). Structural differences and phase separation in alkali silicates. *J. Chem. Phys.* 95: 3634–3642.
72 Huang, C. and Cormack, A.N. (1992). Structure and energetics in mixed-alkali-metal silicate glasses. *J. Mater. Chem.* 2: 281–287.
73 Vessal, B., Leslie, M., and Catlow, C.R.A. (1989). Molecular dynamics simulation of silica glass. *Mol. Simul.* 3: 123–136.
74 Greaves, G.N. (1985). EXAFS and the structure of glass. *J. Non-Cryst. Solids* 71: 203–217.

75 Warren, B.E. and Loring, A.D. (1935). X-ray diffraction study of the structure of soda-silica glass. *J. Am. Ceram. Soc.* 18: 269–276.

76 Warren, B.E. and Biscoe, J. (1938). Fourier analysis of X-ray patterns of soda-silica glass. *J. Am. Ceram. Soc.* 21: 259–265.

77 Huang, C. and Cormack, A.N. (1991). Computer simulation studies of the structure and transport properties of alkali silicate glasses. In: *Physics of Non-crystalline Solids* (ed. L.D. Pye, W.C. LaCourse and H.J. Stevens), 31–35. Sheffield: Taylor and Francis.

78 Stebbins, J.F. (1998). Cation sites in mixed-alkali oxide glasses: correlations of NMR chemical shift data with site size and bond distance. *Solid State Ionics* 112: 137–141.

79 Stebbins, J.F. (1988). Effects of temperature and composition on silicate glass structure and dynamics: Si-29 NMR results. *J. Non-Cryst. Solids* 106: 359–369.

80 Cormack, A.N. and Cao, Y. (1996). Molecular dynamics simulations of silicate glasses. *Mol. Eng.* 6: 183–227.

81 Lewis, G.V. and Catlow, C.R.A. (1985). Potential models for ionic oxides. *J. Phys. C: Solid State Phys.* 18: 1149–1162.

82 Yuan, X. and Cormack, A.N. (2001). Local structure of md-modeled vitreous silica and sodium silicate glasses. *J. Non-Cryst. Solids* 283: 69–87.

83 Teter, D.M. (2021). Personal communication to A.N. Cormack.

84 Matsui, M. (1994). A transferable interatomic potential model for crystals and melts in the system $CaO–MgO–Al_2O_3–SiO_2$. *Mineral. Mag.* 58A: 571–572.

85 Matsui, M. (1996). Molecular dynamics simulation of structures, bulk moduli, and volume thermal expansivities of silicate liquids in the system $CaO–MgO–Al_2O_3–SiO_2$. *Geophys. Res. Lett.* 23: 395–398.

86 Pedone, A., Malavasi, G., Menziani, M.C. et al. (2006). A new self-consistent empirical interatomic potential model for oxides, silicates, and silicas-based glasses. *J. Phys. Chem. B* 110: 11780–11795.

87 Pedone, A., Malavasi, G., Cristina Menziani, M. et al. (2008). Molecular dynamics studies of stress–strain behavior of silica glass under a tensile load. *Chem. Mater.* 20: 4356–4366.

88 Huang, L. and Kieffer, J. (2004). Amorphous–amorphous transitions in silica glass. I. Reversible transitions and thermomechanical anomalies. *Phys. Rev. B* 69: 224203.

89 Guillot, B. and Sator, N. (2007). A computer simulation study of natural silicate melts. Part I: low pressure properties. *Geochim. Cosmochim. Acta* 71: 1249–1265.

90 Micoulaut, M., Guissani, Y., and Guillot, B. (2006). Simulated structural and thermal properties of glassy and liquid germania. *Phys. Rev. E* 73: 031504.

91 Soules, T.F. (1982). Molecular dynamic calculations of glass structure and diffusion in glass. *J. Non-Cryst. Solids* 49: 29–52.

92 Sangster, M.J.L. and Dixon, M. (1976). Interionic potentials in alkali halides and their use in simulations of the molten salts. *Adv. Phys.* 25: 247–342.

93 Walker, A.B., Dixon, M., and Gillan, M.J. (1982). Computer simulation of ionic disorder in high-temperature $PbF_2$. *J. Phys. C: Solid State Phys.* 15: 4061–4073.

94 Cormier, G., Peres, T., and Capobianco, J.A. (1996). Molecular dynamics simulation of the structure of undoped and $Yb^{3+}$-doped lead silicate glass. *J. Non-Cryst. Solids* 195: 125–137.

95 Cormier, G., Capobianco, J.A., and Monteil, A. (1994). Molecular dynamics simulation of lead metaphosphate $Pb(PO_3)_2$ glass. *J. Non-Cryst. Solids* 168: 115–124.

**96** Cormier, G., Capobianco, J.A., and Monteil, A. (1993). Molecular dynamics simulation of the trivalent europium ion doped in silica and sodium disilicate glasses. *J. Non-Cryst. Solids* 152: 225–236.

**97** Peres, T., Litton, D.A., Capobianco, J.A., and Garofalini, S.H. (1997). Three-body potential modeling of undoped and $Er^{3+}$-doped lead silicate glass. *J. Non-Cryst. Solids* 221: 34–46.

**98** Rybicki, J., Rybicka, A., Witkowska, A. et al. (2001). The structure of lead-silicate glasses: molecular dynamics and EXAFS studies. *J. Phys. Condens. Matter* 13: 9781–9797.

**99** Nanba, T., Kieffer, J., and Miura, Y. (2000). Molecular dynamic simulation of the structure of sodium-germanate glasses. *J. Non-Cryst. Solids* 277: 188–206.

**100** Soules, T.F. (1980). A molecular dynamic simulation of the structure of $B_2O_3$ glass. *J. Chem. Phys.* 73: 4032–4036.

**101** Mozzi, R.L. and Warren, B.E. (1970). The structure of vitreous boron oxide. *J. Appl. Crystallogr.* 3: 251–257.

**102** Amini, M., Mitra, S.K., and Hockney, R.W. (1981). Molecular dynamics study of boron trioxide glass. *J. Phys. C: Solid State Phys.* 14: 3689–3700.

**103** Hirao, K. and Soga, N. (1985). Molecular dynamic simulation of $Eu^{3+}$-doped sodium borate glasses and their fluorescence spectra. *J. Am. Ceram. Soc.* 68: 515–521.

**104** Inoue, H., Aoki, N., and Yasui, I. (1987). Molecular dynamics simulation of the structure of borate glasses. *J. Am. Ceram. Soc.* 70: 622–627.

**105** Xu, Q., Kawamura, K., and Yokokawa, T. (1988). Molecular dynamics calculations for boron oxide and sodium borate glasses. *J. Non-Cryst. Solids* 104: 261–272.

**106** Abramo, M.C., Carini, G., and Pizzimenti, G. (1988). $Ag_2O$–$B_2O_3$ glassy systems: network coherence behaviour as seen from molecular dynamics calculations. *J. Phys. C: Solid State Phys.* 21: 527–536.

**107** Soppe, W., van der Marel, C., van Gunsteren, W.F., and den Hartog, H.W. (1988). New insights into the structure of $B_2O_3$ glass. *J. Non-Cryst. Solids* 103: 201–209.

**108** Soppe, W. and den Hartog, H.W. (1989). A molecular dynamics study of $(B_2O_3)_{1-x-y}(Li_2O)_x(Li_2Cl_2)_y$ and $(B_2O_3)_{1-x-y}(Li_2O)_x(Cs_2O)_y$. *J. Non-Cryst. Solids* 108: 260–268.

**109** Verhoef, A.H. and den Hartog, H.W. (1992). A molecular dynamics study of $B_2O_3$ glass using different interaction potentials. *J. Non-Cryst. Solids* 146: 267–278.

**110** Verhoef, A.H. and den Hartog, H.W. (1994). Isotope substitution in $B_2O_3$ glass: a molecular dynamics study. *J. Non-Cryst. Solids* 180: 102–105.

**111** Takada, A., Catlow, C.R.A., and Price, G.D. (1995). Computer modelling of $B_2O_3$. I. New interatomic potentials, crystalline phases and predicted polymorphs. *J. Phys. Condens. Matter* 7: 8659–8692.

**112** Takada, A., Catlow, C.R.A., and Price, G.D. (1995). Computer modelling of $B_2O_3$. II. Molecular dynamics simulations of vitreous structures. *J. Phys. Condens. Matter* 7: 8693–8722.

**113** Takada, A., Catlow, C.R.A., and Price, G.D. (2003). 'Computer synthesis' of $B_2O_3$ polymorphs. *Phys. Chem. Glasses* 44: 147–149.

**114** Takada, A. (2004). Molecular dynamics study of pressure induced structural changes in $B_2O_3$. *Phys. Chem. Glasses* 45: 156–159.

**115** Fernández-Perea, R., Bermejo, F.J., and Senent, M.L. (1996). Importance of rings on the microscopic properties of a strong glass. *Phys. Rev. B* 54: 6039–6042.

**116** Swenson, J. and Börjesson, L. (1997). Fraction of boroxol rings in vitreous boron trioxide. *Phys. Rev. B* 55: 11138–11143.

**117** Cormack, A.N. and Park, B. (2000). Molecular dynamics simulations of borate glasses. *Phys. Chem. Glasses* 41: 272–277.

**118** Huang, L. and Kieffer, J. (2006). Thermomechanical anomalies and polyamorphism in $B_2O_3$ glass: a molecular dynamics simulation study. *Phys. Rev. B* 74: 224107.

**119** Tuheen, M.I., Deng, L., and Du, J. (2021). A comparative study of the effectiveness of empirical potentials for molecular dynamics simulations of borosilicate glasses. *J. Non-Cryst. Solids* 553: 120413.

**120** Umari, P. and Pasquarello, A. (2005). Fraction of boroxol rings in vitreous boron oxide from a first-principles analysis of Raman and NMR spectra. *Phys. Rev. Lett.* 95: 137401.

**121** Ferlat, G., Charpentier, T., Seitsonen, A.P. et al. (2008). Boroxol rings in liquid and vitreous $B_2O_3$ from first principles. *Phys. Rev. Lett.* 101: 065504.

**122** Soler, J.M., Artacho, E., Gale, J.D. et al. (2011). The SIESTA method for ab initio order-N materials simulation. *J. Phys. Condens. Matter* 14: 2745–2780.

**123** Garofalini, S.H. (1983). A molecular dynamics simulation of the vitreous silica surface. *J. Chem. Phys.* 78: 2069–2072.

**124** Garofalini, S.H. (1990). Molecular dynamics computer simulations of silica surface structure and adsorption of water molecules. *J. Non-Cryst. Solids* 120: 1–12.

**125** Zirl, D.M. and Garofalini, S.H. (1992). Structure of sodium aluminosilicate glass surfaces. *J. Am. Ceram. Soc.* 75: 2353–2362.

**126** Zirl, D.M. and Garofalini, S.H. (1990). Reactions on modified silica surfaces. *J. Non-Cryst. Solids* 122: 111–120.

**127** Yuan X. (1999). Molecular dynamics simulation of silicate glasses and their surfaces. PhD thesis. Alfred University, Alfred, NY.

**128** Yuan, X. and Cormack, A.N. (1998). MD simulated structure of soda-lime-silica glass and its surface. In: *Advances in Fusion and Processing of Glass II. Ceramic Transactions*, vol. 82 (ed. A.G. Clare and L.E. Jones), 281–286. Columbus, OH: The American Ceramic Society.

**129** Swiler, T.P., Simmons, J.H., and Wright, A.C. (1995). Molecular dynamics study of brittle fracture in silica glass and cristobalite. *J. Non-Cryst. Solids* 182: 68–77.

**130** Abbas, A., Delaye, J.-M., Ghaleb, D., and Calas, G. (2003). Molecular dynamics study of the structure and dynamic behavior at surface of a silicate glass. *J. Non-Cryst. Solids* 315: 187–196.

**131** Leed, E.A. and Pantano, C.G. (2003). Computer modeling of water adsorption on silica and silicate glass fracture surfaces. *J. Non-Cryst. Solids* 325: 48–60.

**132** Du, J. and Cormack, A.N. (2005). Molecular dynamics simulation of the structure and hydroxylation of silica glass surfaces. *J. Am. Ceram. Soc.* 88: 2532–2539.

**133** Adkins, L. and Cormack, A.N. (2012). A study of silica glass fiber structure and elastic properties, using molecular dynamics simulations. In: *Processing, Properties, and Applications of Glass and Optical Materials: Ceramic Transactions*, vol. 231 (ed. A.K. Varshneya, H.A. Schaeffer, K.A. Richardson, et al.), 115–123. Columbus, OH: The American Ceramic Society.

**134** Yuan, F. and Huang, L. (2012). Molecular dynamics simulation of amorphous silica under uniaxial tension: from bulk to nanowire. *J. Non-Cryst. Solids* 358: 3481–3487.

**135** Plimpton S., Thompson A., Moore S., et al. Lammps: Large-scale atomic/molecular massively parallel simulator. www.lammps.org (accessed 7 June 2021).

**136** Smith W., Forester T.R., Todorov I.T. DL_POLY. www.ccp5.ac.uk/DL_POLY (accessed 7 June 2021).

**137** Abraham, M.J., Murtola, T., Schulz, R. et al. (2015). GROMACS: high performance molecular simulations through multi-level parallelism from laptops to supercomputers. *SoftwareX* 1, 2: 19–25.

**138** Kresse G. The Vienna Ab Initio Simulation Package: Atomic scale materials modelling from first principles. www.vasp.at (accessed 7 June 2021).

**139** Q. Espresso. Quantum Espresso is an integrated suite of Open-Source computer codes for electronic-structure calculations and materials modeling at the nanoscale. It is based on density-functional theory, plane waves, and pseudopotentials. https://www.quantum-espresso.org (accessed 7 June 2021).

**140** Kühne, T.D., Iannuzzi, M., Ben, M.D. et al. (2020). Cp2k: an electronic structure and molecular dynamics software package – quickstep: efficient and accurate electronic structure calculations. *J. Chem. Phys.* 152: 194103.

**141** Clark, S.J., Segall, M.D., Pickard, C.J. et al. (2005). First principles methods using CASTEP. *Zeitschrift fuer Kristallographie* 220: 567–570.

**142** Dovesi, R., Erba, A., Orlando, R. et al. (2018). Quantum-mechanical condensed matter simulations with CRYSTAL. *WIREs Comput. Mol. Sci.* 8: e1360.

**143** TACC 2020 TACC: Texas Advanced Computing Center. https://www.tacc.utexas.edu (accessed 7 June 2021).

# 7

# Silica, Silicate, and Aluminosilicate Glasses

*Thiruvilla Mahadevan and Jincheng Du*

*Department of Materials Science and Engineering, University of North Texas, Denton, TX, USA*

## 7.1 Introduction

Glass has been discovered and used in various parts of the world since ancient times. It is now a material of choice for applications ranging from containers, packaging, lab and cookwares, optical materials and architecture window, to high technologies such as biomedicine, nuclear waste disposal, display, and optical fiber communications. According to an earlier ASTM definition [1], glass is an inorganic product of fusion that has been cooled to a rigid condition without crystallizing. Progress in glass research has led to several updates in the older definitions [2] and is now based on the order in local structures [3]. A related and commonly used definition that is generally broader and includes noncrystalline solids views glass as an amorphous material formed through processes such as physical or chemical vapor deposition, sol–gel processing, radiation, and other means that disrupt the long-range order in crystalline materials. In general, glass is a solid material that lacks long-range order, but with certain short- and medium-range order, and is formed through the rapid cooling of the melt to avoid crystallization.

Silica is an archetype of glass formation systems. It is also a common engineering material due to its excellent mechanical, optical, and electrical properties and its high chemical durability. Pure silica glass, however, has high-melting temperature that makes its production difficult. With addition of alkali or alkali earth oxides, the melting temperature is greatly reduced while still maintaining a good glass formability. Not surprisingly, the first manmade glass is a silicate glass with alkali or alkali earth components [4]. The most industrially produced glass in volume is soda lime silicate glasses that also fall into this category. Addition of alumina, even in small amount, was found to considerably improve the chemical durability and the mechanical properties of alkali or soda lime silicate glasses. Silicate glasses form the largest portion of glasses that find wide range of everyday and industrial applications such as optical window, architecture, containers, and pharmaceutical. This is in a large part related to the high abundance of silica, as well as other elements found in these glasses, in earth's crust. These silicate glasses also form the basis of more advanced compositions such

*Atomistic Simulations of Glasses: Fundamentals and Applications*, First Edition.
Edited by Jincheng Du and Alastair N. Cormack.

as those for nuclear waste disposal and biomedicine that are detailed in Chapters 9 and 11, respectively. The studies of silica and silicate glasses by using molecular dynamics (MD) simulations have been greatly advanced in the past few decades and are the focus of this chapter.

Seeking the correlation between material properties and the microstructure or atomic arrangements in materials is one of the main themes in material science and engineering. Advances in X-ray diffraction [5] in the early half of the twentieth century allowed for investigation of the atomic structure of glasses. This, in turn, led to the formulation of the continuous random network (CRN) model for silica glass and a proposal of Zachariasen's rules for glass formation [6]. Experimental investigation into the detailed atomic structure in alkali, alkaline earth, and other doped silicates using extended X-ray absorption fine structure (EXAFS) in early 1970s allowed for the formulation of modified random networks (MRNs) in silicate glass structures [7]. MD simulations and Monte Carlo (MC) have also been used for studies and have provided additional insights into the correlation between bulk properties and atomic structure and compositions in glasses [8, 9]. Thus, advances in glass science have been achieved due to the confluence of physical sciences and computational modeling of atomistic arrangements in glass. In this chapter, we first discuss various aspects of MD simulations of silicate glasses from interatomic potentials to structural analysis. Then summaries of structural information obtained from MD simulations in alkali silicate, alkali–alkaline earth silicate, and aluminosilicate glasses are presented.

## 7.2 Atomistic Simulations of Silicate Glasses: Ingredients and Critical Aspects

The concepts of MD simulations have been elaborated to various levels in other chapters in this book (e.g. Chapter 1) and some recent reviews [10]. In brief, these simulations are done by numerical integration of the equations of motion of a set of interacting atoms or ions, which give the position and velocities (and further derivatives), of these particles over a period of discrete time steps. The information in a set of configurations that are a part of the trajectory can be used to derive various bulk structures and properties of the system [11]. Atomistic simulations are performed for systems ranging from a few hundred atoms to millions of atoms for ionic systems, or even hundreds of millions of atoms for metallic systems, in a range of picoseconds to several microseconds [12]. In typical ergodic systems, where all iso-energy configurations of these atoms are equally probable, the ensemble average of any physical property $A$ is equivalent to the time average and is given as [13]

$$\langle A \rangle = \lim_{t \to \infty} \frac{1}{t} \int_0^t A(t)dt \tag{7.1}$$

While this equation allows for the extraction of physical properties of a small set of atoms, the correspondence of these properties to those of the bulk material is realized through the use of periodic boundary conditions (PBCs). Implementation of three-dimensional PBC is done by allowing atoms at the edge of a simulation box to interact with the atoms at the opposite edge as if they were in another box at the edge. This way, the entire system of interacting particles can be considered equivalent to a small region inside the bulk of a material. Similarly, simulations of surfaces of materials are done by implementation of two-dimensional PBC or with a vacuum gap on the top of the surface in a three-dimensional PBC [14].

The accuracy of the bulk properties and the simulation itself is dependent on the accuracy of the interaction potential that is chosen. The term "interatomic potential" or force field refer, in this context, to a set of analytical functions that represent the interaction and have been used interchangeably in the rest of this chapter. In the most general case, the total potential energy of N interacting atoms can only be expressed as a series of additive, multibody interactions as [11] follows:

$$U(r_1, r_2, \ldots, r_N) = \frac{1}{2!}\sum_i\sum_j U_{ij} + \frac{1}{3!}\sum_i\sum_j\sum_k U_{ijk} \ldots\ldots + \frac{1}{n!}\sum_i\sum_i\sum_j \ldots\ldots \sum_m U_{ijk\ldots m} \tag{7.2}$$

where $r_N$ are the positions of the atoms and the summation is only for distinct values of $i,j,k\ldots m$. The first term is the pair, or two-body, potential and is typically defined relative to the energy between a pair of the corresponding atoms at finite separations. The successive terms in Eq. (7.2) are the three-, four-, and $n$-body potentials. Implementation of all the multibody terms can be cumbersome. For most molecular simulations, particularly of silicate glasses with mixed ionic and covalent bonds, the effective pair potentials, sometimes with addition of the three-body interactions, have been found to be sufficient to reproduce relevant physical properties [14].

For simulations of silicate glasses, most of the available potentials consist primarily of pair potentials to account for two-body interactions and three-body potentials in some cases to regulate the certain bond angles, for example to ensure the $SiO_4$ units in tetrahedral shape by constraining the O–Si–O angles to be close to 109°. The pair potential typically consists of a so-called "long-range charge-based electrostatic contribution," similar to the Madelung constant in crystals, and an exponential or Lennard–Jones (LJ)-based short-range $1/r^n$-dependent repulsive–dispersive interactions [11, 15]. In most cases in recent times, the Coulombic term uses partial charges on the atoms estimated from first principles or Quantum Mechanics (QM)-based calculations [16]. The most popularly used potentials can thus be split into Coulomb and short-range pair of interactions as follows:

$$U(r_{ij}) = U_{Rep}(r_{ij}) + U_{Disp}(r_{ij}) + U_{Coul}(r_{ij}) \tag{7.3}$$

The three-body term is used to regulate the angles between triplets of atoms and usually acts only among the first shell of neighbors for former glass cations such as $Si^{4+}$ and $Al^{3+}$ [17, 18]. Typically, they are much lower than the pair potential energy and serve to increase the configurational energy of a trio of atoms, e.g. O–Si–O or Si–O–Si, if the angle between them strays away from the targeted values.

Figure 7.1a shows an example of pair potentials that are plotted as a function of the separation between the pairs for a generic Si–O potential. The short-range repulsive and dispersive terms are needed, especially in case of opposite charges like Si–O to preserve the bond distances. A major portion of the pair potential is dominated by the electrostatic portion of the potential as the energy has a $1/r$ dependence on these terms. For large system of thousands of atoms, it is computationally efficient to calculate the pair potentials only for atoms within an interaction cut-off distance. As can be seen in the figure, the coulombic interaction energy is nonnegligible at a cut-off distance of 10 Å. This may lead to large fluctuations in energy of the system depending on the choice of the interaction cut-off distance. To avoid such fluctuations in energy, most practical simulations use a long-range modification and split the

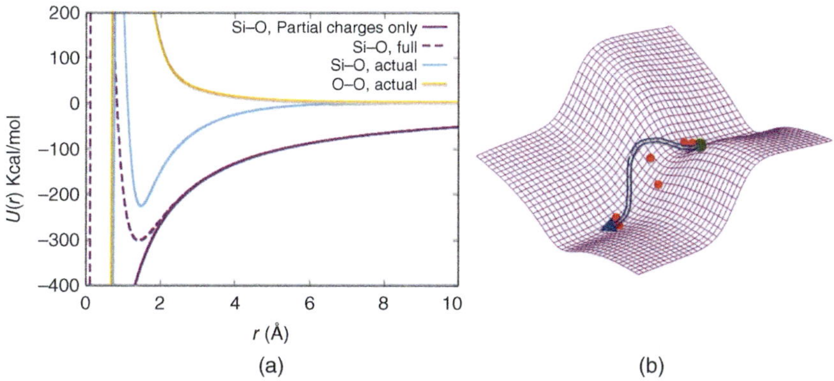


**Figure 7.1** (a) Typical potential energy for interactions between like charges, O–O and unlike charges Si–O. The purely charge-based interactions for Si–O result in high attractive forces at low distances and need to be made repulsive beyond the bond length. This is shown in the broken Si–O curve. Wolf correction was used in the above set of curves to smoothen and shift the interactions beyond 1 nm. (b) A schematic representation of 2D, potential energy surface. The large green dot is the initial configuration. The trajectories obtained by MD are along the blue arrow and the red dots representing trajectories obtained by Monte Carlo methods.

interaction energy into real space and reciprocal (k-space) summations. Common examples of long-range corrections include Ewald [19], Particle–Mesh [20], and the Wolf summation [21] methods. The latter method involves dampening and shifting the pair potentials so that real-space terms are uniformly zero at the cut-off distance for all pair of interactions as shown in the actual Si–O and O–O curves in Figure 7.1a.

The ReaxFF, developed in the early 2000 [22], and the Tersoff potentials [23] are some of the examples of bond order [24]-based potentials that use more than the three-body terms that allow simulations of chemical reactions. Recent development in machine learning (ML)-based potentials [25, 26], usually based on first principles calculations and ab initio molecular dynamics (AIMD) simulations, also offers promises for simulations of complex systems. The interaction potentials based on these twenty-first century developments allow for the simulation of reactions of glass-environment interactions and transport of ions in porous gel structures [27].

In a simulation system consisting of interacting particles such as atoms or ions, the individual energy contributions are added up to give the total configurational energy of the system. The potential energy surface (PES) of a system can thus be defined as a representation of the total potential energy of the system as a function of its atomic configurations. While the PES, by definition is a hyperdimensional surface, it is useful to view the concepts of atomistic simulations of glass as extracting representative atomistic configurations from the PES [28]. Figure 7.1b shows an indicative representation of a PES over a two-dimensional configurational space. Each point on the surface represents the total potential energy of a unique configuration of atom positions and velocities of a system. The valleys in the energy landscape correspond to local stable states and both MC and MD simulations attempt to extract configurations from this energy landscape. The saddle points correspond to energy barriers that need to be overcome for transition of the system between the local stable states. In MC simulations, the starting configuration represented by the large green dot is moved along a random direction on the surface, and the new configuration is either accepted or rejected with a probability directly related to the energy difference between current state and new state. Thus, MC simulations produce configurations over the landscape as shown by the discrete red dots in Figure 7.1b. Depending on how far the configurations are moved from the starting state, MC can thus be used to reach local minima that are far from the initial state, allowing for overcoming energy barriers for reactions. MD simulations, on the other hand, produce configurations that, while are discrete, invariably fall along a continuous path as indicated by the blue arrow. While this has the advantage of producing the underlying temporal information, the system could also likely become confined into to a narrow set of configurations depending on the stability of the local minima. Thus, for a thorough analysis of the glassy system, a combination of MD and MC methods can be used to reach deeper minima of the PES [29].

## 7.3 Characterization and Experimental Validation of Structural and Dynamic Features of Simulated Glasses

As indicated earlier, the primary results of a simulation of an N-atom system are a set of positions $r_N$ and velocities $v_N$. Equilibrium thermodynamic properties such as temperature, pressure, heat, and mass transfer coefficients, dynamics, and spectroscopic properties can eventually be calculated based on the positions and velocities. The formulae used for calculating the physical properties are originated from statistical mechanics and detailed in Chapter 4 in this book, and some of the basic concepts are summarized here. The most easily accessed properties through MD simulations are related to the local order in glassy systems. Hereon, we consider some of the quantitative properties that are used as descriptors for the local structure and how MD simulations are used for accessing this information. We will also briefly describe some of the common experimental techniques that are used for accessing this local structure. More details on characterization of glass structures in general are presented in Chapter 4.

### 7.3.1 Structural Characterizations

One of the most relevant properties that is used as a quantitative description of the atomic structure is the radial pair distribution function (PDF) $g_{\alpha\beta}(r)$. This function and its variants and derivatives (see Chapters 1 and 4 for different forms of distribution functions) can be described as the averaged, normalized probability of encountering an atom of species $\beta$ at a distance $r$ from a central atom $\alpha$ and is given by

$$g_{\alpha\beta}(r) = \frac{1}{4\pi r^2 \rho_\beta} \left( \frac{dN_\beta}{dr} \right) \tag{7.4}$$

where $N_\beta$ is the average number of atoms of type $\beta$ around an ion of type $\alpha$ in a sphere of radius $r$ and $\rho_\beta$ is the number density of atom type $\beta$. The total distribution function is the g($r$) measured for all atomic species without distinction between $\alpha$ and $\beta$. The radial distribution function is a related property that is obtained without normalizing the total PDF to density [30]. These related quantities form a fundamental basis for describing the atomic structure in glassy systems and in materials in general. At the dilution limit, they can be correlated to the pair potential energies and can thus serve as a reference for parameterizing the force field [31]. Experimentally, the distribution function can be obtained as the Fourier transform of the static structure factor ($S(k)$) obtained from X-ray or neutron diffraction:

$$g(r) = 1 + \frac{1}{(2\pi)^3\rho}\int [S(k) - 1)e^{-ik.r}dk \tag{7.5}$$

where $g(r)$ is the total PDF, $k$ is value of the scattering vector, and $\rho$ is average number density. A Lorch-type window function is usually applied to minimize the cut-off effect because maximum $k$ value can be reached depending on the experimental conditions [32, 33]. While in comparison with diffraction studies, the total distribution function $T(r)$ [30], a form of PDF that equals $4\pi r\rho_0 g(r)$, where $\rho_0$ is the average number density, is often used due to the fact that it is broadened symmetrically with $r$ during Fourier transformation. This can be coupled with models to deconvolute the individual PDFs.

Figure 7.2a shows the typical PDFs and coordination charts for a soda lime silicate glass with 15 mol% $Na_2O$ and 10 mol% CaO that was generated using a melt and quench process with the slightly modified Teter potential [10]. As can be seen, the first peak of Si–O indicates the high likeliness of encountering an oxygen around a silicon (or vice versa) at the bond distance of 1.61 Å. The second shell of Si–O neighbors is shown as a much smaller peak at around 4.2 Å. Given the amorphous state of glass, only the short-range order is strictly preserved and, at longer distances, one finds a constant average number density for any species beyond the second shell, as indicated by $g(r)$ values going to 1 at longer distances. The accumulated coordination plot in Figure 7.2b shows the averaged total number of oxygens around a silicon and reaches a plateau of 4.00 in the first shell, indicating that there are four oxygens surrounding each silicon. The Na–O and Ca–O peak around 2.3 Å are much broader indicating a wider distribution in coordination numbers of about 5.5–6, as can be read from the coordination charts. In general, the atomic configurations of the simulated systems can be visually depicted similar to the ball-and-stick model shown in Figure 7.2c, to ascertain uniformity in distribution and amorphous structure.

Another related property that is studied as a measure of the local structure is the coordination number [34] ($CN_{\alpha\beta}$), defined as the number of atoms of $\beta$ found around an average central atom $\alpha$ within the first shell of neighbors. Typically, $\alpha$ and $\beta$ are ions of opposite charge, although it does not need to be so and the cutoff for determining the first shell of neighbors is the location of first minima after the first peak. For network-forming atoms such as Si, Al, and P, one usually finds a predominant coordination number dependent on the valency, whereas network modifiers like alkali cations have a distribution of coordination numbers. In amorphous silica, for instance, the silicon is predominantly coordinated by four oxygens within a first neighbor distance of ~2.2 Å, whereas each oxygen atom is connected to two Si atoms as shown in the integrated PDFs in Figure 7.2b. Closer examination of the trajectory reveals that with a cutoff of 2.2 Å, there are some (<0.1%) three-coordinated and five-coordinated Si and some other defectively coordinated oxygen sites. These defectively coordinated sites correlate to the E' centers and are a factor that affect properties such as optical absorption and the local reactivity of the glass [35, 36].

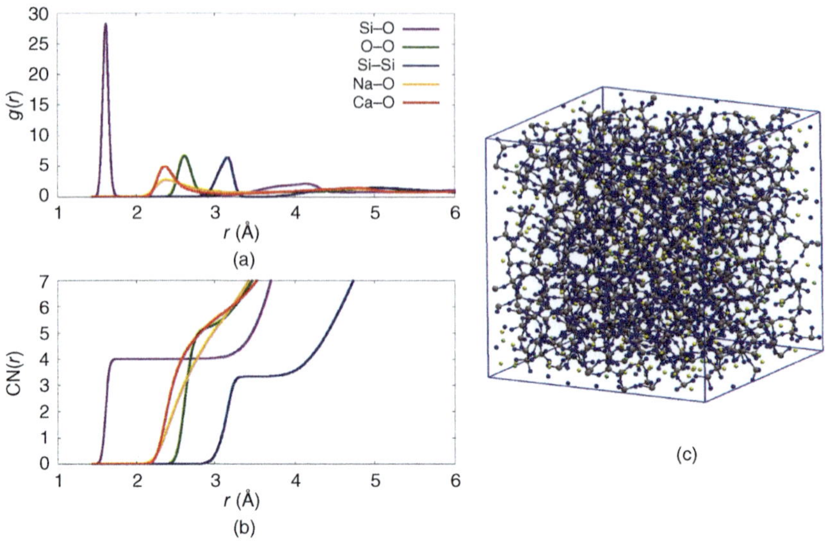


**Figure 7.2** (a) Shows the pair distribution functions of a soda-lime-silicate glass systems with composition $15Na_2O-10CaO-75SiO_2$ (mol%). The first peak of Si–O at 1.615 Å corresponds to the Si—O bond distance in the silica tetrahedra. The Na–O and Ca–O peaks do not have a minimum at 0 after the first peak. This is further demonstrated in the coordination-curve (b) which shows a distinct plateau at 4 indicating the Si–O coordination at 4, whereas the modifiers have no such plateau and show a variation in coordination numbers (c) is an example snapshot of a configuration of equilibrated glass. The gray spheres are Si, the blue are oxygen, and the yellow and cyan spheres are Na and Ca, respectively.

Bond distances measured from the first peak of the PDF curves can also be used to evaluate the bond angle distribution (BAD) of atom triplets. Based on the silicon coordination of four oxygens observed in the integrated PDF and symmetric arrangement, the O–Si–O angle (∢Si–O–Si) can be deduced to be around the tetrahedral angle of 109.47°. Of particular interest is the Si–O–Si angle, which is highly sensitive to the distribution of the Si—Si bond distance [37]. Experimentally, the Si–O–Si angle can be obtained either by calculations based on the Si–Si peak or through extracting a distribution function based on models built from Nuclear Magnetic Resonance (NMR) studies [37–39] of glasses with $^{29}Si$. Such experimental models have predicted a $BAD_{Si\text{-}O\text{-}Si}$ with a peak varying from 140° to 160°, with the most probable on being around 147° for silica glass. Similarly, BADs involving other network-forming species such as B, Al, and P can be investigated experimentally. These bond angles can be used to compare and validate the results from atomistic simulations.

A defining feature of amorphous silicate glass structure is the existence of ring structures with different sizes in a network formed by corner-sharing silica tetrahedra [40]. A ring is formed in the silica network when bridging oxygens (BOs) connecting $SiO_4$ tetrahedra form a closed loop. A primitive ring is a ring that can no longer be divided into smaller rings. Ring size distribution is usually based on the statistics of primitive rings that can

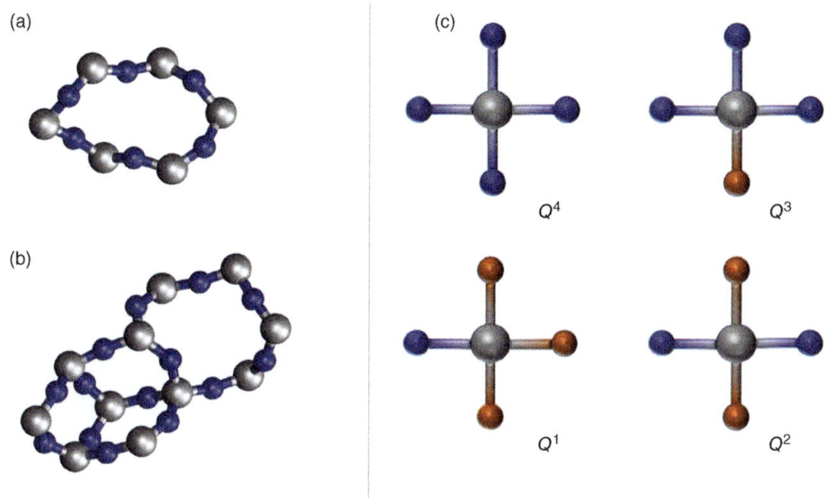


**Figure 7.3** (a) and (b) Examples of ring structures from simulated glass. (a) is a six-membered ring connecting $6SiO_4$ tetrahedra which (b) is a more complex set of interconnected two six-member rings with three, four-member rings within one of them. The gray spheres are silicon, and the blue spheres are oxygen. (c) A schematic depiction of $Q^n$ speciation in glass. The blue spheres are BOs, the orange spheres are NBOs, and the gray spheres are silicon.

be calculated using several different algorithms [41]. Figure 7.3a shows some of the ring structures that could be observed in an atomistic simulation of silica glass. While the nomenclature of rings in general can be based on several different features, the rings in silicates are typically named based on the number of network formers included (Si, Al, B, P) in them. Thus, a six-membered ring in pure silica would thus contain a maximum of six distorted $SiO_4$ tetrahedra connected by six corner-sharing bridging oxygens as shown in Figure 7.3a. Figure 7.3b shows a more complex interconnected ring structure with two edge connected six-member rings, with one of them interlinked to form some four-member rings also. It is easy to see that except for the rarely occurring edge-sharing two-member ring, all higher ring sizes can be accommodated as purely corner-sharing rings. The special case of a two-member ring, which would result in a ∢Si–O–Si of about 70°, is considered a defect structure [42]. While pure silica rarely has rings larger than 10-membered ones, some alkali and multicomponent silicates can have much larger rings. Since the MD trajectories contain all the relevant information on the position of atoms in the system, various algorithms can be used to quantify the ring size distributions [41, 43, 44]. Typically, the silicate glass system is characterized based on the distribution of ring sizes with six- or seven-member rings being the most frequent ones occurring in pure silica [41].

As a corollary to the coordination numbers and ring structures, one can define the network connectivity of a silicate based on BOs and nonbridging oxygens (NBOs). If an oxygen links two network formers, it is considered a BO. NBOs are usually slightly shorter in length compared to the BOs and are linked to only one network former. Figure 7.3c shows a closer view on the types of oxygen in the network with different colors for BO and NBO. The silicon atoms in the network can be further speciated based on the number of oxygen atoms attached to it as shown in Figure 7.3c. Thus, $Q^4$ silicon is one in which all the oxygens that are attached to the central silicon are BOs, $Q^3$ silicon is where it is attached to three BOs and one NBO, and so on. It can thus be seen that a defect-free bulk structure amorphous silica structure will be constituted by all $Q^4$ silicon and BOs only in the bulk. In practice, however, MD simulations of bulk glasses will always result in the formation of some defects, which will be discussed further with the introduction of silica surfaces [45, 46]. The nomenclature of $Q^n$ species has also been referred in a different way with reference to the coordination of the silicon also. In this convention, the silica is classified as $Q^n{}_m$, where the subscript stands for the coordination of silicon and the superscript stands for the number of NBOs [47]. (For aluminosilicates, slightly different $Q^m{}_n$ convention in the nomenclature is also used where $m$ is the total BO and $n$ is the number of BOs linked to Al). $Q^n$ distribution can be determined with high accuracy in solid-state NMR studies [40].

### 7.3.2 Dynamic Properties of Simulated Glasses

The previous sections described some of the properties that were directly associated with the position of atoms in the system, i.e. static structures. Since the results of MD simulations are the trajectories of the atoms, it is also possible to obtain the dynamic properties of the underlying set of particles. This is one of the major advantages of MD simulations as compared to MC or static methods such as energy minimization. While these dynamic properties encompass a wide range of transport and reaction properties, at this juncture, it would be useful to focus on analysis of diffusion and vibrational spectroscopy. Others such as thermal transport, reaction kinetics, and nonequilibrium simulations will depend on the type of system being studied and may require special techniques and modification to the simulation methods [48].

Mean Square Displacement (MSD) of the atoms from an original starting position is always one of the easily available dynamic information that can be calculated on the run during simulations or by postprocessing of the trajectory files. This can be used to calculate the self-diffusion coefficient of the various glass components. In general, because of the stability of the silica network at lower-than glass transition temperatures, the diffusion of silicon and network components is sluggish hence difficult to calculate with high accuracy [8]. At temperatures above the glass transition or close to the melting, diffusion coefficient can be reliably calculated and used to evaluate the activation energy for diffusion. Diffusion coefficients of silicate systems with network-modifying ions are more commonly studied as these modifier cations have higher values of diffusion coefficients through specific paths or diffusion channels embedded in the network [49]. The self-diffusion coefficients can be calculated based on the slope of the MSD curve at longer times, commonly known as the Einstein's equation, or based on the integration of the velocity autocorrelation function [11].

The vibration of atoms and electronic charges in a solid, along different modes at specific frequencies, correlates with the absorption, transmittance, or scattering of radiation of those frequencies. As a result, radiation dispersed by a solid consists of signature spectra with peaks at frequencies determined by the vibrational properties of the material. The measurement of these vibrational spectra of condensed matter provides important insights into the structure, atomic masses, reactivity, and some optical properties of materials. In MD simulations, the vibrational spectrum is measured by Fourier transformation of the autocorrelation function of physical quantities recorded in the trajectories of the atoms. While the spectra can be calculated by Fourier transformation of autocorrelation of velocities or forces [50], a judicious choice of the dynamic variable is required to improve computational efficiency. For instance, use of position as a dynamic variable will require much longer simulations of silicate systems. For amorphous silicates, the velocity autocorrelation is sufficient to cover the important modes of vibration in the silica tetrahedron. These include the rocking mode of Si–O–Si at $400\,cm^{-1}$, the Si–O–Si bending mode at around $800\,cm^{-1}$, and the Si–O stretch mode around [9] $1200\,cm^{-1}$. Ionic components, like alkalis and alkaline earth elements, do not directly affect the primary modes of silica and their effect is primarily seen [49] in the low-frequency modes at around $200–300^{-1}$. While MD simulations do not calculate or incorporate the electronic properties directly, characteristic infrared (IR) and Raman spectrum of materials can still be calculated by a judicious design of the interaction potential or by assigning preselect dipole, quadrupole moments, or charges to the atoms in postprocessing [51].

### 7.3.3 Validation and Experimental Confirmation of Structural and Dynamic Properties

Simulations provide important insights into possible atomistic phenomena that are often not directly accessible through experimental methods. Thus, it is necessary to ensure that the information from the simulations is valid and relevant to real-material systems. Results from MD simulations are generally validated with those from first principles density functional theory (DFT)/QM calculations on the one hand or with experimental results on the other hand. Some of the physical properties such as bulk density, thermodynamic, and chemical properties can be easily validated based on experimental measurements with bulk systems [52]. Spectroscopic methods using electromagnetic radiations have also been commonly used for studying the local compositional variations in glasses [53, 54]. In this section, we consider some of the experimental methods and the principle behind these methods in accessing the atomistic structure. Figure 7.4 shows schematic representations of the experimental setup quantities measured by diffraction and NMR spectroscopy.

#### 7.3.3.1 Diffraction Methods

While the local structure and dynamics have become better accessible in the recent past decades due to development in microscopy [55], direct access to the atomic structure of amorphous or glass materials has been very difficult if possible at all. Often, these methods need to be combined with diffraction studies [56]. Thus, diffraction studies of glasses with either neutrons or X-rays have been one of the most common methods of probing the local atomic structures in glasses [57, 58]. While both these methods are applicable over a wide range of materials systems including crystals and polymers [59], they both follow the same principle of detecting the intensity of scattered beam of neutrons or X-rays as a function of the scattering vector Q.

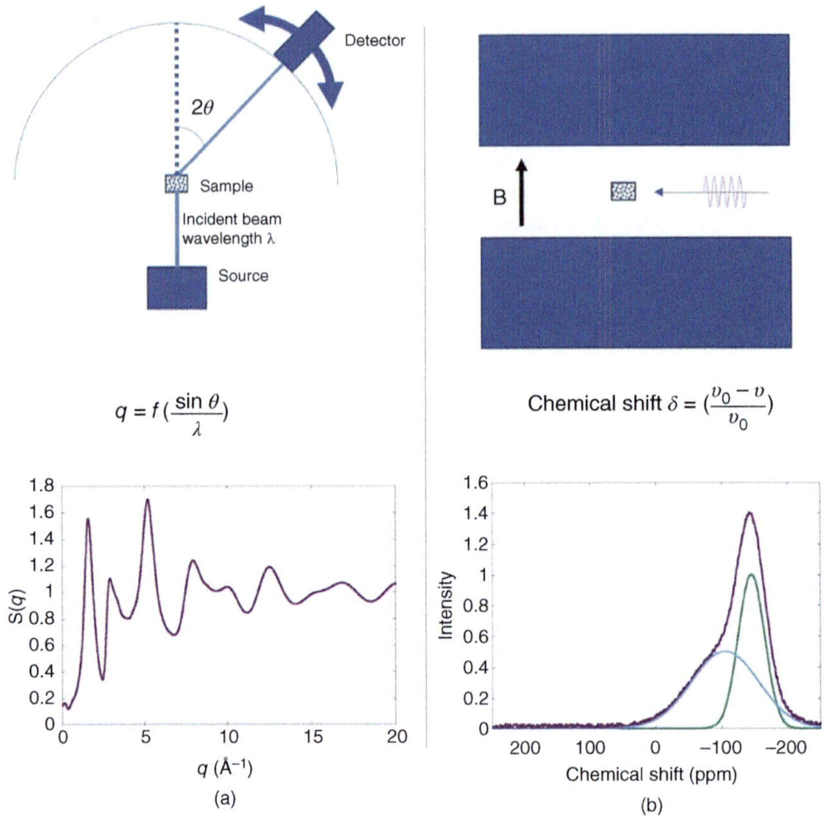


**Figure 7.4** Schematic of the experimental setup for measuring local structure (a) describes diffraction techniques including neutron and X-ray diffraction. A typical result of diffraction studies is the graph of structure factor shown for a $SiO_2$ system. (b) Schematic description of the nuclear magnetic resonance (NMR) technique, which measures the chemical shift in the relaxation frequency relative to a standard system under the same field strength B. The local environments that cause this shift can be deconvoluted and calibrated to standard samples.

X-ray interacts with the electronic charge distribution in an atom and the scattering occurs in all possible directions. In a crystalline system, these scattered beams can interfere constructively and produce high-intensity sharp diffraction peaks (Bragg-diffraction) along the directions determined by the interplanar distance in the crystal. Vitreous systems do not have well-defined scattering planes and hence the diffraction patterns will be a broader peak indicative of the short- and medium-range structure of the material. Scattering of the neutron beam is a consequence of the interaction between the magnetic moments of the neutron and the nucleus as well as with the electronic magnetic moment. Thus, as was the case with the X-rays, the quantity measured in a neutron diffraction study is the intensity of the scattered beam as a function of Q [58, 60].

The intensity of the scattered beam in both the cases is a function of the scattering vector and the scattering cross section of the atoms in the material. While the scattering cross section for X-rays and neutrons can be different for the same sample, since both the radiations are scattered at the location of the nuclei and the charge distribution around it, the information can be used complimentarily to interpret the location of the atoms. Hence, the local structure in both the cases is interpreted based on the Fourier transformation of the diffraction pattern. Without getting into the details of the theoretical considerations, it is sufficient at this point to understand that the structure factor curves (S(Q)) can be obtained from the diffraction patterns. The typical S(Q) curves and experimental setups for neutron and X-ray scattering studies are shown in Figure 7.4a. The Q values in the obtained correspond to the variation in angle, at which the intensity of the beam is measured. While most experimental studies measure the intensity of the scattered radiation at various angles around the sample and express Q as a function of $\sin(\theta)/\lambda$ (units of $Å^{-1}$), variations in Q can also be obtained by varying keeping the angle constant and varying the wavelength of the neutron radiation [61]. Fourier transformation of S(Q) can be used to compare with simulated radial distribution function and also be deconvoluted into the PDF for different species in the glass [7, 30, 58].

A simplistic interpretation of the S(Q) is that the intensity at large Q values correspond to the short-range structure and the low Q intensities correspond to the medium- and long-range structure [62]. Scattering studies can also be further modified to analyze the spectrum at low values of the scattering angle (low-Q) in what is termed as small angle neutron or X-ray scattering (SANS and SAXS) [63–65]. These methods are used for determining the size and shape of particles and pores in glass and are especially useful in the studies of porous silicates and gels.

#### 7.3.3.2 Nuclear Magnetic Resonance

Another important experimental technique for analyzing the local atomic structure in multicomponent silicates is based on the interactions of the nuclei of the material in the presence of a magnetic field. QM establishes that nuclei with an odd number of nucleons are considered spin-half, and such nuclei will possess a net magnetic moment. NMR studies can be applied to materials with a magnetic moment. Applying a magnetic field on a system of nuclei with net magnetic moment will result in the splitting of the allowed energy levels for the nuclei into $2I + 1$ states, where $I$ is the spin of the nucleus. While the energy difference between these allowed levels depends primarily on the applied field, it is also influenced by the effective magnetic field around the nuclei, the presence of surrounding electron clouds and nuclei, chemical environment, anisotropy, and bond strain [40]. Thus, by a precise measurement, the energy difference in terms of the resonant frequency of the nucleus, the local chemical and atomic structure can be interpreted. The frequency is measured relative to a standardized reference frequency and expressed as a fraction of deviation from the resonant frequency of the standard. The standard material is usually chosen as a complex with the same nucleus being studied and in solution so that a sharp peak at the standard frequency is observed. Thus, an NMR spectrum for the sample has the shift measured in ppm on the $x$-axis and an intensity of the shift in the $y$-axis. The location of a given peak is a function of the chemical environment and a broadening of the peak occurs when the nuclei in the sample have multiple orientations as observed in glasses and polycrystalline solids.

The intensity of the NMR peak is an indicator of the quantity of the factor affecting the peak location and thus the NMR spectra can be used in an analysis of the chemical speciation also.

While borosilicate glasses were among the first to be studied with NMR due to the relative abundance of $^{11}$B (80% natural abundance), the $^{29}$Si isotope of silicon with a 4.7% natural abundance is a good candidate for understanding the local structures in silicate glasses as well [38, 66]. The NMR frequency shifts for these $^{11}$B is measured relative to that of B in $BF_3OEt_2$ (Boron trifluoride etherate) and $^{29}$Si is measured relative to $^{29}$Si in tetramethylsilane (TMS) in chloroform-D ($CDCl_3$) solution [67]. For silica and silicate glasses, the NMR spectra can be used to measure the $Q^n$ speciation in the glass as well as the coordination distribution of Si [38]. A specialized technique called magic angle spinning (MAS-NMR) can be used to narrow down the peaks in the spectrum and allows for higher resolution. The spectra from these measurements were fitted to different simulated models of silica to obtain accurate values for the Si–O–Si BAD [37]. While the low abundance of naturally occurring $^{17}$O makes the NMR of oxygen less common, $^{17}$O-enriched glass class can be used for NMR studies. Further specialized techniques such as Multiple-Quantum Magic Angle Spinning (MQ-MAS) and $^{17}$O-enriched MQ-MAS of glass can allow for the speciation of BO [38, 66].

#### 7.3.3.3 Vibrational Spectral Characterization

Among the dynamic properties, the diffusion of silica network is low at room temperature and self-diffusion coefficients become relevant only at temperatures higher than 1000 K. Experimental measures of diffusion in silicates are made by using tracer isotopes. This is useful in understanding the mechanisms of diffusion and activation energy for diffusion at those temperatures [68].

The optical properties and absorption spectra are a more relevant dynamic property considered at all temperatures since many of the applications of glass are dependent on the optical properties. Thus, experimental techniques focus on measuring the wavelength (λ) dependent complex refractive index [69] defined as follows:

$$m(\lambda) = \mu(\lambda) + ik(\lambda) \tag{7.6}$$

where $\mu$ is the refractive index, and the $k$ is the absorption index or extinction coefficient. When light of any given wavelength is incident on a glass, it is transmitted, reflected, or absorbed. The real part of the refractive index corresponds to the refracted portion of light. In highly silicate glasses (as well as in crystalline silicates), this can be easily measured using Snell's law with refractometers. In general, all refractometers work or by allowing light to be refracted through the sample and measuring the critical angle of the light exiting the sample. The refractive index can also be measured by using ellipsometers that measure the polarization state of the refracted light and application of Fresnel equations on these.

The imaginary portion corresponds to the attenuation or extension coefficient, i.e. the extent to which an incident beam loses intensity. This can thus be measured by measuring the intensity of the transmitted beam from a sample of given thickness and obtaining the exponential coefficient [69, 70]. Optical filters can be used to selectively source a given wavelength for the experiment and thus the absorption coefficient over a range of wavelengths can be obtained as utilized in UV–vis spectroscopy. Alternatively, instead of monochromatic source, a broadband radiation source that contains the selected spectrum of wavelengths to be studied can be used and the absorption measured over a period of time, and the absorption spectrum can be back calculated as in Fourier Transformed Infrared (FTIR) spectroscopy [71, 72]. While the specific techniques may be more complex depending on the portion of spectrum being studied, the final result of the study of optical properties is to measure the absorption spectrum of the material.

In general, all experimental characterization of the short- and medium-range structure of silicate glasses involve the application of electromagnetic radiation to the sample and studying the effect of the interaction between the local structure and the radiation. In the case of silicate glasses, these short- and medium-range structures play a major role in their physical properties and in the next few sections, we shall elaborate on these properties of various compositions of silicate glasses as revealed from MD simulations.

## 7.4 MD Simulations of Silica Glasses

A standard method of generating glass structures through MD is by a simulated melt-and-quench starting from a random or a crystal structure. The melting is done by raising the kinetic energy of a set of atoms to much higher than the melting temperature of crystal to remove any memory of the initial structure. The melt is then cooled to room temperature under different temperature profiles (line or nonlinear) and different thermodynamic ensembles (constant pressure or constant volume) [8, 73]. MD-based melt-and-quench, even with the best of computational resources result in a quench rate of about 0.01 K/ps, which is orders of magnitude higher compared to what can be achieved experimentally [74]. Despite these high cooling rates and nonlinear quench rates [75], the local structure and properties of simulated glass approach that of experimentally studied structures asymptotically with cooling rates slower than 5 K/ps and significant deviations in these properties are observable only at cooling rates higher than 20 K/ps [74, 76].

Densities, $T(r)$s, and PDFs evaluated from silica glass simulations are consistent across many different potentials and can be used as a benchmark to evaluate the effectiveness of the potential and the melt-quench procedure [8, 74, 77–79]. Figure 7.5 shows all these features from a typical silica simulation including the scaled $T(r)$ and the deconvoluted partial contributions for specific pairs. Various simulations have confirmed that a reasonable glass structure would produce a $T(r)$ with the first peak at 1.61 Å corresponding to the Si—O bond and the next two peaks at around 2.64 and 3.12 Å corresponding to the O–O and Si–Si first neighbor shells. The primary method of validating the $T(r)$s is through comparison with structure factors obtained from scattering experiments. Most of the simulations have excellent compliance with experimentally obtained structure factors at high Q values, which correspond to good compliance of the short-range structures in real space. The first sharp diffraction peak (FSDP) in the structure factor is an important indicator of the medium-range structural features in the glass [62, 80]. The total variance of a simulated $T(r)$ in comparison to the Fourier transformation of experimental S(Q) is expressed as the $R_\chi$ factor [78, 81]. Figure 7.5a shows the comparison of the experimental and simulated structure factors, Figure 7.5b shows the simulated total and partial structure factors, Figure 7.5c shows the calculated $T(r)$s based on Fourier transformation of the

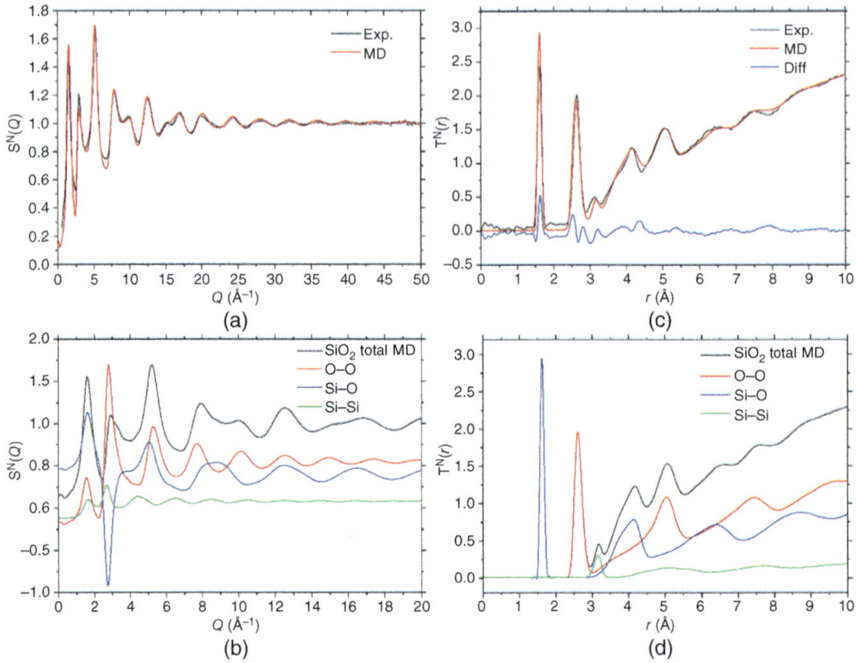


**Figure 7.5** Structure factor and total correlation function of silica glass from MD simulations and validation with experiments. (a) Comparison of experimental neutron structure factor and that from MD simulations for silica glass. Experimental neutron diffraction data was from A. Hannon et al. up to around 50 Å$^{-1}$. MD simulations were performed for a system with 9000 atoms using a cooling rate of 5 K/ps from the melting temperature of 6000 to 300 K under NVT ensemble (b) shows the total and partial O–O, Si–O, and Si–Si structure factors (c) shows the comparison of real space total correlation functions from experiments and simulations. The blue line shows the difference of the two. The $R_x$ factor, a measure of the difference, is 7.2% (d) shows the total partial and correlation functions from MD simulations. The partial correlation function shows that the first peak at 1.62 Å is due to Si–O contribution and the second peak at 2.6 Å is due to O–O contribution and the third peak at 3.1 Å is due to Si–Si contribution. Comparison in both reciprocal (Q) space and real (R) space is important to validate the simulation results. The simulation data such as the partial contributions, on the other hand, can help to assign the peak contributions of experimental results.

experimental and simulated structure factors, and Figure 7.5d shows the partial $T(r)$s. The $R_x$ factor, a quantitative measure of the difference of experimental and simulated total correlation functions, is 7.2% and is also shown in Figure 7.5c. A $R_x$ value under 10% can be deemed an acceptable benchmark for the any silicate simulation and accurate insights can be gained on how the medium-range order contributes to the FSDP.

While the first peak distances can themselves be used to calculate the average bond angles in the silica tetrahedra, a more accurate picture of the angular structure can be obtained by calculating the BAD directly from the trajectories. As can be seen in Figure 7.3, in contrast to what is expected in crystalline forms of silica, the actual shape of silica clusters is more of connected distorted tetrahedra with variations in the angle from the ideal tetrahedral angle of 109.5° [18, 82]. The Si–O–Si BAD has been deduced through by fitting simulated models to experimental data [37, 83]. In simulated models, the range of Si–O–Si BAD has been found to be dependent on the potential with the average varying from 140° to 155°. While experimental measurements suggest that the average bond angle is closer to 142°, MD models have used three-body potentials to regulate the variance in these [18, 79].

The coordination number of atoms in pure amorphous silica is one of the few features that does not vary across experiments or simulations. Depending on the choice of the cut-off, more than 99% of the silicons will be tetrahedrally coordinated with oxygen and on an average, every oxygen will be a linking vertex between two such tetrahedra. This is the significant part of the description of amorphous silica in the CRN model [40, 84, 85]. Further implication of this clear coordination is also that >99% of the oxygen in bulk silica is a BO which also means that most of the $Si^{4+}$ are $Q^4$ species. The ring size distribution in pure silica is also consistently well predicted across various models and shows a maximum of six to seven membered rings with a symmetric ring size distribution curve [18, 62, 78].

The largest deviation in the fraction of four-coordinated silicon occurs in simulations of silica surface. Silica surfaces are simulated by a process of melt quench, followed by cleaving the block of atoms and inducing a vacuum between the cleaved layers. After cleaving, a region of the silica block is "frozen" to prevent the system from floating perpendicular to the axis of cleavage. A subsequent annealing process will allow for diffusion and repair of some of the defective species created at the surface. Generation of silica surface structures is important in the analysis of atomistic phenomena at the interfaces. Silica surfaces are generally oxygen-terminated and there is a large concentration of undercoordinated oxygens at the surface. This leads to the formation defect species, the concentration of which will be highest at the surface [73, 77, 86, 87]. The formation of two- and three-member ring defects at the surface also leads to the presence of low O–Si–O angles of around 80° and Si–O–Si angles of around 100° at the surface [82, 87].

The dynamic optical properties of silica are a consequence of the absorption of incident radiation that activate the different modes of vibration in the local structure of silica. Their modes can be described as Si—O—Si bond rocking, Si–O–Si angle bending, and the Si—O bond stretching at −400, −800, and −1100 $cm^{-1}$, respectively [9, 79, 88, 89]. Besides the IR spectra, the Raman spectra of silica is also analyzed to measure the presence of defect species. The presence of three and four-membered rings results in the glass structures and causes defect bands also known as D1 and D2 peaks in the Raman spectra at around 495 and 606 $cm^{-1}$ [40, 90]. While the analysis of classical MD trajectories themselves can show the presence of the defect species, classical MD simulations with nonpolarizable potentials cannot directly measure the Raman spectra. Hence, by incorporating interacting nuclear polarizations in the postprocessing [91] or through the use of semiclassical ab-inito approaches [51, 92], the Raman spectra of silicates have been evaluated to show the presence of these defects.

In this section, we delved into approaches of MD simulations of pure silica and the analysis of the local structure in MD-simulated silica glasses and surfaces. It is worthwhile to note at this point that other MC-based methods like bond switching in crystalline materials [93, 94] also exist for the generation of structures of silicate glasses. While these can be used for generating a wider variety of configurations, these have not been developed to the extent that MD methods have in the recent past, especially with the advent of various accurate and reactive force fields.

## 7.5 MD Simulations of Alkali Silicate and Alkaline Earth Silicate Glasses

While the network structure in pure silica is well studied in MD simulations, these types of glasses are a minor fraction of those that find industrial applications. Most of the commercially produced glasses have major components ranging from alkalis, alkali earth, to transition and rare-earth elements. The most common additional element in commercial glass is sodium followed by calcium and potassium. Thus, the study of alkali and alkaline earth silicates is essential in the field of glass science. MD simulations for silica involve developing parameter set only for two elements, and each additional element has to be parameterized to be paired with every other element. Depending on the force field employed, this may involve refining dozens of parameters for simple multicomponent systems. Nevertheless, several of the force fields are available to simulate many of the commercially relevant multicomponent silicates [49, 95–97].

### 7.5.1 Local Environments and Distribution of Alkali Ions

While commercial glasses are most likely multicomponent containing at least two or three other oxides than silica, from a simulation and parameter development standpoint, it is easier to consider dopant categories in an appropriate order. In this section, we consider the local structural properties of alkali silicate glasses, including lithium, sodium, and potassium silicates. These are some of the most commercially viable alkalis with available MD [49, 98, 99] parameters for alkali silicate glasses with general composition: $x\mathrm{M_2O}\cdot(1-x)\mathrm{SiO_2}$: while other alkali silicates are possible, they are less investigated [100, 101].

The overall effect of the presence of alkali ions is to disrupt the silicate network and results in the formation of the MRN. The conceptualization of this MRN based on X-ray absorption spectroscopy [54] has been one of the key aspects of the investigation of the structure in alkali silicates and forms the basis for verifying the structure of MD simulations. Similar to pure silica glass structures discussed in the previous section, the FSDP in the structure factors obtained from diffraction studies has also been matched MD simulations with different potentials and have been used to thus verify the local structures [102]. In contrast to $Si^{4+}$, which forms a covalent bond with $O^{2-}$, alkali ions form a heteropolar (ionic) bond with the more electronegative oxygen, in the process of forming a bond, losing an electron from the outermost shell. As a result, the alkali ion attains a positive charge and is not attached to any single oxygen and instead is coordinated by several oxygen atoms depending on the excess charge in the region. Hence, the pair distribution for oxygen around alkali has a first peak at a location dependent on the radius of the alkali ion and does not have a clear cutoff for the first neighbor as was the case with silica [98, 102–104]. Thus, the alkali atoms are surrounded by a variable number of oxygens with partial negative charge. The coordination number of these first neighbor oxygens and the distance between the alkali and oxygen depend on the alkali size.

Sodium silicates can be considered as a representative silicate glass for analyzing the finer details of the structure and its dependence on composition. The extent of alkali content also determines the average bond distances of Si–O in the network. Due to the differences in charge distribution between silicon and oxygen BOs are pulled closer to the Si to form a shorter Si–BO distance. With the addition of more alkali, however, the Si–BO distance decreases and the Si–NBO distance increases [105]. In the network, Si–BO pairs are a major contributor to the overall Si—O bond lengths. Thus, due to the interplay of contributions from BOs to NBOs, the overall Si–O distance can decrease slightly with addition of alkalis [49, 106]. More detailed analysis of the extent of contribution to the Si–O peak has revealed that the Si—O bond distance initially increases and reaches a maximum around the disilicate composition for sodium silicates [105]. Alkali-oxygen bond distances also vary with composition, although different potentials have shown differences in the trends and extents of the variations. For instance, while Du et al. show increasing alkali-oxygen bond lengths with increasing compositions for Na and Li, the same potentials demonstrate that increasing $K_2O$ content results in slight lowering of the alkali-oxygen bond distance similar to the trend shown with other potentials [49, 106, 107]. These small variations in the trends for bond distances are not verifiable by experiment and hence have not shown to affect the overall bulk properties. The most consistently observable trend in all these simulations is the increase in the NBO content in the silicates on addition of the alkali oxide.

The oxygen-alkali-oxygen BADs are also characterized by the oxygen speciation of first coordination shell. In general, the coordination shell around the alkali is dominated by the presence of NBOs which form closer to a 90° angle around the alkali [105], BOs, in contrast to their arrangement around Si. The ratio of NBO to BO around modifier cations in the first coordination shell depends on the type of modifiers and strong affected by the field strength rather than the radius of the cation [102, 121]. A small amount of BOs also can be slightly farther away from the alkali while still being a part of the first shell of neighbors. Due to this farther distance, the BOs are more likely to form more acute angles than the NBOs.

The alkali-alkali PDFs show a greater variation with increase in alkali content and can be used to explain the structure of the alkali silicate glass. The first pair of $g_{\text{Na-Na}}(r)$, for instance, decreases in intensity and length with increasing Na content [49, 105]. This is because the intensity of the first peak in the PDF is inversely proportional to the overall concentration of the alkali. That the first peak is observed at all concentrations indicate that the alkali atoms are not uniformly distributed at low concentrations. Rather, clusters of alkali that are separated by the first neighbor distance are found at low concentrations. As the alkali concentration increases, the number of these clusters also increases which result in the joining of these alkali-rich zones to form alkali channels as shown in Figure 7.6a. The interconnected alkali channels are embedded in the silica network resulting in the formation of the MRN [49, 54, 108]. These alkali channels are themselves stable over long simulation times and serve as pathways for the diffusion transport of the alkali ions [49, 109]. The simulations that contributed to Figure 7.6a qualitatively demonstrated that alkali ions seldom diffused beyond these alkali channels and the overall MRN is stable over the course of the simulation. Analysis of MSD curves for all the components in alkali silicates indicates that the alkali ion has a magnitude of higher diffusivity among all the components [110, 111] and long-time diffusion has a nonlinear dependence of the MSD curve. Analysis of the vibration dynamics

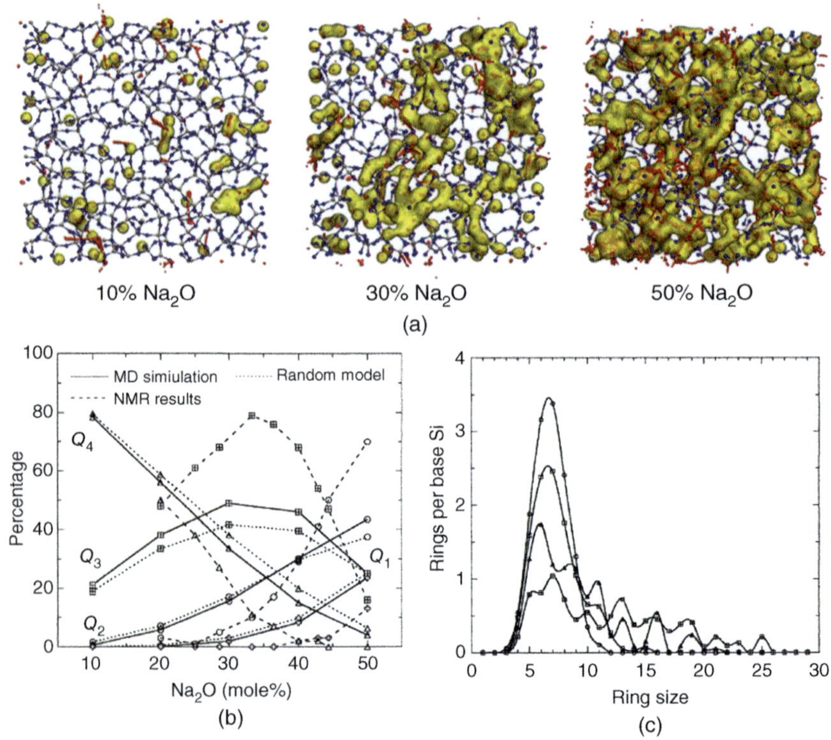


**Figure 7.6** (a) Snapshots of simulated sodium silicate glasses. The yellow blobs are the final position of the sodium ions and the red spheres are the trace of migrating sodium ions over the final 50 ps of simulation time. As can be seen, most of the red tracks are very close to or within the yellow areas indicating that the sodium diffuses along the modified paths in the silicates. (b) shows the $Q^n$ distribution in sodium silicates as a function of composition using random model, computational results, and NMR results. (c) shows the ring size distribution for $SiO_2$ (circle), 10%$Na_2O$ (square), and 20%$Na_2O$ (triangle)and 30%$Na_2O$ glasses (crossed square). Source: (a) Reprinted from Mahadevan et al. [49] (b and c) reprinted with permission from Du and Cormack [105].

of alkali silicates indicates that the Si–O frequencies are not affected by the disruption in the network and that the addition of alkalis results in the increase of sub 200 $cm^{-1}$ vibrations.

The overall atomistic structure of alkali silicate glasses can thus be described based on the MRN. The $Q^n$ speciation and ring structure studies have in various alkali silicates also have been analyzed by MD simulations and are in good agreement with experimental studies [40]. The disruption of the silica network by the alkali channels results in the breakup of some of the regular six to seven membered rings which cut off the existing connections in the network and formation of some large rings. The number of large rings is also dependent on the size of the alkali and larger alkali's could result in the formation of more large rings [102, 105]. The variation of the fraction of $Q^n$ species with increasing alkali content finds consistent fractions of $Q^1$–$Q^4$ species at almost all compositions as shown in Figure 7.6b. This indicates that the formation of $Q^n$ species occurs randomly in the network. A consistent lower value of the peak $Q^3$ fraction compared to experimental values has been noted in many simulations. This lower value has been demonstrated and attributed to the much higher cooling rates that is typically employed in MD simulations [76, 102].

### 7.5.2 The Mixed Alkali Effect

The previous section discussed the available MD studies of a combination of alkali oxide addition to silicate glasses. In general, commercial glasses are usually multicomponent and can comprise of more than one type of alkali. In this context, it is useful to discuss the effect of adding multiple alkali oxides and the mixed-alkali effect in silicates. In general, the addition of a single alkali to silicates results in an almost linear variation in the properties. Instead, if the initial alkali is substituted by another, properties that derive from atomic diffusion will undergo deviation from this linear behavior while some of the other properties show only a slight deviation from the linear behavior [112]. This effect has been analyzed for decades and is the subject of several ongoing studies [112–116]. Initial theories to explain this effect were focused on understanding the diffusivity and coulombic charge-based interactions in the silicates [112, 115, 117]. Since the local environment and structure plays an important role in the diffusion of alkalis, a thorough explanation of the mixed alkali effect will entail understanding of the local structure and dynamics. MD simulations have been used to provide intricate details of the topology, structure, and dynamics and have been used to complement experimental results to understand the mixed alkali effect [116, 118].

### 7.5.3 Alkaline Earth Addition to Alkali Silicates

Binary silicates with alkaline earth oxides of the form $x\text{MO}–(1-x)\cdot\text{SiO}_2$ are analyzed as their respective hydrates in the clayey minerals, cements, and other geochemical materials [119, 120]. In the context of glass, soda-lime silicates are among the most used in commercial glasses, while magnesium and barium silicates are also studied in other multicomponent glass systems.

Several MD simulation studies of soda-lime silicates [121–124] indicate that calcium oxides have nearly the same effect of disrupting the random glass network structure as those of the alkali-oxides. As in sodium silicates, the first shell of O neighbors of Ca does not have a clear cutoff distance at long ranges and the $g_{\text{Ca–Ca}}(r)$ also exhibits a similar decrease of the first peak with increasing calcium concentration [124] indicating clustering and channel formation effect, and these structural features are confirmed based on experimental studies of binary calcium silicates also [125, 126]. Due to the interplay between slightly larger size and higher field strength of Ca, the Ca—O bond distance is slightly lower than that of Na–O. The introduction of Ca to the silicate seems to indicate that it substitutes randomly for Na, albeit, due to the higher field strength of Ca, the O-coordination of Ca is more structured than that of Na [121, 124]. Car–Parrinello MD studies indicate that due to the higher mass and rigid structure around Ca, the diffusivity of Ca and even Na in soda-lime silicates is expected to be lower than those in alkali silicates [127, 128]. This is also validated by MD simulations where Ca substituting for either Na or Si leads to an increase in the activation energy for Na diffusion with the increase being higher when substituted for Na [103].

Likewise, the vibrational density of states (VDOSs) is also not affected at the Si–O stretch and bend frequencies due to the addition of either of these modifiers [124]. A core-shell MD potential has been demonstrated to show a better density and composition-dependent $Q^n$ speciation in soda-lime-silicates at a justified computational cost [123].

Addition of MgO, another alkaline earth oxide, to soda-lime silicates results in improved mechanical properties and chemical durability of the glass. Unlike in the case of lime, $Mg^{2+}$ has a slightly higher field strength than $Ca^{2+}$ and forms a shorter Mg–O first shell neighbor peak around 2.01 Å in the PDF [119, 129, 130]. As a consequence, substitution of CaO by MgO results in the formation of four-coordinated Mg–O clusters and the participation of Mg in the network formation. As a part of the network, Mg also does not obstruct the diffusion pathways of Na ions and thus serves to improve the diffusivity in the alkali channels [129].

## 7.6 MD Simulations of Aluminosilicate Glasses

The previous sections dealt with alkali and alkaline-earth glasses both of which have the effect of disrupting the silicate network, resulting in the formation of MRNs. Alumina additions, especially for the peralkali compositions where the concentrations of alumina is lower than alkali or alkali earth oxide, result in improving the silicate network connectivity and acts as a stabilizer and strengthener for the glass structure [40]. While binary aluminosilicates are of scientific interest [131, 132], MD simulation studies typically emphasize on multicomponent glasses with at least an alkali [133] or alkaline [134] earth oxide in addition to other components.

The condition for alumina to enter the silicate network is that it needs to be charge compensated by positively charged modifier cations. This is because $Al^{3+}$ enters the network by forming the $AlO_4$ tetrahedron, which carries a net negative charge unlike the neutral $SiO_4$ tetrahedron. This is because Al has a charge of +3 and is attached to shared oxygens of charge −2, any coordination higher than three oxygens will result in a net negative charge on $AlO_4$. Assuming even distribution of this negative charge among all the oxygens, it can be easily seen that the oxygens in $Q^4$ Al will hold a negative charge. Hence, positively charged modifier ions in the system can act as charge compensators near these oxygens or they can function like regular modifiers near Si–NBOs or Al–NBOs. These charge-compensating modifier ions result in a higher degree of ionicity of the type of bonding compared to the alkali-(NBO—Si) bond. The extent to which the type of modifier ions become charge compensators is determined by the relative amounts of modifier and Al. Thus, in the case of alkali-alumino silicates, one can define a ratio R = [Al]/[Alkali] on the basis of which the structure and properties can be categorized. Because of the changing role of the alkali ions from charge compensator to network modifier, the properties at the local level near $AlO_4$ will depend on the type of bonding. In general, a linear variation of properties with R has been reported in contrast to the significant deviations seen with mixed-alkali effects [96]. On a related note, the ratio R has been variously defined as [Al]/[modifier] or [modifier]/Al in different texts, with a general preference being to the former.

A majority of Al form a tetrahedrally coordinated [$AlO_4$] structure with the oxygens shared with another Si or Al tetrahedron and maintaining an Al—O bond distance of 1.74 Å. A small fraction of Al, tend to exist in five- or sixfold coordinated forms also, especially at higher Al content [132]. In amorphous aluminosilicates, where there are no charge compensators, charge compensation of $AlO_4$ is accomplished by the formation of three coordinated oxygens [132], also known as tri-clusters or tri-bonded oxygens (TBOs).

Several MD studies of sodium aluminosilicates (NAS) with peraluminous ($R > 1$) to peralkaline ($R < 1$) have studied the trends of variation in local structure and chemical speciation [96, 133, 135, 136]. The formation of TBO is observed to increase in alkali alumino silicates with increase of R. In MD simulations of NASs, this also results in the increase of Na and Si bonded to BOs and the corresponding decrease in Na- and Si-bonded NBOs at high R [133]. The initial addition of alkalis result in them becoming charge compensators at low alkali content [103]. While sixfold coordinated Al is observed in the crystalline alumino silicates, and in Li-alumino silicates [137, 138], they only exist in insignificant quantities at low alumina and alkali content [103, 132]. The BADs and ring structure in alumino silicates also indicate that although the Al–O–Al combination is rare, it is observed in glasses with higher Al content and results in the formation of smaller angles and two member rings.

Similar to alkali aluminosilicates, calcium aluminosilicates (CAS) glasses can also be systematically categorized based on the ratio of modifier to former ($R = [Al]/[Ca]$). These series of glasses present $AlO_5$ and $AlO_6$ units [52, 134, 139], tri-clusters, and NBOs in their local structure. CAS glasses have also been extensively studied by using MD simulations [52, 134]. For example, Lu and Du have recently studied over 100 glass compositions covering the ternary glass formation range by using MD simulations [140]. Comparison with experimental neutron diffraction data shows the employed partial charge potentials that were able to reproduce the experimental structure factor and distribution function in high fidelity, suggesting accuracy of the potential. The structural information such as cation coordination numbers, $Q^n$ distribution, and network connectivity from MD were used to construct descriptors for quantitative structure property relation (QSPR) analysis of various properties ranging from glass transition temperature, elastic moduli, hardness to thermal expansion coefficient.

In terms of dynamic properties, the presence of $AlO_5$ units results in a peak at around 660 $cm^{-1}$ along with Al–O stretch and bend modes in the 700–800 $cm^{-1}$ range. Overall, in all the available literature on soda-lime-aluminosilicate glasses, Al has been demonstrated to act as a network former that requires charge compensates through addition of modifiers or through formation of nonconventional coordination and ring structures. This series of glasses also present an excellent opportunity for further systematic studies with fine variations in the compositions.

## 7.7 MD Simulations of Nanoporous Silica and Silicate Glasses

Nanoporous, amorphous structures, as the name indicates, are characterized by pores that are much larger than the typical first neighbor distance in silicates and include silicate gels, geo-polymers, and also crystalline materials such as zeolites. This section elaborates on the methods and recent developments in the MD simulation of silicate gels.

One of the important characteristics of these nanoporous materials is the presence of large amount of internal surfaces. While MD methods has been used for the study of plane silica surfaces [77, 141], the use of MD for generating the irregular surfaces found in silicate gels is a more recent effort [53, 142]. In general, however, mathematical modeling has been used to study porous, gel-like structures [143–146].

Several methods have been used to generate nanoporous structures in silicates that maintain the local atomistic structures of bulk silicate glasses while introducing pores into the system. Some of the establish methods for generating these structures are as follows:

(1) The volume scaling method [147], where an existing bulk silica structure, is ruptured by gradually increasing the volume which results in the formation of nanoscale cracks.
(2) The charge scaling method [142] in which the charges of atoms in an expanded system is gradually increased, resulting in a low temperature melt-quench of the glass.
(3) Depolymerization of bulk silicate glasses by removing soluble cations or network formers from dense glasses followed by reconstruction to allow for local charges to be neutralized [27].
(4) Coagulation of nanosized chunks of silica and allowing for joining of these in the presence of solvents or through colloidal sintering [148].
(5) Oligomerization of silicic acid [15] in the presence of water, which can be speeded up by the application of accelerated MD/MC [149] or high-temperature processing [150].

Besides the atomic structure characterizations mentioned in the earlier sections, the resultant nanoporous structure from any of the above procedures requires its nanostructural features to be characterized. Porosity measurements based on comparing the density of the gel with that of the starting bulk silicate provide a simple way to understand the efficiency of the method used for gel formation. Since the reactivity of the gels is dependent on the existence of large amounts of internal surfaces in the pores, the solvent accessible surface area, or the Connelly surface can also be measured by assigning suitable radii to the atoms in the gel [151, 152]. Besides the surface area and porosity, an indicative property of the nanoporous structure is measured by its fractal dimension [144, 145], which is useful in the characterization of highly porous aero-gels. Figure 7.7 shows the evolution of the fractal dimension as a function of the processing steps according to the charge scaling method. One of the primary advantages of this method of preparation is that by varying the temperature of processing, the average pore sizes and porosities can be regulated, and tailored structures can be generated. In the case of gels with porosities less than 50%, the pore size distribution can be evaluated [153, 154] to correlate with the reactivities. Experimentally, small angle X-ray scattering experiments have been used to elucidate the gel structures [155].

The nature and connectivity of the pores that are formed are a function of the methodology used for forming the nanoporous structues. For instance, depolymerizing by removing silicon atoms and reorganizing results in the formation of long connected pores, while removing soluble components results in distributed smaller pores [27]. Pore sizes can be controlled by charge scaling at higher temperatures and the generated structures have been demonstrated to be qualitatively similar to experimentally observed structures.

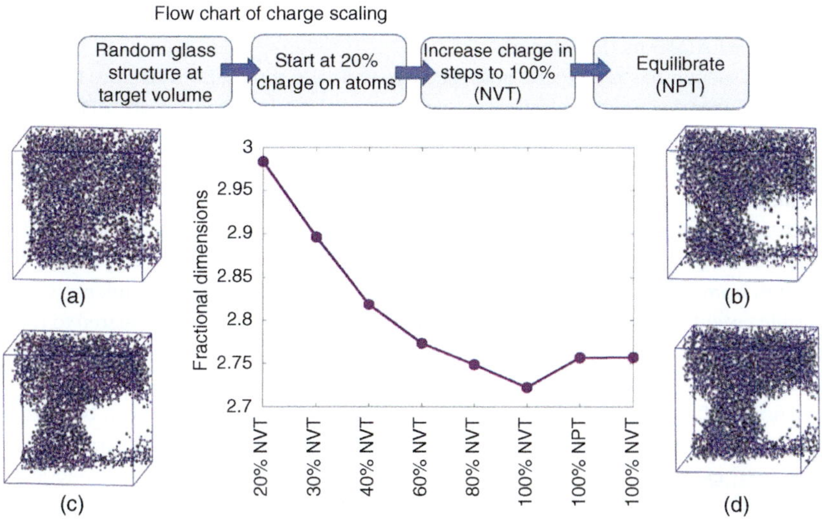


**Figure 7.7** Stagewise evolution of the gel morphology in amorphous silicate: (a) At the end of 20% charge NVT, the atoms are almost evenly distributed giving a fractal dimension close to 3. The structure is still not completely formed at the end of 40% charge NVT (b) and the lowest fractal dimension is achieved after 100% charge equilibration (c). NPT equilibration after this stage preserves the structure, however the overall density is increased as the simulation box shrinks slightly (d).

The local atomistic structures in nanoporous systems are very similar to those of the corresponding dense bulk silicates. As referenced earlier, the presence of large amount of internal surfaces results in the formation of excessive defect structures like undercoordinated network formers and oxygens. MD studies of silicate surfaces have demonstrated that network modifiers tend to migrate toward the surface resulting in a higher NBO concentration and oxygen terminated surfaces [45, 86, 141]. MD studies of nanoporous gel structures also indicate such characteristics, with the extent of the defect formation being highly correlated to the surface area. In general, a large number of small pores results in a much higher surface area and carries far more defective species compared to bulk [151].

## 7.8 AIMD Simulations of Silica and Silicate Glasses

Most of the results presented in the earlier sections are from empirical force field-based MD simulations. Another important method of MD simulation where interatomic forces were calculated from ab initio, so often named AIMD, instead of the empirical parameter development to predefined analytical formulae, which in comparison known as classical MD. AIMD is useful in generating structures based on electronic and atomistic interactions [156–158]. In this method, a system of N atoms is also assigned $N_e$ electrons and the total Hamiltonian is evaluated based on the interactions of the atoms as well as electrons. The electronic interactions and energies can be considered based on the Density Functional

Theory (DFT). The Car–Parrinello [158, 159] and Born–Oppenheimer [160] methods are frequently used AIMD methodologies.

An earlier section has referenced how the Carr-Parinello molecular dynamics (CPMD) method was used for studying diffusion pathways of $Na^+$ in alkali silicates [128]. Du et al. [98] used Born–Oppenheimer molecular dynamics (BOMD), a flavor of AIMD, simulations to study lithium disilicate melt and glass structures. It was found that AIMD can significantly improve the description of Si $Q^n$ distributions in comparison to experiments [98], as compared to classical MD simulations that usually show more broader distributions of $Q^n$ species. This was also observed by Tilocca et al. who showed that $Q^n$ distribution from AIMD for silicate glasses are in closer agreement with experiments, while polarizable potential-based classical simulations show improvements as compared to rigid ion potential-based classical MD simulations [161]. The VDOS were also calculated using two approaches: Fourier transformation of the velocity autocorrelation (VACs) functions and the finite difference method. The two gave similar results and features can be related to the vibration spectral of these glasses. AIMD also has access to electronic structures of the simulated glasses. The electronic density of states (EDOS) of lithium disilicate glass was calculated and compared to the melt, silica glass and a lithium disilicate crystal. A narrowing of bandgap of lithium disilicate glass as compared to silica and corresponding crystal was observed [98]. Furthermore, the atomic charges of the glass and the melt were calculated by using the Bader and Hershfeld methods. Charge values from the two approaches differ significantly. Importantly, the Hershfeld method was found to show the higher negative charges on NBO as compared to BO [98]. The results from these AIMD calculations can be further used for designing and validating the results of classical MD especially in terms of the local structures and dynamic properties. A similar comparison of classical and AIMD results that show incremental improvements in the structure of sodium silicates can also be found in [162].

Using the trajectories from a classical MD system as an input for AIMD is another way of validating the fidelity of the classical potentials. In general, classical potentials have tended to overestimate the average angles subtended by BOs. Refinement of the simulation with CPMD for calcium aluminosilicates allows for this angle to settle to lower values and allows for refinement of other structural attributes and allows the system to "forget" its initial configurations [163].

While these methods allow for an accurate depiction of the atomic phenomena, due to the large number of calculations involved, they are computationally more intense compared to the classical MD simulations that employ empirical potentials, which limits the size and time scale of the systems that can be studied to few hundreds of atoms and over hundreds of picoseconds. Nevertheless, AIMD can be used in conjunction with classical MD to generate silicate structures and understand atomic phenomena.

## 7.9 Summary and Outlook

An important characteristic of amorphous materials like silicate glasses is the lack of long-range order despite of a relatively strict adherence to a short-range local order especially among the first shell of atomic neighbors that closely resemble their crystalline counterparts. Hence, some of the bulk properties of silicate glasses are determined by the short-range order while others properties are affected by medium- and longer-range orders. Atomistic MD simulations are a useful tool for studying silicate glasses as this method can provide insights of both short- and medium-range structures. In this chapter, some of the work on atomistic simulations of silica and silicate glasses were reviewed. Some of the primary means of evaluating the local structure in simulated trajectories of glasses, such as the pair correlation functions, the BADs, coordination and atomic speciation, and medium-range structure features such as $Q^n$ distributions and ring size distributions, are introduced. While some of these quantities are not directly accessible to experimental measurements, characterization techniques like X-ray or neutron diffraction structure factors, NMR, and frequency spectrum measurements can be used, often in combination among each other, to arrive at and validate the simulated structures. For amorphous silica, alkali and alkaline earth silicates, and alumino silicates, a combination of these experimental and simulation techniques has allowed for a detailed understanding of the local structures in these glasses. This understanding has allowed for verification of the CRN in pure silica glass, the MRN in alkali and alkaline earth silicates, and the compensated continuous random network (CCRN) in alumino silicates (Figure 7.8 shows a schematic 2D representation of these networks). As indicated in Section 7.2, as well as elaborated in detail in other chapters, there are several different types of force fields that

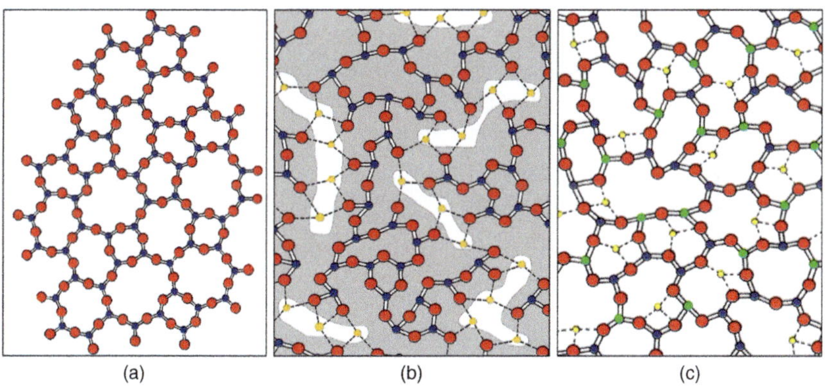


**Figure 7.8** Two-dimensional representation of the network glass structures of silica and silicate glasses. (a) Continuous Random Network (CRN) model of the structure of $SiO_2$. (b) Modified Random Network (MRN) of alkali silicates and alkaline earth silicates. The modifiers, shown as yellow circles, coordinate the NBOs and form channels that disrupt the glass matrix shown as the shaded area. (c) Compensated continuous random network (CCRN) of alkali alumino silicates with an R = [Al]/[Na] > 1. The small green circles represent $Al^{3+}$ that are a part of the silica network. The yellow modifier ions, unlike in the MRN, are coordinated mostly by BOs associated with the network forming $Al^{3+}$. All cations are represented as threefold coordinated. Source: The model of (a) was redrawn from the original image by Zachariasen [6], Wright and Thorpe [164]. Models (b) and (c) were a reprint from Greaves and Nagai [165], modified with color rendering. Reprinted with permission.

can be used for generating trajectories of silicates. Most of the published potentials are designed, parameterized, and tested on how faithfully the local structure and dynamics are reproduced. Such interatomic potentials that have successfully reproduced the local structure can be used to study various bulk phenomena and also evaluate bulk properties of silicates. For example, various methods have been used to measure the bulk mechanical properties and the compliance matrix, ion diffusion and activation energies, and vibrational properties, among others. With the advent of reactive potentials, MD simulations have also been used to investigate the reaction mechanisms and kinetics at the atomistic level.

Advances in machine learning and artificial intelligence approaches to predict the glass properties are growing rapidly and find great interest in glass science. These include methods to correlate the structures and composition with bulk properties and atomistic phenomena. MD simulations have served as a useful tool to validate topological constraint theory [166] which attempts to explain the properties of glass based on degrees of freedom and constraints on the average coordination in glass. Similarly, the QSPR [167, 168] methods have been [169] used to understand the correlations between various bulk properties and the structure of glasses, using structural information based on MD simulations.

Overall, MD simulations have served as a useful method in understanding atomistic phenomena in glasses, including those in regimes that are experimentally unreachable. Recent active development of more accurate and transferable simulations, including some reactive, potentials, enable more and more accurate description of bulk, surface and porous structures, as well as the reactions and interactions of glass with the environments. With the improvement of computational efficiency of algorithms and ever-increasing computation power, larger system sizes, slower cooling rate and more complex glass compositions can be modeled. These developments make atomistic simulations of silicate glasses go beyond university research laboratories to become an important tool in industrial research and developments.

## Acknowledgments

The authors are grateful for funding support National Science Foundation DMR Ceramics program (Project # 1508001).

## References

1 Winter, A. (1957). Glass formation. *J. Am. Ceram. Soc.* 40: 54–58. https://doi.org/10.1111/j.1151-2916.1957.tb12574.x.
2 Brill, R. (1962). A note on the scientist's definition of glass. *J. Glass Stud.* 4: 127–138.
3 Salmon, P.S. (2002). Amorphous materials: order within disorder. *Nat. Mater.* 1 (2): 87–88.
4 Morey, G.W. (1931). Glass, its composition and properties. *J. Chem. Educ.* 8: 421–441. https://doi.org/10.1021/ed008p421.

5 Randall, J.T., Rooksby, H.P., and Cooper, B.S. (1930). 13. X-ray diffraction and the structure of vitreous solids – I. *Zeitschrift Für Krist. Cryst. Mater.* 75: https://doi.org/10.1515/zkri-1930-0114.

6 Zachariasen, W.H. (1932). The atomic arrangement in glass. *J. Am. Chem. Soc.* 54: 3841–3851. https://doi.org/10.1021/ja01349a006.

7 Mozzi, R.L. and Warren, B.E. (1969). The structure of vitreous silica. *J. Appl. Crystallogr.* 2: 164–172. https://doi.org/10.1107/S0021889869006868.

8 Woodcock, L.V., Angell, C.A., and Cheeseman, P. (1976). Molecular dynamics studies of the vitreous state: simple ionic systems and silica. *J. Chem. Phys.* 65: 1565–1577. https://doi.org/10.1063/1.433213.

9 Garofalini, S.H. (1982). Molecular dynamics simulation of the frequency spectrum of amorphous silica. *J. Chem. Phys.* 76: 3189–3192. https://doi.org/10.1063/1.443363.

10 Du, J. (2019). Molecular dynamics simulations of oxide glasses. In: *Springer Handbook of Glass* (ed. J.D. Musgraves, J. Hu and L. Calvez), 1131–1155. https://doi.org/10.1007/978-3-319-93728-1_32.

11 Woodcock, L.V. (1975). Molecular dynamics calculations on molten ionic salts. In: *Advances in Molten Salt Chemistry* (ed. J. Braunstein, G. Mamantov and G.P. Smith), 1–74. Boston, MA: Springer https://doi.org/10.1007/978-1-4615-8270-0_1.

12 Sharma, A., Nakano, A., Kalia, R.K. et al. (2003). Immersive and interactive exploration of billion-atom systems. *Presence: Teleoperators Virtual Environ.* 12: 85–95. https://doi.org/10.1162/105474603763835350.

13 Pang, T. (2006). *An Introduction to Computational Physics*, 2e. Cambridge University Press https://doi.org/10.1017/CBO9780511800870.

14 Allen, M.P. and Tildesley, D.J. (2017). *Computer Simulation of Liquids*. Oxford University Press https://doi.org/10.1093/oso/9780198803195.001.0001.

15 Feuston, B.P. and Garofalini, S.H. (1990). Oligomerization in silica sols. *J. Phys. Chem.* 94: 5351–5356. https://doi.org/10.1021/j100376a035.

16 Cygan, R.T., Liang, J.-J., and Kalinichev, A.G. (2004). Molecular models of hydroxide, oxyhydroxide, and clay phases and the development of a general force field. *J. Phys. Chem. B.* 108: 1255–1266. https://doi.org/10.1021/jp0363287.

17 Feuston, B.P., Newell, R.N., and Garofalini, S.H. (1988). Application of a new three-body potential to vitreous silica and sodium silicate glasses. *MRS Proc.* 141: 207. https://doi.org/10.1557/PROC-141-207.

18 Feuston, B.P. and Garofalini, S.H. (1988). Empirical three-body potential for vitreous silica. *J. Chem. Phys.* 89: 5818–5824. https://doi.org/10.1063/1.455531.

19 Kolafa, J. and Perram, J.W. (1992). Cutoff errors in the Ewald summation formulae for point charge systems. *Mol. Simul.* 9: 351–368. https://doi.org/10.1080/08927029208049126.

20 Darden, T., York, D., and Pedersen, L. (1993). Particle mesh Ewald: an $N\cdot\log(N)$ method for Ewald sums in large systems. *J. Chem. Phys.* 98: 10089–10092. https://doi.org/10.1063/1.464397.

21 Ma, Y. and Garofalini, S.H. (2005). Modified wolf electrostatic summation: incorporating an empirical charge overlap. *Mol. Simul.* 31: 739–748. https://doi.org/10.1080/08927020500262598.

**22** van Duin, A.C.T., Dasgupta, S., Lorant, F., and Goddard, W.A. (2001). ReaxFF: a reactive force field for hydrocarbons. *J. Phys. Chem. A.* 105: 9396–9409. https://doi.org/10.1021/jp004368u.

**23** Yu, J., Sinnott, S.B., and Phillpot, S.R. (2007). Charge optimized many-body potential for the Si/$SiO_2$ system. *Phys. Rev. B Condens. Matter Mater. Phys.* 75: 1–13. https://doi.org/10.1103/PhysRevB.75.085311.

**24** Brenner, D.W., Shenderova, O.A., Harrison, J.A. et al. (2002). A second-generation reactive empirical bond order (REBO) potential energy expression for hydrocarbons. *J. Phys. Condens. Matter.* 14: 783–802. https://doi.org/10.1088/0953-8984/14/4/312.

**25** Botu, V., Batra, R., Chapman, J., and Ramprasad, R. (2017). Machine learning force fields: construction, validation, and outlook. *J. Phys. Chem. C* 121: 511–522. https://doi.org/10.1021/acs.jpcc.6b10908.

**26** Singraber, A., Behler, J., and Dellago, C. (2019). Library-based LAMMPS implementation of high-dimensional neural network potentials. *J. Chem. Theory Comput.* 15: 1827–1840. https://doi.org/10.1021/acs.jctc.8b00770.

**27** Rimsza, J.M. and Du, J. (2018). Nanoporous silica gel structures and evolution from reactive force field-based molecular dynamics simulations. *npj Mater. Degrad.* 2: 1–10. https://doi.org/10.1038/s41529-018-0039-0.

**28** Wales, D. (2013). Chapter 1: Glasses and supercooled liquids. In: *Energy Landscapes*, 592–662. Cambridge University Press https://doi.org/10.1017/cbo9780511721724.011.

**29** Neyts, E.C. and Bogaerts, A. (2013). Combining molecular dynamics with Monte Carlo simulations: implementations and applications. *Theor. Chem. Acc.* 132: 1–12. https://doi.org/10.1007/s00214-012-1320-x.

**30** Fischer, H.E., Barnes, A.C., and Salmon, P.S. (2006). Neutron and X-ray diffraction studies of liquids and glasses. *Reports Prog. Phys.* 69: 233–299. https://doi.org/10.1088/0034-4885/69/1/R05.

**31** Chandler, D. (1987). *Introduction to Modern Statistical Mechanics*. Oxford University Press https://www.bibsonomy.org/bibtex/20b76edaadc1bac83932213edfcbcb615/magdusia.

**32** Du, J. (2015). Challenges in molecular dynamics simulations of multicomponent oxide glasses. In: *Molecular Dynamics Simulations of Disordered Materials*, Springer Series in Materials Science, vol. 215 (ed. C. Massobrio, J. Du, M. Bernasconi and P. Salmon), 157–180. Cham: Springer https://doi.org/10.1007/978-3-319-15675-0.

**33** Soper, A.K. and Barney, E.R. (2012). On the use of modification functions when Fourier transforming total scattering data. *J. Appl. Crystallogr.* 45: 1314–1317. https://doi.org/10.1107/S002188981203960X.

**34** Duffy, J.A. and Macphee, D.E. (2007). The coordination number of silicon in silicon-oxygen compounds: the special case of 6-fold coordination in thaumasite. *J. Phys. Chem. B.* 111: 8740–8745. https://doi.org/10.1021/jp071343n.

**35** Anderson, N.L., Vedula, R.P., Schultz, P.A. et al. (2011). First-principles investigation of low energy E′ center precursors in amorphous silica. *Phys. Rev. Lett.* 106: 1–4. https://doi.org/10.1103/PhysRevLett.106.206402.

**36** Sushko, P.V., Mukhopadhyay, S., Mysovsky, A.S. et al. (2005). Structure and properties of defects in amorphous silica: new insights from embedded cluster calculations. *J. Phys. Condens. Matter.* 17: https://doi.org/10.1088/0953-8984/17/21/007.

**37** Dupree, E. and Pettifer, R.F. (1984). Determination of the Si–O–Si bond angle distribution in vitreous silica by magic angle spinning NMR. *Nature* 308: 523–525. https://doi.org/10.1038/308523a0.

**38** Youngman, R. (2018). NMR spectroscopy in glass science: a review of the elements. *Materials (Basel)* 11: 476. https://doi.org/10.3390/ma11040476.

**39** Greaves, G.N., Fontaine, A., Lagarde, P. et al. (1981). Local structure of silicate glasses. *Nature* 293: 611–616. https://doi.org/10.1038/293611a0.

**40** Greaves, G.N. and Sen, S. (2007). Inorganic glasses, glass-forming liquids and amorphizing solids. *Adv. Phys.* 56: 1–166. https://doi.org/10.1080/00018730601147426.

**41** Yuan, X. and Cormack, A.N. (2002). Efficient algorithm for primitive ring statistics in topological networks. *Comput. Mater. Sci.* 24: 343–360. https://doi.org/10.1016/S0927-0256(01)00256-7.

**42** Levine, S.M. and Garofalini, S.H. (1986). Surface structure of silica glasses by molecular dynamics simulations. *Mater. Res. Soc. Symp. Proc.* 61: 29–37. https://doi.org/10.1557/proc-61-29.

**43** Goetzke, K. and Klein, H.J. (1991). Properties and efficient algorithmic determination of different classes of rings in finite and infinite polyhedral networks. *J. Non-Cryst. Solids* 127: 215–220. https://doi.org/10.1016/0022-3093(91)90145-V.

**44** King, S.V. (1967). Ring configurations in a random network model of vitreous silica. *Nature* 213: 1112–1113. https://doi.org/10.1038/2131112a0.

**45** Garofalini, S.H. (1990). Molecular dynamics computer simulations of silica surface structure and adsorption of water molecules. *J. Non-Cryst. Solids* 120: 1–12. https://doi.org/10.1016/0022-3093(90)90184-N.

**46** Levine, S.M. and Garofalini, S.H. (1987). A structural analysis of the vitreous silica surface via a molecular dynamics computer simulation. *J. Chem. Phys.* 86: 2997–3002. https://doi.org/10.1063/1.452054.

**47** Walsh, T.R., Wilson, M., and Sutton, A.P. (2000). Hydrolysis of the amorphous silica surface. II. Calculation of activation barriers and mechanisms. *J. Chem. Phys.* 113: 9191–9201. https://doi.org/10.1063/1.1320057.

**48** Barhemmati-Rajab, N., Mahadevan, T., Du, J., and Zhao, W. (2020). Thermal transport properties enhancement of paraffin via encapsulation into boron nitride nanotube: a molecular dynamics study. *MRS Commun.* 10: 475–481. https://doi.org/10.1557/mrc.2020.49.

**49** Mahadevan, T.S., Sun, W., and Du, J. (2019). Development of water reactive potentials for sodium silicate glasses. *J. Phys. Chem. B.* 123: 4452–4461. https://doi.org/10.1021/acs.jpcb.9b02216.

**50** Noid, D.W., Koszykowski, M.L., and Marcus, R.A. (1977). A spectral analysis method of obtaining molecular spectra from classical trajectories. *J. Chem. Phys.* 67: 404–408. https://doi.org/10.1063/1.434901.

**51** Fredkin, D.R., Komomicki, A., White, S.R., and Wilson, K.R. (1982). Ab initio infrared and Raman spectra. *J. Chem. Phys.* 78: 7077–7092. https://doi.org/10.1063/1.444751.

**52** Takahashi, S., Neuville, D.R., and Takebe, H. (2015). Thermal properties, density and structure of percalcic and peraluminus $CaO–Al_2O_3–SiO_2$ glasses. *J. Non-Cryst. Solids* 411: 5–12. https://doi.org/10.1016/j.jnoncrysol.2014.12.019.

**53** Hopf, J., Eskelsen, J.R., Chiu, M. et al. (2018). Toward an understanding of surface layer formation, growth, and transformation at the glass–fluid interface. *Geochim. Cosmochim. Acta* 229: 65–84. https://doi.org/10.1016/j.gca.2018.01.035.

**54** Greaves, G.N. (1985). EXAFS and the structure of glass. *J. Non-Cryst. Solids* 71: 203–217. https://doi.org/10.1016/0022-3093(85)90289-3.

**55** García-Heras, M., Carmona, N., Ruiz-Conde, A. et al. (2005). Application of atomic force microscopy to the study of glass decay. *Mater. Charact.* 55: 272–280. https://doi.org/10.1016/j.matchar.2005.07.001.

**56** Hirata, A. and Chen, M. (2014). Angstrom-beam electron diffraction of amorphous materials. *J. Non-Cryst. Solids* 383: 52–58. https://doi.org/10.1016/j.jnoncrysol.2013.03.010.

**57** Warren, B.E. (1940). X-ray diffraction study of the structure of glass. *Chem. Rev.* 26: 237–255. https://doi.org/10.1021/cr60084a007.

**58** Wright, A.C. (1988). Neutron and X-ray amorphography. *J. Non-Cryst. Solids* 106: 1–16. https://doi.org/10.1016/0022-3093(88)90218-9.

**59** Arbe, A., Alvarez, F., and Colmenero, J. (2012). Neutron scattering and molecular dynamics simulations: synergetic tools to unravel structure and dynamics in polymers. *Soft Matter.* 8: 8257–8270. https://doi.org/10.1039/c2sm26061a.

**60** Kittel, C. (1999). *Introduction to Solid State Physics*, 7e. Wiley.

**61** Grimley, D.I., Wright, A.C., and Sinclair, R.N. (1990). Neutron scattering from vitreous silica IV. Time-of-flight diffraction. *J. Non-Cryst. Solids* 119: 49–64. https://doi.org/10.1016/0022-3093(90)90240-M.

**62** Du, J. and Corrales, L.R.R. (2005). First sharp diffraction peak in silicate glasses: structure and scattering length dependence. *Phys. Rev. B Condens. Matter Mater. Phys.* 72: 1–4. https://doi.org/10.1103/PhysRevB.72.092201.

**63** Wong, P.Z., Howard, J., and Lin, J.S. (1986). Surface roughening and the fractal nature of rocks. *Phys. Rev. Lett.* 57: 637–640. https://doi.org/10.1103/PhysRevLett.57.637.

**64** Boldon, L., Laliberte, F., and Liu, L. (2015). Review of the fundamental theories behind small angle X-ray scattering, molecular dynamics simulations, and relevant integrated application. *Nano Rev.* 6: 25661. https://doi.org/10.3402/nano.v6.25661.

**65** Kimanos, F. (2014) SAXS-Guided Metadynamics, http://www.diva-portal.se/smash/get/diva2:767616/FULLTEXT01.pdf.

**66** Bray, P.J. (1985). Nuclear magnetic resonance studies of glass structure. *J. Non-Cryst. Solids* 73: 19–45. https://doi.org/10.1016/0022-3093(85)90335-7.

**67** Becker, E.D. (2000). Chemical Shifts. In: *High Resolution NMR*, 83–117. Elsevier https://doi.org/10.1016/b978-012084662-7/50048-x.

**68** Rodríguez-Viejo, J., Sibieude, F., Clavaguera-Mora, M.T., and Monty, C. (1993). 18O diffusion through amorphous $SiO_2$ and cristobalite. *Appl. Phys. Lett.* 63: 1906–1908. https://doi.org/10.1063/1.110644.

**69** Kitamura, R., Pilon, L., and Jonasz, M. (2007). Optical constants of silica glass from extreme ultraviolet to far infrared at near room temperature. *Appl. Opt.* 46: 8118. https://doi.org/10.1364/AO.46.008118.

**70** Philipp, H.R. (1979). The infrared optical properties of $SiO_2$ and $SiO_2$ layers on silicon. *J. Appl. Phys.* 50: 1053–1057. https://doi.org/10.1063/1.326080.

**71** Battisha, I.K., El Beyally, A., El Mongy, S.A., and Nahrawi, A.M. (2007). Development of the FTIR properties of nano-structure silica gel doped with different rare earth elements, prepared by sol-gel route. *J. Sol-Gel Sci. Technol.* 41: 129–137. https://doi.org/10.1007/s10971-006-0520-z.

**72** D'Souza, A.S., Pantano, C.G., and Kallury, K.M.R. (1997). Determination of the surface silanol concentration of amorphous silica surfaces using static secondary ion mass spectroscopy. *J. Vac. Sci. Technol., A* 15: 526–531. https://doi.org/10.1116/1.580678.

**73** Garofalini, S.H. (2001). Molecular dynamics simulations of silicate glasses and glass surfaces. *Rev. Mineral. Geochem.* 42: 130–168. https://doi.org/10.2138/rmg.2001.42.5.

**74** Vollmayr, K., Kob, W., and Binder, K. (1996). Cooling-rate effects in amorphous silica: a computer-simulation study. *Phys. Rev. B.* 54: 15808–15827. https://doi.org/10.1103/PhysRevB.54.15808.

**75** Corrales, L.R. and Du, J. (2005). Thermal kinetics of glass simulations. *Phys. Chem. Glasses* 46: 420–424.

**76** Li, X., Song, W., Yang, K. et al. (2017). Cooling rate effects in sodium silicate glasses: bridging the gap between molecular dynamics simulations and experiments. *J. Chem. Phys.* 147: 074501. https://doi.org/10.1063/1.4998611.

**77** Garofalini, S.H. (1983). A molecular dynamics simulation of the vitreous silica surface. *J. Chem. Phys.* 78: 2069–2072. https://doi.org/10.1063/1.444927.

**78** Yuan, X. and Cormack, A.N. (2001). Local structures of MD-modeled vitreous silica and sodium silicate glasses. *J. Non-Cryst. Solids* 283: 69–87. https://doi.org/10.1016/S0022-3093(01)00363-5.

**79** Huang, L. and Kieffer, J. (2003). Molecular dynamics study of cristobalite silica using a charge transfer three-body potential: phase transformation and structural disorder. *J. Chem. Phys.* 118: 1487–1498. https://doi.org/10.1063/1.1529684.

**80** Elliott, S.R. (1991). Origin of the first sharp diffraction peak in the structure factor of covalent glasses. *Phys. Rev. Lett.* 67: 711–714. https://doi.org/10.1103/PhysRevLett.67.711.

**81** Wright, A.C. (1993). The comparison of molecular dynamics simulations with diffraction experiments. *J. Non-Cryst. Solids* 159: 264–268. https://doi.org/10.1016/0022-3093(93)90232-M.

**82** Du, J. and Cormack, A.N. (2005). Molecular dynamics simulation of the structure and hydroxylation of silica glass surfaces. *J. Am. Ceram. Soc.* 88: 2532–2539. https://doi.org/10.1111/j.1551‐2916.2005.00352.x.

**83** Yuan, X. and Cormack, A.N. (2003). Si-O-Si bond angle and torsion angle distribution in vitreous silica and sodium silicate glasses. *J. Non-Cryst. Solids* 319: 31–43. https://doi.org/10.1016/S0022-3093(02)01960-9.

**84** A. Varshneya, J. Mauro, *Fundamentals of Inorganic Glasses*, 2019. https://doi.org/10.1016/C2017-0-04281-7.

**85** Eckert, H. (1992). Structural characterization of noncrystalline solids and glasses using solid state NMR. *Prog. Nucl. Magn. Reson. Spectrosc.* https://doi.org/10.1016/0079-6565(92)80001-V.

**86** Mahadevan, T.S. and Du, J. (2018). Evaluating water reactivity at silica surfaces using reactive potentials. *J. Phys. Chem. C* 122: 9875–9885. https://doi.org/10.1021/acs.jpcc.7b12653.

**87** Feuston, B.P. and Garofalini, S.H. (1989). Topological and bonding defects in vitreous silica surfaces. *J. Chem. Phys.* 91: 564–570. https://doi.org/10.1063/1.457440.

**88** Machacek, J., Gedeon, O., and Liska, M. (2013). Normal mode analysis of $Na_2O{\cdot}3SiO_2$ glass. *Physics Procedia* 48: 85–88. https://doi.org/10.1016/j.phpro.2013.07.014.

**89** Haworth, R., Mountjoy, G., Corno, M. et al. (2010). Probing vibrational modes in silica glass using inelastic neutron scattering with mass contrast. *Phys. Rev. B Condens. Matter Mater. Phys.* 81: 98–101. https://doi.org/10.1103/PhysRevB.81.060301.

**90** Galeener, F.L. (1985). Raman and ESR studies of the thermal history of amorphous $SiO_2$. *J. Non-Cryst. Solids* 71: 373–386. https://doi.org/10.1016/0022-3093(85)90308-4.

**91** Shcheblanov, N.S., Mantisi, B., Umari, P., and Tanguy, A. (2015). Detailed analysis of plastic shear in the Raman spectra of $SiO_2$ glass. *J. Non-Cryst. Solids* 428: 6–19. https://doi.org/10.1016/j.jnoncrysol.2015.07.035.

**92** Micoulaut, M. and Phillips, J.C. (2003). Rings and rigidity transitions in network glasses. *Phys. Rev. B Condens. Matter Mater. Phys.* 67: 9. https://doi.org/10.1103/PhysRevB.67.104204.

**93** Du, J. and Kroll, P. (2010). Electronic structure and interfacial properties of Ge nanoclusters embedded in amorphous silica. *J. Non-Cryst. Solids* 356: 2448–2453. https://doi.org/10.1016/j.jnoncrysol.2010.03.040.

**94** Tu, Y., Tersoff, J., Grinstein, G., and Vanderbilt, D. (1998). Properties of a continuous-random-network model for amorphous systems. *Phys. Rev. Lett.* 81: 4899–4902. https://doi.org/10.1103/PhysRevLett.81.4899.

**95** Ha, M.-T. and Garofalini, S.H. (2017). Local structure of network modifier to network former ions in soda-lime alumino-borosilicate glasses. *J. Am. Ceram. Soc.* 100: 563–573. https://doi.org/10.1111/jace.14565.

**96** Xiang, Y., Du, J., Smedskjaer, M.M., and Mauro, J.C. (2013). Structure and properties of sodium aluminosilicate glasses from molecular dynamics simulations. *J. Chem. Phys.* 139: https://doi.org/10.1063/1.4816378.

**97** Yu, Y., Wang, B., Wang, M. et al. (2017). Reactive molecular dynamics simulations of sodium silicate glasses – toward an improved understanding of the structure. *Int. J. Appl. Glass Sci.* 8: 276–284. https://doi.org/10.1111/ijag.12248.

**98** Du, J. and Corrales, L.R.R. (2006). Structure, dynamics, and electronic properties of lithium disilicate melt and glass. *J. Chem. Phys.* 125: 1–12. https://doi.org/10.1063/1.2345060.

**99** Maekawa, H., Maekawa, T., Kawamura, K., and Yokokawa, T. (1991). The structural groups of alkali silicate glasses determined from $^{29}Si$ MAS-NMR. *J. Non-Cryst. Solids* 127: 53–64. https://doi.org/10.1016/0022-3093(91)90400-Z.

**100** Weldes, H.H. and Lange, K.R. (1969). Properties of soluble silicates. *Ind. Eng. Chem.* 61: 29–44. https://doi.org/10.1021/ie50712a008.

**101** Osipov, A., Osipova, L., and Zainullina, R. (2015). Raman spectroscopy and statistical analysis of the silicate species and group connectivity in cesium silicate glass forming system. *Int. J. Spectrosc.* 2015: 1–15. https://doi.org/10.1155/2015/572840.

**102** Du, J. and Corrales, L.R.R. (2006). Compositional dependence of the first sharp diffraction peaks in alkali silicate glasses: a molecular dynamics study. *J. Non-Cryst. Solids* 352: 3255–3269. https://doi.org/10.1016/j.jnoncrysol.2006.05.025.

**103** Cormack, A.N. and Cao, Y. (1996). Molecular dynamics simulation of silicate glasses. *Mol. Eng.* 6: 415–458. https://doi.org/10.1007/BF00161727.

104 Jabraoui, H., Achhal, E.M., Hasnaoui, A. et al. (2016). Molecular dynamics simulation of thermodynamic and structural properties of silicate glass: effect of the alkali oxide modifiers. *J. Non-Cryst. Solids* 448: 16–26. https://doi.org/10.1016/j.jnoncrysol.2016.06.030.

105 Du, J. and Cormack, A.N. (2004). The medium range structure of sodium silicate glasses: a molecular dynamics simulation. *J. Non-Cryst. Solids* 349: 66–79. https://doi.org/10.1016/j.jnoncrysol.2004.08.264.

106 Melman, H. and Garofalini, S.H.H. (1991). Microstructural evaluation of simulated sodium silicate glasses. *J. Non-Cryst. Solids* 134: 107–115. https://doi.org/10.1016/0022-3093(91)90017-Z.

107 Jabraoui, H., Vaills, Y., Hasnaoui, A. et al. (2016). Effect of sodium oxide modifier on structural and elastic properties of silicate glass. *J. Phys. Chem. B.* 120: 13193–13205. https://doi.org/10.1021/acs.jpcb.6b09664.

108 Meyer, A., Horbach, J., Kob, W. et al. (2004). Channel formation and intermediate range order in sodium silicate melts and glasses. *Phys. Rev. Lett.* 93: 027801–027801. https://doi.org/10.1103/PhysRevLett.93.027801.

109 Cormack, A.N., Du, J., and Zeitler, T.R. (2003). Sodium ion migration mechanisms in silicate glasses probed by molecular dynamics simulations. *J. Non-Cryst. Solids* 323: 147–154. https://doi.org/10.1016/S0022-3093(03)00280-1.

110 Hahn, S.H., Rimsza, J., Criscenti, L. et al. (2018). Development of a ReaxFF reactive force field for $NaSiO_x$/water systems and its application to sodium and proton self-diffusion. *J. Phys. Chem. C* 122: 19613–19624. https://doi.org/10.1021/acs.jpcc.8b05852.

111 Habasaki, J., Leon, C., and Ngai, K.L. (2017). *Dynamics of Glassy, Crystalline and Liquid Ionic Conductors*. Cham: Springer International Publishing https://doi.org/10.1007/978-3-319-42391-3.

112 Day, D.E. (1976). Mixed alkali glasses – their properties and uses. *J. Non-Cryst. Solids* 21: 343–372. https://doi.org/10.1016/0022-3093(76)90026-0.

113 Charles, R.J. (1965). The mixed alkali effect in glasses. *J. Am. Ceram. Soc.* 48: 432–433. https://doi.org/10.1111/j.1151-2916.1965.tb14784.x.

114 Isard, J.O. (1969). The mixed alkali effect in glass. *J. Non-Cryst. Solids* 1: 235–261. https://doi.org/10.1016/0022-3093(69)90003-9.

115 Maass, P. (1999). Towards a theory for the mixed alkali effect in glasses. *J. Non-Cryst. Solids* 255: 35–46. https://doi.org/10.1016/S0022-3093(99)00422-6.

116 Onodera, Y., Takimoto, Y., Hijiya, H. et al. (2019). Origin of the mixed alkali effect in silicate glass. *NPG Asia Mater.* 11: 75. https://doi.org/10.1038/s41427-019-0180-4.

117 Vessal, B., Greaves, G.N., Marten, P.T. et al. (1992). Cation microsegregation and ionic mobility in mixed alkali glasses. *Nature* 356: 504–506. https://doi.org/10.1038/356504a0.

118 Hiraoka, Y., Nakamura, T., Hirata, A. et al. (2016). Hierarchical structures of amorphous solids characterized by persistent homology. *Proc. Natl. Acad. Sci. U.S.A.* 113 (26): 7035–7040. https://doi.org/10.1073/pnas.1520877113.

119 Wilding, M.C., Benmore, C.J., Tangeman, J.A., and Sampath, S. (2004). Coordination changes in magnesium silicate glasses. *Europhys. Lett.* 67: 212–218. https://doi.org/10.1209/epl/i2003-10286-8.

**120** Hou, D., Lu, C., Zhao, T. et al. (2015). Structural, dynamic and mechanical evolution of water confined in the nanopores of disordered calcium silicate sheets. *Microfluid. Nanofluid.* 19: 1309–1323. https://doi.org/10.1007/s10404-015-1646-5.

**121** Cormack, A.N. and Du, J. (2001). Molecular dynamics simulations of soda–lime–silicate glasses. *J. Non-Cryst. Solids* 293–295: 283–289. https://doi.org/10.1016/S0022-3093(01)00831-6.

**122** Pedone, A. (2009). Properties calculations of silica-based glasses by atomistic simulations techniques: a review. *J. Phys. Chem. C* 113: 20773–20784. https://doi.org/10.1021/jp9071263.

**123** Tilocca, A., De Leeuw, N.H., and Cormack, A.N. (2006). Shell-model molecular dynamics calculations of modified silicate glasses. *Phys. Rev. B Condens. Matter Mater. Phys.* 73: 104209. https://doi.org/10.1103/PhysRevB.73.104209.

**124** Laurent, O., Mantisi, B., and Micoulaut, M. (2014). Structure and topology of soda-lime silicate glasses: implications for window glass. *J. Phys. Chem. B.* 118: 12750–12762. https://doi.org/10.1021/jp506155p.

**125** Kaseman, D.C., Retsinas, A., Kalampounias, A.G. et al. (2015). Q-speciation and network structure evolution in invert calcium silicate glasses. *J. Phys. Chem. B.* 119: 8440–8445. https://doi.org/10.1021/acs.jpcb.5b02469.

**126** Eckersley, M.C., Gaskell, P.H., Barnes, A.C., and Chieux, P. (1988). Structural ordering in a calcium silicate glass. *Nature* 335: 525–527. https://doi.org/10.1038/335525a0.

**127** Tilocca, A. and De Leeuw, N.H. (2006). Structural and electronic properties of modified sodium and soda-lime silicate glasses by Car–Parrinello molecular dynamics. *J. Mater. Chem.* 16: 1950–1955. https://doi.org/10.1039/b517362k.

**128** Tilocca, A. (2010). Sodium migration pathways in multicomponent silicate glasses: Car–Parrinello molecular dynamics simulations. *J. Chem. Phys.* 133: 1–10. https://doi.org/10.1063/1.3456712.

**129** Pedone, A., Malavasi, G., Cristina Menziani, M. et al. (2008). Role of magnesium in soda-lime glasses: insight into structural, transport, and mechanical properties through computer simulations. *J. Phys. Chem. C* 112: 11034–11041. https://doi.org/10.1021/jp8016776.

**130** Son, N.H., Anh, N.H., Kien, P.H. et al. (2020). Topology of $SiO_x$-units and glassy network of magnesium silicate glass under densification: correlation between radial distribution function and bond angle distribution. *Modell. Simul. Mater. Sci. Eng.* 28: 065007. https://doi.org/10.1088/1361‐651X/ab9bb4.

**131** Perrott, K.W. (1977). Surface charge characteristics of amorphous aluminosilicates. *Clays Clay Miner.* 25: 417–421. https://doi.org/10.1346/CCMN.1977.0250607.

**132** Winkler, A., Horbach, J., Kob, W., and Binder, K. (2004). Structure and diffusion in amorphous aluminum silicate: a molecular dynamics computer simulation. *J. Chem. Phys.* 120: 384–393. https://doi.org/10.1063/1.1630562.

**133** Montorsi, M., Menziani, M.C., Leonelli, C., and Cormack, A.N. (1999). The sodium-alumino silicate glasses: a molecular dynamic study. *Mol. Eng.* 8: 427–445. https://doi.org/10.1023/A:1008355407431.

**134** Hennet, L., Drewitt, J.W.E., Neuville, D.R. et al. (2016). Neutron diffraction of calcium aluminosilicate glasses and melts. *J. Non-Cryst. Solids* 451: 89–93. https://doi.org/10.1016/j.jnoncrysol.2016.05.018.

**135** Zirl, D.M. and Garofalini, S.H. (1990). Structure of sodium aluminosilicate glasses. *J. Am. Ceram. Soc.* 73: 2848–2856. https://doi.org/10.1111/j.1151‐2916.1990.tb06685.x.

**136** Ren, M., Cheng, J.Y., Jaccani, S.P. et al. (2019). Composition – structure – property relationships in alkali aluminosilicate glasses: a combined experimental – computational approach towards designing functional glasses. *J. Non-Cryst. Solids* 505: 144–153. https://doi.org/10.1016/j.jnoncrysol.2018.10.053.

**137** Ren, M. and Du, J. (2016). Structural origin of the thermal and diffusion behaviors of lithium aluminosilicate crystal polymorphs and glasses. *J. Am. Ceram. Soc.* 99: 2823–2833. https://doi.org/10.1111/jace.14292.

**138** Li, W. and Garofalini, S.H. (2004). Molecular dynamics simulation of lithium diffusion in $Li_2O$–$Al_2O_3$–$SiO_2$ glasses. *Solid State Ionics* 166: 365–373. https://doi.org/10.1016/j.ssi.2003.11.015.

**139** Jakse, N., Bouhadja, M., Kozaily, J. et al. (2012). Interplay between non-bridging oxygen, triclusters, and fivefold Al coordination in low silica content calcium aluminosilicate melts. *Appl. Phys. Lett.* 101: 201903. https://doi.org/10.1063/1.4766920.

**140** Lu, X. and Du, J. (2020). Quantitative structure-property relationship (QSPR) analysis of calcium aluminosilicate glasses based on molecular dynamics simulations. *J. Non-Cryst. Solids* 530: 119772. https://doi.org/10.1016/j.jnoncrysol.2019.119772.

**141** Corrales, L.R. and Du, J. (2006). Characterization of ion distributions near the surface of sodium-containing and sodium-depleted calcium aluminosilicate melts. *J. Am. Ceram. Soc.* 89: 36–41. https://doi.org/10.1111/j.1551-2916.2005.00695.x.

**142** Beckers, J.V.L. and De Leeuw, S.W. (2000). Molecular dynamics simulation of nanoporous silica. *J. Non-Cryst. Solids* 261: 87–100. https://doi.org/10.1016/S0022-3093(99)00607-9.

**143** Provis, J.L., Duxson, P., Van Deventer, J.S.J., and Lukey, G.C. (2005). The role of mathematical modelling and gel chemistry in advancing geopolymer technology. *Chem. Eng. Res. Des.* 83: 853–860. https://doi.org/10.1205/cherd.04329.

**144** Bhattacharya, S. and Kieffer, J. (2005). Fractal dimensions of silica gels generated using reactive molecular dynamics simulations. *J. Chem. Phys.* 122: https://doi.org/10.1063/1.1857522.

**145** Lattuada, M., Wu, H., Hasmy, A., and Morbidelli, M. (2003). Estimation of fractal dimension in colloidal gels. *Langmuir* 19: 6312–6316. https://doi.org/10.1021/la034043f.

**146** Connolly, M.L. (1983). Solvent-accessible surfaces of proteins and nucleic acids. *Science* 221: 709–713.

**147** Kieffer, J. and Angell, C.A. (1989). Generation of fractal silicas by negative pressure rupturing of $SiO_2$ glass. *Mol. Simul.* 3: 137–154. https://doi.org/10.1080/08927028908034624.

**148** Habasaki, J. and Ishikawa, M. (2014). Molecular dynamics study of coagulation in silica–nanocolloid–water–NaCl systems based on the atomistic model. *Phys. Chem. Chem. Phys.* 16: 24000–24017. https://doi.org/10.1039/C4CP02984D.

**149** Dupuis, R., Béland, L.K., and Pellenq, R.J.-M. (2019). Molecular simulation of silica gels: formation, dilution, and drying. *Phys. Rev. Mater.* 3: 075603. https://doi.org/10.1103/PhysRevMaterials.3.075603.

**150** Du, T., Li, H., Sant, G., and Bauchy, M. (2018). New insights into the sol-gel condensation of silica by reactive molecular dynamics simulations. *J. Chem. Phys.* 148: https://doi.org/10.1063/1.5027583.

**151** Thiruvilla, M.S., Du, J., Mahadevan, T.S., and Du, J. (2021). Atomic and micro-structure features of nanoporous aluminosilicate glasses from reactive molecular dynamics simulations. *J. Am. Ceram. Soc.* 104: 229–242. https://doi.org/10.1111/jace.17465.

**152** Rimsza, J.M. and Du, J. (2014). Structural and mechanical properties of nanoporous silica. *J. Am. Ceram. Soc.* 97: 772–781. https://doi.org/10.1111/jace.12707.

**153** Gelb, L.D. and Gubbins, K.E. (1999). Pore size distributions in porous glasses: a computer simulation study. *Langmuir* 15: 305–308. https://doi.org/10.1021/la9808418.

**154** Bhattacharya, S. and Gubbins, K.E. (2006). Fast method for computing pore size distributions of model materials. *Langmuir* 22: 7726–7731. https://doi.org/10.1021/la052651k.

**155** Ingham, B., Li, H., Allen, E.L., and Toney, M.F. (2011). SAXSMorph: a program for generating representative morphologies for two-phase materials from small-angle X-ray and neutron scattering data. *J. Appl. Crystallogr.* 44: 221–224. https://doi.org/10.1107/S0021889810048557.

**156** Iftimie, R., Minary, P., and Tuckerman, M.E. (2005). Ab initio molecular dynamics: concepts, recent developments, and future trends. *Proc. Natl. Acad. Sci. U.S.A.* 102: 6654–6659. https://doi.org/10.1073/pnas.0500193102.

**157** Frenkel, D. and Smit, B. (2002). *Understanding Molecular Simulation*, 2e. Academic Press. ISBN: 9780122673511 https://doi.org/10.1016/b978-0-12-267351-1.x5000-7.

**158** Benoit, M., Ispas, S., Jund, P., and Jullien, R. (2000). Model of silica glass from combined classical and ab initio molecular-dynamics simulations. *Eur. Phys. J. B* 13: 631–636. https://doi.org/10.1007/s100510050079.

**159** Tilocca, A. and Cormack, A.N. (2009). Modeling the water-bioglass interface by ab initio molecular dynamics simulations. *ACS Appl. Mater. Interfaces* 1: 1324–1333. https://doi.org/10.1021/am900198t.

**160** Cheng, H.P., Barnett, R.N., and Landman, U. (2002). Structure, collective hydrogen transfer, and formation of $Si(OH)_4$ in $SiO_2$–$(H_2O)_n$ clusters. *J. Chem. Phys.* 116: 9300–9304. https://doi.org/10.1063/1.1473808.

**161** Tilocca, A. (2008). Short- and medium-range structure of multicomponent bioactive glasses and melts: an assessment of the performances of shell-model and rigid-ion potentials. *J. Chem. Phys.* 129: https://doi.org/10.1063/1.2972146.

**162** Ispas, S., Benoit, M., Jund, P., and Jullien, R. (2001). Structural and electronic properties of the sodium tetrasilicate glass (formula presented) from classical and ab initio molecular dynamics simulations. *Phys. Rev. B Condens. Matter Mater. Phys.* 64: 18–21. https://doi.org/10.1103/PhysRevB.64.214206.

**163** Ganster, P., Benoit, M., Delaye, J.M., and Kob, W. (2007). Structural and vibrational properties of a calcium aluminosilicate glass: classical force-fields vs. first-principles. *Mol. Simul.* 33: 1093–1103. https://doi.org/10.1080/08927020701541006.

**164** Wright, A.C. and Thorpe, M.F. (2013). Eighty years of random networks. *Phys. Status Solidi Basic Res.* 250: 931–936. https://doi.org/10.1002/pssb.201248500.

**165** Greaves, G.N. and Ngai, K.L. (1995). Reconciling ionic-transport properties with atomic structure in oxide glasses. *Phys. Rev. B.* 52: 6358–6380. https://doi.org/10.1103/PhysRevB.52.6358.

**166** Mauro, J.C. (2011). Topological constraint theory. *Am. Ceram. Soc. Bull.* 90: 31–37. https://www.scopus.com/record/display.uri?eid=2-s2.0-79955691282&origin=inward&txGid=b60345b7b7133138085cd25c07ac0972.

**167** Lu, X., Deng, L., Gin, S., and Du, J. (2019). Quantitative structure – property relationship (QSPR) analysis of $ZrO_2$ – containing soda-lime borosilicate glasses. *J. Phys. Chem. B* 123 (6): 1412–1422. https://doi.org/10.1021/acs.jpcb.8b11108.

**168** Pedone, A. and Menziani, M.C. (2015). Computational modeling of silicate glasses: a quantitative structure-property relationship perspective. *Springer Ser. Mater. Sci.* 215: 113–135. https://doi.org/10.1007/978-3-319-15675-0_5.

**169** Du, J., Lu, X., Gin, S. et al. (2021). Predicting the dissolution rate of borosilicate glasses using QSPR analysis based on molecular dynamics simulations. *J. Am. Ceram. Soc.* jace.17857. https://doi.org/10.1111/jace.17857.

# 8

# Borosilicate and Boroaluminosilicate Glasses

*Lu Deng and Jincheng Du*

*University of North Texas, Department of Materials Science and Engineering, Denton, Texas, USA*

## 8.1 Introduction

Borosilicate glass is an archetype of various multicomponent oxide glasses with both industrial and technological interests, ranging from cookware, labware, to optical lenses, bioactive coating, and nuclear disposal [1–3]. Their properties, especially the low thermal expansion coefficient and high chemical resistance, enable the applications ranging from our daily lives to those in extreme conditions [4]. Many properties of borate and borosilicate glasses follow a trend of nonlinear change: first increase and then decrease with additional network modifier. This so-called "boron" anomaly phenomenon was initially observed in binary borate glass for structure features and properties such as boron coordination in relation to thermal expansion, density, infrared absorption, and thermal conductivity [5–7]. Later experiments found the boron anomaly also exist in borosilicate and boroaluminosilicate glass systems and even more complicated boron-containing multicomponent oxide glasses, while properties that show non-linear behaviors include ionic diffusion, heat capacity, hardness, and glass transition temperature [1, 8–10]. The origin of this phenomenon is found to be directly correlated with the boron coordination change, particularly the fraction of four-coordinated boron ($N_4$), when composition changes [11]. In addition, superstructural units which consist of multiple boron oxide units, for instance, the boroxol ring, diborate, triborate, and pentaborate groups also contribute to the complexity of the structures of these glasses [11].

Although intensive experimental studies have contributed to the understanding of silicate and borate glasses, their structures and properties are still far from being fully understood. Computer simulations, especially molecular dynamics (MD)-based atomistic simulations, have become a highly valuable tool to investigate the short- and medium-range structural characteristics of these and other glass systems. Both ab initio molecular dynamics (AIMD) and classical molecular dynamics (classical MD) simulations have been employed to study $B_2O_3$ containing glasses. The AIMD methods have higher accuracy than classical MD simulations, as the interatomic interactions are calculated from first principles with various levels of approximations. However, the AIMD simulations of glass systems can only be practically

*Atomistic Simulations of Glasses: Fundamentals and Applications*, First Edition.
Edited by Jincheng Du and Alastair N. Cormack.

performed on a relatively small system (commonly with a few hundreds of atoms) and the cooling rate used for glass-forming process is quite high ($10^{13}$–$10^{15}$ K/s [12, 13]), due to the high computational cost. On the other hand, simulations using classical MD methods can easily simulate large systems from tens of thousands to millions of atoms, and the cooling rate can be orders of magnitude slower than AIMD. Classical MD method plays a more important role in simulating structures of multicomponent silicate and borosilicate glasses due to its higher computational efficiency. To open the door of microscopic world of glass materials using classical MD, availability of effective and transferable interatomic potentials is the key. Particularly, this is true for boron oxide containing glasses that have much fewer available potentials due to the structural complexities mentioned above. Accurate description of boron coordination state, more specifically, the fraction of four-coordinated boron (also known as the $N_4$ value), is one of the most important challenges in simulations of boron-containing oxide glasses. Fortunately, recent active developments of effective interatomic potentials have made classical MD simulations of borate, borosilicate, and more complex multicomponent silicate glasses possible. After reviewing the experimental determination and various theoretical models of boron $N_4$, AIMD and classical MD simulations borosilicate glasses are presented. Particularly, recent development of various interatomic potentials are introduced. Then the applications of MD simulations to borosilicate, boroaluminosilicate, and boron containing multicomponent glasses are presented. A collection of the available interatomic potential forms and parameters are shown in the Appendix.

## 8.2 Experimental Determination and Theoretical Models of Boron $N_4$ Values in Borosilicate Glass

### 8.2.1 Experimental Results on Boron Coordination Number

The property change along with composition in boron oxide-containing glasses has been related to boron coordination, as represented by the $N_4$ value or the fraction of fourfold coordinated boron [14–16], and boron $N_4$ is commonly considered as a characteristic structural indicator of borate and borosilicate glasses. Bray and his coworkers are the pioneers in investigating the boron coordination number using $^{11}B$ nuclear magnetic resonance (NMR) technic dating back to late 1950s [17]. Systematic studies on various borate and borosilicate glass compositions have been performed, and four-coordinated boron numbers are obtained from the NMR spectra [5, 18–20]. Yun and Bray performed one of the earliest $^{11}B$ solid-state NMR study to obtain boron $N_4$ for the sodium borosilicate ternary glass systems [18]. In this work, $N_4$ values of 36 ternary glass compositions were investigated within the experimental glass-forming region of the $Na_2O$–$B_2O_3$–$SiO_2$ ternary system [21]. In addition to the measurement of the $N_4$ values, fourfold boron units are found to be preferred to be associated with the silicon units, while the threefold boron units tend to form the boroxol rings or to be randomly mixed with other glass former units [22]. Fourfold and nonring threefold boron units tend to mix with the silicon tetrahedra, while fourfold boron units prefer to associate with silicon tetrahedra, then threefold boron units, last the fourfold boron groups, instead of fully boron avoidance [22, 23]. It was also found that boron coordination conversion, from three to four, is more favorable than the formation of nonbridging oxygen (NBO) on silicon tetrahedral units in borosilicate glasses with high boron oxide concentration, while the latter one is preferred in the borosilicate glasses with high silica content due to the enthalpic effect [1].

The addition of other glass former oxides such as alumina and phosphorus oxide in multicomponent boron-containing glasses, and the situation of boron coordination change with composition can be more complicated. For example, the addition of alumina in the boroaluminosilicate glasses makes the trend of $N_4$ value change with composition quite different from the one in the borosilicate systems [24]. That is, the sodium content in the composition with the maximum of the $N_4$ values ($N_{4max}$) in the boroaluminosilicate glasses is much higher than the one in borosilicate glasses, which suggests that sodium ions are first used to stabilize and charge-compensate fourfold coordinated aluminum, then the remaining sodium is used for the boron coordination conversion. In addition, the cation field strength (related to the charge and ionic radius) difference of the modifier cations is found to have significant effects on both mixing behavior of formers and the stabilization of NBOs [25]. Similar to the ternary systems, certain level of the avoidance of both boron and aluminum oxygen tetrahedra was noticed and shown by the higher $N_4$ values in higher silica content glasses due to the avoidance [25].

## 8.2.2 Theoretical Models in Predicting Boron $N_4$ Value

Based on $^{11}$B NMR determination of boron coordination in wide ranges of glasses, several theoretical models have been proposed to describe or predict the $N_4$ values from the glass compositions.

### 8.2.2.1 The Yun Bray Model and Variations

Based on the boron coordination with composition determined from $^{11}$B NMR, several theoretical models which enable direct calculation of the $N_4$ value from glass compositions have been proposed for borosilicate glasses since the first work by Yun and Bray in 1978 [18].

The initial model proposed by Yun and Bray [18] (the Y&B model) for the $Na_2O–B_2O_3–SiO_2$ ternary systems are based on two composition ratios: $R$ ($[Na_2O]/[B_2O_3]$) and $K$ ($[SiO_2]/[B_2O_3]$) to represent the glass composition. Models with usage of the super structural units containing boron and/or silicon were employed to fit the experimental $N_4$ values. Two composition regions are defined based on the silica to boron oxide concentration ratio $K$, e.g. $K \geq 8$ and $K \leq 8$. For each region, it also consists of two stages with different trends of the $N_4$ value change separated by $R = R_{max}$. The value of $R_{max}$ is defined as the sodium oxide to boron oxide concentration ratio $R$ of the composition with the maximum of the $N_4$ value, and its value is equal to $K/16 + 0.5$. For the glass compositions with $R \leq R_{max}$, the $N_4$ value are equal to $R$. On the other hand, two different equations are used to predict the $N_4$ values for different $K$ regions with $R \geq R_{max}$: for the compositions with $K \geq 8$, the estimated $N_4$ is equal to 1, while the predicted $N_4$ value of the compositions with $K \leq 8$ can be calculated through the following equation:

$$N_4 = N_{4max} - \frac{0.25}{1+K}(R - R_{max}) \tag{8.1}$$

where $N_{4max}$ is the maximum of the $N_4$ value with same $K$ value. In later work by Yun, Feller, and Bray [19], this model is improved for the compositions with $K \leq 8$, since the diborate units are destroyed to form other super structures, including metaborate and loose units.

This two-stage model provides a simple way to calculate the $N_4$ value directly from the glass compositions without other input information from experiments. However, this model is limited in the composition region with relatively high sodium content (e.g. $R > 4$) due to the limited data of this region.

Dell, Bray, and Xiao [20] later developed another model (the DBX model) for sodium borosilicate glass systems based on the NMR data from Yun and Bray's work [18] as well as later added ones for high sodium content region. The DBX model for compositions with high silicon content ($K \geq 8$) is same to the one in the Y&B model, while for the low silicon compositions ($K \leq 8$), it consists of three stages. The $N_4$ value predicted by this model shows a linear increase when $R$ is smaller than $R_{max}$, and the $N_4$ value is equal to the $R$ value in this region. Then it reaches a plateau until $R = R_{d1}(K/4 + 0.5)$. The last stage in this model is a linear decrease starting from $R_{d1}$ to $R = K + 2$. The $N_4$ value in this stage can be calculated by the following equation:

$$N_4 = \frac{8+K}{12} - R \times \frac{8+K}{12(2+K)} \tag{8.2}$$

This model not only solves the issue in high sodium content region but also gives clear assumptions on the possible reactions processed in each stage. Hence, the DBX model is widely used in calculating theoretical $N_4$ values of various borosilicate glass systems.

In addition to the original DBX model, a modification was proposed by Manara et al. [26] to take both the re-edmergnerite and danburite-like units formation into consideration. The $R_{max}$ value is then adjusted to $R^* = 0.5 + \frac{K}{2N}$, where the $N$ value represents for the average number of the $SiO_2$ units in the borosilicate super units (either reedmergnerite or danburite-like unit) and can be any number between five and six based on the portion of the super units.

#### 8.2.2.2 The Du and Stebbins Model

Systematic studies using $^{11}B$ NMR technic have also been performed on boroaluminosilicate glass and the $N_4$ values are obtained for these systems [22–24, 27]. However, additional alumina to the borosilicate led to the failure of the DBX model which was originally proposed for the sodium borosilicate ternary systems. For instance, the fraction of four-coordinated boron in the boroaluminosilicate glass is overestimated by the DBX model [24]. Du and Stebbins therefore proposed a modified model [25] (the D&S model) based on available NMR data of potassium and sodium boroaluminosilicate systems as well as their own data. The two composition ratios in the DBX model, $R$ and $K$, are modified based on the experimental observations that the four-coordinated aluminum has similar network behavior to the four-coordinated boron units. The new ratios are thus defined as $R'([Na_2O]/([B_2O_3]+[Al_2O_3]))$ and $K'([SiO_2]/([B_2O_3]+[Al_2O_3]))$. As a result, the total four-coordinated boron and aluminum percentage ($N_4'$) can be predicted by using the DBX model with $R'$ and $K'$. The fraction of four-coordinated boron then can be derived from the following equation:

$$N_4 = \frac{N_4' \times (B_2O_3 + Al_2O_3) - Al_2O_3}{B_2O_3} \tag{8.3}$$

With this model, the fraction of four-coordinated boron in boroaluminosilicate glass can be estimated, and the values are in good agreement with experimental NMR data.

#### 8.2.2.3 The Two-State Model and Extension

For glass compositions with high silicon content ($K \geq 8$), the $N_4$ values predicted by both Y&B and DBX models are equal to 1, which are not in accordance with what observed in experiments. Therefore, Smedskjaer et al. [1] proposed another model for $Na_2O$–CaO–$B_2O_3$–$SiO_2$ glass systems based on $^{11}$B and $^{29}$Si magic angle spinning (MAS) NMR experimental results. This so-called "two-state statistical mechanical model" can describe the relationship between the boron coordination change and the thermal history, by employing additional parameters such as fictive temperature and enthalpy differences of boron coordination sates (detailed below). A competition between the boron coordination conversion and formation of silicon NBOs has been observed from NMR results; in addition, NBOs associating with silicon tetrahedra are found to be more favorable than [$BO_3$] units. Hence, the enthalpy difference $\Delta H$ between the boron coordination conversion (from three- to four-coordinated) and NBO formation energy on silicon tetrahedra is employed to calculate the $N_4$ values. For the compositions with a formula of $x$Na$_2$O–$y$CaO–$z$B$_2$O$_3$–$(1-x-y-z)$SiO$_2$, the $N_4$ value can be obtained from the following equation:

$$N_4 = \left( \min \left[ \frac{2(x+y)z}{z + (1-x-y-z)\exp[-\Delta H/kT_f]}, z + \frac{2}{3}(1-x-y-z), 2z \right] \right) /2z \tag{8.4}$$

where $T_f$ is the fictive temperature at which the glass structure is frozen, and it can be taken as the glass transition temperature ($T_g$) for a regular cooled glass. This model was used to predict the liquid fragility index and the hardness of these glass compositions at room temperature, as well as the $T_g$ value from the value of a reference composition with low modifier content. However, this model was not applied to compositions with high modifier content, and it needs input of experimental $T_g$ value.

#### 8.2.2.4 Lu, Deng, Du, and Vienna Model

Although theoretical models have been developed for the borate, borosilicate, and boroaluminosilicate glass systems, there is still a lack of a model which can predict the $N_4$ values for multicomponent boron-containing oxide glasses. Most recently, Lu et al. [28] proposed and compared several sets of formula-based and machine learning-based models. A dataset with 657 compositions from literature was used to obtain the models. Two functional-formula based models, modified D&S model, and modified Bernstein model, have been proposed; moreover, three machine learning-based models, namely the $K$-nearest neighbor, artificial neural network, and Gaussian process regression were developed. The results show that models using functional forms can provide quite accurate prediction of the $N_4$ values in the whole selected composition range, with the $R^2$ value of the linear fitting higher than 0.83 (results of modified Bernstein model is shown in Figure 8.1). On the other hand, the machine learning-based models can achieve even higher accuracy (around 0.91), but the physical meaning of the relationship between $N_4$ and the composition is not clear.

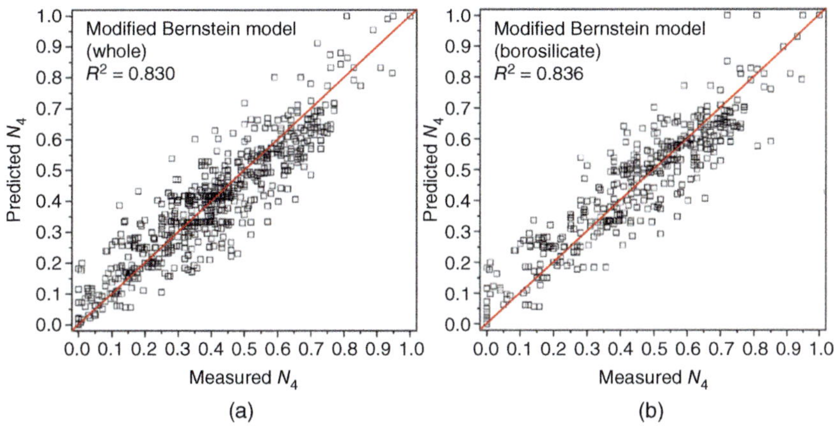


**Figure 8.1** Predicted $N_4$ values using modified Bernstein model versus experimental values using (a) the whole dataset and (b) only borosilicate glasses, respectively. Source: Reproduced from Lu et al. [28] with permission.

## 8.3 Ab Initio Versus Classical MD Simulations of Borosilicate Glasses

The AIMD simulations, also called first principles molecular dynamics, is a computational chemistry method to study various systems based on interatomic forces calculated from quantum mechanics with different levels of approximations. This method was developed in early 1950s and can produce more accurate structures and properties than the empirical or semiempirical methods (more details of AIMD simulations can be found in Chapter 2); however, AIMD simulations are limited to relatively small system size and short time-scale due to the high computational costs. Therefore, this poses some challenges in simulations of some glass structures because some glass systems require large enough number of atoms to fully describe the medium-range order structural features, and the melt-quench process of glass formation requires as long as possible of simulation time. At the earlier time of the AIMD simulations on glass, the glass structures were usually generated through the classical MD methods with relatively small system size and then relaxed with AIMD at the final temperature. For example, Geneste et al. [29] studied the hydrogen-sodium interdiffusion behavior in borosilicate glass system from first principles. Ion exchange, hydronium dissociation, relative acidity of anionic sites, and other structural changes caused by the interdiffusion were investigated. However, the glass structure was generated by classical MD simulations, and the system was with only 200 atoms. Similar method was used later by Soleilhavoup et al. [30] that multinuclear NMR analysis of borosilicate glass were performed; however, the system size was still 200 atoms, and the glass structure was also created by classical MD simulations. With recent development of the computer science, the calculation ability of the super computational clusters has been significantly increased, and the AIMD simulations of glass systems become possible. Pedesseau et al. [12, 31] studied both liquid and glass states of a sodium borosilicate system fully using AIMD simulation method. The system size of the simulated configurations is 320 atoms, and the whole melt-quench process of the glass formation is simulated using AIMD simulation. However, the cooling rate used in this study (around $10^{15}$ K/s) is orders of magnitude higher than those normally used in classical MD simulations (around $10^{12}$ K/s) due to the efficiency consideration. Various structural features and properties such as coordination number, bond distance, bond angle distribution, sodium distribution, vibrational density of states, and dielectric constants were obtained. Due to the high accuracy, this kind of work which fully use the AIMD simulation method to generate glass structures that can be useful for classical MD simulations to compare with, especially for those features which are difficult to obtain through experiments. However, more powerful computers are needed for the glass simulations which need both large system size and long computational time for a melt-cooling process.

Classical MD simulations are a method of computational modeling to investigate the dynamic evolution of the atoms and molecules (please refer to Chapter 1 of this book for more details of classical MD simulation methods.). The movements of the particles in the simulation box are determined by solving the equations of motion following Newton's Law, and the force and potential energy between the particle pairs are described by a form of force field usually called interatomic potentials. As a critical part of the classical MD simulations, the interatomic potential sets describe the interactions between ions or atoms in the systems; in another word, it is the key to open the door of microscopic world of glasses. Most of the potential sets are obtained through a huge number of fittings to the experimental structural features and properties for several compositions, thus these potential sets are also called empirical potential sets. Structures and properties of other compositions can be predicted by using the developed potential sets and are expected to agree well with the experimental ones. Potential sets for various oxide glass systems such as vitreous silica, alkali and alkali earth silicate, and aluminosilicates have been proposed, and structures and properties are well studied using the classical MD simulations [32–35]. However, for the boron-containing oxide glasses, the addition of boron oxide induces structural units such as boroxol rings, three- to four-coordinated boron conversion and other large units involving boron. Addition of oxides such as $Al_2O_3$ and $ZrO_2$ can also lead to other structural complexities such as coordination change and modifier-former preference. These complex structural features lead to a challenge to the potential fitting process and are important for evaluating a potential set developed for these glass systems. With the latest works, several potential sets for boron-containing multicomponent oxide glass systems have been developed.

## 8.4 Empirical Potentials for Borate and Borosilicate Glasses

Dating back to 1960s, the rigid ion potentials, also known as the rigid ion models and based the Born model of solids, were first used to simulate the structures alkali halides, and later this kind of potentials were proved to be successful in the simulations of silica and silicate glasses as well [36, 37]. The pioneer works on MD simulations of borate glass using rigid ion potentials were also conducted by Soules [38], Inoue et al. [39], and Xu et al. [40]. Later, Cormack and Park [41] proposed a potential which allows the coordination change of boron for vitreous $B_2O_3$, alkali borates, and borosilicate glass systems. In this

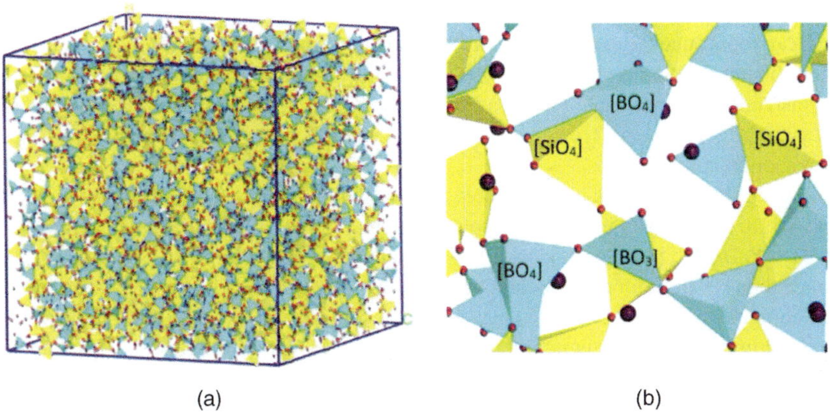


Figure 8.2 Snapshot of the MD simulated glass structure of SBN-B130 glass using Deng and Du potential (a) full image of the simulation cell with 10 000 atoms and (b) a zoomed in snapshot showing $[SiO_4]$ (yellow), $[BO_x]$ (cyan), oxygen (pink), and sodium (purple). Source: Reproduced from Tuheen et al. [42] with permission.

section, we will introduce several recently developed borate and borosilicate potentials that enabled MD simulations of a wide range of borate-containing glasses (Figure 8.2). We will first introduce several rigid ion potentials, either with fixed charge/parameters or with composition-dependent parameters. In addition, polarizable potentials for borosilicate glasses are also introduced. Advantages and potential shortcomings of each of the potentials will be commented.

### 8.4.1 Recent Development of Rigid Ion Potentials for Borosilicate Glasses

Due to various applications in both industrial and technological areas, there are renewed interests in understanding the structure and structure–property relations of borosilicate glasses and boron-containing multicomponent oxide glasses by using MD simulations in the past decade. In this period, several new empirical potentials for these systems have been developed. These developments are reviewed in this section. In addition, to help readers to understand and use these newly developed potentials, several well tested parameters for boron oxide containing glass systems are provided as supplemental material of this chapter.

Kieu et al. [43] successfully developed a pairwise composition-dependent potential set based on an early work of Guillot and Sator [35]. Two composition ratios, $R$ and $K$, are employed which made the charge and boron parameters in the potential composition-dependent. The main potential form includes a Coulombic long-range interaction term and a Buckingham short-range interaction term:

$$\phi(r_{ij}) = \frac{q_i q_j}{r_{ij}} + A_{ij} \exp\left(-\frac{r_{ij}}{\rho_{ij}}\right) - \frac{C_{ij}}{r_{ij}^6} \tag{8.5}$$

where the $A_{ij}$, $\rho_{ij}$, and $C_{ij}$ values are empirical parameters, $r_{ij}$ is the interatomic distance, and $q_i$ and $q_j$ are the charges of atoms $i$ and $j$, respectively. An important feature of this potential set is that the charges of each element are composition-dependent values. Details of the description of the potential and parameters can be found in the Appendix 8.A. This potential set can reproduce the experimental or theoretical predicted boron coordination number for certain composition region ($R < \frac{K}{4} + 0.5$). In addition, other structure features such as bond distance and bond angle as well as the mechanical properties are also in good agreement with the experiments. However, the results need to be improved for compositions with higher sodium content [14].

Following this framework and to enable the simulations of boroaluminosilicate glasses, Deng and Du [14] further extended the potential to include the alumina-related parameters. In this work, Du and Stebbins model [25] was used to replace the Yun and Bray model [18] to estimate the $N_4$ values for a given composition. Subsequent modifications have been made to the equations which to calculate element charge and parameters of B–O interaction for the boroaluminosilicate systems. Details of the equations, modifications, and parameters can be found in the Appendix 8.A. The results, including structure features, mechanical properties, and vibrational density of states, show good agreement with the available results from experiments, theoretical model, as well as ab initio calculations. However, the nature of the framework limited the potential in predicting accurate $N_4$ values for compositions with higher sodium content.

Wang et al. [44] also developed a potential set based on the original Guillot and Sator potential [35] with their additional fitting of boron-related parameters. However, they chose to use fixed charge and fixed parameters, as compared to those composition-dependent strategy used by Kieu et al. [43] and Deng and Du [14] mentioned above, with an aim to improve transferability of the potential for various compositions. Details of the parameterization of the potential are discussed in the Appendix 8.A. Shear viscosity and structure features, e.g. density, coordination number, and neutron structure factor, have been calculated and are in good agreement with experimental data. The fitted parameters enable the simulations of Na–Ca–B–Si–O systems, and this potential can be easily extended with the original Guillot–Sator parameters [35]. With constant values of boron-related parameters, this potential has fewer fitting parameters when adding other elements. This feature makes the framework more transferable compared with the Kieu's framework; however, the lack of composition dependent makes this potential fail to predict the $N_4$ value, thus, other related structure features and properties in a wide range of compositions [42].

Inoue et al. [45] proposed a simple two-body potential set using the Born–Mayer type form:

$$\Phi_{ij}(r_{ij}) = \frac{e^2}{4\pi\varepsilon_0}\frac{Z_i Z_j}{r_{ij}} + B_{ij}\exp\left(-\frac{r_{ij}}{\rho_{ij}}\right) \tag{8.6}$$

where the $\Phi_{ij}$ represents the interaction energy between ions $i$ and $j$, $e$ is the charge of an electron, $\varepsilon_0$ is the vacuum permittivity, $r_{ij}$ is the interatomic distance, and $\rho_{ij}$ is the parameter to describe softness. The $Z_i$ and $Z_j$ are effective charge values of ions $i$ and $j$, respectively. The $B_{ij}$ value is determined at a minimum value of the distance between ions $i$ and $j$. Details of the parameterization of this potential are also included in the Appendix 8.A. Systematical studies on sodium borosilicate glass systems with a wide composition range were performed. The boron coordination number change as a function of $R$ value at different $K$ values were investigated and the results agree well with theoretical model. The relationship between fraction of NBO and the compositions also agrees well with the theoretical predicted one.

Peaks in the silicon $Q_n$ species distribution curves have similar shape comparing with the experimental ones; however, the intensities of the peaks are different. Furthermore, it was found that although the potential works for constant volume simulations, there are issues in constant pressure simulations.

Ha and Garofalini [46] proposed a potential set which also uses a form combining the Born–Mayer–Huggins (BMH) type two-body interaction term but with additional three-body terms to regulate bond angles. The BMH term is described by the following equation:

$$V_{ij}^{BMH} = A_{ij} \exp\left(\frac{-r_{ij}}{\rho_{ij}}\right) + \left(\frac{z_i z_j e^2}{r_{ij}}\right) \operatorname{erfc}\left(\frac{r_{ij}}{\beta_{ij}}\right) \tag{8.7}$$

where $A_{ij}$, $\beta_{ij}$, and $\rho_{ij}$ values are empirical parameters for pair $i$–$j$, $z_i$ and $z_j$ are the full charge of the ions $i$ and $j$, respectively. $e$ is the charge of single electron and $r_{ij}$ value is the interatomic distance. The error function used in the Coulomb term is zero when the interatomic distance is larger than the selected cutoff. Additional angular constrains for Si/Al/B–O–Si/Al/B and O–Si/B–O are applied to better define the angles and thus local structures. Details of the potential are discussed in the Appendix 8.A. This potential set is tested in alumina, calcium aluminosilicate, and soda lime aluminoborosilicate glass systems, and many of the experimental structure features such as the bond distance and silicon coordination numbers can be reproduced. However, boron and aluminum coordination numbers need further improvement. In addition, the potential of the three-body terms for certain triplet angles can well predict some of the glass structure features; however, due to the high constrains induced by the three-body terms, several other features might be quite different from the experimental values. Employing three-body terms may also cause increase the computational cost, although this is a lesser concern as the computational capability increases.

Du and Cormack [33, 47, 48] tested and refined a set of pairwise empirical potential which employs partial charge to describe the partial covalence bond between Si and O that was originally proposed by D. M. Teter. The potential energy is described by a formula combining both the Coulomb term for long-range interaction and the Buckingham potential term for the short-range interaction:

$$V(r_{ij}) = \frac{Z_i Z_j e^2}{4\pi\varepsilon_0 r_{ij}} + A_{ij} \exp(-r_{ij}/\rho_{ij}) - C_{ij}/r_{ij}^{\ 6} \tag{8.8}$$

where $Z_i$ and $Z_j$ are the reduced charges for atoms $i$ and $j$, respectively. These are also called partial charges which are employed to model the partial covalent bonds. $e$ is the charge of a single electron, $\varepsilon_0$ is the vacuum permittivity, and $r_{ij}$ is the interatomic distance between atom $i$ and $j$. The $A_{ij}$, $\rho_{ij}$, and $C_{ij}$ values are empirical parameters for Buckingham term. The parameters for interactions of Al–O, Ca–O, P–O, and rare-earth ion-related pairs were proposed by Du and coworkers [49–51]. This potential sets are well tested in various systems such as silicate, aluminosilicate, and phosphate glass systems. Structure features such as pair distribution function, bond angle distribution function, coordination number, and ring size distribution are well studied and agree well with experimental findings. Properties such as vibrational density of states, mechanical properties, and diffusion coefficient

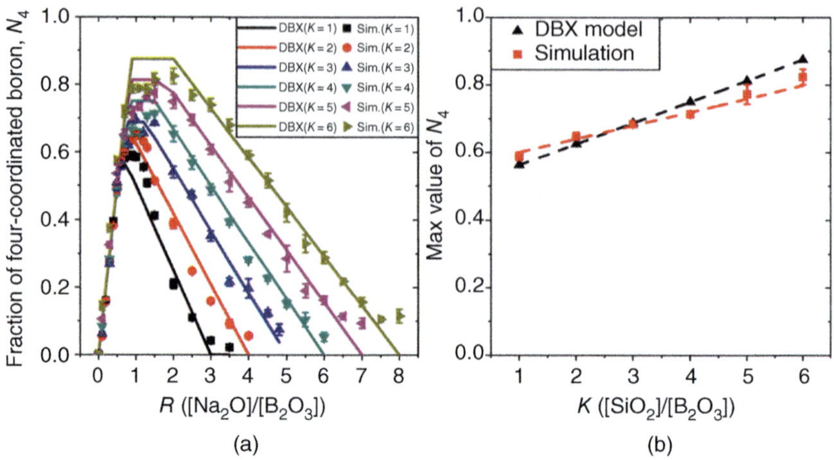


**Figure 8.3** (a) Comparison of four-coordinated boron in sodium borosilicate glasses as function of $R$ with different $K$ values from DBX model and simulation, and (b) the max values of $N_4$ with different $K$ values from DBX model and simulation. The dashed lines in (b) are from linear fitting. Source: Reproduced from Deng and Du [52] with permission.

are also well predicted. Deng and Du [52] recently added boron-related parameters to the framework which enables the simulations of the multicomponent boron-containing oxide glasses including bioactive glass [15] and nuclear waste glass [53]. This potential is also composition-dependent which is similar to the Kieu's framework; however, the $A_{B\text{–}O}$ parameter is the only parameter that changes with composition in this potential. This feature allows the potential to maintain the composition-dependent feature, which is essential to predict boron $N_4$ values in wide composition range (Figure 8.3). On the other hand, the fitting parameters are significantly reduced compared to the Kieu's framework, since only one parameter is composition-dependent and only cation–anion interaction parameters are required to the fitting process. In addition, this potential set is relatively simple since only the two-body term is employed; therefore, it has high computational efficiency. Details of the determination and parameterization of the potential, especially for the B–O interaction, are discussed in the Appendix 8.A. The potential has been tested in various crystal structures, borate glass, borosilicate glasses as well as multicomponent boron-containing oxide glasses. Systematic comparisons of density, coordination number, $Q_n$ species, oxygen species, as well as the mechanical properties between experiments and simulation have been performed and the results are in good agreement [15, 52–56].

Most recently, Sundararaman et al. proposed a set of potentials for alkali and alkaline-earth aluminosilicate glasses [57], and later it is extended to alkali borates, boroaluminate, and borosilicate glasses by adding boron-related parameters [58]. This potential was fitted to the radial distribution function of the liquid at high temperature from AIMD simulations and the density and elastic modulus of glass at room temperature from experiments. The main potential form consists of a short-range interaction described by Buckingham term and the long-range Coulombic interaction using the Wolf truncation method:

$$V^{Buck}(r_{\alpha\beta}) = A_{\alpha\beta}\exp(-B_{\alpha\beta}r_{\alpha\beta}) - \frac{C_{\alpha\beta}}{r_{\alpha\beta}^{6}} + \frac{D_{\alpha\beta}}{r_{\alpha\beta}^{24}} + V^{W}(r_{\alpha\beta}) \tag{8.9}$$

in which $\alpha,\beta \in \{O, Si, Li, Na, K, Al, B\}$, and $V^{W}(r_{\alpha\beta})$ is the Wolf truncation term. Both cation–anion and cation–cation interactions are taken into consideration. Details of the potential description and parameters are discussed in the Appendix 8.A. This potential can reproduce structure features and properties from experiments and AIMD calculations in the tested compositions in terms of density, structure factor, coordination number, vibrational density of states, and mechanical properties. The fitting to high-temperature AIMD results and room temperature experimental data ensures this potential to be valid in whole temperature range. However, the additional cation–cation interactions lead to more parameters for both the fitting process and the simulations, which could limit its transferability to more complicated systems.

### 8.4.2 Development of Polarizable Potentials for Borate and Borosilicate Glasses

Polarizable potentials are another type of empirical potentials used to simulate the glass systems, for instance, a core–shell model polarizable potential was developed by Tilocca et al. [59] for silicate glass systems. The results show that the polarizable potential can reproduce the experimental and ab initio data in terms of density, coordination environment, and network connectivity; in addition, the inclusion of the polarization effects improves the inter-tetrahedral units, $Q_n$ distribution, and local environment around the modifiers for the sodium silicate glass. This framework is then extended by Yu et al. [60] to additional boron and phosphorus-related parameters which enable the simulation of phosphorus-bearing borosilicate glasses. The full functional form is slightly different from the one in the previous work [59], that a repulsive term $D_{\alpha\beta}r^{-12}$ was added as fully presented in a later work [61]:

$$U_{\alpha\beta}(r_{\alpha\beta}) = A_{\alpha\beta}\exp\left(-\frac{r_{\alpha\beta}}{\rho_{\alpha\beta}}\right) - \frac{C_{\alpha\beta}}{r_{\alpha\beta}^{6}}\left\{1-\exp\left[-\left(\frac{r_{\alpha\beta}}{4.3\rho_{\alpha\beta}}\right)^{6}\right]\right\} + \frac{D_{\alpha\beta}}{r_{\alpha\beta}^{12}} \tag{8.10}$$

in which $A_{\alpha\beta}$, $\rho_{\alpha\beta}$, $C_{\alpha\beta}$, and $D_{\alpha\beta}$ are empirical parameters, and the $r_{\alpha\beta}$ is the interatomic distance between element $\alpha$ and $\beta$, respectively. The potential parameters are listed in the Appendix 8.A. In addition to the ability to reproduce the short-range order structure features, the polarizable feature helps the potential to better predict the network connection due to more accurate $Q_n$ species. However, stability of the system during the simulations, especially at high temperatures, is still one of the shortcomings that limit wide applications of the polarizable potentials. On the other hand, more complicated potential form and additional parameters in the three-body terms make it more difficult to extend the potential to more complicated systems than the rigid ion potentials with only two-body terms.

## 8.5 Evaluation of the Potentials

As various potentials have been developed for the borosilicate and boron-containing multicomponent oxide glasses, validation and evaluation of the available potentials become important before further studies on the MD simulations using these potentials. In addition to the results published together with the potential development, direct comparisons of different potential sets are also important to provide a basic understanding of the advantage and shortcomings of the potentials for the same compositions.

For example, Pedone and coworkers [62] investigated a series of sodium borosilicate glasses (five different compositions) with constant $Na_2O/(B_2O_3+SiO_2)$ value of 0.33, using two rigid ion potentials (from Wang et al. [44] and from Deng and Du [52]) and one polarizable potential (Yu et al. [60]). Structure models from all the three potentials are generated and then relaxed in density functional theory (DFT) calculations and the simulated NMR spectra are calculated and compared to the experimental data. Both rigid ion potentials produce reliable boron coordination environment and local environment around sodium; however, second coordination shell of silicon and the Si/B—O—Si/B bond angle distribution were not accurately reproduced by these two potentials. On the other hand, the polarizable potential underestimates the fraction of the fourfold boron units for one of the compositions; however, it reproduces much better results in terms of silicon $Q_n$ species and the inter-tetrahedra Si/B–O–Si/B angles than the two rigid ion potentials. All the three potentials reproduce the B–O distances from the DFT calculations [12], while potential from Deng and Du produce slightly better difference between $B^{[3]}$–O and $B^{[4]}$–O distance possibly due to its composition dependence on the $A$ parameter of Buckingham function.

Later, Lee et al. [63] studied two different major commercial borosilicate glasses (Borofloat®33 and N-BK7®) to compare the potentials from Wang et al. [44] and from Deng and Du [52], and both use a potential form combining short-range Buckingham term and long-range Coulomb term. The potential from Deng and Du has an additional composition-dependent B–O interaction, while the Wang potential employed additional pairs of all the cation–cation interactions. Wang potential is found to give a better description of the short-range structure, while the potential from Deng and Du gives a better prediction of the medium-range structure. Results from the later potential agree well with the experimental $N_4$ value of N-BK7 composition but overestimate it in the Boro-33 glass. On the other hand, Wang potential underestimated the $N_4$ values in both compositions. Both potentials failed to reproduce the experimental elastic modulus for the two compositions, that the Deng and Du potential overestimates the value, while Wang potential underestimates it.

Recently, Tuheen et al. [42] performed a systematic comparison of three rigid ion potentials (named as Kieu [14, 43], Du [52], and Bauchy [44]) by simulating two model glasses, a series of sodium borosilicate compositions with $K = 2$ and $R$ ranging from 0.1 to 4.0 (total number of 18), and a series of sodium boroaluminosilicate compositions (total number of four). Figure 8.4 shows a comparison of the $N_4$ value as a function of

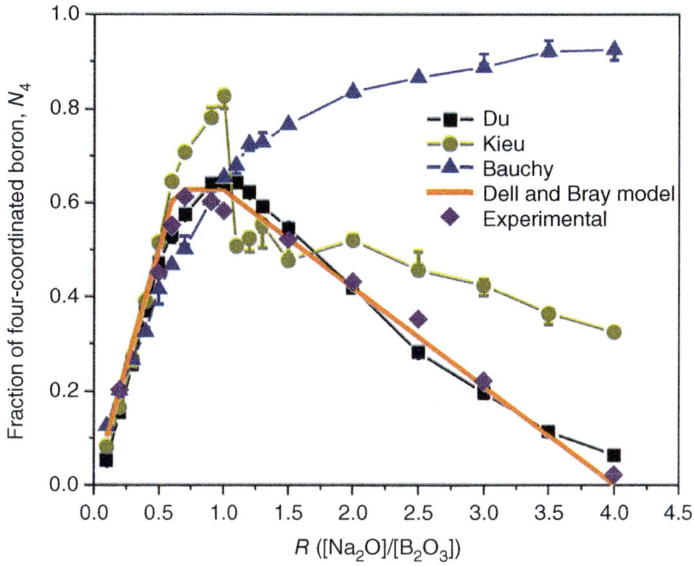


**Figure 8.4** Fraction of four-coordinated boron, $N_4$, as a function of $R$ ($[Na_2O]/[B_2O_3]$) at $K$ ($[SiO_2]/[B_2O_3]$) = 2 obtained from three different potentials (Deng and Du [52], Kieu et al. [43], and Wang et al. [44]), DBX model [20] and experiments in $Na_2O-B_2O_3-SiO_2$ ternary glass system. Source: Reproduced from Tuheen et al. [42] with permission.

$R$ from MD simulations with the three potentials and those from Dell and Bray model and experiments. Among the three potentials, Kieu potential is most complicated one in terms of the parameter numbers, e.g. the composition-dependent charges and the B–O interaction, as well as the cation–cation interactions. Both Du and Bauchy potentials are using fixed charges, but the composition-dependent B–O interaction is applied in Du potential. Figure 8.5 shows a comparison of B–O pair distribution function and O—B—O bond angle distribution of a sodium silicate glass with $K = 2$ and $R = 1.3$. The contributions from three- and four-coordinated boron were also plotted. The three potentials provide quite different predictions of boron coordination in the same composition.

These lead to the differences in the structures and properties in the glass generated. Du potential gives the best agreement with experiments for the ternary systems in terms of the $N_4$ values, density, and oxygen species. Kieu potential can reproduce both structures and bulk modulus well in the low $R$ region; it fails in the high $R$ region due to the limitation of its framework. Bauchy potential also succeeds in the low $R$ region in terms of the structure and shear and Young's moduli; however, it fails in the high $R$ region possibly because of the constant parameters which lacks the composition-dependent feature.

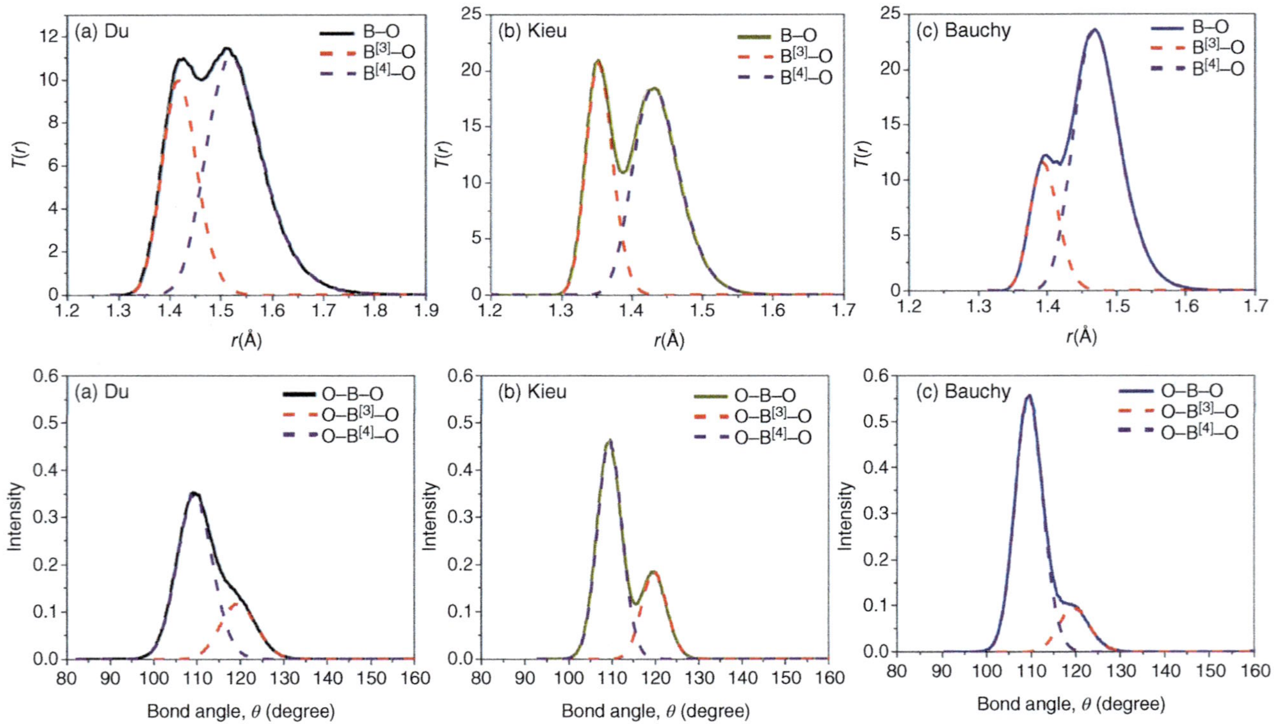


**Figure 8.5** Comparison of boron-oxygen total correlation function ($T(r)$) (top) and O−B−O bond angle distributions (bottom) from the three potentials (Du [52], Kieu [43], and Bauchy [44]) for a sodium borosilicate glass composition with $R = 1.3$ and $K = 2$. Deconvolution of the contributions from $B^{[3]}$ and $B^{[4]}$ were performed. The differences in the shapes of the distribution represent boron $N_4$ value differences from the three potentials. Source: Reproduced from Tuheen et al. [42] with permission.

## 8.6 Effects of Cooling Rate and System Size on Simulated Borosilicate Glass Structures

In addition to the empirical potentials which are essential to the MD simulations, there are other simulation details that would impact the final simulation results. Among all the settings of each simulation, cooling rate and the system size play an important role of generating reasonable results comparing with the experiments [64–67]. Experimentally, thermal history has great impact on the glass formation thus the final glass structures and properties. The commonly used cooling rate in classical MD simulations is between 1 and 10 K/ps, which is orders of magnitude higher than the experimental one, e.g. 0.1 K/s [65]. There are two ways to overcome the gap, extrapolate the data from the cooling rate test [65], or directly validate the results with the macroscopic glass structure and properties regardless of the formation procedure. Most of the MD simulations follow the later method not only because of the computational cost of cooling rate tests, especially for slow cooling one, but also because of their focus on the final glass structure and properties. Without further annealing process, the glass structure has been already established. If the structure features and properties of the final glasses are validated during the potential fitting processes, reasonable results can be expected when using the same protocol to form glass. However, when the thermal history related properties, such as diffusion rate [64] and glass transition temperature [65], are the targets of the simulations, the cooling rate effect then plays an important role to compare with experimental data. In addition, boron-containing oxide glasses, for example the borosilicate glass, have been found to be more sensitive to the cooling rate and the system size used in the MD simulations [68]. The study investigates the cooling rate and system size effects on a sodium borosilicate glass using the potential developed by Kieu et al. [43]. The results of

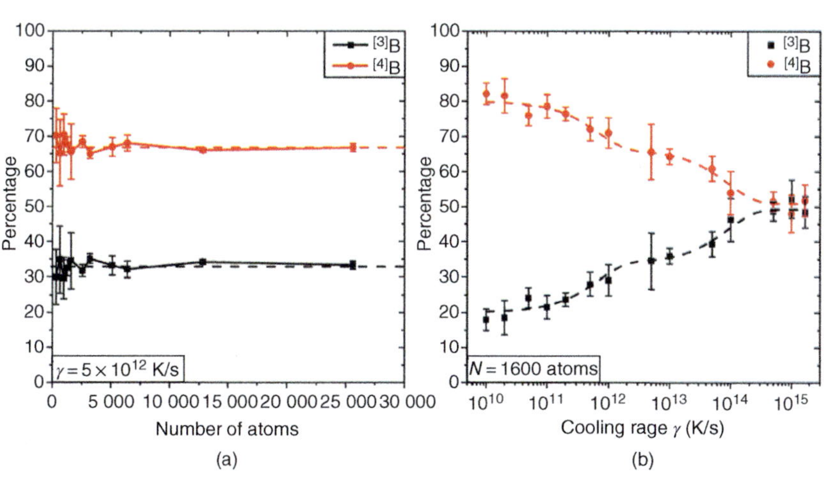


**Figure 8.6** Effects of system size and cooling rate effect on boron $N_4$ in sodium borosilicate glasses: (a) the system size effect and (b) the cooling rate effect, respectively. Source: Reproduced from Deng and Du [68] with permission.

the system size test suggest a system size of 1000 atoms are essential to prevent the impact of further increase of system size for the selected composition (Figure 8.6a). The cooling rate test indicates that structures and properties including density, boron coordination, $Q_n$ speciation, network connectivity, first peak intensity of ring size distribution, vibrational density of state, and mechanical properties all tend to converge to a value with slower cooling rate. This suggests that the cooling rate effect in the borosilicate glass is significant which is mainly due to the cooling rate-dependent boron coordination change (Figure 8.6b), for instance a strong relationship between the $N_4$ and the mechanical properties are observed. In addition to those room temperature structures and properties which rely on the accuracy in the development of the empirical potentials, the thermal-history-based property, $T_g$, is also estimated using the extrapolation method by fitting to a functional form originally used to describe the viscosity of the liquids as a function of temperature. The $T_g$ value at experimental cooling rate extrapolated from the MD cooling rate results (924.3–949.9 K) is slightly higher than the experimental data (763 K), indicating the extrapolation is a valid method to obtain the thermal-history-related properties.

## 8.7 Applications of MD Simulations of Borosilicate Glasses

With the recent active development of empirical potentials presented above, borate and $B_2O_3$-containing multicomponent oxide glasses have been studied more and more by using MD simulations. Compositions studied include those with industrial and technological applications such as commercial borosilicate glasses, borate-containing bioactive glasses, and the nuclear waste glasses. In this section, we review these simulation works based on the glass compositions: borosilicate, boroaluminosilicate, and multicomponent glasses. In addition to the bulk glass structures, glass surface has also been investigated. Some studies also extend to study the properties and structure–property relations.

### 8.7.1 Borosilicate Glass

Meikhail [69] studied the structures of a series of sodium borosilicate glasses differed by the sodium oxide content. Boron coordination conversion from three to four was observed to consume the additional sodium oxide. Sudden increase of the NBO concentration is noticed at the composition with high sodium oxide content. Sodium self-diffusion in the system is found to depend on the temperature, composition, and the cooling rate used in the simulations. By monitoring the number of sodium ions near the interfacial region of the boron and silicon network, a possible phase separation mechanism in the studied composition is discussed. A decrease in the sodium concentration in the interfacial region suggests a lower sodium content that will lead to a higher probability of forming isolated droplets, indicating an increase of the phase separation, e.g. isolated silica-like and borate-like regions.

Gou et al. [70] applied the MD simulations to investigate the structure–composition relationship in the sodium borosilicate glasses. Partial separation of silica-rich and boron-rich regions is observed in the simulated glass structure for the studied compositions, and it is due to the size difference of the borate and silicate polyhedra (~1.4 and 1.6 Å, respectively). The coordination environment of silicon is stable with the boron content change, while the fraction of threefold boron increases with additional boron content. Sodium ions are found to prefer associating with additional boron content.

Kilymis et al. [71] studied the change of local- and medium-range structures and properties of sodium borosilicate glasses under compression. Both the density and the boron coordination numbers are found to increase after the compression tests (Figure 8.7), while the mechanical properties change depends on the glass composition. The atomic packing is found to play an important role of defining the amount of densification and generating the permanent densification, and Na-rich regions are found to have high compressibility based on the shortened Na–Na and Na–O distances.

Delaye et al. [72] applied MD simulations to investigate the radiation damage in the silica and sodium borosilicate glasses using the displacement cascade method. Bond angles between network formers are found to decrease after the simulated radiation for both systems. In addition, silica glass is found to be denser, while the borosilicate glass expands after the radiation. Medium-range structural change and subsequent increase of the internal energy due to local thermal quenching contribute to the radiation-induced change of the macroscopic properties of the glasses.

Ren et al. [16] performed simulations on a series of sodium borosilicate glasses to study the surface structures at atomistic level. $Z$-density profile of the elements is also calculated for the simulated structures in the direction perpendicular to the surface (Figure 8.8). A decreased concentration of all the elements at the surface is observed. Sodium enrichment is observed at the surface, and the surface structure has less fourfold boron and smaller network connectivity than the bulk glass. In addition, structure defects such as two-membered rings and undercoordinated silicon units are also observed at the surface. Interestingly, it is found that the surface of borosilicate silicate glass shows a different boron $N_4$ as compared to the bulk, which is found to be due to the composition variation at the surface glass layer.

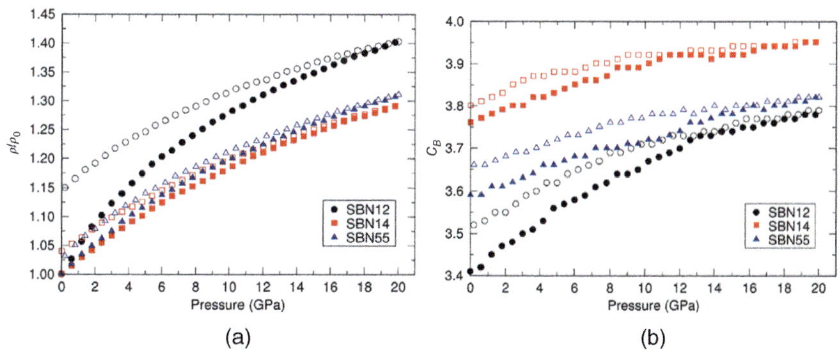


**Figure 8.7** (a) Relative densification, and (b) change of boron coordination of the simulated glasses with compression tests, respectively. The filled and empty symbols are for compression and decompression tests, respectively. Source: Reproduced from Kilymis et al. [71] with permission.

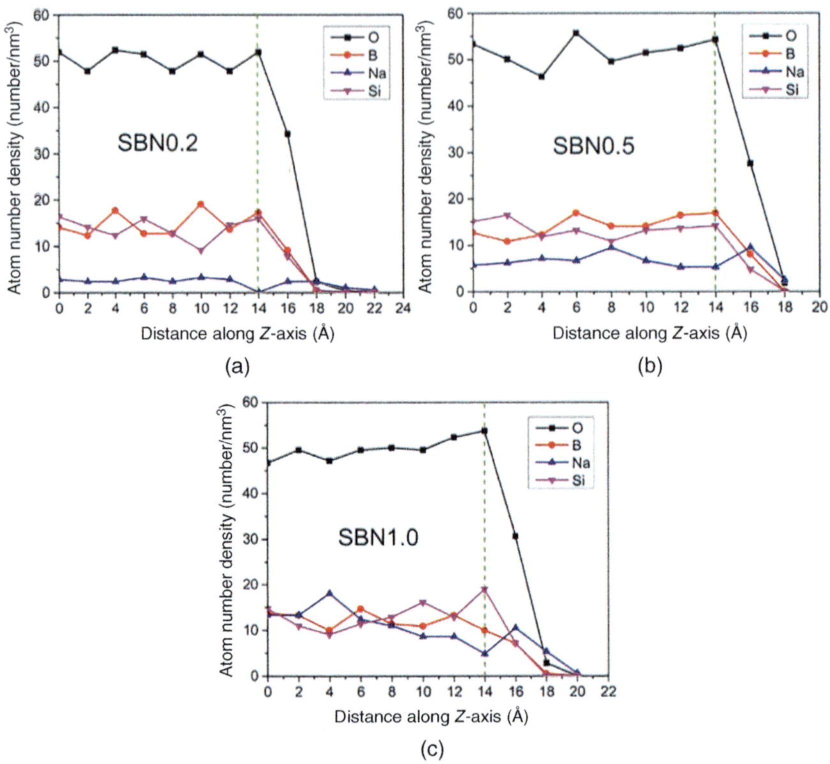


**Figure 8.8** $Z$-density profiles of three sodium borosilicate glasses: (a) SBN0.2, (b) SBN0.5, and (c) SBN1.0, respectively. Source: Reproduced from Ren et al. [16] with permission.

## 8.7.2 Boroaluminosilicate Glasses

Ren et al. [73] investigated the bulk and surface structures as well as the ionic diffusion and mechanical properties of the sodium borosilicate and boroaluminosilicate glasses with the compositions close to the International Simple Glass (ISG) [2]. Sodium enrichment on the glass surface is observed by comparing the bulk glass and surface structures. In addition, boron $N_4$ value and overall network connectivity of the studied borosilicate composition increase with the additional alumina content. Relatively smaller probabilities of the $AlO_x$–$AlO_x$ and $BO_x$–$AlO_x$ connections than $SiO_4$–$AlO_x$ indicate certain avoidance exists between $AlO_x$ units as well as between $AlO_x$ and $BO_x$ units (Figure 8.9a). Figure 8.9b shows a snapshot of the surface structure model from the top and side view of the boroaluminosilicate glass.

Ha and Garofalini [46] studied the local structures of the glass former and modifiers in soda-lime boroaluminosilicate glasses. Majority of the coordination number of silicon and aluminum is four, while the boron coordination number changes from three to four with increasing modifier content. Instead of the idea group that was formed by one Al/B unit

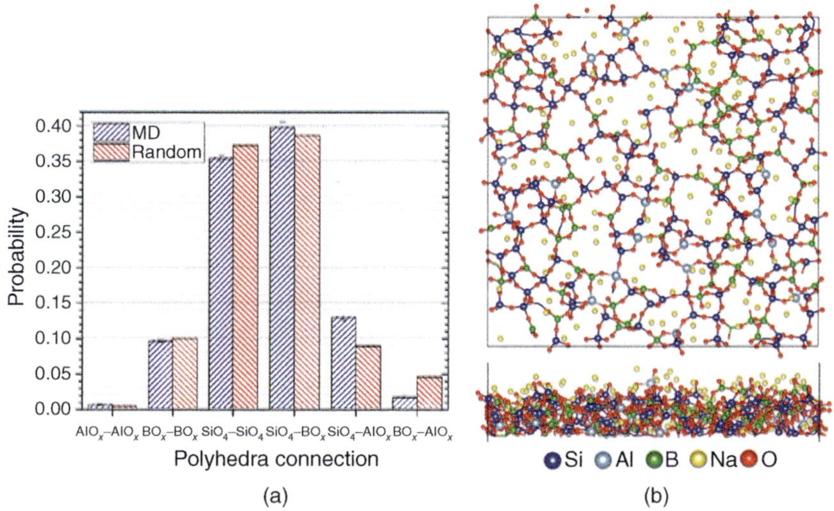


**Figure 8.9** (a) Shows boron coordination ($N_4$) as a function of $R$ for the bulk and surface from MD, and (b) shows a representative surface structure model from MD: top and side views of the relaxed glass surface of sodium boroaluminosilicate glass. Source: Reproduced from Ren et al. [73] with permission.

and one Na, the Al/B unit is found to be able to have more than one Na around; on the other hand, the Na ion can also be surrounded by multiple Al/B units.

Garofalini et al. [74] further investigated the structure change of the surface of soda-lime boroaluminosilicate glasses after being exposed to water. The boron oxygen bonds in the surface area are more likely to get attacked and the network of the surface is thus hydroxylated. Both boron and sodium are found to lose the sites around NBOs, which agree with experimental observations that B and Na are dissolved in the initial stage of the glass corrosion.

### 8.7.3 Boron Oxide-Containing Multicomponent Glass

With the developed B–O parameters that are compatible with other silicate glasses [14, 43, 45, 46, 52, 58, 60], the structure and properties of multicomponent borosilicate and boroaluminosilicate glasses have been studied using MD simulations. For example, boron oxide to silica substitution effects on the structure, properties, and bioactivities of several bioactive glasses have been studied by using MD simulations. Lu et al. investigated the boron effect in the 45S5 ($xB_2O_3$–$[46.1-x]SiO_2$–$24.4Na_2O$–$26.9CaO$, $x = 0, 15.4, 30.7, 46.1$, in mol%) and SrO doped 55S4.3 ($[56.5-x]SiO_2$–$xB_2O_3$–$19.4Na_2O$–$16.5CaO$–$5.0SrO$–$2.6P_2O_5$, $x = 0, 14.1, 28.2, 42.3, 56.5$, in mol%) [15, 75] bioactive glasses with different levels of $B_2O_3$ to $SiO_2$ substitution by combining experiments and MD simulations. Together with the Raman spectrometry, X-ray diffraction (XRD), and Fourier-transform infrared (FTIR) techniques, the structural change caused by the additional boron and the formation of the hydroxyapatite (HAp) are characterized. Additional boron oxide in these systems is

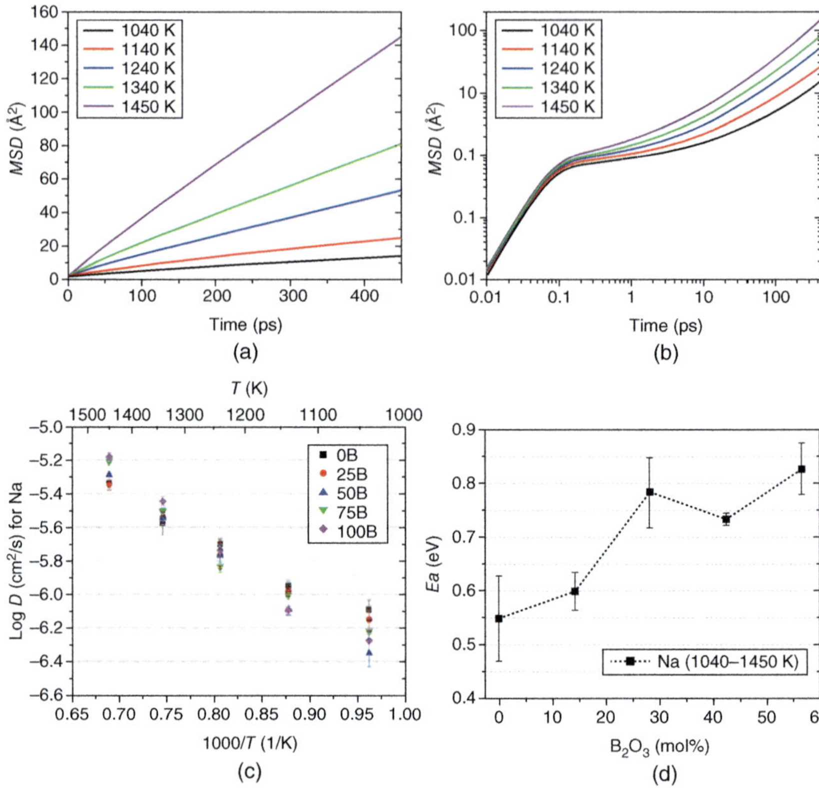


**Figure 8.10** Sodium ion diffusion behaviors of $B_2O_3/SiO_2$ substituted SrO-doped 55S4.3 ($[56.5 - x]SiO_2 - xB_2O_3 - 19.4Na_2O - 16.5CaO - 5.0SrO - 2.6P_2O_5$, $x = 0$, 14.1, 28.2, 42.3, 56.5, in mol%) bioactive glasses. (a) MSD versus time in linear scale, (b) MSD versus time in logarithmic scale, (c) Arrhenius plot of diffusion coefficients (log *D*) versus 1000/*T*, and (d) diffusion energy barriers for Na (error bar from three parallel samples) as a function of $B_2O_3$ concentration (mol%). Source: Reproduced from Ren et al. [56] with permission.

found to increase the network connectivity of the glass structure and prevent or delayed the formation of the HAp. Ren et al. [56] investigated $B_2O_3$ to $SiO_2$ substitution effect on the structure and dynamic properties of SrO-doped 55S4.3 bioactive glasses by using MD simulations. Figure 8.10 shows the calculated diffusion coefficients and energy barrier of sodium ions as a function of $B_2O_3$ concentration. It has been found that the diffusion energy barrier increases with $B_2O_3$, which is related to the increase of the overall network connectivity with the level of substitution. The vibrational density of states has also been calculated, and it shows that $B^{[3]}$ spectra has two peaks: one stronger and sharp peak at $1400\,cm^{-1}$ and a broad peak at $800\,cm^{-1}$, while $B^{[4]}$ spectra only has a broad peak centered at around $900\,cm^{-1}$.

Stevensson et al. [61] investigated the composition–structure relationship in the borate, borosilicate, and phosphorus-containing borosilicate glasses. The local coordination environment of the glass former polyhedra are found to be independent of the coexisting formers, while the major changes of the bond length and angles reflect the change in the oxygen species and subsequent network polymerization. In addition, the glass modifier cations are found to prefer associating with silicon tetrahedra or threefold boron units, while the significant amount of sodium ions are only observed in the boron-rich region. For the NBO-rich glass networks, however, significant amount of NBOs is found to associate with the boron tetrahedra.

Lu et al. [55] compared the structural results from MD simulations and the high-energy X-ray diffraction data for the ISG composition ($60.2SiO_2$–$16.0B_2O_3$–$12.6Na_2O$–$3.8Al_2O_3$–$5.7CaO$–$1.7ZrO_2$). Both the X-ray structure factor and the total correlation function (total distribution function [TDF]) calculated from the simulated ISG structure are in great agreement with the experimental results. The $R_x$ factor that depicts the difference of the simulated and experimental total correlation function in comparison to high-energy X-ray diffraction was found to be 5.5%, indicating a good agreement of the two hence validate the simulated glass structure model of this multicomponent glass. Linkages between silicon tetrahedra as well as the silicon-boron intertetrahedral linkages are significantly observed, while the zirconia octahedra are more likely to form a linkage with the silicon or boron units than itself. Structural features such as the small-sized rings, undercoordinated silicon units, high fraction of threefold boron are observed in the glass surface; in addition, sodium enrichment is also observed at the surface which agrees well with experimental observations.

Lu et al. [76] also performed the quantitative structure–property relationship (QSPR) analysis using the structures from MD simulations to predict the zirconia effect on the chemical durability of the soda-lime borosilicate glasses. A descriptor $F_{net}$ that combines the short-range structure features from MD simulation and cation-oxygen single bond strength provide excellent linear correlations in terms of density and the initial dissolution rate for the selected glass compositions.

## 8.8 Conclusion

As one of the important ways to understanding the atomic information of the glass structures, computational modeling using both classical MD and AIMD methods have been applied to study various borate and borosilicate glass structures. AIMD method has the advantage in the accuracy of the produced results but with significantly higher computational cost; classical MD simulations, on the other hand, are more preferred in the simulations of complicated systems like multicomponent oxide glass systems, given that reliable potentials are available. We have provided a comprehensive review of the borate potential development in the past decade or so. Potential evaluation and simulation parameter effect on borosilicate glass structures are also presented. With these recent active development of potentials for the boron-containing multicomponent oxide glasses, studies of various borate containing glass compositions for various industrial and technological applications are then presented, with the results indicating MD simulation as a powerful tool to understand the structure and properties of borate and borosilicate glasses.

## Acknowledgments

The authors are grateful for support of the Center for Performance and Design of Nuclear Waste Forms and Containers (WastePD), an Energy Frontier Research Center funded by the US Department of Energy (DOE), Office of Science, Basic Energy Sciences under Award # DESC0016584 and U.S. National Science Foundation under awards 1508001 and 1662288.

## 8.A Available Empirical Potentials for Boron-Containing Systems

### 8.A.1 Borosilicate and Boroaluminosilicate Potentials – Kieu et al. and Deng and Du

Kieu et al. [43] successfully developed a pairwise composition-dependent potential set for the sodium borosilicate systems based on an early work of Guillot and Sator [35]. The main potential form includes a Coulombic long-range interaction term and a Buckingham short-range interaction term:

$$\phi(r_{ij}) = \frac{q_i q_j}{r_{ij}} + A_{ij} \exp\left(-\frac{r_{ij}}{\rho_{ij}}\right) - \frac{C_{ij}}{r_{ij}^6} \tag{8.A.1}$$

where $A_{ij}$, $\rho_{ij}$, and $C_{ij}$ are empirical parameters of the Buckingham term, $r_{ij}$ is the interatomic distance, and $q_i$ and $q_j$ are the charges of atoms $i$ and $j$, respectively. Additional empirical parameters are needed to determine the charges of each element and B–O parameters based on the compositions. The composition-dependent charges for each element can be calculated by solving the following two equations:

$$q'_{\mathrm{B}} = -q'_{\mathrm{O}}\left(C_6 K^2 + \sum_{i=1}^{5} C_i R^i + C_0\right) \tag{8.A.2}$$

and

$$q'_i = q_i - N_{\mathrm{B}} \frac{q'_{\mathrm{B}} - q_{\mathrm{B}}}{N_{\mathrm{Si}} + N_{\mathrm{O}} + N_{\mathrm{Na}}} \tag{8.A.3}$$

where $C_i$ ($i = 0, 1, 2, \ldots, 6$) are empirical parameters from Table 8.A.3, $N_i$ ($i$ = B, Na, Si, O) are numbers of atoms for each element, and $q_i$ ($i$ = B, Na, Si, O) are the initial charge values from Table 8.A.1. In addition, the $Aij$ parameter in Eq. (8.A.1) for B–O pair can be calculated using the composition ratios $R$ and $K$, $\frac{Na_2O}{B_2O_3}$ and $\frac{SiO_2}{B_2O_3}$, through the following equation:

$$A_{\mathrm{B-O}} = \sum_{i=1}^{5} a_i R^{*i} + a_0 \tag{8.A.4}$$

in which $a_i$ ($i = 0, 1, 2, \ldots, 5$) values are empirical parameters from Table 8.A.3, and $R^*$ can be obtained through the following equation:

$$R^* = \begin{cases} \min\left(R, \frac{K}{16} + 0.5\right), & R < \frac{K}{4} + 0.5 \\ 0, & R \in \left(\frac{K}{4} + 0.5, K + 2\right) \end{cases} \tag{8.A.5}$$

**Table 8.A.1** Initial charges of elements used in Kieu et al. and Deng and Du potentials [14, 43].

| Element | O | Si | Na | B | Al |
|---|---|---|---|---|---|
| Charge ($q_i$) | −0.945 | +1.89 | +0.4725 | +1.4175 | +1.4175 |

This potential set was then extended to the sodium boroaluminosilicate systems by Deng and Du [14] using the same framework. The additional alumina content added another composition ratio $H$, $\frac{Al_2O_3}{B_2O_3}$; therefore, the charges of elements as well as the $R$, $K$, $R^*$ values in Eqs. (8.A.2)–(8.A.5) are modified correspondingly as the following equations:

$$q'_B = -q'_O\left(C_6K'^2 + \sum_{i=1}^{5} C_i R_e'^{\,i} + C_0\right) \tag{8.A.6}$$

$$q'_i = q_i - N_B \frac{q'_B - q_B}{N_{Si} + N_O + N_{Na} + N_{Al}} \tag{8.A.7}$$

$$A_{B-O} = \sum_{i=1}^{5} a_i \left(R_e^{*\prime}\right)^i + a_0 \tag{8.A.8}$$

and,

$$R_e^* = \begin{cases} \frac{[R^{*\prime}\times(Al_2O_3+B_2O_3)-Al_2O_3]}{B_2O_3} & R \geq H, R^{*\prime} \geq H' \\ 0 & R \geq H, R^{*\prime} < H' \\ 0 & R < H \end{cases} \tag{8.A.9}$$

in which

$$R'_e = R' - H' = \frac{Na_2O - Al_2O_3}{B_2O_3 + Al_2O_3} \tag{8.A.10}$$

and

$$R^{*\prime} = \begin{cases} \min(R', K'/16 + 0.5), & R' \leq K'/4 + 0.5, \quad K' \leq 8 \\ 0, & K'/4 + 0.5 \leq R' < K' + 2, \quad K' \leq 8 \\ R', & R' \leq 1, K' > 8 \\ 0, & R' > 1, K' > 8 \end{cases} \tag{8.A.11}$$

The $R$, $K$, $H$ are the composition ratios of $Na_2O$, $SiO_2$, and $Al_2O_3$ over $B_2O_3$, while the $R'$, $K'$, and $H'$ are ratios of $Na_2O$, $SiO_2$, and $Al_2O_3$ over total number of $Al_2O_3$ and $B_2O_3$, respectively. The empirical parameters used in the above equations can be found in Tables 8.A.1–8.A.3.

**Table 8.A.2** Potential parameters for sodium borosilicate and boroaluminosilicate glass systems in Kieu et al. and Deng and Du potentials [14, 43].

| Pair | $A_{ij}$ (eV) | $\rho_{ij}$ (Å) | $C_{ij}$ (eV·Å$^6$) |
|---|---|---|---|
| O–O[a)] | 9 027.03 | 0.265 | 85.0321 |
| O–Si[a)] | 45 296.72 | 0.161 | 46.1395 |
| O–B | $A_{B-O}$ (from Eq. (8.A.8))[a)] | 0.124[b)] | 35.0019[b)] |
| O–Na[a)] | 120 360.22 | 0.170 | 0.0000 |
| O–Al[b)] | 28 287.00 | 0.172 | 34.7600 |
| Si–Si[a)] | 834.40 | 0.290 | 0.0000 |
| Si–B[a)] | 337.70 | 0.290 | 0.0000 |
| Si–Al[b)] | 646.67 | 0.120 | 0.0000 |
| B–B[a)] | 121.10 | 0.350 | 0.0000 |
| B–Al[b)] | 137.58 | 0.479 | 0.0000 |
| Na–Al[b)] | 175.21 | 0.130 | 0.0000 |
| Al–Al[b)] | 351.94 | 0.360 | 0.0000 |

a) Parameters listed in Kieu et al. [43].
b) Parameters listed in Deng and Du [14].

**Table 8.A.3** Empirical parameters in Kieu et al. and Deng and Du potentials [14, 43]. The original split point of $C_6$, $R_{separate}$, for borosilicate system in Kieu et al. [43] is $R = 0.55$, and the modified one in Deng and Du [14] is $R'_e = 0.5 + K'/16$ for boroaluminosilicate systems, respectively.

| $i$ | $c_i$ | $a_i$ |
|---|---|---|
| 0 | 1.496 43 | 18 390.53 |
| 1 | 0.295 04 | 47 166.67 |
| 2 | −0.256 50 | −43 827.65 |
| 3 | 0.087 21 | 210 268.55 |
| 4 | −0.013 23 | −52 520.42 |
| 5 | 0.000 73 | −139 041.69 |
| 6 | 0.0 ($R \le R_{separate}$) | — |
| | 0.003 15 ($R > R_{separate}$) | |

**Table 8.A.4** Partial charges used in Wang et al. potential [44]. Partial charges for Ti, Al, $Fe^{3+}$, $Fe^{2+}$, Mg, and K from Guillot–Sator potential [35] have also been provided in Wang et al. paper for reference but are not validated to be compatible; therefore, they are not listed here.

| Element | O | Si | B | Ca | Na |
|---|---|---|---|---|---|
| Charge ($q_i$) | −0.945 | +1.89 | +1.4175 | +0.945 | +0.4725 |

**Table 8.A.5** Empirical parameters in Wang et al. potential [44]. Potential parameters for Ti–O, Al–O, $Fe^{3+}$–O, $Fe^{2+}$–O, Mg–O, and K–O from Guillot–Sator potential [35] have also been provided in Wang et al. paper for reference but are not validated to be compatible; therefore, they are not listed here.

| Pair | $A_{ij}$ (eV) | $\rho_{ij}$ (Å) | $C_{ij}$ (eV·Å$^6$) |
|---|---|---|---|
| O–O | 9 022.79 | 0.265 | 85.0921 |
| O–Si | 50 306.10 | 0.161 | 46.2978 |
| O–B | 206 941.81 | 0.124 | 35.0018 |
| O–Na | 120 303.80 | 0.170 | 0.0000 |
| O–Ca | 155 667.70 | 0.178 | 42.2597 |
| Si–B | 337.70 | 0.290 | 0.0000 |
| B–B | 484.40 | 0.350 | 0.0000 |

## 8.A.2 Borosilicate Potential – Wang et al.

Wang et al. developed a partial charge potential set also based on the Guillot and Sator potential with new fitted boron-related parameters for soda-lime borosilicate systems; however, the charges of elements and new boron parameters are constants instead of the composition-dependent values in Kieu et al. potential. The potential form is same to the Kieu et al. potential above (Eq. (8.A.1)), and the charges and potential parameters ($A$, $\rho$, and $C$) values can be found in Tables 8.A.4 and 8.A.5.

## 8.A.3 Borosilicate Potential – Inoue et al.

Inoue et al. [45] developed a set of two-body potential for sodium borate and borosilicate systems with a Born–Mayer type potential form, while the coefficient of the repulsion term is determined at the minimum value of the distance between the two ions of the pair. The potential form follows the Born–Mayer type:

$$\Phi_{ij}(r_{ij}) = \frac{e^2}{4\pi\varepsilon_0}\frac{Z_i Z_j}{r_{ij}} + B_{ij}\left(-\frac{r_{ij}}{\rho_{ij}}\right) \tag{8.A.12}$$

**Table 8.A.6** Potential parameters for the sodium borosilicate systems [45].

| M | $Z_M$ | $B_{M-O}$ ($10^{-16}$ J) | $\rho_{M-O}$ (Å) |
|---|---|---|---|
| Na | +1.0 | 2.300 | 0.300 |
| B | +1.8 | 7.040 | 0.170 |
| Si | +2.4 | 19.600 | 0.174 |
| O | −1.20 to −1.45 | 0.625 | 0.362 |

in which $\Phi_{ij}$ represents for the interaction energy between ions $i$ and $j$, $e$ is the charge of an electron, $\varepsilon_0$ is the vacuum permittivity, $r_{ij}$ is the interatomic distance, and $\rho_{ij}$ is the parameter to describe softness. The $Z_i$ and $Z_j$ are effective charge values of ions $i$ and $j$, respectively. The $B_{ij}$ value is determined at a minimum value of the distance between ions $i$ and $j$, $R_{ij}$, by the following equation:

$$B_{ij} = -\frac{e^2}{4\pi\varepsilon_0}\frac{Z_iZ_j}{R_{ij}^{\ 2}}\rho_{ij}\exp\left(\frac{R_{ij}}{\rho_{ij}}\right) \tag{8.A.13}$$

The charges and potential parameters can be found in Table 8.A.6.

## 8.A.4 Boroaluminosilicate Potential – Ha and Garofalini

Ha and Garofalini [46] proposed a potential using a combination of BMH type two-body term and additional three-body terms to constrain certain bond angles. The two-body potential form is described by the following equation:

$$V_{ij}^{BMH} = A_{ij}\exp\left(\frac{-r_{ij}}{\rho_{ij}}\right) + \left(\frac{z_iz_je^2}{r_{ij}}\right)\operatorname{erfc}\left(\frac{r_{ij}}{\beta_{ij}}\right) \tag{8.A.14}$$

where $A_{ij}$, $\beta_{ij}$, and $\rho_{ij}$ values are empirical parameters for pair $i$–$j$, $z_i$ and $z_j$ are the full charge of the ions $i$ and $j$, respectively. $e$ is the charge of single electron, and $r_{ij}$ value is the interatomic distance. The error function used in the Coulomb term is to zero this part when the interatomic distance is larger than the selected cutoff. In addition, three-body term is used to constrain the selected angles:

$$V_{jik}^{3\text{-body}} = \lambda_{ij}^{\frac{1}{2}}\lambda_{ik}^{\frac{1}{2}} \times \exp\left[\frac{\gamma_{ij}}{(r_{ij}-R_{ij})} + \frac{\gamma_{ik}}{(r_{ik}-R_{ik})}\right] \times \Theta_{jik} \tag{8.A.15}$$

where $\lambda_{ij}$ and $\gamma_{ij}$ values are empirical parameters, $R_{ij}$ value is the cutoff value. The angular component $\Theta_{jik}$ has two different expressions. For angles Si/Al/B–O–Si/Al/B and

**Table 8.A.7** Charges of elements used in Ha and Garofalini potential [46].

| Element | O | Na | Ca | B | Al | Si |
|---|---|---|---|---|---|---|
| Charge ($Z_i$) | −2 | +1 | +2 | +3 | +3 | +4 |

O–Si/B–O, the angular component is given as $\left(\cos\theta_{jik} - \cos\theta^0_{jik}\right)^2$, and for O–Al–O angle, it is given by $\left(\left(\cos\theta_{jik} - \cos\theta^0_{jik}\right)\sin\theta_{jik}\cos\theta_{jik}\right)^2$. The detailed parameters can be found in Tables 8.7–8.9.

## 8.A.5 Borosilicate and Boron-Containing Oxide Glass Potential – Deng and Du

Deng and Du [52] developed a potential for boron-containing multicomponent oxide glass systems based on the widely used partial charge rigid ion potential framework [33, 47]. This potential uses a combination of Coulomb term and Buckingham term to describe the long-range and short-range interactions as shown in the following equation:

$$V(r_{ij}) = \frac{Z_i Z_j e^2}{4\pi\varepsilon_0 r_{ij}} + A_{ij}\exp(-r_{ij}/\rho_{ij}) - C_{ij}/r_{ij}^{\ 6} \tag{8.A.16}$$

where $Z_i$ and $Z_j$ are the partial charges for ions $i$ and $j$, respectively. $e$ is the charge of a single electron, $\varepsilon_0$ is the vacuum permittivity, and $r_{ij}$ is the interatomic distance between atom $i$ and $j$. The $A_{ij}$, $\rho_{ij}$, and $C_{ij}$ values are empirical parameters for the Buckingham term. To solve issues of random or high temperature configurations (unreasonable results caused by the Buckingham type potential when two atoms are too close to each other), a splice function for the low interatomic distance is used to correct:

$$V'(r_{ij}) = B_{ij}/r_{ij}^{\ n} + D_{ij}\cdot r_{ij}^{\ 2} \tag{8.A.17}$$

where $B_{ij}$, $D_{ij}$, and $n_{ij}$ are calculated parameters at the splice point, where the original function, the first, and second derivative values of the Buckingham potential term are equal to those of the splice function. Potential parameters for Al–O, Ca–O, P–O, and rare-earth ion-related pairs were proposed by Du and coworkers [50, 51, 77, 78]. The boron-related parameters [52] employs a composition-dependent $A$ parameter of B–O interaction which is described by the following equation:

$$A_{\text{B–O}} = \begin{cases} A_1 + t_1\times\left(R_{max} - (R_{max} - N_4)^2/R_{max}\right) + t_2\times K^2 & R \le R_{max} \\ A_2 + t_3\times R_{max}\times N_4/R & R \le R_{max} \end{cases} \tag{8.A.18}$$

where, $A_1$, $A_2$, $t_1$, and $t_2$ are empirical parameters. $R$ and $K$ are composition ratios, i.e. $Na_2O/B_2O_3$ and $SiO_2/B_2O_3$ for sodium borosilicate compositions. $N_4$ is the fraction of the four-coordinated boron, and $t_3$ is determined by solving the Eq. (8.A.18) at $R = R_{max}$:

$$t_3 = (A_1 + t_2\times K^2 - A_2)/R_{max} + t_1 \tag{8.A.19}$$

The details of the partial charges, empirical potential parameters, and parameters to determine the $A_{\text{B–O}}$ value are listed in Tables 8.A.10–8.A.12.

## 8.A.6 Borate, Boroaluminate, and Borosilicate Potential – Sundararaman et al.

Sundararaman et al. [58] developed a potential to simulate the alkali borate, boroaluminate, and borosilicate glass systems. The potential form consists of a short-range interaction

**Table 8.A.8** Parameters of the two-body term in Ha and Garofalini potential [46].

| Pair | $A_{ij}$ | $B_{ij}$ | $\rho_{ij}$ (Å) |
|---|---|---|---|
| O–O | 0.0725 | 234 | 29 |
| O–Si | 0.2962 | 234 | 29 |
| O–B | 0.1400 | 234 | 29 |
| O–Al | 0.2490 | 234 | 29 |
| O–Na | 0.3195 | 234 | 29 |
| O–Ca | 0.5700 | 234 | 29 |
| Si–Si | 0.1877 | 230 | 29 |
| Si–B | 0.2000 | 230 | 29 |
| Si–Al | 0.2523 | 233 | 29 |
| Si–Na | 0.2001 | 230 | 29 |
| B–B | 0.1100 | 230 | 29 |
| B–Al | 0.0450 | 230 | 29 |
| B–Na | 0.1740 | 230 | 29 |
| B–Ca | 0.1930 | 230 | 29 |
| Al–Al | 0.0500 | 235 | 29 |
| Al–Na | 0.2178 | 230 | 29 |
| Al–Ca | 0.2420 | 230 | 29 |
| Na–Na | 0.2159 | 230 | 29 |
| Ca–Ca | 0.7000 | 230 | 29 |

**Table 8.A.9** Three-body parameters in Ha and Garofalini potential [46].

| Triplet | $\lambda_{ij}$ (fJ) | $\gamma_{ij}$ (Å) | $R_{ij}$ (Å) | $\theta_{jik}$ (°) |
|---|---|---|---|---|
| Si/Al/B–O–Si/Al/B | 0.001 | 2.0 | 2.6 | 109.5 |
| O–Si/Al–O | 0.024 | 2.8 | 3.0 | 109.5 |
| O–B–O | 0.024 | 2.8 | 3.0 | 120.0 |

**Table 8.A.10** Partial charges of elements used in Deng and Du potential [52].

| Pair | O | Si | Al | B | Zr | P | Li | Na | K | Ca | Sr |
|---|---|---|---|---|---|---|---|---|---|---|---|
| Charge ($e$) | −1.2 | +2.4 | +1.8 | +1.8 | +2.4 | +3.0 | +0.6 | +0.6 | +0.6 | +1.2 | +1.2 |

**Table 8.A.11** Empirical parameters in Deng and Du potential for borosilicate systems together with compatible parameters from previous studies for multicomponent oxide glass [52].

| Pair | $A_{ij}$ (eV) | $\rho_{ij}$ (Å) | $C_{ij}$ (eV·Å$^6$) | $B_{ij}$ (eV·Å$^{n_{ij}}$) | $D_{ij}$ (eV/Å$^2$) | $n_{ij}$ | $r_0$ (Å) |
|---|---|---|---|---|---|---|---|
| B–O | From Eq. (8.A.18) | 0.171 271 | 28.500 | 10.336[a)] | −1.849[a)] | 4.612[a)] | 1.098[a)] |
| O–O | 2 029.2204 | 0.343 645 | 192.58 | 45.868 | −0.333 | 3.402 | 1.903 |
| Si–O | 13 702.9050 | 0.193 817 | 54.681 | 28.950 | −3.059 | 3.932 | 1.168 |
| Al–O | 12 201.4170 | 0.195 628 | 31.997 | 51.728 | −10.197 | 3.179 | 1.024 |
| Li–O | 41 051.9380 | 0.151 160 | 0.000 | — | — | — | — |
| Na–O | 4 383.7555 | 0.243 838 | 30.700 | 48.264 | −4.730 | 2.893 | 1.172 |
| K–O | 20 526.9720 | 0.233 708 | 51.489 | 311.662 | −78.964 | 2.459 | 0.949 |
| Ca–O | 7 747.1834 | 0.252 623 | 93.109 | 74.737 | −3.991 | 3.166 | 1.318 |
| P–O | 26 655.4720 | 0.181 968 | 86.856 | 28.566 | −3.408 | 4.635 | 1.168 |
| Sr–O | 14 566.6370 | 0.245 015 | 81.773 | 189.853 | −23.951 | 2.731 | 1.112 |
| Zr–O | 17 943.3840 | 0.226 627 | 127.650 | 108.655 | −8.481 | 3.305 | 1.224 |

a) Parameters are for selected composition from Deng and Du [52] and will vary for different compositions.

**Table 8.A.12** Empirical parameters used to determine the $A_{\text{B-O}}$ parameter based on Eq. (8.A.19) in Deng and Du potential [52].

| Variable | $A_1$ (eV) | $A_2$ (eV) | $t_1$ (eV) | $t_2$ (eV) |
|---|---|---|---|---|
| Parameter | 11 900.00 | 12 525.00 | 4350.00 | 85.00 |

**Table 8.A.13** Charge of different species of Sundararaman et al. potential [58].

| Pair | Si | Li | Na | K | Al | B |
|---|---|---|---|---|---|---|
| Charge (*e*) | 1.7755 | 0.5727 | 0.6018 | 0.6849 | 1.6334 | 1.6126 |

**Table 8.A.14** Parameters of short-range interactions in Sundararaman et al. potential [58]. Noting that the pairs in this table have been reordered from the original paper.

| Pair | $A_{ij}$ (eV) | $B_{ij}$ (Å$^{-1}$) | $C_{ij}$ (eV·Å$^6$) | $D_{ij}$ (eV·Å$^{24}$) |
|---|---|---|---|---|
| O–O | 1 120.5 | 2.8927 | 26.132 | 16 800.0 |
| O–Si | 23 108.0 | 5.0979 | 139.700 | 66.0 |
| O–Li | 6 745.2 | 4.9120 | 41.221 | 70.0 |
| O–Na | 1 127 566.0 | 6.8986 | 40.562 | 16 800.0 |
| O–K | 258 160.0 | 5.1698 | 130.770 | 16 800.0 |
| O–Al | 21 740.0 | 5.3054 | 65.815 | 66.0 |
| O–B | 16 182.0 | 5.6069 | 59.203 | 32.0 |
| Si–Si | 2 798.0 | 4.4073 | 0.000 | 3 423 204.0 |
| Si–Li | 17 284 | 4.3848 | 0.000 | 16 800.0 |
| Si–Na | 495 653 | 5.4151 | 0.000 | 16 800.0 |
| Si–K | 268 967 | 4.3289 | 0.000 | 16 800.0 |
| Si–B | 4 798.0 | 3.6703 | 207.000 | 16 800.0 |
| Li–Li | 2 323.8 | 3.9129 | 0.000 | 3 240.0 |
| Li–B | 4 148.6 | 3.5726 | 102.360 | 16 800.0 |
| Na–Na | 1 476.9 | 3.4075 | 0.000 | 16 800.0 |
| Na–B | 3 148.5 | 3.6183 | 34.000 | 16 800.0 |
| K–K | 3 648.0 | 4.4207 | 0.000 | 16 800.0 |
| K–B | 1 548.6 | 2.7283 | 201.360 | 16 800.0 |
| B–B | 1 805.5 | 3.8228 | 69.174 | 6 000.0 |
| Al–Al | 1 799.1 | 3.6778 | 100.000 | 16 800.0 |

described by Buckingham term and the long-range Coulombic interaction using the Wolf truncation method:

$$V^{Buck}(r_{\alpha\beta}) = A_{\alpha\beta}\exp(-B_{\alpha\beta}r_{\alpha\beta}) - \frac{C_{\alpha\beta}}{r_{\alpha\beta}^{6}} + \frac{D_{\alpha\beta}}{r_{\alpha\beta}^{24}} + V^{W}(r_{\alpha\beta}) \tag{8.A.20}$$

where $A_{\alpha\beta}$, $\rho_{\alpha\beta}$, $C_{\alpha\beta}$, and $D_{\alpha\beta}$ are empirical potential parameters, and $r_{\alpha\beta}$ is the interatomic distance between ions $\alpha$ and $\beta$, respectively. $\alpha$ and $\beta$ are ions of O, Si, Li, Na, K, Al, and B, while the Wolf truncation term can be calculated by following equation:

$$V^W(r_{\alpha\beta}) = q_\alpha q_\beta \left[ \frac{1}{r_{\alpha\beta}} - \frac{1}{r_{cut}^W} + \frac{\left(r_{\alpha\beta} - r_{cut}^W\right)}{\left(r_{cut}^W\right)^2} \right] \tag{8.A.21}$$

in which $q_\alpha$ and $q_\beta$ are the charges of ions $\alpha$ and $\beta$, and $r_{cut}^W$ is the cutoff value used for the Wolf truncation. The charge values of elements and the empirical parameters of pair interactions can be found in Tables 8.A.13 and 8.A.14.

### 8.A.7 Borate and Borosilicate Polarizable Potential – Yu et al.

Yu et al. [60] proposed a set of potential for phosphorus-bearing borosilicate systems using an established core-shell polarizable potential framework developed by Tilocca et al. [59, 79] by adding additional boron and phosphorus-related parameters. The full functional form is described by the following equation:

$$U_{\alpha\beta}(r_{\alpha\beta}) = A_{\alpha\beta} \exp\left(-\frac{r_{\alpha\beta}}{\rho_{\alpha\beta}}\right) - \frac{C_{\alpha\beta}}{r_{\alpha\beta}^6}\left\{1 - \exp\left[-\left(\frac{r_{\alpha\beta}}{4.3\rho_{\alpha\beta}}\right)^6\right]\right\} + \frac{D_{\alpha\beta}}{r_{\alpha\beta}^{12}} \tag{8.A.22}$$

in which $A_{\alpha\beta}$, $\rho_{\alpha\beta}$, $C_{\alpha\beta}$, and $D_{\alpha\beta}$ are empirical parameters, and the $r_{\alpha\beta}$ is the interatomic distance between element $\alpha$ and $\beta$, respectively. The detailed parameters can be found in Tables 8.A.15 and 8.A.16.

**Table 8.A.15** Empirical potential parameters of pair interactions in Yu et al. potential [60].

| $\alpha-\beta$ | $A_{\alpha\beta}$ (keV) | $\rho_{\alpha\beta}$ (pm) | $C_{\alpha\beta}$ (eV·Å6) | $D_{\alpha\beta}$ (eV·Å$^{12}$) |
|---|---|---|---|---|
| Na–$O_S$ | 56.465 | 19.39 | 0.000 | 0.0 |
| Ca–$O_S$ | 2.152 | 30.92 | 0.099 | 0.0 |
| Si–$O_S$ | 1.284 | 32.05 | 10.662 | 0.0 |
| B–$O_S$ | 0.472 | 33.50 | 0.000 | 5.4 |
| P–$O_S$ | 1.750 | 29.00 | 0.000 | 0.0 |
| $O_S$–$O_S$ | 22.764 | 14.90 | 27.880 | 0.0 |
| $O_C$–$O_S$ | 0.000 | 0.00 | 0.000 | 0.1 |

**Table 8.A.16** Empirical potential parameters of three-body term in Yu et al. potential [60].

| $\alpha-\beta-\alpha$ | $K_{\alpha\beta\alpha}$ (eV/rad$^2$) | $\rho_{\alpha\beta\alpha}$ (Å) | $\theta^0_{\alpha\beta\alpha}$ |
|---|---|---|---|
| $O_S$–Si–$O_S$ | 5.48 | 2.030 | 109.47° |
| $O_S$–P–$O_S$ | 2.74 | 2.030 | 109.47° |

## References


1 Smedskjaer, M.M., Mauro, J.C., Youngman, R.E. et al. (2011). Topological principles of borosilicate glass chemistry. *J. Phys. Chem. B* 115: 12930–12946.

2 Gin, S., Abdelouas, A., Criscenti, L.J. et al. (2013). An international initiative on long-term behavior of high-level nuclear waste glass. *Mater. Today* 16: 243–248. https://doi.org/10.1016/j.mattod.2013.06.008.

3 Hench, L.L., Splinter, R.J., Allen, W.C., and Greenlee, T.K. (1971). Bonding mechanisms at the interface of ceramic prosthetic materials. *J. Biomed. Mater. Res.* 5: 117–141. https://doi.org/10.1002/jbm.820050611.

4 Yue, Y., Tuheen, M.I., and Du, J. Borosilicate glasses. In: *Encyclopedia of Materials: Technical Ceramics and Glasses*, 1e, vol. 2, 519–539. Elsevier https://doi.org/10.1016/B978-0-12-818542-1.00098-9.

5 Bray, P.J. and O'Keefe, J.G. (1963). Nuclear magnetic resonance investigations of the structure of alkali borate glasses. *Phys. Chem. Glasses* 4: 37–46.

6 Boulos, E.N. and Kreidl, N.J. (1971). Structure and properties of silver borate glasses. *J. Am. Ceram. Soc.* 54: 368–375.

7 Sørensen, S.S., Johra, H., Mauro, J.C. et al. (2019). Boron anomaly in the thermal conductivity of lithium borate glasses. *Phys. Rev. Mater.* 3: 1–11. https://doi.org/10.1103/PhysRevMaterials.3.075601.

8 Malchow, P., Johanns, K.E., Möncke, D. et al. (2015). Composition and cooling-rate dependence of plastic deformation, densification, and cracking in sodium borosilicate glasses during pyramidal indentation. *J. Non-Cryst. Solids* 419: 97–109. https://doi.org/10.1016/j.jnoncrysol.2015.03.020.

9 Manara, D., Grandjean, A., and Neuville, D.R. (2009). Advances in understanding the structure of borosilicate glasses: a Raman spectroscopy study. *Am. Mineral.* 94: 777–784. https://doi.org/10.2138/am.2009.3027.

10 Smedskjaer, M.M., Zheng, Q., Mauro, J.C. et al. (2011). Sodium diffusion in boroaluminosilicate glasses. *J. Non-Cryst. Solids* 357: 3744–3750. https://doi.org/10.1016/j.jnoncrysol.2011.07.008.

11 Wright, A.C. (2010). Borate structures: crystalline and vitreous. *Phys. Chem. Glasses* 51: 1–39.

12 Pedesseau, L., Ispas, S., and Kob, W. (2015). First-principles study of a sodium borosilicate glass-former. II. The glass state. *Phys. Rev. B* 91: 134202.

13 Du, J. and Rimsza, J.M. (2017). Atomistic computer simulations of water interactions and dissolution of inorganic glasses. *npj Mater. Degrad.* 1: 16. https://doi.org/10.1038/s41529-017-0017-y.

14 Deng, L. and Du, J. (2016). Development of effective empirical potentials for molecular dynamics simulations of the structures and properties of boroaluminosilicate glasses. *J. Non-Cryst. Solids* 453: 177–194. https://doi.org/10.1016/j.jnoncrysol.2016.09.021.

15 Lu, X., Deng, L., Kuo, P.-H. et al. (2017). Effects of boron oxide substitution on the structure and bioactivity of SrO-containing bioactive glasses. *J. Mater. Sci.* 52: 8793–8811. https://doi.org/10.1007/s10853-017-0836-9.

16 Ren, M., Deng, L., and Du, J. (2017). Surface structures of sodium borosilicate glasses from molecular dynamics simulations. *J. Am. Ceram. Soc.* 100: 2516–2524. https://doi.org/10.1111/jace.14654.

**17** Silver, A.H. and Bray, P.J. (1958). Nuclear magnetic resonance absorption in glass. I. Nuclear quadrupole effects in boron oxide, soda-boric oxide, and borosilicate glasses. *J. Chem. Phys.* 29: 984. https://doi.org/10.1063/1.1744697.

**18** Yun, Y.H. and Bray, P.J. (1978). Nuclear magnetic resonance studies of the glasses in the system $Na_2O$–$B_2O_3$–$SiO_2$. *J. Non-Cryst. Solids* 27: 363–380.

**19** Yun, Y.H., Feller, S.A., and Bray, P.J. (1979). Correction and addendum to "Nuclear magnetic resonance studies of the glasses in the system $Na_2O$–$B_2O_3$–$SiO_2$". *J. Non-Cryst. Solids* 33: 273–277.

**20** Dell, W.J., Bray, P.J., and Xiao, S.Z. (1983). $^{11}B$ NMR studies and structural modeling of $Na_2O$–$B_2O_3$–$SiO_2$ glasses of high soda content. *J. Non-Cryst. Solids* 58: 1–16.

**21** Imaoka, M. (1961). Studies on the glass formation range of borate systems. *J. Ceram. Assoc. Jpn.* 69: 282–306. https://doi.org/10.2109/jcersj1950.69.789_282.

**22** Du, L.S. and Stebbins, J.F. (2003). Nature of silicon-boron mixing in sodium borosilicate glasses: a high-resolution $^{11}B$ and $^{17}O$ NMR study. *J. Phys. Chem. B* 107: 10063–10076. https://doi.org/10.1021/jp034048l.

**23** Du, L.S. and Stebbins, J.F. (2003). Solid-state NMR study of metastable immiscibility in alkali borosilicate glasses. *J. Non-Cryst. Solids* 315: 239–255. https://doi.org/10.1016/S0022-3093(02)01604-6.

**24** El-Damrawi, G., Müller-Warmuth, W., Doweidar, H., and Gohar, I.A. (1993). $^{11}B$, $^{29}Si$ and $^{27}Al$ nuclear magnetic resonance studies of $Na_2O$–$Al_2O_3$–$B_2O_3$–$SiO_2$ glasses. *Phys. Chem. Glasses* 34: 52–57.

**25** Du, L.-S. and Stebbins, J.F. (2005). Network connectivity in aluminoborosilicate glasses: a high-resolution $^{11}B$, $^{27}Al$ and $^{17}O$ NMR study. *J. Non-Cryst. Solids* 351: 3508–3520.

**26** Manara, D., Grandjean, A., and Neuville, D.R. (2009). Structure of borosilicate glasses and melts: a revision of the Yun, Bray and Dell model. *J. Non-Cryst. Solids* 355: 2528–2531. https://doi.org/10.1016/j.jnoncrysol.2009.08.033.

**27** Wu, J. and Stebbins, J.F. (2010). Quench rate and temperature effects on boron coordination in aluminoborosilicate melts. *J. Non-Cryst. Solids* 356: 2097–2108. https://doi.org/10.1016/j.jnoncrysol.2010.08.015.

**28** Lu, X., Deng, L., Du, J., and Vienna, J.D. (2021). Predicting boron coordination in multicomponent borate and borosilicate glasses using analytical models and machine learning. *J. Non-Cryst. Solids* 553: 120490. https://doi.org/10.1016/j.jnoncrysol.2020.120490.

**29** Geneste, G., Bouyer, F., and Gin, S. (2006). Hydrogen-sodium interdiffusion in borosilicate glasses investigated from first principles. *J. Non-Cryst. Solids* 352: 3147–3152. https://doi.org/10.1016/j.jnoncrysol.2006.04.023.

**30** Soleilhavoup, A., Delaye, J., Angeli, F. et al. (2010). Contribution of first-principles calculations to multinuclear NMR analysis of borosilicate glasses. *Magn. Reson. Chem.* 48: S159–S170. https://doi.org/10.1002/mrc.2673.

**31** Pedesseau, L., Ispas, S., and Kob, W. (2015). First-principles study of a sodium borosilicate glass-former. I. The liquid state. *Phys. Rev. B* 91: 134201.

**32** van Beest, B.W.H., Kramer, G.J., and Santen, R.A. (1990). Force fields for silicas and aluminophosphates based on ab initio calculations. *Phys. Rev. Lett.* 64: 1955–1958.

**33** Du, J. and Cormack, A.N. (2004). The medium range structure of sodium silicate glasses: a molecular dynamics simulation. *J. Non-Cryst. Solids* 349: 66–79. https://doi.org/10.1016/j.jnoncrysol.2004.08.264.

**34** Huang, L. and Kieffer, J. (2003). Molecular dynamics study of cristobalite silica using a charge transfer three-body potential: phase transformation and structural disorder. *J. Chem. Phys.* 118: 1487–1498.

**35** Guillot, B. and Sator, N. (2007). A computer simulation study of natural silicate melts. Part II: high pressure properties. *Geochim. Cosmochim. Acta* 71: 4538–4556.

**36** Soules, T.F. (1979). A molecular dynamic calculation of the structure of sodium silicate glass. *J. Chem. Phys.* 71: 4570. https://doi.org/10.1063/1.440632.

**37** Garofalini, S.H. (1982). Molecular dynamics simulation of the frequency spectrum of amorphous silica. *J. Chem. Phys.* 76: 3189–3192. https://doi.org/10.1063/1.443363.

**38** Soules, T.F. (1980). A molecular dynamic calculation of the structure of $B_2O_3$ glass. *J. Chem. Phys.* 73: 4032–4036. https://doi.org/10.1063/1.440632.

**39** Inoue, H., Aoki, N., and Yasui, I. (1987). Molecular dynamics simulation of the structure of borate glasses. *J. Am. Ceram. Soc.* 70: 622–627. https://doi.org/10.1111/j.1151-2916.1987.tb05729.x.

**40** Xu, Q., Kawamura, K., and Yokokawa, T. (1988). Molecular dynamics calculations for boron oxide and sodium borate glasses. *J. Non-Cryst. Solids* 104: 261–272. https://doi.org/10.1016/0022-3093(88)90397-3.

**41** Cormack, A.N. and Park, B. (2000). Molecular dynamics simulations of borate glasses. *Phys. Chem. Glasses* 41: 272–277.

**42** Tuheen, M.I., Deng, L., and Du, J. (2021). A comparative study of the effectiveness of empirical potentials for molecular dynamics simulations of borosilicate glasses. *J. Non-Cryst. Solids* 553: 120413. https://doi.org/10.1016/j.jnoncrysol.2020.120413.

**43** Kieu, L.-H., Delaye, J.-M., Cormier, L., and Stolz, C. (2011). Development of empirical potentials for sodium borosilicate glass systems. *J. Non-Cryst. Solids* 357: 3313–3321.

**44** Wang, M., Anoop Krishnan, N.M., Wang, B. et al. (2018). A new transferable interatomic potential for molecular dynamics simulations of borosilicate glasses. *J. Non-Cryst. Solids* 498: 294–304. https://doi.org/10.1016/j.jnoncrysol.2018.04.063.

**45** Inoue, H., Masuno, A., and Watanabe, Y. (2012). Modeling of the structure of sodium borosilicate glasses using pair potentials. *J. Phys. Chem. B* 116: 12325–12331.

**46** Ha, M.-T. and Garofalini, S.H. (2017). Local structure of network modifier to network former ions in soda-lime alumino-borosilicate glasses. *J. Am. Ceram. Soc.* 100: 563–573. https://doi.org/10.1111/jace.14565.

**47** Cormack, A.N., Du, J., and Zeitler, T.R. (2002). Alkali ion migration mechanisms in silicate glasses probed by molecular dynamics simulations. *Phys. Chem. Chem. Phys.* 4: 3193–3197. https://doi.org/10.1016/S0022-3093(03)00280-1.

**48** Du, J. and Cormack, A.N. (2005). Molecular dynamics simulation of the structure and hydroxylation of silica glass surfaces. *J. Am. Ceram. Soc.* 88: 2532–2539.

**49** Xiang, Y., Du, J., Smedskjaer, M.M., and Mauro, J.C. (2013). Structure and properties of sodium aluminosilicate glasses from molecular dynamics simulations. *J. Chem. Phys.* 139: 44507.

**50** Du, J. and Kokou, L. (2011). Europium environment and clustering in europium doped silica and sodium silicate glasses. *J. Non-Cryst. Solids* 357: 2235–2240. https://doi.org/10.1016/j.jnoncrysol.2010.11.088.

**51** Du, J., Kokou, L., Rygel, J.L. et al. (2011). Structure of cerium phosphate glasses: molecular dynamics simulation. *J. Am. Ceram. Soc.* 94: 2393–2401.

52 Deng, L. and Du, J. (2019). Development of boron oxide potentials for computer simulations of multicomponent oxide glasses. *J. Am. Ceram. Soc.* 102: 2482–2505. https://doi.org/10.1111/jace.16082.

53 Collin, M., Fournier, M., Frugier, P. et al. (2018). Structure of international simple glass and properties of passivating layer formed in circumneutral pH conditions. *npj Mater. Degrad.* 2: 4. https://doi.org/10.1038/s41529-017-0025-y.

54 Lu, X., Deng, L., Kerisit, S., and Du, J. (2018). Structural role of $ZrO_2$ and its impact on properties of boroaluminosilicate nuclear waste glasses. *npj Mater. Degrad.* 2: 19. https://doi.org/10.1038/s41529-018-0041-6.

55 Lu, X., Ren, M., Deng, L. et al. (2019). Structural features of ISG borosilicate nuclear waste glasses revealed from high-energy X-ray diffraction and molecular dynamics simulations. *J. Nucl. Mater.* 515: 284–293. https://doi.org/10.1016/j.jnucmat.2018.12.041.

56 Ren, M., Lu, X., Deng, L. et al. (2018). $B_2O_3/SiO_2$ substitution effect on structure and properties of $Na_2O$–CaO–SrO–$P_2O_5$–$SiO_2$ bioactive glasses from molecular dynamics simulations. *Phys. Chem. Chem. Phys.* 20: 14090–14104. https://doi.org/10.1039/C7CP08358K.

57 Sundararaman, S., Huang, L., Ispas, S., and Kob, W. (2019). New interaction potentials for alkali and alkaline-earth aluminosilicate glasses. *J. Chem. Phys.* 150: https://doi.org/10.1063/1.5079663.

58 Sundararaman, S., Huang, L., Ispas, S., and Kob, W. (2020). New interaction potentials for borate glasses with mixed network formers. *J. Chem. Phys.* 152: 104501. https://doi.org/10.1063/1.5142605.

59 Tilocca, A., De Leeuw, N.H., and Cormack, A.N. (2006). Shell-model molecular dynamics calculations of modified silicate glasses. *Phys. Rev. B: Condens.* 73: 1–14. https://doi.org/10.1103/PhysRevB.73.104209.

60 Yu, Y., Stevensson, B., and Edén, M. (2017). Medium-range structural organization of phosphorus-bearing borosilicate glasses revealed by advanced solid-state NMR experiments and MD simulations: consequences of B/Si substitutions. *J. Phys. Chem. B* 121: 9737–9752. https://doi.org/10.1021/acs.jpcb.7b06654.

61 Stevensson, B., Yu, Y., and Edén, M. (2018). Structure-composition trends in multicomponent borosilicate-based glasses deduced from molecular dynamics simulations with improved B–O and P–O force fields. *Phys. Chem. Chem. Phys.* 20: 8192–8209. https://doi.org/10.1039/c7cp08593a.

62 Fortino, M., Berselli, A., Stone-Weiss, N. et al. (2019). Assessment of interatomic parameters for the reproduction of borosilicate glass structures via DFT-GIPAW calculations. *J. Am. Ceram. Soc.* 102: 7225–7243. https://doi.org/10.1111/jace.16655.

63 Lee, K.H., Yang, Y., Ziebarth, B. et al. (2020). Evaluation of classical interatomic potentials for molecular dynamics simulations of borosilicate glasses. *J. Non-Cryst. Solids* 528: 119736. https://doi.org/10.1016/j.jnoncrysol.2019.119736.

64 Vollmayr, K., Kob, W., and Binder, K. (1996). How do the properties of a glass depend on the cooling rate? A computer simulation study of a Lennard-Jones system. *J. Chem. Phys.* 105: 4714–4728. https://doi.org/10.1063/1.472326.

65 Vollmayr, K., Kob, W., and Binder, K. (1996). Cooling-rate effects in amorphous silica: a computer-simulation study. *Phys. Rev. B* 54: 15808–15827. https://doi.org/10.1103/PhysRevB.54.15808.

66 Lane, J.M.D. (2015). Cooling rate and stress relaxation in silica melts and glasses via microsecond molecular dynamics. *Phys. Rev. E: Stat. Nonlinear Soft Matter Phys.* 92: 1–8. https://doi.org/10.1103/PhysRevE.92.012320.

67 Tilocca, A. (2013). Cooling rate and size effects on the medium-range structure of multicomponent oxide glasses simulated by molecular dynamics. *J. Chem. Phys.* 139: 114501. https://doi.org/10.1063/1.4821150.

68 Deng, L. and Du, J. (2018). Effects of system size and cooling rate on the structure and properties of sodium borosilicate glasses from molecular dynamics simulations. *J. Chem. Phys.* 148: 024504. https://doi.org/10.1063/1.5007083.

69 Meikhail, M.S. (1991). Diffusion of sodium ions in borosilicate glasses by molecular dynamics method. *Zeitschrift Für Phys. B Condens. Matter.* 84: 309–313. https://doi.org/10.1007/BF01313553.

70 Gou, F., Greaves, G.N., Smith, W., and Winter, R. (2001). Molecular dynamics simulation of sodium borosilicate glasses. *J. Non-Cryst. Solids* 293–295: 539–546. https://doi.org/10.1016/S0022-3093(01)00775-X.

71 Kilymis, D.A., Delaye, J.-M., and Ispas, S. (2015). Behavior of sodium borosilicate glasses under compression using molecular dynamics. *J. Chem. Phys.* 143: 094503. https://doi.org/10.1063/1.4929785.

72 Delaye, J.M., Peuget, S., Bureau, G., and Calas, G. (2011). Molecular dynamics simulation of radiation damage in glasses. *J. Non-Cryst. Solids* 357: 2763–2768. https://doi.org/10.1016/j.jnoncrysol.2011.02.026.

73 Ren, M., Deng, L., and Du, J. (2017). Bulk, surface structures and properties of sodium borosilicate and boroaluminosilicate nuclear waste glasses from molecular dynamics simulations. *J. Non-Cryst. Solids* 476: 87–94. https://doi.org/10.1016/j.jnoncrysol.2017.09.030.

74 Garofalini, S.H., Ha, M.T., and Urraca, J. (2018). Simulations of the surfaces of soda lime aluminoborosilicate glasses exposed to water. *J. Am. Ceram. Soc.* 101: 1135–1148. https://doi.org/10.1111/jace.15237.

75 Lu, X., Deng, L., Huntley, C. et al. (2018). Mixed network former effect on structure, physical properties, and bioactivity of 45S5 bioactive glasses: an integrated experimental and molecular dynamics simulation study. *J. Phys. Chem. B* 122: 2564–2577. https://doi.org/10.1021/acs.jpcb.7b12127.

76 Lu, X., Deng, L., Gin, S., and Du, J. (2019). Quantitative structure–property relationship (QSPR) analysis of $ZrO_2$-containing soda-lime borosilicate glasses. *J. Phys. Chem. B* 123: 1412–1422. https://doi.org/10.1021/acs.jpcb.8b11108.

77 Xiang, Y. and Du, J. (2011). Effect of strontium substitution on the structure of 45S5 bioglasses. *Chem. Mater.* 23: 2703–2717. https://doi.org/10.1021/cm102889q.

78 Du, J. (2015). Challenges in molecular dynamics simulations of multicomponent oxide glasses. In: *Molecular Dynamics Simulations of Disordered Materials* (ed. C. Massobrio, J. Du, M. Bernasconi and P.S. Salmon), 157–180. Cham: Springer.

79 Tilocca, A. (2008). Short- and medium-range structure of multicomponent bioactive glasses and melts: an assessment of the performances of shell-model and rigid-ion potentials. *J. Chem. Phys.* 129: https://doi.org/10.1063/1.2972146.

# 9

# Atomistic Simulation of Nuclear Waste Glasses

*Jean-Marc Delaye*

*Département de recherche sur les technologies pour l'Enrichissement, le Démantèlement et les Déchets, Commissariat à l'énergie atomique et aux énergies alternatives, Direction des Energies, Institut pour les Sciences et Technologies pour une Economie, Circulaire des Energies bas carbone, Bagnols-sur-Cèze CEDEX, France*

## 9.1 Preamble

Among the different glass compositions used in industry, nuclear glasses occupy a special place. While on the one hand these glasses are required to contain a very large set of radioelements resulting in complex compositions with more than 30 oxides [1–5], on the other hand, they must be able to resist internal irradiation and external water alteration over very long periods in view of their future geological repository storage, until their radioactivity reaches that of the natural environment.

For these glasses, experiments and simulations are both difficult to manage because it is necessary to work with complex compositions and to extrapolate results just obtained at laboratory scale to much longer geological time.

The approaches generally adopted consist in either conducting integral experiments with the real compositions to identify the global behaviors, or in studying simplified compositions in order to better understand the basic mechanisms [6–10].

This chapter presents a summary of different atomistic simulations (ab initio, classical molecular dynamics, and Monte Carlo) performed on simplified nuclear glasses in order to better understand the effects of irradiation on their structural and macroscopic properties as well as the glass–water alteration mechanisms leading to the formation of a protective layer. Only the case of French borosilicate nuclear glass is discussed here.

## 9.2 Introduction to French Nuclear Glass

### 9.2.1 Chemical Composition

For the treatment of spent nuclear fuels, two options are considered: either direct storage or the incorporation of the minor actinides and fission products (after their separation from

*Atomistic Simulations of Glasses: Fundamentals and Applications*, First Edition.
Edited by Jincheng Du and Alastair N. Cormack.

other fuel elements) into a glassy matrix called nuclear glass. Nuclear glasses are intended to be stored in a deep geological environment to isolate the radioactive elements from the geosphere. This second option has been chosen by several countries (China, France, India, Japan, Russia, United Kingdom, etc.). Different nuclear glass compositions are currently being studied. This chapter focuses on the case of the borosilicate glasses chosen by France to store nonrecoverable radioelements.

In order to select the glass composition, both technical and economic criteria have been taken into account.

First, it is necessary for the glass packages to be manufactured at a reasonable cost. This means that the elaboration temperature has to remain sufficiently low, typically in the range of 1100–1300 °C. The viscosity has to be suitable for the glass to be poured into the steel canisters while in this temperature range. Consequently, the composition has had to be modulated between the glass formers (silica, alumina, borate, etc.) and the glass modifiers (alkaline, alkaline earth, etc.).

Good glass quality means first that the matrix must be homogeneous at microscale, with a limited number of crystalline inclusions or unmelted particles, and able to incorporate a very large set of radioelements (the minor actinides and fission products mainly).

Second, as nuclear glass will be stored in a deep geological repository, it must resist internal radiation effects and external water alteration on a geological timescale, typically several thousands of years, until the residual radioactivity returns to the natural radioactivity level.

To meet all these requirements, the glass composition has to be precisely adjusted, and therefore the resulting composition is quite complex. This work corresponds to the glass formulation. In France, more than 30 different oxides can be found in the nuclear glass, called R7T7. Table 9.1 shows the average French nuclear glass composition. The main components, $SiO_2$, $B_2O_3$, $Na_2O$, $Al_2O_3$, CaO, represent more than 80% in molar percentages. The other components come from the fission product solution or are specifically added to refine the glass properties in the liquid state.

### 9.2.2 About the Long-Term Behavior (Irradiation, Glass Alteration, He Accumulation)

To be accepted, a nuclear glass has to resist the internal disintegrations generated by radioelements. There are several kinds of radioelements, $\alpha$ and $\beta/\gamma$ emitters.

The fission products are responsible for $\beta/\gamma$ disintegrations which induce mainly electronic excitations. Several works have focused on the damage caused by these disintegrations. In particular, a wide variety of point defects can be created [11–15].

The long-lived minor actinides are responsible for $\alpha$-disintegrations. After an $\alpha$ disintegration, one He ion with an energy in the range 4–5 MeV (rapidly neutralized by the surrounding material) and one recoil nucleus with an energy in the range of 80–120 keV are generated. The $\alpha$ particle mainly loses its energy through electronic interaction, while the recoil nucleus provokes a large number of ballistic collisions. On average, one $\alpha$ particle displaces 400 atoms, while the recoil nucleus displaces 2000 atoms.

Hence, as far as the number of displaced atoms is concerned, the recoil nuclei are the more damaging particles.

**Table 9.1** Average composition of the French R7T7 glass in %wt.

| Oxides | Average composition (%wt) |
|---|---|
| $SiO_2$ | 45.6 |
| $B_2O_3$ | 14.1 |
| $Al_2O_3$ | 4.7 |
| $Na_2O$ | 9.9 |
| CaO | 4.0 |
| $Fe_2O_3$ | 1.1 |
| NiO | 0.1 |
| $Cr_2O_3$ | 0.1 |
| $P_2O_5$ | 0.2 |
| $Li_2O$ | 2.0 |
| ZnO | 2.5 |
| Oxides (Fission products, actinides, Zr) + Fine suspensions | 17.0 |
| Minor actinides | 0.6 |

Concerning glass alteration by water, several stages have been identified between the initial alteration rate and the long-term behavior. In the early period of the geological storage, the waste package will be in a closed system, meaning that no liquid water will be in contact with it. After a certain time, the metallic containers will be corroded by the groundwater and the nuclear glass will therefore be directly in contact with water, i.e. in an open system. Thus, it is important to understand the fundamental mechanisms at the origin of glass alteration in order to be able to predict the quantity of radioelements likely to be released into the geological environment.

When a glass is in contact with an aqueous solution, the alteration rate follows the curve shown in Figure 9.1 [16, 17].

At the very beginning of the glass alteration (Stage 0), an interdiffusion occurs between the water molecules and the most easily soluble elements (B, Na). This stage corresponds to a reactive diffusion, as the release of the soluble elements is accompanied by local structural relaxations. Next, the glass is altered at the maximum rate, due to the hydrolysis of the silicate network (Stage I). During this stage, some elements which had passed into solution can be redeposited on the glass surface, leading to the progressive formation of a protective layer (Stage II). Thereafter, the alteration rate decreases. A dynamic equilibrium is reached and the alteration rate stabilizes at the residual rate (Stage III). Depending on the glass composition and the solution pH, some alteration resumptions are possible in the long term, due to the precipitation of secondary phases that can destabilize the protective layer (Stage IV).

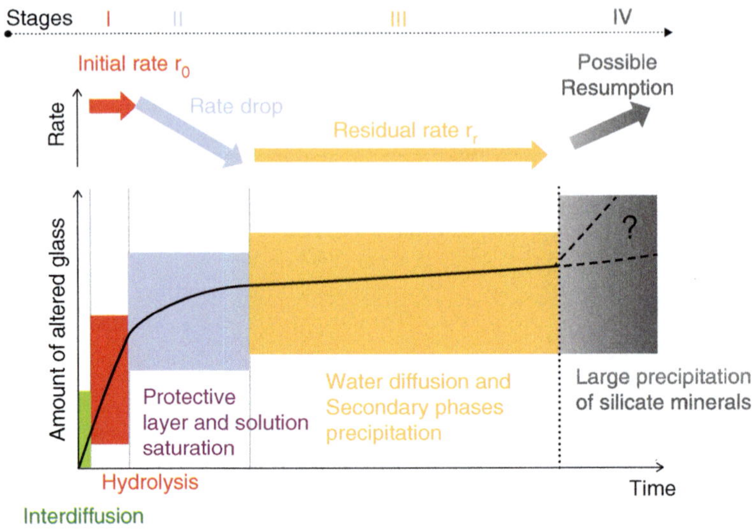


**Figure 9.1** Alteration rate of a glass in contact with an aqueous solution. Different stages are identified. Source: Figure from Gin et al. [16].

In the case of the geological storage planned for France, the pH of the surrounding solution will be lower than 9.5.

Another issue related to the long-term behavior of nuclear glasses in a geological repository concerns gas accumulation within the matrix. As mentioned above, a He molecule is emitted after an α-disintegration, and the quantity of He will increase with the dose. It is therefore important to understand both how the He molecules will behave inside the material and the conditions under which gaseous bubbles will form [18].

### 9.2.3 What Can Atomistic Simulations Contribute?

To better understand the basic mechanisms at the origin of long-term glass behavior, several studies using computational techniques have been developed to complete the experimental approaches. This document is mainly dedicated to the contribution of simulation to a better understanding of the different processes. The reader is invited to consult the following references to obtain a broader overview of experimental approaches [1, 3, 8].

As real nuclear glass contains more than 30 different components, it is impossible to simulate real systems by atomistic simulations. The first step therefore consists in simplifying the real composition to retain only the major components.

For this reason, the $SiO_2$–$B_2O_3$–$Na_2O$ system, possibly completed by $Al_2O_3$, CaO, and $ZrO_2$ oxides, has been used very frequently. This represents a good model for a real nuclear glass, as it contains the main formers ($SiO_2$, $B_2O_3$, $Al_2O_3$), and modifiers ($Na_2O$, CaO) [19–21].

Among the formers, $Al_2O_3$ and $B_2O_3$ oxides play a special role because the Al and B ions in a tetrahedral environment need to be locally charge-compensated by a modifier.

Concerning the B ions, they can be three or four coordinated, and the $^{III}B \leftrightarrow {}^{IV}B$ conversion is a fundamental process in explaining structural and mechanical property changes under irradiation. $ZrO_2$ is regularly considered in water alteration studies because it is insoluble and strongly impacts alteration behavior.

Regarding radiation effects, it has been shown experimentally that the ballistic effects (nuclear energy deposition) are preponderant in explaining structural and mechanical property changes. Several studies have been carried out using light and heavy ion fluxes and in each case, the macroscopic property changes have been plotted versus the electronic or nuclear energy deposited. Figure 9.2 shows an example of the hardness [8]. When the nuclear energy deposited is considered, all the points align along the same regular curve (Figure 9.2a). This is not the case when the electronic energy is considered (Figure 9.2b). In this case, the points are much more dispersed and no clear trend can be identified. Thus, when the level of nuclear energy deposited is known, it is possible to extrapolate the hardness change. However, this is not possible with the electronic energy deposited.

Another point proving the preponderance of the ballistic over the electronic effects can be seen in fracture toughness behavior. It has been shown that there is no fracture toughness change when nuclear glass is subjected to irradiation by light ions (responsible for only electronic energy deposition), while there is a large fracture toughness change when the glass is subjected to heavy ion irradiation (responsible for electronic and nuclear energy depositions) [8].

The analogies between real and simplified glass behaviors under irradiation or water alteration justify using simplified compositions to understand the basic mechanisms at

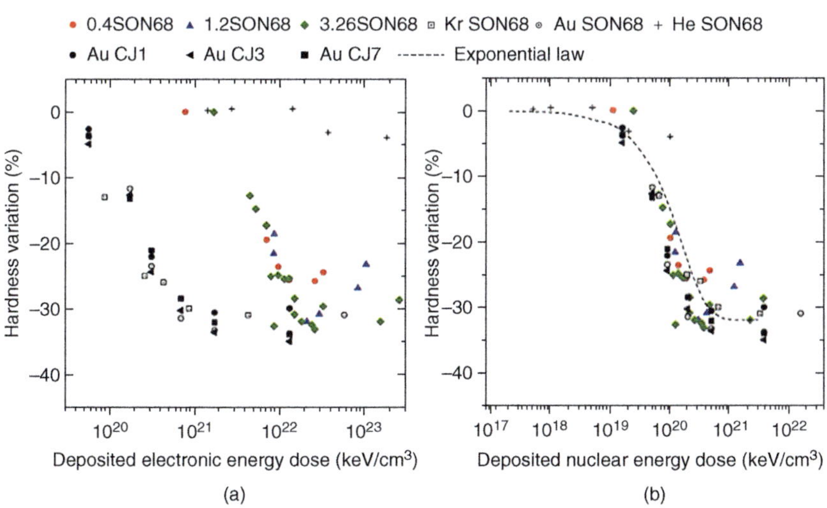


**Figure 9.2** Effects of the deposited electronic (a) and nuclear (b) energies on the hardness change in real and simplified nuclear glasses. Different radiation conditions are considered (Kr, Au, He, and doped glasses). Simplified glasses (SBN14 = CJ1, CJ3, CJ7) and real glass (SON68) were used. Source: Based on Peuget et al. [8].

the origin of long-term glass behavior. It can be seen in Figure 9.2 that under irradiation by heavy ions, the simplified glasses (SBN14 = CJ1: 67.7%$SiO_2$–18.0%$B_2O_3$–14.2%$Na_2O$; CJ3: 61.2%$SiO_2$–16.3%$B_2O_3$–12.9%$Na_2O$–3.0%$Al_2O_3$–5.8%CaO; CJ7: 63.8%$SiO_2$–17.0%$B_2O_3$ –13.4%$Na_2O$–4.1%$Al_2O_3$–1.8%$ZrO_2$) behave qualitatively in the same manner as real glass. In the following references, it can be seen that there are also considerable analogies between the behaviors of simplified and real glasses under water alteration [16, 22, 23].

The rest of this chapter is organized as follows: paragraph 3 describes the computational methodology used to simulate simplified nuclear glasses. The radiation effect simulations are described in paragraph 4. Paragraph 5 is dedicated to glass–water interaction studies using ab initio or Monte Carlo techniques. Finally, paragraph 6 describes how He molecules can be incorporated inside the glass matrix and how radiation effects modify the He solubility. A short conclusion is given in paragraph 7.

## 9.3 Computational Methodology

Several stages are necessary to simulate simplified nuclear glasses by classical molecular dynamics. The first stage consists in choosing (or fitting) classical potentials to represent the interactions between the atoms. The second stage consists in preparing an atomistic structure representative of the system to simulate. Finally, the third stage consists in studying the property of interest by performing nanoindentation, fracture simulation, accumulation of displacement cascades, etc.

These different stages are described below.

### 9.3.1 Review of Existing Classical Potentials for Borosilicate Glasses

Several classical potentials proposed to simulate borosilicate glasses can be found in the literature, all based on a Buckingham form [24–27]. This means that the energy associated with an atomic pair $(i,j)$ separated by a distance $r_{ij}$ is given by:

$$\phi(r_{ij}) = \frac{q_i q_j}{r_{ij}} + \frac{A_{ij}}{\rho_{ij}} \exp\left(-\frac{r_{ij}}{\rho_{ij}}\right) - \frac{c_{ij}}{r_{ij}^6} \tag{9.1}$$

In this equation, $q_i$ and $q_j$ are the ionic charges of the ions $i$ and $j$, and $A_{ij}$ and $\rho_{ij}$ are adjustable parameters.

To better represent the strong local order around the Si and B atoms, some three-body potentials are sometimes added. One possible form is the following:

$$\emptyset_3(r_{ij}, r_{ik}, \theta) = \lambda_i \exp\left(\frac{\gamma_i}{r_{ij} - r_{c1}} + \frac{\gamma_i}{r_{ik} - r_{c2}}\right) \times (\cos\theta - \cos\theta_0)^2 \tag{9.2}$$

In this Eq. (9.2), $(j,i,k)$ represents a triplet of ions with the ion $i$ in the central position. The terms $r_{ij}$ and $r_{ik}$ represent, respectively, the $i$–$j$ and $i$–$k$ distances, and the term $\theta$ represents the <jik> angle. The other parameters are adjustable parameters.

In general, the adjustable parameters of the pair and three-body potentials are chosen in order to reproduce a set of structural, elastic, or mechanical experimental data.

The details of the performances of the first classical potentials we have used to simulate borosilicate glasses can be consulted in the following references [20, 28].

Recently, in order to simulate the mechanical property changes under irradiation of different $SiO_2$–$B_2O_3$–$Na_2O$ glasses, Kieu et al. [24] developed a new version of classical potentials with charges depending on the chemical composition. These potentials were able to reproduce the nonlinear evolution of the B coordination versus the R ($R = [Na_2O]/[B_2O_3]$) and K ($K = [SiO_2]/[B_2O_3]$) ratios as stated by Yun, Dell, and Bray [29, 30], and different mechanical and elastic properties of crystalline and amorphous $SiO_2$–$B_2O_3$–$Na_2O$ systems.

The results presenting in the following have been obtained using either the first potentials (displacement cascades) or Kieu's potentials (displacement cascades, mechanical properties, and He incorporation).

### 9.3.2 Preparation of a Glass

A glass configuration is generally prepared using the following algorithm. First, a liquid is equilibrated at a temperature in the range of (4000–6000 K) typically for 100 ps or more. This stage enables the initial memory of the structure to be completely lost. The liquid is then quenched at a rate in the range of ($10^{11}$–$10^{14}$ K/s) to solidify a glass at ambient temperature. The quench can be done at constant volume or constant pressure. When it is performed at constant volume, the glass contraction as the temperature decreases leads to a decrease in the pressure. In order to determine the final equilibrium volume, a relaxation at constant pressure is applied.

Figure 9.3 summarizes the different stages needed to prepare a glass by classical molecular dynamics.

The final glass configuration can be validated by comparison with different experimental data, for example short-range orders (local coordination, first neighbor distances,

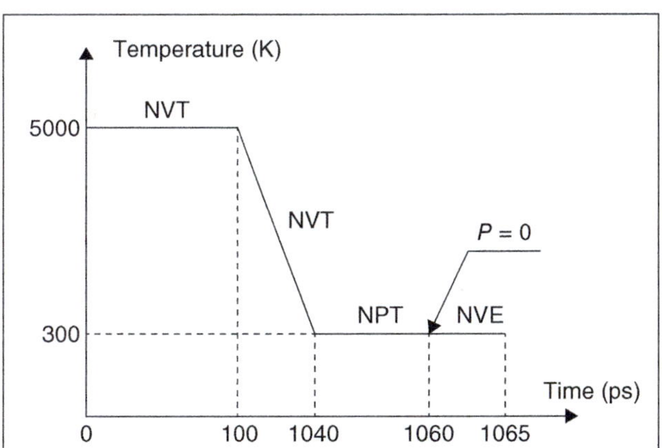


**Figure 9.3** Preparation of a glass. A liquid is prepared in the NVT ensemble, and then quenched to ambient temperature in the NVT ensemble. The equilibrium volume of the final configuration is determined by a relaxation in the NPT ensemble. The glass is finally relaxed in the NVE ensemble.

and angular distribution) or medium-range orders (structural factor, radial distribution function, and ring size distribution, etc.). It is also possible to compare its elastic and mechanical properties to experimental data [31]. Details are not given here because each author proposes his own validation between experimental and simulated data.

The time step used to perform the glass preparation is typically equal to 1 fs.

### 9.3.3 Displacement Cascade Simulations

Displacement cascades can be accumulated in a glass structure in order to study the ballistic effects [32, 33]. A displacement cascade is initiated by accelerating one particle inside a simulation box, thus inducing a collision cascade. Inside the core of the cascade, the distance between atoms can be shortened, and the classical potentials are no longer valid to represent such interactions. To address this issue, special pair potentials have been developed by Ziegler, Biersack, and Littmarck (ZBL) [34] based on ab initio calculations. These potentials are described by the following analytical form:

$$\Phi_{ZBL}(r_{ij}) = \frac{q_i q_j}{r_{ij}} \sum_{k=1}^{4} c_k \exp\left(-\frac{b_k r_{ij}}{a}\right) \tag{9.3}$$

The parameters $a$, $b_k$, and $c_k$ are constants.

The different adjustable parameters of the ZBL potentials have been fitted to best represent the interaction energy when two atoms approach each other.

The ZBL potentials, generally applied for interatomic distances lower than 0.1 nm, are connected to the classical potentials using polynomial expressions. These polynomial expressions are fitted to ensure the continuity of the energies, forces, and force derivatives.

At the beginning of a displacement cascade, few atoms can have very large kinetic energies. As long as the collision cascade continues to extend, the kinetic energy is more and more widely shared among a larger number of atoms and the largest atomic velocity decreases. To avoid simulation instability, a small time step is applied at the beginning (typically $10^{-18}$ s) and this is progressively increased as the largest atomic velocity decreases. Such a methodology is described in the following reference [35]. At the end, when the irradiated structure finally equilibrates, the time step returns to 1 fs.

Accelerating one heavy projectile corresponds to introducing an additional energy to the simulation box, and the temperature of the structure suddenly increases. In order to avoid a divergence of the temperature when a large number of displacement cascades accumulate, one solution is to thermalize the external layer of the simulation box to maintain it at ambient temperature. This can be done by regularly adjusting the velocities of the atoms in the external layer. In this case, any drift of the temperature due to the accumulation of displacement cascades is impeded.

### 9.3.4 Short Bibliography About Simplified Nuclear Glass Structure Studies

This document is mainly dedicated to radiation effect investigation but first a short synthesis of the simplified nuclear glass structural studies performed by classical molecular dynamics is proposed. This section is to indicate a selection of papers about such studies.

There are much fewer structural studies about borosilicate glasses than other silicate compositions, because the boron oxide is difficult to simulate due to the coexistence of its three and four coordinated local environments. The relative proportions of three and four coordinated B atoms are linked to the R and K ratios [29, 30, 36], and precisely simulating the relative interactions between the three $SiO_2$, $B_2O_3$, and $Na_2O$ oxides is a difficult task.

Systems containing between four and six oxides have also been studied by a combination of classical molecular dynamics and Reverse Monte Carlo, leading to a better understanding of the atomic organization inside the glass [19, 20, 37]. Also, coupling nuclear magnetic resonance (NMR) and classical molecular dynamics (MD) enables a better interpretation of the local order around the Na and Ca atoms [38, 39].

## 9.4 Simulation of Radiation Effects in Simplified Nuclear Glasses

### 9.4.1 Accumulation of Displacement Cascades and the Thermal Quench Model

In order to understand how a simplified nuclear glass reacts to a nuclear energy deposition, some series of 600 eV displacement cascades have been accumulated in a simplified nuclear glass (SBN14: 67.7%$SiO_2$–18.1%$B_2O_3$–14.2%$Na_2O$). Simulation boxes contain 4000 atoms [40].

All the cascades were accumulated without modifying the volume of the simulation box but regularly, after each five cascades, a relaxation in the NPT, and then in the NVE, ensemble was performed to estimate the new equilibrium volume of the intermediate configurations.

In the initial box subjected to a series of 100 displacement cascades, the internal pressure gradually increases with the number of cascades, which indicates a trend toward swelling. When the new equilibrium volume is calculated after each five cascades, the actual swelling can be drawn versus the nuclear energy deposited (Figure 9.4). The swelling rate is maximal at the beginning and progressively decreases until a saturation plateau is reached where the density stabilizes.

The behavior of the simplified nuclear glass subjected to a series of displacement cascades mimics those of complex and simplified nuclear glasses experimentally subjected to internal and external irradiations [8, 9]. Qualitatively, the form of the density decrease (or of the volume increase) until a saturation plateau is reached is the same in the simulation as in the experiments. Quantitatively, the global swelling of the simplified nuclear glass is roughly equal to 4%, not far from the experimental value, as the swelling measured by an optical interferometry technique in SBN14 glass subjected to an external multienergy Au implantation (1, 3.5, and 7 MeV) was not far from 4% [41].

In both computational and experimental cases, a Marples model [42] can be fitted with the curves.

The origin of this model is as follows. The model hypothesized that a particle irradiating an already irradiated zone has no additional effects. If $F$ is the volume fraction already

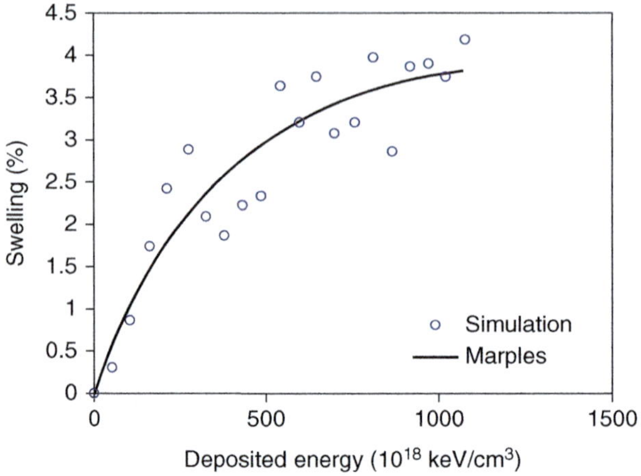


**Figure 9.4** Volume change of a simplified nuclear glass subjected to one hundred 600 eV – displacement cascades. The solid line corresponds to a fit using the Marples model. Source: Figure from Delaye et al. [40].

impacted by the projectiles, $R$ the dose rate (in particle/cm$^3$/s), and $V_c$ the elementary volume disturbed by one irradiating particle (in cm$^3$), this means the volume fraction impacted by the projectiles evolves according to the following equation:

$$\frac{dF}{dt} = V_c R(1 - F) \tag{9.4}$$

Integrating this equation leads to the following formulae:

$$F = 1 - \exp(-V_c Rt) \tag{9.5}$$

According to the Marples model, the swelling is proportional to the volume fraction impacted by the projectiles:

$$\frac{\Delta V}{V_{tot}} = A_c F = A_c[1 - \exp(-V_c Rt)] \tag{9.6}$$

The Marples model contains two adjustable parameters, $A_c$ and $V_c$ that can be fitted on the experimental or simulated curves. In both the experimental and computational cases, this model can be fitted with the data with good accuracy [8].

One strong hypothesis at the origin of this model is that just one projectile is sufficient to completely change one elementary volume. The simulation enables the pertinence of this hypothesis to be verified. The following analysis has been performed. In the simulation box, the numbers of oxygen atoms displaced once, twice, or three times have been identified and plotted versus the nuclear energy deposited (Figure 9.5). It can be seen that the swelling begins to saturate when all the O atoms have been displaced once. The correspondence with Figure 9.4 is less with the oxygen atoms displaced twice or three times. The hypothesis at the origin of the Marples model is thus justified.

The origin of the swelling can be identified at the atomistic level. Parallel to the swelling, decreases of the B and Na local coordinations are observed, as shown in Figure 9.6.

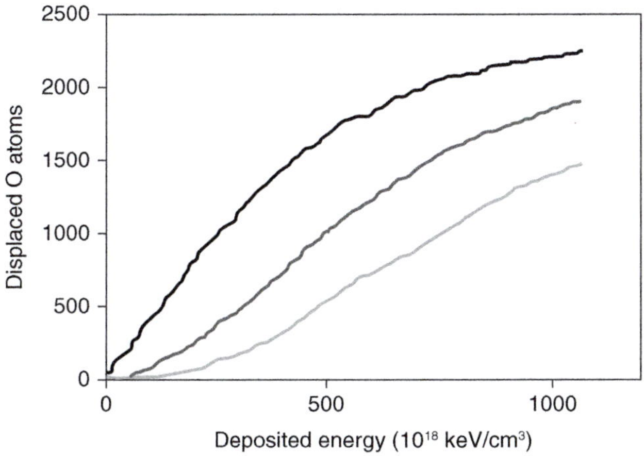


**Figure 9.5** Number of oxygen atoms displaced at least once (black), twice (dark gray), or three times (light gray) versus the deposited energy in a series of 600 eV displacement cascades in SBN14 glass. Source: Figure from Delaye et al. [40].

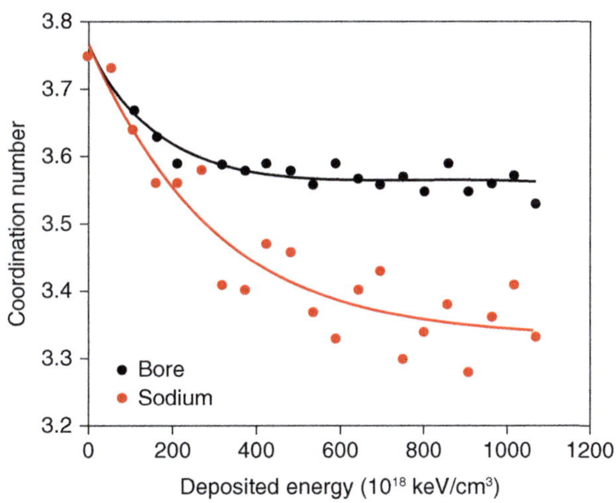


**Figure 9.6** Evolution of the mean coordination number of boron and sodium atoms in SBN14 glass subjected to a series of 600 eV cascades. Trend lines were fitted using exponential functions. The Na coordination is shifted to be visualized on the same scale. Source: Figure from Delaye et al. [40].

As previously mentioned, each B atom can be in a three or a four coordinated state, and the decrease of the B coordination corresponds to transitions from four coordinated to three coordinated local environments. Some Na atoms originally mobilized to charge and to compensate $BO_4$ entities are freed to create nonbridging oxygens. Thus, the B coordination decrease is accompanied by a depolymerization of the network.

As the atomic density of a $BO_3$ entity is lower than that of a $BO_4$ entity, this global process can explain a part of the glass swelling.

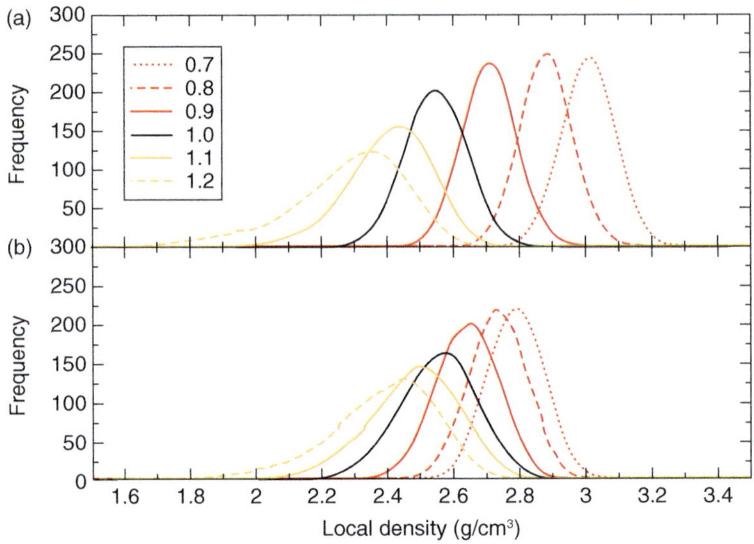


**Figure 9.7** Local density distributions for the (a) pristine and (b) irradiated glass prepared with different initial volumes. Source: Figure from Kilymis et al. [43].

Interestingly, after the accumulation of the displacement cascades, an increase in the level of disorder can be observed in the glass. This can be seen notably in the distribution of the local density around the different atoms. In a study considering several SBN14 glasses prepared with different initial densities and subjected to some series of 800 eV displacement cascades, the local densities were measured around each atom considering a radius of 8 Å [43]. It can be seen that after the accumulation of the displacement cascades, a systematic widening of the local density distribution occurs (Figure 9.7). This effect should not be confused with the standardization of the curves after the accumulation of the displacement cascades.

The increase in disorder can also be seen with the systematic widening of the angular or ring distributions or with the broadening of the first peaks of the radial distribution functions [40]. The increase in the disorder comes with an increase in the potential energy [43].

The origin of the swelling is then associated with the depolymerization, with the B and Na coordination changes, and with the increase in the disorder.

The Marples model shows that one passage of a projectile through a local region is sufficient to disturb it completely. The temperature of the local region rises suddenly when the projectile crosses it and a constrained liquid region is formed, embedded into the surrounding material still at ambient temperature. In this high-temperature region, depolymerization and disordering occur. This region is then quenched very rapidly at ambient temperature (it has been shown that the local quench rate can be larger than $10^{15}$ K/s) because its volume is very low [44]. The entire volume is progressively impacted as the deposited dose increases. When one already disturbed area is crossed again by a new projectile, no further structural changes occur because after the quench, the glass irradiated twice will be identical to the glass irradiated only once. No additional effects can be expected; hence, the saturation observed for the macroscopic properties (density and hardness).

At the end, a new metastable structure is formed as a consequence of the accumulation of these fast local thermal quenches. This explanation is proposed for the saturation of the properties of simplified and real nuclear glasses under irradiation.

Additionally, it can be shown experimentally that the fictive temperature of the irradiated glasses increases [45].

Even if some discrepancies exist, the numerous analogies observed between the quench rate effects and the radiation effects reinforce this explanation [46, 47].

### 9.4.2 Preparation of Disordered and Depolymerized Glasses

In order to study the mechanical property changes under irradiation (hardness and fracture toughness), large simulation boxes are required to simulate nanoindentations or fracture propagations. It would be very time-consuming to accumulate a large number of displacement cascades in large systems to simulate a structure's complete irradiation. To avoid this issue, another approach has been used. It consists in taking advantage of the analogies between the quench rate effects and those of ballistic collisions.

To quickly reproduce the depolymerization and disordering induced by ballistic collisions, a special algorithm has been applied consisting in reheating the initial structure equilibrated at ambient temperature a second time. The heating temperature is fixed at 3000 K. The relaxation lasts 50 ps in the NPT ensemble. The heated structure is then quenched to room temperature at $10^{14}$ K/s in the NVT ensemble. Using this, the new configuration (called the fast quenched structure below) is more depolymerized and disordered than the initial one. The quench rate to prepare the fast quenched structure has been fitted to best reproduce the swelling and depolymerization levels measured after the accumulation of a series of 600 eV displacement cascades in a smaller simulation box.

Table 9.2 compares the structural and macroscopic characteristics of the fast quenched structure and of the glass subjected to a series of displacement cascades. It can be seen that the swelling, depolymerization, and disordering are reproduced, even if some discrepancies

**Table 9.2** Comparison of the effects of accelerated quenching and accumulation of 600 eV displacement cascades in SBN14 glass.

| | Accelerated quenching | Displacement cascades |
|---|---|---|
| Swelling | 7% | 4% |
| Percentage increase in $^{[3]}$B | 10% | 17% |
| Decrease in Si–O–Si angle | 2.2° | 6° |
| Decrease in Si–O–B angle | 2° | 6° |
| Mean ring size | Initial: 4.82 | Initial: 4.95 |
| | Final: 4.72 | Final: 4.79 |

**Table 9.3** Structural evolution of the SBN12 and SBN55 glasses after heat treatment.

| | SBN12 | SBN55 |
|---|---|---|
| Swelling | 2.4% | 6% |
| Percentage increase in tricoordinate boron | 3% | 11% |
| Decrease in Si–O–Si angle | 1.7° | 1.1° |
| Decrease in Si–O–B angle | 1.7° | 1.0° |
| Mean ring size | Initial: 5.09 | Initial: 5.22 |
| | Final: 5.02 | Final: 5.18 |

remain. The fast quenched structure will be considered as a model of a glass subjected to a series of displacement cascades.

The same technique was applied to prepare two other fast quenched glasses, SBN12 glass (59.7%$SiO_2$–28.1%$B_2O_3$–12.2%$Na_2O$, mol%) and SBN55 glass (55.3%$SiO_2$–14.7%$B_2O_3$ -30.0%$Na_2O$, mol%).

Table 9.3 describes the structural changes between the pristine and fast quenched SBN12 and SBN55 glasses.

The equilibrated and fast quenched glasses have been used to manage nanoindentations and fracture propagations in order to understand the origins of the hardness decrease and fracture toughness increase in nuclear glasses.

### 9.4.3 Origin of the Hardness Change Under Irradiation

Hardness has been measured in the pristine and fast quenched SBN12, SBN14, and SBN55 glasses [48, 49]. Each simulation box contains $2 \times 10^6$ atoms.

The periodic boundary conditions were removed along the $Z$ direction and an additional relaxation was performed to relax the freshly created surface. The bottom layer of the simulation box was frozen. The nanoindentations were simulated by pushing an indenter constituted of C atoms down into the glass. The Vickers' indenter geometry was used. Figure 9.8 describes how the nanoindentations were performed.

The C atoms of the indenter were not allowed to move apart from the global translation applied to the tip. The penetration speed was equal to 10 m/s. The interactions between the indenter and the glass atoms were simulated by ZBL potentials [34]. The maximal charge applied on the indenter was such that the penetration depth was 15–20 Å.

When the maximum penetration depth was reached, a 50 ps holding period was applied to cancel most of the collective motions. After this, the indenter was pulled out of the glass. At the end, an imprint had formed at the surface (see the insert in Figure 9.9).

Figure 9.9 compares the indentation imprint profiles in the disordered and pristine glasses. In each case, the same maximum load was applied. Notice that the scale on the $y$-axis is different between the upper and lower curves. For the same load, the indenter

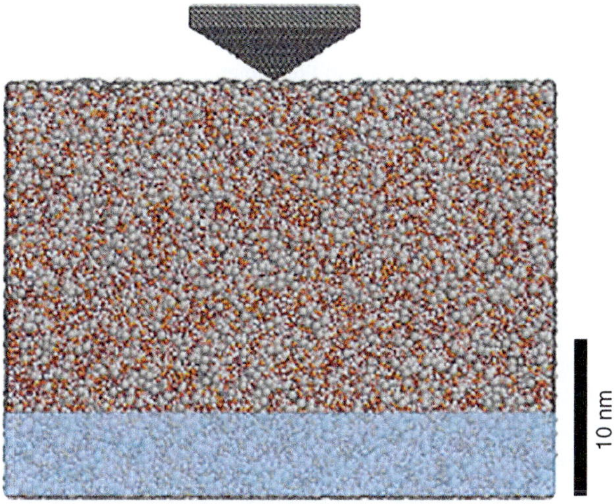


**Figure 9.8** Scheme of a simulation box submitted to nanoindentation. The bottom layer is frozen. A surface is created at the top and the Vickers' indenter is pushed into the structure to form an imprint. Source: Figure from Kilymis et al. [48].

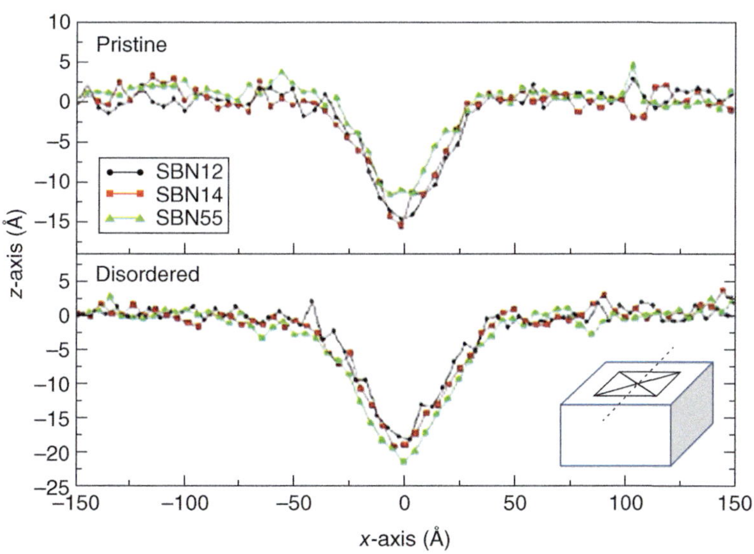


**Figure 9.9** Indentation profiles in the pristine and fast quenched glasses. Source: Figure from Kilymis et al. [48].

penetrates more deeply into the fast quenched glasses than the pristine materials. This means that qualitatively, there is a hardness decrease in the disordered glasses.

To quantitatively measure the hardness, the Oliver and Pharr method was used [50, 51]. This method takes advantage of the loading – unloading curves. The loading – unloading curves for the three glassy compositions are plotted in Figure 9.10. To illustrate the Oliver and Pharr method, a typical loading – unloading curve is also plotted in Figure 9.10 (bottom right).

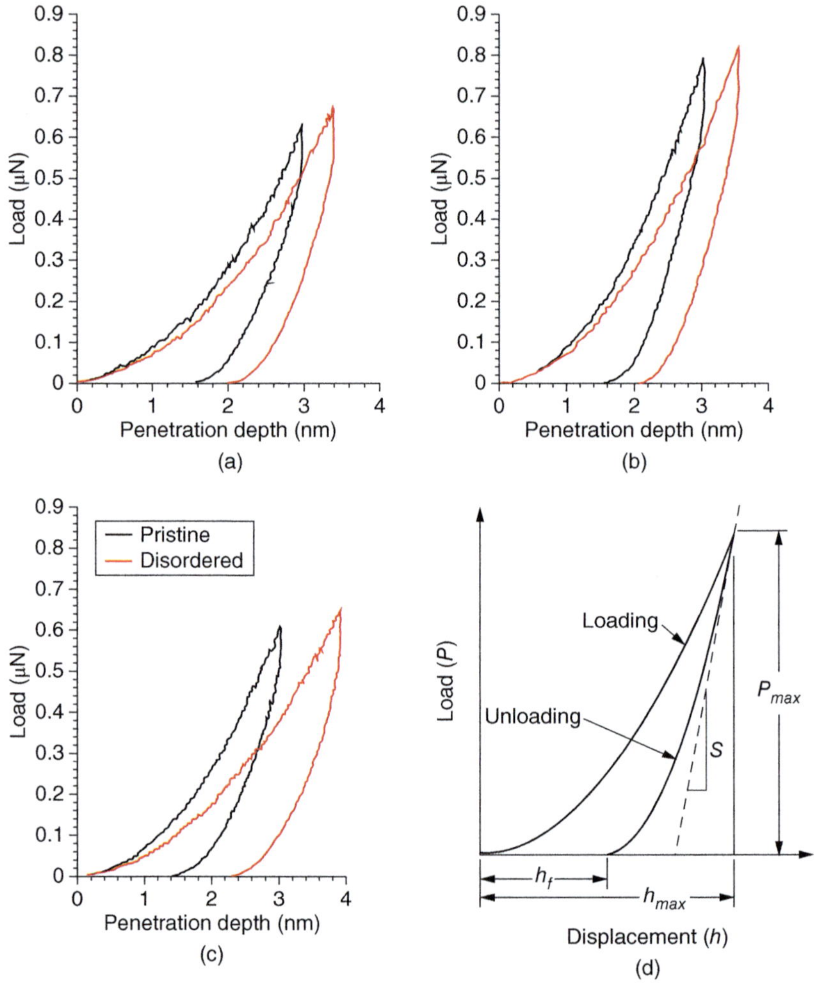


**Figure 9.10** Loading – unloading curves for pristine and fast quenched glasses (a) SBN12, (b) SBN14, (c) SBN55. A typical loading – unloading curve is added in (d) to illustrate the Oliver and Pharr method (see text). Source: Reproduced from Kilymis and Delaye [49] with the permission of AIP Publishing.

The stiffness $S$ is defined as the slope of the curve just at the beginning of the unloading. To determine this, the curve is fitted using the following equation:

$$P = a(h - h_f)^m \tag{9.7}$$

with $h_{max}$ the maximum penetration depth.

In this Eq. (9.7), $h_f$ is the plastic deformation after unloading and $a$ and $m$ are adjustable parameters. Knowing $S$, $h_{max}$, and $P_{max}$, the load at the maximum penetration depth, the contact depth $h_c$ is determined:

$$h_c = h_{max} - 0.75\frac{P_{max}}{S} \tag{9.8}$$

The projected area $A_c$ corresponding to the contact depth can be calculated, knowing the indenter geometry:

$$A_c = 4h_c^2 \tan^2\frac{\varphi}{2} \tag{9.9}$$

with $\varphi$ the angle at the apex of the indenter.

Using this model, the hardness values were calculated for pristine and fast quenched configurations. The values are given in Table 9.4.

A systematic decrease in the hardness can be observed for the fast quenched glasses.

To understand the origin of this decrease, correlations with different structural characteristics were plotted (Figures 9.11 and 9.12). Figure 9.11 shows that the hardness systematically increases with the $SiO_2$ percentage, reaching a maximum for pure silica (the calculations performed on pure silica are described in [52]).

It can be seen that the hardness decreases when the $^{III}B$ or non-bridging oxygen (NBO) percentages increases. Even if it is difficult to separate the structural effects from each other, it appears that the more depolymerized the glass is, the lower the hardness.

Another effect concerning the density cannot be ignored. Systematically, the density decreases in fast quenched glasses. This means that the free volume increases. This could also impact hardness because it can be supposed that the hardness will decrease when the free volume increases.

It appears that a nuclear energy deposition systematically induces a decrease in the B coordination, an increase in the NBO percentage, and probably the formation of free volume. Together with a depolymerization increase favoring shear flow, the tip penetrates more easily. Hence, the hardness has decreased.

**Table 9.4** Hardness values (GPa) for the pristine and fast quenched SBN12, SBN14 and SBN55 glasses.

| | $H_v$ (pristine) | $H_v$ (fast quenched) | $\Delta H_v$ (%) |
|---|---|---|---|
| SBN12 | $4.4 \pm 0.4$ | $3.4 \pm 0.3$ | −22.7 |
| SBN14 | $5.5 \pm 0.5$ | $3.9 \pm 0.6$ | −29.1 |
| SBN55 | $4.3 \pm 0.3$ | $2.5 \pm 0.2$ | −41.9 |

The change in percentages is given in the last column.

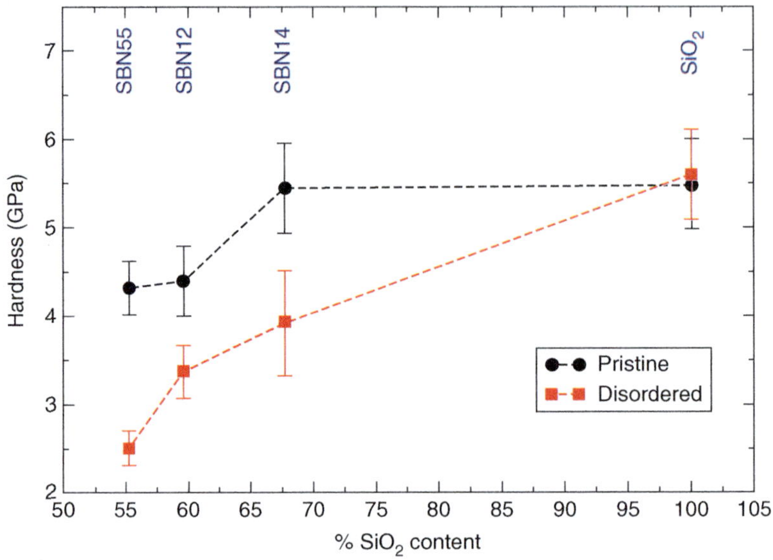


**Figure 9.11** Hardness of the pristine and fast quenched (= disordered) SBN12, SBN14, SBN55, and $SiO_2$ glasses versus the $SiO_2$ concentration. Source: Figure from Kilymis and Delaye [48].

Concerning the simplified nuclear glass SBN14, the order of magnitude of the hardness decrease is not far from the experimental value. So the possible origins of a hardness decrease listed above can explain most of the hardness change in nuclear glasses.

### 9.4.4 Origin of the Fracture Toughness Change Under Irradiation

It has been shown that when real nuclear glasses or simplified nuclear glasses are subjected to heavy ion irradiation, their fracture toughness increases by around 20% or more [8]. In order to better understand the origin of this increase, fracture was simulated in the SBN14 glass [53]. Pristine and fast quenched configurations were considered. A parallelepiped simulation box containing 100 000 atoms was prepared with the desired chemical composition. A notch was drawn on one side of the simulation box, as shown in Figure 9.13. The upper and lower atomic layers were frozen and a tensile stress was applied to initiate the fracture.

The strain rate was equal to $4 \times 10^{-6}$ $fs^{-1}$. Several strain rates were tested without significant differences in the glass behavior.

When the tensile stress was progressively increased, the stress intensity factor increased at the tip of the notch and when the critical threshold was reached, a fracture began to propagate. Several stages can be distinguished during the fracture propagation. They are described in Figure 9.14. At the beginning, nanocavities appeared, not only near the tip of the notch but also in the bulk. These nanocavities progressively swelled and lengthened in the direction of the crack propagation, finishing by coalescing to form larger voids. Finally, the complete decohesion of the structure occurred.

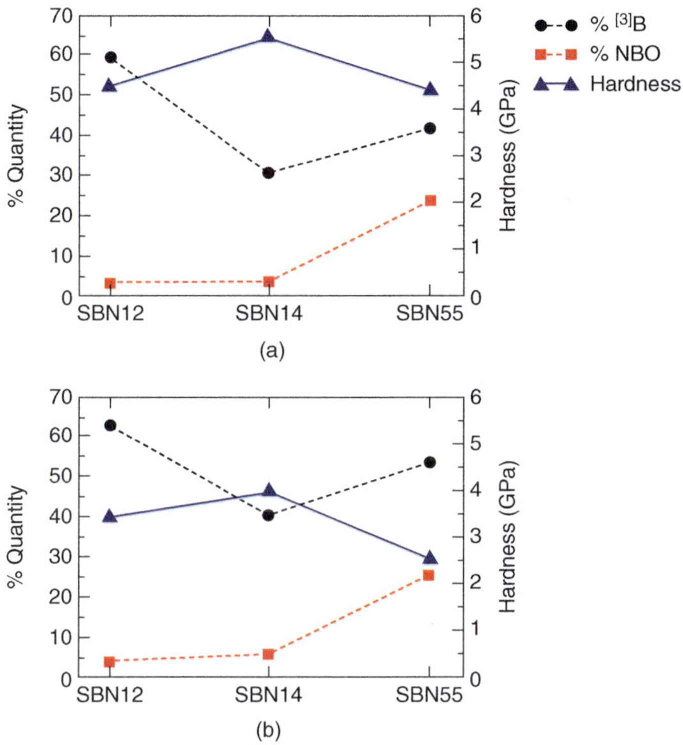


**Figure 9.12** Hardness change in the pristine (a) and fast quenched (b) SBN12, SBN14, and SBN55 glasses. The hardness values are plotted together with the [3]B and NBO percentages. Source: Figure from Kilymis and Delaye [48].

**Figure 9.13** Scheme of a simulation box submitted to a tensile stress. The upper and lower layers are frozen.

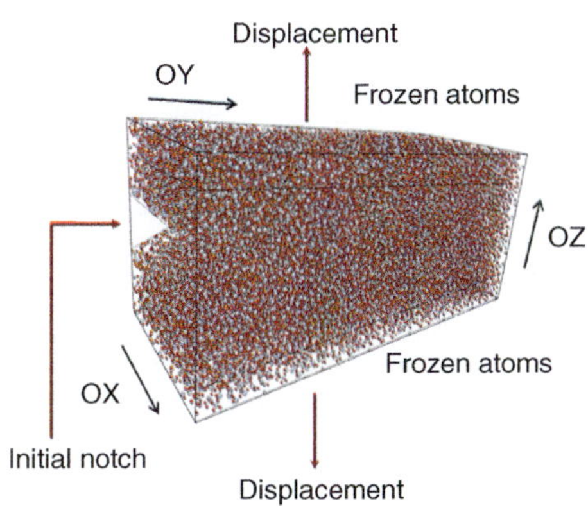

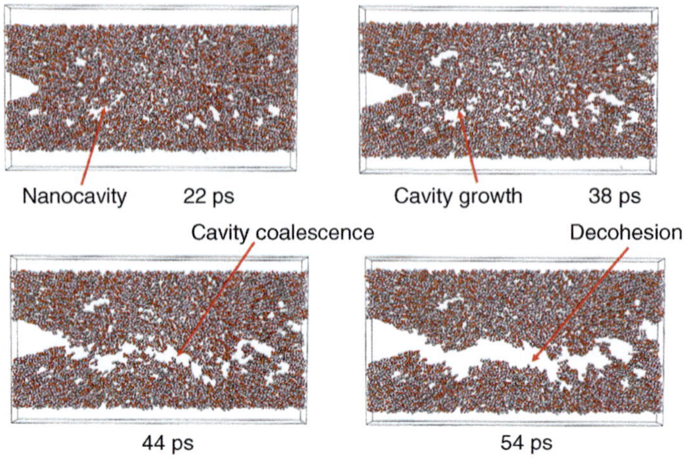


**Figure 9.14** The four stages leading to the complete decohesion of SBN14 glass. Source: Figure from Kieu et al. [53].

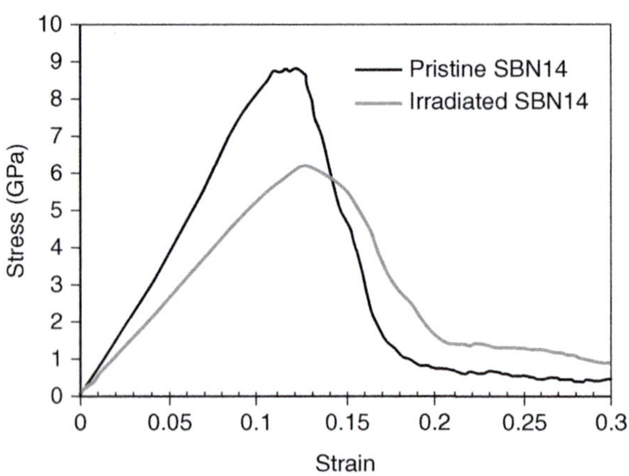


**Figure 9.15** Stress–strain curves for the pristine and fast quenched SBN14 glasses. The term "irradiated" used here means "fast quenched". Source: Figure from Kieu et al. [53].

The stress–strain curves are shown in Figure 9.15.

The curve is linear for the small strains and the slope gives the Young modulus. The Young modulus decreased in the fast quenched glass as expected. The decrease was equal to 30%. A region corresponding to a more plastic deformation was then reached and the strain continued to increase while the stress remained roughly constant. In this region, plastic processes occurred, allowing local stress relaxation. After this, a sudden drop in the stress took place. This stage corresponds to the complete decohesion of the glass and to the formation of new internal surfaces. The local stresses were then completely relaxed. The stress intensity was lower in the plastic region in the fast quenched glass compared to the pristine material. This means that a larger disorder and depolymerization make plastic processes easier. Also,

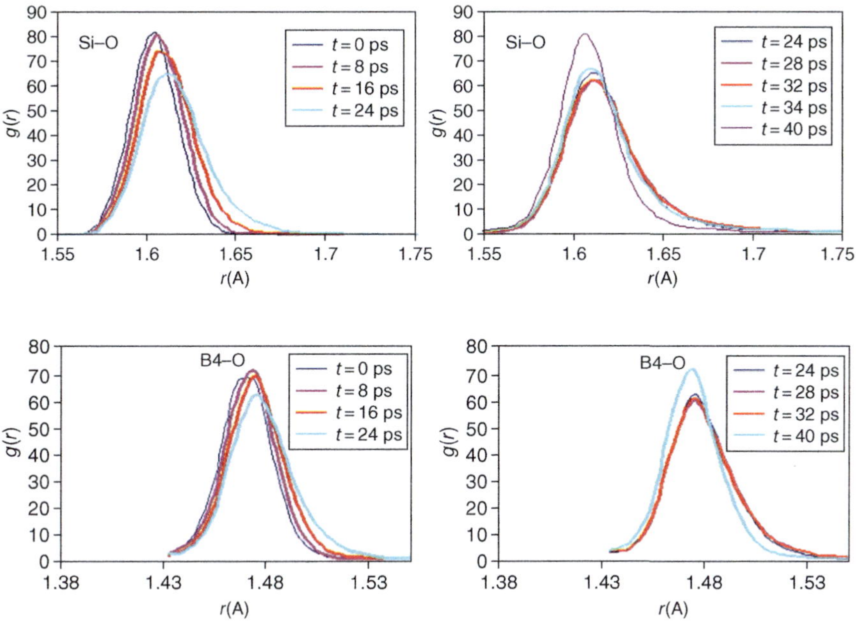


**Figure 9.16** First peaks of Si–O and $^{IV}$B–O radial distribution functions at different instants during fracture propagation in SBN14 glass.

the plastic region extended on a larger strain range in the fast quenched glass. The plastic processes are therefore more numerous in this glass.

The change in the local atomic environments was studied to better understand the origin of the plastic processes. The first peaks of the different radial distribution functions were followed at different instants during the fracture propagation (Figures 9.16 and 9.17). The B atoms are separated depending on their local coordination.

Two categories of elements can be distinguished, the Si and $^{IV}$B atoms on one side, and the $^{III}$B and Na atoms on the other. Concerning the Si and $^{IV}$B atoms, the first peak of their radial distribution functions began to shift toward larger distances during the elastic phase. After, the peaks came back to their initial positions. A decrease in the intensity of the first peak for the $^{IV}$B–O radial distribution function is observed because of the $^{IV}B \rightarrow {}^{III}B$ conversions during the fracture.

As the Si–O and $^{IV}$B–O first neighbor distances widened during the elastic phase and then returned to their initial values, the Si and $^{IV}$B local environments were subjected to elastic deformations.

The behavior is different when the $^{III}$B and Na atoms are considered. In this case, there were practically no changes of the first peak positions. This means that the local stresses were easily accommodated by the $^{III}$B and Na local environments. Hence, these environments favor plastic rather than elastic deformations.

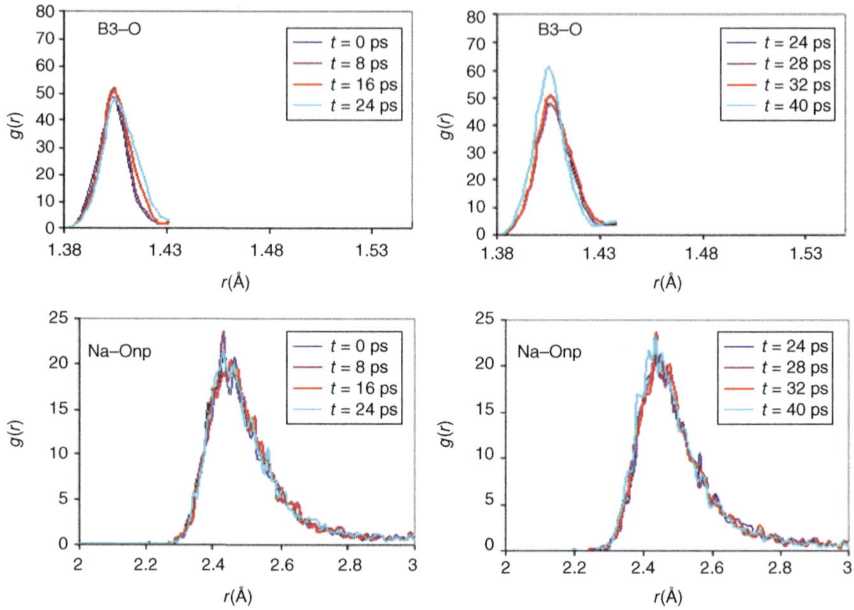


Figure 9.17 First peaks of the $^{III}$B–O and Na–O radial distribution functions at different instants during fracture propagation in SBN14 glass.

In the fast quenched SBN14 glass, there are more $^{III}$B and less $^{IV}$B than in the pristine glass, and this explains why the plastic processes are more numerous, given that plastic processes are favored by the $^{III}$B atoms' local environments.

In conclusion, a natural explanation can be proposed to understand the increase in fracture toughness in simplified nuclear glass under irradiation. The accumulation of ballistic collisions induces a decrease in B coordination. Consequently, the plastic processes are easier and consume a greater energy, hence, the increase in fracture toughness.

Experimentally, the increase in fracture toughness for SBN14 glass after neutron irradiation has also been measured by Vickers' nanoindentation, and a 16% increase has been observed [54], confirming the qualitative behavior revealed by molecular dynamics.

## 9.5 Simulation of Glass Alteration by Water

### 9.5.1 Contribution from Ab Initio Calculations

Ab initio calculations have been performed to study the interactions between amorphous silicate systems (silica, aluminosilicate, borosilicate) and water. Due to the impossibility of treating a large number of atoms, the studies are limited to small clusters.

To begin, the studies performed on pure silica and aluminosilicate can be summarized.

The activation energy for the hydrolysis of a Si—O—Si bond has been calculated using the ab initio method and different degrees of polymerization have been considered for the

Si atoms constituting the Si—O—Si bond. In order that most resemble the experimental case, the following study [55] took into account the solvation of a hydronium ion. Despite some discrepancies compared to previous results concerning activation energies [56], the main steps of the Si—O—Si bond hydrolysis were described. First, the bridging oxygen is protonated. Then a five-coordinated Si atom is formed just before the breaking of the Si—O—Si bond.

The activation energy for Si–O–Si hydrolysis rises as the Si local polymerization degree becomes higher.

The case of aluminosilicate glasses has also been treated. Different clusters containing Si and Al atoms were simulated [56, 57]. It was shown that the $Na^+$ ions stabilized preferentially the O–Si–O–Al–O–Si–O–Al rings rather than the O–Si–O–Si–O–Al–O–Al rings. On the contrary, the $Ca^{2+}$ ions stabilized preferentially the O–Si–O–Si–O–Al–O–Al–O rings. It has been shown that Si–O–Al hydrolysis is easier than Si–O–Si hydrolysis. This result has been recently confirmed by an ab initio study in an aluminosilicate mineral [58, 59].

To most resemble the real nuclear glass case, a gel layer model (67%$SiO_2$–12%$Al_2O_3$ –21%CaO) has been considered [60]. Systems containing around 100 atoms were stabilized by ab initio calculations (SIESTA code, Density Functional Theory) and each Si—O—Si and Si—O—Al bond was broken and put in contact with a $H_2O$ molecule.

Concerning Si—O—Si bonds, it appears that there is a wide distribution of hydrolysis energies depending on the local degree of disorder, evidencing the influence of the mid-range order. The hydrolysis of some bonds is exothermic, while the hydrolysis of others is endothermic. The authors propose an explanation for the fact that the silanol group concentration saturates in a silicate glass when the $H_2O$ concentration increases. They suggest that only the weakest bonds will be hydrolyzed by the water molecules, and once all the weakest bonds have been broken, the concentration of nondissociated water molecules begins to increase.

Similar hydrolysis mechanisms have been observed for the Si—O—Al bonds with the possibility of forming a new entity after the bond breakage, an $AlO_3$–$H_2O$ group. The authors also confirmed the possibility of the Burnham mechanism occurring. This mechanism supposes the migration of a proton after a Si–O–Si breakage to form a Si—OH—Al bond. Then the proton can migrate between different Si—O—Al bonds.

The same kind of work has been performed with a borosilicate glass [58]. In this case, a 70%$SiO_2$–16.7%$B_2O_3$–13.3%$Na_2O$ glass was generated by classical molecular dynamics and stabilized by an ab initio method (SIESTA code, Density Functional Theory). When a $Na^+$ ion in a modifier position (i.e. associated with a Si atom by a non-bridging O) was substituted for by a $H_3O^+$ hydronium ion, a Si–OH silanol group was formed, with a hydrogen bond to a $H_2O$ molecule. There were no deep structural rearrangements.

When a Na in a charge compensating role (near a $BO_4$ group involved in a $^{IV}B$—O—Si or $^{IV}B$—O—$^{IV}B$ bond) was substituted for by a $H_3O^+$ ion, one $^{IV}B$ was converted into a $^{III}B$, and one silanol Si–OH group or boranol B–OH group was formed. The remaining $H_2O$ molecule was linked to the network by three hydrogen bonds.

When the $^{IV}B$ is involved in a $^{IV}B$—O—$^{III}B$ bond, the $^{IV}B$ converts into a $^{III}B$ and a B–OH group forms on the $^{III}B$.

This study showed that after interdiffusion between hydronium and $Na^+$ ions, the hydronium ion has a large probability of dissociating into a proton and a water molecule. The authors propose that near the glass–water surface, the $Na^+$ – water exchange involves hydronium ions, but deeper in the glass, the protons will dominate the exchanges with the $Na^+$ ion.

Another recent work using ab initio calculations has treated the question of the Si—O—B and B—O—B bond hydrolysis under three different conditions: protonated (acidic environment), neutral, and deprotonated (basic environment). It was shown that the energy barriers increase when the conditions change from acidic to basic. Hydrolysis is easier for the B—O—B than Si—O—B bonds in acidic conditions. In the deprotonated (basic) conditions, the reaction energies for the hydrolysis of the B—O—B and Si—O—B bonds become closer to each other. This result can explain the experimental observations. It has in fact been observed [61, 62] that at low pH, borosilicate glass dissolution is incongruent, with a preferential dissolution of the Na and B species. Conversely, at high pH the glass dissolves congruently.

In the same spirit as the work performed by Xiao and Lasaga [56], clusters around the $^{IV}$Si-O-$^{IV}$Si, $^{IV}$Si-O-$^{III}$B, and $^{III}$B-O-$^{III}$B have been simulated using the Gaussian98 code [63]. The enthalpies of hydrolysis were determined for these clusters, confirming the better stability of the $^{IV}$Si-O-$^{IV}$Si cluster. The $^{III}$B—O—$^{III}$B bond presented the weakest stability against hydrolysis. The enthalpy of hydrolysis and the stability of the $^{IV}$Si—O—$^{III}$B bond were intermediate.

### 9.5.2 Contribution from Monte Carlo Simulations

A Monte Carlo simulation has been developed to represent the glass–water alteration [64–66]. This method enabled the representation of $SiO_2$–$B_2O_3$–$Na_2O$–$Al_2O_3$–CaO–$ZrO_2$ systems containing several millions of atoms. First, a network representative of a glass was built. A cubic network was used, and the nodes were occupied by the atoms or water molecules. Only the formers were represented (Si, B, and Al). Some bonds were broken to simulate the mixing of tetra-coordinated and tri-coordinated atoms, as well as the presence of nonbridging oxygens created by the modifiers (Na and Ca). Such a network is represented in Figure 9.18.

The Zr atoms are six coordinated and cannot be located on a site of the network (the local coordination of all the sites is equal to three or four). The technique adopted is represented in Figure 9.19. New bonds are artificially created between an interstitial site and six "normal" network sites.

To simulate the glass–water alteration, an "ocean" of water molecules is placed in contact with the glass. Three categories of atoms are defined; while boron and sodium atoms are immediately dissolved when they are in contact with water, the Si and Al atoms are dissolved with defined probabilities, and the Zr atoms are considered insoluble, i.e. they stay within the network.

When a Si or Al atom is in contact with water, it is connected with a maximum of three other bonds to the network. Then for the Si atoms, three different probabilities ($w_3$, $w_2$, and $w_1$) have been defined depending on the number of bonds connecting the Si atom to the network. The number of bonds between the Si atom and B and Al atoms are taken into

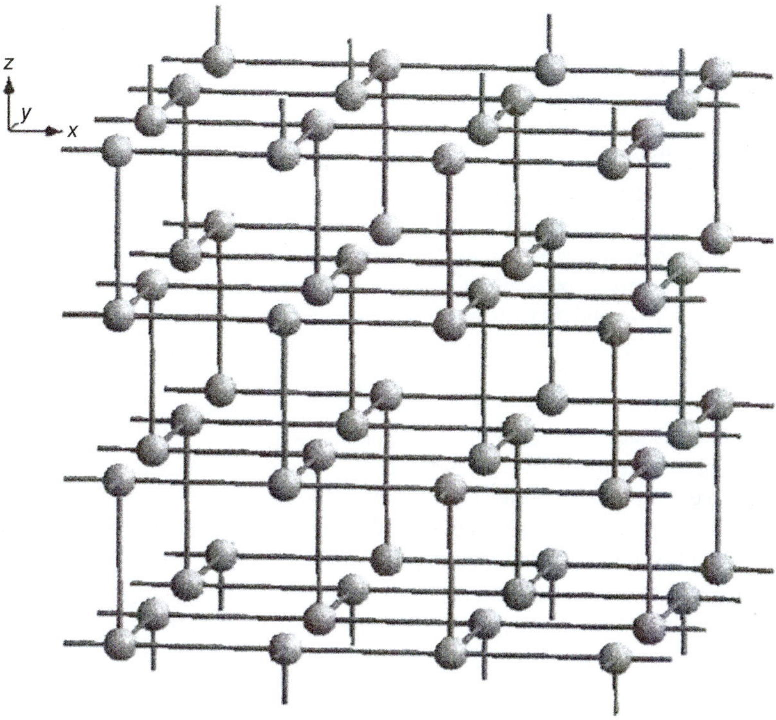


Figure 9.18 Example of a Monte Carlo network used to simulate glass–water alteration.

Figure 9.19 Addition of a six coordinated Zr atom in the Monte Carlo network.

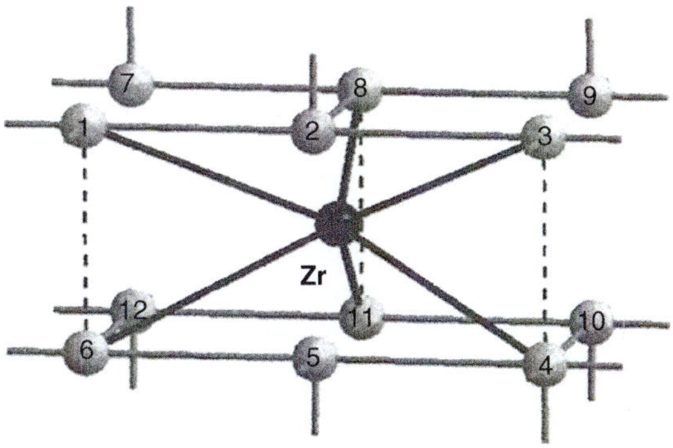


account to modulate the dissolution probability:

$$w_{\mathrm{Si}(i,j,k)} = f_{\mathrm{Al}}^{k} f_{\mathrm{B}}^{j} w_{(i+j+k)} \tag{9.10}$$

$w_{\mathrm{Si}(i,j,k)}$ is the dissolution probability for a Si atom connected to the network by $i+j+k$ bonds, with $i$ Si—O—Si bonds, $j$ Si—O—B bonds, and $k$ Si—O—Al bonds.

The $f_B$ parameter is taken to be greater than one, to favor dissolution when the Si is connected to B atoms, and the $f_{Al}$ parameter is taken as less than one, to take into account the decrease in the dissolution rate when the $Al_2O_3$ concentration increases.

The probability for Al dissolution is taken as equal to:

$$w_{Al(i,j,k)} = f_{Al} w_{Si(i,j,k)} \tag{9.11}$$

No additional empirical parameters are added to treat the Al dissolution except $f_{Al}$.

Condensation probabilities are also considered to allow for the recondensation of the dissolved Si and Al atoms on the glass surface. The diffusion of the dissolved Si and Al atoms within the solution is considered as infinitely fast. The Si and Al concentrations in solution are noted $c_{Si}$ and $c_{Al}$. Two condensation probabilities are defined for the Si and Al atoms, $w_c^{Si}$ and $w_c^{Al}$. On a site occupied by water at the glass–solution interface, the probability for a Si or Al atom to condense is equal, respectively, to $w_c^{Si} c_{Si}$ and $w_c^{Al} c_{Al}$.

The different parameters have been fitted to reproduce the alteration rate and gel morphology of several glasses dissolved in controlled conditions [64].

Before starting a calculation, it is first necessary to choose the surface-to-volume ratio (S/V: glass surface to water volume) and to define the different adjustable parameters. Then the calculation can be run. It corresponds to the repetition of a loop consisting of the different stages listed below, until the alteration stops:

- *Dissolution*: The atoms in contact with the solution are dissolved depending on their dissolution probabilities. A Metropolis algorithm is used to apply the probabilities.
- *Condensation*: Depending on the condensation probabilities, the Si and Al atoms can be redeposited at the glass–solution interface.
- The new connectivity of the structure is determined. During the dissolution and condensation steps, closed pores can form in the glass and solid islands can form in the solution. When a Si or Al atom is dissolved at the surface of a closed pore, it is redeposited on the surface of the pore after a random walk on its surface.

The diffusion of water molecules into the glass is not taken into account, preventing simulation of the residual alteration rate.

The first studies were performed with $SiO_2$–$B_2O_3$–$Na_2O$ systems. In these systems, the accumulation of the hydrolysis and condensation phenomena leads to the progressive formation of a protective layer. First, a porous region forms because of the leaching of the B and Na atoms and the hydrolysis of the Si atoms. When the Si atoms begin to redeposit on the surface, pores are formed. With the Si condensation increase, closed pores can form and the glass alteration stops when the porosity is completely closed.

In the first Monte Carlo model, the diffusion of the Si atoms into the solution was considered to be infinite, i.e. the Si concentration in solution was instantaneously homogenized. This hypothesis had an impact on the gel layer morphology. The alteration layer was denser near the solution and less dense near the pristine glass because more condensation phenomena occur near the solution. This observation was not coherent with the experimental results.

To better resemble the experimental case, a Si concentration profile at the glass–water interface was introduced. It enabled the consideration of a larger Si concentration near the pristine glass and a decreasing concentration profile between the pristine glass and the solution. With this modification, the alteration layer formed becomes denser near the pristine glass and less dense near the solution, in good agreement with the experiments.

More complex glasses have been studied by adding $Al_2O_3$, CaO, and $ZrO_2$. When $Al_2O_3$ is present, the initial alteration rate decreases because it becomes more difficult for the Si atoms to dissolve. Consequently, the Si concentration in the solution increases more slowly, and the formation of the protective layer is delayed [67].

The same kind of mechanism has been observed for glasses containing increasing concentrations of $ZrO_2$ [22, 68]. For small $ZrO_2$ concentrations, a porous alteration layer with a closed porosity was formed. Because of the diffusion of the water molecule in the alteration layer, the glass was not taken into account, and the glass alteration stopped. When the $ZrO_2$ concentration increased, the Si hydrolysis and condensation slowed down, preventing the protective layer formation. In this case, the glass alteration did not stop and the glass was completely altered.

The same kinds of study have also been performed at Pacific Northwest National Laboratory (PNNL) [69–72] using the same algorithm but without considering the condensation processes in the first instance. Results complementary to the previous ones were obtained. It was shown that glass inhomogeneity has an impact on the dissolution rate. In fact, the presence of boroxol rings increases the dissolution rate because the surface of the glass–water interface increases. For the same reason, the dissolution rate increases when the Al atoms are not homogeneously distributed [69].

In later studies, the condensation processes were taken into account, and two cases were considered: a static case where the Si concentration in solution was calculated from all the dissolved Si atoms, and a renewal condition in which some of the dissolved Si atoms were removed from the solution to simulate a renewal rate [70]. Borosilicate glasses were investigated.

In the static case, a protective layer enriched in Si systematically formed at the glass–water interface and the glass dissolution stopped. In the renewal conditions, if the renewal rate was high enough, the protective layer could not form and the dissolution remained congruent until the end of the simulation.

Alumino borosilicate glasses have also been considered [71]. When the $Al_2O_3$ concentration increased, the Boron initial dissolution rate decreased and at the end, the B concentration in solution reached a plateau whose level evolved nonlinearly with the $Al_2O_3$ concentration (see Figure 9.20). For the lowest $Al_2O_3$ concentration, it took more time for the protective layer to form, inducing an increase in the saturation level. For larger $Al_2O_3$ concentrations, the dissolution rate continued to decrease, but in this case, the protective layer was formed nearer the surface, explaining the decrease in the saturation level.

Work has been also performed to improve the diffusion model [72]. Three different models were tested: an immediate homogenization of the solution (as in the previous studies), a linear gradient for the Si concentration profile between the glass and the solution, and a more realistic Si diffusion in the solution according to Fick's law.

Compared to the immediate homogenization of the solution, when a linear concentration profile was considered, the protective layer density changed and the Si distribution in the protective layer became more homogeneous.

When the diffusion model based on Fick's law is applied, the protective layer forms more slowly and the porosity of this layer can reopen over a longer period because of local

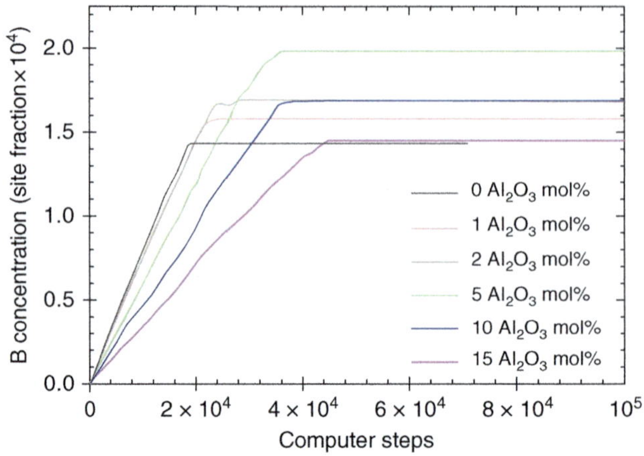


**Figure 9.20** B concentration in solution versus computer steps for several $Al_2O_3$ concentrations. Source: Figure from Kerisit et al. [71].

concentration fluctuations. As a consequence, the definitive formation of the protective layer is delayed.

In conclusion, Monte Carlo approaches constitute a good tool for investigating what happens at the water–glass interface and to evidence the different stages leading to the end of alteration. One of the current drawbacks of this method is that water diffusion inside the pristine glass upstream from the hydrolysis front is not taken into account. It has been suggested recently that the mechanisms occurring in this area could impact the alteration slowdown [23].

## 9.6 Gas Incorporation: Radiation Effects on He Solubility

### 9.6.1 Solubility Model

The notion of free volume in a glass has been linked to the solubility of rare gases. This is due to the fact that the incorporation of rare gases inside the glass occurs via the interstitial sites available to host the gas molecules [73]. The solubility $S$ of a gas is defined by the following equation:

$$C = S \times P \tag{9.12}$$

with $C$ the gas concentration inside the material, and $P$ the pressure of the gas outside the material.

More or less complex models exist to correlate the gas solubility to the free volume. One model widely used nowadays is that of van't Hoff. The dependence between the solubility and the temperature is taken into account. In this model, the solubility is described by the relation:

$$S = S_0 \times \exp\left(-\frac{\Delta H}{RT}\right) \tag{9.13}$$

Other more recent authors [74, 75] have provided a meaning for the parameter $S_0$. The solubility obeys the following equation:

$$S = \left(\frac{h^2}{2\pi m k_B T}\right)^{3/2} \left(\frac{1}{k_B T}\right) N_s \left[\frac{e^{-h\nu/2k_B T}}{1 - e^{-h\nu/k_B T}}\right] e^{-E(0)/RT} \tag{9.14}$$

$N_s$ corresponds to the density of interstitial sites accessible to the rare gas molecules (in at/m$^3$), $\nu$ (s$^{-1}$) represents the characteristic oscillation frequency of the gas molecule, and $E(0)$ is the interaction energy between the gas molecule and the material.

This model includes dependence between the solubility and the glass structure on the one hand, via the parameter $N_s$, and between the solubility and the temperature on the other.

Estimating the density of interstitial sites inside a glass is therefore a crucial point in order to understand the gas solubility.

### 9.6.2 Interstitial Sites in $SiO_2$–$B_2O_3$–$Na_2O$ Glasses

Classical molecular dynamics has been used to quantify the number of interstitial sites in different borosilicate glasses [76]. Twelve different glass compositions were considered (Table 9.5).

Pure silica was also studied, as a reference material. To simulate the borosilicate glasses, the empirical potentials developed by L.H. Kieu were used [24], while pure silica was simulated using the classical BKS potentials [77]. For each glass, the size distribution of

**Table 9.5** Glass compositions (molar fractions) used to quantify interstitial site concentrations.

| Glass | $SiO_2$ | $B_2O_3$ | $Na_2O$ |
|---|---|---|---|
| SBN12 | 59.66 | 28.14 | 12.20 |
| SBN15 | 57.63 | 27.13 | 15.24 |
| SBN20 | 54.19 | 25.51 | 20.30 |
| SBN25 | 50.76 | 23.89 | 25.35 |
| SBN30 | 47.33 | 22.28 | 30.39 |
| SBN35 | 43.95 | 20.63 | 35.42 |
| SBN40 | 40.47 | 19.08 | 40.45 |
| SBN55 | 55.30 | 14.71 | 29.99 |
| SBN59 | 59.24 | 15.76 | 25.00 |
| SBN63 | 63.20 | 16.80 | 20.00 |
| SBN67 | 67.10 | 17.90 | 15.00 |
| SBN14 | 67.73 | 18.04 | 14.23 |

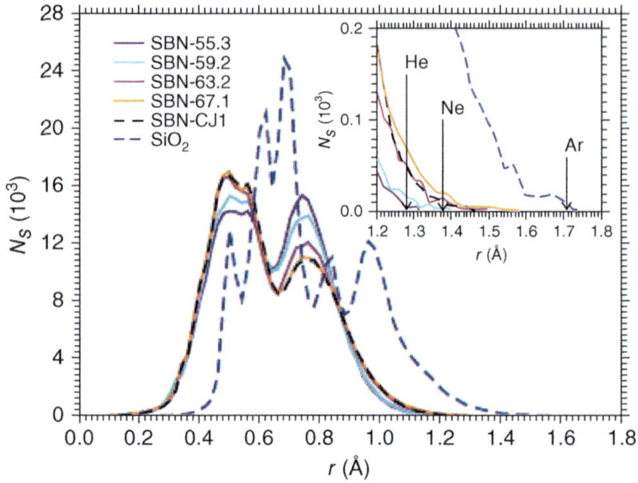


**Figure 9.21** Size distributions of interstitial sites in the series of borosilicate glasses listed in Table 9.5 and in pure silica. Not all the glasses are represented. Source: Figure from Kerrache and Delaye [76].

the interstitial sites was calculated using the Delaunay tessellation and estimating the free volume inside each Delaunay tetrahedron. The atoms were considered as perfect spheres with a radius equal to the ionic radius. This method is described in detail in the following reference [76].

Figure 9.21 shows the size distribution of the interstitial sites for some glasses. All the glasses are not represented here because the distributions are very similar between some glasses. Compared with pure silica, the size distributions of interstitial sites for the borosilicate glasses are shifted to lower values. This is because Na ions fill the free volumes. The $N_s$ quantity can be estimated by considering the interstitial sites as sufficiently large to host a He molecule. A zoom on the distribution tails is shown in Figure 9.21.

Each glass can be characterized by their ratios R (R = $[Na_2O]/[B_2O_3]$) and K (K = $[SiO_2]/[B_2O_3]$). Figure 9.22 shows the dependence between the interstitial site concentration (the sites able to host a He molecule) and the ratio R. The glasses are separated depending on their $K$ ratio ($K = 2.12$ or 3.76).

Two striking features should be underlined. First, there is a decrease of the $N_s$ value when $R$ increases. This means that when the $Na_2O$ concentration increases, there are fewer and fewer sites available for the He atoms because the Na atoms fill the free volume. Second, for a fixed value of $R$, $N_s$ increases when $K$ increases, i.e. when $SiO_2$ replaces $B_2O_3$, the structure becomes more open. This is explained by the fact that the Si–O first neighbor distance is larger than the B–O first neighbor distance.

In another calculation, the SBN14 glass density was modified by modulating the relaxation temperature (called the disordering temperature). This method was used to mimic the disordering and swelling induced by a series of displacement cascades. The $N_s$ values were plotted conjointly with the density versus the disordering temperature (Figure 9.23).

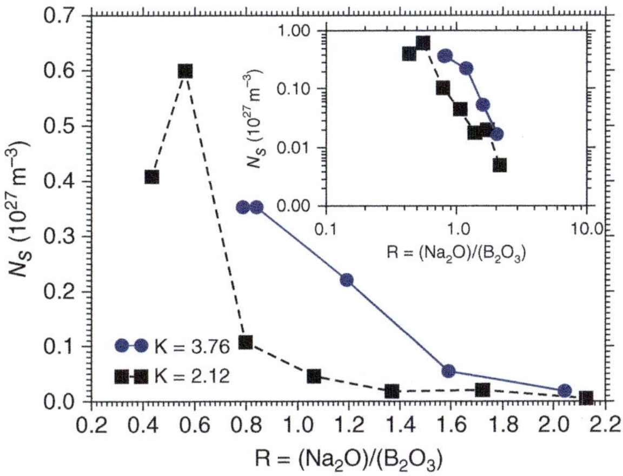


**Figure 9.22** $N_s$ versus R (=[$Na_2O$]/[$B_2O_3$]). The glasses are separated depending on their K ratio. Source: Figure from Kerrache and Delaye [76].

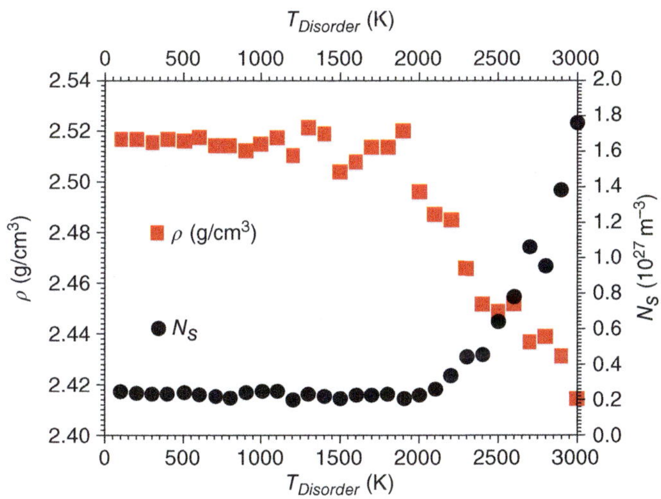


**Figure 9.23** Density and $N_s$ values versus the disordering temperature for the SBN14 glass. Source: Figure from Kerrache and Delaye [76].

Logically, when the density decreases, the $N_s$ value increases. A decrease of 4% of the density is associated with a multiplication by 7 of the $N_s$ value.

### 9.6.3 Discussion About He Solubility in Relation to the Radiation Effects

It is important to go back to the case of real nuclear glass to extrapolate the He behavior and the impact of the radiation effects in the real case. After 100 000 years, the He concentration will be 0.1% at in the French nuclear glass (diffusion is ignored) [18].

In the SBN14 simplified nuclear glass, atomistic simulation has shown that the interstitial site concentration able to host He molecules is $0.35 \times 10^{27}/m^3$. This quantity corresponds to roughly 0.5% at. This is clearly larger than the quantity of Helium generated within the glass by α-disintegrations.

Experimentally, in R7T7 nuclear glass, the interstitial site concentration able to host He molecules has been estimated to be $2.2 \times 10^{27}/m^3$, i.e. 3% at. Here again, this value is clearly larger than the quantity of Helium generated in the glass.

When pristine glasses are under consideration, the number of interstitial sites able to accommodate He molecules is larger than the quantity of gas generated.

Now, the irradiation effect can be considered. Section 9.6.2 described how the concentration of interstitial sites able to accommodate He molecules increases when the glass density decreases. It can be deduced that the ballistic effects will enhance the He solubility, limiting the risk of forming He bubbles. This reasoning only concerns the ballistic effects.

## 9.7 Conclusion

It has been shown that atomistic simulations can provide a better understanding of the effects of irradiation on the structure and macroscopic properties of simplified nuclear glasses. The accumulation of ballistic collisions on the glassy network induces depolymerization ($^{IV}B \rightarrow {}^{III}B$ conversions, NBO formation) and disordering. These structural changes are responsible for swelling and more generally lead to a more plastic material, hence a decrease in hardness and an increase in fracture toughness.

The number of interstitial sites available to host He molecules is sufficiently large in pristine nuclear glass and continues to increase after ballistic collisions, which is a reassuring result for the accommodation of He generated by α-disintegrations.

Even if the subject of glass alteration is complex and still much debated, it has been shown that a Monte Carlo approach can provide clarifications regarding the structure and mechanisms leading to a protective layer formation.

Simulations at different scales (ab initio, classical molecular dynamics, Monte Carlo) are currently used to study simplified nuclear glass behaviors under different internal or external disturbances. With today's innovations and increasingly accurate methods, it is probable that in the future, ever more complex compositions will be developed to refine results and enhance our understanding.

## Acknowledgments

The author is grateful to CEA, Orano, and EDF that have supported a large part of this research work and to all the researchers whose results enable to prepare this summary document.

## References


1 Weber, W.J., Ewing, R.C., Angell, C.A. et al. (1997). *J. Mater. Res.* 12: 1946–1978.
2 Weber, W.J. and Roberts, F.P. (1983). *Nucl. Technol.* 60: 178–198.
3 Matzke, H.J. and Vernaz, E. (1993). *J. Nucl. Mater.* 201: 295.
4 CEA/Valrho summer session. Glass scientific research for high performance containment. Proceedings of the "Summer session, Glass scientific research for high performance containment; Université d'été, CEA/Valrho", Mejannes-Le-Clap, France (31 August–7 September 1997).
5 2nd International Summer School on nuclear glass wasteform: structure, properties and long-term behavior, SumGLASS 2013. *Proc. Mater. Sci.*, 7, 2014,
6 Inagaki, Y., Furuya, H., Ono, Y., and Idemitsu, K. (1993). Scientific basis for nuclear waste management XVI. *Mater. Res. Soc. Symp. Proc.* 294: 191–198.
7 Peuget, S., Cachia, J.-N., Jegou, C. et al. (2006). *J. Nucl. Mater.* 354: 1–13.
8 Peuget, S., Delaye, J.-M., and Jegou, C. (2014). *J. Nucl. Mater.* 444: 76–91.
9 de Bonfils, J., Peuget, S., Panczer, G. et al. (2010). *J. Non-Cryst. Solids* 356: 388–393.
10 Mendoza, C., Peuget, S., Bouty, O. et al. (2012). *Proc. Chem.* 7: 581–586.
11 Boizot, B., Petite, G., Ghaleb, D. et al. (1999). *J. Non-Cryst. Solids* 243: 268–272.
12 Boizot, B., Petite, G., Ghaleb, D. et al. (2000). *Nucl. Instr. Methods Phys. Res.* 166: 500–504.
13 Boizot, B., Ollier, N., Olivier, F. et al. (2005). *Nucl. Instr. Methods Phys. Res.* 240: 146–151.
14 Ollier, N., Charpentier, T., Boizot, B., and Petite, G. (2004). *J. Phys. Condens. Matter* 16: 7625–7635.
15 Griscom, D. (2011). *J. Non-Cryst. Solids* 357: 1945–1962.
16 Gin, S., Beaudoux, X., Angeli, F. et al. (2012). *J. Non-Cryst. Solids* 358: 2559–2570.
17 Poinssot, C. and Gin, S. (2012). *J. Nucl. Mater.* 420: 182–192.
18 Gutierrez, G., Peuget, S., Hinks, J.A. et al. (2014). *J. Nucl. Mater.* 452: 565–568.
19 Cormier, L., Ghaleb, D., Delaye, J.-M., and Calas, G. (2000). *Phys. Rev. B* 61: 14495–14499.
20 Delaye, J.-M., Cormier, L., Ghaleb, D., and Calas, G. (2001). *J. Non-Cryst. Solids* 293: 290–296.
21 Cormier, L., Ghaleb, D., Neuville, D.R. et al. (2003). *J. Non-Cryst. Solids* 332: 255–270.
22 Cailleteau, C., Angeli, F., Devreux, F. et al. (2008). *Nat. Mater.* 7: 978–983.
23 Gin, S., Jollivet, P., Fournier, M. et al. (2015). *Nat. Commun.* 6: 6360.
24 Kieu, L.H., Delaye, J.-M., Cormier, L., and Stolz, C. (2011). *J. Non-Cryst. Solids* 357: 3313.
25 Jolley, K., Smith, R., and Joseph, K. (2015). *Nucl. Instr. Methods Phys. Res.* 352: 140.
26 Stoch, P. and Stoch, A. (2015). *J. Non-Cryst. Solids* 411: 106.
27 Inoue, H., Masuno, A., and Watanabe, Y. (2012). *J. Phys. Chem. B* 116: 12325.
28 Delaye, J.-M., Louis-Achille, V., and Ghaleb, D. (1997). *J. Non-Cryst. Solids* 210: 232.
29 Yun, Y.H. and Bray, P.J. (1978). *J. Non-Cryst. Solids* 27: 363.
30 Dell, W.J., Bray, P.J., and Xiao, S.Z. (1983). *J. Non-Cryst. Solids* 58: 1.
31 Pedone, A. (2009). *J. Phys. Chem. C* 113: 20773.
32 Delaye, J.-M. and Ghaleb, D. (2000). *Phys. Rev. B* 61: 14481.
33 Abbas, A., Serruys, Y., Ghaleb, D. et al. (2000). *Nucl. Instr. Methods Phys. Res.* 166: 445.
34 Ziegler, J.F., Biersack, J.P., and Littmark, U. (1985). *The Stopping and Range of Ions in Matter*. New York: Pergamon.

**35** Delaye, J.-M. and Ghaleb, D. (2003). *J. Non-Cryst. Solids* 330: 106.
**36** Manara, D., Grandjean, A., and Neuville, D.R. (2009). *J. Non-Cryst. Solids* 355: 2528.
**37** Bouty, O., Delaye, J.-M., Beuneu, B., and Charpentier, T. (2014). *J. Non-Cryst. Solids* 401: 27.
**38** Angeli, F., Delaye, J.-M., Charpentier, T. et al. (2000). *J. Non-Cryst. Solids* 276: 132.
**39** Angeli, F., Delaye, J.-M., Charpentier, T. et al. (2000). *Chem. Phys. Lett.* 320: 681.
**40** Delaye, J.-M., Peuget, S., Bureau, G., and Calas, G. (2011). *J. Non-Cryst. Solids* 357: 2763.
**41** de Bonfils, J. (2007). PhD thesis. University Claude Bernard, Lyon.
**42** Marples, J.A.C. (1988). *Nucl. Instr. Methods Phys. Res.* 32: 480.
**43** Kilymis, D.A., Delaye, J.-M., and Ispas, S. (2016). *J. Non-Cryst. Solids* 432: 354.
**44** Delaye, J.-M. and Ghaleb, D. (2005). *Phys. Rev. B* 71: 224204.
**45** Maugeri, E.A., Peuget, S., Staicu, D. et al. (2012). *J. Am. Ceram. Soc.* 95: 2869.
**46** Peuget, S., Maugeri, E.A., Charpentier, T. et al. (2013). *J. Non-Cryst. Solids* 378: 201.
**47** Delaye, J.-M., Peuget, S., Calas, G., and Galoisy, L. (2014). *Nucl. Instr. Methods Phys. Res.* 326: 256.
**48** Kilymis, D.A. and Delaye, J.-M. (2014). *J. Non-Cryst. Solids* 401: 147.
**49** Kilymis, D.A. and Delaye, J.-M. (2014). *J. Chem. Phys.* 141: 014504.
**50** Oliver, W.C. and Pharr, G.M. (1992). *J. Mater. Res.* 7: 1564.
**51** Oliver, W.C. and Pharr, G.M. (2004). *J. Mater. Res.* 19: 3.
**52** Kilymis, D.A. and Delaye, J.-M. (2013). *J. Non-Cryst. Solids* 382: 87.
**53** Kieu, L.H., Delaye, J.-M., and Stolz, C. (2012). *J. Non-Cryst. Solids* 358: 3268.
**54** Kieu, L.H. (2011). PhD thesis. Ecole Polytechnique.
**55** Criscenti, L.J., Kubicki, J.D., and Brantley, S.L. (2006). *J. Phys. Chem. A* 110: 198.
**56** Xiao, Y. and Lasaga, A.C. (1994). *Geochim. Cosmochim. Acta* 58: 5379.
**57** Tossell, J.A. and Saghi-Szabo, G. (1997). *Geochim. Cosmochim. Acta* 61: 1171.
**58** Zapol, P., He, H., Kwon, K.D., and Criscenti, L.J. (2013). *Int. J. Appl. Glass Sci.* 4: 395.
**59** Fenter, P., Zapol, P., He, H., and Sturchio, N.C. (2014). *Geochim. Cosmochim. Acta* 141: 598.
**60** Bouyer, F., Geneste, G., Ispas, S. et al. (2010). *J. Solid State Chem.* 183: 2786.
**61** McGrail, B.P., Ebert, W.L., Bakel, A.J., and Peeler, D.K. (1997). *J. Nucl. Mater.* 249: 175.
**62** Abraitis, P.K., Livens, F.R., Monteith, J.E. et al. (2000). *Appl. Geochem.* 15: 1399.
**63** Lee, S.K., Musgrave, C.B., Zhao, P., and Stebbins, J.F. (2001). *J. Phys. Chem. B* 105: 12583.
**64** Devreux, F., Ledieu, A., Barboux, P., and Minet, Y. (2004). *J. Non-Cryst. Solids* 343: 13.
**65** Ledieu, A., Devreux, F., and Barboux, P. (2004). *J. Non-Cryst. Solids* 345: 715.
**66** Arab, M., Cailleteau, C., Angeli, F. et al. (2008). *J. Non-Cryst. Solids* 354: 155.
**67** Ledieu, A., Devreux, F., and Barboux, P. (2005). *Phys. Chem. Glasses* 46: 12.
**68** Cailleteau, C., Devreux, F., Spalla, O. et al. (2011). *J. Phys. Chem. C* 115: 5846.
**69** Kerisit, S. and Pierce, E.M. (2011). *Geochim. Cosmochim. Acta* 75: 5296.
**70** Kerisit, S. and Pierce, E.M. (2012). *J. Non-Cryst. Solids* 358: 1324.
**71** Kerisit, S., Ryan, J.V., and Pierce, E.M. (2013). *J. Non-Cryst. Solids* 378: 273.
**72** Kerisit, S., Pierce, E.M., and Ryan, J.V. (2015). *J. Non-Cryst. Solids* 408: 142.
**73** Doremus, R.H. (1966). *J. Am. Ceram. Soc.* 49: 461.
**74** Studt, P.L., Shackelford, J.F., and Fulrath, R.M. (1970). *J. Appl. Phys.* 41: 2777.
**75** Shackelford, J.F., Studt, P.L., and Fulrath, R.M. (1972). *J. Appl. Phys.* 43: 1619.
**76** Kerrache, A. and Delaye, J.-M. (2014). *Nucl. Instr. Methods Phys. Res.* 326: 269.
**77** Van Beest, B.W.H., Kramer, G.J., and van Santen, R.A. (1990). *Phys. Rev. Lett.* 64: 1955.

# 10

# Phosphate Glasses

*Gavin Mountjoy*

*School of Physical Sciences, University of Kent, Canterbury, Kent, UK*

## 10.1 Introduction to Phosphate Glasses

### 10.1.1 Applications of Phosphate Glasses

Phosphate glasses constitute one of the major families of oxide glasses based on recognized oxide glass forming cations $B^{3+}$, $Si^{4+}$, $P^{5+}$, and $Ge^{4+}$. Although the production and study of oxide glasses are overwhelmingly oriented toward silicate and borosilicate glass families, the phosphate glass family is nevertheless of great scientific and technical interest for niche applications. These include phosphate glasses for lasers [1], biomedical applications [2], and vitrification of nuclear waste [3]. Based on this interest, an international conference series on phosphate glasses was recently started. The first conference was in Pardubice, Czech Republic, 2014, and the second in Oxford, UK, 2017. (This chapter uses content from a presentation at the first conference.)

### 10.1.2 Synthesis of Phosphate Glasses

Phosphorous oxide, $P_2O_5$, is a pure glass-forming oxide, meaning that melt quenching can be used to make pure $P_2O_5$ glass, i.e. without additives. Melt quenching is by far the most common method to make phosphate glasses, which tend to have lower melting temperatures (and lower glass transition temperatures) compared to silicate glasses. However, $P_2O_5$ is an extremely hygroscopic compound, and historically, this has hindered the production of phosphate glasses. Suitable addition of cations is key to being able to synthesize stable phosphate glasses. In fact, the phosphate network proves to be a particularly suitable host for certain cations, leading to some spectacular results. Phosphate glasses can be made with high rare earth concentrations, and this leads to interesting properties. Binary iron phosphate glasses are extremely durable. In contrast, biomedical applications of (other) phosphate glasses are based on the ease with which they can be dissolved.

*Atomistic Simulations of Glasses: Fundamentals and Applications*, First Edition.
Edited by Jincheng Du and Alastair N. Cormack.

### 10.1.3 The Modified Random Network Model Applied to Phosphate Glasses

As a framework for discussing the modeling of phosphate glasses, it is useful to first cover some important elementary details of phosphate glass structures. This is based on a large body of knowledge about the structure of phosphate glasses obtained via experimental studies, including excellent reviews [4, 5].

The study of the structure of phosphate glasses, like other oxide glasses, benefits from the conceptual framework known as the modified random network (MRN) model. One component of the glass structure is due to the "network former" cation(s), i.e. $P^{5+}$, which forms well-defined structural units ($PO_4$ tetrahedra), with strong links to one another via corners (see Figure 10.1). Other component(s) of the glass structure are "network modifier" cations (e.g. alkalis and alkaline earths) which tend to be "interstitial" to the phosphate glass network, form less well-defined structural units, with weaker links to each other and to $PO_4$ tetrahedra. (Not all cations can be classified so easily in terms of these two categories.) The MRN model also provides a categorization of oxygen anions as "bridging" (BO) or "non-bridging" (NBO) according to whether they do or do not link the corners of $PO_4$ tetrahedra.

### 10.1.4 The Tetrahedral Phosphate Glass Network

The network former cation $P^{5+}$, like $Si^{4+}$ and unlike $B^{3+}$ or $Ge^{4+}$, always takes the coordination number ($N_{PO}$) of 4 and forms a tetrahedral structural unit. Furthermore, the high cationic field strength of $P^{5+}$ means that all oxygen anions are likely to be BO or NBO. This in turn means that the average connectivity of the phosphate network is predictable from the ratio of O to P. The notation $Q^n$ is used, where $Q$ represents a $PO_4$ tetrahedra, and the connectivity $n$ is the number of corner links to other $PO_4$ tetrahedra. A key terminology for phosphate compositions is based on the average connectivity as follows: ultraphosphates have $2 < n < 3$, metaphosphates have $n = 2$, polyphosphates have $n < 2$. (The latter includes pyrophosphates with $n = 1$, and orthophosphates with $n = 0$.) A key experimental technique for studying connectivity is nuclear magnetic resonance (NMR) of P-31, as shown in Figure 10.2 for the case of Zn phosphate glasses.

As a first approximation, a corner linked tetrahedral phosphate glass network can be described in analogy to a silicate glass network (see Figure 10.3), and the key effect of adding

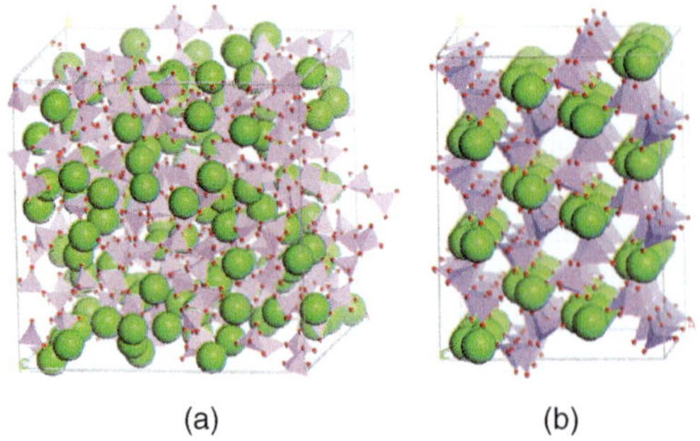


Figure 10.1 Models of (a) calcium metaphosphate glass and (b) calcium metaphosphate crystal showing the common structural features of tetrahedral phosphate network and interstitial calcium modifier ions. Source: Reproduced from Mountjoy et al. [6] with permission of Elsevier.

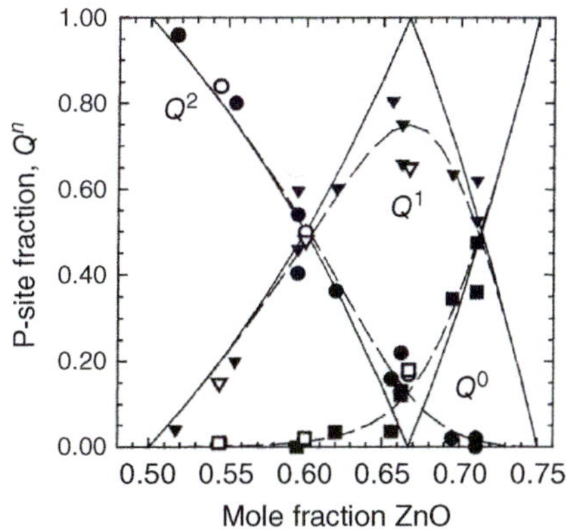


**Figure 10.2** The distribution of $Q^n$ groups in zinc phosphate glasses from (symbols) experimental studies of NMR of P-31, and (solid line) prediction based on the O:P ratio. Source: Reproduced from Brow [4] with permission of Elsevier.

modifier cations is to reduce the average network connectivity $n$ (see Figure 10.2). In this context, the extra valence electron of $P^{5+}$ (compared to $Si^{4+}$) causes the connectivity to be lower. This is easily recognized because pure $P_2O_5$ glass has $Q^3$ connectivity in contrast to pure $SiO_2$ glass which has $Q^4$ connectivity. Associated with this, there can be NBO that are only bonded only to $P^{5+}$ and are not bonded to other cations, i.e. double bonded P=O. In the literature, these are labeled as double-bonded oxygen (DBO) or terminally bonded oxygen (TBO).

An additional difference between tetrahedral phosphate and silicate glass networks is that the charge balancing of NBOs in phosphate glasses requires fewer cations compared to silicate glasses. This means that cations are not as closely packed in phosphate glasses compared to silicate glasses (see Figure 10.3). Also that it is easier to incorporate cations of higher oxidation state in phosphate glasses compared to silicate glasses. In fact, phosphate glasses have attracted a great deal of interest for their ability to incorporate much larger amounts of $Fe^{3+}$ and trivalent rare earths compared to silicate glasses.

An additional complexity with the $P^{5+}$ ion is that whereas the extra valence electron (compared to $Si^{4+}$) is localized to the P=O bond in $Q^3$ tetrahedra in pure $P_2O_5$ glass, it becomes delocalized in $Q^2$, $Q^1$, and $Q^0$ tetrahedra that are present in phosphate glasses. This key, distinctive, feature of phosphate glass structures is illustrated in Figure 10.4 from Brow [4]. A consequence of this is that in phosphate glasses, the P—NBO bond lengths are significantly shorter than the P—BO bond lengths, and the bond lengths vary significantly with composition. (In contrast, the Si—BO and Si—NBO bond lengths in silicate glasses show much smaller variations.)

### 10.1.5 Modifier Cations in Phosphate Glasses

Network modifier cations are essential for applications because they are responsible for useful properties, e.g. chemical durability and luminescence. The short-range order (SRO) around network modifier cations (M) is expressed first in terms of coordination number ($N_{MO}$). To a good approximation, the value of $N_{MO}$ can be viewed as a characteristic of the cation which is similar in phosphate glasses to silicate and other oxide glasses.

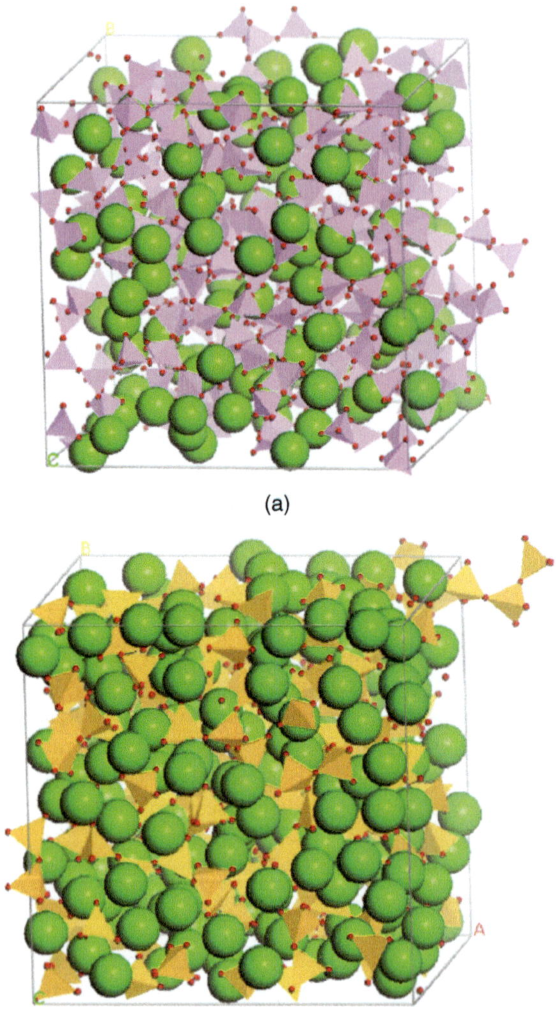


Figure 10.3 Models of (a) calcium metaphosphate glass and (b) calcium metasilicate glass showing the common structural features of a tetrahedral glass network with average connectivity of $n = 2$, and interstitial modifier cations. Source: Reproduced from Mountjoy et al. [6] with permission of Elsevier.

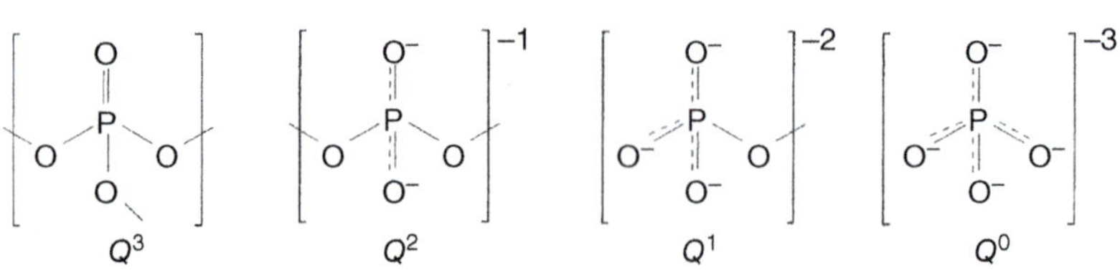


Figure 10.4 The variety of $PO_4$ tetrahedral units found in phosphate glasses labeled with the $Q^n$ notation. Source: Reproduced from Brow [4] with permission of Elsevier.

**Table 10.1** Elementary features of phosphate glass structures.

| Structural feature | Parameter | Value |
|---|---|---|
| Tetrahedral units | P–O $N_{PO}$ | 100% $N_{PO} = 4$ |
| | P—O bond length | Average of 1.55 Å |
| | | Shorter P—NBO than P—BO bonds |
| | O—P—O bond angle | Average of 109° |
| Tetrahedral network | P—O—P bond angle | Average of ~140° |
| | Connectivity $Q^n$ | Average $n$ based on O–P ratio |
| | | Distribution of $n$ from NMR of P-31 |
| Modifier cations | M–O $N_{MO}$ | Values from diffraction and X-ray absorption spectroscopy |
| | M—O bond lengths | |

The medium-range order in terms of correlations among modifier cations is qualitatively described as interstitial modifier cations forming "channels" at high enough concentrations (e.g. metaphosphate glasses).

In the previous paragraphs, we have discussed elementary features of phosphate glass structures. It is suggested that those who are unfamiliar with the structure of phosphate glasses would do well to approach the topic within the framework of the MRN model, and by analogy to the more familiar topic of silicate glass structures. Furthermore, when modeling phosphate glasses, the elementary features of phosphate glass structures provide a useful benchmark with which to evaluate models (see Table 10.1).

## 10.2 Modeling Methods for Phosphate Glasses

### 10.2.1 Configurations of Atomic Coordinates

Fundamentally, a modeling method (of whichever variety) aims to produce a set of atomic coordinates which is representative of the atomic structure of the experimental glass. Starting positions are usually randomly selected to avoid any prior bias. Care should be taken when using crystal or molecular structures for convenience, as the finished model may contain artifacts (see Figure 10.5). Boundary conditions are nearly always periodic to represent a continuous structure. For very small models, periodic boundary conditions would be too artificial, and (nonperiodic) clusters have been used.

The noncrystalline nature of glass structure means that no two parts are the same, and all structural quantities are an average over atomic coordinates. Models typically contain the order of 1000s of atoms, with enough averaging to reliably calculate the elementary details of structure (e.g. bond lengths, $N_{PO}$, $N_{MO}$, and $Q^n$ distribution). The statistical averaging will

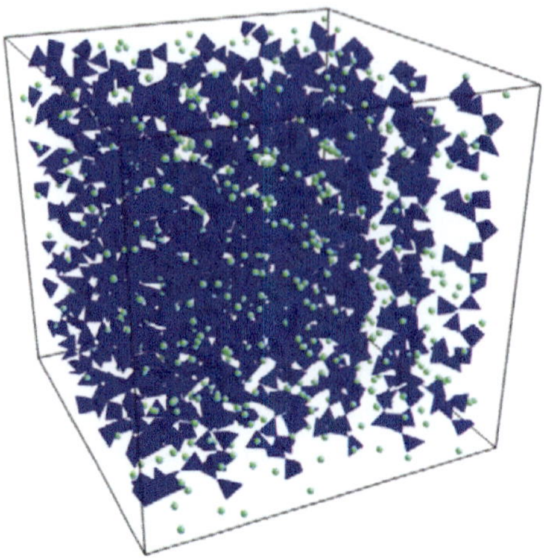

**Figure 10.5** Model of a sodium borophosphate glass in which traces of the crystalline starting configuration is apparent (i.e. a lattice-like distribution of dark blue phosphate tetrahedra and cyan sodium ions). Source: Reproduced from Le Roux et al. [7] with permission of IOP Publishing.

be deteriorated or enhanced (respectively) in much smaller or larger models. In addition, it is a very common average over small differences in a structure corresponding to thermal disorder. Much less common is the more time-consuming approach of averaging over completely independent configurations.

### 10.2.2 Molecular Modeling Versus Reverse Monte Carlo Modeling

In "molecular modeling," the simulated interaction between atomic species provides a criteria for iterating the atomic coordinates from the initial to the final configuration. One approach is to search the (interatomic potential) energy landscape for a minimum energy using a Monte Carlo method. Another approach, molecular dynamics, is to simulate the melting and quenching process used to make an experimental glass.

An alternative modeling method that does not depend on calculating interatomic potentials is called "reverse Monte Carlo" modeling because the criteria for iterating the atomic coordinates is not based on interatomic interactions but is instead based on comparison with pair distribution functions (PDFs) from diffraction experiments. (The PDF shows the different bond lengths as a series of peaks along an axis of interatomic distance, where the coordination numbers are given by the areas of the peaks.) Both reverse Monte Carlo and molecular dynamics methods are capable of producing useful models of phosphate glasses.

### 10.2.3 Classical Versus Ab Initio Molecular Modeling

In classical modeling methods, atoms are represented as whole ions. The interatomic potential energy will always include the long-range electrostatic interaction between charged cations and anions. Charges corresponding to formal valence are simple to assign, whereas partial charges better reflect incomplete transfer of valence electrons. In addition, the interatomic potential energy includes empirical short-range interactions which are (i) repulsive

**Table 10.2** A widely used set of rigid ion potential parameters for phosphate glasses.

| | | $Ae^{-\frac{r}{\rho}} - Cr^{-6}$ | | |
|---|---|---|---|---|
| | $q$ (e) | $A$ (eV) | $\rho$ (Å) | $C$ (eV Å$^6$) |
| P–O | 3.0 | 27 722 | 0.1819 | 86.86 |
| Tb–O | 1.8 | 13 137 | 0.2287 | 36.33 |
| O–O | −1.2 | 1844 | 0.3436 | 192.58 |

| | $\frac{1}{2}k_3(\theta - \theta_0)^2$ | |
|---|---|---|
| | $k_3$ (eV rad$^{-2}$) | $\theta_0$ (°) |
| O–P–O | 3.5 | 109.47 |
| P–O–P | 3.0 | 135.5 |

Source: Clark et al. [8].

due overlap of electron clouds, and (ii) attractive due to induced dipoles of electron clouds. These are referred to as "rigid ion" two body potentials and have been the most common way to model oxide glasses in general, and phosphate glasses in particular (see Table 10.2). A valuable addition has been three-body potentials to better represent O—P—O bond angles in tetrahedral units, and P—O—P bond angles in the linkages of the phosphate network. A more sophisticated approach is to explicitly represent polarization of the anionic electron shell, by using a "polarizable ion" or "shell model" potential, and an interatomic potential of this kind has been successful for phosphate glasses (see Table 10.3).

Modeling methods which specifically include individual electrons are referred to as "ab initio" or "quantum mechanical" due to the necessity to use wavefunctions to describe electrons. The use of these methods is computationally demanding but is becoming increasingly feasible and is expected to produce more accurate models. Despite the name "ab initio," such methods still require assumptions about how to parameterize electron–electron interactions.

### 10.2.4 Evaluating the Simulation of Interatomic Interactions

In the rest of this chapter, the term "potential" is used to refer to the empirical parameters which are required as inputs to calculate interatomic forces in classical molecular modeling. It is a very good practice to evaluate the performance of these potential parameters prior to using them to model glasses. A standard way to do this is to check if simulations can reproduce known crystalline phosphate structures (see Table 10.4). This is part of checking for the attribute of transferability of potential parameters, meaning that the potential parameters are valid for a range of states and compositions. These comparisons should also be made for ab initio modeling (see Table 10.5).

**Table 10.3** A set of shell model potential parameters for phosphate glasses.

| | $Ae^{-\frac{r}{\rho}} - Cr^{-6}$ | | |
|---|---|---|---|
| | $A$ (eV) | $\rho$ (Å) | $C$ (eV Å$^6$) |
| P–$O_s$ | 1 020.0000 | 0.343 22 | 0.030 0 |
| $O_s$–$O_s$ | 22 764.30 | 0.149 00 | 27.88 |
| Na–$O_s$ | 56 465.3453 | 0.193 931 | 0.0 |
| Ca–$O_s$ | 2 152.3566 | 0.309 227 | 0.099 440 |

| | $\frac{1}{2}k_{3b}(\theta - \theta_0)^2$ | |
|---|---|---|
| | $k_{3b}$ (eV rad$^{-2}$) | $\theta_0$ (°) |
| $O_s$–P–$O_s$ | 3.3588 | 109.470 000 |
| P–$O_s$–P | 7.6346 | 141.179 333 |

| | $\frac{1}{2}k_{cs}r^2$ | | |
|---|---|---|---|
| | $k_{cs}$ (eV Å$^{-2}$) | $O_c$ (e) | $O_s$ (e) |
| $O_c$–$O_s$ | 74.92 | +0.8482 | −2.8482 |

Source: Reproduced from Ainsworth et al. [9] with permission of AIP Publishing.

**Table 10.4** Comparison of o′-$P_2O_5$ lattice parameters from ("This work") shell model ("Teter") rigid ion molecular modeling studies with experimental data from X-ray diffraction (XRD) (deviations Δ are in %).

| | *a* | *b* | *c* | *V* |
|---|---|---|---|---|
| o′$(P_2O_5)_\infty$ | | | | |
| XRD | 9.193 | 4.890 | 7.162 | 322 |
| This work | 8.693 | 4.945 | 7.184 | 309 |
| Δ | −5.44 | +1.12 | +0.31 | −4.07 |
| Teter | 8.623 | 5.073 | 6.921 | 302 |
| Δ | −6.20 | +3.74 | −3.36 | −6.21 |

Source: Reproduced from Ainsworth et al. [9] with permission of AIP Publishing.

**Table 10.5** Comparison of o′-$P_2O_5$ lattice parameters from a variety of density functional theory (DFT) simulations with experimental data from X-ray diffraction (deviations Δ are in %).

| | | | | o′$(P_2O_5)_\infty$ | | | | |
|---|---|---|---|---|---|---|---|---|
| **Method** | ***a*** | **Δ*a*** | ***b*** | **Δ*b*** | ***c*** | **Δ*c*** | **Cell vol.** | **Δ vol.** |
| DFT/GGA/PAW | 9.692 | +5.44 | 4.943 | +1.08 | 7.362 | +2.79 | 352.697 | +9.55 |
| DFT/GGA/US-PP | 9.625 | +4.70 | 4.958 | +1.39 | 7.383 | +3.08 | 352.319 | +9.43 |
| DFT-D/GGA/US-PP | 9.245 | +0.57 | 4.950 | +1.24 | 7.244 | +1.15 | 331.557 | +2.98 |
| X-ray diffraction | 9.193 | | 4.890 | | 7.162 | | 321.959 | |
| DFT/LDA/US-PP | 8.984 | −2.28 | 4.881 | −0.17 | 7.072 | −1.25 | 310.156 | −3.67 |
| DFT-D/LDA/US-PP | 8.704 | −5.32 | 4.881 | −0.19 | 6.838 | −4.52 | 290.500 | −9.77 |

Source: Reproduced from Ainsworth et al. [10] with permission of AIP Publishing.

### 10.2.5 Evaluating Models of Glasses by Comparison with Experimental Data

As mentioned in Section 10.1, it is highly advisable to first evaluate models in terms of elementary structural features of phosphate glasses (see Table 10.1) as revealed by experimental methods of structural characterization. An obvious first step is to choose the configuration size and composition to match experimental density. In a molecular dynamics simulation, the initial simulation temperatures should significantly exceed the known melting temperature. Reliable determination of average bond lengths and $N_{MO}$ of cations is possible from careful X-ray diffraction (XRD) and neutron diffraction (ND) experiments and X-ray absorption spectroscopy experiments. Table 10.6 shows an example of evaluating a variety of models of one sodium calcium phosphate-based glass (denoted "P45C30N25") by comparison with experimental data on bond lengths.

Modeling of dynamic structure, i.e. vibrational modes, is more challenging. This is because it depends not only on the minimum in the (potential) energy landscape but also on the curvature around the minimum. In addition, it depends on the coupling strength between the vibrational mode and the probes used in vibrational spectroscopy experiments. Vibrational spectroscopy (i.e. infrared and Raman spectroscopy) is able to identify different $Q^n$ units via the frequencies of their vibrational modes.

## 10.3 Modeling Pure Vitreous $P_2O_5$

### 10.3.1 Modeling of Crystalline $P_2O_5$

Many of the potentials needed for phosphate glasses, e.g. to describe Na–O interactions, have already been determined in modeling studies of silicate glasses, and via the principle of transferability they can be used for phosphate glasses. A challenge in modeling of

**Table 10.6** Comparison of bond lengths in sodium calcium phosphate-based bioglass (denoted "P45C30N25") from ab initio (AI), shell model (SM), and rigid ion modeling (RI) with experimental data (Exp.).

| | AI | SM | RI | Exp. |
|---|---|---|---|---|
| *r*(P–BO) | 1.63 | 1.63 | 1.54 | 1.60 ± 0.02 |
| *r*(P–NBO) | 1.51 | 1.48 | 1.48 | 1.52 ± 0.02 |
| *r*(Na–BO) | 2.51 | 2.41 | 2.53 | 2.41 ± 0.02 |
| *r*(Na–NBO) | 2.35 | 2.32 | 2.35 | |
| *r*(Ca–BO) | – | 2.65 | 2.47 | 2.40 ± 0.02 |
| *r*(Ca–NBO) | 2.35 | 2.29 | 2.41 | |

Source: Di Tommaso et al. [11] CC BY-SA 3.0.

phosphate glasses is to accurately model P—O bonds. This is because of the variety of $PO_4$ tetrahedral units which can form (see Figure 10.3). Crystalline $P_2O_5$ provides an excellent test for modeling P—O bonds. There are three crystal polymorphs of $P_2O_5$. The least stable is a molecular solid with $P_4O_{10}$ molecules which is not relevant. The other two are infinite, corner-linked, networks of $Q^3$ units. They differ in whether the linkages are arranged as two-dimensional sheets (polymorph o′-$P_2O_5$, more stable), or a three-dimensional network (polymorph o-$P_2O_5$, less stable).

Molecular modeling of crystalline $P_2O_5$ using rigid ion potentials has previously been reported in studies of phosphate glasses. Clark et al. [8] reported this as part of a MD modeling study of Tb metaphosphate glass. The potential parameters are shown in Table 10.2, and the comparison with experimental data for crystalline $P_2O_5$ is shown in Table 10.4. The two-body potential parameters were from Teter (see Cygan [12]) and use partial charges which are 60% of formal charges. The three-body potential parameters were derived from Liang et al. [13].

Comprehensive ab initio molecular modeling of crystalline $P_2O_5$ was carried out by Ainsworth et al. [10] as part of the work to develop new shell model potentials for phosphate glasses. This included a Density Functional Theory (DFT) study o-$P_2O_5$ and o′-$P_2O_5$ which reproduced experimental lattice constants to within 1%. A charge analysis indicated that effective charges are approximately 40% of formal charges, and it was concluded that P–O interactions have a mixed ionic/covalent character. The comparison with experimental data for crystalline $P_2O_5$ is shown in Table 10.5.

Ainsworth et al. [9] continued the work by developing a shell model potential for P–O interactions. Via the principle of transferability, this made use of an existing shell model for the oxygen anion from Sanders et al. [14] with formal charges. The potential was developed by using the o′-$P_2O_5$ crystal with infinite layers. The potential parameters are shown in Table 10.3, and the comparison with experimental data for crystalline $P_2O_5$ is shown in Table 10.4. Good results were obtained with smaller deviations from experimental values than in Clark et al. [8].

### 10.3.2 Modeling of Vitreous $P_2O_5$

Phosphorous oxide is one of the four unconditional glass-forming oxides, meaning that it can be prepared as a glass, known as vitreous $P_2O_5$ (v-$P_2O_5$), by melt quenching of a single pure oxide component. Studies of v-$P_2O_5$ are important for a close examination of the phosphate network in its most polymerized state, i.e. 100% $Q^3$ units, and without the complexity of additional cations. Perhaps counter-intuitively, this has proved to be a greater challenge for modeling methods than to model (e.g.) alkali phosphate glasses.

Reverse Monte Carlo modeling was used in Hoppe [15] to create a 1400 atom model of v-$P_2O_5$. This reverse Monte Carlo (RMC) study benefitted from high-quality neutron and XRD data sets, and the agreement with experimental diffraction data is excellent, as is often the case with RMC modeling. The model reproduced several expected features of SRO: average $N_{PO} = 3.9$ (expected 4.0), average O—P—O bond angle of 109° (tetrahedral), and shorter P—NBO than P—BO bonds. The average P—O—P bond angle is 140° which is realistic. The main limitation of the model is that only 82% of $PO_4$ tetrahedra are $Q^3$ rather than 100% expected. Consequently, there are defects of P with $N_{PO} = 3$ and $Q^2$ units. The MRO of the model was analyzed and described as a layer-like network (see Figure 10.6). This is interpreted as being different to either of the crystalline forms of $P_2O_5$.

Suzuki et al. [16] used rigid ion potentials inspired by the earlier work of Speghini et al. [17], with newly estimated effective charges equal to 40% of formal charges, to make a 1050 atom model with experimental density. This produced the expected SRO with 98% of P having $N_{PO} = 4$, shorter P—NBO than P—BO bond lengths, and tetrahedral O—P—O bond angles. However, the P—O—P angle of 170° is much larger than the value obtained from analysis of diffraction data.

Liang et al. [13] and Alam et al. [18] used MD with a Lennard-Jones potential having partial charges equal to 37% of formal charges, and harmonic three-body potentials, to make

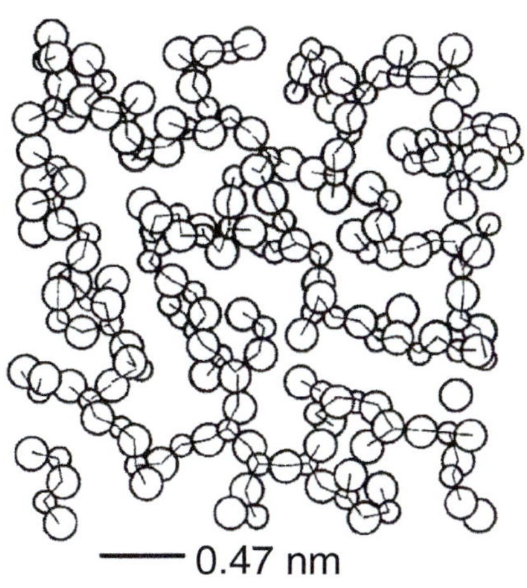


**Figure 10.6** A slice of a RMC model of v-$P_2O_5$ showing (small spheres) P and (large spheres) BO with NBOs not shown for clarity. Source: Reproduced from Hoppe [15] with permission of IOP Publishing.

up to 400 atom models with experimental density. This had good SRO with average $N_{PO} = 4$, shorter P—NBO than P—BO bonds, tetrahedral O—P—O bond angles, and P—O—P bond angles close to 140° as expected from experiment. However, the average connectivity of $Q^n = 2.6$ is less than the 3 expected, with at most 60% $Q^3$ rather than 100% expected.

### 10.3.3 Cluster Models of Vitreous $P_2O_5$

Uchino and Ogata [19] used ab initio molecular orbital theory to model the local structure of v-$P_2O_5$. The oxygens were in the following configurations: P—O—P (i.e. bridging), P—O—H (terminating the cluster as if bridging), and P=O (i.e. double bonded). The motivation was to study the bonding character. Realistic bond lengths of 1.44 and 1.58 A

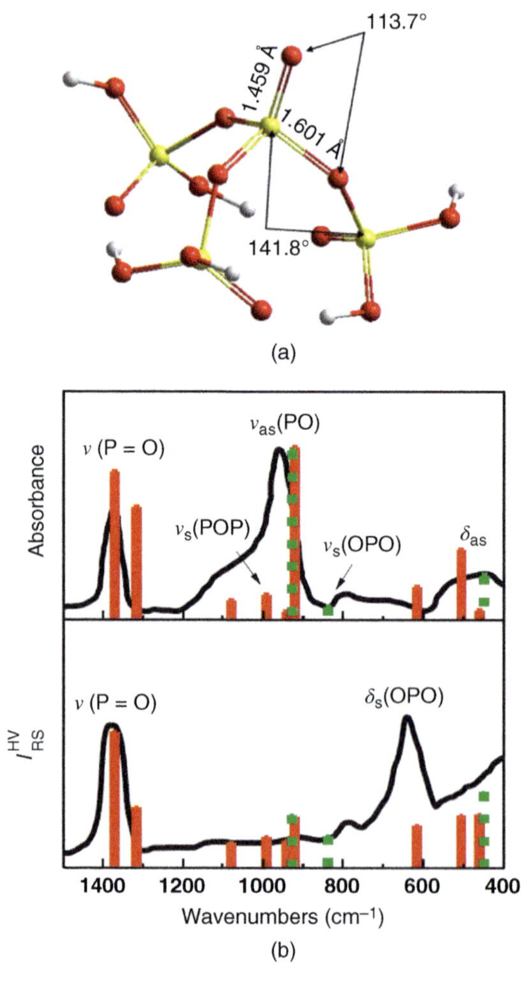


**Figure 10.7** (a) A cluster model of v-$P_2O_5$, and (b) Vibrational modes in v-$P_2O_5$ from (black) experiment and (red) ab initio modeling of the cluster. ("Absorbance" refers to infrared spectroscopy and "RS" refers to Raman spectroscopy.) Source: Reproduced from Guimbretiere et al. [20] with permission of Elsevier.

(respectively) were obtained for P=O and P—BO bonds in v-$P_2O_5$, and Mulliken population analysis gave charges of +1.85 on P and −0.8 on O (which is approximately 40% of formal charges).

Guimbretiere et al. [20] used ab initio modeling of a cluster to study vibrational modes in v-$P_2O_5$. The motivation was to compare the model with experimental results from infrared and Raman spectroscopy. The smallest cluster consisted of single $Q^3$ unit with one P=O bond, and three bridging oxygens terminated by H (i.e. P–O–H). A larger cluster (see Figure 10.7a) consisted of a $Q^3$ unit with one P=O bond, and three bridging oxygens to the smaller clusters. The two higher-frequency vibrational bands (see Figure 10.7b) are assigned to a symmetric stretch of P=O at 1300 $cm^{-1}$, and an asymmetric stretch of $Q^3$ at 1000 $cm^{-1}$. Bending modes occur at lower frequency around 600 $cm^{-1}$.

## 10.4 Modeling Phosphate Glasses with Monovalent Cations

### 10.4.1 Modeling Lithium Phosphate Glasses

Alkali ions are archetypal network modifier cations and are among the most common additives in oxide glasses, including phosphate glasses. It is natural to begin the study of binary phosphate glasses with alkali phosphates, i.e. $M_2O$–$P_2O_5$ systems with M = Li, Na, etc. Note that the addition of modifiers increases the NBOs in the glass and hence depolymerizes the glass network. This effect depends on the valence of modifier, so other monovalent cations have the same effect as alkali ions.

Karthikeyan et al. [21] used classical MD with Born–Mayer potentials and partial charges (60% of formal charge) to produce approximately 800 atom models of $x\text{Li}_2\text{O}$–$(100-x)\text{P}_2\text{O}_5$ glasses with $x = 33$, 50, and 66. The models had experimental densities and the pressure was reported as ~2 GPa. (This is noted to make the observation that in classical molecular modeling, it is frequently difficult to reproduce both experimental density and pressure.) These compositions correspond to average connectivity of $Q^{2.5}$ (ultraphosphate), $Q^2$ (metaphosphate), and $Q^1$ (pyrophosphate). The models have the expected SRO with average $N_{PO} = 4$, shorter P—NBO than P—BO bonds, and tetrahedral O—P—O bond angles. The P—O—P bond angle at ~160° is larger than expected from experiment. For $x = 33$ the value $N_{LiO} = 3.8$ is realistic, as is the average connectivity of $Q^{2.5}$. However, with increasing $x$, there is a decrease in $N_{LiO}$ and the phosphate network becomes insufficiently depolymerized, being average $Q^{1.6}$ at the pyrophosphate composition instead of expected $Q^1$. The models were used to study Li ion mobility which is of interest for applications involving ionic conductivity, such as battery materials. The mean square deviation was used to simulate conductivity for $x = 50$ and the value of $10^{-3}$ S/cm at 300 K was an order of magnitude higher than the experimental value.

Liang et al. [13] and Alam et al. [18] used MD with Morse potentials, partial charges (37% of formal charge), and harmonic three-body potentials to make 300–400 atom models of $x\text{Li}_2\text{O}$–$(100-x)\text{P}_2\text{O}_5$ glasses. These had experimental density and ambient pressure. They produced realistic SRO with $N_{PO} = 4$, shorter P—NBO than P—BO bonds, and average

$N_{LiO} = 4–5$. Models were used to simulate NMR of Li-6. The average connectivity was realistic with $n = 2.5$ for $x = 30$ (expected $n = 2.6$) and $n = 1.95$ for $x = 50$ (expected $n = 2$). The metaphosphate model has a wide spread of $Q^n$ with only 54% $Q^2$, whereas NMR of P-31 indicates closer to 100% $Q^2$ in metaphosphate glasses. The study included an analysis of rings in the phosphate network. As expected, the increasing addition of modifier cations causes the breakup of large rings and a trend toward chains. The glass transition temperature, $T_g$, can be found from a change in slope of energy or density during quenching (see Figure 10.8a). The simulated $T_g$ has a good correspondence with experiment including a minimum around $x = 20$ (see Figure 10.8b). The models are thus useful to illustrate the changes in glass structure that accompany changes in $T_g$ (see Figure 10.9).

Sistla et al. [22] modeled lithium phosphate glasses using MD with Born–Mayer potentials, partial charges of 40% of formal charges, and Vessel type three-body potentials. The model has good SRO with $N_{PO} = 4$ and $N_{LiO} = 4$, and the average connectivity decreases to the expected $Q^2$ for $x = 50$. The study included simulating $T_g$ and Li ion conductivity, but the results were less accurate than the studies discussed previously. Rao et al. [23] continued this work with adjustments to the Born–Mayer potential parameters. A model with $x = 48$ very close to metaphosphate had $N_{PO} = 4.1$ and $N_{LiO} = 3.5$. The model was repeated 50 times, and the results showed an average chain length of 16 $PO_4$ units. This type of analysis is rarely reported.

Hall et al. [24] used RMC modeling to make models of mixed alkali $Li_xRb_{1-x}PO_3$ metaphosphate glasses with 4000 atoms. Good-quality ND and XRD data were used and excellent agreement with experiment was obtained. Note that RMC modeling of metaphosphate glasses is advantageous due to the presence of one type of $Q^n$ unit, and the models were constructed to have all $Q^2$. The models have realistic SRO of $PO_4$ units, although analysis of the finished models is not reported. The main aim was to analyze medium-range order to test the hypothesis that "ionic conduction through a solid should follow a path of lowest bond valence mismatch." The models were analyzed to identify the part of the volume with lowest bond valence mismatch and to check if it is correlated with the activation energy for ionic diffusion. In addition, the analysis looked at whether the part of volume corresponding to small bond valence mismatch becomes percolating. The fractal dimension $d$ of this part of the volume was found to rise from $d \sim 2$ at typical Li–Li nearest neighbor distances of 3 Å to $d \sim 3$ at distances of 5 Å.

### 10.4.2 Modeling Sodium Phosphate Glasses

Speghini et al. [17] used MD with an $r^6$ repulsive term, and partial charges not proportional to formal charges, to make a 4000 atom model of Na metaphosphate glass with experimental density. This has good SRO with average $N_{PO} = 4$, average $N_{NaO} = 5.7$, and tetrahedral O—P—O bond angles. The P—O—P bond angles around 155° are larger than from experiment. The average connectivity was $Q^2$ with significant amounts of $Q^1$ and $Q^3$, as usual for MD modeling with rigid ion potentials. This is one of few MD studies to compare the PDF of the model with diffraction data (see Figure 10.10).

Al Hasni et al. [25] used MD with a Buckingham potential and partial charges (60% of formal charges), and harmonic three-body potentials for O—P—O and P—O—P bond angles

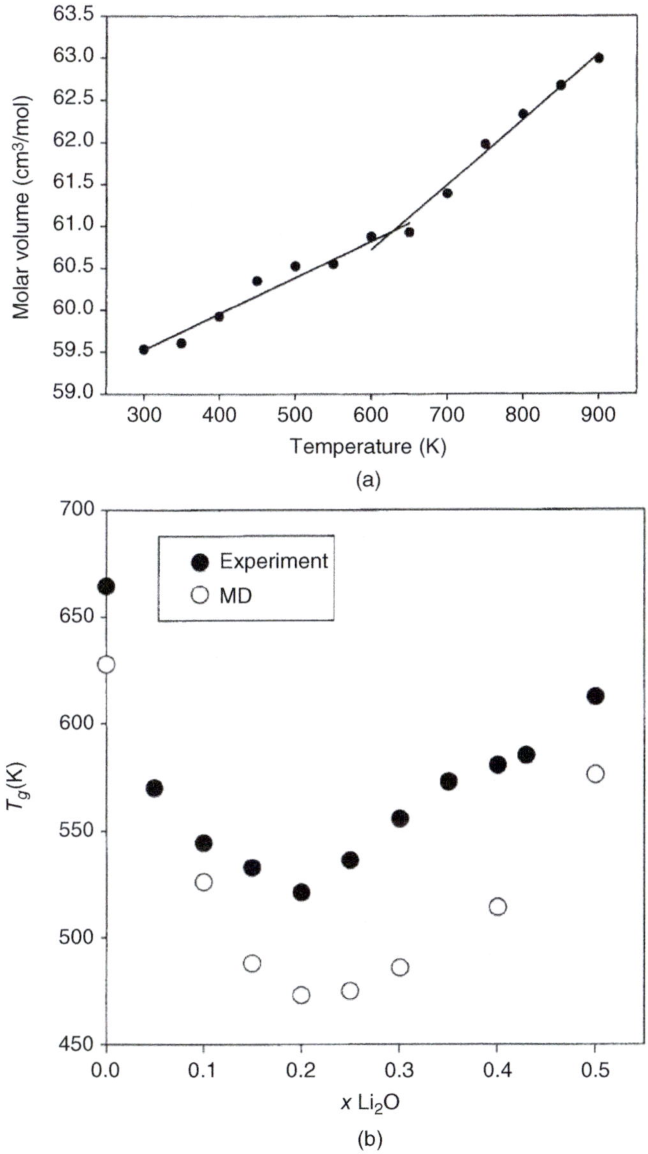


**Figure 10.8** (a) Obtaining $T_g$ from a change in slope of molar volume during a simulated quench of pure phosphate glass. Source: Reproduced from Liang et al. [13] with permission of Elsevier. (b) Comparison of simulated and experimental $T_g$ for lithium phosphate glasses. Source: Reproduced from Alam et al. [18] with permission of Royal Society of Chemistry.

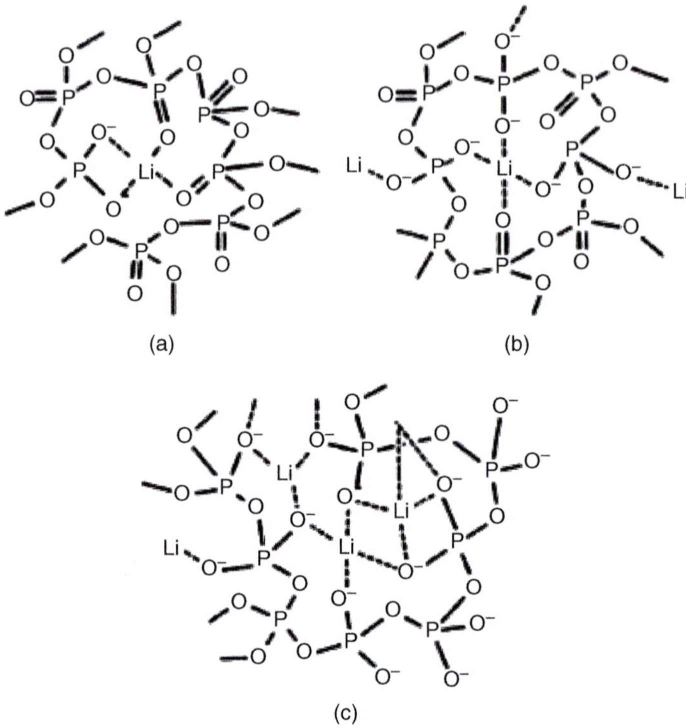


**Figure 10.9** A schematic illustration of the changes in the Li environment in lithium phosphate glasses for (a) low, (b) medium, and (c) high $Li_2O$ concentration. Source: Reproduced from Alam et al. [18] with permission of Royal Society of Chemistry.

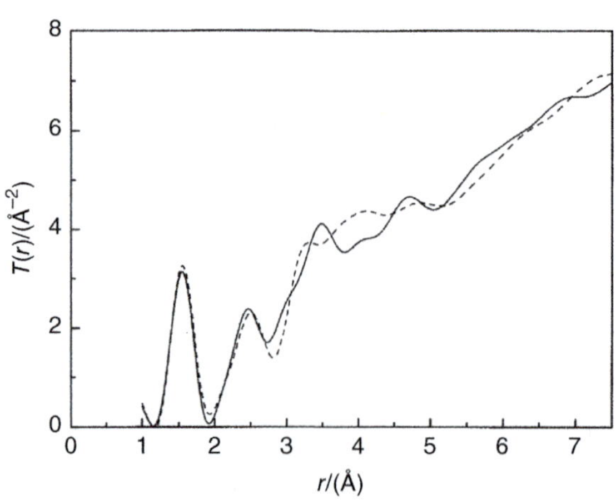


**Figure 10.10** Comparison of the PDF of sodium metaphosphate glass obtained from (solid line) XRD experiment and (dotted line) MD modeling. The peaks at 1.5 and 2.4 Å (respectively) represent P—O and Na—O bonds. Source: Reproduced from Speghini et al. [17] with permission of Royal Society of Chemistry.

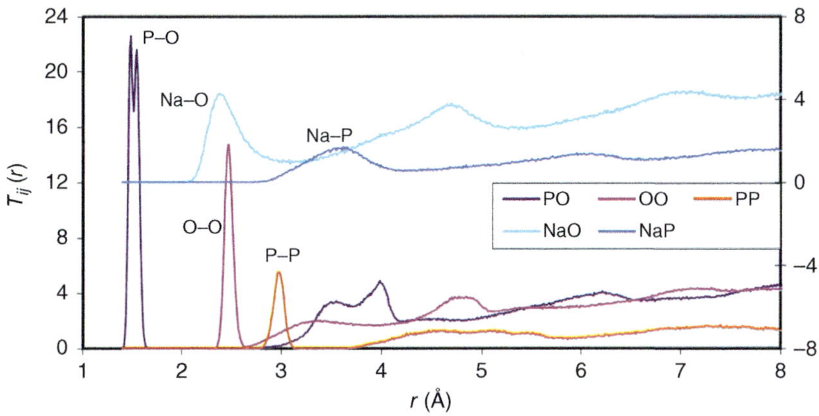


**Figure 10.11** Partial PDFs in a MD model of sodium metaphosphate glass. Source: Reproduced from Al Hasni et al. [25] with permission of Society of Glass Technology.

(see Table 10.2) to make a 1250 atom model of Na metaphosphate glass at experimental density. The model had good SRO with all $N_{PO} = 4$, average $N_{NaO} = 5$, and shorter P—NBO than P—BO bonds. This characteristic difference between P—O bonds can be seen in the partial PDFs shown in Figure 10.11. The model had 50% $Q^2$ with significant amounts of $Q^1$ and $Q^3$. Analysis showed that each Na—NBO bond involves a different $PO_4$ group. Also that the average $N_{NaO} = 5$ requires different Na to bond to the same NBO, i.e. Na—NBO—Na configurations. (This is because in a sodium metaphosphate glass, there are only two NBO for each Na.)

Parsons et al. [26] used ab initio MD to make a 150 atom model of Na metaphosphate glass. Apart from the small model size, the quench rate used of $10^{14}$ K/s is an order of magnitude higher than in classical MD. (These are both consequences of the higher computational demands of ab initio compared to classical MD methods.) The model has good SRO with shorter P—NBO than P—BO bonds, tetrahedral O—P—O bond angles, and P—O—P bond angles around 135° as seen in experiment.

Hall et al. [27] used RMC to model $Ag_xNa_{1-x}PO_3$ metaphosphate glasses with 4000 atoms using ND data and obtaining excellent agreement. The models have chains of all $Q^2$ units; however, there are O–O distances down to 2 Å which is much shorter than from diffraction experiments. In the $x = 0.5$ model, Na has slightly long Na–O distance of 2.5 Å and a somewhat low $N_{NaO} = 3.6$ compared to diffraction experiments. Hall et al. [28] continues this study by also using XRD data, and again obtaining excellent agreement. As in the previous work of Hall et al. [24] on $Li_xRb_{1-x}PO_3$ metaphosphate glasses, the models are analyzed in terms of the part of volume corresponding to small bond valence mismatch. Interestingly, it is found that 5% percolating volume is sufficient for diffusion of Na.

Schuch et al. [29] used RMC to model $35Na_2O$–$65P_2O_5$ glass which is close to the ultraphosphate composition ($x = 33$). An elaborate series of constraints included the experimental $Q^n$ distribution, which makes this the most ambitious RMC model of a nonmetaphosphate composition. However, there were O–O distances down to 2 Å which is unrealistic. Although average $N_{NaO} = 5$ is realistic, detailed results show a spread of $N_{NaO}$ from 1 to 9 which is unrealistic. (This is an example of the challenge of producing models which agree with all details of SRO.)

Parsons et al. [26] used ab initio MD to make models of Na metaphosphate glass. As is often the case with ab initio modeling, these models are very small, containing 150 atoms, although they do match experimental density and ambient pressure. The SRO is briefly reported as having the expected shorter P—NBO than P—BO bonds, tetrahedral O—P—O bond angles, and largely $Q^2$ chains. However, the Na—O bond length of 2.2 Å is slightly shorter than in diffraction results.

Hoppe et al. [5] used RMC to model Na metaphosphate glass. Models of 2000 atoms were fitted with ND and XRD data and obtained very good agreement. The RMC model building process included steps to ensure $PO_4$ units with dominance of $Q^2$ chains, and a realistic average $N_{NaO} = 4.8$ was obtained. The Na–Na partial PDF reported for the model highlights the large number of 5.1 Na–Na nearest neighbors which indicate that most NBO are bonded to 2 Na, i.e. Na–NBO–Na configurations.

### 10.4.3 Modeling Phosphate Glasses with Other Monovalent Cations

Hoppe et al. [5] also used RMC to model K metaphosphate glass with similar results to those obtained for Na metaphosphate glass (see Section 10.4.2). The main difference is larger coordination numbers of K–O and K–K nearest neighbors as expected due to K being a larger alkali than Na.

As discussed in Section 10.4.1, Hall et al. [24] used RMC to make models of mixed alkali $Li_xRb_{1-x}PO_3$ metaphosphate glass with 4000 atoms. This included $RbPO_3$ glass. The average $N_{RbO}$ was not reported. The motivation was to study the mixed alkali effect in phosphate glasses, whereby properties such as ionic conductivity are nonlinear with respect to the mixture composition $x$ (see Figure 10.12). Interestingly, Li and Rb have separate percolating volumes and each of these becomes nonpercolating in the fully mixed composition, i.e. $Li_{0.5}Rb_{0.5}PO_3$ (see Figure 10.13).

As discussed in Section 10.4.2, Hall et al. [27] used RMC to model $Ag_xNa_{1-x}PO_3$ metaphosphate glasses with 4000 atoms using ND data and obtaining excellent agreement. This included $AgPO_3$ glass for which average $N_{AgO} = 2.5$, i.e. smaller than Na. Interestingly, in Hall et al. [28], it is reported that Na and Ag have similar percolating volumes (see Figure 10.13) and there is a much reduced effect of mixing (see Figure 10.12) compared to $Li_xRb_{1-x}PO_3$ glasses.

### 10.4.4 Modeling Phosphate Glasses with Monovalent Cations and Addition of Halides

Interest in ionic diffusivity (which determines ionic conductivity) has motivated some modeling studies of phosphate glasses with mobile cations such as Li, Na, and Ag in which halides have been added to the glass composition. Rao et al. [23] modeled $LiCl–Li_2O–P_2O_5$ glasses as an extension of the study of $x = 52$ Li phosphate glass. The models reproduce the effect of Cl in lowering $T_g$. Instead, the simulated Li diffusion constant does not show a

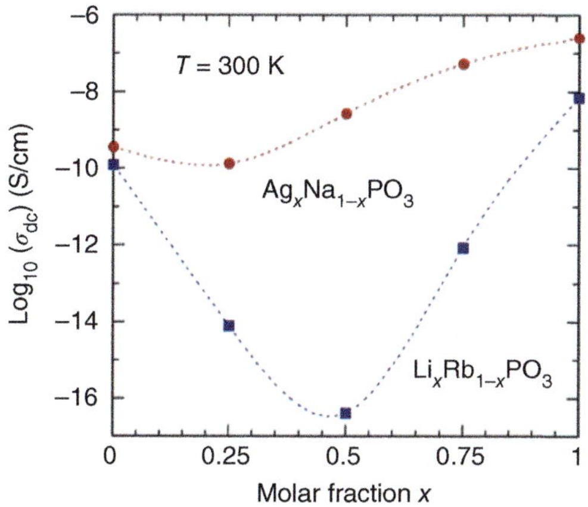


**Figure 10.12** Experimental ionic conductivity in metaphosphate glasses showing the mixed alkali effect in $Li_xRb_{1-x}PO_3$ glasses. Source: Reproduced from Hall et al. [27] with permission of IOP Publishing.

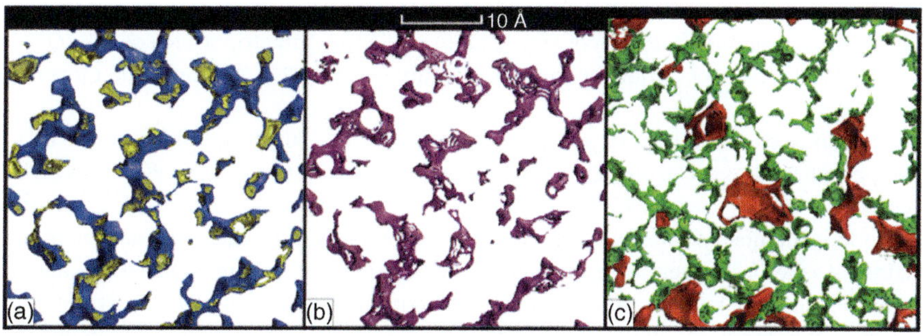


**Figure 10.13** Models of $Ag_{0.5}Na_{0.5}PO_3$ and $Li_{0.5}Rb_{0.5}PO_3$ metaphosphate glasses analyzed in terms of the part of volume corresponding to small bond valence mismatch for (a) (yellow) Ag, (blue) Na, (b) (magenta) both Ag and Na, and (c) (green) Li, and (red) Rb. Source: Reproduced from Hall et al. [28] with permission of American Physical Society.

clear trend with Cl content. Duc et al. [30] modeled $LiBr–Li_2O–P_2O_5$. The approach involved defining the part of volume corresponding to small bond valence mismatch and analyzing under what circumstances it is percolating. The simulated conductivity increases with Br content as seen in experiment, although there is not a significant change in the percolating volume. An analysis of the dimensionality $d$ of percolating volume, like that used by Hall et al. [24], shows $d < 3$ at Li–Li nearest neighbor distances increasing to $d \sim 3$ by 5 Å.

### 10.4.5 Cluster Models of Alkali Phosphate Glasses

Uchino and Ogata [31] used ab initio molecular orbital theory to model clusters representing the local structure of sodium phosphate glasses. These were made to correspond with ultra- and metaphosphate compositions, i.e. $xNa_2O–P_2O_5$ with $x = 0.5$ and 1. The clusters contained one or two Na atoms. The oxygens were in the following configurations: P–O–P (i.e. bridging), P–O–Na (i.e. nonbridging), P–O–H (terminating the cluster), and P=O (double bonded). These models are unrealistic for Na because they give $N_{NaO} = 3$ which is too low (and Na is bonded only to a single $PO_4$ group which is also unrealistic). The motivation was to study the bonding character. The P—O and P=O bond lengths were seen to lengthen (compared to those in $P_2O_5$) with increasing Na content as is also seen in experimental diffraction (see Hoppe et al. [5] for an explanation of this). The covalent character of all types of P—O bond is weakened by the addition of Na, and consequently, the negative charge on oxygen increases. Mulliken population analysis gave charges of +1.85 on P, −0.8 on oxygen and +0.6 on Na (i.e. varying over approximately 40–60% of formal charge).

Uchino and Yoko [32] have continued to use clusters to investigate the vibrational spectra of Na and Li metaphosphate glasses. The clusters contained four composition units (eight $PO_4$ units), in which there are two four-membered rings of $PO_4$ units with two Li interstitial to the chains (and the other Li terminate the exteriors of the chains). The SRO of P—O bonds is realistic. The Na and Li coordination numbers range from realistic values of four to unrealistic values of two. These simple cluster models are able to reproduce vibrational modes in Raman spectra. The highest-frequency band due to P–O stretching is split between asymmetric stretch of $PO_4$ at 1260 $cm^{-1}$ and symmetric stretch of $PO_2$ ($Q^2$) at 1100–1200 $cm^{-1}$. Lower-frequency bands are bending of P–O–P at 700 $cm^{-1}$, and torsion of $Q^2$ at 300 $cm^{-1}$.

Simandiras et al. [33] continued to use clusters to investigate infrared spectra of Li metaphosphate glass. The clusters contain four composition units (eight $PO_4$ units). The cluster differs from that used by Uchino and Yoko [32] because it has four Li interstitial to two four-membered phosphate chains (and the other Li terminate the exteriors of the chains). As well as presenting a molecular orbital calculation, there is also a simulation using DFT, and the best agreement with experiment is obtained using DFT. Interpretation of vibrational bands is different to Uchino and Yoko [32]. The split in high-frequency P–O stretching bands is due to P–NBO (asymmetric stretch at 1372 $cm^{-1}$ and symmetric stretch at 1172 $cm^{-1}$) and P–BO (asymmetric stretch at 1127 $cm^{-1}$ and symmetric stretch at 1069 $cm^{-1}$). The lower-frequency bands are P–O–P bending at 541 $cm^{-1}$ and torsion of $Q^2$ at 340 $cm^{-1}$. Liakos et al. [34] presented analysis of SRO in the clusters with expected features of shorter P—NBO than P—BO bonds, and tetrahedral O—P—O bond angles. However, the P–P nearest neighbor distances of 3.3 Å are longer than expected from diffraction data.

## 10.5 Modeling Phosphate Glasses with Divalent Cations

### 10.5.1 Modeling Zinc Phosphate Glasses

Among binary phosphate glasses with divalent cations, the ZnO–$P_2O_5$ system has attracted much attention. First, Zn phosphate glasses are found to have good properties for applications, including in optoelectronics. Second, because an $M^{2+}$ cation will introduce four NBOs to the phosphate glass network, so for $Zn^{2+}$ with $N_{ZnO} = 4$ there is enough NBO that there is no need for Zn–NBO–Zn configurations. (Instead, Zn metaphosphate crystal has $Zn^{2+}$ with $N_{ZnO} = 6$ and Zn–NBO–Zn present.)

It was seen in Section 10.4 that the metaphosphate composition is popular for RMC modeling, and this is because it can be well approximated as having 100% $Q^2$ units. Bionducci et al. [35] used RMC to model Zn metaphosphate glass. The details of the RMC method were reported in Navarra et al. [36]. This RMC study is very impressive because it used six sets of experimental data: ND, XRD from laboratory and synchrotron sources, anomalous X-ray scattering at two energies close to the Zn K-edge, and X-ray absorption spectroscopy of Zn. RMC models with 6000 atoms were constructed from random starting configurations, and very good agreement was obtained with experimental data. The RMC model has good SRO in terms of having $N_{PO} = 4$ and connectivity of $Q^2$. The phosphate network was found to be overwhelmingly composed of long chains with 6% of terminal $Q^1$. However, P—O bond lengths are not split into shorter P—NBO and longer P—BO bonds as would be seen in MD modeling. The O—P—O bond angle is centered on tetrahedral, and the P—O—P angle is peaked around 145°, similar to the value from diffraction data. These bond angle distributions obtained from RMC are also broader than obtained from MD models. (This is a typical differences between SRO in MD and RMC models, being more disordered in the later.) The average $N_{ZnO} = 3.9$ is in agreement with experimental results. The Zn–P $N_{ZnP} = 3.9$ is consistent with Zn being coordinated to four NBO from four different $PO_4$ units. Nevertheless, the number of Zn–Zn nearest neighbors is 0.9 (not zero) which means that some Zn–NBO–Zn is present.

Hoppe et al. [5] used RMC to model Zn metaphosphate glasses, and Hoppe et al. [37] also used RMC to model Zn phosphate glasses with ultraphosphate, metaphosphate, and (close to) pyrophosphate composition. In Hoppe et al. [38], this work was extended past the orthophosphate composition. The models are in good agreement with high-quality ND and XRD data. The modeling attempted to control $Q^n$ units by prohibiting $Q^3$ beyond the metaphosphate composition, and prohibiting $Q^2$ beyond the pyrophosphate composition. It was reported to be difficult to eliminate defects of $N_{PO} = 3$ and of O not bonded to P. Due to significant $Q^1$, the metaphosphate has lower than expected connectivity of $n = 1.65$. The other glasses have good average $N_{PO} = 4$ and reasonable average connectivity and $Q^n$ distributions. Nevertheless, it is found that only ~50% of Zn have $N_{ZnO} = 4$, and the remainder has $N_{ZnO} = 3$ or 5. Remaking of the models in Hoppe et al. [38] gave an improvement to 90% of Zn having $N_{ZnO} = 4$. Further details of connectivity showed that BO is not normally bonded to Zn, and NBO are normally bonded to only one Zn (up to the metaphosphate composition), as expected. Interestingly, Hoppe et al. [5] comment that a Zn metaphosphate glass with only $PO_4$ and $ZnO_4$ units linked by corners would be topologically similar to $SiO_2$ glass, and in fact, there is a strong similarity in the diffraction data from these two glasses. For higher Zn concentrations, there begins to be Zn–NBO–Zn configurations.

Hoppe et al. [37] and Hoppe et al. [38] carefully examined MRO in terms of P–P and Zn–Zn correlations in PDFs. They also included an RMC modeling protocol to maximize Zn–P correlations, to avoid too high P–P connectivity, and to minimize Zn–Zn nearest neighbors. This was successful in producing a model of Zn metaphosphate glass with no Zn–Zn nearest neighbors, whereas in the RMC model of Navarra et al. [36] these were present. The RMC models of Hoppe show a progression in Zn–Zn partial PDF with marked decrease in Zn–Zn nearest neighbor distance as the ZnO concentration increases past metaphosphate.

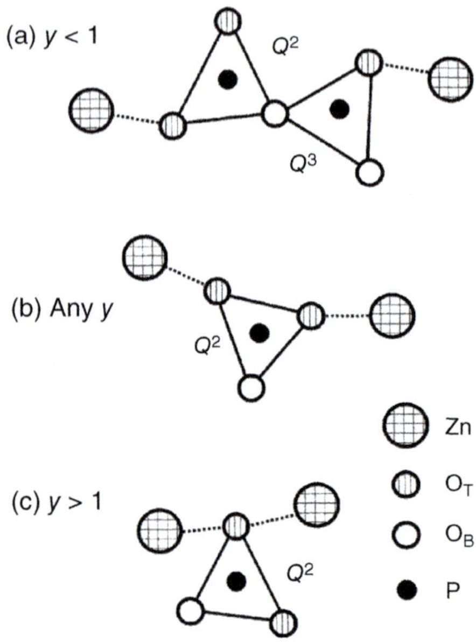


**Figure 10.14** Schematic to explain the decrease in typical Zn–Zn nearest neighbor distance in $x$ZnO–$(1-x)P_2O_5$ glasses with increasing ZnO content, i.e. (a) $y < 1$, (b) $y = 1$, and (c) $y > 1$, where $y = x/(1-x)$. Source: Reproduced from Hoppe et al. [37] with permission of Elsevier.

This provides an explanation for changes in the diffraction data with ZnO concentration. The results are illustrated in the useful schematic shown in Figure 10.14.

Boiko et al. [39] used MD to make a model of Zn pyrophosphate glass, with same protocols as for Na pyrophosphate glass as in Section 10.5.2. The model had good SRO with average $N_{PO} = 4$, including shorter P—NBO than P—BO bonds, and tetrahedral O—P—O bond angles. This model had average connectivity of $n = 1$ as expected, but with a spread in the $Q^n$ distribution including $Q^0$ and $Q^2$. The model has a realistic average $N_{ZnO} = 4.3$ with nearly all Zn having $N_{ZnO} = 4$ or 5. The Zn–Zn partial PDF shows similar features to those reported by Hoppe et al. [38] in the previous paragraph.

Tischendorf et al. [40] used MD to make models of Zn phosphate glasses from below metaphosphate to above pyrophosphate composition. They used rigid ions with Lennard-Jones potentials and partial charges (of 37% of formal charges), and harmonic three-body potentials for O—P—O and P—O—P bond angles. The potential parameters were obtained empirically by modeling of Zn ultra-, meta-, and two forms of orthophosphate crystals, although the analysis is not reported. Models of 500 atoms were made with experimental density, and then relaxed at ambient pressure. After relaxation, there is a good agreement with experimental density up to 55% ZnO, but the density is 10% below experiment around the pyrophosphate composition. The study reports average $N_{ZnO} = 3.8$–$4.0$ and average O—Zn—O bond angle is tetrahedral. Typically, 50–60% of Zn have $N_{ZnO} = 4$ and the others have $N_{ZnO} = 3$ or 5. The Zn–Zn partial PDF for the MD model of metaphosphate shows a shorter nearest neighbor distance compared to the RMC model of Zn metaphosphate by Hoppe et al. [37, 38]. The simulation of $T_g$ shows good agreement with experiment and reproduces the minimum in $T_g$ around 60% ZnO (see Figure 10.15).

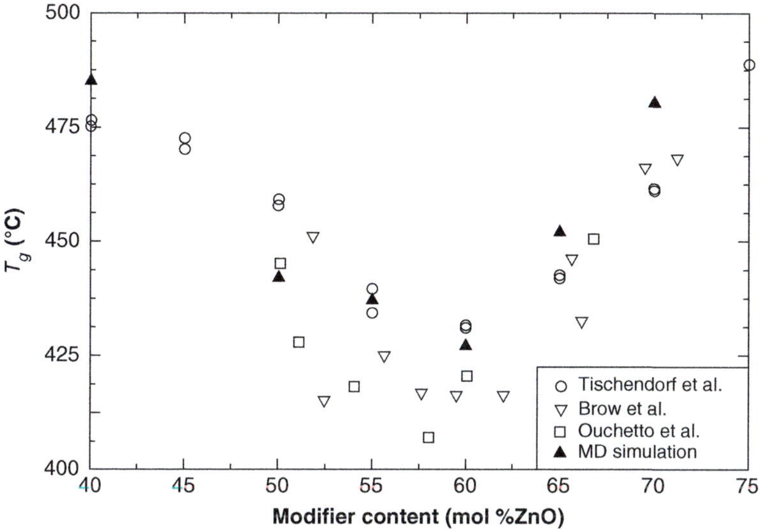


**Figure 10.15** $T_g$ from (solid symbols) MD simulation and (open symbols) experimental values reported in the literature. Source: Reproduced from Tischendorf et al. [40] with permission of Elsevier.

This is between meta- and pyrophosphate compositions and corresponds to maximum mixing of $Q^2$ and $Q^1$ groups.

### 10.5.2 Modeling Zinc Phosphate Glasses with Additional Cations

Boiko et al. [39] modeled Na and Zn pyrophosphate glasses as a prequel to modeling mixed pyrophosphate compositions $33ZnO–33Na_2O–33P_2O_5$ and $44ZnO–22Na_2O–33P_2O_5$, the latter being around the eutectic in the mixed Zn–Na pyrophosphate system. (Note that these are mixed pyrophosphates with Na:Zn ratios of 2 : 1 and 1 : 1, respectively.) The mixed models had similar SRO to the Na pyrophosphate and Zn pyrophosphate models (discussed previously). The results for mixed models suggest that mixing appears to discourage the presence of close Zn–Zn neighbors, whereas it encourages the presence of close Na–Na neighbors. This can be interpreted as $Zn^{2+}$ being more competitive in bonding to NBOs compared to $Na^{1+}$.

Navarra et al. [41] carried out RMC modeling of $Eu_2O_3–9ZnO–12P_2O_5$ glass which is a mixed metaphosphate with a Eu:Zn ratio of 1 : 4.5. This built on the previous work of Navarra et al. [36] in using RMC to model Zn metaphosphate glass with 6000 atom models. The method involved randomly replacing 96 Zn cations with 64 Eu cations and 32 charge-compensating vacancies. The RMC method was then applied utilizing seven experimental data sets as for the previous work on Zn metaphosphate (discussed previously). Good SRO was maintained and the new information concerned the environment of Eu cations. Due to the reduced statistical accuracy of only 64 Eu cations, the main result is restricted to

Eu–O correlations. It was found that the average $N_{EuO} = 6.2$, ranging from $N_{EuO} = 5–8$, and the Eu—O bond length was 2.35 Å, both being realistic. Inspection of the models showed typically two, three, or four phosphate chains that are twisted around each $Eu^{3+}$ site, making it one of the few modeling studies to examine this aspect of phosphate glass structure.

### 10.5.3 Modeling Alkaline Earth Phosphate Glasses

Divalent cations include alkaline earths which are among most common additives to oxide glasses. Alkaline earths are also the most common group of modifier cations after alkalis. Hence, addition of alkaline earths will depolymerize phosphate glass network. It is expected that alkaline earths will provide stronger cross-linking between phosphate chains compared to alkalis due to their higher cationic field strength.

Sourial et al. [42] used MD to make a model of Mg metaphosphate glass using rigid ion potentials with partial charges (57% of formal charge). The model had 4000 atoms and experimental density. The models have some good features of SRO, such as average $N_{PO} = 4$ and average $N_{PP} = 2$, with close to expected proportions of BO and NBO. The longest phosphate chain in model was 25 $PO_4$ units. The model has mostly Mg with $N_{MgO} = 4$ or 5, but the average $N_{MgO} = 4.3$ is slightly lower than from diffraction data. Mg is entirely bonded to NBO as expected. (This confirms that BO bonded to two P and one Mg are unfavorable.)

Karakasides et al. [43] used MD to model Mg and Ca phosphate glasses from meta- to around orthophosphate composition. Initially, ab initio calculations on Mg and Ca phosphate compounds were used to determine partial charges, which were ~35% of formal charge (except 50% for Mg). However, these charges have different proportions of the formal charges for each ion and for each different stoichiometry, and so are not transferable. Two body potentials of the Lennard-Jones form were used to make models of 300–800 atoms, although it is not clearly stated if the densities correspond to experiment. The SRO appears good with average $N_{PO} = 4.0$, tetrahedral O—P—O bond angles, and average P—O—P bond angles of 140°. Modifiers have $N_{CaO} = 6$ and $N_{MgO} = 5.2$ which is realistic. However, the model has an unrealistically short Ca—O bond length of 2.16 Å. Interestingly, the analysis shows that Ca–NBO–Ca configurations are more common than Mg–NBO–Mg configurations. The models with high alkaline earth content have some "free" oxygen not bonded to P, and Mg–O–Mg configurations become more common than Ca–O–Ca configurations. The study included modeling an equimolar mixed Mg–Ca metaphosphate glass. This had similar SRO to the Mg metaphosphate and Ca metaphosphate glass models. Unfortunately, M–M correlations were not reported so it is difficult to evaluate mixing.

Parsons et al. [26] used ab initio MD modeling to make models of Mg and Ca metaphosphate glasses. Due to the use of ab initio MD modeling, these models are very small, containing only 135 atoms, but the models were at experimental density and ambient pressure. The SRO in the models is briefly reported to have the expected shorter P—NBO than P—BO bonds, tetrahedral $PO_4$ units, and largely $Q^2$ chains. This study presents a very interesting analysis of the P–P partial PDF which focuses on the P–P second neighbor distance which is either between chains or in loops of the same chain. The analysis shows the effect of the $Mg^{2+}$ modifier to contract the P–P second neighbor distance compared to $Na^{1+}$ (see

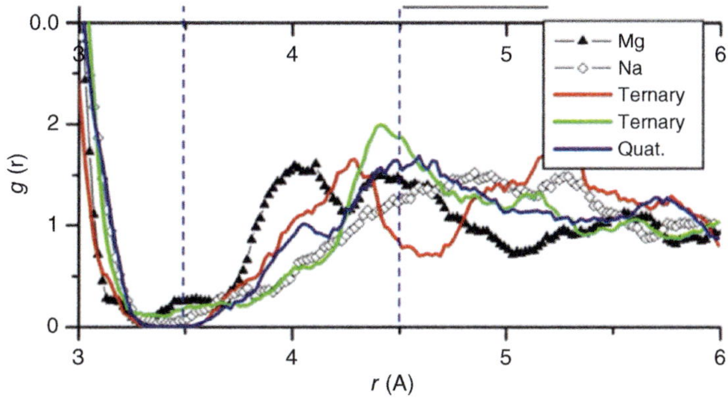


**Figure 10.16** P–P partial PDF in metaphosphate glasses with different modifier cations ("ternary" refers to Na–Mg metaphosphates, and "quarternary" refers to Na–Mg–Ca metaphosphate). Source: Reproduced from Parsons et al. [26] with permission of IOP Publishing.

Figure 10.16). This is understood as an effect of cation field strength, and the effect goes in the sequence $Mg^{2+} > Ca^{2+} > Na^{1+}$.

Hoppe et al. [44] used RMC to model Mg metaphosphate glass using XRD and ND data. This included formation of largely $Q^2$ chains. Models had 2000 atoms and experimental density. The models showed average $N_{MgO} = 4.6$, slightly lower than the experimental value of 5.3. The number of Mg–P nearest neighbors is $N_{MgP} = 4.6$, consistent with each Mg–NBO belonging a different $PO_4$ unit (i.e. NBO are not bonded to two P and one Mg). Note that $N_{MgO} = 4$ is the maximum possible value without NBO being bonded to two Mg, and the Mg–Mg partial PDF shows a small nearest neighbor peak due to sharing with average $N_{MgMg} = 0.5$ (i.e. there are some NBO bonded to one P and two Mg).

The RMC method of Hoppe et al. [44] was previously used to model a series of alkaline earth metaphosphate glasses including Ca, Sr, and Ba. Hoppe et al. [44] makes a very informative presentation of the collective results which show a sequence of metaphosphate glasses with increasing size of alkaline earth ion, i.e. M = Mg, Ca, Sr, and Ba. This is also a sequence of increasing cation coordination number (see Table 10.7). The $N_{MP}$ is observed to be closely similar to the $N_{MO}$, indicating that each M–NBO belongs to a different $PO_4$ unit. The M–M partial PDFs show increase in size of nearest neighbor peak (see Figure 10.17), corresponding to an increase in the number of M–M nearest neighbors increase. This happens due to $N_{MO} > 4$ requiring NBOs to be bonded to two M, hence, increasing close M–M nearest neighbors.

Wetherall et al. [45] used RMC to model Ca metaphosphate glass. A model of 1080 atoms and experimental density was made using ND and XRD data. The model construction procedure ensured almost 100% $Q^2$ units in chains. The model had good SRO with tetrahedral $PO_4$ units and P—O—P bond angle of 130° close to the value from diffraction data. The model also had fairly realistic $N_{CaO} = 5.6$ consisting largely of $N_{CaO} = 5$, 6, or 7. The model topology was shown to consist of a phosphate "backbone" and interstitial Ca ion channels,

**Table 10.7** The effect of increasing size of divalent modifier cations (Me) on numbers of nearest neighbors (*N*) in RMC models of $MeP_2O_6$ metaphosphate glasses.

| Glass | $N_{MeO}$ | $N_{MeP}$ | $N_{MeMe}$ |
|---|---|---|---|
| $ZnP_2O_6$ | 4.2 | 4.2 | 0.1 |
| $MgP_2O_6$ | 4.6 | 4.6 | 0.5 |
| $CaP_2O_6$ | 6.9 | 7.0 | 1.7 |
| $SrP_2O_6$ | 8.2 | 7.5 | 2.4 |
| $PbP_2O_6$ | 8.2 | 7.6 | 3.5 |
| $BaP_2O_6$ | 8.9 | 8.8 | 3.7 |

Source: Reproduced from Hoppe et al. [44] with permission of Elsevier.

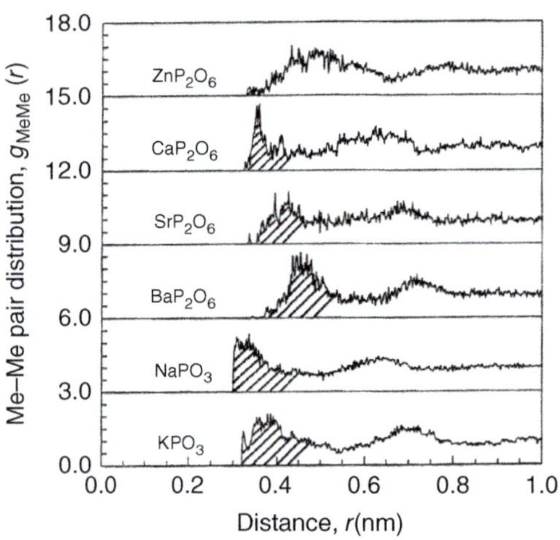


**Figure 10.17** M–M partial PDFs from RMC models of metaphosphate glasses, where M = modifier cation. Source: Reproduced from Hoppe et al. [5] with permission of Elsevier.

as shown in Figure 10.18. This motivation for this illustration is to provide a comparison with the concepts in the MRN model (see Section 10.1).

Mountjoy et al. [6] used MD to model Ca metaphosphate glass with rigid ion potentials as given in Table 10.2. A model of 1000 atoms and experimental density was made. Potentials used were two body potentials of Buckingham form with partial charges (of 60% of formal charges), and harmonic three body potentials for O—P—O and P—O—P bond angles. The model has realistic SRO with 98% of P having $N_{PO} = 4$ (the remaining $N_{PO} = 5$ being a defect). The model also has fairly realistic average $N_{CaO} = 5.8$ consisting of largely $N_{CaO} = 5$, 6, or 7. The connectivity showed average $n = 2$ as expected; however, as usual for rigid ion MD, the connectivity showed a spread with 50% $Q^2$ (as compared to 96% $Q^2$ from NMR of P-31) and remainder $Q^1$ and $Q^3$. Thus, network connectivity is the key difference compared to the RMC model by Wetherall et al. [45] which had 100% $Q^2$. Figure 10.18 shows a comparison of the MD and RMC models which emphasize the MRN concept by highlighting the phosphate backbone and interstitial $Ca^{2+}$ ions.

Figure 10.18 Models of calcium metaphosphate glass made using (a) RMC and (b) MD methods. In both cases, the "backbone" of the phosphate network is shown in red and purple, and the "channels" of calcium modifier ions are shown in green. Source: Reproduced from Wetherall et al. [45] with permission of IOP Publishing.

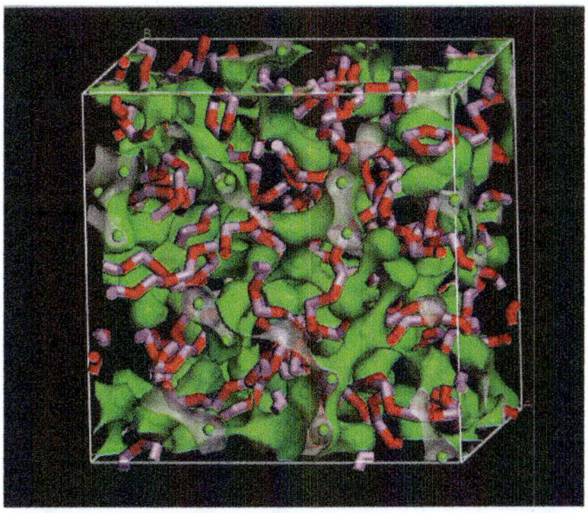

(a)

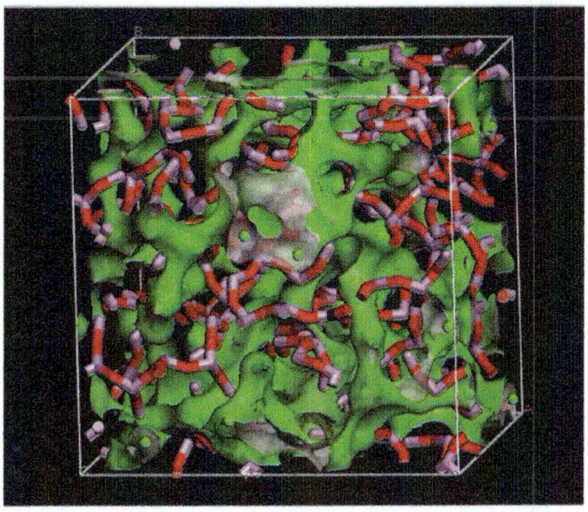

(b)

## 10.5.4 Modeling Lead Phosphate Glasses

Divalent cations include $Pb^{2+}$ which is not necessarily a modifier cation. That is because the attributes of the $Pb^{2+}$ cation depend on its lone pair of electrons. When the lone pair is not stereochemically active, $Pb^{2+}$ has a high $N_{PbO}$ and behaves like a modifier cation. When the lone pair is stereochemically active, $Pb^{2+}$ takes on a small $N_{PbO}$ of 3 or 4 similar to network former cations. For this reason, $Pb^{2+}$ is referred to as an intermediate cation and can sometimes play the role of strengthening the glass network.

Cormier et al. [46] used MD to model Pb metaphosphate glass. An unusual Mitra potential with partial charges (of 57% of formal charges) was used to make a 630 atom model with experimental density. The model has good SRO with all $N_{PO} = 4$, and tetrahedral O—P—O bond angle. The P—O—P bond angle of 165° is significantly larger than experiment. Although average connectivity is $Q^2$ as expected for a metaphosphate, this includes $Q^1$ and $Q^3$ as common in rigid ion MD. The study is notable for a very detailed analysis of topology (see Figure 10.19). This form of analysis is rarely reported. Attempts to compare this with high-performance liquid chromatography (which identifies the size of

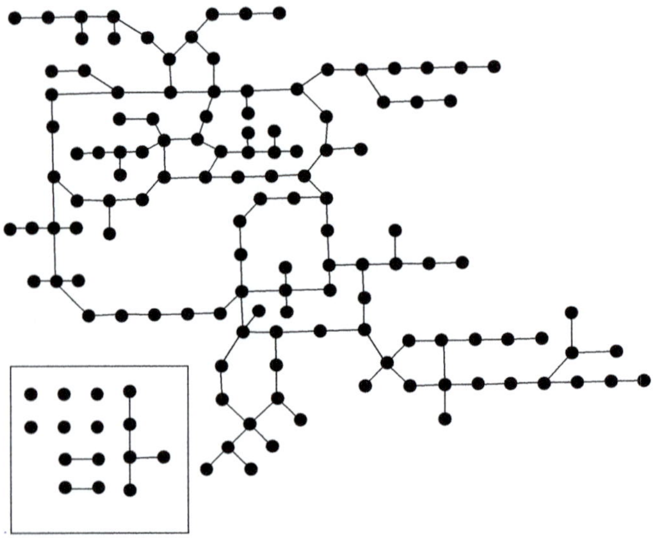

**Figure 10.19** Two-dimensional schematic representation of the phosphate network connectivity in a MD model of Pb metaphosphate glass, indicating (circles) $PO_4$ tetrahedra and (lines) BO. Source: Reproduced from Cormier et al. [46] with permission of Elsevier.

molecular fragments) requires too many assumptions about how the network would break up. The accuracy is also limited by finite size of the model. The Pb–O distance of 2.43 Å is in good agreement with experiment. The average $N_{PbO} = 6.9$ is consistent with Pb acting as a network modifier. As expected, to achieve this $N_{PbO} > 4$, it is necessary for NBOs to have a mixture of 1 or 2 bonds to Pb. This leads to the average number of Pb–Pb nearest neighbors being about 1.3.

Hoppe et al. [44] used RMC to model Pb metaphosphate glass using the same procedure as the previously used one for a series of alkali and alkaline earth metaphosphate glasses, including the formation of largely $Q^2$ chains. Models had 2000 atoms and experimental density. The models had average $N_{PbO} = 8.2$, and the number of Pb–P nearest neighbors is $N_{PbP} = 7.6$, consistent with most P-bonds involving a different $PO_4$ unit. This confirms that BO with two bonds to P and one bond to Pb is unfavorable. A value of $N_{PbO} = 4$ is the maximum possible without NBOs having two bonds to Pb. In fact, the Pb–Pb partial PDF shows a nearest neighbor peak due to Pb–NBO–Pb configurations with average $N_{PbPb} = 3.5$. This confirms that NBO bonded to one P and two Pb are favorable.

Sourial et al. [42] used MD with rigid ion potentials and partial charges (of 57% of formal charge) to make a model of Pb metaphosphate glass. The model had 4000 atoms and experimental density. The model has some good features of SRO such as average $N_{PO} = 4$ and average $N_{PP} = 2$, and expected proportions of BO and NBO. The longest phosphate chain in the model was 19 $PO_4$ units. The average $N_{PbO} = 5.5$, consisting mostly of Pb with $N_{PbO} = 4$–7. The Pb is entirely bonded to NBO as expected.

## 10.6 Modeling Phosphate-Based Glasses for Biomaterials Applications

### 10.6.1 Modeling $Na_2O$–CaO–$P_2O_5$ Glasses with 45 mol% $P_2O_5$

The $Na_2O$–CaO–$P_2O_5$ glass family is presented separately from the preceding sections on alkali and alkaline earth phosphates because of its importance in studies of biomaterials. This has given rise to the term "phosphate-based glass" or PBG. The most frequently modeled PBG compositions have 45 mol% $P_2O_5$, with a connectivity of $n = 1.8$, i.e. less than metaphosphate, which is chosen to aid dissolution.

Tang et al. [47] used ab initio MD to model PBGs of composition $(55 - x)Na_2O$–$x$CaO – $45P_2O_5$ with $x = 30$, 35, and 40. As usual for ab initio MD the models were small, with approximately 90 atoms of which there were less than 10 Na and 10 Ca. Modeling was based on experimental density values, but found to have a pressure of 2 GPa. This was a valuable study due to the detailed report on SRO compared to the similar study by Parsons et al. [26] as discussed in Section 10.6.3. There is a good SRO with a well-defined tetrahedral phosphate network including shorter P—NBO than P—BO bonds. Interestingly, the O–P–O tetrahedral angle has relatively large full width at half maximum (FWHM) of 15°. The P—O—P bond angles are 130° similar to diffraction data and have a FWHM of around 15°. Note that these compositions have average connectivity of $n = 1.8$, and this is formed from majority $Q^2$ with remainder $Q^1$ as expected. The SRO of modifiers shows realistic bond lengths, and realistic values of $N_{MO}$ of 6 for Na and 7 for Ca. Importantly, it is one of the few studies which analyses modifier bonding to NBO and BO (rather than simply total O). As expected, the bonding of Ca—BO is minor, for which typically 0.5 of Ca nearest neighbors are BO. It is more significant for Na, for which typically 0.9 of nearest neighbors are BO.

Ainsworth et al. [9] used classical MD with a shell model potential (see Table 10.3) to model $(55 - x)Na_2O$–$x$CaO–$45P_2O_5$ glasses with $x = 30$, 35, and 40. Models were made with 3000 atoms at experimental density. The models showed the expected SRO with shorter P—NBO than P—BO bonds, close to experimental bond lengths, and a good tetrahedral network with tetrahedral O—P—O bond angles and O—P—O bond angles around 135° close to that expected from diffraction data. The models have average $n = 1.8$, a majority of $Q^2$ and the remainder of $Q^1$. There is only a slight amount of $Q^3$ making $Q^n$ distribution slightly broader than from NMR of P-31. The connectivity is found to correspond to chain-like segments and no rings are found. The Na—O and Ca—O bond lengths are close to experimental values, and $N_{MO}$ of 6.5 for Na and 6.8 for Ca are realistic. The analysis of cation bonding to BO was reported, with the contribution being minor for Ca (0.5 of nearest neighbor O) and more noticeable for Na (1.3 of nearest neighbor O). The bond angle distribution of O—M—O bonds (where M = Na, Ca) indicates that most O neighbors of M are on different $PO_4$ groups.

Di Tommaso et al. [11] reported both ab initio and classical MD modeling of $(55 - x)Na_2O$–$x$CaO–$45P_2O_5$ glasses with $x = 30$, 35, and 40. In fact, the ab initio work has the same details for models as Tang et al., including approximately 100 atom models, and the results are the same. The same glasses were also modeled by classical MD

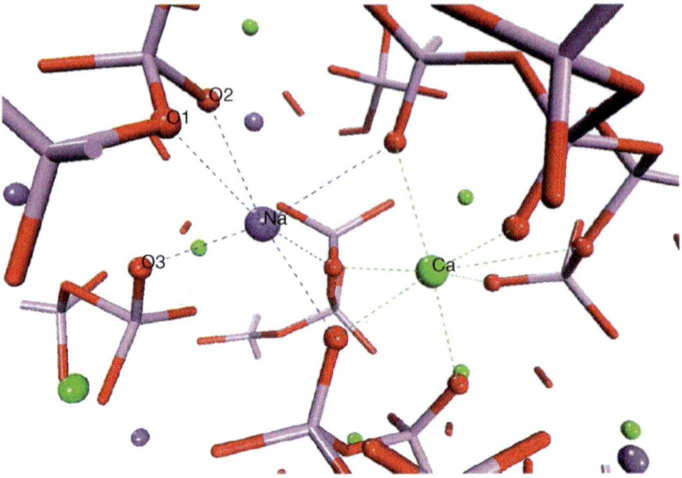


**Figure 10.20** Snapshot from ab initio modeling of PBG with composition $25Na_2O–30CaO–45P_2O_5$ showing local environment of (purple) Na and (green) Ca. Source: Di Tommaso et al. [11] CC BY-SA 3.0.

using the same shell model potential developed by Ainsworth et al. [9]. The results for the bond lengths and $Q^n$ distribution are slightly different to Ainsworth et al. [9] possibly due to the small model size. A very useful comparison is given showing that modeling with the shell model potential produced very similar results to that from ab initio work. A limited comparison was given with classical MD using rigid ion potentials (see Table 10.2). Better agreement with experimental P—O bond lengths are obtained from ab initio and shell model potentials compared to rigid ion potentials. For modifier cation bond lengths, there are not significant differences. Larger Na and Ca coordination numbers are obtained from ab initio and shell model potentials compared to rigid ion potentials. In this study, it was concluded that Ca has a greater cross-linking effect on phosphate network due to its higher field strength compared to Na (see Figure 10.20).

Christie et al. [48] continued the work of Ainsworth et al. [9] and Di Tomasso et al. [11] in using a shell model potential to make classical MD models of $(55 - x)Na_2O–xCaO–45P_2O_5$ glasses and extend the compositional range from $x = 25$ to 45. Models of 3000 atoms were made at experimental density. Although not reported, the SRO is expected to be similar to that in the previous studies using the same potentials. Of great significance is the presentation of a detailed analysis of modifier cation bonding to phosphate chains. These glasses have average connectivity of $Q^n = 1.8$, and hence the phosphate network mostly consists of terminated chains. The most common chain lengths are 3–7, but the overall average chain length is around 8. The distribution of chain lengths does not seem to show any statistically significant differences with Ca content $x$. As previously discussed, Na and Ca have $N_{MO}$ around 6.5 and are mostly bonded to NBO. Typically, Ca is bonded to 0.5 BO, and Na is bonded to 1.3 BO. Analysis showed that Na is typically bonded to 6.0 $PO_4$ tetrahedra, whereas Ca is typically bonded to 6.3 $PO_4$ tetrahedra. This study included a very insightful analysis of how many different phosphate chains were bonded to each modifier cation. The results were largely independent of $x$. It was found that the typical percentages of Na bonded to 1, 2, 3, 4, 5, and 6 chains (respectively) were 1%, 15%, 48%, 32%, 4%, and 0%, meaning that on average Na bonds to 3.2 chains. In comparison, the typical percentages of Ca bonded to 1, 2, 3, 4, 5, and 6 chains (respectively) were 0%, 5%, 26%, 50%, 21%, and 3%, meaning that on average Ca bonds to 3.9 chains. This result has taken on significance for explaining why Ca decreases the solubility of these glasses.

Shaharyar et al. [49] continued the study of $10Na_2O$–$45CaO$–$45P_2O_5$-based glasses using shell model potential of Ainsworth et al. [9]. They used classical MD with shell model potentials to model a system of fluorinated PBG with compositions $10Na_2O$–$(45 - x)CaO$–$xCaF_2$–$45P_2O_5$, where $x = 0$, 5, and 10. Models of 3000 atoms were made using densities extrapolated from $x = 0$ glass. The $x = 0$ composition corresponds to a glass studied by Christie et al. [48]. The P—O, Na—O, and Ca—O bond lengths were observed to be unchanged by addition of F. With addition of F up to $x = 10$, the ratios of F:P, F:Na, and F:Ca became 1 : 2.5, 1 : 1, and 1 : 2.5, respectively. The F coordination numbers are $N_{FP} = 0.2$ with P, $N_{FNa} = 0.3$ with Na and $N_{FCa} = 2.55$ with Ca. (Note that Na—F and Ca—F bond lengths are very similar to Na—O and Ca—O.) This indicates strong affinity for F bonding to Ca and aversion for F bonding to P. (Although in a related study by Christie et al. [50] which is discussed in Section 10.6.2 there was more P—F bonding observed.) The main effect of replacing CaO with $CaF_2$, where F is not bonded to P, is to decrease the O:P ratio, and hence, increase in polymerization to a connectivity higher than $n = 1.8$.

### 10.6.2 Modeling $Na_2O$–$CaO$–$P_2O_5$ Glasses with 50 mol% $P_2O_5$

The other common PBG compositions have 50 mol% $P_2O_5$ corresponding to metaphosphate (with connectivity $n = 2$). Al Hasni et al. [25] used previously discussed work on MD modeling of Na and Ca metaphospohate glasses as a basis for studying a $20Na_2O$–$30CaO$–$50P_2O_5$ composition. A larger model was made of the mixed composition to ensure adequate statistics for each modifier. The model showed fair agreement with ND and XRD data. The SRO in the mixed Na–Ca metaphosphate model was found to be very similar to that in the Na metaphosphate and Ca metaphosphate glass models. Particular emphasis was placed on evaluating the mixing of Na and Ca. This is a topic which is often not addressed in modeling of mixed-modifier glasses. The analysis based on average number density showed there appears to be random mixing. This is a reasonable result given that the glass is formed by quenching of melt.

Section 10.6.1 discussed some studies of PBG using ab initio MD and classical MD with shell model potential. The same methods have been extended to the study $(50 - x)Na_2O$–$xCaO$–$50P_2O_5$ metaphosphate glasses in Ainsworth et al. [51]. In particular, $x = 40$ glasses were modeled. A 300 atom model was made using ab initio MD and a 6000 atom model was made using classical MD with shell model potential, both at experimental density. Most features of cation–oxygen bond lengths and numbers of nearest neighbors were similar to those discussed in Section 10.6.1 for $(55 - x)Na_2O$–$xCaO$–$45P_2O_5$ glasses. A difference between ab initio and shell model potentials was found in the P—O—P bond angle being smaller at 130° for ab initio compared to 135° for shell model potentials. These metaphosphate models have average $Q^n$ of $n = 2$. Especially, noteworthy is the high proportion of $Q^2$, being 80% for shell model potentials and 96% for ab initio (which is another difference between the two methods). Both are significantly higher compared to models of metaphosphate glasses made using rigid ion potentials such as Mountjoy et al. [6].

Christie et al. [50] used ab initio MD to model $20Na_2O$–30CaO–$50P_2O_5$ metaphosphate glass in which fluorine is added in the form of $CaF_2$ in the compositions $xCaF_2$–$(100 - x)$($20Na_2O$–30CaO–$50P_2O_5$) with $x = 2$ and 6. Ab initio MD was chosen due to the lack of classical potentials for F. Models were made based on densities extrapolated from $x = 0$ glass. Small models were made with 200 atoms for $x = 2$, corresponding to 4 F, and 360 atoms for $x = 6$, corresponding to 6 F. Note that network connectivity is reduced by F because it cannot link $PO_4$ units, i.e. P–F–P does not occur. It is found that F is roughly divided between forming P—F—Na bonds and being surrounded by a mixture of Na and Ca but not P.

### 10.6.3 Modeling $Na_2O$–CaO–$P_2O_5$ Glasses with Additional Cations

Parsons et al. [26] applied the ab initio MD modeling method of Tang et al. [47], described in Section 10.6.1, to model mixed $Na_2O$–MgO–CaO–$P_2O_5$ glasses, still with small models of approximately 150 atoms. The previous work on Na and Mg metaphosphates served as a basis for a modeling mixed $Na_2O$–MgO–$P_2O_5$ glasses. This included a $3Na_2O$–12MgO–$15P_2O_5$ metaphosphate glass with composition equivalent to $10Na_2O$–40MgO–$50P_2O_5$. (Unfortunately, the details of the other glass compositions are confusing due to statements about "charge balancing oxygens.") The goal was to compare the model of this last glass with the experimental glass and the ND shows fair agreement. This study is notable for a sophisticated analysis of cross-linking of phosphate chains by cations which is not easily summarized here.

The work of Ainsworth et al. [51] on $(50 - x)Na_2O$–$x$CaO–$50P_2O_5$ metaphosphate glasses was extended to silver containing glasses having composition $xAg_2O$–$(20 - x)Na_2O$–30CaO–$50P_2O_5$ with $x = 10$, 15, and 20. (This parallels the study by Christie et al. [52] discussed next.) The SRO in terms of P, Ca, and Na bonding to O are similar to those in the $x = 0$ glass. The Ag cations are found to have $N_{AgO} \sim 5.5$ with on average 0.8 bonds to BO. This high value of $N_{AgO}$ is consistent with results from X-ray absorption spectroscopy of Ag.

Christie et al. [52] looked at the effect of replacing Na with Ag in $xAg_2O$–$(25 - x)Na_2O$–30CaO–$45P_2O_5$ glasses with $x = 5$, 10, and 15. Ag is known to be antibiotic, and in 50 mol% $P_2O_5$ glasses replacing Na with Ag causes a decrease in dissolution rate. Models of 2000 atoms were prepared using same methodology as previously discussed with densities extrapolated from the $x = 0$ composition. The SRO around P, Na, and Ca are similar to before, but the $Q^n$ distribution is not discussed. The results for Ag show an average $N_{AgO}$ of around 5.5 of which there is typically about 0.9 BO, slightly less than Na. The results also show that Ag typically links to 3.2 phosphate chains, similar to Na. A clustering analysis was performed using criteria of average number density. Apart from the $x = 5$ model which has only 27 Ag atoms, the $x = 10$ and $x = 15$ models had slight clustering of Ag, similar to Na, whereas Ca does not appear to be clustered.

## 10.7 Modeling Phosphate Glasses with Trivalent Cations

### 10.7.1 Modeling Iron Phosphate Glasses

There is significant interest in iron phosphate glasses because their extraordinarily high chemical durability makes them suitable to use for vitrification of nuclear waste. Iron is a transition metal with two common oxidation states, $Fe^{2+}$ and $Fe^{3+}$. For convenience, this section will discuss phosphate glasses with $Fe^{2+}$, and with $Fe^{3+}$. In fact, iron-containing glasses will frequently have a mixture of both oxidation states.

Al-Hasni and Mountjoy [53] used classical MD to model binary iron phosphate glasses with different oxidation states. The motivation was to compare with experimental data on two sets of glasses: $x$FeO–(100 – $x$)$P_2O_5$ glasses with $x = 30$, 40, and 50 having a majority of $Fe^{2+}$, and $x Fe_2O_3$–(100 – $x$)$P_2O_5$ glasses with $x = 30$ and 40 having majority of $Fe^{3+}$. To simplify the range of compositions, models were made either with 100% of the nominal oxidation state, or with 80% of the nominal oxidation state and 20% of the alternative oxidation state. The work used rigid iron potentials (see Table 10.2). The models had tetrahedral $PO_4$ units, and P—O—P bond angle around 150° which is somewhat larger than from diffraction data. The models were compared with ND and XRD data. The models were able to reproduce the different $Fe^{2+}$—O and $Fe^{3+}$—O bond lengths seen in experiments. The values of $N_{FeO}$ around 4–5 were not found to be very different for $Fe^{2+}$ and $Fe^{3+}$.

The average connectivity in models of FeO–$P_2O_5$ glasses is close to that expected. For models of $Fe_2O_3$–$P_2O_5$ glasses, there is a noticeable trend for the average connectivity to be higher than expected. This is a significant observation and is explained by the presence of oxygens which are not BOs or NBOs (see Figure 10.21). These oxygens are not bonded to P and are only bonded to $Fe^{3+}$. Such oxygens, when they have been observed in models of silicate glasses (see Mountjoy et al. [6]), are referred to as either nonnetwork oxygens, or "free" oxygens (although the latter should not to be confused with gaseous molecular oxygen). More "free" oxygens were observed for the composition 40$Fe_2O_3$–60$P_2O_5$, corresponding to pyrophosphate, which has an O:P ratio of 3.5 : 1. It is worthwhile to make an analogy with MgO–$SiO_2$ glass models around the pyrosilicate composition which were also found to contain free oxygen (see Al Hasni and Mountjoy [54]). In both cases the network former cation, i.e. $P^{5+}$ (or $Si^{4+}$) is competing for oxygens with another cation of relatively high field strength, i.e. $Fe^{3+}$ (or $Mg^{2+}$), and when the concentration of the latter is high enough, some oxygens are removed from the network.

The models included the composition 40$Fe_2O_3$–60$P_2O_5$ which has an extraordinarily low dissolution rate compared to silicate glasses. This is all the more remarkable given that phosphate glasses traditionally have high dissolution rates, i.e. they tend to be hygroscopic. The models of $x$FeO–(100 – $x$)$P_2O_5$ glasses with $x$ = 30, 40, and 50, and $x Fe_2O_3$–(100 – $x$)$P_2O_5$ glasses with $x = 30$ and 40, show that $N_{FeO}$ decreases as $x$ increases, something also seen in experimental results. This is explained by a model of the availability of NBO per modifier cation given in Hoppe et al. [5], which counter-intuitively decreases as modifier cation concentration increases. At the same time that $Fe^{3+}$ coordination number decreases, the bond strength of $Fe^{3+}$—NBO bonds increases. Fortuitously, this matches a

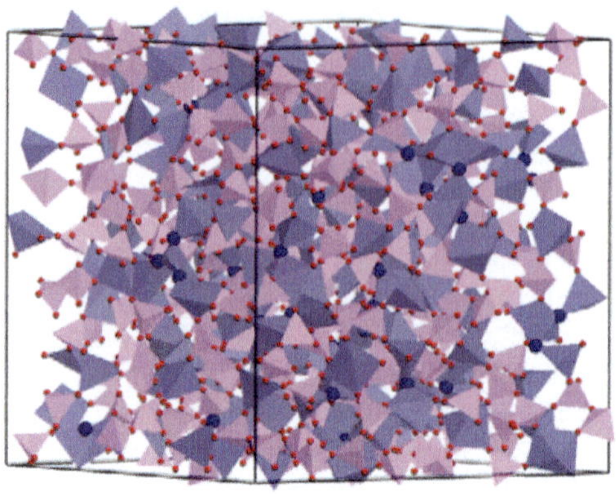

**Figure 10.21** MD model of $40Fe_2O_3-60P_2O_5$ glass with (pink) $PO_4$ tetrahedra, (gray) $FeO_n$ polyhedra, (red) BO and NBO, and (blue) nonnetwork oxygens. Source: Reproduced from Al Hasni and Mountjoy [53] with permission of Elsevier.

| $25Fe_2O_3$–$75P_2O_5$ metaphosphate $Q^2$ | $Fe^{3+}$ coordination |
|---|---|
| BO $\overset{1.0}{—}$ P $\overset{1.5}{—}$NBO$\overset{0.5}{------}$ | $\overset{0.5}{------}$ $Fe^{3+}$ CN = 6 |
| $40Fe_2O_3$–$60P_2O_5$ pyrophosphate $Q^1$ | |
| BO $\overset{1.0}{—}$ P $\overset{1.33}{—}$NBO$\overset{0.67}{------}$ | $\overset{0.67}{------}$ $Fe^{3+}$ CN = 4.5 |
| $50Fe_2O_3$–$50P_2O_5$ orthophosphate $Q^0$ | |
| P $\overset{1.25}{—}$NBO$\overset{0.75}{------}$ | $\overset{0.75}{------}$ $Fe^{3+}$ CN = 4 |

**Figure 10.22** Schematic showing complementary trends in electrostatic bond valence (numbers in blue) with (i) depolymerization of the phosphate network, and (ii) decrease in $Fe^{3+}$ coordination number.

decrease in P—NBO bond strengths as the phosphate network is depolymerized (e.g. from $Q^2$ to $Q^1$). These changes enhance the cross-linking effect of $Fe^{3+}$ with increasing $x$ (see Figure 10.22), and contributes to the $40Fe_2O_3$–$60P_2O_5$ glass having such high chemical durability.

The other interesting analysis in this study concerned the Fe–Fe partial PDF in the model of $40Fe_2O_3$–$60P_2O_5$ glass. An uncommon magnetic ND measurement had been performed to obtain the experimental Fe—Fe partial PDF for $44Fe_2O_3$–$56P_2O_5$ glass. The presence or not of close M–M neighbors has been discussed in Sections 10.4.2, 10.5.1–10.5.3, 10.7.3, in the context of the effect of increasing M concentration, or increasing the M cation size. In the $40Fe_2O_3$–$60P_2O_5$ glass model, the value $N_{FeO} = 4.5$ is possible without close Fe–Fe neighbors (i.e. no Fe-NBO–Fe configurations). However, the high-field strength of $Fe^{3+}$ means that

Fe–O–Fe configurations occur in association with "free" oxygen. The Fe–Fe partial PDF from both model and experiment show some such close Fe–Fe neighbors.

Joseph et al. [55] continued the work on classical MD modeling of iron phosphate glasses started by Al Hasni et al. [53]. The two-body potential from Al Hasni et al. [53] was adapted, and then combined with a three-body potential of the Stillinger Weber kind to be more compatible with techniques for modeling of radiation damage. As discussed previously, the interest in iron phosphate glasses stems from their high chemical durability and ability to vitrify nuclear waste, hence the motivation to model radiation damage. This modeling study looked at idealized compositions of $Fe^{3+}$ metaphosphate and pyrophosphate, i.e. $25Fe_2O_3$–$75P_2O_5$ and $40Fe_2O_3$–$60P_2O_5$ (respectively). In addition, the study looked at a mixed oxidation state composition $25FeO$–$25Fe_2O_3$–$50P_2O_5$, i.e. with 33% of Fe as $Fe^{2+}$ (whereas Al Hasni et al. [53] looked $x = 40$ with 20% of Fe as $Fe^{2+}$). In addition, Joseph et al. made models of $x = 40$ with 4% and 17% $Fe^{2+}$. Small models of approximately 100 atoms and varying densities were made to find densities giving the lowest potential energy. Then the resulting densities were used to make larger models of 1000–1500 atoms. The SRO is not reported in detail, but the reported average values of $N_{PO} = 4$ and $N_{FeO} = 4$–5 are realistic. As seen by Al Hasni et al. [53] the value of $N_{FeO}$ decreases with $x$, being mostly 5 for $x = 25$ and mostly 4 for $x = 40$. This is one of few studies to simulate elastic properties, and good agreements with experimental values were obtained.

The initial study of radiation damage by Joseph et al. [55] was to simulate the threshold displacement energy for a range of ions. This involves randomly injecting kinetic energy into an ion and checking if it is displaced. Typical values average of displacement energies were 20 eV for $Fe^{2+}$, 22 eV for $P^{5+}$, 27 eV or $O^{2-}$, and 34 eV for $Fe^{3+}$. There was not a noticeable effect from coordination number of Fe. This work was continued in Jolley and Smith [56] where models of 2000–3000 atoms were made for $x = 40$ with 4% and 17% $Fe^{2+}$. It included more work on optimizing densities, and interestingly the optimized densities were 5% higher than in the previous study (and in experiment). In addition the simulated elastic constants were 25–35 GPa higher than in previous study (and in experiment). Lastly the melting point in simulations was examined and found to be 800 K higher than experiment. The new study is very valuable because it is one of the only modeling studies of radiation damage in phosphate glasses. The resulting models of 450 000 atoms are believed to be the largest models of phosphate glasses which have ever been reported. The radiation damage is simulated via a primary knock-on atom (PKA) which has 4 keV of kinetic energy injected into it (see Figure 10.23). The displacements (greater than 1.2 Å) of other atoms in the model were then monitored over 1 ps. Typically 800 ions were displaced in total, or 0.2% of model. The most commonly displaced ions were $O^{2-}$, followed by $Fe^{2+}$ and P, and lastly $Fe^{3+}$. The most common cation defects were temporarily broken bonds of $Fe^{3+}$ or P leading to coordination numbers of 3.

### 10.7.2 Cluster Models of Iron Phosphate Glasses

Stoch et al. [57] modeled iron phosphate glasses using ab initio methods based on clusters with H termination. One cluster is based on the $FePO_4$ crystal structure, i.e. $xFe_2O_3$–$(100 - x)P_2O_5$ with $x = 50$, where $PO_4$ and $FeO_4$ tetrahedra form an alternating

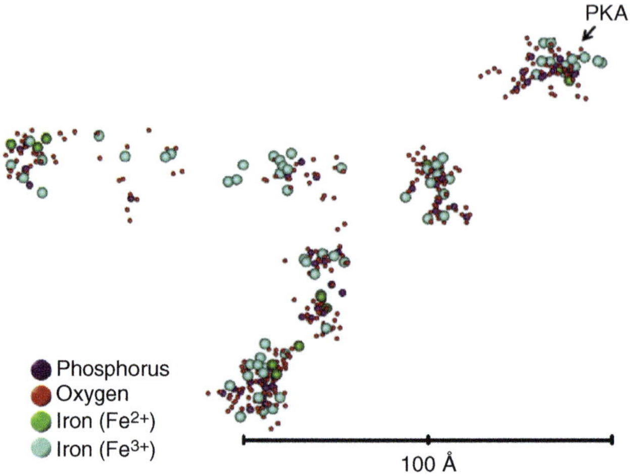


**Figure 10.23** A snapshot image of an iron phosphate glass with 4% $Fe^{2+}$ showing all atoms displaced by at least 1.2 Å at 4 ps after injecting 4 keV of kinetic energy into a $P^{5+}$ ion (see arrow with "PKA"). Source: Reproduced from Jolley and Smith [56] CC BY-SA 4.0.

corner shared network, although such alternation is unrealistic in a disordered material like a glass. The study developed another two cluster models. One has an Fe:P ratio of 2 : 3 with only $Fe^{3+}$ (i.e. $x = 40$) and the other has an Fe:P ratio of 3 : 4 (i.e. $x = 43$) with 33% of Fe as $Fe^{2+}$. However, both clusters are built with $Q^2$ units which is not realistic when the pyrophosphate composition $x = 40$ has a dominance of $Q^1$ units. SRO was good for P—O bonds, reproducing experimental bond lengths. Analysis of ionicity showed a value of 61% for P–O increasing to 71% for $Fe^{3+}$–O and 82% for $Fe^{2+}$–O. Simulation of Raman and IR spectra showed good agreement with experiment. The contribution of asymmetric stretching of $Fe^{3+}$—O—P bonds was identified in the range of 900–1000 $cm^{-1}$.

Stoch et al. [58] used ab initio methods based on clusters to look in more detail at vibrational modes in pyrophosphate $xFe_2O_3$–$(100 - x)P_2O_5$ glass with $x = 40$. Two clusters were built with only $Q^1$ units. One had $N_{FeO} = 4$ via 4 $Q^1$ units, and another had $N_{FeO} = 6$ via six $Q^1$ units. This gave Fe–O distances slightly shorter than expected. The cluster with $N_{FeO} = 4$ gave asymmetric stretch of P–O–Fe around 900 $cm^{-1}$, similar to the previous study, and as expected the cluster with higher $N_{FeO} = 6$ gave lower frequency asymmetric stretch of P–O–Fe around 800 $cm^{-1}$. These identifications of vibrational bands in the cluster models were compared with experimental IR and Raman spectra from $x = 37$ glass.

### 10.7.3 Modeling Trivalent Rare Earth Phosphate Glasses

Rare earth phosphate glasses are of particular interest because of their ability to incorporate large amounts of rare earth ions (R), something which is difficult to achieve in silicate glasses. Mountjoy et al. [59] used RMC to model Eu and Tb metaphosphate glasses, i.e. $25R_2O_3$–$75P_2O_5$ with R = Eu, Tb, for which XRD data was available. The typical modeling procedure for metaphosphate glasses was used, i.e. to construct chains of $Q^2$ units. Models of 550 atoms were made at experimental density with good agreement to XRD. These models have SRO with expected features corresponding to $PO_4$ tetrahedra and R cations with $N_{RO} = 6$.

Clark et al. [8] used classical MD with rigid ion potentials (see Table 10.2) to model Tb metaphosphate glass, i.e. $25Tb_2O_3$–$75P_2O_5$. Models of 1300 atoms were made with experimental density, but nonambient pressure. (In contrast a model made to have ambient pressure had density 8% higher than experiment.) The model has good SRO with shorter P—NBO than P—BO bonds, good tetrahedral shape of $PO_4$ units, but the P—O—P bond angle around 150° is somewhat larger than from diffraction data. The phosphate network has a majority of $Q^2$ but with $Q^1$ and $Q^3$. The value for average $N_{TbO} = 5.7$ and consists of five- and sixfold coordinated Tb. For this glass experimental data on the Tb–Tb partial PDF was available from an uncommon magnetic ND measurement, and the model showed good agreement. The model results show a Tb–Tb nearest neighbor peak at short distance of 4.7 Å, which is due to Tb–O–Tb configurations, but it is not clear if such a peak is present in the results from magnetic ND measurement experimental data.

Hoppe [60] used RMC to make models of a series of rare earth metaphosphate glasses, i.e. $25R_2O_3$–$75P_2O_5$ with R = La, Nd, Er, and Yb in the sequence decreasing ion size. The study also used a methodology to ensure chains of $Q^2$ tetrahedra and to minimize as much as possible close R–R neighbors, i.e. R–O–R configurations. Models of 3900 atoms were made at experimental density and showed good comparison to high-quality XRD and ND data. A particular focus was the R–R partial PDF and whether or not there are close R–R neighbors, i.e. R–O–R configurations are present. Note that a metaphosphate with trivalent modifier cation has six NBOs per modifier cation, so a value of $N_{RO} = 6$ is possible with no sharing of NBOs, i.e. no R–NBO–R configurations. Instead, a $N_{RO} > 6$ requires R–NBO–R and hence close R–R neighbors. It was found that $N_{RO} = 6.9$ for larger ions La and Nd, and $N_{RO} = 6.5$ for smaller ions Er and Yb. Corresponding to this there are significant numbers of close R–R neighbors for La and Nd, and a minor amount for Er and Yb. This size effect is analogous to that seen for close M–M neighbors in alkaline earth metaphosphate glasses (see Figure 10.17).

### 10.7.4 Modeling Aluminophosphate Glasses

The trivalent $Al^{3+}$ cation is considered to be an intermediate cation. This is on the basis that $Al^{3+}$ can take up fourfold tetrahedral coordination, in which regard it mirrors $B^{3+}$, and which makes it compatible with tetrahedral network glasses like phosphates. (In addition, although a glass cannot be made from pure $Al_2O_3$, glass formation becomes possible on the addition of a modifier like Ca.)

Martin et al. [61] used classical MD with rigid ion potentials (see Table 10.2) to model a glass with composition $23Dy_2O_3$–$7Al_2O_3$–$70P_2O_5$. The inclusion of Al in the experimental glass is a consequence of the glass-making method whereby Al from the alumina crucible enters the melt. A model of 2600 atoms was made using experimental density and showed fair agreement with ND data. The connectivity of the phosphate network is $n = 1.7$ as expected for 70 mol% $P_2O_5$ composition. An interesting result from the study concerns role of Al. In the model, Al is found to be 80% fourfold coordinated (the rest being $N_{AlO} = 5$),

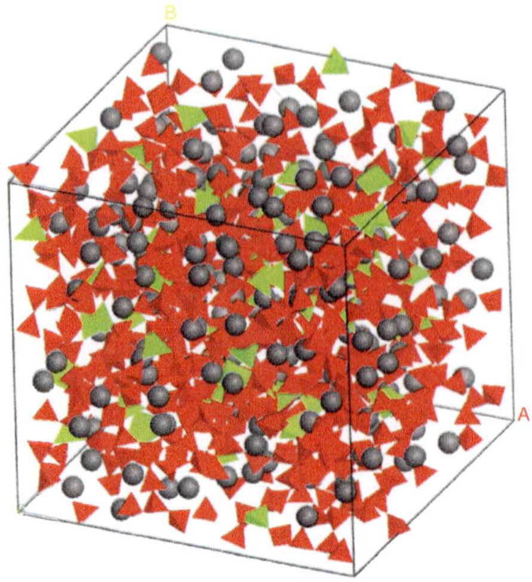


**Figure 10.24** MD model of $23Dy_2O_3$–$7Al_2O_3$–$70P_2O_5$ glass with (gray) Dy ions, (red) $PO_4$ tetrahedra, and (green) $AlO_4$/$AlO_5$ polyhedra. Source: Reproduced from Martin et al. [61] with permission of IOP Publishing.

although the O—Al—O bond angle does not show strong tetrahedral geometry. The model shows that Al is never neighboring to Dy, i.e. Al–O–Dy configurations do not occur. This is apparent in Figure 10.24 where $AlO_4$ tetrahedra are never found close to Dy.

In the study by Martin et al. [61] of $23Dy_2O_3$–$7Al_2O_3$–$70P_2O_5$ glass the choice of Dy is due to the availability of experimental ND measurements which were performed using isomorphic substitution, i.e. replacement of Dy with Ho (two ions with very similar ionic radius). This enabled the analysis of experimental ND data to extract separate partial PDFs representing (i) the matrix–matrix correlations, and (ii) the Dy–matrix correlations. (This is a significant improvement over the total PDF representing all correlations which is normally the result of XRD or ND.) The average $N_{\mathrm{DyO}} = 5.8$ is similar to experiment. Although the metaphosphate composition, i.e. 75 mol% $P_2O_5$, has six NBO per trivalent modifier, which means that Dy–O–Dy configurations would not be necessary, it is found that each Dy typically has 0.8 close Dy–Dy neighbors. This results closely parallels that in the modeling study by Clark et al. [8] for Tb metaphosphate glass which was discussed previously. Nevertheless, the Dy–Dy partial PDF is subject to experimental uncertainty.

Du et al. [62] made a related study of another $R_2O_3$–$Al_2O_3$–$P_2O_5$ glass close to metaphosphate, in particular $5Ce_2O_3$–$19Al_2O_3$–$76P_2O_5$. Note this composition with Ce:Al ratio of approximately 1 : 4 forms a sequence with the previously discussed studies of Martin et al. [61], which has Dy:Al ratio of approximately 3 : 1, and of Clark et al. [8], which has only Tb and no Al. All used classical MD with rigid ion potentials to model compositions close to metaphosphate. Du et al. [62] made a 5000 atom model of $5Ce_2O_3$–$19Al_2O_3$–$76P_2O_5$ glass, and in this model, 5% of the Ce ions are $Ce^{4+}$. The model had the expected SRO with tetrahedral phosphate groups. The Al is 80% fourfold coordinated (the rest being five), and the O—Al—O bond angle does not show strong tetrahedral geometry. The network connectivity analyzed in terms of both Al and P was $n = 3.2$. Ce in both 3+ and 4+ oxidation states is found to have a mix of $N_{\mathrm{CeO}} = 5$ and 6. The Al is overwhelming linked to P, i.e. Al–O–P configurations, and Ce–O–Al configurations are almost never found, consistent with Martin et al. [61]. Du et al. [62] note that there is very little clustering of Ce due to the de-clustering effect of aluminum.

Kokou and Du [63] continued the methodology to study further cerium aluminophosphate glasses with approximately 70 mol% $P_2O_5$ and varying ratios of Ce:Al. These show results similar to those already discussed. This study included a glass with composition $23.8Ce_2O_3$–$4.1Al_2O_3$–$72.1\,P_2O_5$ (and Ce:Al ratio of approximately 5 : 1) which might be considered comparable to the $23Dy_2O_3$–$7Al_2O_3$–$70P_2O_5$ glass studied by Martin et al. [60]. The results are very similar apart from the larger $Ce^{3+}$ ion having larger $N_{CeO} = 6.4$ (mostly $N_{CeO} = 6$ or 7) compared to Dy. In addition, there is greater clustering of Ce with each Ce having almost three close Ce–Ce neighbors, compared to Dy. The other three compositions that were studied are $xCe_2O_3$–$(30.2 - x)Al_2O_3$–$69.8P_2O_5$ with $x = 1.8$, 13.8, and 19.8. Regarding the distribution of Ce, the results for $x = 13.8$ and 19.8 correspond to approximate Ce:Al ratios of 1 : 1 and 2 : 1, respectively, and the models have typically two close Ce–Ce neighbors, i.e. Ce–O–Ce. Instead, the model with $x = 1.8$ and Ce:Al ratio of 1 : 15 has highly declustered Ce ions (as was discussed for Du et al. [62] in the previous paragraph).

## 10.8 Modeling Phosphate Glasses with Tetravalent and Pentavalent Cations

Pickup et al. [64] used RMC to model a titanium pyrophosphate glass, i.e. $50(TiO_2)$–$50(P_2O_5)$ using XRD data. The model was based on a crystalline starting configuration consisting of the $TiP_2O_7$ crystal structure. The crystal contains pairs of $Q^1$ tetrahedra, $P_2O_7$, in which each $Q^1$ tetrahedra provide three P–NBO–Ti linkages. This enables Ti to have octahedral coordination and to cross-link pyrophosphate groups. This study illustrates the use of the RMC method to test a hypothesis, but also the ability of the RMC method to obtain consistency with XRD data of glass even based on a highly ordered crystalline input configuration. It contrasts with the preferred approach of using a random starting configuration.

Broglia et al. [65] used classical MD to model lithium vanadophospohate glasses. Broglia et al. [65] carried out modeling using rigid ions with Morse potentials. This was done by augmenting the set of parameters published by Pedone et al. [66]. Both $V^{4+}$ and $V^{5+}$ oxidation states were included, where the fraction of $V^{4+}$ was based on experimental data and varied from 25% to 40% of V. Models of 5100 atoms were made. The P is overwhelmingly in tetrahedral $PO_4$ groups, although there are a small amount of $N_{PO} = 5$ defects. The value of $N_{LiO}$ is slightly under 4 which is consistent with Li phosphate glasses. The value of $N_{VO}$ is between 5 and 6 for $V^{5+}$ and 4–6 for $V^{4+}$, with a lower average $N_{VO}$ for $V^{4+}$ than $V^{5+}$. Presumably, this explains why V—O bond lengths are found to be shorter for $V^{4+}$ compared to $V^{5+}$.

One glass composition studied by Broglia et al. [65] is $10V_2O_5$–$45Li_2O$–$45P_2O_5$, i.e. a Li metaphosphate, i.e. with 10 mol% $V_2O_5$ added. This might naturally be compared to Li metaphosphate $50Li_2O$–$50P_2O_5$ glass. In the $10V_2O_5$–$45Li_2O$–$45P_2O_5$ model, there is no V with $N_{VO} = 4$ but mixtures V with $N_{VO} = 5$ or 6, with $N_{VO} = 6$ being more common. If V is considered a modifier cation (like Li), then the O:P ratio of 3.55 is close to that in the pyrophosphate composition, i.e. $P_2O_7$ with O:P ratio of 3.5. The latter has only $Q^1$ units. Instead, the $10V_2O_5$–$45Li_2O$–$45P_2O_5$ glass model has an average $Q^n$ with $n = 1.22$. This is due to the role of V in creating 4.8% nonnetwork or "free" oxygen, i.e. O not bonded to P, which repolymerizes the phosphate network. The two other compositions studied both have 10 mol% $Li_2O$. They differ in their vanadophosphate compositions, one being $30V_2O_5$–$60P_2O_5$ with O:P ratio of 3.83, and the other being $50V_2O_5$–$40P_2O_5$ with O:P ratio of 5.75. The results for O:P ratio of 3.83 are very similar to those for O:P ratio of 3.55 except that the free oxygen increases to 9.7%. The results for O:P ratio of 5.75 have a large proportion of $Q^0$ units as expected (noting that the orthophosphate composition $PO_4$ has O:P ratio of 4), but still substantial $Q^1$ because of the larger amount of 32.6% free oxygen.

The alternative approach is to consider V to be an intermediate cation if it is providing low coordination number structural units which are corner-linked to the phosphate network (analogous to Al in aluminophosphate glasses). The vanadophosphate glass models by Broglia et al. [65] have only a minor amount of $V^{4+}$ with $N_{VO} = 4$. The majority of V has $N_{VO} = 5$ or greater. Nevertheless, it could be interpreted that V–O–P and V–O–V linkages form part of the glass network. The study does report that 100% of P–O–P and V–O–P linkages are corner-shared, as are 98% of V–O–V linkages (the exception being ~2% of $V^{4+}$–O–$V^{4+}$ linkages). Taking the $10V_2O_5$–$45Li_2O$–$45P_2O_5$ glass composition as an example, the ratio O to (V + P) is 2.9, i.e. slightly lower than metaphosphate. Then the average $Q^n$ counting both V and P would be expected to have $n$ slightly higher than 2. This type of $Q^n$ analysis was not reported in the study.

## 10.9 Modeling Phosphate Glasses with Mixed Network Formers

### 10.9.1 Modeling Borophosphate Glasses

Le Roux et al. [7] used RMC to model sodium borophosphate glasses starting from $35Na_2O$–$65P_2O_5$ close to ultraphosphate (i.e. with 67 mol% $P_2O_5$), and replacing P with B, i.e. $35Na_2O$–$65(yB_2O_3$–$(1 - y)P_2O_5)$ with $0 \leq y \leq 1$. This RMC study made use of high-quality XRD data, and models of approximately 6000 atoms were made. The modeling method used crystalline starting conditions of sodium metaphosphate for $y = 0$, and sodium diborate for $y = 1$, and replacing P with B, or vice versa, for $0 < y < 1$. Regrettably, the final configurations in the models show very clear traces of the crystalline starting configurations (see Figure 10.5). This serves to remind of the hazard of using crystalline starting configurations.

Follow up work on the same glasses was much improved by Schuch et al. [29] who used RMC with random starting configurations. As usual for the RMC method, a number of constraints were included, such as all O being bonded to network formers $P^{5+}$ or $B^{3+}$ (i.e. as bridging or nonbridging oxygen), all P with $N_{PO} = 4$, and all B with $N_{BO} = 3$ or 4. Experimental data was used to specify the fraction of B with $N_{BO} = 4$ to be equal to $y$ up to $y = 0.43$, and to be equal 0.43 up to $y = 1$. Initial RMC models produced a majority of P with $Q^3$ and $Q^2$ units as expected because $y = 0$ has O:P ratio of 2.8 (i.e. compared to metaphosphate with O:P ratio of 3 and dominance of $Q^2$). However, there is also $Q^1$ and $Q^4$, and this unrealistic feature was removed by use of further constraints during RMC modeling. In the resulting models, the addition of B was observed to produce a direct replacement of $Q^2$ with B having $N_{BO} = 4$. A further weakness of the models was tendency for Na to have average $N_{NaO} \sim 3$, and some Na to have $N_{NaO} = 0$. Further constraints were added to give a model with realistic average $N_{NaO} \sim 5$. Lastly, the model had a significant content of B–O–B involving B with $N_{BO} = 4$ which was deemed unrealistic and removed, again by adding constraints. The resulting models were analyzed in terms of connectivity of network formers. This showed a preference for B–O–P linkages (see Figure 10.25). Interestingly, this comes from the avoidance of B–O–B linkages in the RMC modeling, whereas in the MD model discussed next this comes from avoidance of P–O–P linkages.

Tho et al. [67] used MD to model lithium borophosphate glasses from $45Li_2O–P_2O_5$ close to Li metaphosophate, with increasing replacement of P by B, i.e. $45Li_2O–55(yB_2O_3–(1-y)P_2O_5)$, with $0.09 \leq y \leq 0.64$. The MD modeling used rigid ion potentials of Buckingham form with partial charges (of 60% of formal) and Vessel type three-body potentials for O—P—O, P—O—P, and O—B—O bond angles. Models were made with experimental density and had pressures around ambient. Taking $y = 0.64$ model as an example, it is seen to have reasonable SRO with average bond lengths and coordination numbers as expected for P, B, and Li. The model is found to have a mixture of B with $N_{BO} = 3$ and 4. The analysis reports the proportions of P–O–P, P–O–B, and B–O–B linkages. These can be compared to those expected according to a theory of random linkages, and the comparison shows clear preference for P–O–B linkages instead of P–O–P linkages. This is understandable as it prevents close neighbors of $P^{5+}$ cations. The modeling study also looked at diffusion of Li and the simulations reproduce trend of increasing diffusivity of Li with increasing y (B concentration), then a plateau after $y = 0.36$. As per the studies reported in Sections 10.4.1–10.4.4, the analysis looked at the part of volume corresponding to small bond valence mismatch and to what extent this can explain the trend in conductivity (see Figure 10.26).

## 10.9.2 Modeling Phosphosilicate Glasses

Malavasi et al. [68] used classical MD to model phosphosilicate glasses starting from tetrasilicate $20CaO–80SiO_2$ and replacing Ca with P, i.e. $(20-x)CaO–xP_2O_5–80SiO_2$. Values of

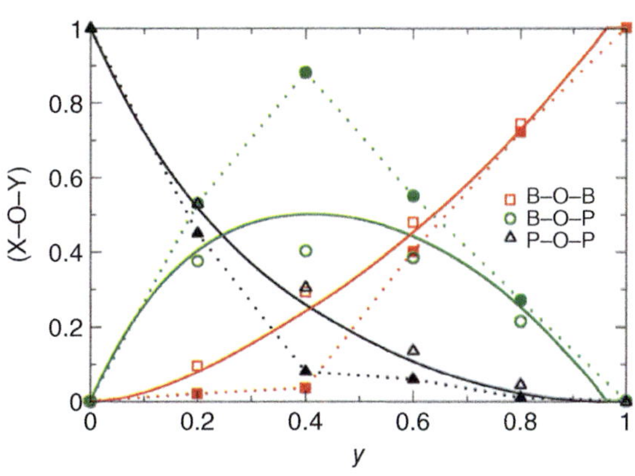


**Figure 10.25** Fraction of bridging oxygens connecting different network former cations from (symbols) RMC modeling and (solid lines) for random linkages in sodium borophosphate glasses $35Na_2O–65(yB_2O_3–(1-y)P_2O_5)$. Source: Reproduced from Schuch et al. [29] with permission of American Chemical Society.

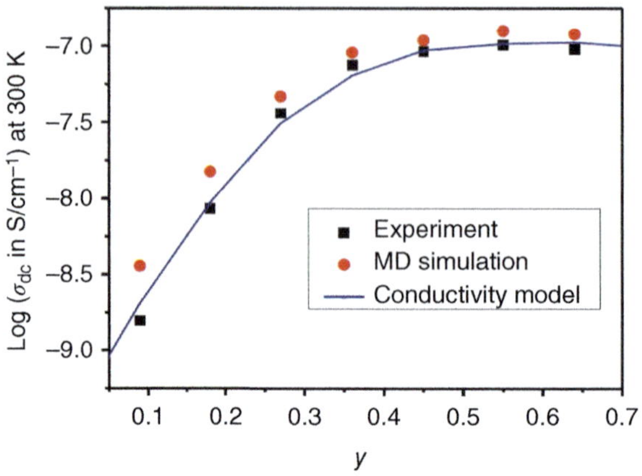


**Figure 10.26** Li ion conductivity in $45Li_2O$–$55(yB_2O_3-(1-y)P_2O_5)$ lithium borophosphate glasses. Source: Reproduced from Tho et al. [67] with permission of Springer Nature.

$x = 0$, 3, 5, 10, 15, 17, and 20 gave P:Si ratios of 0, 0.075, 0.125, 0.25, 0.375, 0.425, and 0.5. Models have at least 2000 atoms and were made with experimental density. The potentials used were those developed by Pedone et al. [66]. The analysis reported good average bond lengths and coordination numbers for P, Si, and Ca, and tetrahedral O—P—O and O—Si—O bond angles. Note that the endpoint compositions have 78% bridging oxygen for $x = 0$ (i.e. $20CaO$–$80SiO_2$), and 85% bridging oxygen for $x = 20$ (i.e. $20P_2O_5$–$80SiO_2$). Connectivity analysis shows a great aversion to P–O–P linkages. For $x = 20$, there are 3% of linkages as P–O–P when the expected proportion for random linkages would be 11%. Instead, there is a larger than expected amount of Si–O–P linkages. The associated $Q^n$ analysis shows the expected tendency for Si to be $Q^4$ and P to be $Q^3$ at $x = 20$. For $x = 20$, there is a spread in the $Q^n$ distribution of P that includes $Q^2$ and $Q^4$ (such a spread in $Q^n$ is usual for rigid ion MD modeling). The results also show a clear preference for Ca to be a close neighbor to P which is understandable due to NBOs being bonded to P.

Linati et al. [69] used classical MD to model phosphosilicate glasses with compositions $24.35Na_2O$–$26.91CaO$–$48.73(yP_2O_5-(1-y)SiO_2)$, where $y = 0.05$, 0.10, and 0.13. The value $y = 0.05$ corresponds to bioglass "45S" which is discussed in Section 10.10. The values $y > 0$ correspond to replacing Si with P, and the values of $y = 0.05$, 0.10, and 0.13 (respectively) give a P:Si ratio of 0.105, 0.22, and 0.30. Models of 2800 atoms were made with experimental density using rigid ion potentials developed by Pedone et al. [66]. There is 51 mol% content of (total) $Na_2O$ and CaO, and so these compositions are very close to a mix of metasilicate and metaphosphate. An analysis of $Q^n$ is presented and gives an average (Si + P) connectivity of $n \sim 1.9$ as expected for O:Si + P ratio of 3.05. It is very noteworthy that P has a majority of ~60% $Q^0$, i.e. far from $Q^2$ expected for a metaphosphate. This causes an increased polymerization of the silicate network to average connectivity $n = 2.2$, which is greater than $n = 2.0$ expected for a metasilicate. Increasing replacement of Si with P causes a slight increase in $Q^n$ of both Si and P.

Kapoor et al. [70] used classical MD to model glasses in the mixed diopside ($25CaO$–$25MgO$–$50SiO_2$) and tricalcium phosphate ($75CaO$–$25P_2O_5$) system with compositions $y(25CaO–25MgO–50SiO_2)–(1-y)(75CaO–25P_2O_5)$ with $0 \leq y \leq 1$. Experimentally, such glasses could only be obtained for $0.5 < y \leq 1$. Rigid ion potentials of the Buckingham form were used to make models of 5000 atoms with experimental density. All the cations were seen to have realistic bond lengths. The modifiers Mg and Ca (respectively) had realistic $N_{MO} = 4.5$ and 7 which did not change much over this composition range. As expected the $x = 100$ (diopside) metasilicate glass had $Q^n$ with average n close to 2. (It was reported that 2–3% of oxygen were "free" oxygen, consistent with results by Al Hasni and Mountjoy [54] for Mg metasilicate glass.) For $y = 0.9$, where the P:Si ratio is 1 : 9 it is found that 90% of P is $Q^0$, in agreement with results from NMR of P-31. For $y = 0.7$, where the P:Si ratio is 1 : 4.7, it is found that there is still a majority of 70% P as $Q^0$. This is less than the result of 90% of P as $Q^0$ from NMR of P-31 of the $y = 0.7$ glass.

## 10.10 Modeling Bioglass 45S and Related Glasses

### 10.10.1 Modeling Bioglass 45S and Related Glasses from the Same System

One of the most modeled glass systems containing $P_2O_5$ is the famous "Bioglass" system. The most frequently cited of these bioglasses is so-called "45S," having the composition $24.35Na_2O–26.91CaO–2.6P_2O_5–46.13SiO_2$ (equivalent to 6 wt% $P_2O_5$), which optimizes the bioactivity within the $CaO–SiO_2–Na_2O$ phase diagram (with fixed 6 wt% $P_2O_5$). There have been many studies of glasses with related compositions. This section has been included in the chapter because the 45S glass is technically a phosphosilicate glass. However, $P_2O_5$ is the most minor component in 45S glass, and accordingly, this section will be of limited scope.

Tilocca and De Leeuw [71] used ab initio MD to model a glass composition close to 45S glass. For ab initio, they used a small system size of 116 atoms, i.e. $19SiO_2–10Na_2O–11CaO–1P_2O_5$. (Note this corresponds to the composition $24.4Na_2O–26.8CaO–2.4P_2O_5–46.3SiO_2$.) Models were made initially using classical MD with shell model potentials. Ab initio modeling was then used at room temperature to relax the models. The results found good SRO with 100% $N_{Si} = 4$ and $N_P = 4$, realistic bond lengths for all cations, and average $N_{MO}$ of 4.5 for Na and 5.4 for Ca. Detailed study shows a higher contribution to $N_{NaO}$ from BO and from intratetrahedral NBO, compared to $N_{CaO}$. The $Q^n$ distribution counting both Si and P as network-forming units has overall average $n = 2.0$ which is consistent with the O:(Si + P) ratio of 3.05 being very close to metasilicate. While Si is majority $Q^2$ with the usual spread from $Q^0$–$Q^4$, a significant finding is that P is 50% $Q^1$ and 50% $Q^0$. This indicates a tendency of P to form $Q^0$ isolated from the silicate network when there is a supply of modifier cations. This study also performed a detailed analysis of vibrational density of states (VDOS) and electronic density of states (DOS).

Tilocca et al. [72] reported a further MD modeling of 45S-based glass with increasing $SiO_2$ content. This work focused on glasses with compositions $(1 - y)(24.4Na_2O–26.9CaO)–2.6P_2O_5–(1 + y)46.1SiO_2$ with $y = 0$, 0.23, and 0.45. (The value $y = 0$ is very close to 45S glass, and the other values of y correspond to 55S and 65S glasses.) In this study, models of approximately 1500 atoms were made with experimental densities for $y = 0$ and 0.23, and estimated density for $y = 0.45$. Given the large number of atoms, this study used only

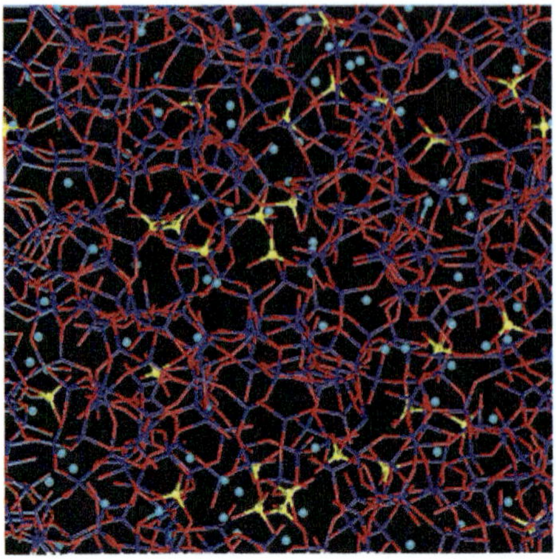

**Figure 10.27** Phosphate-rich regions in the BG65 model with (blue) silicon, (yellow) phosphorous, (cyan) calcium, and (red) oxygen (for clarity sodium are not shown). Source: Reproduced from Tilocca et al. [72] with permission of American Chemical Society.

classical MD with the same details as in the previous study by Tilocca and De Leeuw [71]. Some SRO details for the 45S glass model are the same as the previous model. The study found average $N_{NaO} = 5.9$ and $N_{CaO} = 6.1$ for the 45S glass model, which is slightly higher than in the previous model. In the 45S glass model, the connectivity with regards to Si was $n = 2.07$ like the previous model, but with a narrower distribution of $Q^1$–$Q^3$. Instead, P was found to be majority 65% $Q^0$ with 33% $Q^1$, i.e. more $Q^0$ than the previous model. (These differences could be due to the previous model having a much smaller size, or being relaxed by using ab initio modeling.) With increasing y from 0 to 0.45, the connectivity increased as expected due to decrease in O:(Si + P) ratio from 3.05 to 2.45. This was accompanied by a shift in the average $Q^n$ to $n = 3.23$ for Si and $n = 1.26$ for P. This study provides a detailed analysis which reveals a tendency for Na and Si to be neighboring to Si, and for Ca and P to be neighboring to P (see Figure 10.27). (Note this is associated with higher average $Q^n$ for Si, and low average $Q^n$ for P.)

In Tilocca and Cormack [73], the previous work was extended by using MD modeling to study surfaces of $y = 0$ (45S) and $y = 0.45$ (65S) glasses. Given the application of these glasses as biomaterials, where surface reactions are crucial to the process by which bone minerals are eventually deposited, there is a very important motivation for this kind of study. As expected, the models obtained showed enhancement of NBO at the surface corresponding to termination of glass network. This is accompanied by enhancement of Na concentration, which is understandable as Na is the more mobile of charge compensating cations.

Malavasi et al. [74] used ab initio MD to model the structure of 45S glass. Due to the use of ab initio modeling, the model size was small at 78 atoms, having the composition $6Na_2O$–$7CaO$–$1P_2O_5$–$13SiO_2$ (or approximately, $25.9CaO$–$22.2Na_2O$–$3.7P_2O_5$–$48.1SiO_2$). The model was initially made at experimental density using classical MD with the Morse potentials developed by Pedone et al. [66]. This model was then relaxated using ab initio MD, which allowed the initially cubic model to become triclinic, and also increased density by 3.5% above experimental density. The SRO details in terms of bond lengths and coordination numbers are quite reasonable, except there is a Si coordination defect with 1 out of 13 Si having $N_{SiO} = 3$. The $N_{NaO} = 5.3$ and $N_{CaO} = 6.0$ are similar to those seen by

Tilocca and De Leeuw [71] using ab initio MD. The $Q^n$ distribution for Si is majority of $Q^2$ as expected for the O:(Si + P) ratio, with narrow spread from $Q^1$–$Q^3$. The 2 P have $Q^0$ and $Q^1$ which is consistent with previous models.

Corno et al. [75] continued the work of Malavasi et al. [74] using the ab initio model produced to simulate vibrational density of states. This showed good correspondence with the experimental infrared spectrum. It confirms assignment of P modes, but such modes are difficult to identify in experimental infrared spectra due to minority of P compared to Si.

Berardo et al. [76] continued the work of Malavasi et al. [74] on 45S glass by making a small 80 atom model of so-called "77S glass" with composition $16CaO$–$4P_2O_5$–$80SiO_2$. The SRO in the 77S glass model was similar to that in 45S model (with the obvious exception that Na is absent in the 77S model). The 77S glass model has O:(P + Si) ratio of 2.23 and as expected has a majority of Si as $Q^4$. Interestingly, two of the eight P in the 77S model have $Q^2$ and $Q^3$, very different to the 45S glass model, and also different to the 65S glass model studied by Tilocca et al. [72]. These models were adapted to study the surfaces of the glasses. That was done by cutting the models and passivating broken bridging oxygen bonds by adding $OH^-$ or $H^+$. The surfaces were then relaxed using ab initio modeling. (This differs from the work of Tilocca and Cormack [73] due to the presence of hydroxyls.) However, the small statistics do not reveal anything about P in relation to the surface.

### 10.10.2 Modeling Bioglass 45S and Related Glasses with Additional Components

Pedone et al. [77] used classical MD modeling to look at the effect of replacing Ca with Mg. The glasses had compositions $46.2SiO_2$–$(26.9 - x)CaO$–$xMgO$–$24.3Na_2O$–$2.6P_2O_5$ with $x = 0$, 5, 10, 15, 20, and 26.9. The study used shell model potentials previously used by Tilocca and De Leeuw [71] which were extended to include Mg. The potentials were shown to have good performance on related crystal structures. Models of 2800 atoms were made with estimated glass densities. A relatively new property calculation was to simulate elastic constants. The bulk modulus decreases with increasing replacement of Ca with Mg, something seen in soda-lime silicate glasses. The SRO is like that previously seen in models of 45S glass, and the Mg bond length of 1.98 Å and $N_{MgO} = 5$ are realistic. As previously discussed for models of 45S glass, there appears to be a preference for Na to be neighboring to Si, and for Ca to be neighboring to P. There does not appear to be a preference in the case of Mg. The connectivity shows a majority of $Q^2$ for Si and a majority of $Q^0$ for P, with little effect from Mg content.

Malavasi et al. [78] used the classical MD modeling methods to look at the addition of 1 mol% $Ga_2O_3$ or $Al_2O_3$ to 45S glass. Two different potentials were used, the first being the Morse potentials from Pedone et al. [66], and the second being the shell model potentials from Tilocca and De Leeuw [71]. In both the cases, parameters for Al–O interactions were taken from the literature, and parameters for Ga–O parameters were derived. (The fitted Ga–O parameters work well for Ga in crystals and sodium gallium silicate glass.) The SRO for Si, P, Na, and Ca is reported to be similar to that in previous models of 45S glass. The Al is found to be in tetrahedral $AlO_4$ units with realistic bond length of 1.75 Å. Ga is found to have $N_{GaO}$ somewhat higher than 4, and the O—Ga—O bond angle distribution shows disordered Ga sites. This is thought to be due to Ga being a larger ion than Al.

Christie et al. [79] used ab initio MD modeling to look at the effects of F addition to 45S bioglass. This was done for a single composition of 45S with 10 mol% of CaO replaced with 10 mol% of $CaF_2$, according to the composition $10Na_2O$–7CaO–$4CaF_2$–$1P_2O_5$–$19SiO_2$ (or approximately $24.4Na_2O$–17.1CaO–$9.8CaF_2$–$2.4P_2O_5$–$46.4SiO_2$). Models were made with experimental density and 120 atoms, i.e. eight F atoms, and results were based on three simulations. The glass composition has ratios of F:Na, F:Ca, F:P, and F:Si (respectively) equal to, 1 : 2.50, 1 : 1.33, 1 : 0.25, and 1 : 2.38. The coordination numbers for F–Na, F–Ca, F–P, and F–Si (respectively) were found to be 2.41, 1.69, 0.00, and 0.17. This indicates an aversion to F—Si and F—P bonds, and a roughly equal preference for F bonds to Ca and Na. The small statistics do not provide clear information about $Q^n$.

## 10.11 Summary

The earliest modeling study of phosphate glasses that this chapter references is from 1994. This chapter has referenced roughly 80 modeling studies carried out in the 20 years since. The studies have most commonly involved RMC, classical, and ab initio MD methods. There is a scarcity of modeling studies using the empirical potential structure refinement method. The most popular experimental phosphate glass compositions have now been modeled in multiple studies, and with significant success in reproducing experimental data and providing insights into the structures and properties. Hence, the modeling of phosphate glasses has become an established topic.

The literature provides widely used rigid ion and shell model potentials for classical MD modeling of phosphate glasses. The studies show the typical differences between rigid ion, shell model, and ab initio MD modeling methods. The differences between these methods are not significant for SRO in terms of cation bond lengths and $N_{MO}$, shorter P—NBO than P—BO bonds, and tetrahedral $PO_4$ units. The differences start to be noticeable for how closely P—O—P bond angles and the $Q^n$ distribution in models agree with experimental data. On this note, the simulation of NMR spectra from models is scarce in contrast to silicate glasses. Perhaps counterintuitively, a big challenge in modeling of phosphate glasses is (pure) vitreous $P_2O_5$.

The range of modeling studies encompass all of the most well-known families of experimental phosphate glasses. One emphasis has been on compositions such as alkali and alkaline earth metaphosphates which represent archetypal glasses (e.g. with phosphate chains only). In these systems, the models have been analyzed to quantitatively evaluate theories of phosphate glass structure. The more challenging part of the MRN model is to quantify the distribution of modifier cations (M). Models of phosphate glasses have clarified much about why the numbers and distances of M–M nearest numbers depend on the oxidation state, size, and concentration of modifier cations.

Another emphasis has been on glasses for applications. These include phosphate-based glasses as biomaterials, rare earth phosphate glasses for applications involving high rare earth concentrations, iron phosphate glasses for vitrification of nuclear waste, and phosphate glasses with mobile ions for high ionic conductivity. The interpretation of ionic conductivity in terms of structure has so far only been done via the concept of volumes of minimum change in bond valence. There is a scarcity of studies which simulate optical spectra of phosphate glasses. Another scarcity is of models of the surfaces of phosphate glasses. There is a sound basis for many future modeling studies to provide important advances, and so to help to realize further applications of phosphate glasses.

## References


1 Boetti, N. G., Pugliese, D., Ceci-Ginistrelli, E., Lousteau, J., Janner, D., and Milanese, D. (2017). Highly doped phosphate glass fibers for compact lasers and amplifiers: a review. *Appl. Sci.*, 7(12), 1295. doi:https://doi.org/10.3390/app7121295

2 Knowles, J. C. (2003). Phosphate based glasses for biomedical applications. *J. Mater. Chem.*, 13(10), 2395–2401. doi:https://doi.org/10.1039/b307119g

3 Brow, R. K., and Schmitt, M. L. (2009). A survey of energy and environmental applications of glass. *J. Eur. Ceram. Soc.*, 29(7), 1193–1201. doi:https://doi.org/10.1016/j.jeurceramsoc.2008.08.011

4 Brow, R. K. (2000). Review: the structure of simple phosphate glasses. *J. Non-Cryst. Solids*, 263(1–4), 1–28. doi:https://doi.org/10.1016/S0022-3093(99)00620-1

5 Hoppe, U., Walter, G., Kranold, R., and Stachel, D. (2000). Structural specifics of phosphate glasses probed by diffraction methods: a review. *J. Non-Cryst. Solids*, 263(1–4), 29–47. doi:https://doi.org/10.1016/S0022-3093(99)00621-3

6 Mountjoy, G., Al-Hasni, B. M., and Storey, C. (2011). Structural organisation in oxide glasses from molecular dynamics modelling. *J. Non-Cryst. Solids*, 357(14), 2522–2529. doi:https://doi.org/10.1016/j.jnoncrysol.2011.01.015

7 Le Roux, S., Martin, S., Christensen, R., Ren, Y., and Petkov, V. (2011). Three-dimensional structure of multicomponent $(Na_2O)_{0.35}[(P_2O_5)_{(1-x)}(B_2O_3)_x]_{0.65}$ glasses by high-energy X-ray diffraction and constrained reverse Monte Carlo simulations. *J. Phys. Condens. Matter*, 23(3), 035403. doi:https://doi.org/10.1088/0953-8984/23/3/035403

8 Clark, E. B., Mead, R. N., and Mountjoy, G. (2006). A molecular dynamics model of the atomic structure of Tb metaphosphate glass $(Tb_2O_3)_{0.25}(P_2O_5)_{0.75}$. *J. Phys. Condens. Matter*, 18(29), 6815–6826. doi:https://doi.org/10.1088/0953-8984/18/29/019

9 Ainsworth, R. I., Di Tommaso, D., Christie, J. K., and de Leeuw, N. H. (2012). Polarizable force field development and molecular dynamics study of phosphate-based glasses. *J. Chem. Phys.*, 137(23), 234502. doi:https://doi.org/10.1063/1.4770295

10 Ainsworth, R. I., Di Tommaso, D., and de Leeuw, N. H. (2011). A density functional theory study of structural, mechanical and electronic properties of crystalline phosphorus pentoxide. *J. Chem. Phys.*, 135(23), 234513. doi:https://doi.org/10.1063/1.3666017

11 Di Tommaso, D., Ainsworth, R. I., Tang, E., and de Leeuw, N. H. (2013). Modelling the structural evolution of ternary phosphate glasses from melts to solid amorphous materials. *J. Mater. Chem. B*, 1(38), 5054–5066. doi:https://doi.org/10.1039/c3tb20662a

12 Cygan, R.T. (2001). Molecular modeling in mineralogy and geochemistry. *Rev. Miner. Geochem.*, 42(1), 1–35. doi:https://doi.org/10.2138/rmg.2001.42.1

**13** Liang, J. J., Cygan, R. T., and Alam, T. M. (2000). Molecular dynamics simulation of the structure and properties of lithium phosphate glasses. *J. Non-Cryst. Solids*, 263(1–4), 167–179. doi:https://doi.org/10.1016/S0022-3093(99)00632-8

**14** Sanders, M. J., Leslie, M., and Catlow, C. R. A. (1984). Interatomic potentials for $SiO_2$. *J. Chem. Soc. Chem. Commun.*, 1984(19), 1271–1273. doi:https://doi.org/10.1039/c39840001271

**15** Hoppe, U. (2000). A refined reverse Monte Carlo analysis of the structure of vitreous $P_2O_5$. *J. Phys. Condens. Matter*, 12(41), 8809–8818. doi:https://doi.org/10.1088/0953-8984/12/41/306

**16** Suzuki, Y., Takase, K., Akiyama, I., Suzuya, K., Umesaki, N., and Ohtori, N. (2001). Short-range structure of vitreous $P_2O_5$ by MD simulation. *Mater. Trans.*, 42(11, SI), 2242–2246. doi:https://doi.org/10.2320/matertrans.42.2242

**17** Speghini, A., Sourial, E., Peres, T., Pinna, G., Bettinelli, M., and Capobianco, J. A. (1999). Structural investigation of $NaPO_3$ glass using molecular dynamics simulation. *Phys. Chem. Chem. Phys.*, 1(1), 173–177. doi:https://doi.org/10.1039/a808829b

**18** Alam, T. M., Liang, J.-J., and Cygan, R. T. (2000). Molecular dynamics simulations of the lithium coordination environment in phosphate glasses. *Phys. Chem. Chem. Phys.*, 2(19), 4427–4432. doi:https://doi.org/10.1039/b004627m

**19** Uchino, T., and Ogata, Y. (1995). Ab initio molecular orbital calculations on the electronic structure of phosphate glasses – binary alkali metaphosphate glasses. *J. Non-Cryst. Solids*, 191(1, 2), 56–70. doi:https://doi.org/10.1016/0022-3093(95)00295-2

**20** Guimbretiere, G., Begue, D., Dussauze, M., and Rodriguez, V. (2012). Combined vibrational multipolar analyses and ab initio cluster description of glasses: toward structural insights in vitreous v-$P_2O_5$. *Vib. Spectrosc.*, 63, 426–431. doi:https://doi.org/10.1016/j.vibspec.2012.09.008

**21** Karthikeyan, A., Vinatier, P., Levasseur, A., and Rao, K. J. (1999). The molecular dynamics study of lithium ion conduction in phosphate glasses and the role of non-bridging oxygen. *J. Phys. Chem. B*, 103(30), 6185–6192. doi:https://doi.org/10.1021/jp990228c

**22** Sistla, R. K., and Seshasayee, M. (2004). Structural study of lithium phosphate glasses by X-ray RDF and computer simulations. *J. Non-Cryst. Solids*, 349, 22–29. doi:https://doi.org/10.1016/j.jnoncrysol.2004.08.257

**23** Rao, R. P., and Seshasayee, M. (2004). Molecular dynamics simulation of ternary glasses $Li_2O$–$P_2O_5$–LiCl. *Solid State Commun.*, 131(8), 537–542. doi:https://doi.org/10.1016/j.ssc.2004.04.019

**24** Hall, A., Adams, S., and Swenson, J. (2004). The nature of conduction pathways in mixed alkali phosphate glasses. *Ionics*, 10(5, 6), 396–404. doi:https://doi.org/10.1007/BF02378000

**25** Al Hasni, B., Martin, R. A., Storey, C., Mountjoy, G., Pickup, D. M., and Newport, R. J. (2016). Molecular dynamics modelling of sodium and calcium metaphosphate glasses for biomaterial applications. *Phys. Chem. Glasses*, 57(6), 245–253. doi:https://doi.org/10.13036/17533562.57.6.080

**26** Parsons, A. J., Ahmed, I., Rudd, C. D., Cuello, G. J., Pellegrini, E., Richard, D., and Johnson, M. R. (2010). Neutron scattering and ab initio molecular dynamics study of cross-linking in biomedical phosphate glasses. *J. Phys. Condens. Matter*, 22(22), 485403. doi:https://doi.org/10.1088/0953-8984/22/48/485403

**27** Hall, A., Swenson, J., Karlsson, C., Adams, S., and Bowron, D. T. (2007). Structure of $Ag_xNa_{1-x}PO_3$ glasses by neutron diffraction and reverse Monte Carlo modelling. *J. Phys. Condens. Matter*, 19(41), 415115. doi:https://doi.org/10.1088/0953-8984/19/41/415115

**28** Hall, A., Swenson, J., Adams, S., and Meneghini, C. (2008). Mixed mobile ion effect and cooperative motions in silver-sodium phosphate glasses. *Phys. Rev. Lett.*, 101(19), 195901. doi:https://doi.org/10.1103/PhysRevLett.101.195901

**29** Schuch, M., Christensen, R., Trott, C., Maass, P., and Martin, S. W. (2012). Investigation of the structures of sodium borophosphate glasses by reverse Monte Carlo modeling to examine the origins of the mixed glass former effect. *J. Phys. Chem. C*, 116(1), 1503–1511. doi:https://doi.org/10.1021/jp2085654

**30** Duc, T. T., Rayavarapu, P. R., and Adams, S. (2010). Mobile ion transport pathways in $(LiBr)_x[(Li_2O)_{0.6}(P_2O_5)_{0.4}]_{(1-x)}$ glasses. *J. Solid State Electrochem.*, 14(10), 1781–1786. doi:https://doi.org/10.1007/s10008-010-1005-0

**31** Uchino, T., and Ogata, Y. (1995). Ab initio molecular orbital calculations on the electronic structure of phosphate glasses – sodium phosphate glasses. *J. Non-Cryst. Solids*, 181(1, 2), 175–188. doi:https://doi.org/10.1016/0022-3093(94)00496-X

**32** Uchino, T., and Yoko, T. (2000). Structure and vibrational properties of alkali phosphate glasses from ab initio molecular orbital calculations. *J. Non-Cryst. Solids*, 263(1–4), 180–188. doi:https://doi.org/10.1016/S0022-3093(99)00633-X

**33** Simandiras, E. D., and Liakos, D. G. (2004). Theoretical study of glass systems using ab initio molecular electronic structure theory. I. Lithium metaphosphate glass. *J. Phys. Chem. A*, 108(17), 3854–3858. doi:https://doi.org/10.1021/jp035819o

**34** Liakos, D. G., and Simandiras, E. D. (2008). Structural properties of lithium metaphosphate glasses by ab initio molecular electronic structure calculations. *J. Non-Cryst. Solids*, 354(14), 1569–1574. doi:https://doi.org/10.1016/j.jnoncrysol.2007.08.077

**35** Bionducci, M., Licheri, G., Musinu, A., Navarra, G., Piccaluga, G., and Pinna, G. (1996). The structure of a Zn(II) metaphosphate glass. 1. The cation coordination by a combination of X-ray and neutron diffraction, EXAFS and X-ray anomalous scattering. *Z. Naturforsch. A: Phys. Sci.*, 51(12), 1209–1215. doi:https://doi.org/10.1515/zna-1996-1211

**36** Navarra, G., Falqui, A., Piccaluga, G., and Pinna, G. (2002). The structure of a zinc metaphosphate glass. A reverse Monte Carlo study. *Phys. Chem. Chem. Phys.*, 4(19), 4817–4822. doi:https://doi.org/10.1039/b204964c

**37** Hoppe, U., Walter, G., Carl, G., Neuefeind, J., and Hannon, A. C. (2005). Structure of zinc phosphate glasses probed by neutron and X-ray diffraction of high resolving power and by reverse Monte Carlo simulations. *J. Non-Cryst. Solids*, 351(12, 13), 1020–1031. doi:https://doi.org/10.1016/j.jnoncrysol.2005.01.013

**38** Hoppe, U., Dimitriev, Y., and Jovari, P. (2005). Structure of zinc phosphate glasses of 75 and 80 mole% ZnO content studied by X-ray diffraction and reverse Monte Carlo simulations. *Z. Naturforsch. A: Phys. Sci.* 60 (7): 517–526.

**39** Boiko, G. G., Andreev, N. S., and Parkachev, A. V. (1998). Structure of pyrophosphate $2ZnO{\cdot}P_2O_5{-}2Na_2O{\cdot}P_2O_5$ glasses according to molecular dynamics simulation. *J. Non-Cryst. Solids*, 238(3), 175–185. doi:https://doi.org/10.1016/S0022-3093(98)00693-0

**40** Tischendorf, B. C., Alam, T. M., Cygan, R. T., and Otaigbe, J. U. (2003). The structure and properties of binary zinc phosphate glasses studied by molecular dynamics simulations. *J. Non-Cryst. Solids*, 316(2, 3), 261–272. doi:https://doi.org/10.1016/S0022-3093(02)01795-7

**41** Navarra, G., Picealuga, G., and Pinna, G. (2004). A reverse Monte Carlo study of a zinc-europium phosphate glass. *Phys. Chem. Chem. Phys.,* 6(13), 3714–3720. doi:https://doi.org/10.1039/b401045k

**42** Sourial, E., Peres, T., Capobianco, J. A., Speghini, A., and Bettinelli, M. (1999). A structural investigation of $Mg(PO_3)_2$, $Zn(PO_3)_2$ and $Pb(PO_3)_2$ glasses using molecular dynamics simulation. *Phys. Chem. Chem. Phys.,* 1(8), 2013–2018. doi:https://doi.org/10.1039/a809926j

**43** Karakassides, M. A., Saranti, A., and Koutselas, I. (2004). Preparation and structural study of binary phosphate glasses with high calcium and/or magnesium content. *J. Non-Cryst. Solids,* 347(1–3), 69–79. doi:https://doi.org/10.1016/j.jnoncrysol.2004.08.111

**44** Hoppe, U., Kranold, R., Barz, A., Stachel, D., Neuefeind, J., and Keen, D. A. (2001). Combined neutron and X-ray scattering study of phosphate glasses. *J. Non-Cryst. Solids,* 293, 158–168. doi:https://doi.org/10.1016/S0022-3093(01)00666-4

**45** Wetherall, K. M., Pickup, D. M., Newport, R. J., and Mountjoy, G. (2009). The structure of calcium metaphosphate glass obtained from X-ray and neutron diffraction and reverse Monte Carlo modelling. *J. Phys. Condens. Matter*, 21(3), 035109. doi:https://doi.org/10.1088/0953-8984/21/3/035109

**46** Cormier, G., Capobianco, J. A., and Monteil, A. (1994). Molecular-dynamics simulation of lead metaphosphate $Pb(PO_3)_2$ glass. *J. Non-Cryst. Solids*, 168(1, 2), 115–124. doi:https://doi.org/10.1016/0022-3093(94)90126-0

**47** Tang, E., Di Tommaso, D., and de Leeuw, N. H. (2010). An ab initio molecular dynamics study of bioactive phosphate glasses. *Adv. Eng. Mater.*, 12(7, SI), B331–B338. doi:https://doi.org/10.1002/adem.201080011

**48** Christie, J. K., Ainsworth, R. I., Di Tommaso, D., and de Leeuw, N. H. (2013). Nanoscale chains control the solubility of phosphate glasses for biomedical applications. *J. Phys. Chem. B*, 117(36), 10652–10657. doi:https://doi.org/10.1021/jp4058115

**49** Shaharyar, Y., Wein, E., Kim, J.-J., Youngman, R. E., Munoz, F., Kim, H.-W., Tilocca, A., and Goel, A. (2015). Structure–solubility relationships in fluoride-containing phosphate based bioactive glasses. *J. Mater. Chem. B*, 3(48), 9360–9373. doi:https://doi.org/10.1039/c5tb01494h

**50** Christie, J. K., Ainsworth, R. I., and de Leeuw, N. H. (2014). Ab initio molecular dynamics simulations of structural changes associated with the incorporation of fluorine in bioactive phosphate glasses. *Biomaterials*, 35(24), 6164–6171. doi:https://doi.org/10.1016/j.biomaterials.2014.04.032

**51** Ainsworth, R. I., Christie, J. K., and de Leeuw, N. H. (2014). On the structure of biomedical silver-doped phosphate-based glasses from molecular dynamics simulations. *Phys. Chem. Chem. Phys.,* 16(39), 21135–21143. doi:https://doi.org/10.1039/c4cp00574k

**52** Christie, J. K., Ainsworth, R. I., and de Leeuw, N. H. (2016). Investigating structural features which control the dissolution of bioactive phosphate glasses: beyond the network connectivity. *J. Non-Cryst. Solids*, 432(A), 31–34. doi:https://doi.org/10.1016/j.jnoncrysol.2015.01.016

**53** Al-Hasni, B., and Mountjoy, G. (2011). Structural investigation of iron phosphate glasses using molecular dynamics simulation. *J. Non-Cryst. Solids*, 357(15), 2775–2779. doi:https://doi.org/10.1016/j.jnoncrysol.2010.10.010

**54** Al-Hasni, B. M., and Mountjoy, G. (2014). A molecular dynamics study of the atomic structure of $x(MgO)\cdot(100-x)(SiO_2)$. *J. Non-Cryst. Solids*, 400, 33–44. doi:https://doi.org/10.1016/j.jnoncrysol.2013.11.011

**55** Joseph, K., Jolley, K., and Smith, R. (2015). Iron phosphate glasses: structure determination and displacement energy thresholds, using a fixed charge potential model. *J. Non-Cryst. Solids*, 411, 137–144. doi:https://doi.org/10.1016/j.jnoncrysol.2014.12.033

**56** Jolley, K., and Smith, R. (2016). Iron phosphate glasses: structure determination and radiation tolerance. *Nucl. Instrum. Methods Phys. Res. Sect. B*, 374, 8–13. doi:https://doi.org/10.1016/j.nimb.2015.09.043

**57** Stoch, P., Szczerba, W., Bodnar, W., Ciecinska, M., Stoch, A., and Burkel, E. (2014). Structural properties of iron-phosphate glasses: spectroscopic studies and ab initio simulations. *Phys. Chem. Chem. Phys.*, 16(37), 19917–19927. doi:https://doi.org/10.1039/c4cp03113j

**58** Stoch, P., Stoch, A., Ciecinska, M., Krakowiak, I., and Sitarz, M. (2016). Structure of phosphate and iron-phosphate glasses by DFT calculations and FTIR/Raman spectroscopy. *J. Non-Cryst. Solids*, 450, 48–60. doi:https://doi.org/10.1016/j.jnoncrysol.2016.07.027

**59** Mountjoy, G., Anderson, R., Bowron, D.T., and Newport, R.J. (1998). Reverse Monte Carlo modelling of Eu and Tb metaphosphate glasses. *J. Non-Cryst. Solids*, 232-4, 227–233. doi:https://doi.org/10.1016/S0022-3093(98)00384-6

**60** Hoppe, U. (2008). Rare-earth site distributions in $R(PO_3)(3)$ (R = La, Nd, Er, Yb) metaphosphate glasses by reverse Monte Carlo simulations. *J. Phys. Condens. Matter*, 20(16), 165206. doi:https://doi.org/10.1088/0953-8984/20/16/165206

**61** Martin, R. A., Mountjoy, G., and Newport, R. J. (2009). A molecular dynamics model of the atomic structure of dysprosium alumino-phosphate glass. *J. Phys. Condens. Matter*, 21(21), 75102–75107. doi:https://doi.org/10.1088/0953-8984/21/7/075102

**62** Du, J., Kokou, L., Rygel, J. L., Chen, Y., Pantano, C. G., Woodman, R., and Belcher, J. (2011). Structure of cerium phosphate glasses: molecular dynamics simulation. *J. Am. Ceram. Soc.*, 94(8) 2393–2401. doi:https://doi.org/10.1111/j.1551-2916.2011.04514.x

**63** Kokou, L., and Du, J. (2014). Short- and medium-range structures of cerium aluminophosphate glasses: a molecular dynamics study. *J. Non-Cryst. Solids*, 403, 67–79. doi:https://doi.org/10.1016/j.jnoncrysol.2014.07.014

**64** Pickup, D. M., Speight, R. J., Knowles, J. C., Smith, M. E., and Newport, R. J. (2008). Sol-gel synthesis and structural characterisation of binary $TiO_2$–$P_2O_5$ glasses. *Mater. Res. Bull.*, 43(2), 333–342. doi:https://doi.org/10.1016/j.materresbull.2007.03.005

**65** Broglia, G., Mugoni, C., Du, J., Siligardi, C., and Montorsi, M. (2014). Lithium vanado-phosphate glasses: structure and dynamics properties studied by molecular dynamics simulations. *J. Non-Cryst. Solids*, 403, 53–61. doi:https://doi.org/10.1016/j.jnoncrysol.2014.07.003

**66** Pedone, A., Malavasi, G., Menziani, M. C., Cormack, A. N., and Segre, U. (2006). A new self-consistent empirical interatomic potential model for oxides, silicates, and silica-based glasses. *J. Phys. Chem. B*, 110(24), 11780–11795. doi:https://doi.org/10.1021/jp0611018

**67** Tho, T. D., Rao, R. P., and Adams, S. (2012). Structure property correlation in lithium borophosphate glasses. *Eur. Phys. J. E*, 35(1), 8. doi:https://doi.org/10.1140/epje/i2012-12008-y

**68** Malavasi, G., Pedone, A., and Menziani, M. C. (2013). Study of the structural role of gallium and aluminum in 45S5 bioactive glasses by molecular dynamics simulations. *J. Phys. Chem. B*, 117(15), 4142–4150. doi:https://doi.org/10.1021/jp400721g

**69** Linati, L., Lusvardi, G., Malavasi, G., Menabue, L., Menziani, M. C., Mustarelli, P., Pedone, A., and Segre, U. (2008). Medium-range order in phospho-silicate bioactive glasses: insights from MAS-NMR spectra, chemical durability experiments and molecular dynamics simulations. *J. Non-Cryst. Solids*, 354(2–9), 84–89. doi:https://doi.org/10.1016/j.jnoncrysol.2007.06.076

**70** Kapoor, S., Semitela, A., Goel, A., Xiang, Y., Du, J., Lourenco, A. H., Sousa, D. M., Granja, P. L., and Ferreira, J. M. F. (2015). Understanding the composition-structure-bioactivity relationships in diopside ($CaO{\cdot}MgO{\cdot}2SiO_2$)–tricalcium phosphate ($3CaO{\cdot}P_2O_5$) glass system. *Acta Biomater.*, 15, 210–226. doi:https://doi.org/10.1016/j.actbio.2015.01.001

**71** Tilocca, A., and de Leeuw, N. H. (2006). Ab initio molecular dynamics study of 45S5 bioactive silicate glass. *J. Phys. Chem. B*, 110(51), 25810–25816. doi:https://doi.org/10.1021/jp065146k

**72** Tilocca, A., Cormack, A. N., and de Leeuw, N. H. (2007). The structure of bioactive silicate glasses: new insight from molecular dynamics simulations. *Chem. Mater.*, 19(1), 95–103. doi:https://doi.org/10.1021/cm061631g

**73** Tilocca, A., and Cormack, A. N. (2010). Surface signatures of bioactivity: MD simulations of 45S and 65S silicate glasses. *Langmuir*, 26(1), 545–551. doi:https://doi.org/10.1021/la902548f

**74** Malavasi, G., Menziani, M. C., Pedone, A., Civalleri, B., Corno, M., and Ugliengo, P. (2007). A computational multiscale strategy to the study of amorphous materials. *Theor. Chem. Acc.*, 117(5, 6), 933–942. doi:https://doi.org/10.1007/s00214-006-0214-1

**75** Corno, M., Pedone, A., Dovesi, R., and Ugliengo, P. (2008). B3LYP simulation of the full vibrational spectrum of 45S5 bioactive silicate glass compared to v-silica. *Chem. Mater.*, 20(17), 5610–5621. doi:https://doi.org/10.1021/cm801164u

**76** Berardo, E., Pedone, A., Ugliengo, P., and Corno, M. (2013). DFT modeling of 45S5 and 77S soda-lime phospho-silicate glass surfaces: clues on different bioactivity mechanism. *Langmuir*, 29(19), 5749–5759. doi:https://doi.org/10.1021/la304795w

**77** Pedone, A., Malavasi, G., and Menziani, M. C. (2009). Computational insight into the effect of CaO/MgO substitution on the structural properties of phospho-silicate bioactive glasses. *J. Phys. Chem. C*, 113(35), 15723–15730. doi:https://doi.org/10.1021/jp904131t

**78** Malavasi, G., Menabue, L., Menziani, M. C., Pedone, A., Salinas, A. J., and Vallet-Regi, M. (2013). New insights into the bioactivity of $SiO_2$–CaO and $SiO_2$–CaO–$P_2O_5$ sol–gel glasses by molecular dynamics simulations. *J. Sol-Gel Sci. Technol.*, 67(1), 208–219. doi:https://doi.org/10.1007/s10971-011-2453-4

**79** Christie, J. K., Pedone, A., Menziani, M. C., and Tilocca, A. (2011). Fluorine environment in bioactive glasses: ab initio molecular dynamics simulations. *J. Phys. Chem. B*, 115(9), 2038–2045. doi:https://doi.org/10.1021/jp110788h

# 11

# Bioactive Glasses

*Jamieson K. Christie*

*Department of Materials, Loughborough University, Loughborough, UK*

## 11.1 Introduction

A material is bioactive if, when implanted into the body as part of a medical device, it undergoes some chemical reaction with its surroundings which is beneficial to its purpose. The canonical bioactive glass composition is 45S5 Bioglass which contains 45 wt% $SiO_2$, 24.5 wt% $Na_2O$, 24.5 wt% CaO, and 6 wt% $P_2O_5$, and which, when implanted, was found to bond directly to human bone [1]. This bonding takes place as the final steps of a complicated series of chemical reactions between the glass and its physiological surroundings [2, 3]. In the 50 years since its discovery, research in bioactive glass has continued to expand into new fields such as the development of porous bioactive scaffolds, and of tough glass/polymer hybrids to allow for implants to be used for bone regeneration and repair [4]. 45S5 Bioglass, and other related compositions, can be synthesized containing a variety of ions and other therapeutic species, which are released into the body as the material degrades; applications related to ionic release include antimicrobials, bone regeneration through osteogenesis, vascularization through angiogenesis, and stem cell differentiation induced toward the formation of osteoblasts [5].

45S5 is not the only bioactive glass composition currently used. In addition to silicate glass compositions such as 13–93 [6] and others [7], phosphate-based glasses, which have faster degradation than silicate-based glasses, have potential for tissue engineering applications [8], and delivery of therapeutic ions for bone repair [9]. Borate and borosilicate glass compositions have also been proposed for hard and soft tissue repair, as they too can be doped with therapeutic substances for, e.g. wound healing [10].

In this chapter, we will concentrate on the computational modeling of Bioglass® and other silicate glass compositions, as these have been most widely studied and used in clinical applications. Computational studies of bioactive phosphate glasses have been recently reviewed [11]. If we are to understand how to optimize the design of bioactive glasses for improved efficacy, it is crucial to understand the structure and reactions of the glass at the atomic scale. Bioactive glasses typically contain five or more atomic species, so elucidation of their

*Atomistic Simulations of Glasses: Fundamentals and Applications*, First Edition.
Edited by Jincheng Du and Alastair N. Cormack.

structure from experiment alone is challenging, even using combinations of diffraction and spectroscopic techniques [12], or advanced diffraction methods [13]. The use of computer modeling, typically through molecular dynamics (MD) simulations, provides the atomic structure of the glass at all length and timescales accessible to the simulation, acting as a valuable complement to experimental techniques [14–16].

## 11.2 Methodology

Both experimentally and in simulation, glass is typically prepared using the melt-quench method, in which a melt of the appropriate composition is prepared and rapidly cooled, resulting in the formation of a disordered, glassy state.

Bioactive glasses, by contrast, are often prepared experimentally using the sol–gel method, in which alkoxide precursors are mixed with a catalyst to produce a solution (*sol*) of nanoparticles which then bond and aggregate through gelation (*gel*) into a disordered network which is then aged, dried, and stabilized [4]. The glasses so created have a wider range of compositions which are bioactive [17], with controllable porosity and mechanical properties [18], and are synthesized at lower temperatures than melt-quenched glasses, allowing for the incorporation of molecules which are less thermally stable [19].

Currently, however, most simulations of glass, bioactive or otherwise, are prepared using the melt-quench method. Because the sol–gel process is markedly more complicated, it has received much less effort from a computational point of view. Sol–gel-prepared glasses often have substantial amounts of water left inside the glass, which increases the complexity of the structure, and hence any investigation of their properties is also more complicated. This point will be discussed further below.

In classical MD, the interatomic forces are represented by an empirical expression called the *potential*. The development of accurate interatomic potentials is crucial to get a reliable structure. This is particularly important for bioactive glass because the connectivity of the $SiO_4$ and/or $PO_4$ tetrahedra (the so-called "network connectivity" [NC]) very strongly affects the bioactivity of the glass [20, 21], and so it is crucial to represent this connectivity correctly in any simulation. We discuss the development of appropriate interatomic potentials in Section 11.3.

## 11.3 Development of Interatomic Potentials

The simplest form of interatomic potentials are so-called "rigid-ion" (RI) potentials, in which the ions are represented as electrically charged point particles. The electrostatic forces are described using Coulomb's Law, typically with partial charges on the ions, and there is an additional short-range term which needs to be parameterized to the system under study.

Two families of RI interatomic potentials have been used for the simulation of bioactive glasses and related compositions. In both families, the electrostatic charge on the ions are 60% of their formal charge. The first group is the Teter potentials [22–24] in which the short-range interatomic energy between two ions $i$ and $j$ a distance $r$ from each other is represented by a Buckingham potential

$$V(r) = A_{ij}e^{-r/\rho_{ij}} - C_{ij}/r^6 \tag{11.1}$$

in which $A_{ij}$, $\rho_{ij}$, and $C_{ij}$ are parameters which need to be fitted to each pair of atomic species $i$ and $j$. Second, a family of potentials developed by Pedone et al. [25] in which the short-range interatomic energy between two ions $i$ and $j$ a distance $r$ from each other is represented by a Morse potential

$$V(r) = D_{ij}\left[\left\{1 - e^{-a_{ij}(r-r_{0,ij})}\right\}^2 - 1\right] + C_{ij}/r^{12} \tag{11.2}$$

in which $D_{ij}$, $a_{ij}$, $r_{0,ij}$, and $C_{ij}$ are parameters which need to be fitted to each pair of atomic species $i$ and $j$. The Morse potential is more normally used to model covalent bonds, e.g. within a molecule, in which the Coulomb term is explicitly subtracted; this is not the case in these simulations, implying that the physical interpretations usually assigned to $D_{ij}$, $a_{ij}$, and $r_{0,ij}$ are not applicable [25].

When developing and parametrizing these potentials, it is usual to assume a force between all cation–anion and anion–anion pairs, but not between any cation–cation pairs. The usual assumption is that cation–cation interactions can be neglected on the grounds that cations are electrostatically shielded from one another by the negative anions.

RI potentials typically give very good representations of the local environment, with bond lengths, coordination numbers, and O–X–O, where X is a cation, with bond-angle distributions in agreement with experimental data. They are less good at representing the medium-range structure of the glass, specifically the connectivity of the tetrahedra as described through the $Q^n$ distribution and NC, than potentials in which the polarizability of the oxide anion is explicitly taken into account [26]. Since the NC is one of the major indicators of the bioactivity, this drove the development of polarizable interatomic potentials.

In these potentials, the polarizability of the oxide anion is incorporated through the shell model (SM), in which the $O^{2-}$ ion is split into a positively charged core and a negatively charged shell. A small mass (usually 0.2 amu) is assigned to the shell, and the core and shell are connected by a harmonic spring. The short-range forces, Buckingham potentials in this case, interact between the shells and other atoms, but both the core and the shell are subjected to Coulombic forces. This potential was developed from a polarizable interatomic potential for $SiO_2$ previously developed by Sanders et al. [27]. The first step was to develop a reliable interatomic potential to take account of the interatomic forces between those species present in bioactive glasses: first, sodium and calcium [24]. The Na–O and Ca–O potential parameters were verified by comparing the lattice structures of sodium calcium silicate crystals simulated with the potential to those observed experimentally, as well as by comparison to first-principles MD simulations [28].

In order to simulate 45S5 and related compositions, this force field was extended to include P–O interatomic forces by fitting to the crystal structures of sodium phosphate, calcium phosphate, and sodium calcium phosphate. Once this was successfully done, classical MD simulations of a truly bioactive composition could be performed [26], and compared to first-principles simulations [29]. Glass structures so generated from both types of simulations were compared to experimental diffraction results and NMR spectra, and the superiority of the polarizable potential over nonpolarizable potentials in reproducing experimental data on the medium-range structure was confirmed. Due to the importance of this part of the structure in bioactive applications, the use of SM potentials for the simulation of bioactive compositions should be preferred over RI, where possible.

The SM is not the only way to introduce polarizable ions into a computer simulation, although it is the only one widely used in the simulation of bioactive glasses. Dipoles and quadrupoles can be explicitly placed on the point positions of the ions, which allows for an alternative description of polarizability [30]; although this has not been applied to bioactive glass, it successfully represented the infrared spectrum of amorphous $SiO_2$ [31].

Care must be taken, particularly if the simulation methodology changes during the simulation. To save computational time, the melt-quench preparation of the glass was often performed using RI potentials, before switching to SM potentials or first-principles methods at room temperature. While this can yield valuable information about, e.g. the electronic structure and density of states from first-principles simulations [28, 29], the $Q^n$ distribution is set during the melt-quench part of the simulation, and does not change significantly when the methodology is changed [32]. This means that if an RI potential is used for the quench, the $Q^n$ distribution obtained will be due to the RI potential, which is known to be inaccurate, irrespective of the use of more accurate methods later on.

Both SM and RI potentials are typically fitted to static structures at equilibrium and are likely to be less suitable for modeling of processes which occur far from equilibrium, such as bond breaking and making. This is particularly relevant in bioactive glass for glass structures made by the sol–gel method, which contains water. Recently, reactive interatomic potentials have been used to investigate precursors of sol–gel bioactive glass [33, 34] and the condensation of a silica network as occurs in the early stages of the sol–gel process [35]. Although these have not yet been extended to the full study of bioactive glass compositions, they remain a promising direction for future research.

## 11.4 Structure of 45S5 Bioglass

We begin by explaining the understanding of the structure of 45S5 Bioglass which has been developed from computer simulation and experimental data. It is the best-studied bioactive glass composition, meaning that there is a wealth of experimental data available against which simulations can be validated, and also a full analysis of its structure highlights with many important aspects of glass structure and their relation to the bioactivity, which will be relevant throughout this chapter.

In 45S5, the silicon and phosphorus atoms act as network formers and are located at the center of $SiO_4$ or $PO_4$ tetrahedra, with four oxygen atoms as nearest neighbors. There are very few (typically <1%) over- or undercoordinated silicon or phosphorus atoms. Si—O bond lengths are about 1.6 Å, while P—O bond lengths are slightly shorter at 1.5–1.55 Å [26]. The oxygen atoms are divided into two types: bridging oxygen (BO) atoms, which are oxygen atoms that are chemically bonded to two or more network formers, and nonbridging oxygen (NBO) atoms, which are not.

Sodium and calcium exist in the glass as network modifiers with typical Na—O and Ca—O bond lengths of 2.3–2.4 Å and coordination numbers close to six [26]. The distribution of bond lengths and coordination numbers is much broader for network modifiers than for network formers. These cations are at the center of distorted polyhedra, often octahedra, with broad bond-angle distributions. Network modifiers prefer to bond chemically to NBO atoms, with the majority of their first coordination shells made up of NBOs; for 45S5, sodium had an average of 4.4 NBO and 1.4 BO atoms in its first coordination shell, while calcium had an average of 5.5 NBO and 0.6 BO atoms in its first coordination shell [26]. The observation that calcium has fewer BO atoms in its first coordination shell than sodium is generally true and is due to the effect by which two cations of different field strengths (defined as $Z/a^2$, where $Z$ is the formal ionic charge, and $a$ is the distance between the cation and the oxygen anion, that is, the sum of the cation and oxygen ionic radii [36]) "compete" for their preferred bonding environment, with the higher-field-strength cation, calcium in this case, being more successful [37].

As mentioned above, the NC is a crucial structural parameter for understanding the bioactivity of glass compositions [20, 21]. The NC is defined as the average number of BO atoms per network former. Although there is not a precise relationship between NC and bioactivity, compositions with smaller NC are more likely to be bioactive; for reference, 45S5 has a NC of 1.9 [21]. The NC can be measured directly from an analysis of the MD trajectory, although it is also related to the ratio $r$ of the number of oxygen atoms to network formers, $NC = 8 - 2r$.

A network former with $n$ BO atoms bonded to it is described as a $Q^n$ cation, and cations with separate values of $n$ in $Q^n$ are distinguishable in MAS-NMR spectra, allowing for a direct comparison of computational structures with experimentally obtained spectroscopic data [38]. Simulation showed that the silicon atoms in 45S5 were mainly $Q^2$ (53%) with almost equal amounts of $Q^3$ (24%) and $Q^1$ (20%) [26]; there were also very small amounts of $Q^0$ (1%) and $Q^4$ (2%) present. By comparison, NMR data gave 67% $Q^2$ with 22% $Q^3$ and 11% $Q^1$ [38]; the NMR spectrum is shown in Figure 11.1. The phosphorus atoms in 45S5 were found from simulation to be mostly orthophosphate, that is, $Q^0$ (65%), with most of the rest $Q^1$ (33%), and a very small amount of $Q^2$ (2%) [26]. The same NMR data implied that all phosphorus was present as orthophosphate, that is, $Q^0$ [38].

A "binary model" had often been assumed in the deconvolution and fitting of NMR spectra, in which ions with only two values of $n$ were assumed to be present [39, 40]. Simulations showed that this is not necessarily true, with at least three, and often more, values of $n$ existing in the glass, e.g. for the 45S5 composition, every single value of $n$ from 0 to 4 is found in the $Q^n$ distribution of silicon [26]. The more extreme values of $n$ are only present at very low quantities, typically 1%–2%, and would be very difficult to deconvolve accurately from an experimental NMR spectrum.

Both first-principles [29] and classical [26] MD simulations of 45S5 have shown that although the majority of the phosphate groups are orthophosphate ($Q^0$), that is, not chemically bonded to the silicate part of the network, there is a significant number of P—O—Si bonds, allowing for phosphate tetrahedra to become part of the connected glass structure, usually as $Q^1$. These bonds become more prevalent at higher-silicate compositions [26]. Although present in all simulations, reporting their existence has been controversial, as they are not observed in NMR [38]. Some NMR spectra show them becoming more common at higher-silicate compositions, although, in many experiments, their presence is below the detectable limit for 45S5 [41]. Solid-state NMR of a sodium-free glass composition with identical silicate and phosphate content to 45S5 showed 8% of the phosphorus atoms to be $Q^1$ [42]. It is clear that simulation is, at least, overstating the prevalence of $Q^1$ phosphorus atoms. This might, at least partially, be due to the use of very fast cooling rates during the simulation of glass. Tilocca [43] showed that a slower cooling rate will reduce the amount of $Q^1$ phosphorus atoms down to about 10%, consistent with experiment [42], for cooling rates of less than 1 K/ps.

Particularly powerful is the direct calculation of the NMR spectrum of a simulated structure in the GIPAW approach [44], as first proposed by Pickard and Mauri [45]. It allows for the computation of the NMR spectra based on various ions and was first applied to glass in 2004 [46]. When applied to 45S5, polarizable potentials were used to generate example structures from which to compute the NMR spectra [38]. This allowed for an accurate fitting of the $Q^n$ distribution, showing that there were three different values of $n$ (= 1–3) present for the $Q^n$ distribution of the silicate ions as in Figure 11.1. Also, as above, the presence of Si—O—P bonds was explicitly ruled out, despite their presence in very many computer simulations of these materials. There was also evidence that the distribution of sodium and calcium was not random around the Si–NBO and P–NBO sites.

MD simulation gives access to the positions of all of the atoms at all times, allowing for the computation of structural properties that might be inaccessible through experiment. For example, the mobility of the ions, as measured through their mean-squared displacement, implies that sodium ions are more mobile than calcium ions [28]. The O—M—O bond-angle distributions, where M is a modifier cation, are broad, with peaks typically centered at about

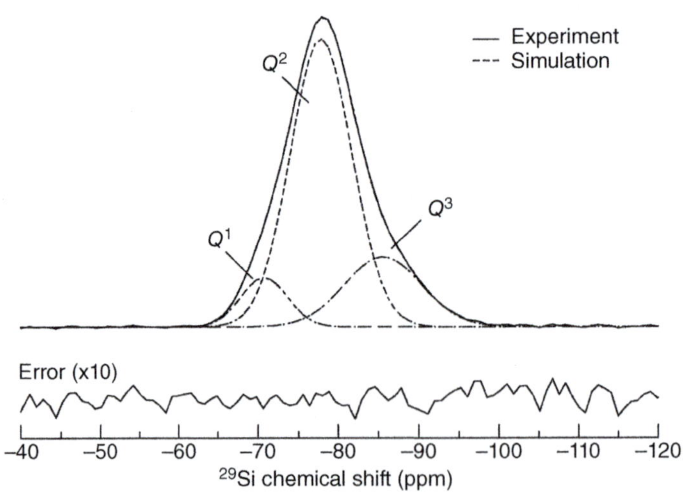


**Figure 11.1** Fit of the experimental $^{29}$Si MAS NMR spectrum of 45S5 Bioglass constrained by using Eq. (2) in ref. [38]. Source: Reprinted with permission from ref. Pedone et al. [38]. Copyright (2010) American Chemical Society.

90° in accordance with the quasi-octahedral structure usually present around these ions [29]. There is also a peak at about 60° corresponding to the case where the two oxygen atoms in the bond angle come from the same $SiO_4$ tetrahedron (so-called "intratetrahedral bonding"). This peak is more prevalent in the O—Na—O bond-angle distribution than in the O—Ca—O distribution, as seen in Figure 11.2.

The structure of the glass is not necessarily homogeneous at length scales larger than the bond lengths; certain types of ions and chemical moieties can show tendencies to cluster together. It is difficult to characterize, either in experiment or in simulation, in part because of the wide range of different possibilities in which the ions can show a nonrandom distribution and the different terms used by different investigators to describe this, as adroitly explained in the discussion in ref. [47].

This clustering is rooted in the local chemical bonding of the various species. For typically bioactive compositions, such as 45S5, it seems that any clustering is rather slight. By computing the ratio $R^{A}_{B/C} = (CN_{A-B}/CN_{A-C})(N_C/N_B)$, where $CN_{A-B}$ is the $A$–$B$ coordination number and $N_A$ is the number of $A$ ions in the structure, Tilocca et al. [26] showed that, in 45S5, sodium and calcium are randomly distributed around the silicate and phosphate groups (the relevant ratios are between 0.96 and 1.01, where 1.0 would represent a perfect random distribution), but that phosphate groups prefer to be located close to other phosphate groups (the ratio $R^{P}_{Si/P} = 0.74$). By setting A in the above to a silicon or phosphorus atom with a specific value of $Q^n$, it was observed that low-$n$ sites prefer to be located close to calcium and high-$n$ sites prefer to be located close to sodium (for silicon, the ratios $R^{Q^0}_{Na/Ca} \sim 0.5$ and $R^{Q^1}_{Na/Ca} \sim 0.6$, while the ratios $R^{Q^3}_{Na/Ca} \sim 1.2$ and $R^{Q^4}_{Na/Ca} \sim 1.2$, and for phosphorus, the ratios $R^{Q^0}_{Na/Ca} \sim 0.8$, $R^{Q^1}_{Na/Ca} \sim 1.2$, and $R^{Q^2}_{Na/Ca} \sim 2.0$). This agrees with the larger amount of NBO atoms in calcium's first coordination shell, as discussed above. These data are in agreement with the nonrandom distribution of modifier ions found through direct computation of NMR spectra and simulation [38], as well as MAS-NMR spectra combined with REDOR and REAPDOR NMR [47].

The tendency of ions to cluster seems larger for glass compositions out of the bioactive range. Glasses with 65 wt% $SiO_2$ show a slight preference for phosphate to aggregate to calcium $\left(R^{P}_{Na/Ca} \sim 0.8\right)$, and silicate to sodium $\left(R^{Si}_{Na/Ca} \sim 1.1\right)$, where no preference is seen for 45 wt% $SiO_2$ glasses [26]. Increasing the amount of phosphate to 12 mol% $P_2O_5$ increases

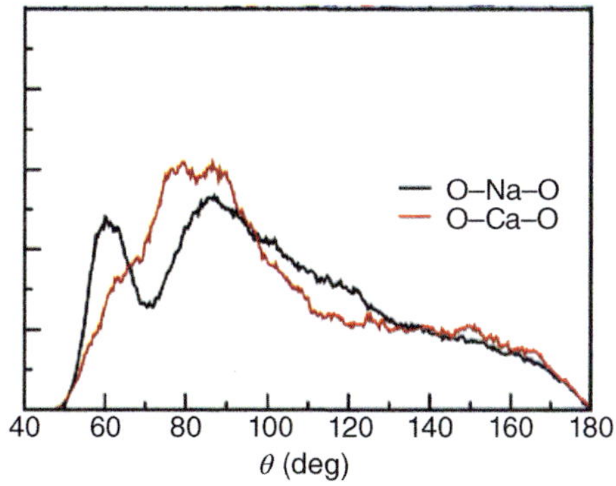


**Figure 11.2** O–Na–O and O–Ca–O bond-angle distribution functions from MD simulations of 45S5. Source: Adapted with permission from ref. Tilocca and de Leeuw [29]. Copyright (2006) American Chemical Society.

phosphate's tendency to be located close to calcium, and, for this composition, there are clear regions of the glass rich in calcium phosphate; experimental investigations at compositions up to 6 mol% $P_2O_5$ showed no difference from random clustering [48].

A thorough comparison of MD simulations, using polarizable potentials, of a range of bioactive and bioinactive glass compositions and their NMR spectra, confirmed and extended some of this previous analysis [41, 47, 48]. The phosphate groups are found to be mostly orthophosphate, although the fraction of $Q^1$ phosphate groups increases with increasing NC and hence bioinactivity. The fraction does not seem to change with phosphate content. Phosphate and silicate species are broadly surrounded randomly by sodium and calcium ions, although there is a preference of $Q^4$ silicate species to bond to sodium, and for $Q^1$ species to bond to calcium, as in simulation. No significant phosphate aggregation was observed. The effect of modifier cations can be semiquantitatively explained by the cation field strength and the strengths of associated bonds.

## 11.5 Inclusion of Ions into Bioactive Glass and the Effect on Structure and Bioactivity

In this section, we consider the effects of adding ions, typically for therapeutic use, on the glass structure and bioactivity. As with many glass compositions, bioactive glasses are able to accommodate a wide range of other ions in their structure. Because of the lack of crystal structure, there are no restrictions on stoichiometry caused by the need to occupy specific lattice sites, so a continuous range of compositions can often be produced, implying the possibility of tuning the glass properties by changes in the composition. The salient property for bioactive glass is of course its bioactivity. The inclusion of other ions, which would be released into the body as the glass matrix dissolves, allows for the possibility of using bioactive glass as a vector for the controlled delivery and release of therapeutically relevant ions. Of course, the presence of these ions will alter the glass structure, and hence the bioactivity, so these changes must be understood in order to optimize the structure properly for a given application.

Additional work is needed to include such ions in computer modeling, especially for classical MD simulations. As mentioned above, the typical potentials used exclude cation–cation interactions, but each new cation included must have a potential parametrized to represent its interaction with oxygen. If a new anion A is included, then the interactions of all cations with A, as well as oxygen–A and A–A interactions, must also be parameterized, a much more complicated task. For first-principles MD, the inclusion of an extra ion is easier, although care must still be taken that the choice of, e.g. pseudopotential and basis set, is appropriate for the new species.

We now examine the several different dopant ions investigated through computer simulation, commenting, where possible, on their effect on the glass's bioactivity.

One of the earliest such ions was fluoride, which has applications in dentistry, where it is currently used commercially. The local structure around fluoride ions was found through simulations, using RI potentials, of 45S5 in which the $Na_2O$ and CaO were, respectively, replaced with $CaF_2$ [49], building on earlier simulations of F-free compositions from the same group [50]. The direct replacement of $Na_2O$ or CaO with $CaF_2$ also alters the NC so there were associated changes in the network polymerization. Very few Si—F and P—F bonds were found, and then only at the compositions with the highest fluoride content. This means that the fluoride ion only very rarely substitutes for an oxide ion in the first coordination shell of either of the network formers (silicon and phosphorus), instead preferring to bond directly to the network modifiers, sodium and calcium in this case. No preference was found for bonding to either one of the modifiers over the other [49]. More accurate first-principles simulations, albeit on smaller models, confirmed these results [51].

This *local* preference for fluoride bonding to modifier ions leads to a segregation on the medium-range length scale [52], with regions of the glass which are rich in fluoride and modifiers, and regions of the glass which are rich in phosphosilicate, as in Figure 11.3. This could have relevance for the bioactivity of the glass, as there are regions of the glass which have starkly different local network connectivities, even if the average NC is deemed suitable for a given application.

Building on the same work, a quantitative structural descriptor was developed [53] to correlate the macroscopic glass transition temperature of fluoridated bioactive glasses with local structural properties such as bond enthalpies and measured coordination numbers with some success ($r^2 = 0.912$ for 18 compositions).

GIPAW was applied to the computation of the NMR spectra of fluoridated 45S5 bioactive glasses, including that of $^{19}$F NMR [54]. Again, no or very few, Si—O—P bonds were found, and, in agreement with previous simulations [49, 51], fluorine was found to be bonded

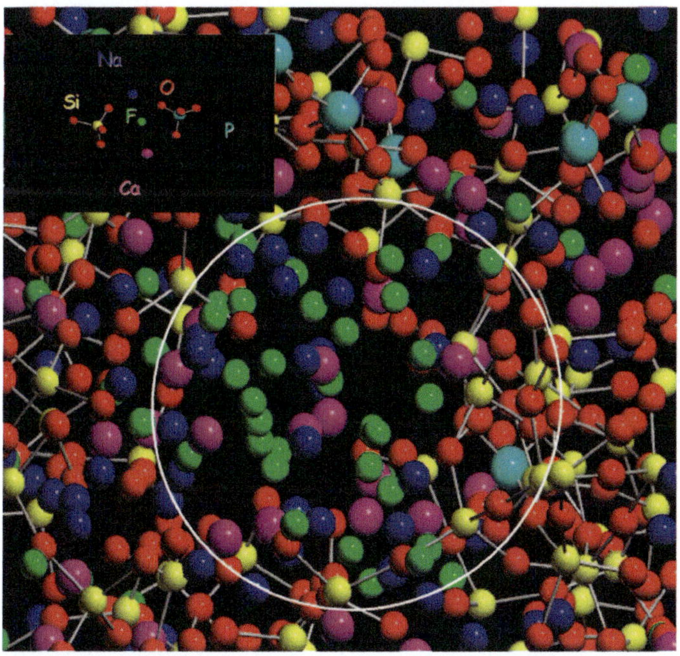


**Figure 11.3** Snapshot of glass structure of a composition based on 45S5 but containing 15 mol% $CaF_2$. The clustering of modifiers (Ca and Na) and fluoride ions is clearly seen within the white circle. Source: Reprinted with permission from ref. Lusvardi et al. [49]. Copyright (2008) American Chemical Society.

entirely to the modifier ions, sodium and calcium. In contrast to the first-principles simulations [51], however, there was some evidence for fluorine atoms exclusively bonded to sodium [54].

The other anion investigated through simulation is chloride. Swansbury et al. [55] used a mixed RI and SM potential approach to investigate the structure of chloride in bioactive calcium silicate glasses, and Pedone et al. [56] developed a SM potential to include chloride and fluoride in sodium-free bioactive phosphosilicate glasses, allowing direct comparison to earlier experimental work from that group [57]. In both works, as for the fluoride-only glasses discussed above, no or very few Si—(F, Cl) and P—(F, Cl) bonds were found, as the fluoride and chloride anions prefer to bond exclusively to the modifiers in the glass, as found experimentally [57]. As in the fluoridated glasses [49], this local bonding preference leads to heterogeneities in the structure at larger length scales, with chloride-rich and chloride-poor regions of the glass. In the phosphosilicate glasses containing both chloride and fluoride, there is evidence to suggest that the fluoride ions aggregate preferentially to the phosphate groups, while the chloride ions prefer to cluster to the silicate groups [56] with $R^{Si}_{F/Cl} \sim 0.8$ and $R^{P}_{F/Cl} \sim 1.2-1.3$. The same quantitative structural descriptor [53] developed for fluoridated glass was extended to these compositions and showed similar positive correlation with the glass transition temperature of the mixed glasses [56].

Magnesium is from the same group in the periodic table as calcium, and it plays a key role in bone tissue redevelopment, among others, and is present in human bone, muscle, and soft tissue, making it potentially very important for inclusion into bioactive glasses [58]. Computer simulation, using an SM potential, showed that magnesium enters the glass as a network modifier with a typical Mg—O bond length of about 2.0 Å, shorter than Na—O or Ca—O bond lengths at 2.3–2.4 Å [59]. Mg–O coordination numbers are typically five, compared to six for Na–O or Ca–O. The first coordination shell of magnesium is almost all NBO atoms, which it has in a higher proportion than calcium or sodium, reflecting the fact that the magnesium has the highest field strength of these three cations and can obtain its preferred bonding environment more easily [37]. This leads, as above, to a preference for low-*n* silicate sites around calcium and magnesium, pushing the sodium toward higher-*n* silicate sites. The peak at 60° in the O—M—O bond-angle distribution, where M is a modifier, corresponding to intratetrahedral bonding, is essentially absent for O–Mg–O, consistent with this observation [59]. The clustering behavior did not vary monotonically with MgO content. For the most biomedically relevant composition, based on 45S5 but with 10 mol% MgO replacing 10 mol% CaO, all three modifier ions are equally likely to be located close to the silicate groups, but magnesium is less likely than the other two ($R^{P}_{Na/Mg} \sim 1.4$, $R^{P}_{Na/Mg} \sim 1.7$) to be close to the phosphate groups.

Strontium is from the same group in the periodic table as calcium and magnesium, and it is often used a part of a treatment for osteoporosis (low bone density). Computer modeling using RI potentials, of 45S5 in which CaO was replaced by SrO, showed a long Sr—O bond length of 2.6 Å and an associated high coordination number of 7.0 [60]. The density and molar volume agreed with experiment across the range of compositions studied. Strontium has a field strength intermediate between sodium and calcium, and this is reflected in the amount of BO atoms in its first coordination shell, and in the size of the peak at 60° in the O—Sr—O bond-angle distribution, which are both intermediate between that seen in O–Na–O and O–Ca–O. Strontium, like calcium, has a slight preference to be located close to the phosphate groups, and this preference is enhanced with increasing strontium content. The diffusion of the ions within the structure was also examined: strontium and calcium have very similar diffusion constants, two (or more, depending on temperature) orders of magnitude lower than of sodium [61, 62]. The split in modifier-oxygen bond lengths between modifiers bonded to NBO or BO atoms, as observed experimentally [12], was not examined in simulation. GIPAW was also used to compare experimental and simulation spectra of $^{29}$Si and $^{87}$Sr NMR in a strontium calcium silicate glass, with some evidence for phase segregation of the glass at the nanoscale [63].

Zinc was also added to strontium-containing bioactive glasses and examined through X-ray diffraction and MD using the Pedone RI potentials [25]. Zinc enters the glass as a so-called "network intermediate." The Zn—O bond length is about 2.0 Å, and the coordination number is typically four, although this increases slightly with increasing zinc content. Zinc has very low numbers of BO atoms in its first coordination shell, lower than sodium, calcium, or strontium, which were also present, reflecting its higher field strength than these three cations. Ionic diffusion rates of zinc were comparable to strontium and calcium. The silicate NC slightly increases with increasing zinc content. Computer simulation and spectroscopy were also used to investigate the structure and properties of Zn/Sr and fluoride-containing alkali-free bioactive glasses, showing that SrO/CaO or ZnO/MgO substitution had very little effect on the silicate NC, despite affecting the dissolution rate [64].

Yttrium can be incorporated into glass for radiotherapy [65] but the glass must not dissolve during the time in which the yttrium ions are radioactive. The structure of yttrium-containing bioactive glass was investigated using SM potentials [66, 67] to see if, despite the bioactivity of the glass, it would be suitable for this application. Yttrium enters as a modifier with a Y—O bond length of 2.2 Å and a coordination number of less than six. The first peak in the Y–O radial distribution function and the distribution of Y–O coordination numbers are both narrower than the equivalent distributions for Ca–O. The yttrium cation has a formal charge of +3, leading to a higher field strength than Ca or Na, and this is clear in the reduced intratetrahedral bonding peak in the O—Y—O bond-angle distribution compared to that of the O–Ca–O [67]. By computing the ratio $r_{X\text{-}X}$

$$r_{X-X} = \frac{N_{X-X,obs}}{N_{X-X,hom}} = \frac{CN_{X-X}+1}{\frac{4}{3}\pi r_c^3 \frac{N_X}{V_{box}}}$$

between the observed $X$–$X$ coordination numbers ($N_{X\text{-}X,obs} = CN_{X\text{-}X} + 1$) and that which would be expected if the X ions were distributed homogeneously ($N_{X\text{-}X,hom} = \frac{4}{3}\pi r_c^3 \frac{N_X}{V_{box}}$), using a cut-off of $r_c = 5$ Å, it could be seen that yttrium and calcium both aggregate into clusters [66]. This preference is stronger for yttrium at low yttrium content (typically $r_{Y\text{-}Y} = 1.4$–$1.6$ for 3–4 mol% $Y_2O_3$) and becomes stronger for calcium at lower calcium content, up to $r_{Ca\text{-}Ca} = 2.6$, in the compositions studied.

An elucidation of the proportions of different amounts of modifier-tetrahedral bonding was made for each modifier; yttrium is able to coordinate, through NBO atoms, to a larger number of individual silicate tetrahedra than calcium or sodium. This is thought to strengthen the glass against dissolution compared to Y-free compositions [67]. This tendency would be opposed by the reduction in NC which could be caused by incorporation of more modifier cations, although this can be corrected by increasing silicate content.

Gallium and aluminum in bioactive glass were considered through the use of RI and SM potentials [68]. Aluminum is mostly tetrahedral and four-coordinated, with some ions having higher coordination numbers of five and six; it acts as a network former with typical Al—O bond lengths of about 1.8 Å. Gallium has a slightly larger Ga—O bond length of about 1.9 Å, and coordination numbers between four and five. One difference is that aluminum ions tend to aggregate with $r_{Al-Al} = 1.14$, an aggregation also found for bioinactive yttrium aluminosilicate glasses [69], whereas there was no evidence for clustering of gallium ions.

One big challenge for computational studies of glass is ions which can enter into the glass in more than one oxidation state. In first-principles simulations, this is taken into account during the calculation of the electronic structure, but, in the more commonly used classical simulations, this has to be taken into account during the parameterization of the potential. As it would be exceptionally difficult to have a single set of potential parameters represent the interactions of an ion across all its oxidation states, the more usual practice is to fit a separate potential for each oxidation state. The disadvantage of this method is that one has to specify in advance which ions will have which oxidation state and that the ratios of such ions is effectively fixed throughout the simulation, because including interconversion between ionic types, with associated charge transfer, is extremely complicated.

For example, cerium can enter into bioactive glass either as $Ce^{3+}$ or $Ce^{4+}$. Gambuzzi and Pedone [70] fitted separate SM potentials to represent the $Ce^{3+}$–O and $Ce^{4+}$–O interactions to investigate the structure of cerium-containing compositions based on glass compositions with very high phosphate contents (35–72 mol% $P_2O_5$), in which the $Ce^{4+}/Ce^{3+}$ ratio was fixed at the start of the simulation. The $Ce^{3+}$—O bond length is about 2.4 Å, and there is a broad range of coordination numbers with the distribution centered on a CN of about 6.5. By contrast, the $Ce^{4+}$—O bond length is shorter, at about 2.3 Å, with a much narrower distribution of coordination numbers, centered on a CN of about 7.5. Increasing the cerium content in these glass compositions leads to the separation of the glass into silica-rich and cerium–phosphate-rich regions, as phosphate tetrahedra are seen to solvate cerium ions. For high cerium content, upward of 18 mol% $Ce_2O_3$, cerium clustering is also observed. Simulation data on the cerium local environment were also found to be in agreement with EXAFS data taken on experimental samples [71].

Boron has been extensively used in glasses, for example, in laboratory glass and glass for nuclear waste encapsulation, as well as its inclusion into bioactive glasses of various compositions. For the computational scientist, boron represents a very difficult challenge, as it can enter the glass as $B^{3+}$ or $B^{4+}$, and these ions have very different structures. $B^{3+}$ will usually bond to three oxygen atoms in a trigonal planar shape, repeated units of which will lead to a layered medium-range structure in the glass, while $B^{4+}$ will usually bond to four oxygen atoms in a tetrahedron, repeated units of which will lead to a three-dimensional disordered glass structure. It follows, therefore, that even small changes in the proportions of $B^{3+}$ and $B^{4+}$ in the glass will lead to very different medium- and long-range structures of the glass. Agreement with experimental data has proven to be very challenging for these systems.

Based on the RI Teter potentials, Deng and Du [72] have developed a set of potential parameters for boron, relevant for its inclusion into bioactive glass systems. The B–O interaction is represented by a single Buckingham potential, in which the A parameter governing the magnitude of the attractive part of the potential depends on the composition. This potential gives a proportion of four-coordinated boron atoms within 7% of the proportion predicted by the Yun-Dell-Bray (YDB) model [73, 74].

This potential was applied to boron-containing glasses based on 45S5 [75] and strontium-containing 55S4 [76, 77], in which $B_2O_3$ was included at the expense of $SiO_2$. A clear distinction was observed between the $B^3$—O bond length at about 1.4 Å, and the $B^4$—O bond length at about 1.5 Å. For these compositions, the B–O coordination number was about 3.2, with the exact number depending on the precise composition. The O–$B^3$–O and O–$B^4$–O peaks in the bond-angle distribution are clear, centered on angles close to the ideal trigonal angle of 120°, and the ideal tetrahedral angle of 109.5°, respectively. Boron enters the glass as a network former, and the overall NC of the glass increases with increasing boron content.

The ongoing difficulty in the development of a reliable interatomic potential for boron is reflected in recent work [78], in which three recent potentials [72, 79, 80] are reviewed and compared by computation of the NMR spectra. No single potential is able to represent all observed features of the first and second coordination shell.

## 11.6 Glass Nanoparticles and Surfaces

The chemical reactions between bioactive glasses and their physiological environment occur, of course, at the surface of the glass. While most simulations of bulk materials are carried out using a supercell repeated through the use of periodic boundary conditions (PBCs), which effectively model the material as an infinite sample, it is possible to turn off PBCs in one or more dimensions and allow for the creation of surfaces in a simulated material.

One way in which this is pertinent to bioactive glasses is that they are often implanted in particulate form, and a nanoparticle can be simulated in this way. Dry nanoparticles of 45S5 [81] and of Ce-containing 45S5 and related compositions [82] have been simulated with similar results, specifically, the NC decreases toward the nanoparticle surface, enhancing its reactivity, and the $Na^+/Ca^{2+}$ ratio increases toward the surface. For Ce-containing structures, the $Ce^{3+}/Ce^{4+}$ ratio increases strongly at the surface with deleterious effects on that material's antioxidant properties.

By removing PBCs in one dimension, a single surface can be prepared, and its interactions with water studied. The preparation of such a surface in a physically and chemically realistic manner is not trivial; a straight cut at a given value of, e.g. $z$-coordinate leads to unrealistically under-coordinated atoms. When preparing these surfaces, it is usual not to permit the formation of two-coordinated (or lower) silicon atoms, or isolated oxygen atoms; NBO atoms created by cutting the surface are saturated with hydrogen atoms [83].

The interaction of water with the surface of a bioactive glass necessarily involves chemical reactions, and the breaking and formation of bonds, so the first studies of this interaction were done with first-principles Car–Parrinello MD; first with a single water molecule interacting with the surface [83]; then water dimers, trimers, and a full water layer [84].

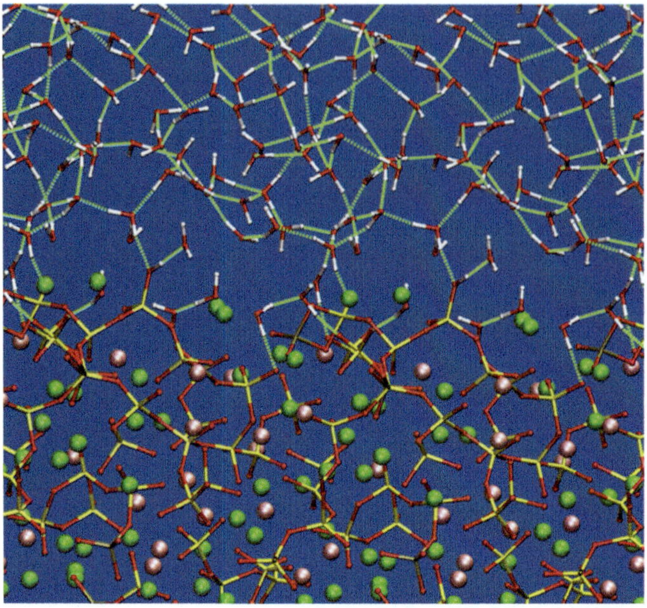

**Figure 11.4** A snapshot of the bioglass–water interface. Source: Extracted from the CPMD trajectory in ref. Tilocca and Cormack [85], where it was originally published under the CC-BY license (https://creativecommons.org/licenses/by/4.0). The phosphosilicate network (bottom) is represented as yellow/red ball-and-stick, with Na and Ca cations as green and brown spheres, respectively. Dashed lines highlight hydrogen bonds.

The very strongest interaction in such cases was between undercoordinated silicon atoms and water molecules, which dissociated immediately provided there was a nearby NBO atom to receive the proton. A snapshot of the glass/water interface is shown in Figure 11.4.

More interesting is the strong interaction between the network modifiers, sodium and calcium, and the oxygen atom on the water molecule, such that about half of the bonds in the first coordination shell of modifiers on the surface are to water molecules [84]. Due to the typically large content of modifier atoms in bioactive glass compositions, there is clear evidence that Na and Ca can act as "channels" to allow water to penetrate below the immediate surface quite quickly.

At the start of the simulation, and hence at the start of the reaction, the interactions between sodium and water dominate, and there is an enrichment of sodium atoms at the surface of the glass, consistent with the $H^+/Na^+$ exchange which is one of the first steps in the mechanism for the degradation of Bioglass [4]. Interactions between calcium and water only dominate once the surface of the material has begun to degrade and significant amounts of sodium have dissolved from the glass [84]. Subsequent very long (40 ps) first-principles MD simulations of the interface between the 45S5 surface and water showed the importance of the interactions between sodium and water, with water dissociating at an $Na^+$…NBO pair with the release of $Na^+$ ions into solution [85].

NBO atoms exposed on the surface can be protonated through water dissociation provided the $OH^-$ can be stabilized by the surrounding structure. This stabilization seems to happen in one of two ways: either through $OH^-$ interaction with a group of (typically three) modifier cations, or by proton hopping across several water molecules shifting the charge defect to another position [84]. As a result, the density of $OH^-$ species is high on the surface, which is effective in attracting calcium cations from solution and promoting apatite formation.

Two-membered silicate rings are found on the glass surface, and despite their intrinsic strain, are typically quite stable, as the other water dissociation mechanisms discussed above are more favorable [84]; this stability was also seen in first-principles simulations of water embedded in the bulk of 45S5 [86].

SM simulations can also be used to look at surface structure and reactivity. Although they do not model the chemical reactions well, they allow for the construction of larger models than first-principles simulations, and so existing structural motifs known to correlate with bioactivity can be examined. SM simulations of the surfaces of the bioactive 45S5 and bioinactive 65S5 compositions showed the sodium enrichment at the surfaces of the glass mentioned above, as well as the fact that there are many more $Na^+$...NBO interactions at the surface than there are $Si^3$...NBO interactions, highlighting again the importance of the sodium–water interactions in the initial stages of bioactive glass dissolution [87]. There are more stable small rings on the surface of the bioinactive composition, but the amount of these rings is known not to correlate strongly with bioactivity.

## 11.7 Discussion and Future Work

There has been extensive use of computer simulation in examining bioactive glasses. Computer simulation is of most use when it acts as a complement to experimental studies by examining properties or finding relationships which are inaccessible or difficult to find experimentally. For bioactive glasses, the property of interest is their bioactivity, and so experiment and simulation have jointly aimed to link their composition and structure to the bioactivity. While, for example, the NC has long been known to be strongly correlated to bioactivity [20, 21], a full understanding of the links between glass composition, structure, and bioactivity has not yet been obtained. The simulations discussed above have highlighted the presence of clustering and nanosegregation of the glass, which leads to regions of the glass with different network connectivities and so presumably different bioactivities. The inclusion of certain dopant ions can enhance this clustering or not, depending on the effects of the local bonding around the ion. The quantity of phosphate tetrahedra bonded into the silicate network will also affect the bioactivity, as orthophosphate will presumably dissolve much more easily. Mechanisms by which water can travel into the glass, among other surface properties, have also been revealed.

Of course, our understanding of these interactions is not complete. Four areas seem to me to be particularly relevant for bioactive glasses, and promising for further work. First, the study of the interaction between the glass and the physiological environment, either at the surface or in the bulk of the glass. This typically requires computationally expensive first-principles simulations but more research into reactive potentials would offer a cheaper way to accelerate research in this area. This is particularly important given the growth in bioactive glasses manufactured using the sol–gel approach, which typically have water retained in their structure. Second, the success of the GIPAW method allowing for the direct computation of NMR spectra offers a direct link between computer simulation and experiment that can be exploited to the benefit of both. Of course, this is relevant beyond bioactive glasses, but is particularly important to bioactive glasses because the medium-range structure, represented by the $Q^n$ distribution so strongly affects the bioactivity, and is accessible to NMR. Third, the development of better ways of representing interatomic interactions through classical potentials, or other means, would expand the range of compositions accessible to these methods. This is particularly important for boron, as this has wide uses in the biomedical field, among others. Finally, more experimental observations of clustering and aggregation, perhaps through small-angle scattering, would allow us to constrain results from simulations and help us to understand the significance of this widely seen structural motif.

## References


1 Hench, L.L., Splinter, R.J., Allen, W.C., and Greenlee, T.K. (1971). Bonding mechanisms at the interface of ceramic prosthetic materials. *J. Biomed. Mater. Res. Symp.* 334: 117–141.

2 Hench, L.L. (1991). Bioceramics: from concept to clinic. *J. Am. Ceram. Soc.* 74: 1487–1510.

3 Hench, L.L. (1998). Bioceramics. *J. Am. Ceram. Soc.* 81: 1705–1728.

4 Jones, J.R. (2013). Review of bioactive glass: from Hench to hybrids. *Acta Biomater.* 9: 4457–4486.

5 Hoppe, A., Guldal, N.S., and Boccaccini, A.R. (2011). A review of the biological response to ionic dissolution products from bioactive glasses and glass-ceramics. *Biomaterials* 32: 2757–2774.

6 Brink, M., Turunen, T., Happonen, R.P., and Yli-Urpo, A. (1997). Compositional dependence of bioactivity of glasses in the system $Na_2O$–$K_2O$–MgO–CaO–$B_2O_3$–$P_2O_5$–$SiO_2$. *J. Biomed. Mater. Res.* 37: 114–121.

7 Groh, D., Doehler, F., and Brauer, D.S. (2014). Bioactive glasses with improved processing. Part 1. Thermal properties, ion release and apatite formation. *Acta Biomater.* 10: 4465–4473.

8 Knowles, J.C. (2003). Phosphate based glasses for biomedical applications. *J. Mater. Chem.* 13: 2395–2403.

9 Lakhkar, N.J., Lee, I.-H., Kim, H.-W. et al. (2013). Bone formation controlled by biologically relevant inorganic ions: role and controlled delivery from phosphate-based glasses. *Adv. Drug Delivery Rev.* 65: 405–420.

10 Rahaman, M.N., Day, D.E., Sonny Bal, B. et al. (2011). Bioactive glass in tissue engineering. *Acta Biomater.* 7: 2355–2373.

11 Christie, J.K., Ainsworth, R.I., Ruiz Hernandez, S.E., and de Leeuw, N.H. (2017). Structures and properties of phosphate-based bioactive glasses from computer simulation: a review. *J. Mater. Chem. B* 5: 5297–5306.

12 Martin, R.A., Twyman, H.L., Rees, G.J. et al. (2012). An examination of the calcium and strontium site distribution in bioactive glasses through isomorphic neutron diffraction, X-ray diffraction, EXAFS and multinuclear solid state NMR. *J. Mater. Chem.* 22: 22212–22223.

**13** Martin, R.A., Moss, R.M., Lakhkar, N.J. et al. (2012). Structural characterization of titanium-doped bioglass using isotopic substitution neutron diffraction. *Phys. Chem. Chem. Phys.* 14: 15807–15815.

**14** Christie, J.K., Cormack, A.N., Hanna, J.V. et al. (2016). Bioactive sol–gel glasses at the atomic scale: the complementary use of advanced probe and computer modeling methods. *Int. J. Appl. Glass Sci.* 7: 147–153.

**15** Tilocca, A. (2009). Structural models of bioactive glasses from molecular dynamics simulations. *Proc. R. Soc. A* 465: 1003–1027.

**16** Tilocca, A. (2010). Models of structure, dynamics and reactivity of bioglass: a review. *J. Mater. Chem.* 20: 6848–6858.

**17** Li, R., Clark, A.E., and Hench, L.L. (1991). An investigation of bioactive glass powders by sol–gel processing. *J. Appl. Biomater.* 2: 231–239.

**18** Arcos, D. and Vallet-Regi, M. (2010). Sol–gel silica-based biomaterials and bone tissue regeneration. *Acta Biomater.* 6: 2874–2888.

**19** Pickup, D.M., Newport, R.J., and Knowles, J.C. (2012). Sol–gel phosphate-based glass for drug delivery applications. *J. Biomater. Appl.* 26: 613–622.

**20** Strnad, Z. (1992). Role of the glass phase in bioactive glass ceramics. *Biomaterials* 13: 317–321.

**21** Hill, R.G. (1996). An alternative view of the degradation of bioglass. *J. Mater. Sci. Lett.* 15: 1122–1125.

**22** Cormack, A.N., Du, J., and Zeitler, T.R. (2002). Alkali ion migration mechanisms in silicate glasses probed by molecular dynamics simulations. *Phys. Chem. Chem. Phys.* 4: 3193–3197.

**23** Du, J. and Cormack, A.N. (2004). The medium range structure of sodium silicate glasses: a molecular dynamics simulation. *J. Non-Cryst. Solids* 349: 66–79.

**24** Tilocca, A., de Leeuw, N.H., and Cormack, A.N. (2006). Shell-model molecular dynamics calculations of modified silicate glasses. *Phys. Rev. B* 73: 104209.

**25** Pedone, A., Malavasi, G., Menziani, M.C. et al. (2006). A new self-consistent empirical interatomic potential model of oxides, silicates and silica-based glasses. *J. Phys. Chem. B* 110: 11780–11795.

**26** Tilocca, A., Cormack, A.N., and de Leeuw, N.H. (2007). The structure of bioactive silicate glasses: new insight from molecular dynamics simulations. *Chem. Mater.* 19: 95–103.

**27** Sanders, M.J., Leslie, M., and Catlow, C.R.A. (1984). Interatomic potentials for $SiO_2$. *J. Chem. Soc., Chem. Commun.* 1271–1273.

**28** Tilocca, A. and de Leeuw, N.H. (2006). Structural and electronic properties of modified sodium and soda-lime silicate glasses by Car–Parrinello molecular dynamics. *J. Mater. Chem.* 16: 1950–1955.

**29** Tilocca, A. and de Leeuw, N.H. (2006). Ab initio molecular dynamics study of 45S5 bioactive silicate glass. *J. Phys. Chem. B* 110: 25810–25816.

**30** Wilson, M., Madden, P.A., and Costa-Cabral, B.J. (1996). Quadrupole polarization in simulations of ionic systems: application to AgCl. *J. Phys. Chem.* 100: 1227–1237.

**31** Wilson, M., Madden, P.A., Hemmati, M., and Angell, C.A. (1996). Polarization effects, network dynamics, and the infrared spectrum of amorphous $SiO_2$. *Phys. Rev. Lett.* 77: 4023–4026.

**32** Tilocca, A. (2007). Structure and dynamics of bioactive phosphosilicate glasses and melts from ab initio molecular dynamics simulations. *Phys. Rev. B* 76: 224202.

**33** Cote, A., Cormack, A.N., and Tilocca, A. (2016). Influence of calcium on the initial stages of the sol–gel synthesis of bioactive glasses. *J. Phys. Chem. B* 120: 11773–11780.
**34** Cote, A., Cormack, A.N., and Tilocca, A. (2017). Reactive molecular dynamics: an effective tool for modelling the sol–gel synthesis of bioglasses. *J. Mater. Sci.* 52: 9006–9012.
**35** Du, T., Li, H., Sant, G., and Bauchy, M. (2018). New insights into the sol–gel condensation of silica by reactive molecular dynamics simulations. *J. Chem. Phys.* 148: 234504.
**36** Dietzel, A. (1948). Glasstruktur und Glaseigenschaften. *Glastechn. Ber.* 22: 41.
**37** Varshneya, A.K. and Mauro, J.C. (2019). *Fundamentals of Inorganic Glasses*. Elsevier.
**38** Pedone, A., Charpentier, T., Malavasi, G., and Menziani, M.C. (2010). New insights into the atomic structure of 45S5 Bioglass by means of solid-state NMR spectroscopy and accurate first-principles simulations. *Chem. Mater.* 22: 5644–5652.
**39** FitzGerald, V., Pickup, D.M., Greenspan, D. et al. (2007). A neutron and X-ray diffraction study of bioglass with reverse Monte Carlo modelling. *Adv. Func. Mater.* 17: 3746–3753.
**40** Angelopoulou, A., Montouillout, V., Massiot, D., and Kordas, G. (2008). Study of the alkaline environment in mixed alkali compositions by multiple-quantum magic angle nuclear magnetic resonance (MQ-MAS NMR). *J. Non-Cryst. Solids* 354: 333–340.
**41** Mathew, R., Stevensson, B., Tilocca, A., and Eden, M. (2014). Toward a rational design of bioactive glasses with optimal structural features: composition-structure correlations unveiled by solid-state NMR and MD simulations. *J. Phys. Chem. B* 118: 833–844.
**42** Fayon, F., Duee, C., Poumeyrol, T. et al. (2013). Evidence of nanometric-sized phosphate clusters in bioactive glasses as revealed by solid-state $^{31}$P NMR. *J. Phys. Chem. C* 117: 2283–2288.
**43** Tilocca, A. (2013). Cooling rate and size effects on the medium-range structure of multicomponent oxide glasses simulated by molecular dynamics. *J. Chem. Phys.* 139: 114501.
**44** Charpentier, T. (2011). The PAW/GIPAW approach for computing NMR parameters: a new dimension added to NMR study of solids. *Solid State Nucl. Magn. Reson.* 40: 1–20.
**45** Pickard, C.J. and Mauri, F. (2001). All-electron magnetic response with pseudopotentials: NMR chemical shifts. *Phys. Rev. B* 63: 245101.
**46** Charpentier, T., Ipsas, S., Profeta, M. et al. (2004). First-principles calculation of $^{17}$O, $^{29}$Si, and $^{23}$Na NMR spectra of sodium silicate crystals and glasses. *J. Phys. Chem. B* 13: 4147–4161.
**47** Mathew, R., Stevensson, B., and Eden, M. (2015). Na/Ca intermixing around silicate and phosphate groups in bioactive phosphosilicate glasses revealed by heteronuclear solid-state NMR and molecular dynamics simulations. *J. Phys. Chem. B* 119: 5701–5715.
**48** Stevensson, B., Mathew, R., and Eden, M. (2014). Assessing the phosphate distribution in bioactive phosphosilicate glasses by $^{31}$P solid-state NMR and molecular dynamics simulations. *J. Phys. Chem. B* 118: 8863–8876.
**49** Lusvardi, G., Malavasi, G., Cortada, M. et al. (2008). Elucidation of the structural role of fluorine in potentially bioactive glasses by experimental and computational investigation. *J. Phys. Chem. B* 112: 12730–12739.
**50** Linati, L., Lusvardi, G., Malavasi, G. et al. (2008). Medium-range order in phospho-silicate bioactive glasses: insights from MAS-NMR spectra, chemical durability experiments and molecular dynamics simulations. *J. Non-Cryst. Solids* 354: 84–89.
**51** Christie, J.K., Pedone, A., Menziani, M.C., and Tilocca, A. (2011). Fluorine environment in bioactive glasses: ab initio molecular dynamics simulations. *J. Phys. Chem. B* 115: 2038–2045.
**52** Christie, J.K. and Brauer, D.S. (2017). The role of fluoride in the nanoheterogeneity of bioactive glasses. *Phys. Chem. Glasses* 58: 180–186.

**53** Lusvardi, G., Malavasi, G., Tarsitano, F. et al. (2009). Quantitative structure-property relationships of potentially bioactive fluoro phospho-silicate glasses. *J. Phys. Chem. B* 113: 10331–10338.

**54** Pedone, A., Charpentier, T., and Menziani, M.C. (2012). The structure of fluoride-containing bioactive glasses: new insights from first-principles calculations and solid state NMR spectroscopy. *J. Mater. Chem.* 22: 12599–12608.

**55** Swansbury, L.A., Mountjoy, G., Chen, X. et al. (2017). Modeling the onset of phase separation in $CaO–SiO_2–CaCl_2$ chlorine-containing silicate glasses. *J. Phys. Chem. B* 121: 5647–5653.

**56** Pedone, A., Chen, X., Hill, R.G., and Karpukhina, N. (2018). Molecular dynamics investigation of halide-containing phospho-silicate bioactive glasses. *J. Phys. Chem. B* 122: 2940–2948.

**57** Chen, X., Chen, X., Pedone, A. et al. (2018). New insight into mixing fluoride and chloride in bioactive silicate glasses. *Sci. Rep.* 8: 1316.

**58** Diba, M., Tapia, F., Boccaccini, A.R., and Strobel, L.A. (2012). Magnesium-containing bioactive glasses for biomedical applications. *Int. J. Appl. Glass Sci.* 3: 221–253.

**59** Pedone, A., Malavasi, G., and Menziani, M.C. (2009). Computational insight into the effect of CaO/MgO substitution on the structural properties of phospho-silicate bioactive glasses. *J. Phys. Chem. C* 113: 15723–15730.

**60** Xiang, Y. and Du, J. (2011). Effect of strontium substitution on the structure of 45S5 Bioglasses. *Chem. Mater.* 23: 2703–2717.

**61** Du, J. and Xiang, Y. (2012). Effect of strontium substitution on the structure, ionic diffusion and dynamic properties of 45S5 bioactive glasses. *J. Non-Cryst. Solids* 358: 1059–1071.

**62** Du, J. and Xiang, Y. (2016). Investigating the structure-diffusion-bioactivity relationship of strontium containing bioactive glasses using molecular dynamics based computer simulations. *J. Non-Cryst. Solids* 432A: 35–40.

**63** Bonhomme, C., Gervais, C., Folliet, N. et al. (2012). $^{87}Sr$ solid-state NMR as a structurally sensitive tool for the investigation of materials: antiosteoporotic pharmaceuticals and bioactive glasses. *J. Am. Chem. Soc.* 134: 12611–12628.

**64** Kapoor, S., Goel, A., Tilocca, A. et al. (2014). Role of glass structure in defining the chemical dissolution behaviour, bioactivity and antioxidant properties of zinc and strontium co-doped alkali-free phosphosilicate glasses. *Acta Biomater.* 10: 3264–3278.

**65** Erbe, E.M. and Day, D.E. (1993). Chemical durability of $Y_2O_3–Al_2O_3–SiO_2$ glasses for the in vivo delivery of beta radiation. *J. Biomed. Mater. Res.* 27: 1301–1308.

**66** Christie, J.K., Malik, J., and Tilocca, A. (2011). Bioactive glasses as potential radioisotope vectors for in situ cancer therapy: investigating the structural effects of yttrium. *Phys. Chem. Chem. Phys.* 13: 17749–17755.

**67** Christie, J.K. and Tilocca, A. (2012). Integrating biological activity into radioisotope vectors: molecular dynamics models of yttrium-doped bioactive glasses. *J. Mater. Chem.* 22: 12023–12031.

**68** Malavasi, G., Pedone, A., and Menziani, M.C. (2013). Study of the structural role of gallium and aluminium in 45S5 bioactive glasses by molecular dynamics simulations. *J. Phys. Chem. B* 117: 4142–4150.

**69** Christie, J.K. and Tilocca, A. (2010). Aluminosilicate glasses as yttrium vectors for in situ radiotherapy: understanding composition-durability effects through molecular dynamics simulations. *Chem. Mater.* 22: 3725–3734.

**70** Gambuzzi, E. and Pedone, A. (2014). On the structure of Ce-containing silicophosphate glasses: a core-shell molecular dynamics investigation. *Phys. Chem. Chem. Phys.* 16: 21645–21656.

**71** Benedetti, F., Luches, P., D'Addato, S. et al. (2017). Structure of active cerium sites within bioactive glasses. *J. Am. Ceram. Soc.* 100: 5086–5095.

**72** Deng, L. and Du, J. (2019). Development of boron oxide potentials for computer simulations of multicomponent oxide glasses. *J. Am. Ceram. Soc.* 102: 2482–2505.

**73** Dell, W.J., Bray, P.J., and Xiao, S.Z. (1983). 11B NMR studies and structural modeling of $Na_2O$–$B_2O_3$–$SiO_2$ glasses of high soda content. *J. Non-Cryst. Solids* 58: 1–16.

**74** Yun, H.Y. and Bray, P.J. (1978). Nuclear magnetic resonance studies of the glasses in the system $Na_2O$–$B_2O_3$–$SiO_2$. *J. Non-Cryst. Solids* 27: 363–380.

**75** Lu, X., Deng, L., Huntley, C. et al. (2018). Mixed network former effect on structure, physical properties and bioactivity of 45S5 bioactive glasses: an integrated experimental and molecular dynamics simulation study. *J. Phys. Chem. B* 122: 2564–2577.

**76** Lu, X., Deng, L., Kuo, P.-H. et al. (2017). Effects of boron oxide substitution on the structure and bioactivity of SrO-containing bioactive glasses. *J. Mater. Sci.* 52: 8793–8811.

**77** Ren, M., Lu, X., Deng, L. et al. (2018). $B_2O_3/SiO_2$ substitution effect on structure and properties of $Na_2O$–CaO–SrO–$P_2O_5$–$SiO_2$ bioactive glasses from molecular dynamics simulations. *Phys. Chem. Chem. Phys.* 20: 14090–14104.

**78** Fortino, A., Berselli, A., Stone-Weiss, N. et al. (2019). Assessment of interatomic parameters for the reproduction of borosilicate glass structures via DFT-GIPAW calculations. *J. Am. Ceram. Soc.* 102: 7225–7243.

**79** Wang, M., Krishnan, N.M.A., Wang, B. et al. (2018). A new transferable interatomic potential for molecular dynamics simulations of borosilicate glasses. *J. Non-Cryst. Solids* 498: 294–304.

**80** Stevensson, B., Yu, Y., and Eden, M. (2018). Structure-composition trends in multicomponent borosilicate-based glasses deduced from molecular dynamics simulations with improved B–O and P–O force fields. *Phys. Chem. Chem. Phys.* 20: 8192–8209.

**81** Tilocca, A. (2011). Molecular dynamics simulations of a bioactive glass nanoparticle. *J. Mater. Chem.* 21: 12660–12667.

**82** Pedone, A., Muniz-Miranda, F., Tilocca, A., and Menziani, M.C. (2016). The antioxidant properties of Ce-containing bioactive glass nanoparticles explained by molecular dynamics simulations. *Biomed. Glass* 2: 19–28.

**83** Tilocca, A. and Cormack, A.N. (2008). Exploring the surface of bioactive glasses: adsorption and reactivity. *J. Phys. Chem. C* 112: 11936–11945.

**84** Tilocca, A. and Cormack, A.N. (2009). Modeling the water-bioglass interface by ab initio molecular dynamics simulations. *ACS Appl. Mater. Interfaces* 1: 1324–1333.

**85** Tilocca, A. and Cormack, A.N. (2011). The initial stages of bioglass dissolution: a Car–Parrinello molecular-dynamics study of the glass–water interface. *Proc. R. Soc. A* 467: 2102–2111.

**86** Bernardo, E., Corno, M., Cormack, A.N. et al. (2014). Probing the fate of interstitial water in bulk bioactive glass by ab initio simulations. *RSC Adv.* 4: 36425–36436.

**87** Tilocca, A. and Cormack, A.N. (2010). Surface signatures of bioactivity: MD simulations of 45S and 65S silicate glasses. *Langmuir* 26: 545–551.

# 12

# Rare Earth and Transition Metal Containing Glasses

*Jincheng Du[1], Monia Montorsi[2], Silvia Barbi[2], and Xiaonan Lu[3]*

[1]*Department of Materials Science and Engineering, University of North Texas, Denton, TX, USA*
[2]*Department of Science and Methods for Engineering, University of Modena and Reggio Emilia, Reggio Emilia, Italy*
[3]*Pacific Northwest National Laboratory, USA*

## 12.1 Introduction

### 12.1.1 TM and RE Oxides in Glasses: Potential Applications and Structural Roles

Transition metal (TM) or rare earth (RE) oxides have been added to oxide glass systems such as silicate, phosphate, or tellurate glasses with the aim to promote peculiar optical, magnetic, and electrical properties [1, 2]. TM and RE oxides are also commonly found in borosilicate or phosphate glasses for nuclear waste disposal. In all these applications, the structural features, such as the local environments of the TM and RE cations, their distribution, and clustering behaviors, as well as their interactions with the network-forming cations, play important roles on the properties and behaviors such as optical fluorescence, electrical conductivity, and chemical durability. One interesting feature of TM and some RE ions is their possibility to present be in the glass materials in multiple oxidation states that can affect the short- and intermediate-range structures and various properties. Applications for this class of materials include magnetic recording, heterogeneous catalysis, solid-state lasers, photonic devices, optical fiber amplifiers for communication, luminescence solar energy concentrator, and nuclear waste management [3–16].

RE ion-containing glasses have desirable optical emission and fluorescence properties that make them candidate material for glass laser, optical amplification, photonic devices, and solid-state lighting [17–21]. The optical properties of these glasses are sensitive to the coordination environment of RE ions and their spatial distributions. Indeed, the luminescences of RE ions have been used as a structural probe in glasses and crystals. Erbium- and europium-ion doped glasses are among the most-studied RE ions in glasses due to their

*Atomistic Simulations of Glasses: Fundamentals and Applications*, First Edition.
Edited by Jincheng Du and Alastair N. Cormack.

applications in optical fiber communications and phosphors, respectively [18, 19]. Experimental techniques such as X-ray and neutron diffraction and extended X-ray absorption fine structures (EXAFS) have been used to determine the local environment of RE ions in glasses and some insights have been gained. However, these methods provide averaged coordination environments hence are often difficult to provide detailed structural information. Molecular dynamics (MD) [22] and Monte Carlo (MC) [23]-based atomistic computer simulations have been applied to gain insights of the structures of RE-doped glasses which are then used to understand the optical emission and other properties [24, 25].

Some TM oxide-containing glasses show semiconductive or mixed-ionic and electronic conductive behaviors that thus can be used as transparent conductors or for the development of the next generation of solid-state batteries [26]. Among the potential glass matrices to which TM oxides could be added, phosphates represent an interesting opportunity due to their strong glass-forming character as well as their easy preparation at lower temperatures [27]. Moreover, relatively low chemical durability of phosphate glasses can be overcome by adding several extra components such as alumina and iron oxide in the basic composition [28]. The peculiar properties induced by typical phosphate network was highlighted in zinc phosphate used for the obtainment of organic/inorganic composites [29] and the improvement in chemical durability and low-processing temperature were reported in literature for phosphate glasses derived from nuclear waste [30]. Phosphate-based glasses were also considered for biomedical applications [28, 29] as well as lithium-containing phosphate compositions and were proposed recently as potential candidate for secondary batteries applications [30, 31]. In this field, it was demonstrated that any compositional change regarding both the type and concentration of the modifiers in glass formulation can strongly affect the final properties of the investigated system [32–34]; therefore, particular attention has to be devoted to the design of glass formulation that could promote the desired final properties.

TM and RE elements commonly exist in nuclear wastes, and they are also used as surrogate of radioactive actinides in nuclear waste studies [4]. Vitrification using borosilicate glasses is a widely accepted nuclear waste disposal method [31, 35, 36]. Hence, the effect of TM and RE ions on glass formability, physical properties, and chemical durability are important considerations for vitrification of these wastes in nuclear waste disposal. The solubility of TM in glasses depends on the type of RE ions, redox condition, and the base glass composition [37, 38]. For example, the solubility limit of $RuO_2$ in borosilicate glasses is only 600 wt ppm [38]. Molybdenum also has low solubility in borosilicate glasses, below 1 wt% of $MoO_6$, above which molybdenum-containing crystal phase, also known as the yellow phase, will precipitate in vitrified borosilicate glasses. It would be important to understand the effect of the glass matrix structures on the solubility of these TMs. Addition of zirconia, even in the small amounts, was found to improve the chemical durability of borosilicate glasses but too high a concentration would deteriorate the long-term chemical durability [32, 33]. Furthermore, it was found that the addition of small amount of TM and RE oxides can change the chemical durability of glasses significantly [34]. To interpret these behaviors and other properties such as chemical durability, we need a deeper understanding of the structures of TM and RE ions containing borosilicate and other related glasses [39, 40].

### 12.1.2 Rare Earth Ions in Glasses: Effects of Local Structure and Clustering Behaviors

RE and TM containing glasses have potential applications for lasers, amplifiers, scintillators, and as model systems for nuclear waste disposal materials [41, 42]. The solubility of RE and TM oxides in the glass matrix differs significantly from one glass system to another [43, 44]. In some systems, such as phosphate, multicomponent silicate, and fluoride glasses, RE ions can be major glass constituents (higher than 5 mol%). While in vitreous silica, or silica glass, the solubility of RE or TM oxides can be as low as fraction of a mole percent, as in the case of erbium oxide [43]. In erbium-doped fiber amplifier (EDFA), which has brought revolutionary changes to optical fiber telecommunications [45], erbium concentration is in the level of 100–1000 erbium wt ppm to avoid the clustering behavior of erbium ions. The difference in the RE and TM solubility can be explained from the rigid three-dimensional network in silica glass as compared to more open and flexible network structures in phosphate and modified silicate glasses [44, 46].

The local environments and distribution of RE and TM ions can play an important role on the optical properties. The local oxygen coordination shell defines the crystal field around these ions hence directly influence the splitting of energy levels and absorption and emission spectra shape and intensity. Furthermore, energy transfer between the clustered RE or TM ions can lead to nonradiative transitions and lower the population of the excited state ions; this reduces population inversion and hence optical efficiency. There are two primary mechanisms involved: energy transfer up-conversion, which leads to losing two excited ions, and energy migration between excited and nonexcited ions. In the latter case, the process itself will not lead to the loss of excited ions; however, it will increase their likelihood of being close to defect sites, such as hydroxyl groups, and hence increase the probability of nonradiative transition. Arai et al. observed a Fast Decay Component (FDC) in time-resolved fluorescence spectra of neodymium-doped silica, which they explained as being due to the clustered RE ions [47]. They also found that addition of $P_2O_5$ or $Al_2O_3$ could eliminate the FDC, presumably by "declustering" the RE ions [47].

Experimental techniques such as neutron and X-ray diffraction, EXAFS, and Nuclear Magnetic Resonance (NMR) have been used to study the local environments and, to a less extend, the clustering of RE and TM ions in glasses. Marcus and Polman studied erbium-doped silica and sodium silicate glasses using EXAFS, and found that the erbium environment in silica was quite different from that in sodium silicate, but observed no clustering [48]. Peters and Houde Walter used EXAFS to study a series of multicomponent oxide and fluoride glasses, and also reported no evidence of clustering [49]. Sen studied Nd-doped and Nd/Al-codoped silicate glasses by using Advanced NMR and EXAFS and reported the formation of Nd—O—Nd bonding; this is the only direct evidence of clustering to our knowledge [50, 51].

### 12.1.3 Redox Reaction and Transition Metal with Multioxidation States in Glasses

As previously mentioned, the peculiarity of TM oxide (and some RE oxide)-containing glasses is the possibility to have TM ions in multiple oxidation states. This obviously corresponds to a wide range of potential local environment of the TM, as well as to different medium-range order of the glass structure that can directly affect the final behavior of material. In particular, the presence of multiple oxidation states of TM in the bulk can promote the phonon-assisted polaron hopping thanks to the concurrent presence of partly reduced and oxidized TM species. This phenomenon leads to the presence of an electronic contribution to total conductivity of material as widely detailed in literature [2, 19–22, 52]. Materials containing both oxidized and reduced TM become an interesting opportunity as concerning the applications in system in which electronic contribution is required, i.e. cathode materials for secondary batteries [26].

One of the major problems in this field is that the relative proportion of multiple oxidation states in the glass strongly depend on glass composition, melting conditions (temperature and time), oxygen partial pressure in furnace atmosphere, and presence of oxidizing/reducing agents [4]. These variables are often not easily controlled during glass obtainment; therefore, quantification of the ratio between oxidized and reduced TM become a key point in design and development of glasses for specifically application in electronic field. Several works concerning how parameters, such as temperature, oxidizing environments, affect the ratio between oxidized and reduced TM in glasses are present in literature. For example, the effects of the increasing melting temperature on the oxidized and reducing ratio of TM were investigated in iron [53] and copper-containing systems. It was demonstrated that the increasing temperature favors the presence of reduced species in the material and this is reflected on its electronic conductivity value [4, 54–57]. It was also documented that the effect of TM oxide concentration on the ratio between reduced and oxidized TM species increased, while chromium, manganese, and vanadium result completely reduced in the system depending on the increasing amount of TM in the composition; copper, molybdenum, and titanium tend to remain at higher valence states [4, 58–60] in higher TM containing systems [26]. It is worth noting that the effect of TM concentration on $R$ value, defined as $TM^{m+}/TM^{(m+1)+}$, did not show an unique trend: while for vanadium [61] the oxidized form is prevalent in a higher vanadium containing system, for copper and iron [56], the reduced form is stabilized at higher transition metal oxide (TMO) concentrations as well as in binary copper (II) phosphate glasses. In fact, for this system, an increase of the copper oxide content lead to the $R$ ratio decrease reaching a minimum at 0.3 mol%, and then an increase was observed, confirming the strong dependence of $R$ on the glass composition [62].

### 12.1.4 Effect of Matrix Composition on Multioxidation States in Glasses Containing TM

In addition to the TM concentration, the type and composition of the glass matrices in which the TM are added also play a key role on the redox ratio of TM ions. In particular, phosphate glasses showed lower basicity with respect to silicate or borate, and this promoted a higher reducing effect on the TM [60, 63]. As an example, higher content of oxidized species ($Cr^{6+}$ and $V^{5+}$) was detected in TM-doped silicate – with respect to phosphate-based systems [54, 64]. Another factor affecting the $R$ value and therefore basicity is the change of the amount of modifier cations introduced in the glass matrix. In particular, it was reported in literature that the $R$ values was affected by the modifiers concentration more in silicate- and borate-based glasses, while in phosphate-based ones $R$ remains practically unaffected by modifiers concentration [26]. It was evident that *R* shows a wide variety of values depending on the combination of the factors detailed above; therefore, a large experimental characterization was promoted on TM containing glasses in order to derive quantitative correlation between all these parameters and the glass structure and final properties. Electron Paramagnetic Resonance (EPR), X-ray Photoelectron Spectroscopy (XPS), and EXAFS and X-ray Absorption near-edge structure (XANES) [7, 65–70] have been used to gain insight of the 3D network and its peculiar aspect in terms of short- and medium-range structure. It was pointed out that the TM coordination (commonly ranging four to six, but higher coordinations possible for larger cations) is affected by its concentration in glasses due both to the potential distortion of the metal local site and the formation of complex ion of TM in phosphate glass [4, 55, 71–74]. The coordination number (CN) of the TM as a function of its concentration in glass formulation define the main role (former or modifier) played by this species in the framework: generally, while low concentration of TM promotes the network modifier role and therefore the glass depolymerization, higher content of TM leads to a reticulated, and interconnected structure according to a former role [4, 74–78].

For potential application that requires electronic conductivity, the generation of a well-defined TM oxide-based network in glasses is essential to guarantee the condition for small polaron hopping. In particular by using NMR spectroscopy and EPR [26, 61], it was possible to confirm that, with increasing the vanadium content in phosphate-based glass formulation, the creation of a vanadate network (connected to isolated phosphate polyhedra) takes place. This was supported by the increase of both the P–O–V and V–O–V species at the expenses of the P–O–P ones in a glass structure [77]. Literatures highlighted also that in phosphate-based matrices, vanadium, chromium, titanium, iron, and others are predominantly arranged in octahedral coordination, and a main modifier character of these TM was associated [79, 80]. Moreover, while at low $V_2O_5$ concentration (c. 5 mol%), vanadium exists only as six-folded $V^{4+}$ to which was associated "modifier" behavior, for vanadium content ranging from 20 to 50 mol%, a mixture of $V^{4+}$ and $V^{5+}$ ions was detected and characterized by four and five coordination, respectively [81]. Also in iron phosphate glasses, both Fe(II) and Fe(III) are present and the Fe(II) fraction increases with the temperature and using a reducing melting atmosphere [53]. When Na or Ba was added to iron phosphate matrices, an average coordination of 4–5 was reported and a mixed tetrahedral–octahedral local environment for iron ion was proposed [82, 83]. Copper is another interesting TM largely investigated in the past [8, 84, 85] due to its potential ability to improve both mixed electronic–ionic conduction and/or fast ionic conductivity (FIC) in glasses [62, 80]. Experimental reports on copper-containing phosphate glasses points out the presence of both $Cu^{2+}$ and $Cu^{+}$ occupying sites characterized by different coordination geometries, 2 + 4 distorted octahedral, and 4 + 2 octahedral, respectively [86]. Moreover, UV absorption and emission studies [87] indicate that $Cu^{+}$ ions in calcium phosphate glasses can exist in trigonally distorted octahedra.

### 12.1.5 The Role of MD in Investigating TM and RE Containing Glasses

The characterization of the glass structure is a key step to correlate the final properties with glass composition and this become particularly complicated in systems with TM and RE ions, especially those with multiple oxidation states. The amorphous nature and inherent complexity make characterization of the glass structures challenging. For RE- and TM-doped glasses, the previously presented experimental techniques have contributed to the understanding of the local environments of these cations in glasses, but there remain many unanswered questions. Atomistic computer simulations have been successfully used to gain insights of RE- and TM-doped glasses [22]. In particular, MD, allowing the evaluation of both the structure and dynamic properties is a powerful technique, while others such as MC simulations have also been used to study RE- and TM-containing glasses. Several MD studies have been conducted in the past on TM- and RE-containing glasses in order to quantify the correlations between RE or TM concentration in glass formulation and final properties such as chemical durability, magnetic, and electrical properties [88–102].

MD simulations have been used as a valuable tool to understand the general structure and particularly the coordination environments of RE and TM in glasses. MD has been widely used to study the structure of oxide glasses and has provided valuable information on different glass systems since 1970s [14, 15]. Brawer and Webber reported the earliest MD study on RE-doped floroberyllate glass in early 1980s [23, 25]. Later, Capobianco and his group have published a series of papers on RE-doped glass using MD [18–21]. Some of the studies have also simulated optical spectra using crystal field theory, based on the configurations obtained from MD simulation [21–24]. Peres et al. have studied erbium-doped lead silicate glasses [20], and Chaussedent has studied erbium-doped titanium silicate glasses [25]. We have also reported RE- and TM-doped silica, silicate, aluminophosphate, and borosilicate glasses. Examples of MD simulations of RE-doped silica and sodium silicate glasses and cerium-doped aluminophosphate glasses will be reviewed in Section 13.3.1. Furthermore, density functional theory (DFT)-based first principles calculations have been used to obtain the electronic structures of TM and RE doped glasses based on structural models generated from MD simulations. More details of the theory is introduced in Section 12.2.5. DFT calculations of charge trapping and electronic structures of cerium-doped glasses are given as an example in Section 12.3.1.3.

Concerning phosphate glasses, just a few works based on MD simulations are reported in literature. Boiko et al. and Tischendorf et al. studied zinc oxide addition (from $40 < x < 70$ mol%) in phosphate glasses belonging to the $x$ZnO–$(100-x)P_2O_5$ system with a focus on glass transition temperature ($T_g$) change dependence on structural rearrangements [103, 104]. The local structure of Zn was studied with the presence of Na ions, and Zn coordination decreases as a function of the modifier concentration. The coordination trend of iron in phosphate glasses are quite similar, but the two valence states lead to different effects on the phosphorus network as detailed by the computational work done by Al-Hasni et al. [105, 106]. The other papers studied the structure of TMO containing glasses derived by computer simulations and characterized in terms of bond distances and angle distributions, in a coordination environment of TM. The result was confirmed by experimental data if available on the same systems, but several problems were encountered in analyzing complex glass formulations. In fact, the absence of experimental references on recent glass compositions limits also the setup of appropriate force field (FF) able to describe the interactions between different species in atomistic modeling. Several FFs were proposed for phosphate glasses [99, 100, 105, 107–110], but the quality of the potential highly depends on the properties that would be analyzed. Amorphous-$V_2O_5$ was largely investigated in the past to evaluate the coordination of V and the presence of corner-sharing (CS) or edge-sharing (ES) distribution that could directly affect the overlapping polarons (OLP) tunneling and therefore the electrical properties [100, 101, 111]. Most available potentials, however, have difficulties to describe the amorphous $V_2O_5$ structures.

Ionic diffusion in TM containing glasses have been studied by using MD simulations. For example, Garofalini et al. [112] studied the interface between intergranular films of $V_2O_5$ and lithium silicate glass by MD, in order to analyze the Li diffusion at the electrode/electrolyte. Recently, Ori et al. include also the valence state 4+ of vanadium in MD simulations of vanadate glasses to investigate the structural environment of vanadium depending on its valence state [113].

In this chapter, we will first introduce the general simulation methodologies ranging from empirical potentials, procedures used to generate glass structures in MD simulations, to dynamic property calculations. DFT-based first principles methods to calculate the electronic structure and charge trapping will be introduced as well. Then we will use three examples to illustrate how MD and related simulations are used to study RE- and TM-doped glasses. The first case study is erbium- and europium-doped silica and sodium silicate glasses where the differences in rigid Si–O network in fully polymerized silica and modified silicate glasses are compared. These structural changes affect both the local structure of the RE ions and their clustering behaviors. In addition, MD simulations of cerium-doped aluminophosphate and phosophosilicate glasses are introduced, where the mixed glass former effect on the RE environment and distribution are explained in detail. The second case study is investigating the structure and transport behavior in sodium and lithium vanadophosphate glasses, a mixed conductor that have potential to be used as anode materials in battery applications. The third case study is zirconium-doped borosilicate glasses for nuclear waste disposal, where structural role in $Zr^{4+}$ and their effect on a series of properties is elucidated by using the recently developed boron potentials for a series of borosilicate glasses.

## 12.2 Simulation Methodology

### 12.2.1 MD Simulation Details

Classical MD simulations of TM and RE oxide containing aluminosilicate, borosilicate, and phosphate glasses are usually simulated using parallel simulation packages such as DL_POLY or the large-scale atomic/molecular massively parallel simulator (LAMMPS) package. Empirical potentials are the most important input in MD simulations, and it is of no exception for TM- and RE-doped glasses. Here we introduced two sets of partial atomic charges pairwise potentials that have been successfully used to describe the interactions in these glass systems. The first set uses the Buckingham short-range interactions, while the other uses a modified Morse potential, with both sets of potentials having the same partial atomic charge assignments. The potential parameters are usually obtained by empirical fitting multiple crystal phases simultaneously. Subsequent retuning to reproduce the glass structures, as well as thermal and mechanical properties are commonly used.

The first set of potential is partial charge pairwise potential developed based on the commonly known Teter potentials, which were first tested and modified by Cormack

**Table 12.1** Atomic charges and Buckingham potential parameters for simulations of RE and TM containing silicate, borosilicate, and aluminophosphate glasses [88, 114–118].

| Pairs | A (eV) | ρ (Å) | C (eV · Å$^6$) |
|---|---|---|---|
| $O^{-1.2}$–$O^{-1.2}$ | 2 029.2204 | 0.343 645 | 192.58 |
| $Si^{2.4}$–$O^{-1.2}$ | 13 702.905 | 0.193 817 | 54.681 |
| $Al^{1.8}$–$O^{-1.2}$ | 12 201.417 | 0.195 628 | 31.997 |
| $B^{1.8}$–$O^{-1.2}$ | 14 877.069[a)] | 0.171 281 | 28.5 |
| $P^{3.0}$–$O^{-1.2}$ | 26 655.472 | 0.181 968 | 86.856 |
| $Na^{0.6}$–$O^{-1.2}$ | 4 383.7555 | 0.243 838 | 30.70 |
| $Ca^{1.2}$–$O^{-1.2}$ | 7 747.1834 | 0.252 623 | 93.109 |
| $Ce^{1.8}$[b)]–$O^{-1.2}$ | 11 476.9522 | 0.242 032 | 46.7604 |
| $Ce^{2.4}$[c)]–$O^{-1.2}$ | 31 697.724 | 0.218 36 | 90.689 |
| $Eu^{2.4}$–$O^{-1.2}$ | 5 950.5287 | 0.253 669 | 27.818 |
| $Er^{2.4}$–$O^{-1.2}$ | 58 934.851 | 0.195 478 | 47.651 |
| $Zr^{4+}$–$O^{-1.2}$ | 17 943.384 | 0.226 627 | 127.65 |

a) B–O A parameter is composition dependent.
b) For $Ce^{3+}$.
c) For $Ce^{4+}$.

and Du [114] and then further developed to introduce more TM and RE species and to add boron-related parameters by Du and coworkers [88, 115–117]. The potentials use partial atomic charges (charge of oxygen is −1.2, and cation charges scale proportionally to maintain charge neutrality) and Buckingham type short-range interactions:

$$V(r_{ij}) = Z_i Z_j / 4\pi\varepsilon r_{ij} + A_{ij}(-r_{ij}/\rho_{ij}) - C_{ij}/r_{ij}^{\ 6} \tag{12.1}$$

where $Z_i$ and $Z_j$ are charges, $A$, $\rho$, and $C$ are parameters. Commonly used parameters are shown in Table 12.1. Atomic charges and short-range interaction correction was commonly introduced to overcome the inherent limitation of the Buckingham potentials.

Another rigid ionic partial charges potential with Morse short-range interaction [110] developed for study of vanadium-doped phosphate glasses are shown in Table 12.2. A $1/r^{12}$ term is added to provide stronger short-range repulsion.

$$V(r_{ij}) = D_{ij}\left[\left\{1 - e^{-a_{ij}(r_{ij}-r_0)}\right\}^2 - 1\right] + C_{ij}/r_{ij}^{12} \tag{12.2}$$

A simulated melt-and-quench process is commonly used to generate the glass structures. The initial glass configuration can be from random generation, from a known crystal structure with the same composition, or from a previous simulation. Typically, cubic simulation boxes with a few thousands to a hundreds of thousands of atoms are used in the simulations. In order to generate statistical meaning structures and properties, multiple samples from

**Table 12.2** Atomic charges and Morse parameters vanadium phosphate glasses [119].

| | $D_{ji}$ (eV) | $a_{ij}$ (Å$^{-2}$) | $r_0$ (Å) | $C_{ij}$(eVÅ$^{+12}$) |
|---|---|---|---|---|
| $Li^{+0.6}$–$O^{-1.2}$ | 0.0011 | 34 295 | 26 814 | 1.0 |
| $P^{+3.0}$–$O^{-1.2}$ | 0.8313 | 25 858 | 18 008 | 1.0 |
| $V^{+3.0}$–$O^{-1.2}$ | 0.0219 | 14 959 | 33 985 | 1.0 |
| $V^{+2.4}$–$O^{-1.2}$ | 0.0033 | 21 093 | 26 636 | 1.0 |
| $O^{-1.2}$–$O^{-1.2}$ | 0.0424 | 13 790 | 36 187 | 22.0 |

different initial configurations are commonly generated. The melting temperature is usually a few thousands (3000–8000) Kelvin to ensure full melting. Then the melt is gradually cooled to room temperature, either under constant pressure or constant volume, with a typical cooling rate in classical simulation of 1–10 K/ps.

One particular challenge in MD simulations of RE- and TM-doped glasses is when studying compositions with low RE or TM concentrations. As in RE-doped silica glasses, the solubility limit is a fraction of a mole percent. In this case, in order to obtain statistically meaning results of structures and properties, large simulation cells are needed to have enough RE or TM ions in the simulation cell. Additionally, when characterizing the clustering behaviors of these cations, simulation cell size was also shown to play a critical role.

### 12.2.2 Structural and Cation Distribution Analysis

In addition to common structural analyses such as pair distribution functions (PDFs), bond angle distributions (BAD), and cation coordination numbers (refer to Chapter 4 of this book for details of these analysis methods), a few specific analyses have been used to understand the local environments and their distributions of RE and TM cations in glass structures.

The environments of modifiers, and especially some TM or RE ions, are not as regular as those of glass former cations, such as $Si^{4+}$, $P^{5+}$, and $B^{3+}$, and therefore require more specific methods to characterize them. Two particular methods have been introduced to study the local environment of TM and RE cations: the dipole and quadrupole moments. The dipole moment of a cation represents central symmetry of the cation-oxygen polyhedron, while the components of quadrupole moments describe the shapes of the polyhedron.

The dipole moment reflects the symmetry of the cation oxygen polyhedron [118]. The dipole moment $\mu$ maybe defined as [118]

$$\mu = \sum_{i=1}^{n} q_i(\vec{r}_i - \vec{r}_{ct}) \tag{12.3}$$

where $q_i$ and $\vec{r}_i$ are the charge and position vectors of the ligand ion $i$, while $\vec{r}_{ct}$ is the position vector of the center cation. A higher dipole moment suggests a more asymmetric environment around the center cation. If a RE ion has a regular octahedral or tetrahedral coordination, the dipole moment becomes zero.

The quadruple moments, or second moments, calculation is another way to characterize the symmetry of the cation oxygen polyhedron [118]. First used by Weber and Brawer [25], the quadruple moment calculation was used to study the crystal field of RE ions in simulated fluoride glasses. The second moment can be expressed as matrix, $M$,

$$M_{ij} = \sum_{L} \frac{q_L r_i(L) r_j(L)}{R_L^5} \tag{12.4}$$

in which $q_L$ is the charge of each ligand ion; $i$ and $j$ represent $x$, $y$, or $z$ in Cartesian coordinates, while $r_i(L)$ and $r_j(L)$ are the $x$, $y$, or $z$ components of the position of ligand ion $L$. The second moments can be diagonalized by rotating the axes. A set of principal axis dipole moments can thus be obtained by

$$\begin{aligned} \lambda_1 &= \sum_{L} q_L X_L^2 / R_L^5 \\ \lambda_2 &= \sum_{L} q_L Y_L^2 / R_L^5 \\ \lambda_3 &= \sum_{L} q_L Z_L^2 / R_L^5 \end{aligned} \tag{12.5}$$

These second moment values indicate the shape of the charge ellipsoid around the center cation. Two large dipole moment values and a small one represent a disc or ellipsoid shape, while one large value and two small value represent a needle shape. The crystal field around the central cation can be calculated based on these values.

To define the first coordination shell of the RE or TM cation-oxygen polyhedron, a cutoff distance of the cation-oxygen pair is commonly used in the calculations of the dipole and quadruple moments. The cutoff distance is usually determined by the first minimum in the corresponding cation-oxygen PDF.

As an example, Table 12.3 shows the dipole and quadruple moments of erbium ions in three erbium-containing crystals [118]. It can be seen that in the crystal structures, erbium ions have relatively small dipole moments and uniform quadruple moments, suggesting regular erbium-oxygen coordination environment [118]. Based on the MD simulations of erbium-doped silica and sodium silicate glasses, the average erbium dipole moment was found to be 6.46 in silica glass, while it ranged from 3.95 to 2.73 in sodium silicate glasses

**Table 12.3** Erbium ion local structures, dipole and quadruple moments in erbium-containing crystals $Er_2O_3$, $Si_2Er_2O_7$, and $NaErO_2$ [118].

| Crystal | Space group | Er–O distance (Å) | Nearest Er–Er distance | Er site | Coordination number | Dipole moment (Debye) | Quadrupole moment $\lambda_1$ | $\lambda_2$ | $\lambda_3$ |
|---|---|---|---|---|---|---|---|---|---|
| $Er_2O_3$ | I a −3 | 2.24–2.31 | 3.49 | 2 | 6 | 0.00 | 0.342 | 0.342 | 0.342 |
| | | | | | 6 | 2.10 | 0.368 | 0.341 | 0.323 |
| $Si_2Er_2O_7$ | C 2/m | 2.23–2.28 | 3.48 | 1 | 6 | 3.57 | 0.431 | 0.361 | 0.258 |
| $NaErO_2$ | C 2/c | 2.21–2.39 | 3.37 | 1 | 6 | 0.09 | 0.388 | 0.357 | 0.296 |

and decreases with increasing sodium oxide concentration. The quadruple moment also shows much larger variations from 0.1 to 0.4 in these glasses. These examples show that dipole and quadrupole moments are effective tools to evaluate the symmetry of shape of the cation-oxygen polyhedron for RE and TM cations.

In addition to the local environment that is usually represented by the first coordination shell environment of TM and RE cations, how these cations distribute in the glass structure are also important to properties such as excitation and emission but difficult to directly characterize. The local environments of RE and TM cations, which are usually accepted to play a role as modifiers in silicate glasses, have significant consequences in the optical properties. Clustering of these cations can lead to nonradiative de-excitation of excited RE ions and thereby lowers the efficiency of devices such as laser and amplifiers. Several techniques can be used to study the clustering of cations in a glass structure:

(1) The cation–cation PDFs provide an initial indication of clustering and correlations of the cations [115].
(2) Magnetic second moments calculations. The second moment is calculated by the summation of $1/r^6$ of all cations in the structure that can give an indication of the clustering behavior, especially when compared with those of the crystal and homogeneous or random models [114].
(3) Cluster size distribution. Based on the criteria of cation–cation distance or cation–oxygen–cation linkages, one can calculate the statistics of how many atoms are isolated, in dimer, trimmer, or $n$-clusters [116, 117]. This is the cluster size distribution and can provide information, based on the clustering criteria selected, how many atoms are isolated, and how many of them are in clusters.
(4) Lastly, the probably of clustering can be calculated and compared with random distribution [115, 116]. The probability of finding neighboring ions in the glass structure can be compared with those from a random distribution. For an atom in a simulation cell with volume $V$, when a second atom is added, the probability of not being seen by the first one atom at distance $R$ is

$$g_1 = \frac{V - V_R}{R} \tag{12.6}$$

where $V_R = \frac{4}{3}\pi R^3$. By adding nth atom to the system, and the probability of it not being *not* within distance $R$ with the $n - 1$ atoms already in the system is

$$g_n = (g_1)^n = \left(\frac{V - V_R}{R}\right)^n \tag{12.7}$$

Thus, based on random distribution, the probability of finding a neighboring atom within distance $R$ is

$$P = 1 - g_n = 1 - \left(\frac{V - V_R}{R}\right)^n \tag{12.8}$$

In the MD generated structural models, we can calculate the actual probability of finding another atom within distance $R$ and the probably will change with $R$. Hence, a plot of actual and theoretical based on random distribution can be compared, from which whether clustering does happen in a glass structure can be evaluated. Due to the shortest atom–atom

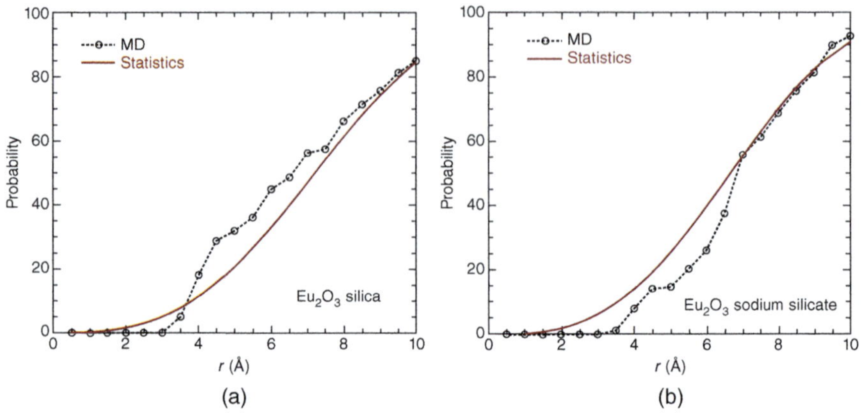


**Figure 12.1** Probability of finding Eu ions as a function of Eu–Eu distance from random distribution (statistics) and that calculated from structural models generated from MD simulations (MD) for (a) $1Eu_2O_3$–$99SiO_2$ (mol%) glass and (b) $1.1Eu_2O_3$–$27.5Na_2O$–$71.4SiO_2$ (mol%). The simulation cell contains 24 160 and 24 176 total atoms, 160 and 170 $Eu^{3+}$ ions, with the side of cubic simulation cell being 71.1 and 67.8 Å for silica and sodium silicate glasses, respectively [115].

distance of RE or TM cations in a glass structure, the probability will be zero before the shortest distance in actual distribution. However, in random distributions, there is no such a limit; hence, there is always a probability no matter how small is the *R* in random distribution. Figure 12.1 shows the comparison of actual distribution from MD and those from random distribution in $Eu^{3+}$-doped silica and sodium silicate glasses, with $Eu_2O_3$ concentration of 1 mol% in both cases. The comparison shows that there is a clear tendency of clustering in silica glass as shown by higher than random distribution values from 4 to 10 Å. On the other hand, in sodium silicate glass, the actual probably from MD is lower than random distribution from 4 to 7 Å and above 7 Å the two are similar. These results demonstrate the capability of this kind of analysis in elucidating the clustering behavior in different types of glasses.

### 12.2.3 Vibrational and Dynamic Property Calculations

The transport properties of mobile ion (e.g. $Li^+$) can be studied by calculating the mean square displacement (MSD) at different temperatures. At each temperature, the systems were equilibrated in NVT and then the production was in NVE ensemble, for a typical 200 000–400 000 timesteps (400 ps). Nosè–Hoover or Berendsen thermostats were used to control the temperature. Trajectories were collected every 20 steps (20 fs), during the NVE runs are collected to calculate the MSD. The MSD evaluates the averaged distance traveled ($r_i(t) - r_i(t_0)$) traveled by atoms as a function of time ($t$), and it was calculated according to Eq. (12.9), with multiple origins, from these trajectories [6, 8]:

$$MSD(t) = \frac{1}{N}\sum_{i=1}^{N}[r_i(t) - r_i(t_0)]^2 \tag{12.9}$$

Indeed, the ions trajectories can be approximated to a random walk that can be analyzed by Einstein equation (Eq. (12.10)) that permits to calculate the self-diffusion coefficient for each particle type and at different temperature [1, 3, 113]:

$$D = \frac{1}{6}\frac{d}{dt}MSD(t) \tag{12.10}$$

where $t$ is the time.

$$D = D_0 e^{\frac{-E_a}{RT}} \tag{12.11}$$

Equation (12.11) allows to obtain the activation energy ($E_a$) value where $D_0$ is the diffusion pre-factor and $R = 8.62 \times 10^{-5}$ eV/atom·K is the gas constant. In particular, the $E_a$ is the energy per atom required to activate the diffusive process, while the $D_0$ corresponds to the self-diffusion coefficient when $T > -\infty$.

Additionally, the velocity auto-correlation function (VACF) shows the intrinsic nature of the motion of the particles and it is defined as the scalar product of the velocity:

$$VACF(t) = < v_i(0).v_i(t) > \tag{12.12}$$

where $v_i$ is the velocity of the particles of the $i$ species. Depending on the role that the ions play inside the glass structure, the VACF curves can be characterized as being of a different behavior [120, 121] that is correlated to the force between these ions. For network formers, their positions are very stable, and they cannot escape from them, so they continue to vibrate around their positions. Differently, the network modifiers move more easily inside the glass structures and the oscillatory motion is rapidly destroyed. Thus, the VACF curves show only one damped oscillation, which correspond to a minimum, and then decay to zero [120, 121]. Integration of VACF would provide another way to calculate the self diffusion coefficient but less commonly used than MSD calculations.

The Van Hove correlation function is also commonly used to study the dynamic behaviors. It considers both the space and time dependence of the particle motion [120, 122], and it is defined as

$$G(r,t) = \langle \frac{1}{N} \sum_{i=1}^{N} \sum_{j=1}^{N} \int \delta[r - r_j(t) + r_i(0)] \rangle \tag{12.13}$$

Physically, the Van Hove function is proportional to the probability density of finding a particle $j$ around a point $r$ at time $t$, given that a particle $i$ is at the origin at time $t = 0$. This function can be divided into two contributions, the first correlated to the probability that a particle $i$ has moved a distance $r$ in time $t$, and the second representing the probability to find at time $t$ a different particle $i$ at a distance $r$ from a place at which at time $t = 0$, there was a particle $j$. Both functions acquire typical curves with the increase of time: the first contribution tends to broaden to a bell-shaped curve, whereas the peaks that characterized the second one disappears.

Thus, simulations at constant temperature were performed in order to obtain trajectories that accurately describe the ions movements. The systems were systematically studied in a range between 300 and 3000 K and for each temperature; they were first equilibrated in NVT while the production was performed in NVE ensemble, for a total of 1 000 000 timesteps (1 ns). Nosè–Hoover thermostat was used to control the temperature with a frictional coefficient of 0.5 and a timestep of 1 fs was chosen during the simulations. Trajectories were collected every 20 steps (20 fs), during the NVE runs and totally 9000 final configurations were obtained. From the trajectories collected during the NVE simulations, the transport properties were analyzed using MSD, VACF, and Van Hove correlation function ($G_s(r, t)$).

### 12.2.4 Quantitative Structure–Property Relationship (QSPR) Analysis

In addition to direct structural comparison and property calculations, short- and medium-range structures of glasses from MD simulations can be used to generate numerical representation called structural descriptors that are related to properties through regression or machine learning approaches. This is classified into to the category of Quantitative Structure–Property Analysis (QSPR) analysis that has already find wide applications in drug discovery and more recently in materials studies. The application of QSPR in glass materials has only developed in the past decade or so but has great potential due to the compositional flexibility and large compositional space of glass-forming systems. QSPR approaches have been successfully and widely utilized to predict various properties of polymers (e.g. refractive indices [123–126], glass transition temperature [127–129], and viscosity [130, 131]). Coupled with MD simulations, QSPR analysis has been performed to find correlations of several glass properties such as density, glass transition temperature, chemical durability, thermal expansion coefficients, and mechanical properties such as elastic moduli and hardness [132–137]. Good reviews of QSPR methods and application can be found by Pedone and Menziani for glass materials [135] and Le et al. for general materials applications [138].

The key in QSPR analysis is the formulation and calculation of the structural descriptors. Lusvardi et al. has proposed the $F_{net}$ descriptor that combines structural features such as cation coordination number and energetics such as single bond strength [39, 139]. The original definition of $F_{net}$ is

$$F_{net} = \frac{1}{N}\sum_{X}^{cations} n_X \cdot CN_{XO} \cdot BS_{XO} \cdot m_X \tag{12.14}$$

where $N$ is the total number of atoms, $n_X$ is the number of type $X$ cations, $CN_{XO}$ is the average coordination number of oxygen around cation $X$, $BS_{XO}$ is the bond strength of the X—O bond calculated from the bond enthalpy of diatomic molecules, $m_X$ is a multiplication factor that is the coordination number for glass former and intermediate cation (e.g. 4 for $Si^{4+}$ and $Al^{3+}$), and valence electrons of modifier cations (e.g. 2 for $Ca^{2+}$ and 1 for $Na^{+}$). Lu and Du have extended the definition of $F_{net}$ to use other parameters such as use Sun's single bond strength for $BS_{XO}$ and average network connectivity for $m_X$ of network-forming cations [137]. These modifications were found to improve the quality of linear regression of properties such as thermal expansion coefficient, initial dissolution rate, and glass transition temperature. The structural descriptor $F_{net}$ thus contains both the structural and energetic input. The structural input, such as cation coordination number, glass former cation network connectivity, can be conveniently calculated from MD simulation results. QSPR analyses of complex behaviors such as the initial dissolution rate and mechanical properties of glasses has been found to be an effective approach in glass research and composition design, which serves as an extension to the commonly adopted MD-based structural analysis and property calculations.

### 12.2.5 Electronic Structures from First Principles Calculations

In order to study the electronic structure of the RE or TM containing glasses, small glass samples (usually with one to a few hundred atoms) can be generated using classical MD simulations. These small systems will make this kind of calculations possible due to the much higher computational cost in first principles calculations. Commonly used first principles approach is the DFT due to its computational efficiency. The initial glass structures from MD simulations are first fully optimized with DFT calculations to relax simulation cell dimension and atomic positions. The fully optimized structures are used for electronic structure calculations to understand properties such as charge trapping mechanisms, and in some cases, the absorption or emission spectra. First principles-based DFT provides more accurate interatomic forces than those derived from interatomic potentials used in MD. Geometry optimization allowed atoms to reach local energy minimum and served good starting point for further charge trapping or spectra calculations.

Plane wave-based periodic DFT total energy calculations are commonly used due to its computational efficiency and possibility to systematically improve the accuracy by increasing the basis set size of plane waves, essentially by increasing the cutoff kinetic energy cutoffs. Generalized gradient corrected exchange and correlation functional are commonly used with the projected augmented wave (PAW) pseudopotentials [26, 27]. The DFT+U correction based on the Hubbard model is commonly used to address the band gap issue of conventional DFT and for strongly correlated systems. DFF+U has widely used for calculations of TM and RE containing materials. For example, Sun and Du used DFT+U to study a series of vanadium phosphate crystals to understand the physical properties and relative stabilities [140]. After obtaining optimized geometry, the simulation of electron or hole trapping in the glasses has been done by adding an electron or hole to the simulation cell, then a further relaxation of the structures is performed with planewave kinetic energy cutoff being 400 eV. The convergence criterion for geometry is for the force acting on each atom being smaller than 0.01 eV/A. DFT calculations were performed using the Vienna ab initio Simulation Package (VASP) using plane wave basis sets. Spin polarization was turned on in all the calculations due to introduction of unpaired electrons and potential bond breaking and formation in the relaxation process. Full geometry optimization including atom position and cell parameters was performed with a convergence criterion of the forces acting on each of the atoms being smaller than 0.01 eV/Å.

## 12.3 Case Studies of MD Simulations of RE and TM Containing Glasses

### 12.3.1 Rare Earth Containing Silicate and Phosphate Glasses

#### 12.3.1.1 Erbium-Doped Silica and Silicate Glasses: From Melt-Quench to Ion Implantation

Due to their application as EDFAs in optical fiber communications and related photonic devices, erbium-doped glasses are of particular interests and there are a number of experimental and simulation studies to understand the structures of these glasses [141]. In EDFA, the base glass is silica where erbium solubility is limited so the doping concentration is usually very low (a fraction of a 1 mol%), and this is compensated by long fibers to achieve high optical amplification efficiency, and codoping with alumina and phosphorus oxide are commonly used to mediate the effect of erbium clustering [18, 141]. For integrated photonics, the length limit requires a much higher erbium concentration to achieve the same efficiency, hence, a higher doping level is needed, and this is achieved by using silicate or other multicomponent glasses [141]. Erbium ion local environment, clustering behavior, mechanism of clustering, and effect to delustering are thus questions of both scientific and technological importance.

Du and Cormack studied erbium ($Er_2O_3$)-doped silica glass using MD simulations [142]. $Er_2O_3$ concentration ranges from 0.25 up to 5 mol%. To ensure statistical meaningful results, large simulation cells with over 10 000 atoms were used. Partial charge pairwise potential with Buckingham form (Table 12.1) was used, and it provides high computational efficiency. Figure 12.2 shows typical PDFs and accumulated coordination number of 1 mol% $Er_2O_3$ doped silica glass. It can be seen that the Er—O bond distance is around 2.19 Å and $Er^{3+}$ has a coordination number of 4.9. The Er–Er PDF shows a strong peak at 3.2 Å suggesting certain level of clustering of $Er^{3+}$ ions at this concentration. $Er^{3+}$ ion environment shows slight changes with $Er_2O_3$ concentration in silica glass. Figure 12.3a shows a snapshot of the erbium ions and their oxygen coordination environments, where it can be seen that erbium ions be isolated, corner- or edge-sharing with another erbium ion, and a few form even larger clusters. Figure 12.3b shows the accumulated coordination number for $Er_2O_3$ concentration of 1, 3, and 5 mol%. The average erbium coordination number increases with $Er_2O_3$ concentration. The increase of erbium coordination number is mainly due to the increase of contribution of nonbridging oxygens (NBOs), which is originated from sharing of NBOs through cluster formation (detailed below). The results suggest that $Er^{3+}$ form clusters to share NBOs to better satisfy their coordination requirements.

To quantify the erbium ion clustering behavior, the probability of finding another erbium ion at different distances was calculated from MD generated structural models and compared with those calculated based on random distribution [142]. The results are shown in Figure 12.4. With increasing erbium concentration, the tendency of clustering as compared to random distribution becomes higher and higher, as shown from the much higher probability from MD than those from random distribution. The statistics of $Er^{3+}$ ions in cluster size 1 (isolated), 2 (dimer), and 3 or higher are shown in Figure 12.5 calculated by using two criteria: direct Er–O–Er linkage formation (shown in Figure 12.5a) and Er–Er cutoff distance of 6 Å (Figure 12.5b), which is based on the effective energy transfer distance of erbium ions. The two shows similar trend. At the lowest concentration (0.25%, or 2500 ppm), 90% or more erbium ions are isolated. This decreases rapidly with increasing $Er_2O_3$ concentration to around 10% for the glass with 5 mol% $Er_2O_3$. At the same time, erbium cluster with size 3 or above increase monotonically while those with dimer reach a plateau and does not change much with further increase of concentration.

Erbium local environment and clustering behavior in sodium silicate glasses was also studied by Du and Cormack using MD simulations [118]. Erbium-doped sodium silicate glasses with constant $Er_2O_3$ but increasing soda concentration (composition $1Er_2O_3$–$x$$Na_2O$–$(99-x)SiO_2$, $x = 0$, 10, 20, 30 mol%, named Ess1 for $x = 0$ and Enssx for other $x$ values) were studied. The main purpose was to see how the modified silicate glass matrix accommodate erbium ions, as compared to pure silica where clear tendency of

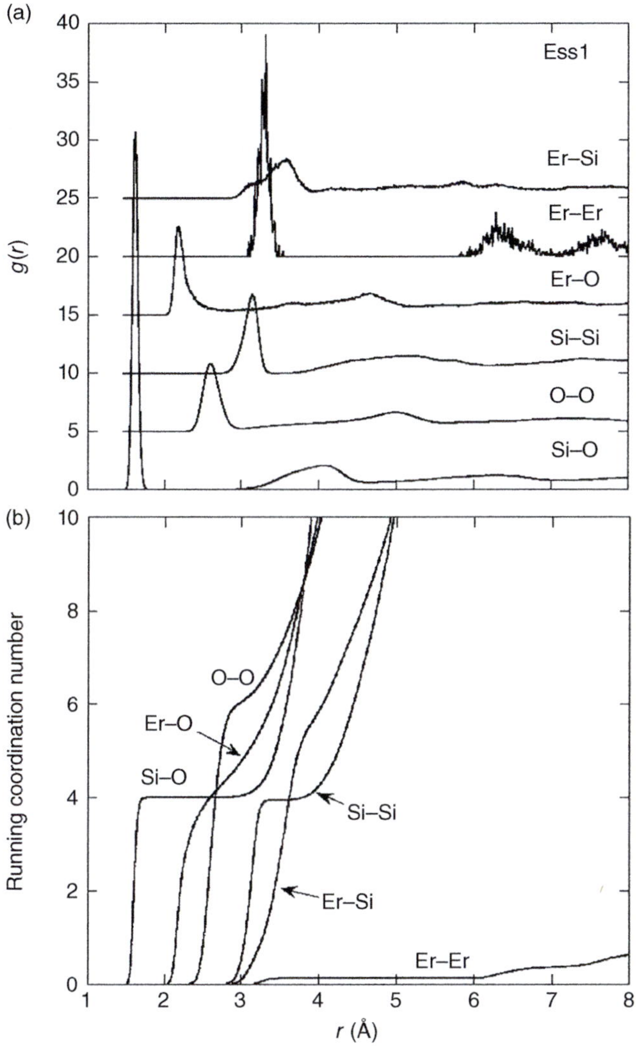


**Figure 12.2** Pair distribution function (a) and coordination number (b) of erbium-doped silica ($1Er_2O_3-99SiO_2$) glass.

clustering was observed [142]. Table 12.4 shows the calculated dipole moment (Eq. (12.3)) and erbium coordination number in the series. There is a significant drop of erbium dipole moment from 6.46 for Ess1 to 3.96 for Enss10. Further increase of soda content leads to more gradual decreases of the dipole moment values. Since the value of the dipole moment is related to level of symmetry of the coordination environment, erbium ions have more symmetric coordination environment in sodium silicate than silica. This is also reflected in erbium coordination number that increases with soda concentration: from 4.88 for Ess1 to 5.06 in Enss10 and 5.81 in Enss30. The increase of coordination number mainly originated

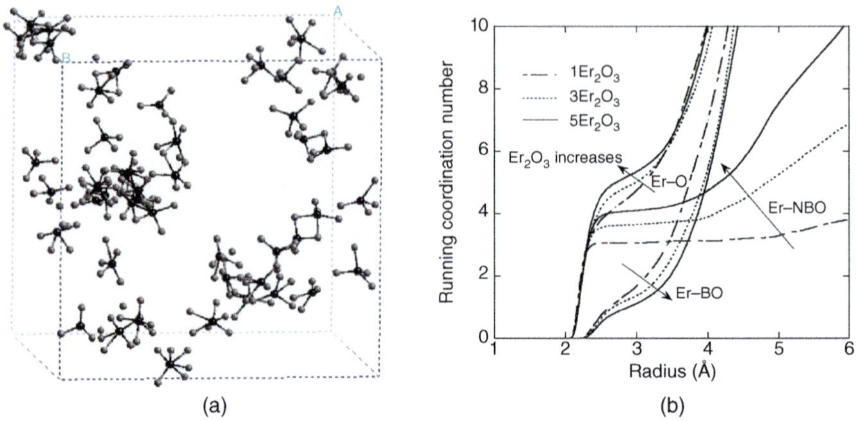


**Figure 12.3** (a) Snapshots of coordination environment of erbium ions and their first coordination shell of $3Er_2O_3$–$97SiO_2$ [142]. All other atoms are not shown. The big black balls are erbium ions, small gray balls are oxygen ions, and the small white balls are free oxygen ions. (b) Oxygen coordination around erbium in $xEr_2O_3$–$(100 - x)SiO_2$ glasses ($x = 1,3,5$).

**Table 12.4** Erbium coordination number and average dipole moment in $1Er_2O_3$–$xNa_2O$–$(99 - x)SiO_2$ glasses.

| Name | Ess1 | Enss10 | Enss20 | Enss30 |
|---|---|---|---|---|
| $x$ | 0 | 10 | 20 | 30 |
| $Er^{3+}$ dipole moment | 6.46 (2.33) | 3.96 (2.22) | 3.41 (1.75) | 3.23 (1.79) |
| $Er^{3+}$ coordination number | 4.88 | 5.06 | 5.81 | 5.81 |
| -NBO | 3.00 | 4.88 | 5.19 | 5.50 |
| -BO | 1.88 | 0.18 | 0.62 | 0.31 |

Dipole moment in parenthesis are standard deviations [118].

from the increase of NBO, which increases from 3.0 in Ess1 to 4.9 in Enss10 and 5.50 in Enss30. The coordination change is reflected in Figure 12.6 where the partial correlation function ($T(r)$) and their NBO and bridging oxygen (BO) contributions are shown. The Er–O correlation function in Enss30 has a stronger first peak than in Ess1 that is related to the coordination number change. This is explained by the fact that erbium ions enter the environment rich in sodium and NBO that better satisfy their coordination requirement in soda silicate glasses.

Quadruple moments of the erbium ions were calculated to further quantify the coordination environment. The three values of the quadruple moment represent the symmetry and shape of the erbium coordination environment. Figure 12.7 shows the calculated erbium quadruple moments sorted from large to small for Ess1 and Enss30. It can be seen that

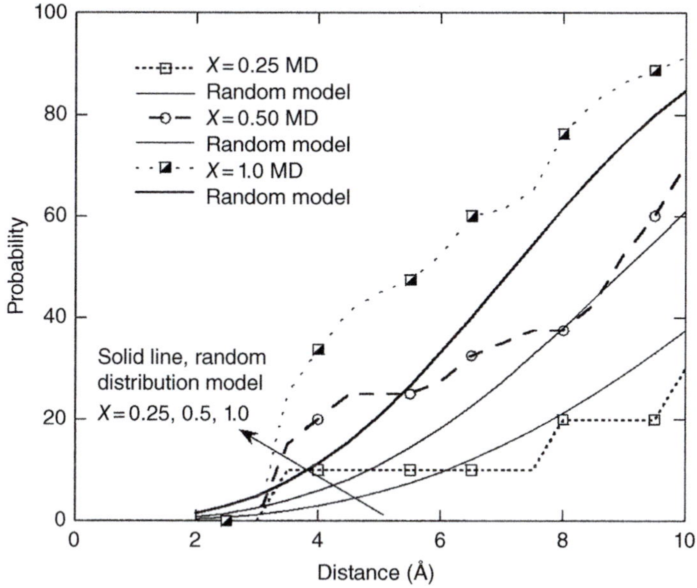


**Figure 12.4** Comparison of MD modeled rare earth distribution with those from random distribution model for erbium-doped silica glasses in $xEr_2O_3$–$(100 - x)SiO_2$ glasses ($x = 0.25, 0.5, 1.0$).

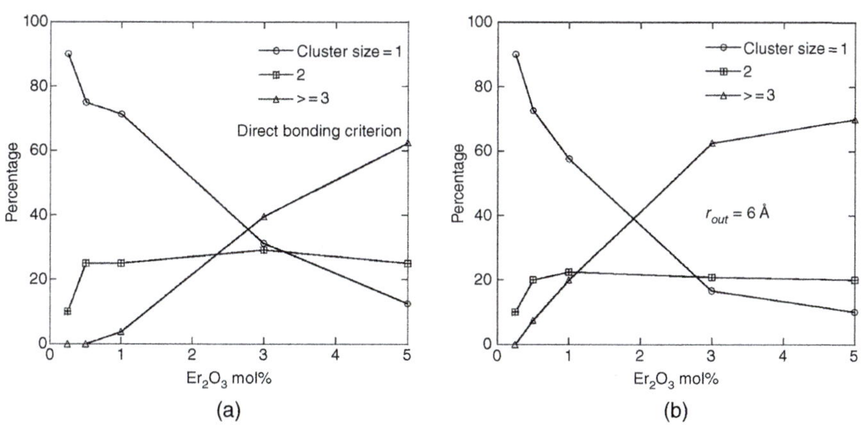


**Figure 12.5** (a) Erbium clustering analyses using Er—O—Er bonding criterion (b) Erbium clustering analyses using Er–Er distance criterion in $xEr_2O_3$–$(100 - x)SiO_2$ glasses ($x = 0.5, 1, 3, 5$) [142].

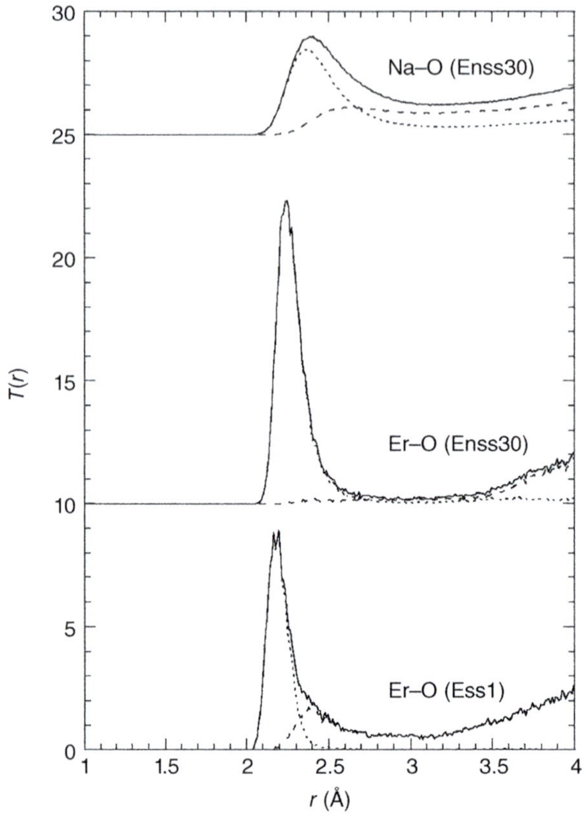


**Figure 12.6** Comparison of Na–O and Er–O partial distribution functions and their BO/NBO components for Enss20 ($1Er_2O_3$–$20Na_2O$–$79SiO_2$) and Er–O distribution in Ess1 ($1Er_2O_3$–$99SiO_2$). Solid line: total $T(r)$, dotted line: NBO contributions, dashed line: BO contributions [118].

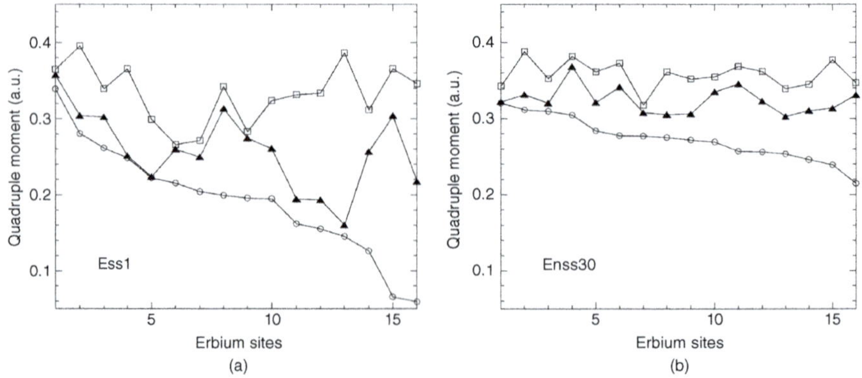


**Figure 12.7** Quadruple moments of the first coordination shell of erbium ions (with a cut off of 3 Å) for (a) Ess1 ($1Er_2O_3$–$99SiO_2$, mol%) (b) Enss30 ($1Er_2O_3$–$30Na_2O$–$69SiO_2$, mol%) [118].

those in Ess1 show diverse values meaning large variation of the shape of the erbium oxygen polyhedra, while that for Enss30 the three values are much closer indicating more uniform coordination environment. Together with lower dipole moment, these results indicate erbium ions in soda silicate glasses are more symmetric and uniform. As the optical emission spectra due to f–f transition of erbium ions is closely related to the local environment, it is expected that there are large differences of the optical emission spectra of the glasses with and without soda.

In addition to the traditional melt-and-quench process of glass processing, ion implantation is a common way to fabricate optical waveguides of RE or TM doped glasses [143]. Energetic ions dissipate their energy through electronic excitation and nuclear collision cascades, with the latter be usually called nuclear stopping and is responsible for most of the damage and defect formation in the substrate material, during ion implantation [144]. MD simulations are an effective tool to understand nuclear collision cascade-induced defect formation in ceramic materials [144, 145]. Du and Corrales have studied erbium implantation-induced structural changes in amorphous and crystalline silica by using MD simulations [146]. Atomic-level structural details of erbium ions implantation in silica glass and cristobalite, as a crystal counterpart to compare amorphous structure versus crystalline lattices, were studied to understand the implantation-induced structural changes in the matrix and the structural features of erbium ions in the glass. Erbium ions were given a kinetic energy from 100 up to 1000 eV and the induced collision cascades were simulated. Due to the high velocity, especially in the initial stage of the collision cascade, a variable timestep algorithm was used in MD simulations and the Buckingham potential (Table 12.1) was modified by splicing with the Ziegler–Biersack–Littmark (ZBL) potentials to provide realistic repulsion at short interatomic distances [146].

Erbium ion was selected as the Primary Knock-on Atom (PKA) and the kinetic energy was added to that atom, usually placed at the edge of the simulation cell, in an equilibrated large simulation cell with 12 000 atoms. Figure 12.8a shows the kinetic energy evolution on the PKA atom after the cascade with initial energy of 1 keV [146]. Within 0.4 ps, the PKA kinetic energy decreases exponentially to be close to the regular lattice vibration energy. At the same time, the number of atoms with 0.5 eV due to the collision cascades show a Gaussian type distribution and reach a maximum of 200 atoms per simulation cell at 0.2 ps. While the number of displaced atoms (by distance larger than 1 Å) shows a maximum at around 1 ps and then decreases to reach a steady value. At the same time, erbium ion (as the PKA atom) shows a large change of coordination evolution as a function of time (Figure 12.8b). In the initial stage of the cascade, erbium ion shows a wide range of coordination distribution, while toward the end of the cascades, its coordination reaches around 6 as a typical value in silica glass.

Atoms are displaced along the collision cascade track caused by the PKA atom and coordination defects are created in the matrix structures. The displaced atoms along the track of the collision cascade of the PKA in a 1 keV cascade in silica glass is shown in Figure 12.9 [146]. It can be seen that along the track of the collision cascade, both silicon and oxygen atoms were displaced. The simulation results also indicate that oxygen ions are more easily displaced than silicon ions. This was explained by the difference of the ionic size and coordination number of the two species. $O^{2-}$ ion has larger ionic radius (~1.4 Å) than $Si^{4+}$

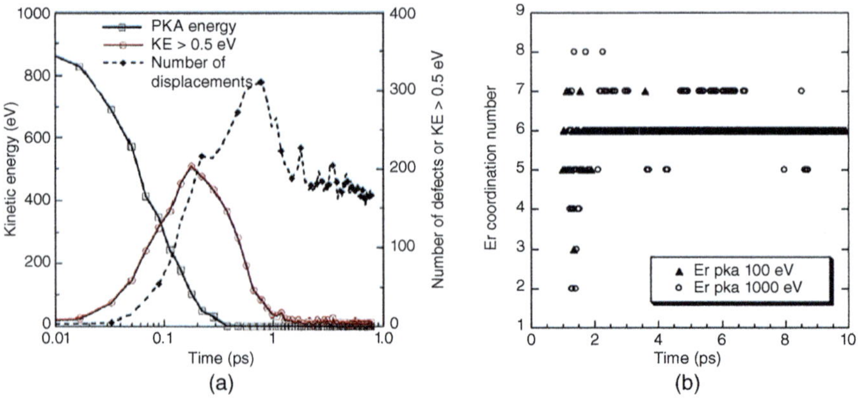


**Figure 12.8** (a) Evolution of PKA kinetic energy, number of energetic ions (>0.5 eV), and displaced atoms (>1 Å) with time in a 1 keV Er cascade (b). Evolution of erbium coordination number with time in silica glass after erbium ion implantation of 100 and 1000 eV in silica glass [146].

(~0.26 Å), while $Si^{4+}$ has a higher coordination number (4) than $O^{2-}$ (2). The results are consistent with other collision cascade simulations in silica and silicate glasses. Figure 12.10a shows the coordination defects such as under- and overcoordinated silicon and oxygen in the original glass after cascade, immediately after 1 keV cascade, and after annealing at 2000 K (below the simulated glass transition temperature of ~2500 K) for 10 ps [146]. The results show that collision cascade increases the percentage of coordination defects of all coordination defect species (except five coordinated silicon which shows a slight decrease). The annealing below $T_g$ can effectively recover the percentages of coordination defects close to prior cascade values, suggesting that thermal annealing is an effective way to decrease the concentration of these defects caused due to collision cascade. Figure 12.10b shows the number of coordination defects formed in silica glass and cristobalite as a function of the energy on the PKA atom [146]. A close to linear increase of the coordination defects from 100 to 1000 eV was observed in both the glass and the crystal. Interestingly, it was found that the coordination defect caused in the glass is consistently higher than in the crystal. This was explained by the fact that the regular crystal matrix helps to restore the displaced atoms back to their original coordination state after the cascade [146]. These results shed light of the atomic-level information of the RE ion environments and the impact of the collision cascades in the glass and crystal matrixes during ion implantation.

#### 12.3.1.2 Europium and Praseodymium-Doped Silicate Glasses

In addition to erbium, other RE ions-doped silicate glasses have been studied by using MD simulations [115, 116, 147]. Du and Kokou investigated europium ($Eu^{3+}$)-doped silica and sodium silicate glasses to understand the difference local environment and clustering behaviors in the two glass matrixes, both with around 1 mol% $Eu_2O_3$ concentration [115]. Much larger simulation cells (with over 24 000 atoms) than earlier MD simulations were

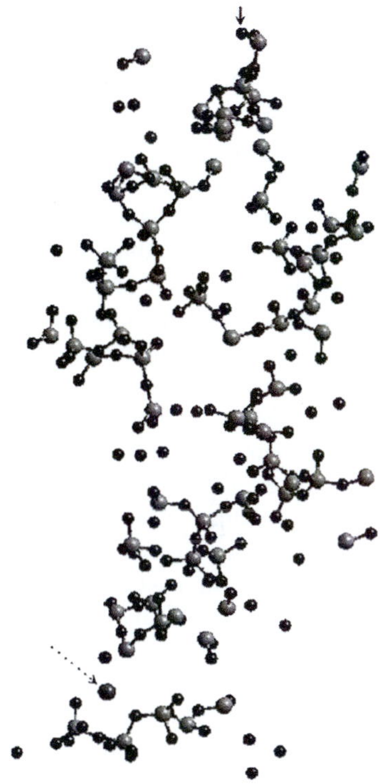

**Figure 12.9** Snapshots of the displaced atoms after 1 keV on Er PKA. The short arrow shows the direction and the initial position of the Er PKA. The long arrow with broken line shows the final position of the erbium ion. Black small balls represent silicon ions and gray large balls represent oxygen ion [146].

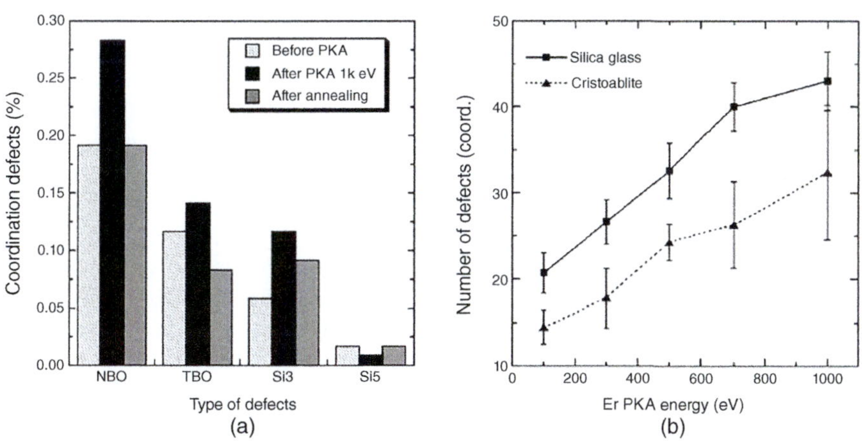


**Figure 12.10** (a) Percentage of Si and O coordination defect sites before and after 1 keV PKA, and after annealing (b) statistics of the number of coordination defects versus PKA energy. Lines are used to guide eyes [146].

used to ensure statistically more accurate results of both the RE environments and distributions. The simulation results revealed a clear difference of the $Eu^{3+}$ environments in silica and sodium silicate glass, similar to what was found for $Er^{3+}$ introduced earlier [114, 118]. Eu–O partial correlation function ($T(r)$) shows very different behaviors for the two glasses (Figure 12.11a,b) [115]. In silica, Eu–O first peak has a lower intensity, less symmetric with higher BO contribution than that in sodium silicate glass. This indicates that $Eu^{3+}$ has a lower average coordination number in silica: on average $Eu^{3+}$ coordination number is 4.8 in silica and 5.9 in sodium silicate glass, about 22% higher [115]. This difference is also shown in Figure 12.11c where coordination distribution of $Eu^{3+}$ is shown [115]. In addition to higher coordination number, the coordination distribution in sodium silicate glass is also narrow suggesting more uniform coordination environments. This was explained by the abundance of NBOs in the sodium silicate glass that helps to satisfy the coordination requirement of $Eu^{3+}$ ions. This fits well into the modified random network model in which the modifiers (both $Eu^{3+}$ and $Na^{+}$) and NBO segregate into modifier rich regions to separate the network-forming regions. In silica, $Eu^{3+}$ ions segregate to share NBOs to satisfy the oxygen coordination requirements. This is also supported by the clustering analysis by comparing with random distributions (Figure 12.1), where it is clearly shown that $Eu^{3+}$ shows clear clustering tendency in silica while not so much in the sodium silicate glass.

Kokou and Du further studied the simulation system size and glass structure effect on the clustering of europium ions [147]. By using simulation cell from around 1500 atoms up to over 24 000 atoms and total $Eu^{3+}$ ions from 10 to 170, $Eu^{3+}$ ion clustering in silica and sodium silicate glasses were studied in detail. It was found that large enough simulation cell is required to get consistent clustering analysis results [147]. For the smallest simulation cell (with ~1500 atoms), only around ten $Eu^{3+}$ ions exist in the system, and this leads to large variations from simulation to simulation of the clustering analysis. Some can even predict the wrong trend of clustering as compared to random distribution as shown in Figure 12.12, where the calculated clustering probability from MD-generated structures and those from random distribution are shown for both silica and sodium silicate glasses. This is also supported by Eu–Eu pair distribution analysis, in which with small simulation cell shows quite different behaviors as compared to those with large simulation cells where more consistent distribution is obtained [147]. These results clearing indicate that the usage of larger simulation cell for compositions with low doping levels is desired, especially if clustering analysis is the goal.

The europium ($Eu^{3+}$) and praseodymium ($Pr^{3+}$) environments and distributions in sodium silicate glasses with different doping levels between 0 and 10 mol% of $RE_2O_3$ were studied by coupling neutron, X-ray diffractions, and MD simulations by Johnson et al. [116]. Figure 12.13a,b shows a comparison of the experimental X-ray and neutron structure factors with those calculated from models generated from MD simulations for the base glass and two europium doped glasses [116]. Generally, good agreement suggests that the simulation models reproduce the structural features of these glasses. The average Eu—O and Pr—O bond lengths are 2.30 and 2.39 Å, respectively, from MD which agree well with 2.36–2.40 Å from the diffraction experiments [116]. Both experimental and simulation data show a slight increase in bond distance with higher RE oxide concentration. To better elucidate the RE–RE distribution from the experimental data, calculations to remove the

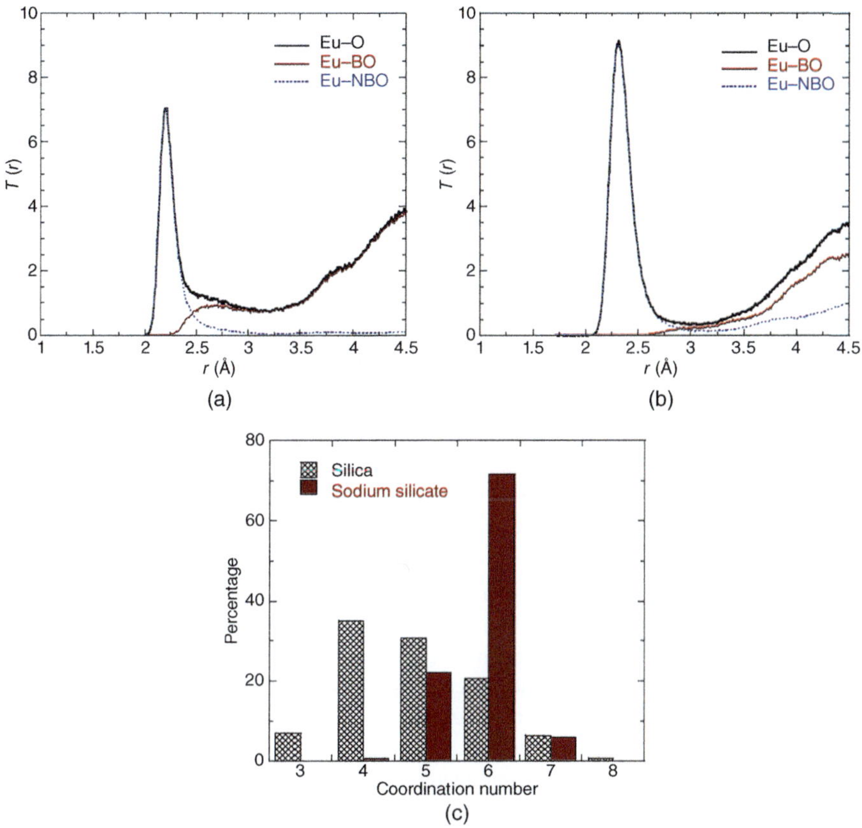


**Figure 12.11** Eu–O pair correlation function and their BO/NBO contributions in Eu-doped silica glass (a) and sodium silicate glass (b). Eu coordination number distribution of europium-doped silica and sodium silicate glasses. (c) Average Eu coordination number is 4.8 in silica and 5.9 in sodium silicate glass. (Eu–O cutoff of 3.1 Å was used for the coordination calculations) [115].

Na–Si correlation in the base glass were performed for the X-ray diffraction total correlation results ($T(r)$). The difference correlation for a praseodymium and a europium-doped glass is shown in Figure 12.13c [116]. The results show the RE–RE distances to be 3.6–3.8 for $Eu^{3+}$ doped glass and 3.8 Å for $Pr^{3+}$ doped glass. These distances are much shorter than that average RE–RE distance of 9 Å predicted from a homogeneous distribution. MD simulations revealed that the RE ions enter the sodium-rich regions in the sodium silicate glasses and act as modifiers. Cluster analysis performed on the model systems indicates that the tendency for clustering is higher in praseodymium-containing glasses than in the europium glasses.

The coordination of $Pr^{3+}$ show a similar trend of $Eu^{3+}$. The RE coordination numbers are 5.9 for Eu in Eu35 ($3.5Eu_2O_3$–$27Na_2O$–$69.5SiO_2$) and 6.0 for Pr in Pr44

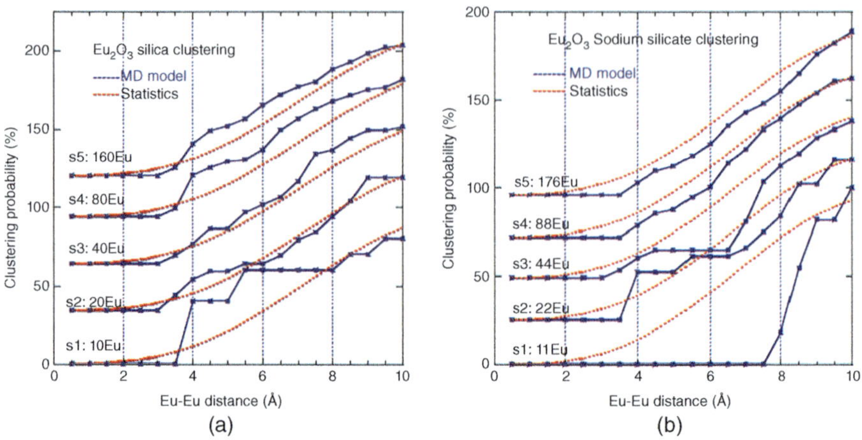


**Figure 12.12** Probabilities of finding neighboring Eu ions as a function of Eu–Eu distance in random distribution (red curves) and Eu ions distribution from MD simulation (blue curves) in $Eu_2O_3$ doped (a) silica and (b) sodium silicate glasses [147].

($4.4Eu_2O_3$–$28Na_2O$–$67SiO_2$) from MD simulations [116]. Figure 12.14a shows the partial correlation function for Pr–O and Figure 12.14b shows associated coordination numbers for Pr44. Similar to $Eu^{3+}$, majority contribution to the first coordination shell of $Pr^{3+}$ is NBOs. The coordination number increase slightly with increasing RE concentration. This is shown in Figure 12.14c [116]. The average coordination number increased from 6.0 in Pr44 to 6.3 in Pr78, in which higher percentage of $Pr^{3+}$ ions takes six- and sevenfold coordination. Figure 12.15 shows europium ions and their first coordination shell of oxygen ions for Eu35 and Eu86 [116]. Some clustering behaviors were observed in Eu35 and the clustering tendency is much more obvious in Eu86. As a result of the clustering of RE ions, the fraction of free oxygen (oxygen ions not bonded to any silicon) increases with RE concentration. These RE ions enter the $Na^+$ and NBO-rich regions and play the role of modifiers as indicated from the calculated fraction NBOs that are consistent with the theoretical model of each RE ion creates three NBOs. Although the clustering behavior can be clearly revealed from visualization in Figure 12.15, comparison with probability of random distributions show that in both Eu35 and Eu86, the clustering tendency is not higher than what the random model predicted suggesting that in sodium silicate glasses there is no significant tendency of clustering than that predicted from random distribution.

#### 12.3.1.3 Cerium-Doped Aluminophosphate and Mixed Former Glasses

Some RE and most of TM ions have multiple oxidation states depending on the oxidation and reduction reaction of these ions have had during the glass formation process. The most notable example might be iron; it can take the ferrous ($Fe^{2+}$) or ferric ($Fe^{3+}$) states depending on the oxidation environments, glass composition, and processing temperatures [148]. One RE ion of particular interest is the cerium ions. Cerium ion can also take two oxidation states ($Ce^{3+}$ and $Ce^{4+}$) in glass and the change of oxidation states can happen not only during

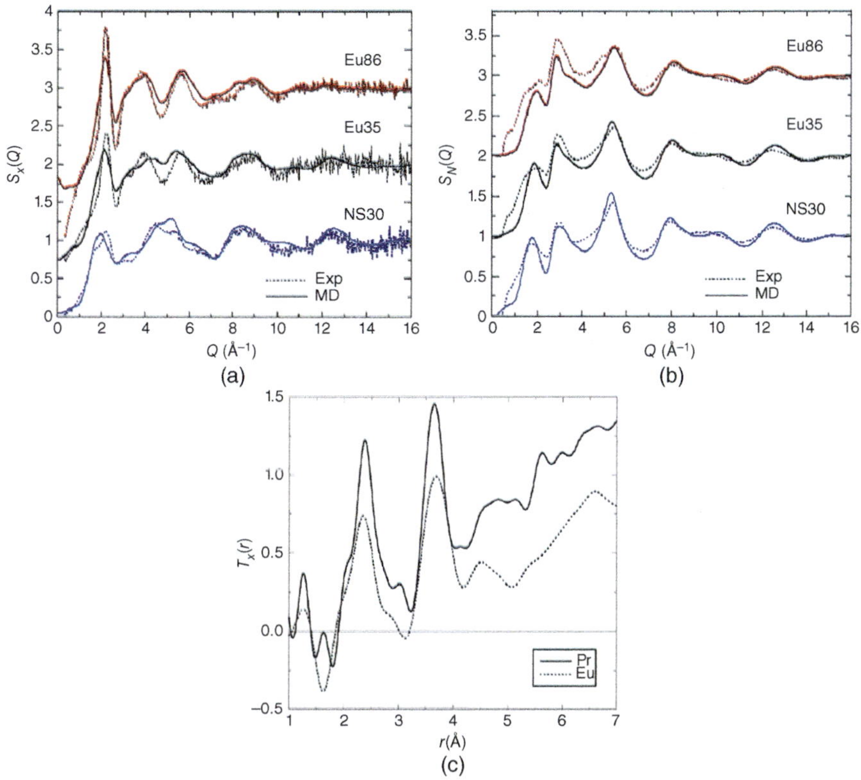


**Figure 12.13** A comparison of experimental and MD calculated X-ray (a) and neutron (b) structure factors of europium-doped sodium silicate glasses (NS30: $31Na_2O-69SiO_2$, Eu35: $3.5Eu_2O_3-27Na_2O-69.5SiO_2$, Eu86: $8.6Eu_2O_3-27.5Na_2O-64SiO_2$, in mol%). (c) First-order difference total correlation functions the Pr ($7.8Eu_2O_3-29Na_2O-63SiO_2$) and Eu ($8.6Eu_2O_3-27.5Na_2O-64SiO_2$) glasses with the base glass subtracted to eliminate the Na–Si correlation [116].

glass formation but also in the formed glass [149]. This means cerium ions can be a donor or acceptor depending on the environment, and the relatively low energy of the transition between the two oxidation states makes cerium doping an effective way to protect glass windows from radiation-induced color center formations [88, 149]. Ionization radiation such as UV light, X-ray, γ ray, neutron, and electron can create color centers in oxide glasses that led to absorption in the visible and infrared ranges [88]. The mechanism of the radiation protection of cerium-doped glasses is mainly due to cerium ions existing in two oxidation states: the trivalent cerium can act as a hole trapper while tetravalent cerium act as an electron trapper to protect the glass matrix from radiation-induced color center formation which leads to darkening and loss of transparency, hence, deteriorate their properties as an optical window either in solar cell or detector applications. Cerium-doped silicate, phosphate, and borate glasses have been studied for these applications [149].

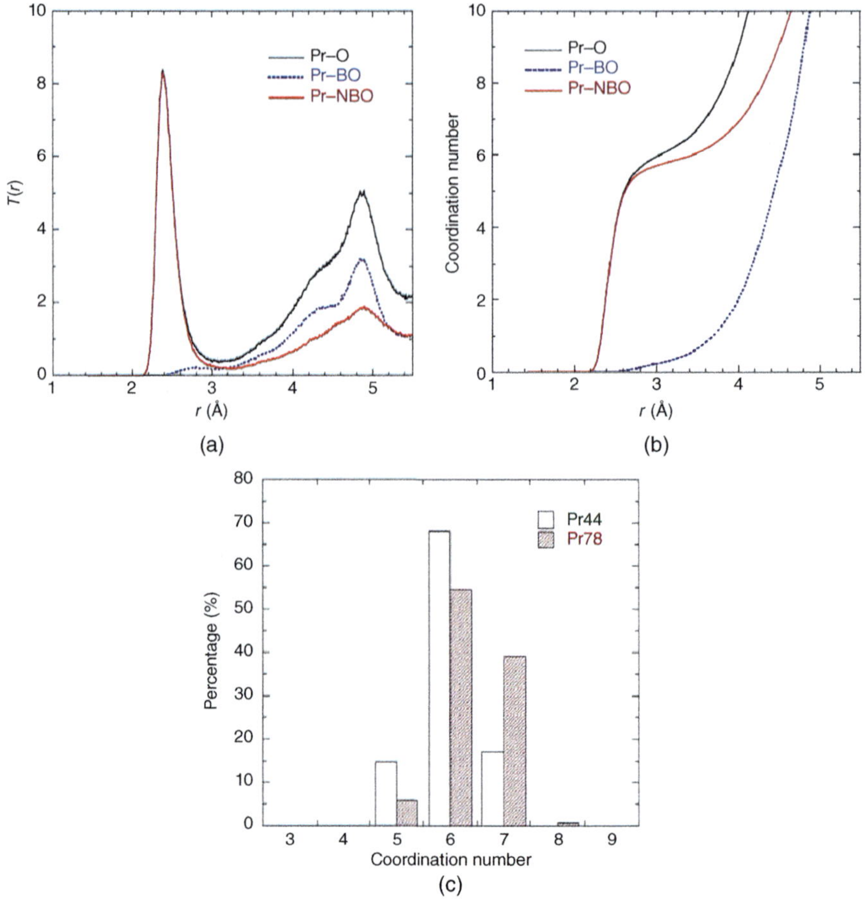


**Figure 12.14** (a) The Pr−O partial correlation function and (b) coordination number in Pr44. (c) Coordination number distributions for praseodymium ions in Pr44 ($4.4Eu_2O_3-28Na_2O-67SiO_2$) and Pr78 ($7.8Eu_2O_3-29Na_2O-63SiO_2$) [116].

MD simulations of cerium-doped glasses with mixing three different kinds of glass formers: $SiO_2$, $P_2O_5$, and $Al_2O_3$, have been studied by Du et al. [88]. The exact redox ratio of the two cerium oxidation states depends on the processing condition and glass composition. Cerium oxidation states were experimentally determined by Ce 3d spectra XPS or Ce $L_3$-edge spectra in XANES [149]. Direct simulation of the redox reactions during glass formation is a great challenge, one will need ab initio molecular dynamics (AIMDs), which is very expensive computationally, or need the empirical potentials to be fully reactive and can accurately describe the redox reactions of the TM or RE ions. However, if the redox ratio is known, one alternative approach is to simulate the glass with correct ratio of these ions in the oxidation states and treat each of them as separate species. Based on this, the

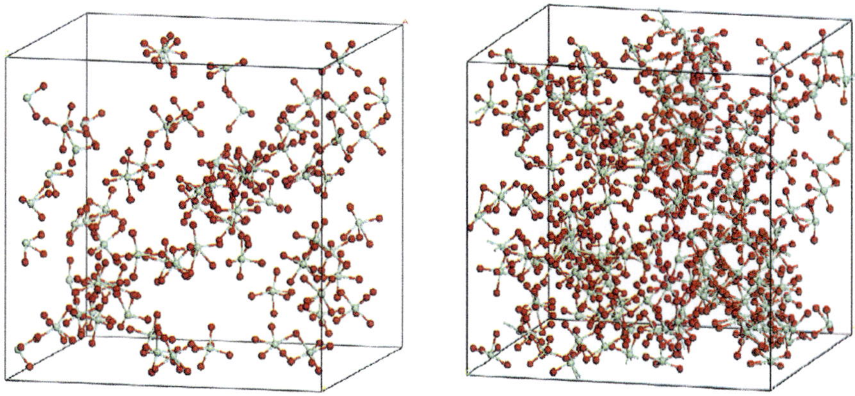


**Figure 12.15** Europium and its first coordination shell oxygen ions in (a) Eu35 ($3.5Eu_2O_3-27Na_2O-69.5SiO_2$) and (b) Eu86 ($8.6Eu_2O_3-27.5Na_2O-64SiO_2$) glasses. Silicon and sodium ions are not shown in the snapshots [116].

oxygen concentration is adjusted to maintain overall charge neutrality. Du et al. adopted the later approach to perform MD simulations of a series of cerium-containing glasses [88]. They first developed parameters for $Ce^{3+}$ and $Ce^{4+}$ with the existing partial charge pairwise potentials that already include $SiO_2$, $Al_2O_3$, and $P_2O_5$. With the newly developed potential parameters for the two cerium oxidation states, the short- and medium-range structures of these glasses, particularly cerium environments have been studied. Figure 12.16a shows the Ce–O PDFs and coordination numbers, while Figure 12.16b shows the BAD of O–Ce–O for the two cerium species [88]. The two oxidation states show different behaviors in their distribution functions. The first peak of $Ce^{4+}$–O distribution has a shorter distance than that of the $Ce^{3+}$–O pair, consistent with the smaller ionic radius of the cerium ion with higher oxidation state. Furthermore, the first peak of $Ce^{4+}$–O is sharper and better defined than $Ce^{3+}$–O, which shows a broader distribution with a longer tail, indicating that the first coordination shell of $Ce^{4+}$ is better defined than $Ce^{3+}$. This difference is reflected in the accumulated coordination number, in which $Ce^{4+}$ has a plateau with around sixfold coordination while that of $Ce^{3+}$ does not have such a plateau and with a slightly higher coordination number (cerium oxygen polyhedra can be visualized in Figure 12.16c,d). The difference is also reflected in the O–Ce–O BADs (Figure 12.16b), where $Ce^{4+}$ shows clearly a peak at 90°, 110°, and 160°, corresponding to a distorted tetrahedron, while $Ce^{3+}$ shows a broader distribution of bond angles.

Beyond the first coordination shell, the species in the second coordination shell is important to understand the clustering behaviors. As discussed earlier, RE or TM cations tend to form clusters in silica glass to share NBOs to better satisfy their coordination environments. In phosphate glasses, cerium ions however may have other choices, e.g. a $[PO_4]$ unit can have a terminal oxygen that can satisfy the coordination requirement of these cations. Du et al. investigated three cerium-containing glasses with mixed glass formers: CPS ($8CeO_2-37P_2O_5-55SiO_2$), PAC1 ($5CeO_2-76P_2O_5-19Al_2O_3$), and PASC2

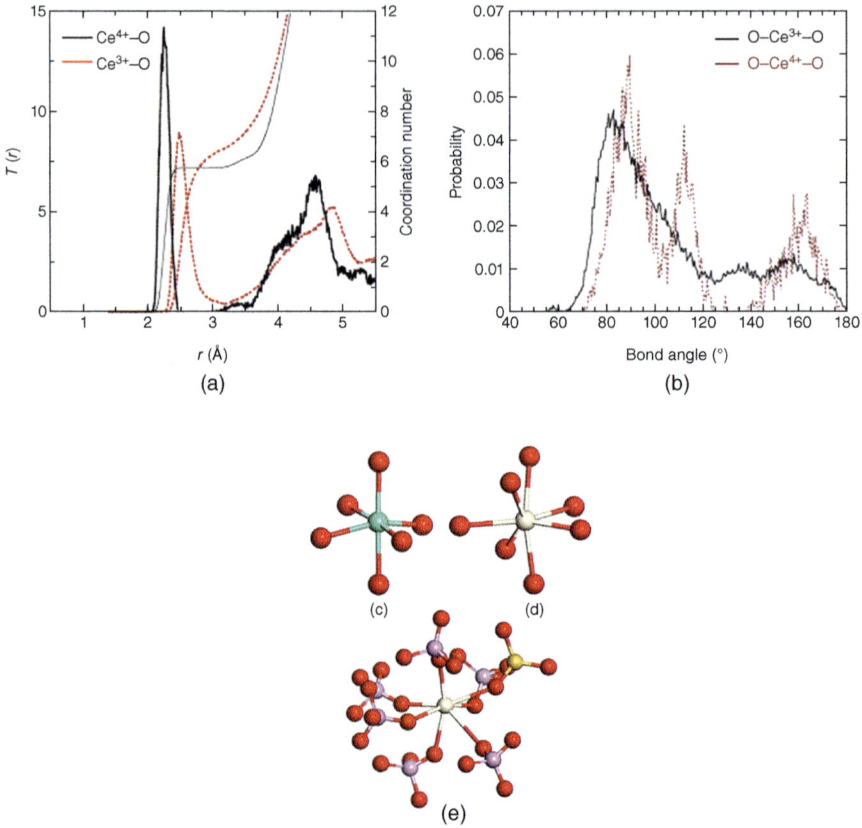


**Figure 12.16** $Ce^{3+}/Ce^{4+}-O$ pair-distribution functions, coordination numbers (a) and $O-Ce^{3+}/Ce^{4+}-O$ (b) bond-angle distribution in cerium phosphosilicate CPS ($8CeO_2-37P_2O_5-55SiO_2$, mol%) glass. Typical cerium oxygen polyhedron for $Ce^{4+}$ (c) and $Ce^{3+}$ (d) and the second coordination shell of a cerium ion ($Ce^{3+}$) in the cerium phosphosilicate glass (e). Red ball: oxygen, green ball: $Ce^{4+}$, yellow ball: $Ce^{3+}$, purple ball: P, orange ball: Si [88].

($5CeO_2$–$71.25P_2O_5$–$14.25Al_2O_3$–$9.5SiO_2$) (all compositions in mol% [88]. Experimentally measured redox ration shows that 95% of cerium ions are $Ce^{3+}$ and 5% are $Ce^{4+}$) [88]. The Ce–Si/P/Al PDF is an effective way to show the correlations of cerium with the glass former cations. Figure 12.17a shows the $Ce^{3+}$–Si/P distributions. It can be clearly seen that $Ce^{3+}$–P first peak intensity is much stronger than $Ce^{3+}$–Si, indicating $Ce^{3+}$ has a preference of $P^{5+}$ than $Si^{4+}$ in the second coordination shell (consistent with visualization shown in Figure 12.16e). Cerium ions are surrounded by the glass former cation oxygen polyhedra, and this is shown in Figure 12.17e, in which a $Ce^{3+}$ ion is linked to six [$PO_4$] units and one [$SiO_4$] unit. In PAC1 (Figure 12.17b), $Ce^{3+}$ also shows a preference for $P^{5+}$ than $Al^{3+}$ in the second coordination shell, and in PASC2 in which contains all three glass formers,

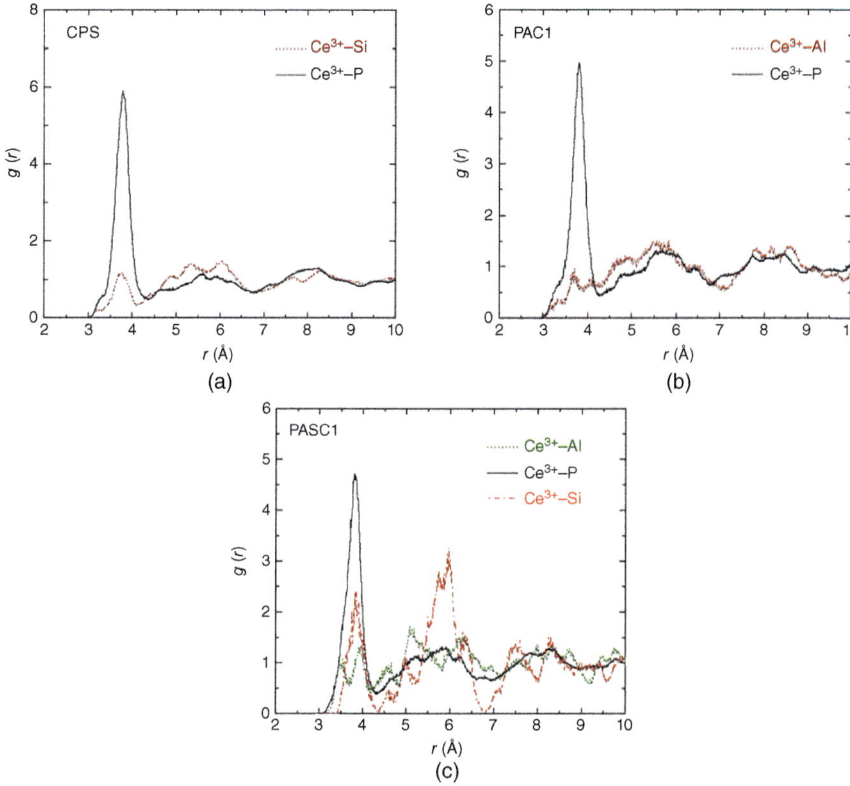


**Figure 12.17** (a) $Ce^{3+}$-Si/P in cerium phosphosilicate glass, (b) $Ce^{3+}$–Si/P in cerium aluminophosphate (CAP) glass, (c) $Ce^{3+}$–Si/P/Al in CAP glasses: CPS ($8CeO_2-37P_2O_5-55SiO_2$), PAC1 ($5CeO_2-76P_2O_5-19Al_2O_3$), and PASC2 ($5CeO_2-71.25P_2O_5-14.25Al_2O_3-9.5SiO_2$) (all compositions in mol%) [88].

the preference of $P^{5+}$ for $Ce^{3+}$ is also clearly observed. Similar behaviors were observed for $Ce^{4+}$ in their preferences of the glass former cations, except in PAC1 and PASC2, $Ce^{4+}$ has exclusive preference for $P^{5+}$ in the second coordination shell. The preference can be quantified by calculating the coordination numbers of the network former cations around the cerium ions. This is summarized in Table 12.5. For example, in PASC2, there is on an average 6.6 P, 0.5 Al, and 0.2 Si in the second coordination shell of $Ce^{3+}$, while that for $Ce^{4+}$ is 6.6 P and no $Al^{3+}$ or Si in the second coordination shell. That is consistent with the distribution functions shown in Figures 12.17 and 12.18 [88]. Based on the investigations of these three glasses with mixed glass formers, it was concluded that the preference sequence of cerium (or other RE or TM cations) to the three common glass former cations is $P^{5+} \gg Al^{3+} > Si^{4+}$ [88]. This finding is very useful to understand the structures of RE-doped glasses and can also help to design glass compositions to lower the clustering of RE or TM ions.

**Table 12.5** Average network former cation ($Si^{4+}$, $P^{5+}$, and $Al^{3+}$) around cerium ions in the three glasses [88].

| | CPS | | Pac1 | | Pasc2 | | |
|---|---|---|---|---|---|---|---|
| | Si | P | Al | P | Si | Al | P |
| $Ce^{3+}$ | 1.2 | 5.7 | 0.5 | 6.5 | 0.2 | 0.5 | 6.6 |
| $Ce^{4+}$ | 1.2 | 4.9 | 0.0 | 6.8 | 0.0 | 0.0 | 6.5 |

Note: The cutoff distances used are 4.37/4.00, 4.27/4.20, 4.36/4.00 Å for $Ce^{3+}/Ce^{4+}$ for CPS, Pac1, Pasc2 glasses, respectively.

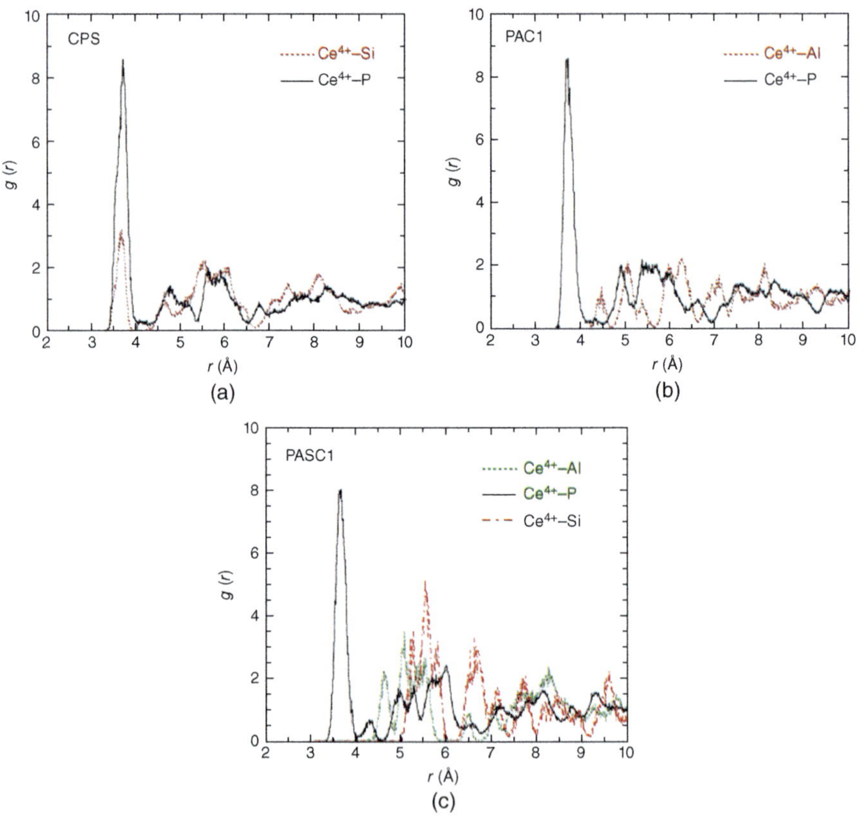


**Figure 12.18** (a) $Ce^{4+}$–Si/P in cerium phosphosilicate glass, (b) $Ce^{4+}$–Si/P in cerium aluminophosphate glass, (c) $Ce^{4+}$–Si/P/Al in cerium aluminophosphosilicate glasses: CPS ($8CeO_2-37P_2O_5-55SiO_2$), PAC1 ($5CeO_2-76P_2O_5-19Al_2O_3$), and PASC2 ($5CeO_2-71.25P_2O_5-14.25Al_2O_3-9.5SiO_2$) (all compositions in mol%) [88].

**Table 12.6** Glass composition, density, and cerium redox, $Al^{3+}$ and $P^{5+}$ coordination numbers of cerium aluminosilicate glasses studied by MD simulations [150].

| Name | $Ce_2O_3$ (mol %) | $P_2O_5$ (mol %) | $Al_2O_3$ (mol %) | $Ce^{4+}$ (mol %) | Density ($g/cm^3$) | $^{IV}Al$ | $^{V}Al$ (%) | $^{VI}Al$ | Al Ave. | P Ave. |
|---|---|---|---|---|---|---|---|---|---|---|
| CAP_3 | 23.8 | 72.1 | 4.1 | 1.3 | 3.105 | 87.8 | 10.98 | 1.22 | 4.1 | 4.0 |
| CAP_5.1 | 19.8 | 69.6 | 10.6 | 1.1 | 3.004 | 88.92 | 11.08 | 0 | 4.1 | 4.0 |
| CAP_6 | 13.8 | 69.8 | 16.4 | 1.6 | 2.793 | 87.35 | 12.65 | 0 | 4.1 | 4.0 |
| CAP_9 | 1.8 | 69.8 | 28.3 | 5.1 | 2.586 | 79.59 | 19.88 | 0.53 | 4.2 | 4.0 |

In addition to the cerium ion environments, the simulation results also show interesting structural features of glass former cation coordination in these glasses. In these three glasses (CPS, PAC1, PASC2), $P^{5+}$ is almost always fourfold coordinated (>99%). While $Si^{4+}$ in CPS was found to have 4.4% fivefold coordination (and remaining 95.6% of $Si^{4+}$ is fourfold coordination) and $Al^{3+}$ in PAC1 has 1.2% sixfold (and 20.1% fivefold coordination [average coordination number 4.23]). Similarly, in PASC2, fivefold coordinated $Si^{4+}$ and five- and sixfold coordinated $Al^{3+}$ were observed. As in silicate and aluminosilicate glasses, $Si^{4+}$almost always takes fourfold coordination, this result suggests that high concentration of $P_2O_5$ forces the other two glass former cations ($Si^{4+}$ and $Al^{3+}$) to take higher coordination in the glasses with mixed glass formers. These results are in good agreement with experimental findings.

Kokou and Du have also studied a series of cerium-doped aluminophosphate glasses by using MD simulations [150]. Glasses with cerium oxide have concentration ranging from 1.8 to 23.8 mol%, while $P_2O_5$ have concentration fixed around 70% and were studied. The oxidation states of cerium ions were determined experimentally by XPS with the percentage of $Ce^{4+}$ being from 1.3% to 5.1%; hence, majority of the cerium ions are $Ce^{3+}$ [150]. Table 12.6 shows the composition, density, and $Ce^{4+}$ percentage in all the studied glasses. Simulation cells with over 13 000 atoms were used to accommodate the glasses with low cerium oxide concentrations. The structures of these glasses were also determined using high-energy X-ray diffractions. Figure 12.19 shows the comparison of experimental and simulated structure factors (Figure 12.19a) and total correlation functions (Figure 12.19b) [150]. It can be seen that the agreement was very good in both reciprocal space structure factors and real space correlation functions. The $R_x$ factor defines that the quality of the fitting was found to be 5.45% for CAP6 and 6.85% for CAP3 [150]. In addition, the position and intensity of the major peaks are all reproduced by those from MD simulations, suggesting that the structure models from MD simulations well represent the real glass structures.

Figure 12.20 shows cerium–Al/P PDFs that show correlations of these cation pairs. It can be seen that for both $Ce^{3+}$ and $Ce^{4+}$, they show a strong correlation with $P^{5+}$ with a peak position at 3.5–4 Å, while Ce–Al first peak is located at a much longer distance and lower intensity. Two glasses with different cerium concentration (23.8% versus 1.8%) show very similar behaviors (Figure 12.20). The inset of Figure 12.20a shows a snapshot of the coordination shell around a $Ce^{3+}$ in CAP3. It can be seen that $Ce^{3+}$ is surrounded by $PO_4$

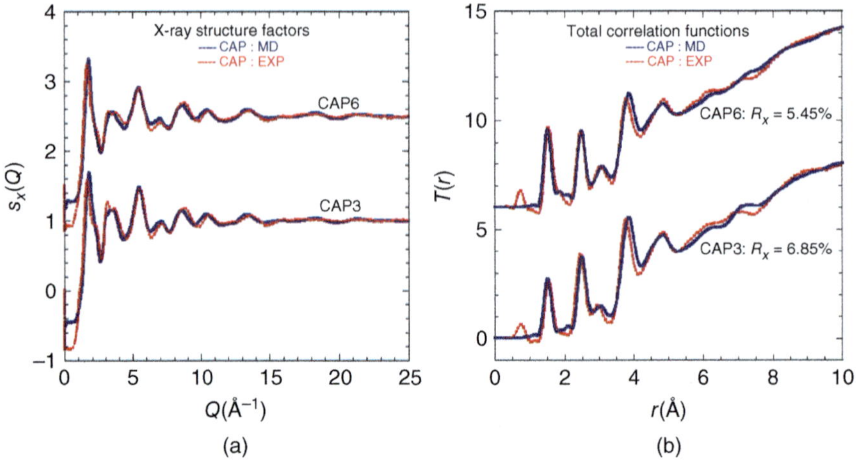


**Figure 12.19** (a) Comparison of MD simulation (blue) and experimental (red) X-ray diffraction structure factors $S(Q)$ comparison and (b) MD simulation (blue) and experimental (red) X-ray total correlation functions $T(r)$ in CAP3 and CAP6 glasses (for glass composition see Table 12.6) [150].

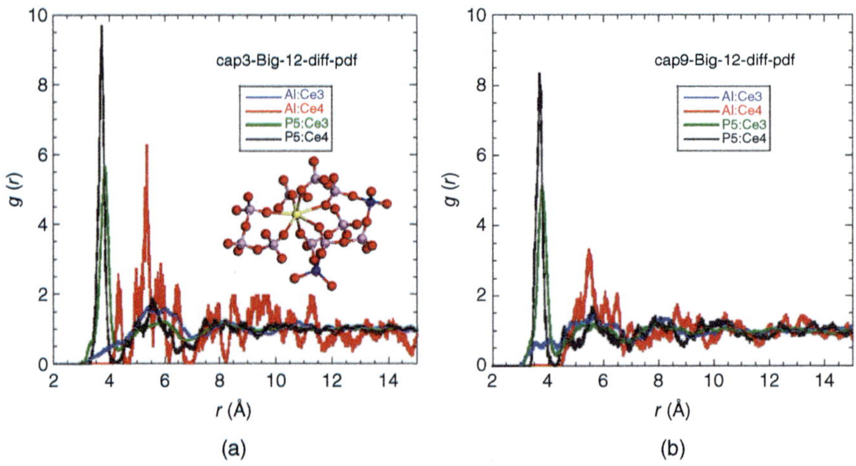


**Figure 12.20** Ce–Al/P partial pair distribution functions in cerium aluminophosphate (CAP) glasses from (a) CAP3 (b) CAP9. Inset of (b) $Ce^{3+}$–O polyhedron and its second coordination shell in CAP3 with O (red), Al (blue), Ce (yellow), and P (purple) (for glass composition see Table 12.6) [150].

units through which they are linked to $AlO_4$ units. Hence, $PO_4$ is in the second coordination shell around cerium ions, while $AlO_4$ is in the next coordination shell. This is consistent with the findings from the earlier study that showed the preference sequence is P > Al > Si in around cerium ions [88].

The clustering behavior was analyzed by studying the cluster size distributions. The cluster size distribution in CAP6 and CAP9 are shown in Figure 12.21a,b [150]. Due to relatively low cerium concentration (1.8 mol%) in CAP9, 100% of $Ce^{4+}$ and 80% of $Ce^{3+}$ are isolated, i.e. they do not bond to another cerium ion, while around 15% of $Ce^{3+}$ ions are in dimmers and 5% in trimmer clusters (Figure 12.21a,b). For CAP6, $Ce^{4+}$ is still 100% isolated, while $Ce^{3+}$ shows a much broader distribution with cluster sizes up to 11 and only about 33% of them are in isolated units.

The average coordination number of $Ce^{3+}$ and $Ce^{4+}$ and the average cluster size is shown in Figure 12.21c. $Ce^{4+}$ has an average coordination number of 6 and increases slightly with cerium oxide concentration, while that of $Ce^{3+}$ is 6.5, and it decrease slightly with cerium oxide concentration. The average cluster size increases almost linearly with the cerium oxide concentration from around 1 for CAP6 to about 2.8 for CAP3. With higher cerium oxide concentration, the clustering of cerium ions become inevitable, as even in random distributions, these ions will form clusters. It was found that both cerium ions prefer phosphorous in the second coordination shell. At low concentration (<2 mol%), most cerium ions are in isolated units and do not cluster with other cerium ions, which is good for the optical emission properties [150]. While at higher concentrations, their segregation through forming Ce–O–Ce linkages to form larger clusters was observed and the cluster sizes increase with cerium oxide concentration.

With carefully developed potentials and experimental validations, MD simulations prove to be an effective tool to understand the coordination environments and distribution of TM and RE ions in oxide glasses. However, in order to understand the charge-trapping behaviors of the cerium ions, first principles-based calculations will be needed where the electronic degree of freedom is treated explicitly. DFT is a commonly used first principles methods for the study of electronic structures in solids due to its computational efficiency. Normally, due to the high computational cost, one cannot afford to calculate the simulation cell of a few thousands to tens of thousands of atoms as we used in the classical MD simulations of RE-doped glasses discussed earlier. Normally, glass structure models with smaller simulation cells (with one to a few hundred atoms) are first generated from classical MD simulations and then the structural models optimized in first principles [151]. Based on the relaxed structures, the electronic properties such as electronic density of states, charge trapping is studied [111].

Du et al. have investigated the electronic structure and charge trapping of cerium-doped aluminophosphate glasses using DFT calculations. Smaller simulation cell with 150–200 atoms were generated using classical MD with potentials and procedures reported earlier [88, 150]. Due to the low concentration of cerium ions in these glasses and small size of simulation cell, usually only one or two cerium ions exist in the simulation glass structures, which is sufficient to understand the electronic structure behaviors. Figure 12.22 shows the relaxed simulation cells of cerium-doped aluminophosphate glasses. Geometry optimization was performed by using periodic DFT calculations with generalized gradient

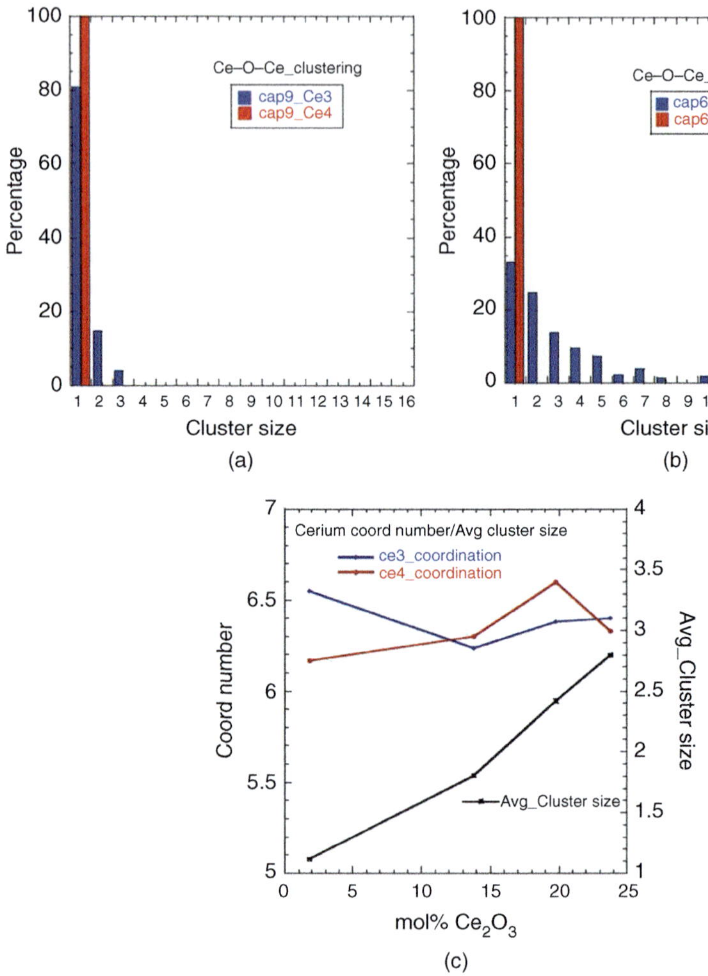


**Figure 12.21** Cerium ion clustering statistics in CAP9 (a) and CAP6 (b) based on Ce–O–Ce linkages using cutoff 3.2 Å. (c) Cerium average coordination number and average cluster size correlation in cerium aluminophosphate glasses (for glass composition see Table 12.6) [150].

approximation (GGA) functionals and plane wave basis sets. PAW pseudo potentials were used to describe the core electrons. Charge trapping calculations were performed by an addition of an electron to a glass structure with $Ce^{4+}$ or subtraction of an electron from a structure with $Ce^{3+}$. The resulting optimized structures were analyzed to see the structural changes (e.g. bond distance and bond angles), as well as electronic structure changes such as charge localization and electronic density of states. Figure 12.22a shows localized charge on a $Ce^{4+}$ containing glass, in which it is shown that the addition of an electron to the glass structure

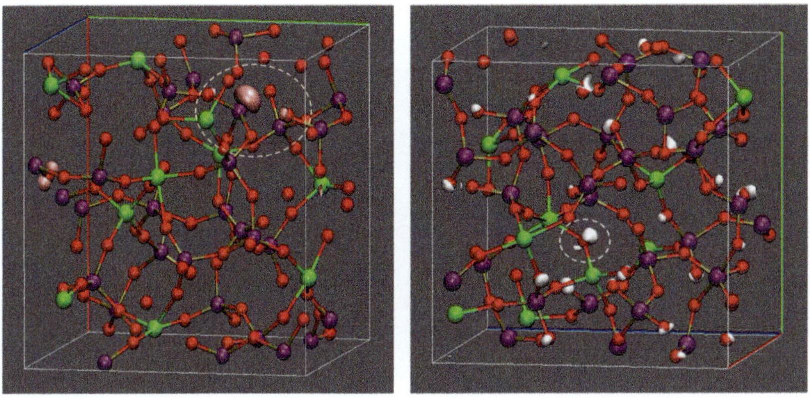

(a) (b)

**Figure 12.22** Structure snapshots and excess charge density distributions after charge trapping in an aluminophosphate glass without cerium doping from DFT calculations. (a) Electron trapping. Excess electron density iso-surface is shown in pink, circled with dashed line. (b) Hope trapping. Excess hole density iso-surface is shown in white, circled in dashed line. (Red ball: oxygen; purple ball: phosphorus; green ball: aluminum; pink shape: electron density; white shape: hole density.)

is localized on $Ce^{4+}$ and convert it to a $Ce^{3+}$. The excess electron density distribution shows a distinct shape of an f electron, supporting that the extra electron is localized on the cerium ion and convert it to a $Ce^{3+}$. Change of bond distances was observed, after conversion from a $Ce^{4+}$ to $Ce^{3+}$, there is a slight increase of Ce—O bond distances. However, due to the transition that happens in the rigid glass matrix, no coordination number change was observed. It also suggests that once there is a hole in the system, the reduced cerium ion can be easily converted back to a $Ce^{4+}$. These results explain the role of cerium ions to absorb the radiation-induced electron hole pairs to prevent the glass matrix to form color centers for its role in radiation hardening of the material.

Irradiation-induced free electron and hole pairs in glass materials can be localized on network structures and create color centers [152]. The commonly known color centers in silica glass after irradiation include the E' center or non-bridging oxygen hole centers (NBOHCs) in silica glass [152]. In $GeO_2$ doped silica, excited electrons were found to be localized on germanium atoms and led to distortion of their coordination from a $[GeO_4]$ tetrahedron shape to a trigonal bipyramid shape with the electron locating on the lateral position based DFT calculations [111]. To understand the ionization-induced charge trapping in the base glass structure, aluminophosphate glass structure without cerium doping was generated from MD simulations. Afterwards, the glass structure model was fully optimized by using DFT-based first principle calculations and then an excess electron or hole was added to the structure, and it was optimized again. The excess charge density distribution was then calculated and plotted in Figure 12.22a for excess electron distribution and Figure 12.22b for excess hole distribution. The excess electron is mainly localized on phosphorous atoms. It leads to the breaking of a P—O bond and converts a $[PO_4]^-$ unit to a $[PO_3]^{2-}$ unit. The pink

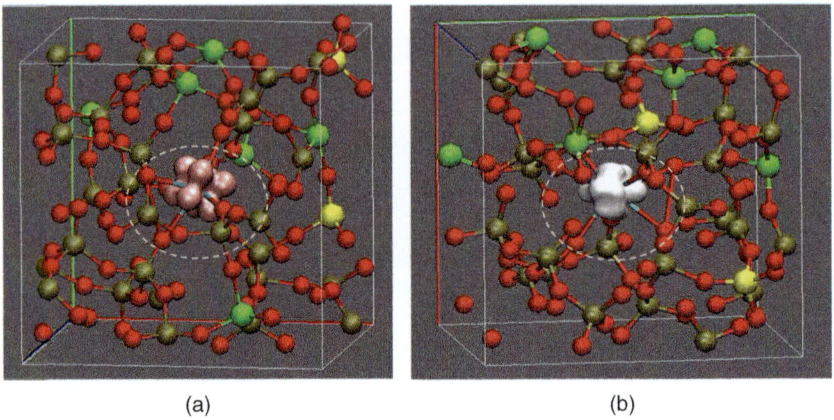


**Figure 12.23** Excess charge trapping in a cerium-doped aluminophosphosilicate glass structure based on DFT calculation. (a) excess electron trapping on a $Ce^{4+}$ site, (b) excess hole trapping on a $Ce^{3+}$ site [154].

round shape in the dashed circle shows the excess electron density distribution localized on a phosphorus atom (Figure 12.20a). Such a center was observed experimentally using EPR in phosphorous-doped silica glass and is considered to defect analog of phosphoryl radicals [153]. For the hole-trapped structure, the excess hole density is distributed among the BO atoms. This is similar to phosphorous oxygen hole center (POHC) [153]. The results are thus encouraging as this type of charge-trapping calculations-based fully optimized glass structures can provide insights of excited electron and hole-induced defect centers in glass structures.

Further calculations were performed on cerium-doped glasses to understand the effect of cerium on charge trapping and defect formation. Structure models of cerium-doped aluminophosphosilicate glasses (around 150 atoms) similar to compositions studied earlier were generated using MD simulations. Fully optimized structures have undergone similar excess charge-trapping calculations to understand which site will be excess electron or hole that will be located with cerium doping. Figure 12.23a shows a structural model-doped with a $Ce^{4+}$ ion. After addition of an extra electron to the system, the excess electron density is clearly located on the $Ce^{4+}$ ion. Importantly, the shape of the excess electron density distribution is similar to the lobes of a 4f electron orbital, suggesting the excess electron enters the f orbital of the Ce ion and convert it from $Ce^{4+}$ to a $Ce^{3+}$. This result thus provides direct evidence that $Ce^{4+}$ in the glass can trap excited electron and be converted to a $Ce^{3+}$. Similarly, excess charge density of a hole-induced structure is localized on a $Ce^{3+}$ site and converted it into a $Ce^{4+}$. This can be visualized in Figure 12.23b. The charge trapping also induced a slight distortion of the original coordination environment of the cerium ions. A slight increase of Ce—O bond distance was observed when $Ce^{4+}$ traps an electron and while slight decrease of Ce–O distance is observed when a $Ce^{3+}$ traps a hole [154]. This is consistent with our MD simulations that shows the bond distance of $Ce^{4+}$–O (around 2.31 Å) and is slightly shorter than that of $Ce^{3+}$–O (around 2.52 Å).

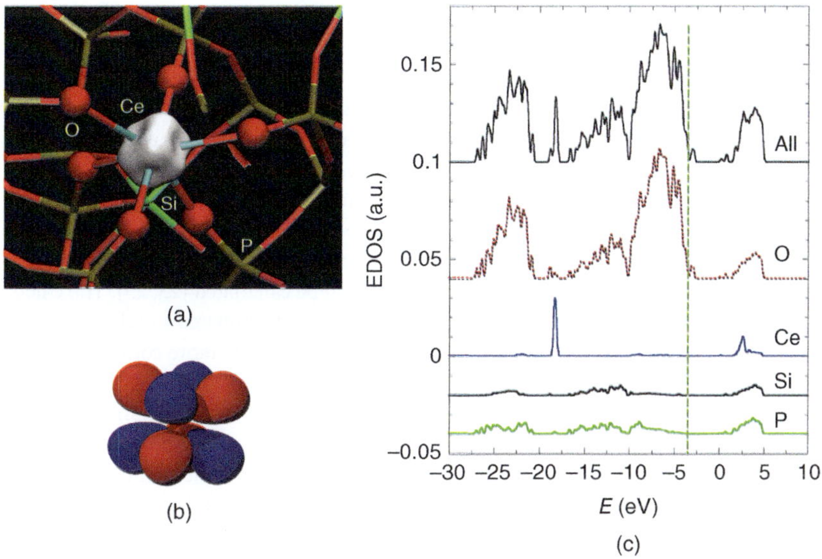


**Figure 12.24** (a) Excess charge density of around a $Ce^{4+}$ ion after trapping an electron in cerium a phosphosilicate glass. White shape, excess electron density iso surface. (b) Shape of electron orbital of a f electron ($f_{xyz}$). (c) Total and partial electronic density of states of a cerium-doped phosphosilicate glass. Vertical dash line Fermi level.

In addition to charge trapping, the calculated electronic density of states can tell the information on electronic structure and influence on the band gap behaviors due to charge trapping. Figure 12.24a shows the excess electron density distribution in a phosphosilicate glass subject to addition of one electron. The shape of the additional one electron is localized on the $Ce^{4+}$, similar to Figure 12.24a, and the shape is similar to the $f_{xyz}$ orbital of an f electron as in Figure 12.24b, suggesting that electron trapping of $Ce^{4+}$ and coverting to a $Ce^{3+}$ can happen in different glass matrix. Figure 12.24c shows the total and partial electronic density of states of a cerium ($Ce^{3+}$)-doped phosphosilicate glass [154]. It can be seen that the valence band top is dominated by O 2p orbitals, while the conduction band bottom is dominated by cerium contributions. Interestingly, there are states in the band gap suggesting the semiconducting behavior of the glass, largely due to the potential change of oxidation states of cerium ions. Further calculations of electron and hole-trapped samples can provide further information of the electronic behaviors of these glasses. It is worth noting that standard DFT approaches are known to have the issues of underestimation of the band gap and tendency of charge delocalization, hybrid functionals in DFT or other first principles methods will be needed to fully understand the charge-trapping-induced behaviors [111].

Based on the structural models from MD simulations, the optical absorption and emission spectra of RE- and TM-doped glasses can be calculated. Standard DFT methods cannot provide accurate description of excited states hence would not able to calculate the electronic excitation such as d–d or f–f transitions. Other methods such as time-dependent density functional theory (TD-DFT) or ligand field density functional theory (LD-DFT) or a higher level of theory is needed for these calculations [155]. For various theories of electronic structure theories of TM complexes, we refer to a nice review by Ziegler [156]. These first principles methods are limited to small systems with a few to tens of atoms. In addition, the calculation of RE absorption and emission spectra has been practiced in the glass simulation community since early 1980s [23]. The RE 4f electron energy levels and splitting can be calculated by crystal field Hamiltonian based on atomic positions from models generated from MD or MC simulations while the transition rates between the energy levels can be obtained according to the Judd–Ofelt theory. Based on the calculated energy levels and transition rates, the absorption and emission spectra of RE ions can be calculated [23, 25]. The calculations based on these theories are much more manageable as compared to fully first principles methods. The applications of these methods have extended to more complex glasses. For example, Inoue et al. have used this approached to study the $Tm^{3+}$ doped ZBLAN, a type of fluoride glass with composition $52ZrF_4$–$20BaF_2$–$3.5LaF_3$–$3AlF_3$–20NaF–$0.5InF_3$–$1TmF_3$. The calculated spectra was found to be in good agreement with experiments [157].

The results presented in this section thus show classical MD simulations can provide significant insights of RE- and TM-doped glasses for optical, photonics, and other applications. From MD simulations, the local coordination environments of RE and TM cations, coordination in the second coordination shell, and their clustering behaviors as a function of glass system and composition can be studied in detail. It has also been shown that the power of combining MD and first principles calculations to understand charge trapping, electronic structure and defects, optical absorption, and emission can be achieved. All of these can significantly advance our understanding of RE- and TM-doped glasses for optical, radiation hardening, and other applications.

### 12.3.2 Alkali Vanadophosphate Glasses as a Mixed Conductor

#### 12.3.2.1 General Features of Vanadophosphate Glasses

Among the TM, Vanado-phosphate glasses (VPGs) were widely investigated in the past because the presence of $V^{4+}$ and $V^{5+}$in the glass bulk could be considered to promote the polaron hopping. In particular, it was highlighted that the different oxidation states (4+ and 5+) of V leads to quite different local and medium environments characterized by edge- and/or corner-sharing interconnected polyhedra that become the preferential path for hopping phenomenon [26]. The $V^{4+}/V_{tot}$ ratio depends on several factors as previously reported such as the vanadium content, the alkali type, and concentration in glass formulation as well the oxidizing or reducing condition in the process obtainment of the final materials [158–160]. It was hypotheses that the presence of $V^{4+}$–O–$V^{5+}$ bridges in the VPGs 3D could affect the electronic conductivity but the presence of these species is often difficult to quantify experimentally, and the data are inconclusive especially for complex VPG formulations that present a quite hard experimental characterization. Literature suggests that the edges-sharing mechanism of V-based polyhedra could favor the polaron hopping [62, 161] but experimental and simulated evidences show that V polyhedra prefer to share corners, which denote longer V–V distances [61, 160, 162–165]. In particular, vanadium shows a coordination comprising of between 4 and 6, to which

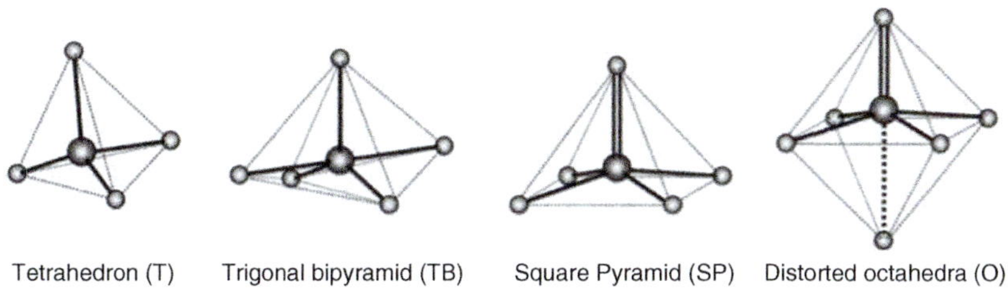


**Figure 12.25** $V^xO_n$ structural units in phosphate glasses.

corresponds tetrahedral, square pyramidal, trigonal pyramidal, and octahedral V-based units detailed in Figure 12.25 [81, 161, 163].

Moreover, the local coordination and the role played by V is strongly dependent by its content in glass formulation [61, 81]: while for low $V_2O_5$ content, vanadium ions present higher coordination number and a stronger modifier character, increasing the $V_2O_5$ content, the coordination decreases, and a former behavior could be associated to vanadium ions. Spatial reorganization of V depending on its coordination clearly influences the presence of $V^{4+}$–O–$V^{5+}$ path in the network and therefore the potential electronic contribution to total conductivity of VPG glasses. Moreover, the presence of extra modifiers leads to a more depolymerized network suggesting a lower electronic contribution, but, on the counterpart, an acquired ionic contribution to total electronic conductivity of these systems [166–169]. The type of modifier affects the final semiconductive behavior of VPG glasses: the V–V distances become larger in the function of the increase of alkali radius, leading to longer hopping paths that justify the decrease of the electronic contribution to total conductivity experimentally observed [19–21]. Furthermore, also the dimension of alkali ions affects the ionic conductivity: larger network modifiers show lower mobility in glass matrix, which produces a decrease in the conductivity and an increase of the activation energy [166–168, 170]. Contemporary presence of more than one modifiers in VPGs results in an anomalous trend of ionic conductivity in a function of glass formulation due to the so-called "mixed alkali effect" (MAE) [169, 171]. The main attracting feature of VPGs is the possibility to promote mixed ionic and electronic conductivity that could open to potential application of these systems in secondary batteries. Nowadays, one of the main problems of the secondary batteries is ascribed to the presence of heterojunctions (solid electrolyte and cathode) in the assembled device. In particular, the presence of different crystalline structures constituting the solid electrolyte or cathode leads to the drastic change in the diffusion properties due to the creation of a potential barrier at the interface that limits the ionic exchange mechanism and therefore the final capacity of the assembled system [26].

The use of the same type of material for electrolyte and cathode could overcome this problem and therefore, it represents a very interesting challenge in the field. In this perspective, glasses represent a very good choice due to advantages that they present if compared to the crystalline-based systems. Glasses are stable, safe, characterized by isotropic properties due to the absence of grain boundaries, they can be produced as film or coating, and the so-called "open structure" can promote ionic conductivity [2, 172–174]. For these reasons, nowadays, $V_2O_5$ containing glass and glass–ceramic systems are extensively investigated for their potential application as cathode materials in secondary batteries [2, 172–174]. Indeed, the presence of V ions in multiple oxidation states (+3, +4, and +5) could favor the small polaron hopping (supported also by the creation of specific geometries) [175, 176] in the glass network, that results in the final electronic conductivity [65, 113, 177–181]. Specific glass compositions belonging to this class of material can be potentially proposed for both cathode material and/or solid electrolyte depending on the main requirements: while for the cathode application, the electronic transport has to be maximized, for the electrolyte high ionic conductivity is needed.

Cathode materials usually work as electron exchanger; therefore, they should be able to intercalate the ions from electrolytes inside their lattice and own enough electrons to balance the redox reactions when modifiers such as Li or Na ions are introduced. Thus, to guarantee both effects, cathodes need to be mixed electronic-conductivity and the addition of TM ions inside the glass matrices that are already ionic conductors, supplies the electronic contribution to create an electron exchanger [1, 52, 182, 183]. For instance, crystalline vanadium oxide ($V_2O_5$) owns a good electronic conductivity and can be fully populated by lithium ions. However, as $LiCoO_2$, $V_2O_5$ transforms into several crystal phase with lithium, and the structural changes could modify the electrical properties and decrease the cycle efficiency. Therefore, the use of glass matrices, for instance lithium $V_2O_5$–$B_2O_3$, could prevent the structural modification because lithium ions generally occupy the free volume that characterize the disordered network without causing severe changes [184].

In this section, we summarize the information obtained on a selected VPGs investigated with the aim to explore the mixed electronic–ionic conductivity phenomenon and the correspondent behavior of these systems in term of conductivity measurements. Moreover, the ionic conduction of $Li^+$ and the effect of the presence of two modifiers in phosphate glasses were analyzed.

MD simulations were used to quantify the V–O–V [163] mixed interlinkages which favor the polaron hopping among V in different oxidation states, leading to changes in experimentally derived conductivity [61]. The relationship between V–O–V species and the electronic contribution to conductivity in vanadophosphate glasses was investigated in the recent past by experimental methods but results are still nonconclusive [5, 6, 52]. Insight in the glass structure and dynamical properties of VPGs were achieved by Montorsi et al. by using classical MD validated by experimental characterization on the glasses [26]. Conductivity measurements confirm the potential application of the VPG for solid-state batteries applications especially as concerning devices that should work in temperature.

In this contribution, we keep into account two series of glasses belonging to $Na_2O$–$V_2O_5$–$P_2O_5$ (NaVP) and to $Li_2O$–$V_2O_5$–$P_2O_5$ (LiVP) systems. Structure of NaVP and LiVP was explored and compared to each other, particularly short- and medium-local environment were analyzed by using bond distances, bond angles, coordination, and $Q_n$ species analysis depending by the V concentration in the glass bulk. The creation of V–O–V linkages was analyzed by MD and then correlated to experimental conductivity in order to explore potential compositions of interest for future applications.

It is worth noting that in this contribution, the results achieved by classical MD were reported, and the main disadvantage of this technique is that the electronic part is not directly accounted in simulations. This could seem misleading if we are interested in electronic conduction, but the main topic of the work of authors was to focus the MD studies on the detachment of the structural features and the dynamic phenomenon that really affects the experimental conductivity. One this information is reached, the attention will split on the details concerning the electronic part (by using ab initio methods), and this will be done only on interesting and promising compositions. This approach allows to save computational effort due to the fact that ab initio studies are more complicated from a computational point of view; therefore, they can be focused only on the glasses presenting the most interesting conductivity.

#### 12.3.2.2 Sodium Vanadophosphate Glasses

Glasses belonging to the $(Na_2O)_{(100-z)/2}$-$(V_xO_y)_z$–$(P_2O_5)_{(100-z)/2}$ system in which z goes from 10 to 60 mol% were investigated by Montorsi et al. [185]. The main advantage in considering this system resides in the wide experimental characterization performed by solid-state $^{31}$P- and $^{51}$V-MAS-NMR that open to the easy validation of the computational procedure and the selected FF in exploring the structure of these glasses [186]. An added value, in comparison with previous work on vanadate glasses, is the experimental quantification of the $V^{4+}/V_{tot}$ ratio (where $V_{tot}$ represents the sum of $V^{4+}$ and $V^{5+}$ present simultaneously in the glass structure) by using the XPS. Beyond the $V^{4+}/V_{tot}$ ratio also, the density was experimentally obtained, and all the values were used to set up the MD input to better reproduce the real system.

In this way, further details on the glass structure and in particular, the coordination of the V in the two-oxidation states can be explored and correlated to the final properties of glass. Coordination environment also defines the role played by V in these glasses, and the data derived by MD can be successfully used to complement experimental information and to clarify if TM acts as a network former or modifier in the glass structure. The data derived by MD, such as PDF, CN BAD, and so on, explain and clarify the role played by V in these glasses as modifier or network former to complete the experimental information. In Table 12.7, composition discussed in this chapter is detailed, as well as $V^{4+}/V_{tot}$ ratio, density, and total atoms. Only for the NaVP-20 and NaVP-60 compositions, the data reported are provided by literature [186].

The coherence among the MD derived bond distances and BAD, coordination number distribution as well as the NBO distribution, as a function of glass composition is confirmed as already reported by Montorsi et al. [185]. The trend of $V^{5+}$-O PDF confirmed the presence of a shoulder at about 1.65 Å, corresponding to the typical vanadyl (V = O) bonds, found in vanadophosphate systems, that was progressively overlapped by the main peak of $V^{5+}$–O for increasing the content of vanadium in glass composition. The creation of V = O· · ·V bonds tends to elongate the vanadyl bonds, emphasizing the disappearance of the shoulder [163].

As already mentioned, V can be present in the glass structure in different local environments characterized by an averaged coordination ranging from four to six as detailed in Figure 12.25. Moreover, the different ionic states can produce quite different environments that can represent a key point in the explanation of some final properties shown by the glassy material. The results demonstrated that the average coordination of $V^{4+}$ and $V^{5+}$ is very different, in fact the $V^{4+}$ ions (Figure 12.26a) show a relative high percentage of

**Table 12.7** Composition (mol%), $V^{4+}/V_{tot}$ ratio, density and total atoms for the NaVP system [185].

| System | $Na_2O$ | $V_2O_5$ | $P_2O_5$ | $V^{4+}/V_{tot}$ (%) | Density (g/cm$^3$) | Total atoms |
|---|---|---|---|---|---|---|
| NaVP-10 | 44.94 | 10.13 | 44.94 | 29.0 | 2.54 | 3515 |
| NaVP-20 | 40.00 | 20.00 | 40.00 | 25.8[a)] | 2.62[a)] | 3520 |
| NaVP-35 | 32.58 | 34.85 | 32.58 | 15.0 | 2.73 | 3520 |
| NaVP-50 | 10.00 | 50.00 | 40.00 | 34.0 | 2.83 | 3518 |
| NaVP-60 | 20.00 | 60.00 | 20.00 | 7.5[a)] | 2.89[a)] | 3520 |

a) Derived from Tricot et al. [186].

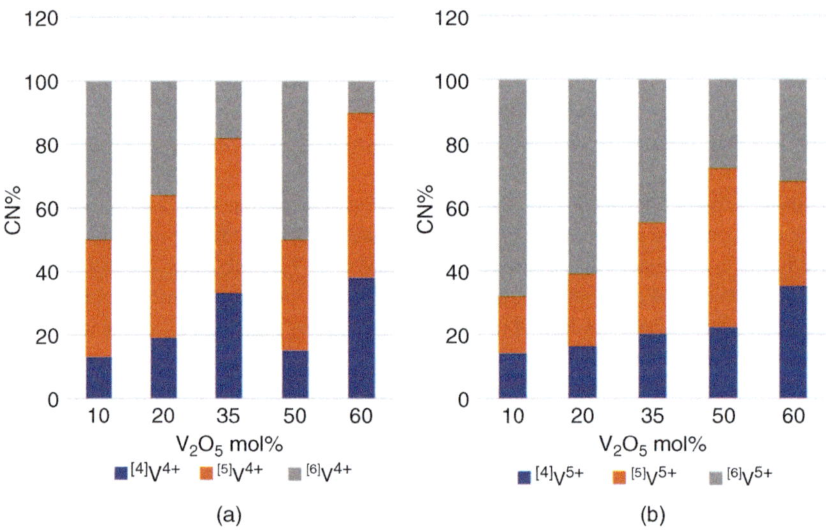


**Figure 12.26** Coordination number (CN) % as a function of $V_2O_5$ content for (a) $V^{4+}$ and (b) $V^{5+}$ (Error within ±0.3%).

octahedral units ($^{[6]}V^{4+}$ ~ 49%) and lower percentages of four- and penta-coordination (~13% and ~37%, respectively) for the lower vanadium containing system (NaVP-10). Regarding the NaVP-50 to NaVP-60, the main contribution derives from penta-coordinated ($^{[5]}V^{4+}$) ions reaching more or less the 50%. On the contrary, Figure 12.26b points out that $V^{5+}$ ions shift from about a 69% of octahedral ($^{[6]}V^{5+}$) sites in low vanadium-containing system (NaVP-10) to a 38% of sixfold coordination in NaVP-60. For NaVP-60, the $V^{5+}$ ions show a quite homogeneous distribution of four-, five-, and six-coordinated structural units.

The BAD (not reported here) confirmed these differences in structural rearrangement of the V local sites as a function of the V content [185]. The information agreed well with

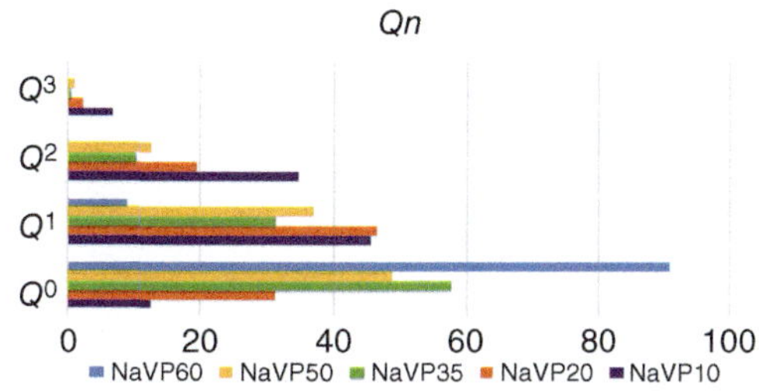


**Figure 12.27** $Q^n$ distribution for P for the NaVP series.

the work of Bianconi et al. in which a clear change from extra-coordination to mainly four-coordination with the increase of the total vanadium content was highlighted [81].

The different local coordination of $V^{4+}$ and $V^{5+}$ presents different features also by considering the type of oxygens linked to the central atom distinguishing in BO or NBO contribution. In this system, MD analysis pointed out the increase of the glass rigidity supported by the creation of BO induced by $V_2O_5$ addition in glass composition at the expenses of the $Na_2O$ content. Both $V^{4+}$ and $V^{5+}$ resulted in connected with BO species for more than the 80% in NaVP60 as detailed by Montorsi et al. [185].

The highlighted transition from phosphate to vanadate-based network was confirmed also by the $Q^n$ distribution derived by MD. As reported in Figure 12.27, the $Q^n$ distribution of P ions in NaVP systems (calculated considering only the phosphorus ions as network former) show high percentage of $Q^0$ species (about 90%) in NaVP60 suggesting that P polyhedra tend progressively to be surrounded by the vanadium ones moving from NaVP10 up to NaVP60. The high concentration of Na ions in the NaVP-10 and NaVP-20 justify the observed high values of $Q^1$ and $Q^2$ species that is congruent with the BO and NBO distribution suggesting a strong depolymerization of the glass network at high Na concentration [113].

Among the medium-range order features, particular attention must be focused on the creation of V–O–V linkages in the glass structure due to the key role that these species cover on the potential small polaron hopping phenomenon between the vanadium ions. In fact, in order to promote the electronic conductivity, the different oxidation states of V ($V^{5+}$ and $V^{4+}$) have to be organized in an interconnected polyhedral as $V^{5+}$–O–$V^{4+}$ linkages in the glass network.

It was demonstrated that at higher V content, the second shell of both $V^{5+}$ and $V^{4+}$ is characterized by the presence of other V ions, according to the shift toward a vanadium-based network as a function of the $V_2O_5$ content, as previously discussed. This is also confirmed in Figure 12.28a where a clear decrease of the P–O–P linkages from 55% (NaVP-10) to 1% (NaVP-60), which corresponds to a progressive increase from 3% to 65% of the V–O–V linkages can be observed. The P–O–V linkages show a maximum of 67.3% in NaVP-20 composition, while for higher amount of $V_2O_5$, the percentage tends to decrease favoring the V–O–V creation.

Moreover, the conductivity was derived for the NaVP10, NaVP35, and NaVP50 by using complex impedance spectroscopy (not reported here). It was verified (Figure 12.28b) that the conductivity values at room temperature for these glasses is quite low, but they show a clear increase as a function of both T and vanadium concentration in the investigated

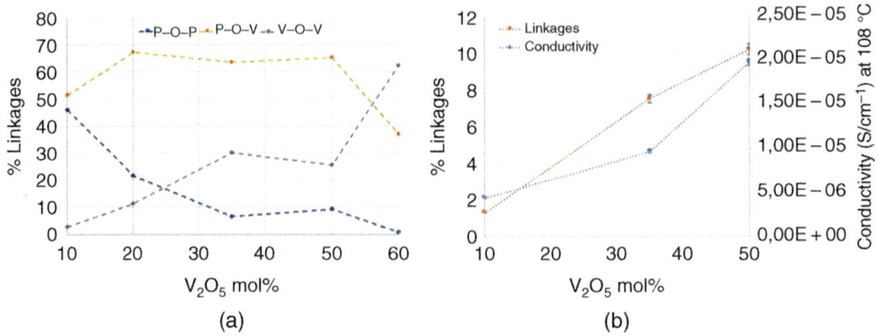


**Figure 12.28** (a) P–O–P, P–O–V, and V–O–V linkages (%) and (b) linkages (%) and conductivity values at 180 °C as a function of $V_2O_5$ content.

glasses. Thus, the observed conductivity can be generally interpreted as being due to small polaron hopping from the lower ($V^{4+}$) to the higher ($V^{5+}$) valence state of vanadium ions [185]. Specifically, Figure 12.28b reports the three bulk conductivity values measured for the NaV10, 35, and 50 at 180 °C, and it is possible to note that at high vanadium content, the conductivity reaches to $10^{-5}$ S/cm. The combined experimental and computational study allows to confirm that V–O–V increase due to the increasing content of V in glasses, resulting in high percentage of $V^{5+}$–O–$V^{4+}$ to which correspond an increase of conductivity at 180 °C.

It can be concluded that NaVP system can be well characterized by MD, confirming the reproducibility of the FF and the computational procedure using the ad hoc derived $V^{4+}$–O and $V^{5+}$–O parameters. MD allows a snapshot of the structural features of these glasses with bond distances and angle as well as medium-range structure well reproduced. V content increase results in more polymerized vanadium-based framework characterized by a high percentage of BO. The different $V^{4+}$ and $V^{5+}$ coordination promotes structural rearrangement in which V-based polyhedral interact forming V–O–V species. Moreover, the $V^{4+}$–O–$V^{5+}$ presence was correlated to the electronic conductivity confirming hypothesis previously proposed by experimentalists that was explored in this work with the atomistic level of detail.

#### 12.3.2.3 Lithium Vanadophosphate Glass

The effect of the network modifier cation on the electrical properties has been extensively investigated, since it can affect the V–V distances in glasses modifying the hopping of polarons along defined paths and thus the electronic contribution to total conductivity [166–168, 187]. Lithium oxide is usually introduced in vanadophosphate glasses to promote the mixed electronic–ionic conductivity for potential application as cathode materials [187–190]. In particular, lithium ions are the major charge carriers of ionic conductivity, while the vanadium ions in different valence states are responsible of the electronic ones. Therefore, the development of this class of materials is mainly driven by the suitable conductivity ($10^{-8}$–$10^{-13}$ cm$^2$/s) and the better stability as compared to the crystalline cathodes during the charge–discharge cycles [183, 191]. It is worth noting that the alkali acting as modifiers promote glass depolymerization creating also useful channels that enhance

**Table 12.8** Composition (mol%), $V^{4+}/V_{tot}$ ratio, density, and total atoms for the LiVP system [191].

| System | $Li_2O$ | $V_2O_5$ | $P_2O_5$ | $V^{4+}/V_{tot}$ | Density (g/cm$^3$) | Total atoms |
|---|---|---|---|---|---|---|
| LiVP10 | 45 | 10 | 45 | 0.2500 | 2.579 | 5175 |
| LiVP30 | 10 | 30 | 60 | 0.4000 | 2.638 | 5184 |
| LiVP50 | 10 | 50 | 40 | 0.3100 | 2.776 | 5156 |

the ionic conductivity. At the same time, this ions breaking the network can destroy the $V^{4+}$–O–$V^{5+}$based path, as discussed for NaVP, and therefore strongly affect the electronic contribution to conductivity [159, 190]. In this section, the analysis of LiVP compositions is reported with the aim to highlight the structural changes induced by vanadium addition in Li containing glasses and at the same time to investigate the dynamical properties such as the $Li^+$ diffusion, which can affect the ionic contribution to conductivity.

Broglia et al. [191] performed MD investigation on LiVP glasses for which the $V^{4+}/V_{tot}$ ratio were experimentally obtained by XPS analysis. As it can be seen in Table 12.8, the compositions investigated presents both high and low content of lithium ions (45 versus 10 mol%) as well as low and high vanadium content (from 10 to 50 mol%) in order to explore the potential ionic contribution to conductivity (higher $Li_2O$ content) and the electronic one (high $V_2O_5$ concentration) [187]. Otherwise, LiVP30 and LiVP50 have the same content of Li ions (10 mol% $Li_2O$) but different V/P ratios, which should permit to a better understand the influence of vanadium content on the final structural properties.

As for the NaVP system, the short-range order of glasses, codified in terms of PDF, well reproduce the structural data experimentally obtained. In particular, Figure 12.29 reports a focus on the V—O bond, where the $V^{5+}$–O PDF is characterized by a shoulder at lower distances (typical of the vanadyl bonds) that tends to disappear by increasing the vanadium content due to the creation of V = O· · ·V bonds that tend to elongate the vanadyl bonds [163] and to the shift of the peak to lower values resulting in the overlapping of the vanadyl bonds by the main V—O bond. The $V^{4+}$-O curves shape is quite broad for the LiVP10 and narrower for the other two compositions according to a limited bond lengths distribution.

It was proposed that in glasses an edge-sharing V-based polyhedral presents averaged $V^{4+}$—$V^{5+}$ bond distances that could promote the polaron hopping phenomenon [164, 167–169]. This trend was verified for the investigated glasses by considering the MD results. It is worth noting that for vanadophosphate glasses, the V–V distances for edge- (ES), and corner-sharing (CS) polyhedra fall in the range between 3.18–3.37 and 3.52–3.60 Å, respectively [163, 164].

The MD data reported in Table 12.9 shows the decrease of the V–V distance moving to higher V containing systems but remain greater than 3.38 Å that suggest a CS distribution. This trend is confirmed by the percentage of CS, always greater than 88%, that allows to conclude that in these glasses the preferential distribution for vanadium atoms is CS [191]. As reported for NaVP system, the averaged coordination of $V^{4+}$ and $V^{5+}$ ranging from 5.75 to 4.78 (depending on $V_2O_5$ content in glasses) shows quite different trend as detailed in

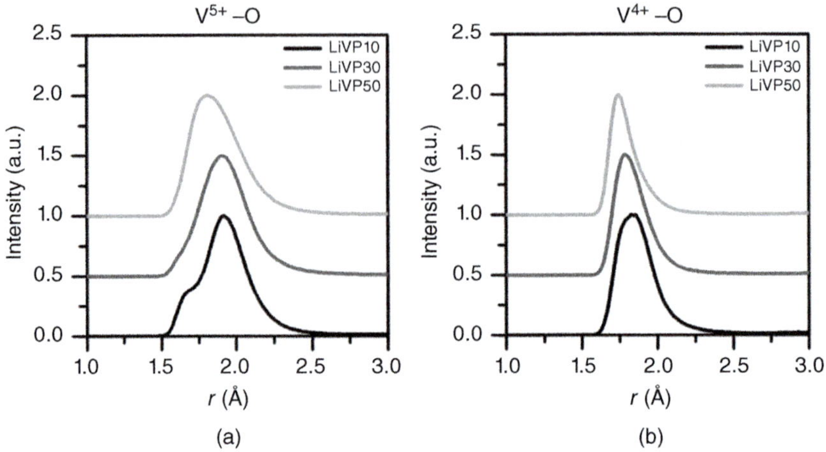


**Figure 12.29** PDF curves of the (a) $V^{5+}$—O and (b) $V^{4+}$—O bonds in the LiVP series (for glass compositions see Table 12.8) [191].

**Table 12.9** Bond distances derived by PDF curves from MD (Å, error within ±0.05).

| | V–V distances | | | %CS | | |
|---|---|---|---|---|---|---|
| | $V^{4+}$–$V^{4+}$ | $V^{4+}$–$V^{5+}$ | $V^{5+}$–$V^{5+}$ | $V^{5+}$–$V^{5+}$ | $V^{5+}$–$V^{4+}$ | $V^{4+}$–$V^{4+}$ |
| LiVP10 | — | 3.49 | 3.63 | 100.0 | 98.6 | 88.9 |
| LiVP30 | 3.39 | 3.51 | 3.63 | 100.0 | 99,5 | 96.0 |
| LiVP50 | 3.39 | 3.48 | 3.61 | 99.5 | 98.4 | 97.7 |

Figure 12.30 pointing out a different local environment as a function of the considered oxidation state. The tetrahedral, square pyramidal and trigonal pyramidal, octahedral structural units identified as $V^{X+}O_4$, $V^{X+}O_5$, $V^{X+}O_6$, where $X$ is the valence state of vanadium ions are showed for both the oxidation state of V as well as the corresponding O—V—O bond angle trend in function of glass composition.

The $V^{4+}O_4$ units increase with the increase of $V_2O_5$ content, moving from around 8% to 40% for LiVP50 to which also correspond a broader O–V–O BAD, where two peaks at 90 and about 110° can be noted according to the typical tetrahedral and square pyramidal units. The percentage of $V^{4+}O_6$ units decreases, while the $V^{4+}O_5$ units become the predominant contribution to the average coordination in LiVP50. Square and trigonal pyramids were detected as the most stable geometries for $V^{4+}$ ions [75, 165, 192, 193], but even sixfold coordination [73, 163, 165, 193] can be found, and it could be also described as a square pyramidal with a sixth long bond. A different trend can be observed for the of $V^{5+}O_4$, $V^{5+}O_5$, and $V^{5+}O_6$ which show the increase of the $V^{5+}O_4$ and $V^{5+}O_5$ units in function of V content in glasses. However, the $V^{5+}O_6$ units remain the predominant environment for the $V^{5+}$. In LiVP, the

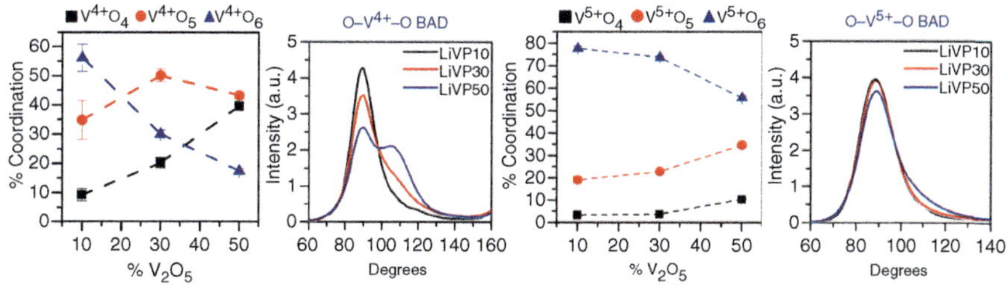


**Figure 12.30** Contributions to the averaged CN of $V^{5+}$ and $V^{4+}$ ions and bond angle distribution (BAD) for $V^{4+}$ and $V^{5+}$ ions in LiVP series (for glass compositions see Table 12.8) [191].

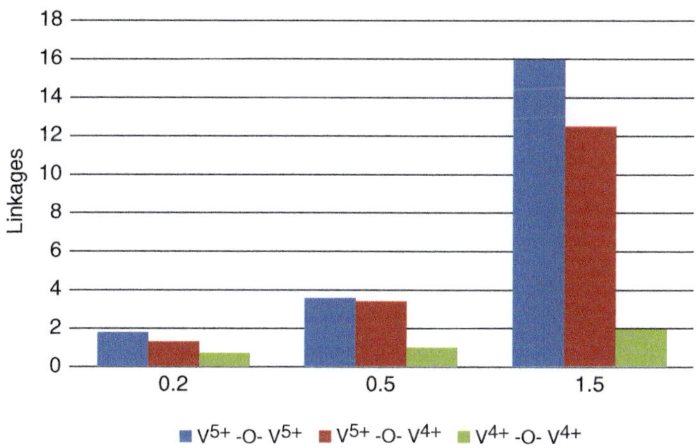


**Figure 12.31** V–O–V Linkages as a function of V content.

$Q^n$ distribution is the same observed for the NaVP and previously discussed highlighting the transition from phosphate-based network to a vanadate one. This structural change was promoted both by the decrease of the Li content that act as a modifier open the glass network favoring the $Q^1$ and $Q^2$ local configuration of P and by the increase of $V_2O_5$ that play a former role leading to a more interconnected structure in which $Q^0$ (P isolated) tend to prevail. Moving from LiVP10 to LiVP50, it was observed the linear increase of the V–O–V linkages at the expenses of the P–O–P depending by the $V_2O_5/P_2O_5$ ratio (Figure 12.31).

The presence of $Li^+$ in the VPG can affect the conductivity as a function of the Li transport properties on which the ionic contribution depends. For this reason, Broglia et al. [191] focused the attention on the dynamic phenomenon in glass systems by using MD simulations and in particular by analyzing the MDS and the diffusion coefficient of $Li^+$ as a function of the V content in glass formulations. The diffusion coefficient as a function of temperature ($1/T$) is shown in Figure 12.32 for the three glasses. The plot in Figure 12.32 highlights a clear change of slope between 610 and 680 K in these systems. However, it can be observed

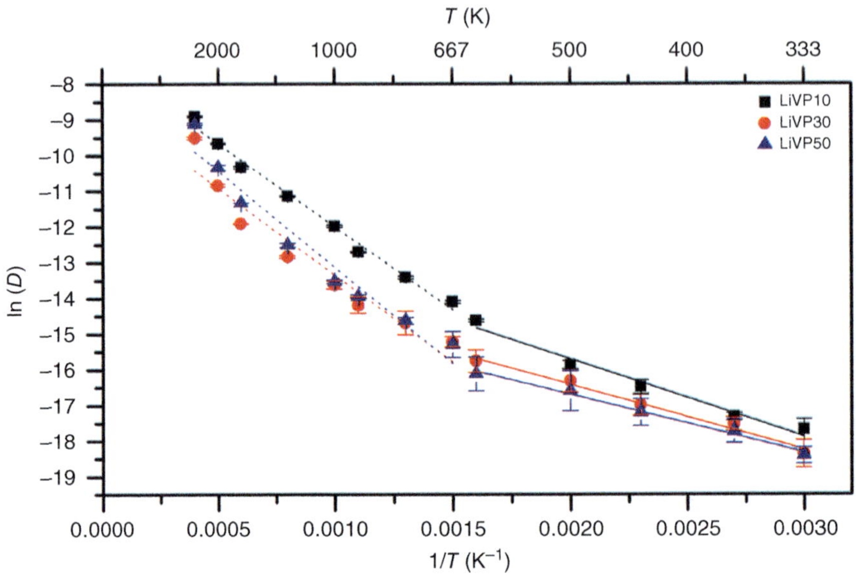


**Figure 12.32** Diffusion coefficient as a function of 1/temperature for three lithium vanadophosphate glasses with different vanadium oxide concentrations.

**Table 12.10** Activation energy ($E_a$) of Li ions and the diffusion prefactor ($D_0$) at low and high temperatures.

| | Low temperature | | High temperature | |
|---|---|---|---|---|
| | $E_a$ | $D_0$ | $E_a$ | $D_0$ |
| LiVP10 | 0.19 (2) | $10^{-5}$ | 0.41 (0) | $10^{-3}$ |
| LiVP30 | 0.16 (3) | $10^{-6}$ | 0.42 (1) | $10^{-4}$ |
| LiVP50 | 0.16 (3) | $10^{-6}$ | 0.47 (2) | $10^{-4}$ |

that there is a good linearity at higher and lower temperatures. The activation energy ($E_a$) of Li ions and the diffusion prefactor ($D_0$) derived by the application of the Arrhenius equation to the MSD data clearly shows two sets of $E_a$ and $D_0$ values, one for high temperature (above 680 K) and low temperature (below 680 K) regimes (Table 12.10).

It was pointed out that the self-diffusion coefficient of Li ions in LiVP10 is higher than the one detected for both LiVP30 and LiVP50 systems, in the entire temperature range. The $D_0$ whose value for LiVP10 is one order of magnitude higher with respect to those calculated for the others two systems, it could be explained considering the different amount of lithium ions in the compositions (45 mol% for LiVP10 and 10 mol% for both LiVP30 and LiVP50), since it is well known that the ionic conductivity is strictly dependent on the mobility and on the number of mobile ions [61]. For the $E_a$, two set of values were considered: one at high and one at low temperature. While for the set at lower temperature, no significant changes were showed, at highest temperature, small discrepancies were noted. The LiVP10 and LiVP30 systems showed slight lower (0.41) activation energy than LiVP50 (0.47), and this trend was related to the shift from a phosphate to a predominant vanadate glass network that change the interactions between the alkali and the vanadium ions. Literature data reports activation energy values ranging from 0.32 to 0.88 eV [166, 188], and the $E_a$ obtained by using the MSD data previously reported well agreed to the one proposed by Assem et al. [168] that tried to split the ionic and electronic contribution (0.09 and 0.38 eV for electronic and ionic contribution, respectively).

## 12.3.3 Zirconium Containing Silicate and Borosilicate Glasses for Nuclear Waste Disposal

### 12.3.3.1 Structures of Zirconia Containing Silicate and Borosilicate Glasses

$ZrO_2$ is an important component in glass technology. It is commonly added in silicate or borosilicate glasses to enhance properties such as chemical durability [32, 33], mechanical strength and crystal precipitation [194–196]. $ZrO_2$ is also a well-known component of nuclear waste glasses as it can be introduced in glasses from the cladding and the walls of glass melter with zirconia-based refractory material, as well as the fines formed after spent fuel reprocessing [197–199].

Experimental studies of $ZrO_2$-containing silicate and borosilicate glasses have explored their structural information [195, 200–207], physical properties [198, 208, 209], and dissolution behavior [32, 33, 198, 208–214]. $ZrO_2$ has been found to increase density, increase glass transition temperature, and decrease thermal expansion coefficient, all consistent with an increase in average glass network strength [198, 208]. Computer simulations such as MD simulation and MC simulation have also been utilized to study the effect of $ZrO_2$ on the structural properties of silicate and borosilicate glasses [32, 197, 203, 215–217]. With the developments of potential parameters for B–O pairs added to existent parameters for aluminosilicate glasses [117, 218], atomistic structural models of multicomponent borosilicate glasses can be generated [39, 40, 219–224]. The atomic-level structural model of a simplified nuclear waste model glass, International Simple Glass (ISG), with a composition of $60.10SiO_2–12.65Na_2O–5.73CaO–15.96B_2O_3–3.84Al_2O_3–1.72ZrO_2$ (mol%) was successfully generated using the composition-dependent potential parameters for boron–oxygen interaction [224, 225], with excellent agreement to results of high-energy X-ray diffraction (as shown in Figure 12.33 NMR, and EXAFS experiments) [117, 224, 225].

Changes to glass properties with the addition of $ZrO_2$ are commonly explained by the strengthening of the glass network due to corner-sharing of $[ZrO_6]$ octahedra with $[SiO_4]$ tetrahedra in the glass-forming networks [203, 206, 226]. Figure 12.34 shows snapshots of a simulated $15Na_2O–10CaO–68SiO_2–7ZrO_2$ glass by MD, presenting the connection between $[ZrO_6]$ and $[SiO_4]$. In silicate and borosilicate glasses, Zr ions exist mainly as sixfolded octahedra with a Zr–O distance of 2.09 Å, and charge compensated preferentially by Na ions [33, 203, 205, 211, 215, 227]. Figure 12.35 shows the PDFs of cation–oxygen pairs in a simulated $ZrO_2$-containing glass. Moreover, MD simulations revealed that amorphization of

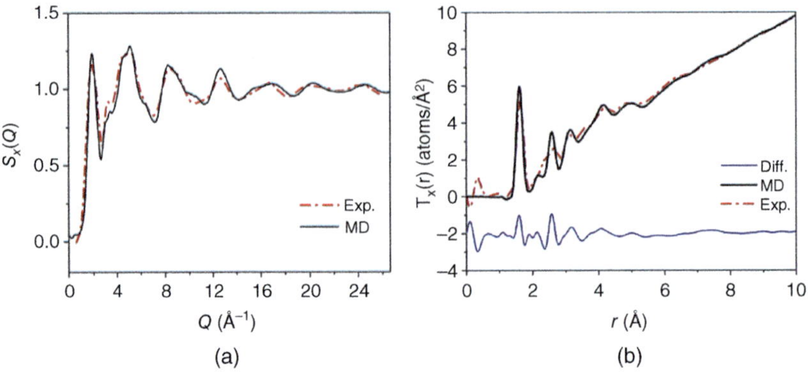


**Figure 12.33** Comparisons of experimental and simulated X-ray diffraction spectra of ISG ($60.2SiO_2-12.6Na_2O-5.7CaO-16.0B_2O_3-3.8Al_2O_3-1.7ZrO_2$ (mol%)). (a) X-ray structure factors. (b) Total correlation functions with a $R_x$ value of 5.5% [224].

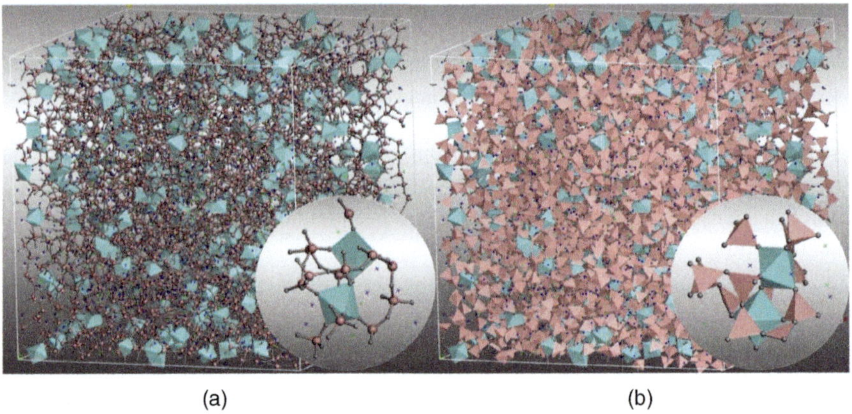


**Figure 12.34** Snapshots of simulated $15Na_2O-10CaO-68SiO_2-7ZrO_2$ glass by MD. Part (a) shows [$ZrO_6$] distribution: Zr, blue octahedra; Si, large pink balls; O, small gray balls; Na, blue crosses; Ca, green crosses. Part (b) shows the connection between [$ZrO_6$] and [$SiO_4$]: Zr, blue octahedra; Si, pink tetrahedra; O, small gray balls; Na, blue crosses; Ca, green crosses [217].

crystalline zircon ($ZrSiO_4$) leads to polymerization of the silicon-oxygen network [216]. Similarly, the addition of $ZrO_2$ was found to be polymerized the glass network of sodium-lithium borosilicate glasses by attracting alkali cations to charge compensate sixfold coordinated Zr ion [197].

However, the structural role of zirconia in oxide glasses is still controversial, for example the preference of charge compensators around negatively charged $[ZrO_6]^{2-}$ is unclear and difficult to elucidate from experiments; zirconia is sometimes categorized as a modifier [203, 209] or an intermediate [202, 206], instead of a glass former [215, 217].

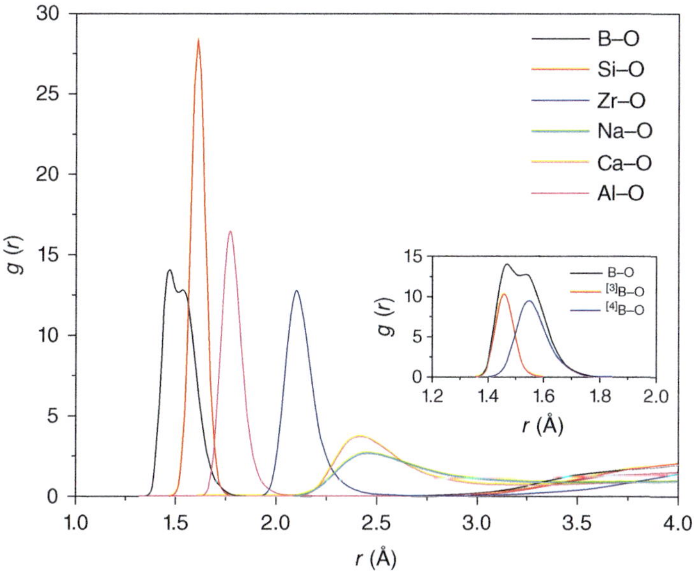


**Figure 12.35** PDFs of cation–oxygen pairs in a simulated $57.9SiO_2$–$12.6Na_2O$–$5.7CaO$–$16.0B_2O_3$–$3.8Al_2O_3$–$4ZrO_2$ glass [40].

The effect of $ZrO_2$ replacing $SiO_2$ on glass structure and properties in silicate and borosilicate glasses was systematically studied by MD simulation [39, 40, 217]. It was found that local environments (e.g. bond distance, bond angle, and coordination numbers) of cation-oxygen pairs were insensitive with $ZrO_2/SiO_2$ substitution in silicate glasses ($(75-x)SiO_2$–$15Na_2O$–$10CaO$–$xZrO_2$, $x = 0, 1, 3, 5, 7$) [217], borosilicate glasses ($(61-x)SiO_2$–$17B_2O_3$–$18Na_2O$–$4CaO$–$xZrO_2$ with $x = 0, 1, 2, 4, 6$, and 8) [39], and a simplified nuclear waste model glass ($(61.9-x)SiO_2$–$12.6Na_2O$–$5.7CaO$–$16.0B_2O_3$–$3.8Al_2O_3$–$xZrO_2$, $x = 0, 1.7, 3, 4$) [40]. In these three glass systems, Zr ions are in slightly distorted $[ZrO_6]$ octahedral environments and have a higher preference of Ca ions as charge compensators in comparison with Na ions [40, 217]. Replacing $SiO_2$ with $ZrO_2$ results in a higher network connectivity, fewer NBO species, and smaller-sized rings, which decreases the initial dissolution rate by strengthening the glass network. Method of calculating Zr K-edge EXAFS spectra was developed (as shown in Figure 12.36), providing more information than traditional shell-by-shell fitting, which opens the door to characterization of local structural environment around Zr and other glass components [40].

#### 12.3.3.2 Effect of Zirconia on Physical Properties and Chemical Durability

Diffusion and mechanical properties of $ZrO_2$-containing glasses have been studied by MD as well. Diffusion and dynamic properties of the ions provide a better understanding of the chemical durability and ionic conductivity of these glasses. Figure 12.37 shows self-diffusion coefficients of B, Ca, Na, Al, Zr, and Si in ISG at different temperatures. Self-diffusion activation energy of all the ions slightly increases with more $ZrO_2$ added, especially for Zr ions,

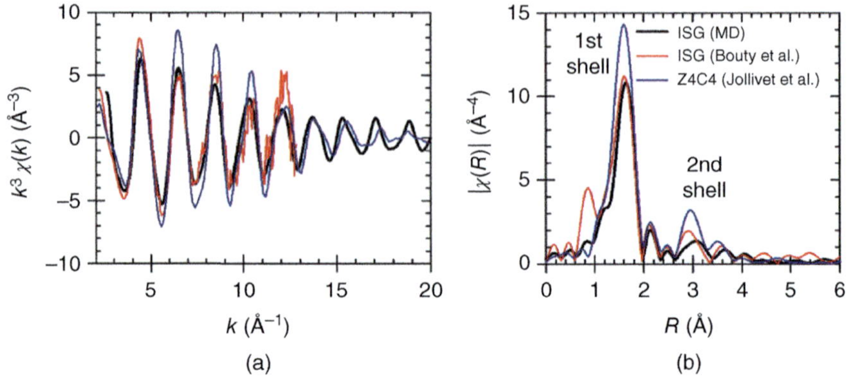


**Figure 12.36** Experimental and calculated ($\Delta E_0 = 10$ eV) Zr K-edge EXAFS spectra (a) and corresponding Fourier transform magnitudes (b) for Zr in ISG (300 K) or Z4C4 (35 K) [40].

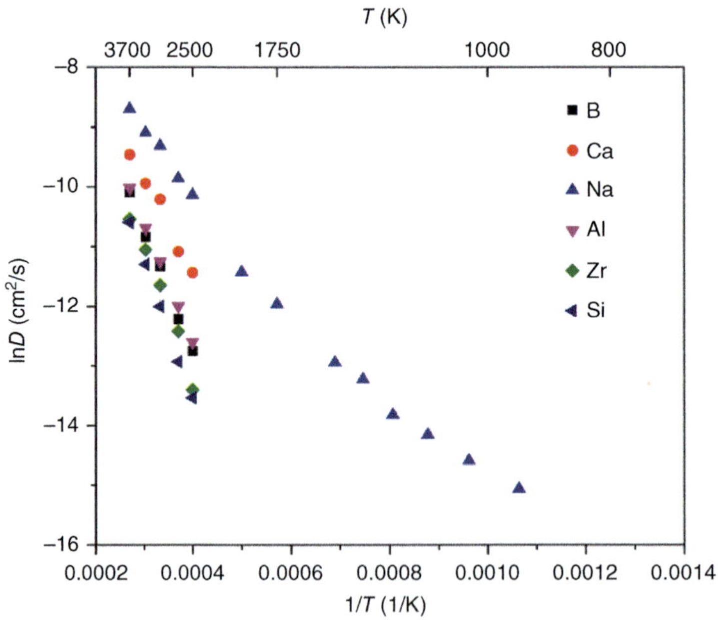


**Figure 12.37** Self-diffusion coefficients of B, Ca, Na, Al, Zr, and Si in ISG at different temperatures [40].

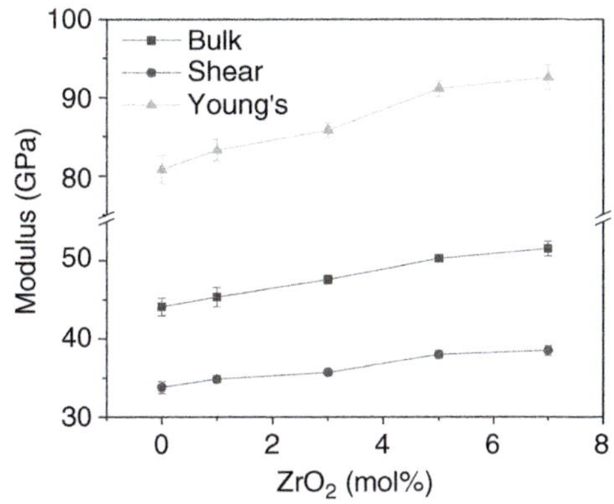


**Figure 12.38** Bulk, shear, and Young's modulus (GPa) obtained by MD simulation of $(75 - x)SiO_2$–$15Na_2O$–$10CaO$–$xZrO_2$, $x = 0, 1, 3, 5, 7$, glasses [217].

suggesting that $ZrO_2$ addition strengthens the glass structure and immobilizes ions for diffusion [40], which is consistent with the increased viscosity observed experimentally [208]. The strengthening of the network structure also explains the increase of Young's, shear and bulk moduli obtained by MD in the glasses with increasing zirconia content, as shown in Figure 12.38.

$ZrO_2$ has been reported to enhance the chemical durability of glasses not only in neutral pH condition but also in alkali-rich environments [198, 208, 210]. Fisher et al. [198] found that addition of up to 4 mol% of $ZrO_2$ in soda-lime silicate glasses dramatically decreases the leaching rates (tested for 28 days) of glasses in distilled water, while this effect saturates for glasses with higher $ZrO_2$ concentrations (up to 12 mol%), where no additional decrease of leaching rates was observed. However, even though the initial dissolution rate drops significantly with increasing $ZrO_2$ concentration, the inhibiting effect from $ZrO_2$ on the dynamics of reorganization of the alteration layer ultimately results in a greater degree of corrosion in the long term [32, 33, 213]. Cailleteau et al. [213] combined advanced experimental characterization with MC simulations (as shown in Figure 12.39) and found that glasses with a fast dissolution rate undergo a fast restructuring of the corroded layer into a passivating film, which rapidly stops corrosion by porosity closure. A study by Jollivet et al. [214] observed a presence of two layers in the gels of the $ZrO_2$-containing soda-lime borosilicate glass series altered at pH 1 and pH 7, where significant fraction of 7- and 8-coordinated Zr appear in bulk gel. This study highly suggests that glass dissolution is being driven by in situ condensation and dissolution-precipitation, where the pH has a greater influence on the structure of gels as compared to Zr content [214].

QSPR analysis for zirconia-containing borosilicate glasses was performed, and it was found that for glasses with different levels of zirconia content, properties such as density and initial dissolution rate have excellent linear correlations ($R^2 > 0.99$) with $F_{net}$ using an updated definition [39]. The initial dissolution rate measured by two separate groups with different conditions were correlated with $F_{net}$ and a linear negative correlation was found for both sets of experimental results (shown in Figure 12.40). Dissolution rate is a complex property, and it depends on both glass composition, glass structure, and external test conditions such as solution pH, solution composition, and temperature. Direct calculation of dissolution rate at atomic scale is not feasible due to the time scale involved. QSPR

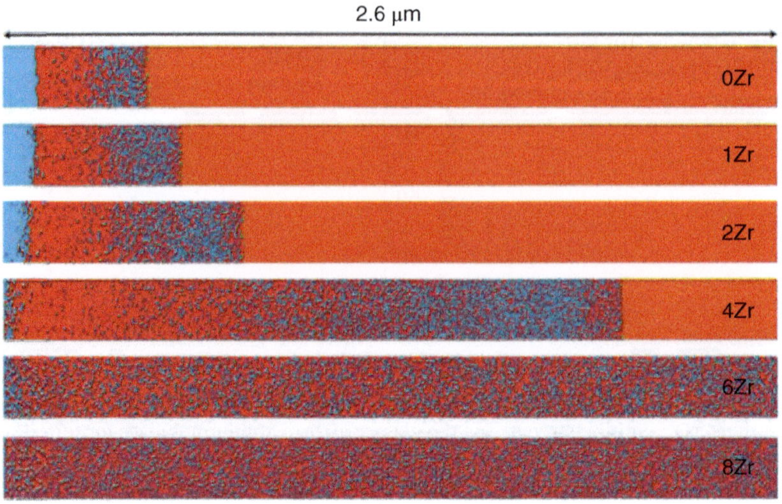


**Figure 12.39** Two-dimensional longitudinal cross sections of the simulated corrosion layer after $3 \times 10^5$ computer steps for $(61-x)SiO_2-17B_2O_3-18Na_2O-4CaO-xZrO_2$ with $x = 0, 1, 2, 4, 6$, and 8 glasses by Monte Carlo simulations. The silicon and boron lattice nodes are shown in red and yellow, respectively, and water in blue [213].

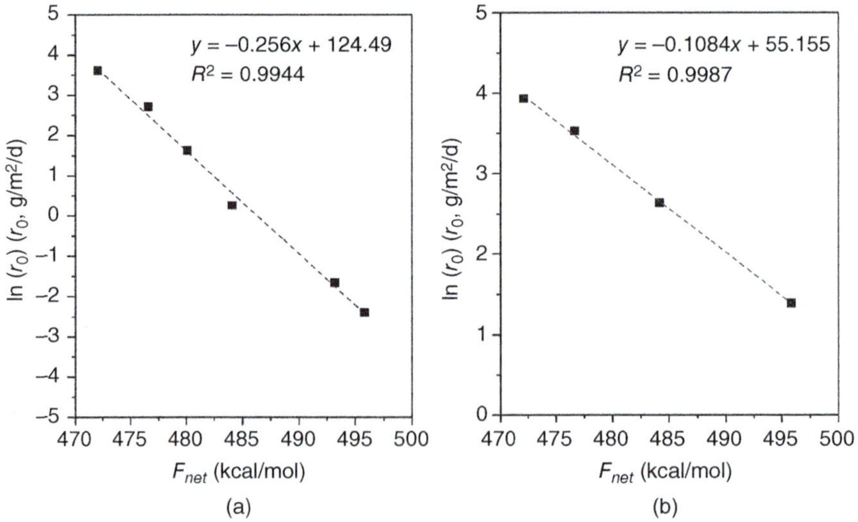


**Figure 12.40** QSPR analysis of initial dissolution rate of soda lime borosilicate glasses with different zirconia concentration. Correlation between the $F_{net}$ descriptor and two types of measurements of experimental initial dissolution rate ($r_0$) of $(61-x)SiO_2-17B_2O_3-18Na_2O-4CaO-xZrO_2$ with $x = 0, 1, 2, 4, 6$, and 8 glasses [39].

approach serves as an effective bridge between the property and structural descriptor ($F_{net}$). This structural descriptor can be modified and refined for different properties and glass systems [137]. Overall, coupling QSPR analysis and MD simulations are a promising approach to connect structural features with experimentally measured glass properties that can benefit both the understanding of composition–structure–property relations and design of new glass compositions for various technological applications.

## 12.4 Conclusions

In this chapter, we have reviewed the application of MD and related atomistic simulation methods to study TM and RE dope glasses. After summarizing the general structural features of these glasses revealed from experiments and glass chemistry, the simulation methodologies such as typical types of interatomic potentials and glass simulation procedures are introduced. The analysis approaches that are specific to the understanding of the local environment and clustering behaviors of these ions were then introduced. Methods for dynamic property calculations such as MSD, van Hove function are introduced while other properties such as electronic structure, optical absorption are introduced in the sections where relevant applications are discussed. Three case studies representing wide range of applications of these glasses are then introduced. The first one is the RE-doped silicate, phosphate, and mixed former glasses for optical and photonics applications. In this section, MD simulations silicate and phosphate glasses doped with RE ions such as $Er^{3+}$, $Eu^{3+}$, $Ce^{3+}$, $Ce^{4+}$, and $Pr^{3+}$ are reviewed. A detailed analysis of RE ions local environments and clustering of RE ions in these glasses from MD simulations are introduced. The effect of system size on clustering behaviors is discussed. Electronic structure and charge trapping in cerium-doped glass from first principles calculations is also introduced. The second case study is TM (vanadium)-containing alkali phosphate glasses as a mixed ionic and electronic conductor. Two vanadium oxidation states ($V^{4+}$ and $V^{5+}$) are investigated, and ionic transport of lithium and sodium in these glasses are studied by using MD simulations. The third case study focuses on TM (zirconium)-doped silicate and borosilicate glasses for nuclear waste disposal. The effect of zirconium addition to silicate glasses and borosilicate glasses is studied by using MD simulations. A comparison with experimental diffraction and EXAFS (Zr) is used to validate the structural models. Related techniques such as MD-based QSPR to correlation structural descriptor and initial correlation rate and MC simulations of glass dissolution are also introduced.

## Acknowledgments

JD and XL acknowledge the support of Center for Performance and Design of Nuclear Waste Forms and Containers (WastePD), an Energy Frontier Research Center funded by the US Department of Energy (DOE), Office of Science, Basic Energy Sciences under Award # DESC0016584. JD also acknowledges support of U.S. National Science Foundation (project #1662288).

## References


1 Mott, N.F. (1972). Introductory talk: Conduction in non-crystalline materials. *J. Non-Cryst. Solids* 8–10: 1–18. https://doi.org/10.1016/0022-3093(72)90112-3.

2 Moawad, H.M.M., Jain, H., and El-Mallawany, R. (2009). DC conductivity of silver vanadium tellurite glasses. *J. Phys. Chem. Solids* 70: 224–233. https://doi.org/10.1016/j.jpcs.2008.10.009.

3 Mott, N.F. (1990). *Metal Insulator Transistor*. London: Taylor & Francis.

4 Mercier, C., Palavit, G., Montagne, L., and Follet-Houttemane, C. (2002). A survey of transition-metal-containing phosphate glasses. *Comptes Rendus Chim.* 5: 693–703. https://doi.org/10.1016/S1631-0748(02)01437-6.

5 Khattak, G.D., Tabet, N., and Wenger, L.E. (2005). Structural properties of glasses in the series $(SrO)_x(V_2O_5)_{1-x}$, $(SrO)_{0.5-y}(B_2O_3)_y(V_2O_5)_{0.5}$, and $(SrO)_{0.2}(B_2O_3)_z(V_2O_5)_{0.8-z}$. *Phys. Rev. B: Condens. Matter Mater. Phys.* 72: 104203.1–104203.12.

6 Faiz, M., Mekki, A., Mun, B.S., and Hussain, Z. (2007). Investigation of vanadium–sodium silicate glasses using XANES spectroscopy. *J. Electron Spectros. Relat. Phenomena.* 154: 60–62. https://doi.org/10.1016/j.elspec.2006.11.006.

7 Mekki, A., Khattak, G.D., Holland, D. et al. (2003). Structure and magnetic properties of vanadium–sodium silicate glasses. *J. Non-Cryst. Solids* 318: 193–201. https://doi.org/10.1016/S0022-3093(02)01883-5.

8 Mugoni, C., Montorsi, M., Siligardi, C., and Jain, H. (2014). Electrical conductivity of copper lithium phosphate glasses. *J. Non-Cryst. Solids* 383: 137–140. https://doi.org/10.1016/j.jnoncrysol.2013.04.048.

9 Bausa, L.E., Jaque, F., Garcia Sole, J., and Duran, A. (1988). Photoluminescence of $Ti^{3+}$ in $P_2O_5$-$Na_2O$-$Al_2O_3$ glass. *J. Mater. Sci.* 23: 1921–1922. https://doi.org/10.1007/BF01115751.

10 Jamnick, M., Znfigik, P., Tunega, D., and Ingram, M.D. (1995). Glass formation and structure in the system $Cu_{20}$–$P_2O_5$–$MoO_3$. *J. Non-Cryst. Solids* 185: 151–158.

11 Satyanarayana, N., Govindaraj, G., and Karthikeyan, A. (1991). Effects of differing ratios of network modifier ($Ag_2O$) to network formers ($MoO_3$ + $V_2O_5$) and dopant salt (AgI) concentrations in silver-based superionic glassy compounds. *J. Non-Cryst. Solids* 136: 219–226.

12 Selvaraj, U. and Rao, K.J. (1985). Characterization studies of molybdophosphate glasses and a model of structural defects. *J. Non-Cryst. Solids* 72: 315–334. https://doi.org/10.1016/0022-3093(85)90187-5.

13 Minami, T., Imazawa, K., and Tanaka, M. (1980). Formation region and characterization of superionic conducting glasses in the systems AgI–$Ag_2O$–$M_xO_y$. *J. Non-Cryst. Solids* 42: 469–476. https://doi.org/10.1016/0022-3093(80)90045-9.

14 Kim, S.-H. and Yoko, T. (1995). Nonlinear optical properties of $TeO_2$-based glasses: $MO_x$–$TeO_2$ (M = Sc, Ti, V, Nb, Mo, Ta, and W) binary glasses. *J. Am. Ceram. Soc.* 78: 1061–1065. https://doi.org/10.1111/j.1151-2916.1995.tb08437.x.

15 Boolchand, P. (2000). *Insulating and Semiconducting Glasses*. World Scientific Publishing Co. Pte. Ltd. https://doi.org/10.1142/9789812813619.

16 Bhattacharya, S. and Ghosh, A. (2002). Polaron transport in semiconducting silver vanadate glasses. *Phys. Rev. B* 66: 132203. https://doi.org/10.1103/PhysRevB.66.132203.

17 Digonnet, M.J.F. (ed.) (2001). *Rare-Earth-Doped Fiber Lasers and Amplifiers, Revised and Expanded*. CRC Press https://doi.org/10.1201/9780203904657.

18 Urquhart, P. (1988). Review of rare earth doped fibre lasers and amplifiers. *IEE Proc. J. Optoelectron.* 135: 385. https://doi.org/10.1049/ip-j.1988.0071.

19 He, M., Jia, J., Zhao, J. et al. (2020). Glass-ceramic phosphors for solid state lighting: a review. *Ceram. Int.* 47: 2963–2980. https://doi.org/10.1016/j.ceramint.2020.09.227.

20 Peres, T., Litton, D.A., Capobinanco, J.A., and Garofalini, S.H. (1997). *J. Non-Cryst. Solids* 221 (1): 34.

21 Barbi, S., Mugoni, C., Montorsi, M. et al. (2017). Structural and optical properties of rare-earths doped barium bismuth borate glasses. *J. Non-Cryst. Solids* 481: 239–247. https://doi.org/10.1016/j.jnoncrysol.2017.10.043.

22 Chaussedent, S., Teboul, V., and Monteil, A. (2003). Molecular dynamics simulations of rare-earth-doped glasses. *Curr. Opin. Solid State Mater. Sci.* 7: 111–116.

23 Brawer, S.A. and Weber, M.J. (1980). Monte Carlo simulation of $Eu^{3+}$-doped $BeF_2$ Glass. *J. Chem. Inf. Model.* 53: 1689–1699. https://doi.org/10.1017/CBO9781107415324.004.

24 Soga, K., Inoue, H., and Makishima, A. (2000). Calculation and simulation of spectroscopic properties for rare earth ions in chloro-fluorozirconate glasses. *J. Non-Cryst. Solids* 274: 69–74. https://doi.org/10.1016/S0022-3093(00)00203-9.

25 Chaussedent, S., Bernard, C., Monteil, A. et al. (2000). Structural properties of erbium-activated silica-titania glasses: modeling by molecular dynamics method. In: *Proceedings of SPIE: Rare-Earth Doped Materials and Devices* (ed. S. Jiang), 243. Bellingham, WA: SPIE.

26 Montorsi, M., Broglia, G., and Mugoni, C. (2015). Structural insight into transition metal oxide containing glasses by molecular dynamic simulations BT – molecular dynamics simulations of disordered materials: from network glasses to phase-change memory alloys. In: *Molecular Dynamics Simulations of Disordered Materials: from Network Glasses to Phase-Change Memory Alloys* (ed. C. Massobrio, J. Du, M. Bernasconi and P.S. Salmon), 181–213. Cham: Springer International Publishing https://doi.org/10.1007/978-3-319-15675-0_8.

27 Ravikumar, R.V.S.S.N., Chandrasekhar, A.V., Ramamoorthy, L. et al. (2004). Spectroscopic studies of transition metal doped sodium phosphate glasses. *J. Alloys Compd.* 364: 176–179. https://doi.org/10.1016/S0925-8388(03)00498-5.

28 Brow, R.K. (2000). Review: the structure of simple phosphate glasses. *J. Non-Cryst. Solids* 263, 264: 1–28. https://doi.org/10.1016/S0022-3093(99)00620-1.

29 Quinn, C.J., Bell, G.H., and Dickinson, J.E. (1992). Alkali zinc pyrophosphate glasses for polymer blends. In: *Proceeding of XVI International Congress on Glass*, Madrid, Spain (4–9 October 1992), vol. 4 (ed. A. Durán), 79–82. Sociedad Española de Cerámica y Vidrio (Spanish Ceramic and Glass Society).

30 Day, D.E., Wu, Z., Ray, C.S., and Hrma, P. (1998). Chemically durable iron phosphate glass wasteforms. *J. Non-Cryst. Solids* 241: 1–12. https://doi.org/10.1016/S0022-3093(98)00759-5.

31 Vienna, J.D., Ryan, J.V., Gin, S., and Inagaki, Y. (2013). Current understanding and remaining challenges in modeling long-term degradation of borosilicate nuclear waste glasses. *Int. J. Appl. Glass Sci.* 4: 283–294. https://doi.org/10.1111/ijag.12050.

32 Cailleteau, C., Algeli, F., Devreux, F. et al. (2008). Insight into silicate-glass corrosion mechanisms. *Nat. Mater.* 7: 978–983. https://doi.org/10.1038/nmat2301.

**33** Bergeron, B., Galoisy, L., Jollivet, P. et al. (2010). First investigations of the influence of IVB elements (Ti, Zr, and Hf) on the chemical durability of soda-lime borosilicate glasses. *J. Non-Cryst. Solids* 356: 2315–2322. http://dx.doi.org/10.1016/j.jnoncrysol.2010.07.065.

**34** Lu, X., Neeway, J.J., Ryan, J.V., and Du, J. (2016). Influence of low concentration V and Co oxide doping on the dissolution behaviors of simplified nuclear waste glasses. *J. Non-Cryst. Solids* 452: https://doi.org/10.1016/j.jnoncrysol.2016.08.026.

**35** Vienna, J.D. (2010). Nuclear waste vitrification in the United States: recent developments and future options. *Int. J. Appl. Glass Sci.* 3: 309–321. https://doi.org/10.1111/j.20411294.2010.00023.x.

**36** Gin, S., Abdelouas, A., Criscenti, L.J. et al. (2013). An international initiative on long-term behavior of high-level nuclear waste glass. *Mater. Today* 16: 243–248. https://doi.org/10.1016/j.mattod.2013.06.008.

**37** Short, R.J., Hand, R.J., and Hyatt, N.C. (2002). Molybdenum in nuclear waste glasses – incorporation and redox state. *MRS Proc.* 757: II5.4. https://doi.org/10.1557/PROC-757-II5.4.

**38** Kamat, H., Arias-Serrano, B.I., Yaremchenko, A., and Goel, A. (2020). Ruthenium solubility and its impact on the crystallization behavior and electrical conductivity of $MoO_3$-containing borosilicate-based model high-level nuclear waste glasses. *J. Non-Cryst. Solids* 549: 120356. https://doi.org/10.1016/j.jnoncrysol.2020.120356.

**39** Lu, X., Deng, L., Gin, S., and Du, J. (2019). Quantitative structure-property relationship (QSPR) analysis of $ZrO_2$-containing soda-lime borosilicate glasses. *J. Phys. Chem. B* 123: 1412–1422. https://doi.org/10.1021/acs.jpcb.8b11108.

**40** Lu, X., Deng, L., Kerisit, S., and Du, J. (2018). Structural role of $ZrO_2$ and its impact on properties of boroaluminosilicate nuclear waste glasses. *npj Mater. Degrad.* 2: https://doi.org/10.1038/s41529-018-0041-6.

**41** Bruce, A.J., Grodkiewicz, W.H., and "Lanthanide-containing Oxide Glass," pp. (1996). *21 in Ceramic Transaction*, vol. 67 (ed. B.G. Potter Jr., and A.J. Bruce). Cincinnati, OH: American Ceramic Society.

**42** Clare, A.G. (1994). Rare earth in glasses for laser applications. In: *Rare Elements in Glasses* (ed. J.E. Shelby), 161. Uetikon-Zuerich, Switzerland: Trans Tech Publications.

**43** Miniscalco, W.J. (1991). Erbium doped glasses for fiber amplifiers at 1500 nm. *J. Lightwave. Technol.* 9 (2): 234–250.

**44** Neeves, A.E., Bruce, A.J., Reed, W.A. et al. (1992). Effect of concentration, host glass composition, and processing on the absorption and emission of $Er^{3+}$ ions in sodium silicate glasses. In: *Solid State Optical Materials* (ed. A.J. Bruce and B.V. Hiremath), 353. Westerville, OH: American Ceramic Society.

**45** France, P.W. (1991). *Optical Fiber Lasers and Amplifiers*, 51. Heidelberg, Netherlands: Springer.

**46** Quimby, R.S., Miniscalco, W.J., and Thompson, B. (1994). Clustering in erbium-doped silica glass fibers analysed using 980 nm excited-state absorption. *J. Appl. Phys.* 76 (8): 4472–4478.

**47** Arai, K., Namikawa, H., Kumata, K., and Honda, T. (1985). Aluminum or phosphorus co-doping effects on the fluorescence and structural properties of neodymium-doped silica glass. *J. Appl. Phys.* 59 (10): 3430–3436.

**48** Marcus, M.A. and Polman, A. (1991). Local structure around Er in silica and sodium silicate glasses. *J. Non-Cryst. Solids* 136 (3): 260–265.

**49** Peters, P.M. and Houde-Walter, S.N. (1997). X-ray absorption fine structure determined of the local environment of $Er^{3+}$ in glass. *Appl. Phys. Lett.* 70 (5): 541–543.

**50** Sen, S. and Stebbins, J.F. (1995). Structural role of $Nd^{3+}$ and $Al^{3+}$ cations in $SiO_2$ glass: a $^{29}$Si MAS-NMR spin-lattice relaxation, $^{27}$Al NNR and EPR study. *J. Non-Cryst. Solids* 188 (1-2): 54–62.

**51** Sen, S. (2000). Atomic environment of high-field strength Nd and Al cations as dopants and major components in silicate glasses: a Nd LIII-edge and Al K-edge X-ray absorption spectroscopic study. *J. Non-Cryst. Solids* 261 (1-3): 226–236.

**52** Moawad, H.M.M., Jain, H., El-Mallawany, R. et al. (2004). Electrical conductivity of silver vanadium tellurite glasses. *J. Am. Ceram. Soc.* 85: 2655–2659. https://doi.org/10.1111/j.1151-2916.2002.tb00511.x.

**53** Karabulut, M., Marasinghe, G.K., Ray, C.S. et al. (2002). An investigation of the local iron environment in iron phosphate glasses having different Fe(II) concentrations. *J. Non-Cryst. Solids* 306: 182–192. https://doi.org/10.1016/S0022-3093(02)01053-0.

**54** Ehrt, D., Leister, M., Matthai, C. et al. (1997). Determination of the redox states of vanadium in glasses and melts by different methods. *Proc. Fundam. Glass Sci. Technol. Conf.* 204–211.

**55** Cozar, O. and Ardelean, I. (n.d.). The local symmetry of $Cu^{2+}$ ions in phosphate glasses. *J. Non-Cryst. Solids* 92: 278–281.

**56** Bae, B.-S. and Weinberg, M.C. (1993). Ultraviolet optical absorptions of semiconducting copper phosphate glasses. *J. Appl. Phys.* 73: 7760. https://doi.org/10.1063/1.353975.

**57** Fang, X., Ray, C.S., Moguš-Milanković, A., and Day, D.E. (2001). Iron redox equilibrium, structure and properties of iron phosphate glasses. *J. Non-Cryst. Solids* 283: 162–172. https://doi.org/10.1016/S0022-3093(01)00416-1.

**58** Majumdar, R. and Lahiri, D. (1972). Redox equilibrium and spectral absorption of some colouring oxides in sodium metaphosphate glass. *Trans. Ind. Ceram. Soc.* XXXI: 121–126.

**59** Calas, G. and Petiau, J. (1983). *Structure of Non-crystalline Materials II*. London: Taylor & Francis.

**60** Calas, G. and Petiau, J. (1983). Structure of oxide glasses: spectroscopic studies of local order and crystallochemistry. Geochemical implications. *Bull Miner.* 106.

**61** Tricot, G. and Vezin, H. (2013). Description of the intermediate length scale structural motifs in sodium vanado-phosphate glasses by magnetic resonance spectroscopies. *J. Phys. Chem. C* 117: 1421–1427.

**62** Metwalli, E., Karabulut, M., Sidebottom, D.L. et al. (2004). Properties and structure of copper ultraphosphate glasses. *J. Non-Cryst. Solids* 344: 128–134. https://doi.org/10.1016/j.jnoncrysol.2004.07.058.

**63** Moringa, K., Yoshida, H., and Takebe, H. (1994). Compositional dependence of absorption spectra of $Ti^{3+}$ in silicate, borate, and phosphate glasses. *J. Am. Ceram. Soc.* 77: 3113–3118. https://doi.org/10.1111/j.1151-2916.1994.tb04557.x.

**64** Murata, T., Torisaka, M., Takebe, H., and Morinaga, K. (1997). Compositional dependence of the valency state of Cr ions in oxide glasses. *J. Non-Cryst. Solids* 220: 139–146. https://doi.org/10.1016/S0022-3093(97)00264-0.

**65** Chaurand, P., Rose, J., Briois, V. et al. (2007). New methodological approach for the vanadium K-edge X-ray absorption near-edge structure interpretation: application to the speciation of vanadium in oxide phases from steel slag. *J. Phys. Chem. B* 111: 5101–5110. https://doi.org/10.1021/jp063186i.

**66** Bogomolova, L.D., Glassova, M.P., Dubatovko, O.E. et al. (1983). The study of interactions between iron and vanadium ions in semiconducting barium-vanadate glasses doped with $Fe_2O_3$. *J. Non-Cryst. Solids* 58: 71–89. https://doi.org/10.1016/0022-3093(83)90104-7.

**67** Bae, B.-S. and Weinberg, M.C. (1991). Oxidation-reduction equilibrium in copper phosphate glass melted in air. *J. Am. Ceram. Soc.* 74: 3039–3045. https://doi.org/10.1111/j.1151-2916.1991.tb04299.x.

**68** Brown, G.E., Calas, G., Waychunas, G.A., and Petiau, J. (1988). X-ray absorption spectroscopy; applications in mineralogy and geochemistry. *Rev. Mineral. Geochem.* 18: 431–512.

**69** Wong, J., Lytle, F., Messmer, R., and Maylotte, D. (1984). K-edge absorption spectra of selected vanadium compounds. *Phys. Rev. B* 30: 5596–5610.

**70** Farges, F., Lefrère, Y., Rossano, S. et al. (2004). The effect of redox state on the local structural environment of iron in silicate glasses: a combined XAFS spectroscopy, molecular dynamics, and bond valence study. *J. Non-Cryst. Solids* 344: 176–188. https://doi.org/10.1016/j.jnoncrysol.2004.07.050.

**71** Musinu, A. and Piccaluga, G. (1995). An X-ray diffraction study of the short-range order around Ni(II) Zn(II) and Cu(II) in pyrophosphate glasses. *J. Non-Cryst. Solids* 193: 32–35.

**72** Cormier, L., Galoisy, L., Delaye, J.-M. et al. (2001). Short- and medium-range structural order around cations in glasses: a multidisciplinary approach. *C.R. Acad. Sci., Ser. Ser. IV Phys. Astrophys.* 2: 249–262. https://doi.org/10.1016/S1296-2147(01)01168-4.

**73** Ravikumar, R.V.S.S.N., Reddy, V.R., Chandrasekhar, A.V. et al. (2002). Tetragonal site of transition metal ions doped sodium phosphate glasses. *J. Alloys Compd.* 337: 272–276. https://doi.org/10.1016/S0925-8388(01)01963-6.

**74** Farrow, L.A. and Vogel, E.M. (1992). Raman spectra of phosphate and silicate glasses doped with the cations Ti, Nb and Bi. *J. Non-Cryst. Solids* 143: 59–64. https://doi.org/10.1016/S0022-3093(05)80553-8.

**75** Cozar, O., Ardelean, I., Simon, V., and David, L. (1999). The local structure and interactions between $V^{4+}$ ions in soda-phosphate glasses. *Appl. Magn. Reson.* 537: 529–537.

**76** Bruni, S., Cariati, F., Corrias, A. et al. (1995). Short range order of sodium-zinc, sodium-copper, and sodium-nickel pyrophosphate glasses by diffractometric and spectroscopic techniques. *J. Phys. Chem.* 99: 15229–15235. https://doi.org/10.1021/j100041a044.

**77** Khattak, G.D., Mekki, A., and Wenger, L.E. (2009). X-ray photoelectron spectroscopy (XPS) and magnetic susceptibility studies of vanadium phosphate glasses. *J. Non-Cryst. Solids* 355: 2148–2155. https://doi.org/10.1016/j.jnoncrysol.2009.06.042.

**78** Majjane, A., Chahine, A., Et-tabirou, M. et al. (2014). X-ray photoelectron spectroscopy (XPS) and FTIR studies of vanadium barium phosphate glasses. *Mater. Chem. Phys.* 143: 779–787. https://doi.org/10.1016/j.matchemphys.2013.10.013.

**79** Albon, C., Muresan, D., Vandenberghe, R.E., and Simon, S. (2008). Iron environment in calcium-soda-phosphate glasses and vitroceramics. *J. Non-Cryst. Solids* 354: 4603–4608. https://doi.org/10.1016/j.jnoncrysol.2008.06.008.

**80** Shih, P.Y., Ding, J.Y., and Lee, S.Y. (2003). 31P MAS-NMR and FTIR analyses on the structure of CuO-containing sodium poly- and meta-phosphate glasses. *Mater. Chem. Phys.* 80: 391–396. https://doi.org/10.1016/S0254-0584(03)00098-1.

**81** Bianconi, A., Giovannelli, A., Dovoli, I. et al. (1982). Xanes (X-ray absorption near edge structure) of V in vanadium-iron phosphate glasses. *Solid State Commun.* 42: 547–551. https://doi.org/10.1016/0038-1098(82)90605-6.

**82** Concas, G., Congiu, F., Manca, E. et al. (1995). Mössbauer spectroscopic investigation of some iron-containing sodium phosphate glasses. *J. Non-Cryst. Solids* 192, 193: 175–178. https://doi.org/10.1016/0022-3093(95)00347-9.

**83** Bingham, P.A. and Barney, E.R. (2012). Structure of iron phosphate glasses modified by alkali and alkaline earth additions: neutron and X-ray diffraction studies. *J. Phys. Condens. Matter.* 24: 175403. https://doi.org/10.1088/0953-8984/24/17/175403.

**84** Chandra, S. (1981). *Superionic Solids: Principles and Applications*. Amsterdam: North-Holland.

**85** Singh, K. (1999). Ion conducting glasses for solid state electrochemical applications. *Indian J. Pure Appl. Phys.* 37: 266–271.

**86** Musinu, A., Piccaluga, G., Pinna, G. et al. (1991). Coordination of zinc and copper in phosphate glasses by EXAFS. *J. Non-Cryst. Solids* 136: 198–204. https://doi.org/10.1016/0022-3093(91)90490-W.

**87** Debnath, R., Chaudhury, J., and Bera, S.C. (1990). Optical properties and nature of coordination of $Cu^+$ ions in calcium metaphosphate glass. *Phys. Status Solidi.* 157: 723–733. https://doi.org/10.1002/pssb.2221570225.

**88** Du, J., Kokou, L., Rygel, J.L. et al. (2011). Structure of cerium phosphate glasses: molecular dynamics simulation. *J. Am. Ceram. Soc.* 94: 2393–2401. https://doi.org/10.1111/j.1551-2916.2011.04514.x.

**89** Martin, R.A., Mountjoy, G., and Newport, R.J. (2009). A molecular dynamics model of the atomic structure of dysprosium alumino-phosphate glass. *J. Phys. Condens. Matter.* 21: 075102. https://doi.org/10.1088/0953-8984/21/7/075102.

**90** Pedone, A., Malavasi, G., and Menziani, M.C. (2009). Computational insight into the effect of CaO/MgO substitution on the structural properties of phospho-silicate bioactive glasses. *J. Phys. Chem. C* 113: 15723–15730. https://doi.org/10.1021/jp904131t.

**91** Malavasi, G., Pedone, A., and Menziani, M.C. (2013). Study of the structural role of gallium and aluminum in 45S5 bioactive glasses by molecular dynamics simulations. *J. Phys. Chem. B* 117: 4142–4150. https://doi.org/10.1021/jp400721g.

**92** Lusvardi, G., Malavasi, G., Menabue, L. et al. (2005). A computational tool for the prediction of crystalline phases obtained from controlled crystallization of glasses. *J. Phys. Chem. B* 109: 21586–21592. https://doi.org/10.1021/jp0546857.

**93** Tilocca, A., Cormack, A.N., and De Leeuw, N.H. (2007). The structure of bioactive silicate glasses: new insight from molecular dynamics simulations. *Chem. Mater.* 95–103.

**94** Lin, C.-C., Chen, S.-F., Leung, K.S., and Shen, P. (2012). Effects of $CaO/P_2O_5$ ratio on the structure and elastic properties of $SiO_2$–CaO–$Na_2O$–$P_2O_5$ bioglasses. *J. Mater. Sci. Mater. Med.* 23: 245–258. https://doi.org/10.1007/s10856-011-4504-3.

**95** Clark, E.B., Mead, R.N., and Mountjoy, G. (2006). A molecular dynamics model of the atomic structure of Tb metaphosphate glass $(Tb_2O_3)_{0.25}(P_2O_5)_{0.75}$. *J. Phys. Condens. Matter.* 18: 6815–6826. https://doi.org/10.1088/0953-8984/18/29/019.

**96** Lusvardi, G., Malavasi, G., Cortada, M. et al. (2008). Elucidation of the structural role of fluorine in potentially bioactive glasses by experimental and computational investigation. *J. Phys. Chem. B* 112: 12730–12739. https://doi.org/10.1021/jp803031z.

**97** Karakassides, M.A., Saranti, A., and Koutselas, I. (2004). Preparation and structural study of binary phosphate glasses with high calcium and/or magnesium content. *J. Non-Cryst. Solids* 347: 69–79. https://doi.org/10.1016/j.jnoncrysol.2004.08.111.

**98** Linati, L., Lusvardi, G., Malavasi, G. et al. (2008). Medium-range order in phospho-silicate bioactive glasses: insights from MAS-NMR spectra, chemical durability experiments and molecular dynamics simulations. *J. Non-Cryst. Solids* 354: 84–89. https://doi.org/10.1016/j.jnoncrysol.2007.06.076.

**99** Tilocca, A. and Cormack, A.N. (2007). Structural effects of phosphorus inclusion in bioactive silicate glasses. *J. Phys. Chem. B* 111: 14256–14264. https://doi.org/10.1021/jp075677o.

**100** Xiang, Y. and Du, J. (2011). Effect of strontium substitution on the structure of 45S5 bioglasses. *Chem. Mater.* 23: 2703–2717. https://doi.org/10.1021/cm102889q.

**101** Tilocca, A., Cormack, A., and de Leeuw, N. (2007). The formation of nanoscale structures in soluble phosphosilicate glasses for biomedical applications: MD simulations. *Faraday Discuss.* 136: 44–55.

**102** Di Tommaso, D., Ainsworth, R.I., Tang, E., and de Leeuw, N.H. (2013). Modelling the structural evolution of ternary phosphate glasses from melts to solid amorphous materials. *J. Mater. Chem. B* 1: 5054. https://doi.org/10.1039/c3tb20662a.

**103** Boiko, G.G., Andreev, N.S., and Parkachev, A.V. (1998). Structure of pyrophosphate $2ZnO{\cdot}P_2O_5$–$2Na_2O{\cdot}P_2O_5$ glasses according to molecular dynamics simulation. *J. Non-Cryst. Solids* 238: 175–185. https://doi.org/10.1016/S0022-3093(98)00693-0.

**104** Tischendorf, B.C., Alam, T.M., Cygan, R.T., and Otaigbe, J.U. (2003). The structure and properties of binary zinc phosphate glasses studied by molecular dynamics simulations. *J. Non-Cryst. Solids* 316: 261–272. https://doi.org/10.1016/S0022-3093(02)01795-7.

**105** Al-Hasni, B. and Mountjoy, G. (2011). Structural investigation of iron phosphate glasses using molecular dynamics simulation. *J. Non-Cryst. Solids* 357: 2775–2779. https://doi.org/10.1016/j.jnoncrysol.2010.10.010.

**106** Al-Hasni, B.M., Mountjoy, G., and Barney, E. (2013). A complete study of amorphous iron phosphate structure. *J. Non-Cryst. Solids* 380: 141–152. https://doi.org/10.1016/j.jnoncrysol.2013.08.031.

**107** Vessal, B. (1994). Simulation studies of silicates and phosphates. *J. Non-Cryst. Solids* 177: 103–124. https://doi.org/10.1016/0022-3093(94)90520-7.

**108** Liang, J.-J., Cygan, R., and Alam, T. (2000). Molecular dynamics simulation of the structure and properties of lithium phosphate glasses. *J. Non-Cryst. Solids* 263, 264: 167–179. https://doi.org/10.1016/S0022-3093(99)00632-8.

**109** Karthikeyan, A., Vinatier, P., Levasseur, A., and Rao, K.J. (1999). The molecular dynamics study of lithium ion conduction in phosphate glasses and the role of non-bridging oxygen. *J. Phys. Chem. B* 103: 6185–6192. https://doi.org/10.1021/jp990228c.

**110** Pedone, A., Malavasi, G., Menziani, M.C. et al. (2006). A new self-consistent empirical interatomic potential model for oxides, silicates, and silica-based glasses. *J. Phys. Chem. B* 110: 11780–11795. https://doi.org/10.1021/jp0611018.

**111** Du, J., René Corrales, L., Tsemekhman, K., and Bylaska, E.J. (2007). Electron, hole and exciton self-trapping in germanium doped silica glass from DFT calculations with self-interaction correction. *Nucl. Instrum. Methods Phys. Res., Sect. B* https://doi.org/10.1016/j.nimb.2006.11.066.

**112** Garofalini, S. (2002). Molecular dynamics simulations of Li transport between cathode crystals. *J. Power Sources* 110: 412–415. https://doi.org/10.1016/S0378-7753(02)00205-7.

**113** Ori, G., Montorsi, M., Pedone, A., and Siligardi, C. (2011). Insight into the structure of vanadium containing glasses: a molecular dynamics study. *J. Non-Cryst. Solids* 357: 2571–2579. https://doi.org/10.1016/j.jnoncrysol.2011.02.002.

**114** Du, J. and Cormack, A.N. (2004). The medium range structure of sodium silicate glasses: a molecular dynamics simulation. *J. Non-Cryst. Solids* 349: 66–79. https://doi.org/10.1016/j.jnoncrysol.2004.08.264.

**115** Du, J. and Kokou, L. (2011). Europium environment and clustering in europium doped silica and sodium silicate glasses. *J. Non-Cryst. Solids* 357: 2235–2240. https://doi.org/10.1016/j.jnoncrysol.2010.11.088.

**116** Johnson, J.A., Benmore, C.J., Holland, D. et al. (2011). https://doi.org/10.1088/0953-8984/23/6/065404). Influence of rare-earth ions on $SiO_2$–$Na_2O$–$RE_2O_3$ glass structure. *J. Phys. Condens. Matter.* 065404.

**117** Deng, L. and Du, J. (2019). Development of boron oxide potentials for computer simulations of multicomponent oxide glasses. *J. Am. Ceram. Soc.* 102: 2482–2505. https://doi.org/10.1111/jace.16082.

**118** Du, J. and Cormack, A.N. (2005). The structure of erbium doped sodium silicate glasses. *J. Non-Cryst. Solids* 351: 2263–2276. https://doi.org/10.1016/j.jnoncrysol.2005.05.018.

**119** Broglia, G., Ori, G., Larcher, L., and Montorsi, M. (2014). Molecular dynamics simulation of amorphous $HfO_2$ for resistive RAM applications. *Model. Simul. Mater. Sci. Eng.* 22: https://doi.org/10.1088/0965-0393/22/6/065006.

**120** Hansen, J.-P. and McDonald, I.R. (1990). *Theory of Simple Liquids*. Academic Press.

**121** Côté, A.S., Smith, B., and Lindan, P.J.D. (1995). *Democritus* 51: 4626–4641.

**122** Kob, W. (1995). Testing mode-coupling theory for a supercooled binary Lennard-Jones mixture I: the van Hove correlation function. *Phys. Rev. E* 51: 4626–4641. https://doi.org/10.1103/PhysRevE.51.4626.

**123** Katritzky, A.R., Sild, S., and Karelson, M. (1998). Correlation and prediction of the refractive indices of polymers by QSPR. *J. Chem. Inf. Comput. Sci.* 38: 1171–1176. https://doi.org/10.1021/ci980087w.

**124** Xu, J., Chen, B., Zhang, Q., and Guo, B. (2004). Prediction of refractive indices of linear polymers by a four-descriptor QSPR model. *Polymer (Guildf).* 45: 8651–8659. https://doi.org/10.1016/j.polymer.2004.10.057.

**125** Xu, J., Liang, H., Chen, B. et al. (2008). Linear and nonlinear QSPR models to predict refractive indices of polymers from cyclic dimer structures. *Chemom. Intell. Lab. Syst.* 92: 152–156. https://doi.org/10.1016/j.chemolab.2008.02.006.

**126** Duchowicz, P.R., Fioressi, S.E., Bacelo, D.E. et al. (2015). QSPR studies on refractive indices of structurally heterogeneous polymers. *Chemom. Intell. Lab. Syst.* 140: 86–91. https://doi.org/10.1016/j.chemolab.2014.11.008.

**127** Katritzky, A.R., Sild, S., Lobanov, V., and Karelson, M. (1998). Quantitative structure - property relationship (QSPR) correlation of glass transition temperatures of high molecular weight polymers. *J. Chem. Inf. Model.* 38: 300–304. https://doi.org/10.1021/ci9700687.

**128** Afantitis, A., Melagraki, G., Makridima, K. et al. (2005). Prediction of high weight polymers glass transition temperature using RBF neural networks. *J. Mol. Struct. Theochem.* 716: 193–198. https://doi.org/10.1016/j.theochem.2004.11.021.

**129** Yu, X., Yi, B., Wang, X., and Xie, Z. (2007). Correlation between the glass transition temperatures and multipole moments for polymers. *Chem. Phys.* 332: 115–118. https://doi.org/10.1016/j.chemphys.2006.11.029.

**130** Afantitis, A., Melagraki, G., Sarimveis, H. et al. (2006). Prediction of intrinsic viscosity in polymer-solvent combinations using a QSPR model. *Polymer (Guildf).* 47: 3240–3248. https://doi.org/10.1016/j.polymer.2006.02.060.

**131** Gharagheizi, F. (2007). QSPR analysis for intrinsic viscosity of polymer solutions by means of GA-MLR and RBFNN. *Comput. Mater. Sci.* 40: 159–167. https://doi.org/10.1016/j.commatsci.2006.11.010.

**132** Voigt, U., Lammert, H., Eckert, H., and Heuer, A. (2005). Cation clustering in lithium silicate glasses: quantitative description by solid-state NMR and molecular dynamics simulations. *Phys. Rev. B: Condens. Matter Mater. Phys.* 72: 1–11. https://doi.org/10.1103/PhysRevB.72.064207.

**133** Malavasi, G., Pedone, A., and Menziani, M.C. (2006). Towards a quantitative rationalization of multicomponent glass properties by means of molecular dynamics simulations. *Mol. Simul.* 32: 1045–1055. https://doi.org/10.1080/08927020600932793.

**134** Christie, J.K. and Tilocca, A. (2012). Molecular dynamics simulations and structural descriptors of radioisotope glass vectors for in situ radiotherapy. *J. Phys. Chem. B* 116: 12614–12620. https://doi.org/10.1021/jp304200f.

**135** Pedone, A. and Menziani, M.C. (2015). Computational modeling of silicate glasses: a quantitative structure-property relationship perspective. *Springer Ser. Mater. Sci.* 215: 113–135. https://doi.org/10.1007/978-3-319-15675-0_5.

**136** Pedone, A., Chen, X., Hill, R.G., and Karpukhina, N. (2018). Molecular dynamics investigation of halide-containing phospho-silicate bioactive glasses. *J. Phys. Chem. B* 122: 2940–2948. https://doi.org/10.1021/acs.jpcb.8b00547.

**137** Lu, X. and Du, J. (2020). Quantitative structure-property relationship (QSPR) analysis of calcium aluminosilicate glasses based on molecular dynamics simulations. *J. Non-Cryst. Solids* 530: https://doi.org/10.1016/j.jnoncrysol.2019.119772.

**138** Le, T., Epa, V.C., Burden, F.R., and Winkler, D.A. (2012). Quantitative structure-property relationship modeling of diverse materials properties. *Chem. Rev.* 112: 2889–2919. https://doi.org/10.1021/cr200066h.

**139** Lusvardi, G., Malavasi, G., Tarsitano, F. et al. (2009). Quantitative structure-property relationships of potentially bioactive fluoro phospho-silicate glasses. *J. Phys. Chem. B* 113: 10331–10338. https://doi.org/10.1021/jp809805z.

**140** Sun, W. and Du, J. (2017). Structural stability, electronic and thermodynamic properties of VOPO4 polymorphs from DFT+U calculations. *Comput. Mater. Sci.* 126: 326–335. https://doi.org/10.1016/j.commatsci.2016.

**141** Quimby, R.S., Miniscalco, W.J., and Thompson, B. (1994). Clustering in erbium-doped silica glass fibers analyzed using 980 nm excited-state absorption. *J. Appl. Phys.* 76: 4472–4478. https://doi.org/10.1063/1.357278.

**142** Du, J. and Cormack, A.N. (2002). Structure study of rare earth doped vitreous silica by molecular dynamics simulation. *Radiat. Eff. Defects Solids* 157: 789–794. https://doi.org/10.1080/10420150215841.

**143** Polman, A., Snoeks, E., van den Hoven, G.N. et al. (1995). Ion beam synthesis of planar opto-electronic devices. *Nucl. Inst. Methods Phys. Res. B* 106: 393–399. https://doi.org/10.1016/0168-583X(95)00738-5.

**144** Devanathan, R., Weber, W.J., and Gale, J.D. (2010). Radiation tolerance of ceramics – insights from atomistic simulation of damage accumulation in pyrochlores. *Energy Environ. Sci.* 3: 1551. https://doi.org/10.1039/c0ee00066c.

**145** Devanathan, R., Durham, P., Du, J. et al. (2007). Molecular dynamics simulation of amorphization in forsterite by cosmic rays. *Nucl. Instrum. Methods Phys. Res., Sect. B* https://doi.org/10.1016/j.nimb.2006.11.021.

**146** Du, J. and Corrales, L.R. (2007). Erbium implantation in silica studied by molecular dynamics simulations. *Nucl. Instrum. Methods Phys. Res., Sect. B* 255: https://doi.org/10.1016/j.nimb.2006.11.065.

**147** Kokou, L. and Du, J. (2012). Rare earth ion clustering behavior in europium doped silicate glasses: simulation size and glass structure effect. *J. Non-Cryst. Solids* 358: 3408–3417. https://doi.org/10.1016/j.jnoncrysol.2012.01.068.

**148** Nienhuis, E.T., Tuheen, M., Du, J., and McCloy, J.S. (2021). In situ pair distribution function analysis of crystallizing Fe-silicate melts. *J. Mater. Sci.* 56: 5637–5657. https://doi.org/10.1007/s10853-020-05643-x.

**149** Rygel, J.L., Chen, Y., Pantano, C.G. et al. (2011). Local structure of cerium in aluminophosphate and silicophosphate glasses. *J. Am. Ceram. Soc.* 94: 2442–2451. https://doi.org/10.1111/j.1551-2916.2011.04596.x.

**150** Kokou, L. and Du, J. (2014). Short- and medium-range structures of cerium aluminophosphate glasses: a molecular dynamics study. *J. Non-Cryst. Solids* 403: 67–79. https://doi.org/10.1016/j.jnoncrysol.2014.07.014.

**151** Deskins, N.A., Du, J., and Rao, P. (2017). The structural and electronic properties of reduced amorphous titania. *Phys. Chem. Chem. Phys.* 19: https://doi.org/10.1039/c7cp02940c.

**152** Griscom, D.L. (1978). Defects in amorphous insulators. *J. Non-Cryst. Solids* 31: 241–266. https://doi.org/10.1016/0022-3093(78)90107-2.

**153** Griscom, D.L., Friebele, E.J., Long, K.J., and Fleming, J.W. (1983). Fundamental defect centers in glass: electron spin resonance and optical absorption studies of irradiated phosphorus-doped silica glass and optical fibers. *J. Appl. Phys.* 54: 3743–3762. https://doi.org/10.1063/1.332591.

**154** Du, J., Kokou, L., Li, Y. (2010). Physical Modeling of Cerium Doped Cover Glasses. Technical Report.

**155** Vlahović, F., Perić, M., Gruden-Pavlović, M., and Zlatar, M. (2015). Assessment of TD-DFT and LF-DFT for study of d–d transitions in first row transition metal hexaaqua complexes. *J. Chem. Phys.* 142: 214111. https://doi.org/10.1063/1.4922111.

**156** Ziegler, T. (2012). A chronicle about the development of electronic structure theories for transition metal complexes. *Struct. Bond.* 143: 1–38. https://doi.org/10.1007/430_2011_47.

**157** Inoue, H., Soga, K., and Makishima, A. (2002). Simulation of the optical properties of Tm:ZBLAN glass. *J. Non-Cryst. Solids* 306: 17–29. https://doi.org/10.1016/S0022-3093(02)01059-1.

**158** Gupta, S., Khanijo, N., and Mansingh, A. (1995). The influence of $V^{4+}$ ion concentration on the EPR spectra of vanadate glasses. *J. Non-Cryst. Solids* 181: 58–63. https://doi.org/10.1016/0022-3093(94)00485-4.

**159** Roling, B. and Funke, K. (1997). Polaronic transport in vanadium phosphate glasses. *J. Non-Cryst. Solids* 212: 1–10. https://doi.org/10.1016/S0022-3093(96)00557-1.

**160** Nabavi, M., Sanchez, C., and Livage, J. (1991). Structure and properties of amorphous $V_2O_5$. *Philos. Mag. B* 63: 941–953. https://doi.org/10.1080/13642819108205549.

**161** Jordan, B. and Calvo, C. (1977). Transport properties and the structure of vanadium phosphate glasses. *Can. J. Phys.* 55: 436–441.

**162** Hoppe, U. and Kranold, R. (1999). A reverse Monte Carlo study of the structure of vitreous $V_2O_5$. *Solid State Commun.* 109: 625–630. https://doi.org/10.1016/S0038-1098(98)00630-9.

**163** Hoppe, U., Wyckoff, N.P., Schmitt, M.L. et al. (2012). Structure of $V_2O_5$–$P_2O_5$ glasses by X-ray and neutron diffraction. *J. Non-Cryst. Solids* 358: 328–336. https://doi.org/10.1016/j.jnoncrysol.2011.09.038.

**164** Murawski, L., Barczynski, R.J., and Rybicka, A. (1997). $V_2O_5$–$P_5O_5$ glass and its polaron transport properties derived from molecular dynamic simulations of structure. In: *Dielectric and Related Phenomena: Materials Physico-Chemistry, Spectrometric Investigations, and Applications* (ed. A. Wlochowicz), 136–141. International Society for Optics and Photonics https://doi.org/10.1117/12.276267.

**165** Mosset, A., Lecante, P., Galy, J., and Livage, J. (1982). Structural analysis of amorphous $V_2O_5$ by large-angle X-ray scattering. *Philos. Mag. B* 46: 137–149.

**166** Barczyński, R.J., Król, P., and Murawski, L. (2010). Ac and dc conductivities in $V_2O_5$-$P_2O_5$ glasses containing alkaline ions. *J. Non-Cryst. Solids* 356: 1965–1967. https://doi.org/10.1016/j.jnoncrysol.2010.07.001.

**167** Ungureanu, M.C., Lévy, M., and Souquet, J.L. (1998). Mixed conductivity of glasses in the $P_2O_5$–$V_2O_5$–$Na_2O$ system. *Ionics* 4: 200–206. https://doi.org/10.1007/BF02375946.

**168** Assem, E.E. and Elmehasseb, I. (2011). Structure, magnetic, and electrical studies on vanadium phosphate glasses containing different oxides. *J. Mater. Sci.* 46: 2071–2076. https://doi.org/10.1007/s10853-010-5040-0.

**169** Devidas, G.B., Sankarappa, T., Chougule, B.K., and Prasad, G. (2007). DC conductivity in single and mixed alkali vanadophosphate glasses. *J. Non-Cryst. Solids* 353: 426–434. https://doi.org/10.1016/j.jnoncrysol.2006.12.011.

**170** Jozwiak, P. and Garbarczyk, J. (2005). Mixed electronic–ionic conductivity in the glasses of the LiO–VO–PO system. *Solid State Ionics* 176: 2163–2166. https://doi.org/10.1016/j.ssi.2004.06.028.

**171** Sankarappa, T., Devidas, G.B., Prashant Kumar, M. et al. (2009). AC conductivity studies in single and mixed alkali vanadophosphate glasses. *J. Alloys Compd.* 469: 576–579. https://doi.org/10.1016/j.jallcom.2008.02.012.

**172** Bhattacharya, S. and Ghosh, A. (2003). AC relaxation in silver vanadate glasses. *Phys. Rev. B* 68: 224202. https://doi.org/10.1103/PhysRevB.68.224202.

**173** Ghosh, A. (1989). Temperature-dependent thermoelectric power of semiconducting bismuth-vanadate glass. *J. Appl. Phys.* 65: 227. https://doi.org/10.1063/1.342575.

**174** Nagamine, K., Honma, T., and Komatsu, T. (2008). Selective synthesis of lithium ion-conductive β-$LiVOPO_4$ crystals via glass-ceramic processing. *J. Am. Ceram. Soc.* 91: 3920–3925. https://doi.org/10.1111/j.1551-2916.2008.02812.x.

175 Schindler, M., Hawthorne, F., and Baur, W. (2000). Crystal chemical aspects of vanadium: polyhedral geometries, characteristic bond valences, and polymerization of ($VO_n$) polyhedra. *Chem. Mater.* 12: 1248–1259.

176 Zavalij, P.Y. and Whittingham, M.S. (1999). Structural chemistry of vanadium oxides with open frameworks. *Acta Crystallogr. Sect. B: Struct. Sci.* 55: 627–663. https://doi.org/10.1107/S0108768199004000.

177 Denton, E., Rawson, H., and Stanworth, J. (1954). Vanadate glasses. *Nature* 173: 1030–1032. https://doi.org/10.1038/1731030b0.

178 Feltz, A. and Unger, B. (1985). Redox reactions in condensed oxide systems II. Variation of the structure of vanadium phosphate glasses in dependence on the oxidation state of vanadium. *J. Non-Cryst. Solids* 72: 335–343. https://doi.org/10.1016/0022-3093(85)90188-7.

179 Frazier, L.L. and France, P.W. (1977). Compositional dependence of the electrical conductivity of vanadium phosphate glass. *J. Phys. Chem. Solids* 38: 801–808. https://doi.org/10.1016/0022-3697(77)90075-0.

180 Giuli, G., Paris, E., Mungall, J. et al. (2004). V oxidation state and coordination number in silicate glasses by XAS. *Am. Mineral.* 89: 1640–1646.

181 Leister, M., Ehrt, D., von der Gönna, G. et al. (1999). Redox states and coordination of vanadium in sodium silicates melted at high temperatures. *Phys. Chem. Glasses* 40 (6): 319–325.

182 Balkanski, M. (2000). Solid-state microbatteries for electronics in the 21st century. *Sol. Energy Mater. Sol. Cells* 62.

183 Duclot, M. and Souquet, J.L. (2001). Glassy materials for lithium batteries: electrochemical properties and devices performances. *J. Power Sources* 610–615. https://doi.org/10.1016/S0378-7753(01)00641-3.

184 Afyon, S., Krumeich, F., Mensing, C. et al. (2014). New high capacity cathode materials for rechargeable Li-ion batteries: vanadate-borate glasses. *Sci. Rep.* 4: 7113. https://doi.org/10.1038/srep07113.

185 Montorsi, M., Broglia, G., and Mugoni, C. (2015). Structural insight into transition metal oxide containing glasses by molecular dynamic simulations. In: *Molecular Dynamics Simulations of Disordered Materials*, Springer Series in Materials Science, vol. 215 (ed. C. Massobrio, J. Du, M. Bernasconi and P. Salmon). Cham: Springer https://doi.org/10.1007/978-3-319-15675-0_8.

186 Tricot, G., Montagne, L., Delevoye, L. et al. (2004). Redox and structure of sodium-vanadophosphate glasses. *J. Non-Cryst. Solids* 345, 346: 56–60. https://doi.org/10.1016/j.jnoncrysol.2004.07.043.

187 Jozwiak, P. and Garbarczyk, J.E. (2005). Mixed electronic-ionic conductivity in the glasses of the $Li_2O$–$V_2O_5$–$P_2O_5$ system. *Solid State Ionics* 2163–2166. https://doi.org/10.1016/j.ssi.2004.06.028.

188 Takahashi, H., Karasawa, T., Sakuma, T., and Garbarczyk, J.E. (2010). Electrical conduction in the vitreous and crystallized $Li_2O$–$V_2O_5$–$P_2O_5$ system. *Solid State Ionics* 181: 27–32. https://doi.org/10.1016/j.ssi.2009.12.001.

189 Tykarski, L., Nowinski, J.L., Garbarczyk, J.E. et al. (2002). Studies of $Li_2O$–$V_2O_5$–$P_2O_5$ glasses by DSC, EPR and impedance spectroscopy. *Solid State Ionics* 154, 155: 367–373.

**190** Behzad, H., Hekmatshoar, M.H., Mirzayi, M., and Azmoonfar, M. (2009). Activation energy and conductivity of glasses in the $P_2O_5$–$V_2O_5$–$Li_2O$ system. *Ionics* 15: 647–650. https://doi.org/10.1007/s11581-009-0319-3.

**191** Broglia, G., Mugoni, C., Du, J. et al. (2014). Lithium vanado-phosphate glasses: structure and dynamics properties studied by molecular dynamics simulations. *J. Non-Cryst. Solids* 403: 53–61. https://doi.org/10.1016/j.jnoncrysol.2014.07.003.

**192** Hoppe, U., Kranold, R., and Gattef, E. (1998). An X-ray diffraction study of the structure of vitreous $V_2O_5$. *Solid State Commun.* 108: 71–76.

**193** Ardelean, I., Cozar, O., Vedeanu, N. et al. (2007). EPR study of $V_2O_5$–$P_2O_5$–$Li_2O$ glass system. *J. Mater. Sci. Mater. Electron.* 18: 963–966. https://doi.org/10.1007/s10854-006-9107-2.

**194** Guo, Y., Liu, C., Wang, J. et al. (2020). Effect of $ZrO_2$ crystallization on ion exchange properties in aluminosilicate glass. *J. Eur. Ceram. Soc.* 40: 2179–2184. https://doi.org/10.1016/j.jeurceramsoc.2020.01.001.

**195** Cormier, L., Dargaud, O., Calas, G. et al. (2015). Zr environment and nucleation role in aluminosilicate glasses. *Mater. Chem. Phys.* 152: 41–47. https://doi.org/10.1016/j.matchemphys.2014.12.008.

**196** Dargaud, O., Cormier, L., Menguy, N. et al. (2011). Mesoscopic scale description of nucleation processes in glasses. *Appl. Phys. Lett.* 99: https://doi.org/10.1063/1.3610557.

**197** Connelly, A.J., Travis, K.P., Hand, R.J. et al. (2011). Composition-structure relationships in simplified nuclear waste glasses: 2. the effect of $ZrO_2$ additions. *J. Am. Ceram. Soc.* 94: 137–144. https://doi.org/10.1111/j.1551-2916.2010.04036.x.

**198** Fisher, J.G., James, P.F., and Parker, J.M. (2005). Soda lime zirconia silicate glasses as prospective hosts for zirconia-containing radioactive wastes. *J. Non-Cryst. Solids* 351: 623–631. http://dx.doi.org/10.1016/j.jnoncrysol.2005.01.064.

**199** Achigar, S., Caurant, D., Régnier, E., and Majérus, O. (2021). Dismantling nuclear waste rich in $P_2O_5$, $MoO_3$ and $ZrO_2$: how do these oxides incorporate in aluminoborosilicate glasses? *J. Nucl. Mater.* 544: 152731. https://doi.org/10.1016/j.jnucmat.2020.152731.

**200** Galoisy, L., Pelegrin, E., Arrio, M. et al. (1999). Evidence for 6-coordinated zirconium in inactive unclear waste glasses. *J. Am. Ceram. Soc.* 82: 2219–2224.

**201** McKeown, D.A., Muller, I.S., Buechele, A.C., and Pegg, I.L. (1999). X-ray absorption studies of the local environment of Zr in high-zirconia borosilicate glasses. *J. Non-Cryst. Solids* 258: 98–109. https://doi.org/10.1016/S0022-3093(99)00517-7.

**202** Meneghini, C., Gualtieri, A.F., and Siligardi, C. (1999). Differential anomalous wide-angle X-ray scattering and X-ray absorption experiments to investigate the formation of glass ceramics in the CaO–$SiO_2$–$ZrO_2$ system. *J. Appl. Crystallogr.* 32: 1090–1099. https://doi.org/10.1107/S0021889899010675.

**203** Ferlat, G., Cormier, L., Thibault, M.H. et al. (2006). Evidence for symmetric cationic sites in zirconium-bearing oxide glasses. *Phys. Rev. B* 73: https://doi.org/10.1103/PhysRevB.73.214207.

**204** Connelly, A.J., Hyatt, N.C., Travis, K.P. et al. (2011). The structural role of Zr within alkali borosilicate glasses for nuclear waste immobilisation. *J. Non-Cryst. Solids* 357: 1647–1656. http://dx.doi.org/10.1016/j.jnoncrysol.2011.01.005.

**205** Jollivet, P., Calas, G., Galoisy, L. et al. (2013). An enhanced resolution of the structural environment of zirconium in borosilicate glasses. *J. Non-Cryst. Solids* 381: 40–47. http://dx.doi.org/10.1016/j.jnoncrysol.2013.09.013.

**206** Calas, G., Galoisy, L., Cormier, L. et al. (2014). The structural properties of cations in nuclear glasses. *Proc. Mater. Sci.* 7: 23–31. https://doi.org/10.1016/j.mspro.2014.10.005.

**207** Angeli, F., Charpentier, T., De Ligny, D., and Cailleteau, C. (2010). Boron speciation in soda-lime borosilicate glasses containing zirconium. *J. Am. Ceram. Soc.* 93: 2693–2704. https://doi.org/10.1111/j.1551-2916.2010.03771.x.

**208** Karell, R., Kraxner, J., and Chromčíková, M. (2006). Properties of selected zirconia containing silicate glasses. *Ceram. Silikaty* 50: 78–82.

**209** Karell, R., Kraxner, J., Chromčíková, M., and Liška, M. (2007). Properties of selected zirconia containing silicate glasses II. *Ceram. Silikaty* 51: 125–130.

**210** Nkurunziza, G., Debaiky, A., Cousin, P., and Benmokrane, B. (2005). Durability of GFRP bars: a critical review of the literature. *Prog. Struct. Mater. Eng.* 7: 194–209. https://doi.org/10.1002/pse.205.

**211** Angeli, F., Charpentier, T., Gaillard, M., and Jollivet, P. (2008). Influence of zirconium on the structure of pristine and leached soda-lime borosilicate glasses: towards a quantitative approach by $^{17}O$ MQMAS NMR. *J. Non-Cryst. Solids* 354: 3713–3722. http://dx.doi.org/10.1016/j.jnoncrysol.2008.03.046.

**212** Pèlegrin, E., Calas, G., Ildefonse, P. et al. (2010). Structural evolution of glass surface during alteration: application to nuclear waste glasses. *J. Non-Cryst. Solids* 356: 2497–2508. http://dx.doi.org/10.1016/j.jnoncrysol.2010.02.022.

**213** Cailleteau, C.C., Devreux, F.F., Spalla, O. et al. (2011). Why do certain glasses with a high dissolution rate undergo a low degree of corrosion? *J. Phys. Chem. C* 115: 5846–5855. https://doi.org/10.1021/jp111458f.

**214** Jollivet, P., Galoisy, L., Calas, G. et al. (2018). Zirconium local environment in simplified nuclear glasses altered in basic, neutral or acidic conditions: evidence of a double-layered gel. *J. Non-Cryst. Solids* https://doi.org/10.1016/j.jnoncrysol.2018.10.002.

**215** Montorsi, M., Leonelli, C., Menziani, M.C. et al. (2002). Molecular dynamics study of zirconia containing glasses. *Phys. Chem. Glasses* 43: 137–142.

**216** Du, J., Devanathan, R., Corrales, L.R. et al. (2006). Short- and medium-range structure of amorphous zircon from molecular dynamics simulations. *Phys. Rev. B: Condens. Matter Mater. Phys.* 74: https://doi.org/10.1103/PhysRevB.74.214204.

**217** Lu, X., Deng, L., and Du, J. (2018). Effect of $ZrO_2$ on the structure and properties of soda-lime silicate glasses from molecular dynamics simulations. *J. Non-Cryst. Solids* 491: 141–150. https://doi.org/10.1016/j.jnoncrysol.2018.04.013.

**218** Deng, L. and Du, J. (2016). Development of effective empirical potentials for molecular dynamics simulations of the structures and properties of boroaluminosilicate glasses. *J. Non-Cryst. Solids* 453: 177–194. https://doi.org/10.1016/j.jnoncrysol.2016.09.021.

**219** Lu, X., Deng, L., Kuo, P.-H. et al. (2017). Effects of boron oxide substitution on the structure and bioactivity of SrO-containing bioactive glasses. *J. Mater. Sci.* 5: https://doi.org/10.1007/s10853-017-0836-9.

**220** Yu, Y., Stevensson, B., and Edén, M. (2019). The structural role of sodium in borosilicate, phosphosilicate, and borophosphosilicate glasses unveiled by solid-state NMR and MD simulations. *J. Phys. Chem. C* https://doi.org/10.1021/acs.jpcc.9b06823.

221 Ren, M., Lu, X., Deng, L. et al. (2018). $B_2O_3/SiO_2$ substitution effect on structure and properties of $Na_2O$–CaO–SrO–$P_2O_5$–$SiO_2$ bioactive glasses from molecular dynamics simulations. *Phys. Chem. Chem. Phys.* 20: 14090–14104. https://doi.org/10.1039/c7cp08358k.

222 Lu, X., Deng, L., Huntley, C. et al. (2018). Mixed network former effect on structure, physical properties, and bioactivity of 45S5 bioactive glasses: an integrated experimental and molecular dynamics simulation study. *J. Phys. Chem. B* 122: 2564–2577. https://doi.org/10.1021/acs.jpcb.7b12127.

223 Bhaskar, P., Kumar, R., Maurya, Y. et al. (2020). Cooling rate effects on the structure of 45S5 bioglass: Insights from experiments and simulations. *J. Non-Cryst. Solids* 534: https://doi.org/10.1016/j.jnoncrysol.2020.119952.

224 Lu, X., Ren, M., Deng, L. et al. (2019). Structural features of ISG borosilicate nuclear waste glasses revealed from high-energy X-ray diffraction and molecular dynamics simulations. *J. Nucl. Mater.* 515: 284–293.

225 Collin, M., Fournier, M., Frugier, P. et al. (2018). Structure of international simple glass and properties of passivating layer formed in circumneutral pH conditions. *npj Mater. Degrad.* 2: 4.

226 Galoisy, L., Pélegrin, E., Arrio, M.-A. et al. (1999). Evidence for 6-coordinated zirconium in inactive nuclear waste glasses. *J. Am. Ceram. Soc.* 82: 2219–2224. https://doi.org/10.1111/j.1151-2916.1999.tb02065.x.

227 Quintas, A., Caurant, D., Majérus, O. et al. (2017). $ZrO_2$ addition in soda-lime aluminoborosilicate glasses containing rare earths: impact on rare earths environment and crystallization. *J. Alloys Compd.* 719: 383–391. https://doi.org/10.1016/j.jallcom.2017.05.211.

# 13

# Fluoride and Oxyfluoride Glasses

*Junjie Zhao[1,2], Xiuxia Xu[2], Xusheng Qiao[2], and Jincheng Du[1]*

[1] *Department of Materials Science and Engineering, University of North Texas, Denton, TX, USA*
[2] *School of Materials Science and Engineering, Zhejiang University, Hangzhou, China*

## 13.1 Introduction

Fluoride glasses are an important category of inorganic glasses that find a wide range of applications in optics and photonics such as optical fibers, laser host, optical window materials because of their overall high transparency from near UV to middle infrared (IR) wavelength ranges and their good solubility of transition and rare earth dopants [1–4]. In addition, these glasses have high fluorine ionic conductivity that enables potential applications in sensors or batteries [5–7]. Oxyfluoride glasses and glass–ceramics have also attracted lots of attention in recent developments in lasers, light emitting diodes (LEDs), optical amplifiers, and solar cells, owing to the combined benefits from both oxide and fluoride rich phases: the oxide glass matrix provides mechanical and chemical stability, while the fluoride phase (either glass or crystal) provides low phonon energy and high rare earth solubility. In particular, transparent oxyfluoride glass–ceramics, in which, through controlling heat treatment, the fluoride-rich phase is converted into fluoride nanocrystals, are ideal hosts of rare earth ions and hence can be used in both downconversion and upconversion optical emissions.

Significant progresses in understanding fluoride glass structures have been made due to advancement of characterization techniques such as neutron and X-ray diffraction, solid-state nuclear magnetic resonance (NMR), Raman and Fourier-transform infrared (FTIR) spectroscopy, and other methods. Atomistic computer simulations, on the other hand, have also greatly advanced the understanding of fluoride glass structures. Fluoride glasses actually were one of the first systems that have been successfully studied by using atomistic simulations, which will be discussed in detail below. In this section, we first provide a brief history of atomistic simulations, both molecular dynamics (MD) and Monte Carlo (MC) methods, of fluoride and other halide glasses, and then explain the empirical potentials, one of the most important inputs in these simulations.

Atomistic computer simulations for fluoride glasses started in the early 1970s. Rahman et al. reported the first MD simulation of fluoride liquids in the compositions of $BeF_2$ and

*Atomistic Simulations of Glasses: Fundamentals and Applications*, First Edition.
Edited by Jincheng Du and Alastair N. Cormack.

$BeLiF_3$ in 1972 [8]. It was found that the melt structure consists of fourfold coordinated $BeF_4$ units that form three-dimensional network through corner-sharing, consistent with the random network model of network glasses by Zachariasen [9]. These encouraging results from MD simulations, although with only point charge, simple repulsive potential and small simulation box with less than hundred atoms, have initiated the field of MD simulations of network glasses [10]. Later, Brawer and Weber reported MC simulations of europium ($Eu^{3+}$)-doped beryline fluoride ($BeF_2$) glasses in 1980, with the goal to understand the local environments and distribution of the rare earth ion and to interpret the optical spectroscopic behaviors [11]. Based on the simulation results, it was found that there exist large variations of the local environments of $Eu^{3+}$ ions and no single coordination number can be assigned [12]. Brawer and Weber continued to use MD simulations to simulate fluoroberyllate glasses with alkali and alkaline earth fluoride glasses [12]. Neodymium ($Nd^{3+}$) was the targeted rare earth ion in this work. It was found that the coordination number of rare earth ions increases from 6.5 (in pure beryline fluoride glasses) to 8 (in beryline fluoride glasses modified with alkali or alkali earth fluorides) [12]. The atom positions from MD simulations were used to predict the energy levels of the $^4F_{3/2}$ state of $Nd^{3+}$ and were found to be in good agreement with experiments, which helped to understand the splitting in absorption and emission spectra [12]. Since then, a number of simulations of fluoride, chloride, and mixed anion glasses have been reported. For example, fluoride and chloride-containing bioactive glasses from MD simulations have been reported [13–15]. We have also reported a series of studies about structures and phase separation behaviors in aluminosilicate oxyfluoride glasses [16–19]. Ab initio molecular dynamics (AIMD) simulations have also been recently applied to halide containing glasses [20, 21]. Due to high computational cost of AIMD and quite successful application of empirical potential-based classical MD simulations, reports of AIMD of halide-containing glasses are relatively rare.

Owing to high ionicity of the chemical bonding in fluoride glasses as compared to oxide glasses, early simulations used the Born Model with point charges and simple repulsive potentials [8]. Later simulations used more sophisticated potential models such as those developed by Tosi and Fumi with the generalized Huggins–Mayer form [10, 22]. Figure 13.1 shows representative potential curves for fluoride, chloride, and oxide systems. These potentials were shown to be able to reproduce both the structure and cohesive energies of crystalline halides [10]. Following these early works, more parameters in the Born–Meyer–Huggins framework of potential have been developed and applied to a wide range of fluoride glasses such as fluorozirconate, fluoroaluminate, or the more complex ZBLAN systems [23]. Later simulations used partial charge instead of full charges and the Buckingham form of short-range interaction. Later developments have included mixed interactions of F–F, O–O, and O–F pairs, which enable the simulations of oxyfluoride glasses [24, 25]. Other potentials that employ the shell model to account for the ion polarizability have also been introduced and enabled the simulations of halides and mixed anion glasses [14].

In this chapter, we aim to cover how MD simulations are used to understand the structure features (including both atomic structures and micro structures such as phase separation), and properties of fluoride and oxyfluoride glasses. The general structure information of these glasses from experiments are first summarized in Section 13.2. The structures and properties of fluorozirconate glasses by MD simulation are discussed in Section 13.3. The phase-separation features of fluoroaluminosilicate glasses from MD

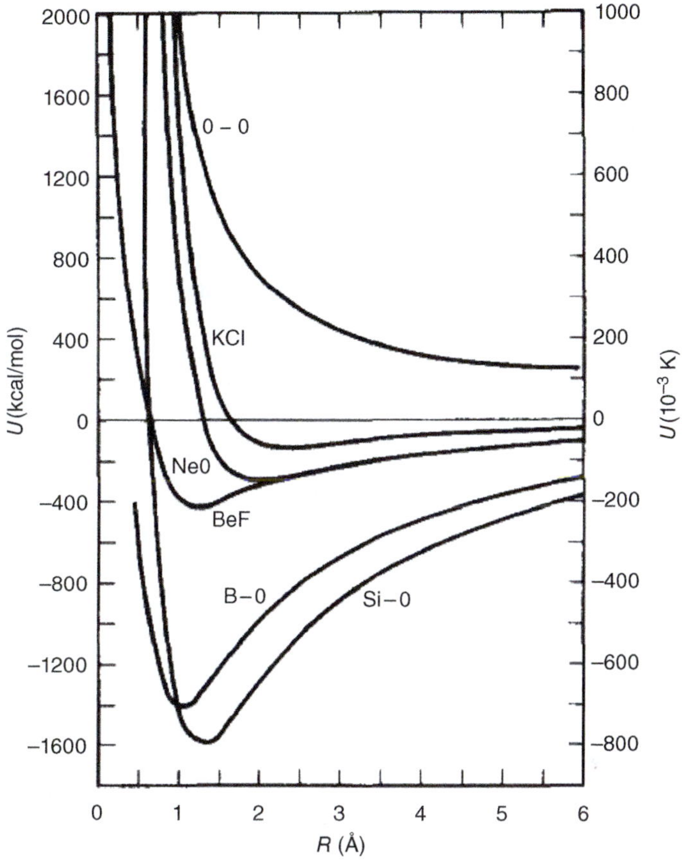


**Figure 13.1** Representative Born–Meyer pair potentials for oxide, fluoride, and chloride glasses [10].

simulations are presented in Section 13.4, with the a special focus on understanding the structural originates of nanocrystallization and therefore help to design the transparent glass-ceramics for optical applications. In Section 13.5, AIMD simulations of oxyfluoride glasses are presented. This is followed by a conclusion of the chapter.

## 13.2 General Structure Features of Fluoride and Oxyfluoride Glasses

### 13.2.1 Structure Features of Fluoride Glasses

Although the structure model for network glass with covalent bonds was proposed by Zachariasen in his seminal paper in 1932 [9], there's no suitable theory for fluoride glasses until Baldwin and Mackenzie proposed a structural model of these glasses in 1979 [26]. Later on, Poulain proposed another theory of fluoride glass structures in 1981 [27]. Baldwin's model was based on Sun's single bond strength approach [28] to define the fluoride glass formers, intermediates, and modifiers by calculating the single bond strength for fluoride [26]. Poulain's model, on the other hand, defined glass forming fluorides (or glass progenitors) as those with cation to anion field strength ratio between 2.5 and 10, while too high or too low a ratio not leading to glass formation [27]. Based on these two theories, $Be^{2+}$, $Al^{3+}$, $Zr^{4+}$, $Hf^{4+}$, and a few other cations can be identified as glass formers in fluoride glasses.

In general, $BeF_2$ is the only fluoride that can form vitreous state without another additional compound, which is similar to $SiO_2$ or $ZnCl_2$. Hence, the three-dimensional network in $BeF_2$ glass is built up by the corner-sharing $[BeF_4]$ tetrahedra [29], which is the same as that in silica glass and can be described by Zachariasen's model. This structural features in $BeF_2$ glass were also observed from MD simulations [8]. In $BeF_2$ glasses with alkali or alkali earth ions modifiers, some researchers observed fivefold $[BeF_5]$ polyhedra, half of which are actually $[BeF_4]$ tetrahedra with an extra $F^-$ at a larger distance than other four [30].

Group IIIA cations such as $Al^{3+}$, $Ga^{3+}$, and $In^{3+}$ can take the role as glass formers with $[XF_6]$ octahedral coordination geometry. Some studies argued the existence of $[AlF_4]$ in fluoroaluminate glasses, but it was explained that the observation of $[AlF_4]$ may be due partly to the rapid quenching of melts (there are $[AlF_6]$ and $[AlF_4]$ in the melts) and partly to metastability inherent to the glassy state [31]. It is also worthy to note that the degree of the glass polymerization decreases in the sequence of Al, Ga, and In [32]. For instance, in $40InF_3–20ZnF_2–20SrF_2–20BaF_2$ (mol%) glass, the discrete $[InF_6]^{3-}$ anions bonded together by $Ba^{2+}$ without significant amount of bridging fluorine [33].

In fluorozirconate glass, the three-dimensional glass network consists with $[ZrF_n]$ units, with modifiers ($Ba^{2+}$, $Na^+$, etc.) donating their fluorine and to be randomly distributed throughout the network [34]. The $Zr^{4+}$ coordination number of 6 and the $[ZrF_6]$ octahedra forming zigzag chains as the basic structure of fluorozirconate glasses were proposed by Almeida and Mackenzie [35, 36]. However, this picture of fluorozirconate glasses was found to be uncommon and likely to be due to analogy made with fluorozirconate crystals [37]. Most of the $[ZrF_n]$ polyhedra were found to consist of seven or eight fluorine, confirmed by the following IR and Raman studies [38]. Initially, researchers estimated the fluorozirconate glass structure from those of the crystals precipitated from the glass [39]. For example, by analyzing the X-ray diffraction of $Ba^{2+}$ or $Pb^{2+}$ containing meta-fluorozirconate glass, and comparing with the corresponding fluoride crystals, the authors suggested that the cations coordination environment is similar to that in β-$BaZrF_6$ and α-$PbZrF_6$ crystals, where $Zr^{4+}$ surrounded by $8F^-$, and $Ba^{2+}$, $Pb^{2+}$ coordinated by 10 $F^-$, respectively [40]. Based on the thermal analyses and Raman studies, it was found that small amount of seven-fold coordinated $Zr^{4+}$ may take place in $BaZrF_6$ glass, and the glass structure of $SrZrF_6$ have the same structural feature with $PbZrF_6$ glass [39].

In addition, $HfF_4$ is chemically similar to and structurally isomorphous to $ZrF_4$ in fluorozirconate glasses, and hence it can partially or totally substitute $ZrF_4$ in any fluorozirconate glasses [41, 42]. Due to potential applications of fluorozirconate glasses in optical devices, e.g. as optical fiber, a number of simulation work has been performed on the fluorozirconate glass system and detailed examples will be presented in Section 13.3.

### 13.2.2 Structure Features of Oxyfluoride Glasses

Oxyfluoride glasses are mixed anion system with $O^{2-}$ and $F^-$, while $P^{5+}$, $B^{3+}$, and $Si^{4+}$ are regarded as classical glass formers [43]. In some oxyfluoride glasses, it is said that $F^-$ can enter the network by substituting $O^{2-}$, forming $G(O,F)_x$ (G: glass former) units [44–46]. For other systems, fluoride phase separation can be observed from as-quench glasses [47, 48]. However, unlike oxide or fluoride glasses, there has not been a general structural model to describe the oxyfluoride glasses.

Fluorine-containing phosphate glasses can be categorized into two types depending on whether $F^-$ substitutes $O^{2-}$ in the [$PO_4$] units. The ones with the presence of P-F bonding are named as fluorophosphate glasses (with classical composition $NaPO_3$–$AlF_3$ [NAPF]) [49, 50], while the ones without P-F bonding referred as fluoride phosphate glasses (with classical composition $AlF_3$–$P_2O_5$ [FP]) [51, 52]. Therefore, the glass network of NAPF glasses consists of metaphosphate chains or rings along with minor pyrophosphate, while the one in FP glasses is made up of pyro- and orthophosphate tetrahedra [53].

Similar to the fluorophosphate glasses, the glass network of fluoroborate glasses is built up with [$BO_4$] or [$BO_3$] units with one or more $O^{2-}$ site being substituted by $F^-$ [46, 54]. Because the coordination number of boron strongly depends upon the amount of modifiers (such as $Na^+$, $Ba^{2+}$, etc.), and on the preferred coordination between fluorine and modifiers, the addition of fluorine in fluoroborate glass could promote more [$BO_4$] transform into [$BO_3$] [55–57]. In addition, unlike fluorine-containing phosphate glass, it is possible to observe droplet-like fluoride phase separation in fluoroborate glasses [47].

Different from fluorophosphate and fluoroborate glass, fluorine is unlikely to coordinate with Si in fluorosilicate glasses [58], especially when with $Al^{3+}$ in the composition [59, 60]. Therefore, the possibility of observing fluoride phase separation in fluorosilicate glass is much higher than that in fluorophosphate or fluoroborate glass. Details of fluoride-phase separation in oxyfluoride glasses will be presented in Section 13.2.3.

### 13.2.3 Phase Separation in Fluoride and Oxyfluoride Glasses

Phase separation has been observed experimentally in different glass systems (silicate and borate glasses), and those with mixed glass formers (borosilicate glass, phosphosilicate glass, etc.) or with mixed anion species (fluorochlorozirconate glass [61], oxyfluoride glass [62], etc.). Phase separation can stem from the liquid phase immiscibility or be developed after heat treatment of the homogeneous glass formed from rapid quenching. Phase separations in these glasses can show different morphologies ranging from spherical second phase (nucleation and growth) to intermixing of the two phases (spinodal decomposition). Figure 13.2 shows typical morphologies of phase separation in fluoride and oxyfluoride glasses.

Direct observations from electron microscopy in combination with thermal analysis have been used to exam the phase separation in fluoride glass systems. Phase separation was observed in fluorozirconate and fluorolead fluoride glass systems [4, 65, 66]. In the fluorozirconate glass system, for example in $60ZrF_4$–$31BaF_2$–$4YF_3$–$2.5AlF_3$–$1.0LaF_3$–$1.5PbF_2$ glass [67] or $52.0ZrF_4$–$18.5BaF_2$–$5.0LaF_3$–$3AlF_3$–$20.0LiF$–$1.5PbF_2$ glass [63], after quenching, opaque glass samples with 1-μm sized droplet phase separation with $Pb^{2+}$ enrichment were

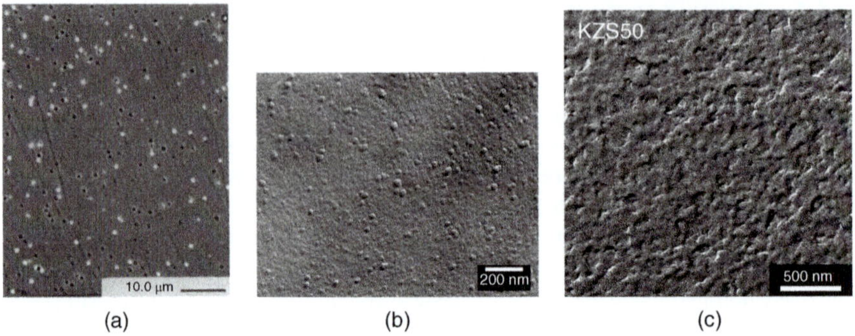


**Figure 13.2** (a) Scanning electron micrograph (backscatter image) of opaque white glass for ZBLAYPb fluoride glass where the white spots are the $Pb^{2+}$ enriched droplet phase separation. Source: Boehm et al. [63], Fig 04 (p. 153)/Esevier. (b) TEM replica of droplet like nano-scale phase separation in $60.3SiO_2-3.7Al_2O_3-12.3SrO-5.3K_2O-8.0Na_2O-10.4SrF_2$ glass. Source: Bocker et al. [64], Fig 03b (p. 1739)/Esevier. (c) TEM replica of interpenetrating fluoride phase separation in $50SiO_2-25KF-25ZnF_2$ glass. Source: Lin et al. [48], Fig 07 (p. 0E)/American Chemical Society.

obtained (Figure 13.2a). In addition, fluoride phase separation with two interconnected phases with a volume ratio of about 40 : 60 was observed with replica transmission electron microscopy (TEM) in $51.1ZrF_4-25.2BaF_2-16.4ThF_4-3.2AlF_3-3.5LiF$ glass [68]. Thermal analysis has also been used to provide indication of phase separation, and those glasses with two or more glass transition temperatures is usually an indication of phase separation. For example, two glass transition temperatures at 262 and 294 °C were observed, indicating that liquid–liquid immiscibility exists in the above fluoride glass system [68]. Phase separation in this glass system was suggested to be due to the high enthalpy of mixing between the fluorozirconate and the $BaF_2-ThF_4$ base glass compositions, each of which gives a different $T_g$, due to the difference in glass composition.

In oxyfluoride glass, fluoride phase separation has been observed and extensively studied. It has been suggested that fluoride phase separation is the initial stage of crystallization [69, 70] and can help to achieve transparent glass–ceramics for optical devices [71, 72]. From this aspect, understanding the phase separation behavior is critical for controllable crystallization that can obtain certain crystal types and sizes with thermal treatment. It has been shown that fluorine and network modifiers (alkali, alkali-earth ions, etc.) can enrich in one phase while glass formers and oxygen built up the glass network in another phase [64, 73, 74]. When with thermal treatment above $T_g$, because of the low viscosity in the fluoride phase, fluoride crystals can be promoted to precipitate from these regions [75, 76]. Two types of fluoride phase separation have been discovered based on the morphology, the droplet like one (Figure 13.2b) and the interpenetrating one (Figure 13.2c) [48, 74]. Multiple nanocrystals can precipitate from a droplet-like fluoride phase separation region [64, 77, 78]. On the contrary, crystals that precipitated from the regions with interpenetrating phase separation is larger in size because of such regions can provide faster transport channels for fluorine and modifiers [48, 79]. In addition, the interface between oxide and fluoride phases can not only provide sites that benefit the crystallization [77, 80] but also help to achieve self-limiting growth of nanocrystals [81, 82]. Owing to the depletion of fluorine and modifiers, the residual glass matrix around these regions is enriched in glass formers, which can result in the increase of the viscosity up to $10^{12}$ Pa·s [80] and therefore build up a barrier preventing ions diffusion [83, 84]. This can effectively hinder the growth of fluoride crystals and therefore help to realize controllable nanocrystallization and obtain transparent oxyfluoride glass–ceramics.

## 13.3 Structures and Properties of Fluoride Glasses from MD Simulations

Since their first applications in the 1970s, atomistic simulations have been successfully used to obtain structural information of a number of fluoride glass systems. Compared to MC simulations, MD simulations can not only provide global and local structural features but also calculate dynamic properties such as diffusion coefficients as a function of temperature and pressure. Owing to their wide IR optical applications, fluorozirconate glasses have been extensively studied in experiments and are considered as one of the best understood among the fluoride glass systems, which laid a good foundation for the simulation studies. Hence, in this section, we will review of structural features and property calculations obtained from MD simulations of fluorozirconate glasses.

### 13.3.1 General Structures from MD Simulations

The structure of zirconia fluoride glass is considered as a three-dimensional network built up by [$ZrF_n$] polyhedral with glass modifiers (e.g. alkali or alkali earth ions) distributed irregularly in the space between [$ZrF_n$] polyhedral [85] and glass intermediate (e.g. rare earth or transition metal ions) that take octahedral sites (Figure 13.3) [3, 87]. The glass modifiers donate fluorine and charge compensate to the network when possible, being surrounded by $F^-$ and bonding ionically to [$ZrF_n$] polyhedra [88]. The [$ZrF_n$] polyhedra connect with each other through corner- or edge-sharing [86]. In barium fluorozirconate glass, edge-sharing [$ZrF_n$] are usually surrounded and stabilized by $Ba^{2+}$ [86]. Among all the pairs in pair distribution function (PDF), the peaks for Zr–Zr and Zr–F are very narrow and well defined as compared to others, which suggests that the regularity of [$ZrF_n$] polyhedral as well as intermediate range order of arrangements among [$ZrF_n$] [85]. In contrast, modifier cations such as $Ba^{2+}$ and $Na^+$ have weaker and broader cation–cation PDF as compared to $Zr^{4+}$ (Figure 13.4) [61], suggesting a more random distribution in the glass structure.

Before demonstrating the detailed researches on cation and fluorine coordination environment and structure role, it needs to be pointed out that due to the limitation of computer resources back in 1980s, only up to a few hundreds of atoms were used in these MD simulations. Despite many useful general structural features obtained, more quantitative description of both the structural and property features might be challenging from these early studies, especially considering the multicomponent and often complex compositions of these glasses.

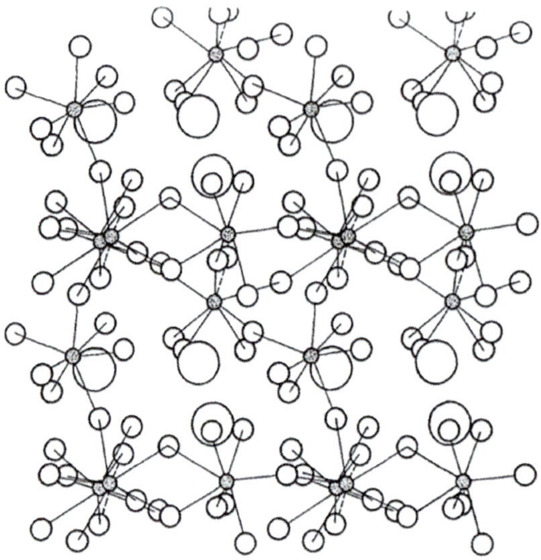

**Figure 13.3** The structure model of $3ZrF_4$–$2BaF_2$ glass derived from the crystal structure of $\alpha$-$ZrF_4$ with Born–Mayer potential [86].

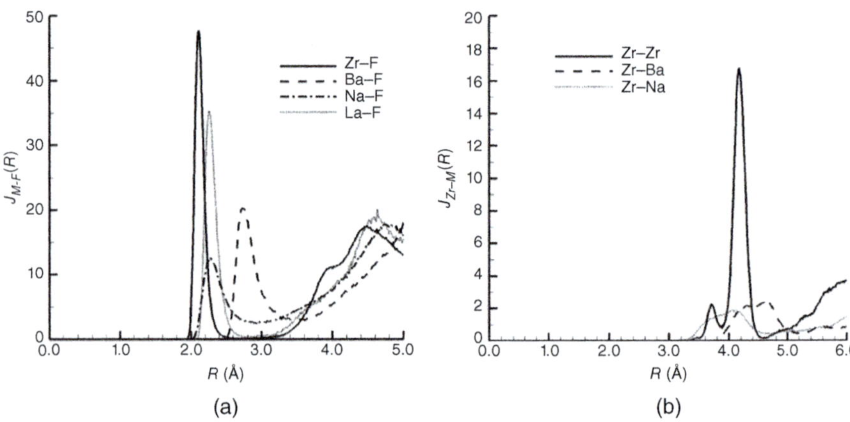


**Figure 13.4** M–F (a) and Zr–M (b) radial distribution function in $55ZrF_4$–$20BaF_2$–$20NaF$–$5LaF_3$ MD simulated glass using Born–Mayer–Huggins potential and containing 790 anions [61].

## 13.3.2 Cation Coordination and Structural Roles

Most of the studies showed that $Zr^{4+}$ is normally 7- or 8-coordinated, indicating that the glass structure consists of $[ZrF_8]$ dodecahedral and $[ZrF_7]$ monocapped trigonal prismatic [86, 88–90]. Though the proportion of the two units depends on the glass composition, it was proposed that $[ZrF_8]$ units dominate, which is consistent with the radial distribution function (RDF) from X-ray scattering [91] and diffraction [41]. Some studies showed that about 75% of the $Zr^{4+}$ are 8-coordinated with $F^-$ [61], with the average coordination number of 7.7 [90, 92]. With a total coordination number of eight fluorine ions, $Zr^{4+}$ in barium fluorozirconate glass are coordinated on an average by about six bridging-fluorines and two nonbridging-fluorines [93]. Similarly, $AlF_3$ is also a glass former in fluoride glasses, but each $Al^{3+}$ ion coordinates with six fluorines and form $[AlF_6]$ octahedral [89, 90], which contrasts to its fourfold coordination when $Al^{3+}$ ions enter the silicate network in oxide glasses.

The structural role of cations in fluoride glass can be defined by the shape of the first peak of the cation-fluorine PDFs. For instance, the strong and narrow first peak at about $R = 2.1$ Å in Zr–F PDF indicates that $Zr^{4+}$ is surrounded by a well-defined coordination shell of $F^-$ ions. Two additional well-defined peaks at $R = 4.4$ and 6.0 Å imply that the $[ZrF_n]$ polyhedral are linked to form a network structure [92]. The simulated bond distance of Zr–F is close to that found in diffraction and extended X-ray absorption fine structure (EXAFS) experiments of $ZrF_4$–$BaF_2$ glass [94].

Lanthanum fluorides ($LnF_3$) are commonly introduced to the fluoride compositions to improve the glass formability due to the fact that they can increase the coordination number of glass forming cation such as $Zr^{4+}$ and the variety of possible $[ZrF_n]$ polyhedral linkages [95]. $Ln^{3+}$ ions were suggested to take a coordination environment in fluorozirconate glass similar to that in $ZrLnF_7$ crystals, in which the Ln–F coordination number is $8 \sim 9$ and depends on the lanthanide [3]. For instance, the coordination number decreases from 9.2 for $La^{3+}$ to 8.3 for $Lu^{3+}$ due to lanthanum contraction. In the samples containing $LnF_3$ (i.e. $LaF_3$, $CeF_3$, $PrF_3$, $NdF_3$, or $PmF_3$) with hexagonal crystal structure, ninefold coordination of $Ln^{3+}$ predominates. The glass samples with other $LnF_3$ contain a mixture of $[LnF_8]$ and $[LnF_9]$ units. In particular, the glass with $YF_3$ was found to contain a significant amount of $[YF_7]$ [95]. From the PDF of Ln-F, the first peak is generally well defined and the peak position shifts to shorter bond length and becomes more asymmetrical with increasing atomic number, which can be explained by the lanthanum contraction. The second coordination shell presents a broad double peak while the third coordination shell is located at about 6 Å [95]. In some studies, $Eu^{3+}$ and $Y^{3+}$ were proposed as glass formers [96, 97]. The average Eu–F distances for the 7-, 8-, and 9- coordination are 2.265, 2.335, and 2.373 Å, respectively, which correspond well with the crystalline analogues [94].

In fluorozirconate glass, $Ba^{2+}$ is the prime network modifier because of the similar ionic radius between $Ba^{2+}$ (~0.135 nm) and $F^-$ (~0.136 nm). $Pb^{2+}$ is a good substitution for $Ba^{2+}$ in terms of the glass-forming ability, while in contrast, $Sr^{2+}$ was found to lead to less stable glass than $Ba^{2+}$ [35]. In barium fluorozirconate glass, as compared to strontium fluorozirconate glass, the bond distances of Zr–F and Zr–Zr are longer, and both the first peak of Zr–F PDF and the distribution of $Zr^{4+}$ coordination number are broader [98, 99]. Coupled with the observation that the percentage of edge-sharing situation between $[ZrF_n]$ polyhedral decreased with the increase of $Ba^{2+}$ content [98], it was implied that $Ba^{2+}$ can loosen and add more $F^-$ to the coordination sphere of $Zr^{4+}$, increasing the disorder in the glass structure. Hence, with high $BaF_2$ content, it is possible to observe free fluorine [100]. With more $BaF_2$ addition into $ZrF_4$ glass, the coordination number of $Ba^{2+}$ can decrease from 10.8 [88, 101] to 8.5 [23, 94] (about 8), while the distances between $[ZrF_n]$ and $[BaF_n]$ units increase [99]. In rare earth-doped fluorozirconate glasses, it was found that $Ba^{2+}$ ions have the tendency to coordinate with $[EuF_n]$ than with $[ZrF_n]$ [102].

Alkali ions, $Na^+$ and $Li^+$ for example, can be introduced as glass modifiers that break the fluoride glass structure and therefore create nonbridging fluorine. For instance, in structure models of 2LiF–40$ZrF_4$ from simulations, each $Li^+$ ion coordinates with 7.4 $F^-$ on average, 73% of which are the nonbridging fluorine. The coordination number of $Na^+$ and $Li^+$ drops slightly with increasing concentration of alkali fluoride, which can be partially explained by the increase of available nonbridging fluorines [1]. The average $F^-$ coordination number of $Na^+$ is around 7 [99]. The addition of $Na^+$ or $Li^+$ leads to a reduction in Zr coordination number as well. With increasing $Li^+$ concentration, the coordination number of $Li^+$ falls from 7.4 to 5.6, which is lower than $Na^+$ with a similar concentration due to smaller ionic radius of $Li^+$ [93].

### 13.3.3 Fluorine Environments

In fluorozirconate glass, there are three typical fluorine species, (i) free fluorine that originated from high modifier concentration [93], (ii) nonbridging fluorine ($F_{NB}$) that breaks the linkage of the Zr–F network, and (iii) bridging fluorine ($F_B$), which can be further divided into two types: edge-sharing fluorine ($F_{Bi}$) and corner-sharing fluorine ($F_B$) (Figure 13.5) [34, 85, 88, 103]. Nonbridging fluorine has a much lower activation energy for diffusion than the bridging ones, which means the nonbridging fluorine are more mobile than the bridging ones and contribute more to electrical conductivity in fluoride glass [104, 105]. With the increase of $ZrF_4$, free and nonbridging fluorines will be converted into bridging fluorines [100]. Some studies even observed the face-sharing that two [$ZrF_n$] polyhedral are connected by more than two bridging fluorine [99]. However, the appearance of face-sharing is controversial, and some argue that face-sharing fluorine only exist under high pressure and they would disappear after decompression [96].

The fluorine environment and the percentage between corner- and edge-sharing fluorine in fluorozirconate glass can be understood from the F–F PDFs. Based on the calculated F–F distances, [$ZrF_n$] polyhedron should have a shape of triangular dodecahedra, close to that in crystalline α-$ZrF_4$ [89]. Interestingly, by detailed examination of the F–F PDFs, splitting

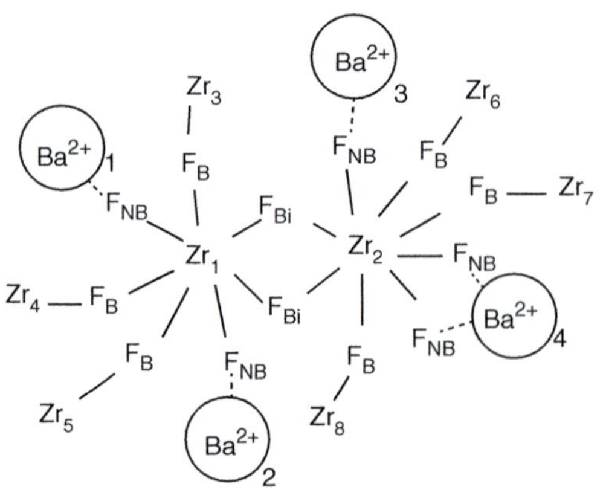


**Figure 13.5** Typical structural motif with $Zr_2F_{13}$ bipolyhedron unit proposed for the structure of $BaZr_2F_{10}$ glass. $F_{NB}$ is the non-bridging fluorine, $F_B$ is the corner-sharing fluorine, $F_{Bi}$ is the edge-sharing fluorine [103].

of the first peak can be observed, which can be related to the corner- and edge-sharing of [$ZrF_n$] polyhedra. Corner-sharing of [$ZrF_n$] dominates in most of the fluorozirconate glass compositions [61, 93]. Only about 17%–20% [88] of fluorine bridges are edge-sharing [34, 99]. In another study by combining X-ray scattering and MD simulations, it was proposed that about one out of four bridges between [$ZrF_n$] polyhedra is edge-sharing [23].

The fluorine coordination environment can also be understood indirectly from the peaks in Zr–Zr PDFs. A smaller peak at about 3.7 Å can be attributed to the edge-sharing fluorine which compares favorably with 3.6 Å in corresponding crystalline phases, while the major peak at about $r = 4.2$ Å can be attributed to the distance of corner-sharing of [$ZrF_n$] units [23, 94]. The number of peaks in Zr–Zr PDFs varied, indicating various types of inter-polyhedral connection among [$ZrF_n$]. A few studies observed only a single peak in Zr–Zr PDFs [99] while most showed a splitting of the first Zr–Zr peak [23, 88, 94, 106]. This discrepancy might be due to the potentials and the number of the atoms used in the simulation cell.

### 13.3.4 Property Calculations Based on Structures from MD

The vibrational spectra of the simulated fluorozirconate glass can be obtained via Fourier transformation of the autocorrelation functions. For example, the IR spectrum can be calculated from the Fourier transformation of the dipole moment correlation function, and the Raman spectrum can be generated from the polarizability tensor of the Zr–F bonds (Figure 13.6) [107]. Generally speaking, the vibrational spectra from MD simulations show similar features as compared to those from experiments, but disagreement is also commonly observed. These disagreement are attributed to the empirical potentials used in MD simulations, unrealistic quenching rate, and the limitation of system size (especially for those used in earlier studies). For example, the calculated spectrum of the high-symmetry stretching mode has similar shape and frequency as compared to the high-frequency band around 500–620 $cm^{-1}$ in the experimental Raman spectrum [107]. In another work, the high-frequency Raman spectrum only involve bond stretching and was calculated by Fourier transformation of the total bond stretching autocorrelation function [108]. The simulated Raman spectra well reproduced the band frequency of fluorozirconate glass but missed the second band at 485 $cm^{-1}$ that is related to the edge-sharing fluorine [108]. In addition, the optical absorption spectra of $Eu^{3+}$ were calculated from simulated glass structures using a point charge approximation [94, 102].

The peculiar permanent densification behavior of fluoride glass has attracted attention owing to its potential application in the control of optical, mechanical, and electrical properties without changing glass composition. It was found experimentally that the density of a fluorozirconate glass (55ZrF4–5$EuF_3$–17$BaF_2$–23NaF) increases with applied pressure until around 3 GPa, while with further increase, the pressure decreases the density (see Figure 13.7) [109, 110]. A maximum of density was observed as a function of applied pressure. This is different from oxide or chalcogenide glasses whose densities increase with the treatment pressure. MD simulations provide an alternative way to high pressure experiments to study the pressure effect on glass structure since it can provide detailed structural change during and after the densification process. As shown in Figure 13.7, MD simulation results reveal that the pressure at which the maximum density is obtained changes with the temperature during densification [96].

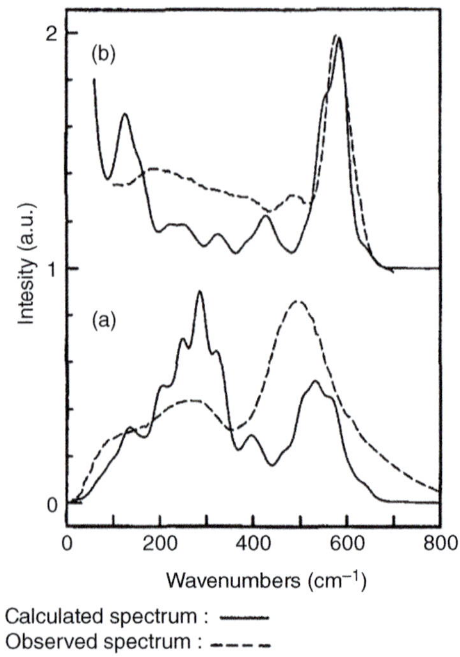


**Figure 13.6** The calculated and observed infrared (a) and Raman (b) spectra of $53ZrF_4$–$20BaF_2$–$20NaF$–$4LaF_3$ glass [107].

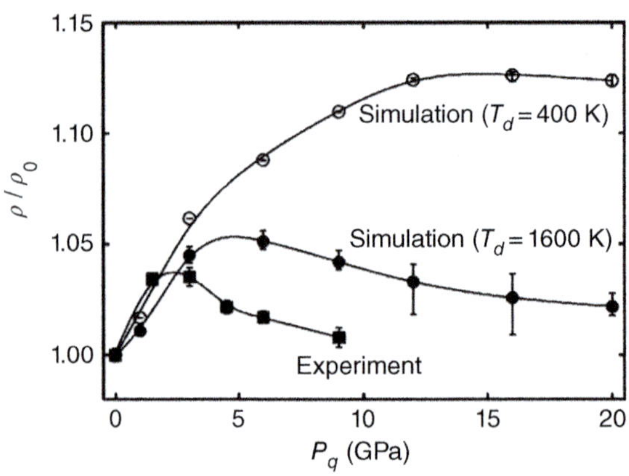


**Figure 13.7** Dependence of relative density $\rho/\rho_0$ on quenching-pressure $P_q$, where density $\rho$ is measured under $P_0$ after decompression from $P_q$ and density $\rho_0$ is measured under $P_0$ without high-pressure treatments. Based on the experiment by Miyauchi et al. [109], the glass composition is 55ZrF4–$5EuF_3$–$17BaF_2$–23NaF [96].

MD simulation results also showed that the connectivity between glass former polyhedral ($[ZrF_n]$ and $[EuF_n]$) changes under pressure [96, 111]. Some of the corner-sharing bridging fluorine species change to the edge-sharing ones. Hence, the coordination cages around glass modifiers ($Ba^{2+}$ and $Na^+$) shrink and the bond distances of Ba–F and Na–F decrease. Strong correlations were found between the density maxima and the $F^-$ coordination number around $Ba^{2+}$ and $Na^+$ ions [112]. After decompression, due to the difference between glass former polyhedral ($[ZrF_n]$ and $[EuF_n]$) and the cages around glass modifiers ($Ba^{2+}$ and

$Na^{+}$), the high-pressure structure has the tendency to relax toward the original structure before pressure applied. The extent of the relaxation was also found to depend on temperature and pressure [113], which is responsible for the peculiar permanent densification behavior.

## 13.4 MD Simulations of Fluoroaluminosilicate Oxyfluoride Glasses

Recent potential developments enabled MD simulations of the oxyfluoride mixed anion glasses [24, 114, 115]. These partial charge potentials include cation-oxygen, cation-fluorine, oxygen–oxygen, fluorine–fluorine, and oxygen–fluorine short-range interactions, while no additional cation–cation terms needed in addition to the Coulombic interactions [24, 114, 115]. Successful applications of MD simulations with these potentials include homogeneous fluorine-containing phosphosilicate bioactive glasses [114], alkali earth aluminosilicate oxyfluoride glasses [24], and fluorine-containing soda lime aluminosilicate glasses [116]. Importantly, for some compositions, clear phase separation into oxide-rich phases and fluoride-rich phase was observed from direct MD simulations using normal cooling rate (e.g. 5 or 1 K/ps) [24, 117–119]. On the other hand, direct observation of phase separation in pure oxide or fluoride glasses are very difficult from MD simulations [120]. The focus of this section is to review the observed phase separation behaviors and interfacial structural features between the oxide and fluoride phases in oxyfluoride glasses from MD simulations.

A structure model of the fluoroaluminosilicate glass system has been proposed by Zhao et al. via MD simulations [24, 120]. In their MD results, owing to the bonding preference between cations and anions, immiscible fluoride and oxide phases can be clearly observed in the $SiO_2$–$Al_2O_3$–$BaF_2$ (BAFS) glass systems. The snapshot of a phase-separated region is shown in Figure 13.8, which can be seen that the fluoride glass phase mainly consists of [$BaF_n$] polyhedra and a few [$AlF_6$] octahedra, while the oxide phase network is made up of [$SiO_4$] and [$AlO_4$] tetrahedra [120]. Hence, in the proposed structure model, they suggest that the fluoroaluminosilicate glass is an immiscible system made up by fluoride-rich phase and aluminosilicate-rich matrix, and these two phases have analogous structural features that can be compared with fluoride glass and aluminosilicate glass, respectively [120]. Furthermore, the two phases are connected by a certain network of intermediate cations, such as $Al^{3+}$, to maintain stability of the glass [24, 120]. In addition, coupled with various experiment results, the separated fluoride phase has been revealed as the structural origins for fluoride nanocrystals precipitating from oxyfluoride glass [24, 64, 75, 114].

### 13.4.1 Oxide and Fluoride Glass Phase Separation Observed from MD Simulations

In fluoroaluminosilicate glass systems, the occurrence of phase separation results from the immiscibility between the glass former tetrahedra (e.g. [$SiO_4$]), and the glass modifier polyhedral (e.g. [$BaF_n$]). According to Zachariasen's random network model [9], oxide glass is constructed with glass-forming tetrahedra coordinating with three or four

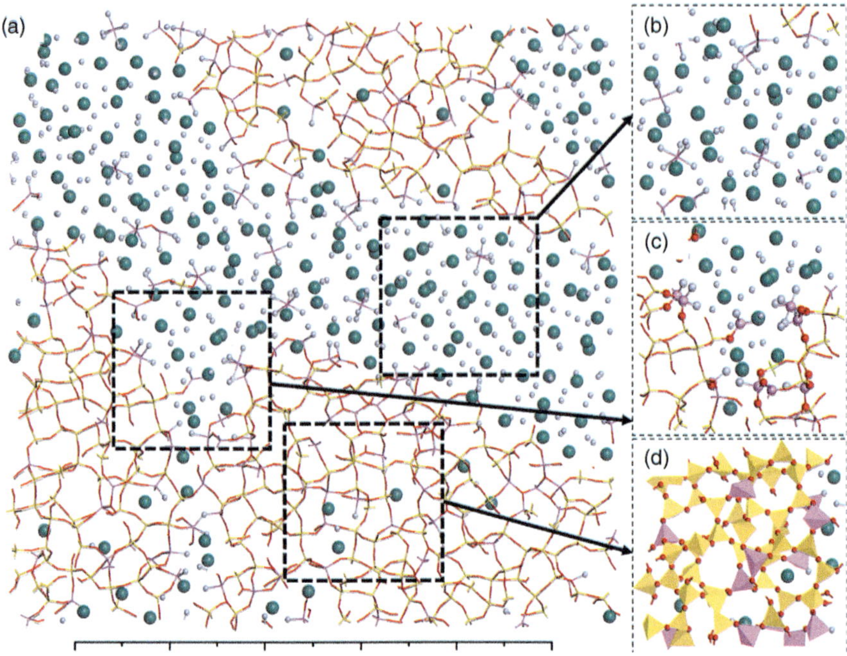


**Figure 13.8** Fluoride phase separation observed in $50SiO_2-15Al_2O_3-35BaF_2$ glass. Snapshot of about 0.5 nm thick layer in the model glass (a) with magnified fluoride phase (b), phase interface (c) and oxide phase (d). Color indications: red-O, yellow-Si, pink-Al, gray-F, bluegreen-Ba. Source: This figure is recreated from Ref. Zhao et al. [120].

oxygens, such as $[SiO_4]$, $[AlO_4]$, or $[BO_3]$. These tetrahedra are connected to form a three-dimension continuous network only by corner-sharing. On the other hand, according to Poulain's structure model for ionic system [27], the network of fluoride glass is made up of cation-fluorine polyhedra with coordination number greater than 4. The connections between fluoride polyhedral can be corner-, edge-, or even face-sharing. Therefore, the cation-anion bonding and cation coordination number differences make it easier for the existence of phase separation in oxyfluoride glass.

The structural features of the aluminosilicate matrix are similar to the aluminosilicate glass in both short- and middle-ranges, with $[SiO_4]$ and $[AlO_4]$ tetrahedra building up a three-dimensional network through corner-sharing (Figure 13.9a,b,d,e). However, with phase separation, some of the modifiers are enriched in the fluoride phase, which lead to a lack of the glass modifier in the oxide phase. As a result, the edge-sharing $[AlO_4]$ tetrahedra can be observed in the modeled glass as a mechanism to maintain local charge neutrality, which is seen less common than the edge-shared ones in oxide glass [120, 121]. In addition, the lack of glass modifiers in oxide phase can also result in the increased level of aggregation of the $[SiO_4]$ network in the BAFS glass, as indicated by the higher network connectivity [120]. Therefore, with the increased introduction of fluoride, the oxide network becomes more compact instead of generating more nonbridging oxygen as one would expect with fluoride addition.

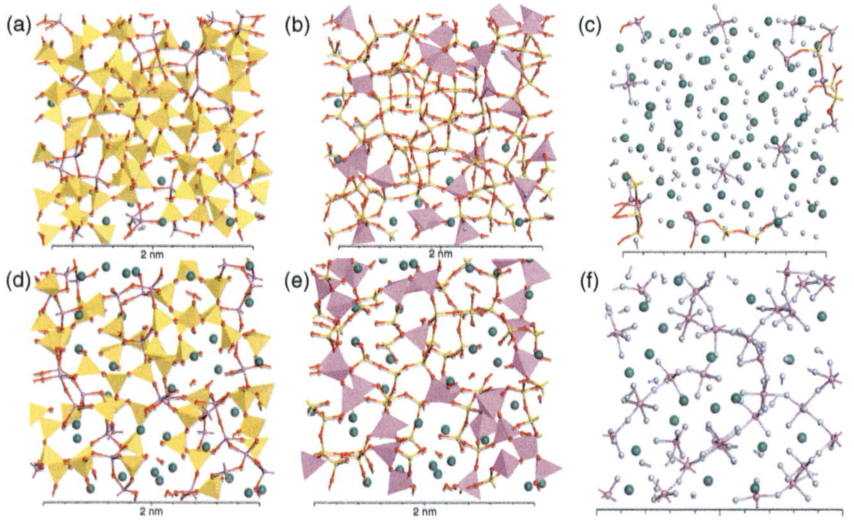


**Figure 13.9** Glass structure comparison among oxide, fluoride and oxyfluoride glass. (a), (b), and (c) show snapshots of silicate, aluminate and fluoride rich regions in structure of MD simulated glass with composition $50SiO_2-15Al_2O_3-35BaF_2$, respectively. (d) and (e) are the structural features with emphasis from silicate and aluminate networks in glass $50SiO_2-15Al_2O_3-35BaO$, respectively. (f) demonstrates the structure of $65AlF_3-35BaF_2$ glass. Color indications: red-O, yellow-Si, pink-Al, gray-F, bluegreen-Ba. Source: This graph is recreated from Ref. Zhao et al. [120].

The structural features of the separated fluoride phase regions can be compared to the fluoride glass, i.e. through comparing total distribution function (TDF) and bond angle distribution function (BAD) [120]. The short range structures are similar, but there are significant differences in the middle-range structure (Figure 13.9c,f). In fluoroaluminosilicate glasses, because of the anion bonding preference that Al–O and Ba–F bonds are more preferable than Al–F and Ba–O, it is more likely for the formation of $[AlO_4]$ in oxide phase and $[BaF_n]$ polyhedral in fluoride phase [120]. As a result, the separated fluoride phase consists of mainly $[BaF_n]$ polyhedra and minor $[AlF_6]$ polyhedra, with both $Ba^{2+}$ and $Al^{3+}$ acting as glass formers to build up the structure of fluoride phase [24, 120]. On the contrary, in the fluoride rich phase, without the competition between $O^{2-}$ and $F^-$, $[AlF_6]$ polyhedra are the network forming units while $Ba^{2+}$ ions are the modifiers [31, 104].

### 13.4.2 Oxide-Fluoride Interfacial Structure Features from MD Simulations

Some cations such as $Al^{3+}$ and $Ba^{2+}$ can bond with both $O^{2-}$ and $F^-$ simultaneously, being in a mixed anion environment, which makes it possible for $[AlO_xF_y]$ and $[BaO_xF_y]$ polyhedral to become the main structure on the interface between the fluoride and oxide phase (Figure 13.10a,b), and to maintain the stability of the immiscible BAFS glass [120]. However,

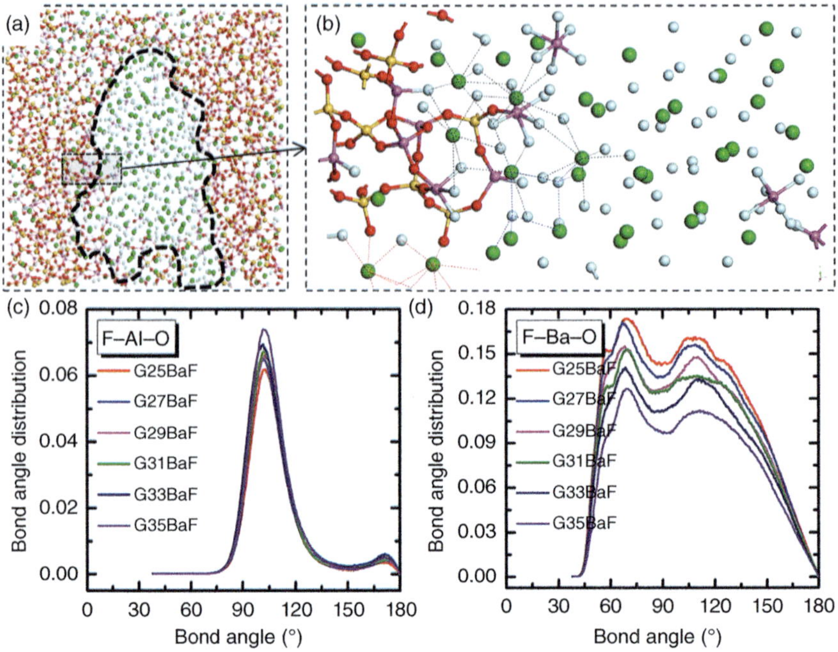


**Figure 13.10** Oxide and fluoride phase interface in $50SiO_2-15Al_2O_3-35BaF_2$ glass. (a) and (b) demonstrate the location and features of two-phase interface. (c) and (d) is the bond angle distribution of $Al^{3+}$ and $Ba^{2+}$ in mixed anion environment, respectively. Color indications: red-O, yellow-Si, pink-Al, cyan-F, green-Ba. Source: (a) and (b) are from Ref. Zhao et al. [24], while (c) and (d) are from Ref. Zhao et al. [120].

$[AlO_xF_y]$ and $[BaO_xF_y]$ play different roles at the phase interface. For one thing, in experiments, after the precipitation of fluoride crystals and with the decreasing concentration of modifiers in fluoride phase, the $[AlO_xF_y]$ polyhedral at the two-phase interface could form a shell-like diffusion barrier around the fluoride phase that can limit the infinite growth of fluoride crystals [81, 122]. For another, compared with F–Al–O bond, the F–Ba–O one has broader distribution (Figure 13.10c,d), indicating that $Ba^{2+}$ locates in a more flexible mixed coordination environment than $Al^{3+}$ [120]. Hence, $[BaO_xF_y]$ polyhedron can be regarded as the intermediate form of $[BaO_6]$ to $[BaF_n]$ that helps $Ba^{2+}$ to diffuse from the oxide phase to the fluoride phase and helps the further growth of the crystal nuclei [120].

### 13.4.3 Correlation of Structural Features Between MD and Crystallization

With the above observations and when coupled with the experiments, the results of MD simulations can lead to deeper understanding of the relationship between the structures and properties of oxyfluoride glass and glass–ceramics, and to provide insights for the design of transparent oxyfluoride glass and glass–ceramics.

Based on the results from MD simulations and experiments, Zhao et al. further proposed strategies of phase separation induced fluoride nanocrystallization, which helps to develop the rare earth or transition metal doped oxyfluoride glass–ceramics with good photoluminescence performances [24, 119]. Aiming at various photonic applications, MD simulations of oxyfluoride glasses also showed the potential to reveal the mechanisms of nanocrystallization, crystalline phase evolution, lattice type conversion, element selective enrichment as well as ionic diffusion routes [117, 118, 123]. The followings are some examples. Qiao and coworkers revealed elemental diffusion principles via MD simulations to achieve the enhanced silver solubility and thermal stability in oxyfluoride glass–ceramic, which could help to improve white light emitting diode performances [123, 124]. Dong et al. and Ballato et al. employed MD simulations to investigate the local environments of rare earth ions in fluorine containing optical fibers, which helped to explain the laser emission enhancement in oxyfluoride glass–ceramic cores and understand the origins of their spectroscopic properties [16, 19]. Chen et al. combined MD simulations and experimental results to discover the selective nanocrystallization mechanism between the cubic and hexagonal $NaLnF_4$ in fluoroaluminosilicate glass–ceramics, which can provide guidance to develop glass–ceramic compositions with desired fluoride nanocrystals and advance optical and other applications [17, 18]. Qiu et al. employed MD simulations to understand the enhancement in up-conversion luminescence of $Yb^{3+}/Er^{3+}$ in $CaF_2$ containing oxyfluoride glass–ceramics that can be attributed to the long-range structural ordering and the change of site symmetry surrounding rare earth ions [25]. In summary, coupled with experiments, MD simulations can provide significant insights of phase separation, nanocrystallization, and their corresponding properties relationship in oxyfluoride glass and glass–ceramics.

## 13.5 Ab Initio MD Simulations of Oxyfluoride Glasses

Compared to many classical MD simulation studies, only a handful studies of AIMDs of halide containing glasses have been reported, and almost all of them are on bioactive glasses with the aim to understand the relationship between fluorine environment and the bioactivity. Because of the high computational cost of AIMD, only relatively small model (up to a few hundred atoms) can be used in those AIMD simulations. These simulations can provide accurate local structure features of these glasses without being affected by the force field and, importantly, some other structural features such as NMR spectra can be calculated and compared with experiments.

In the work by Pedone and coauthors, structure models with 120 atoms in the simulation cell for bioactive glass with composition $19SiO_2$–$10Na_2O$–$1P_2O_5$–$4CaF_2$–$7CaO$ were generated using Car–Parrinello molecular dynamics (CPMD), a type of AIMD simulation [125, 126]. Based on the structure models generated from AIMD, gauge including projected augmented wave (GIPAW)-based algorithm was used to calculate the $^{29}Si$, $^{19}F$, $^{23}Na$, and $^{31}P$ NMR spectra which were then compared with experiments to gain deeper insights of the structural features of these glasses [126]. It was found that fluorine is predominantly coordinated to glass modifiers, $Na^+$ and $Ca^{2+}$, with similar probabilities. A small percentage (2% of the total silicon related bond) of F–Si bonds were observed. The existence of Si–F bonds and high percentage of Si–O–P linkages were considered to be "defects" arisen from the high quenching rate and small box sizes in the simulations. Despite these, the alkali-fluorine rich regions that attracted modifiers from the silicate network were observed, suggesting the existence of phase separation between alkali-fluoride rich regions and phosphosilicate rich regions, which was observed for glass compositions with only 10 mol% of $CaF_2$.

In addition, AIMD was also applied in the study of bioactive fluorophosphate glass $P_2O_5$–$Na_2O$–CaO–$CaF_2$ with 363 and 197 atoms, respectively [20]. It was found that fluorine ions are incorporated into the phosphate network by replacing one oxygen in [$PO_4$] tetrahedra or coordinating with glass modifiers, $Na^+$ and $Ca^{2+}$ in this case. However, owing to the non-preference for fluorine to coordinate with $P^{5+}$, $Na^+$, or $Ca^{2+}$, segregation as being observed in AIMD of silicate-based bioactive glass is not likely to happen in fluorophosphate glass, which is beneficial to the bioactivity [127].

Although AIMD could not provide direct observation and statics of fluoride phase separation as in classical MD due to simulation cell size and simulation time limitations, these simulations provide accurate local structures of cations and anions hence can complement, as well as used to validate, classical MD simulation results.

## 13.6 Conclusions

This chapter has provided an overview for MD simulations of fluoride and oxyfluoride glasses. The history of fluoride glass simulations could be dated back to when MD just began to be applied to study melt and glass structures, with various forms of empirical potentials being proposed and employed, and much progress has been made since then. After introducing the general features of fluoride and oxyfluoride glasses, the applications of MD simulations on fluorozirconate glass structure and their properties calculations were presented. Latest developments and investigations of MD simulations on fluoride phase separations in fluoroaluminosilicate glasses were summarized. Mixed anion oxyhalide glass presents one of the few examples that phase separations can be directly observed from MD simulations with typical cooling rates of a few K/ps. The recent progresses of AIMD simulations of fluoride containing phosphosilicate glasses were also briefly discussed. In summary, MD simulations have been shown to be a very effective tool to elucidate the structure, phase separation and properties of halide and mixed anion glasses. The field has regained considerable attention and will likely continue to grow in the future.

## Acknowledgments

JD acknowledges support of the US National Science Foundation (project number 1662288). JZ would like to thank the sponsorship of the China Scholarship Council.

## References


1 Almeida, R.M. (2012). *Halide Glasses for Infrared Fiberoptics*, vol. 123. Springer Science & Business Media.

2 Tran, D., Sigel, G., and Bendow, B. (1984). Heavy metal fluoride glasses and fibers: a review. *J. Lightwave Technol.* 2 (5): 566–586.

3 Lucas, J. (1985). Rare earths in fluoride glasses. *J. Less-Common Met.* 112 (1, 2): 27–40.

4 Kanamori, T. (1987). Recent advances in fluoride glass fiber optics in Japan. *Mater. Sci. Forum*, Trans Tech Publ 19: 363–374.

5 Lucas, J. (1989). Fluoride glasses. *J. Mater. Sci.* 24 (1): 1–13.

6 Parker, J. (1989). Fluoride glasses. *Annu. Rev. Mater. Sci.* 19 (1): 21–41.

7 Almeida, R.M. (1991). Fluoride glasses. In: *Handbook on the Physics and Chemistry of Rare Earths*, vol. 15 (ed. K.A. Gschneidner Jr., and L.R. Eyring), 287–346. Elsevier Science.

8 Rahman, A., Fowler, R., and Narten, A. (1972). Structure and motion in liquid $BeF_2$, $LiBeF_3$, and LiF from molecular dynamics calculations. *J. Chem. Phys.* 57 (7): 3010–3011.

9 Zachariasen, W.H. (1932). The atomic arrangement in glass. *J. Am. Chem. Soc.* 54 (10): 3841–3851.

10 Soules, T.F. (1982). Molecular dynamic calculations of glass structure and diffusion in glass. *J. Non-Cryst. Solids* 49 (1–3): 29–52.

11 Brawer, S. and Weber, M. (1980). Monte Carlo simulation of $Eu^{3+}$-doped $BeF_2$ glass. *Phys. Rev. Lett.* 45 (6): 460.

12 Brawer, S. and Weber, M. (1981). Neodymium fluorescence in glass: comparison with computer simulations of glass structure. *J. Lumines.* 24: 115–118.

13 Lusvardi, G., Malavasi, G., Tarsitano, F. et al. (2009). Quantitative structure-property relationships of potentially bioactive fluoro phospho-silicate glasses. *J. Phys. Chem. B* 113 (30): 10331–10338.

14 Pedone, A., Chen, X., Hill, R.G., and Karpukhina, N. (2018). Molecular dynamics investigation of halide-containing phospho-silicate bioactive glasses. *J. Phys. Chem. B* 122: 2940–2948.

15 Swansbury, L.A., Mountjoy, G., Chen, X. et al. (2017). Modeling the onset of phase separation in $CaO–SiO_2–CaCl_2$ chlorine-containing silicate glasses. *J. Phys. Chem. B* 121: 5647–5653.

16 Cavillon, M., Faugas, B., Zhao, J. et al. (2019). Investigation of the structural environment and chemical bonding of fluorine in Yb-doped fluorosilicate glass optical fibres. *J. Chem. Thermodyn.* 128: 119–126.

17 Li, X., Chen, D., Huang, F. et al. (2018). Phase-selective nanocrystallization of $NaLnF_4$ in aluminosilicate glass for random laser and 940 nm LED-excitable upconverted luminescence. *Laser Photonics Rev.* 12: 1800030.

18 Peng, Y., Zhong, J., Li, X. et al. (2018). Controllable competitive nanocrystallization of $La^{3+}$-based fluorides in aluminosilicate glasses and optical spectroscopy. *J. Eur. Ceram. Soc.* 39: 1420–1427.

**19** Kang, S., Huang, Z., Lin, W. et al. (2019). Enhanced single-mode fiber laser emission by nano-crystallization of oxyfluoride glass-ceramic cores. *J. Mater. Chem. C* 7 (17): 5155–5162.

**20** Christie, J.K., Ainsworth, R.I., and de Leeuw, N.H. (2014). Ab initio molecular dynamics simulations of structural changes associated with the incorporation of fluorine in bioactive phosphate glasses. *Biomaterials* 35 (24): 6164–6171.

**21** Machacek, J., Kostka, P., Liska, M. et al. (2011). Calculation and analysis of vibrational spectra of $PbCl_2$–$Sb_2O_3$–$TeO_2$ glass from first principles. *J. Non-Cryst. Solids* 357 (14): 2562–2570.

**22** Tosi, M. and Fumi, F. (1964). Ionic sizes and born repulsive parameters in the NaCl-type alkali halides – II: the generalized Huggins–Mayer form. *J. Phys. Chem. Solids* 25 (1): 45–52.

**23** Lucas, J., Angell, C.A., and Tamaddon, S. (1984). Fluoride bridging modes in fluorozirconate glasses by X-ray and computer simulation studies. *Mater. Res. Bull.* 19 (7): 945–951.

**24** Zhao, J., Ma, R., Chen, X. et al. (2016). From phase separation to nanocrystallization in fluorosilicate glasses: structural design of highly luminescent glass-ceramics. *J. Phys. Chem. C* 120: 17726–17732.

**25** Ali, M.A., Ren, J., Liu, X. et al. (2017). Understanding enhanced upconversion luminescence in oxyfluoride glass-ceramics based on local structure characterizations and molecular dynamics simulations. *J. Phys. Chem. C* 121: 15384–15391.

**26** Baldwin, C. and Mackenzie, J. (1979). Fundamental condition for glass formation in fluoride systems. *J. Am. Ceram. Soc.* 62 (9, 10): 537–538.

**27** Poulain, M. (1981). Glass formation in ionic systems. *Nature* 293: 279.

**28** Sun, K.H. (1947). Fundamental condition of glass formation. *J. Am. Ceram. Soc.* 30 (9): 277–281.

**29** Wright, A.C. (1987). Diffraction studies of halide glasses. In: *Halide Glasses for Infrared Fiberoptics* (ed. R.M. Almeida), 75–117. Springer.

**30** Brawer, S.A. and Weber, M.J. (1981). Molecular dynamics simulations of the structure of rare-earth-doped beryllium fluoride glasses. *J. Chem. Phys.* 75 (7): 3522–3541.

**31** Kawamoto, Y. and Kono, A. (1986). Raman spectroscopic study of $AlF_3$–$CaF_2$–$BaF_2$ glasses. *J. Non-Cryst. Solids* 85 (3): 335–345.

**32** Fedorov, P.P., Zakalyukin, R.M., Ignat'eva, L.N., and Bouznik, V.M. (2000). Fluoroindate glasses. *Russ. Chem. Rev.* 69 (8): 705.

**33** Almeida, R., Pereira, J., and Alcala, R. (1992). Vibrational spectra and structure of glasses based on $ZnF_2$–$CdF_2$. *Phys. Chem. Glasses* 33 (5): 187–190.

**34** Phifer, C.C. and Lucas, J. (1987). A new perspective on fluorozirconate glass structure. *Mater. Sci. Forum*, Trans Tech Publ 19: 111–120.

**35** Almeida, R.M. and Mackenzie, J.D. (1981). Vibrational spectra and structure of fluorozirconate glasses. *J. Chem. Phys.* 74 (11): 5954–5961.

**36** Almeida, R.M. and Mackenzie, J.D. (1983). A structural interpretation of the vibrational spectra of binary fluorohafnate glasses. *J. Chem. Phys.* 78 (11): 6502–6511.

**37** Phifer, C.C., Gosztola, D.J., Kieffer, J., and Austen Angell, C. (1991). Effects of coordination environment on the Zr–F symmetric stretching frequency of fluorozirconate glasses, crystals, and melts. *J. Chem. Phys.* 94 (5): 3440–3450.

**38** Kawamoto, Y. (1984). The Raman-spectra of barium fluorozirconate glasses and their interpretation. *Phys. Chem. Glasses* 25 (3): 88–91.

**39** Kawamoto, Y. and Sakaguchi, F. (1983). Thermal properties and Raman spectra of crystalline and vitreous $BaZrF_6$, $PbZrF_6$, and $SrZrF_6$. *Bull. Chem. Soc. Jpn.* 56 (7): 2138–2141.

**40** Kawamoto, Y. and Horisaka, T. (1983). Short-range structures of barium, lead, and strontium meta-fluorozirconate glasses. *J. Non-Cryst. Solids* 56 (1–3): 39–44.

**41** Etherington, G., Keller, L., Lee, A. et al. (1984). An X-ray diffraction study of the structure of barium fluorozirconate and fluorohafnate glasses. *J. Non-Cryst. Solids* 69 (1): 69–80.

**42** Almeida, R.M. and Mackenzie, J.D. (1984). A vibrational spectroscopy study of the structure of binary thorium fluorohafnate glasses. *J. Non-Cryst. Solids* 68 (2, 3): 203–217.

**43** Kiprianov, A. and Karpukhina, N. (2006). Oxyhalide silicate glasses. *Glass Phys. Chem* 32: 1–27.

**44** Bradtmüller, H., Zhang, L., de Araujo, C.C. et al. (2018). Structural studies of $NaPO_3$–$AlF_3$ glasses by high-resolution double-resonance nuclear magnetic resonance spectroscopy. *J. Phys. Chem. C* 122 (37): 21579–21588.

**45** Brow, R.K., Osborne, Z., and Kirkpatrick, R.J. (1992). A multinuclear MAS NMR study of the short-range structure of fluorophosphate glass. *J. Mater. Res.* 7 (7): 1892–1899.

**46** Boussard-Plédel, C., Le Floch, M., Fonteneau, G. et al. (1997). The structure of a boron oxyfluoride glass, an inorganic cross-linked chain polymer. *J. Non-Cryst. Solids* 209 (3): 247–256.

**47** Rodriguez Chialanza, M., Schneider, J.F., Keuchkerian, R. et al. (2020). Structural analysis of oxyfluoride borate glass and $BaF_2$ crystallization from phase separation. *J. Am. Ceram. Soc.* 103 (5): 3126–3137.

**48** Lin, C., Bocker, C., and Rüssel, C. (2015). Nanocrystallization in oxyfluoride glasses controlled by amorphous phase separation. *Nano Lett.* 15: 6764–6769.

**49** Möncke, D. and Eckert, H. (2019). Review on the structural analysis of fluoride-phosphate and fluoro-phosphate glasses. *J. Non-Cryst. Solids X* 3: 100026.

**50** Velli, L., Varsamis, C.-P.E., Kamitsos, E. et al. (2005). Structural investigation of metaphosphate glasses. *Phys. Chem. Glasses* 46 (2): 178–181.

**51** Videau, J.-J., Portier, J., and Piriou, B. (1982). Raman spectroscopic studies of fluorophosphate glasses. *J. Non-Cryst. Solids* 48 (2, 3): 385–392.

**52** Möncke, D. and Ehrt, D. (2005). Structure and properties of mixed phosphate and fluoride glasses. *Phys. Chem. Glasses* 46 (2): 67–71.

**53** Möncke, D., Ehrt, D., Velli, L. et al. (2007). Comparative spectroscopic investigation of different types of fluoride phosphate glasses. *Phys. Chem. Glasses-Eur. J. Glass Sci. Technol. Part B* 48 (6): 399–402.

**54** Jager, C. and Haubenreisser, U. (1985). A reexamination of studies of the structure of $NaF$–$Na_2O$–$B_2O_3$ glasses. *Phys. Chem. Glasses* 26 (5): 152–156.

**55** Mishra, R., Sudarsan, V., Kaushik, C. et al. (2009). Effect of fluoride ion incorporation on the structural aspects of barium–sodium borosilicate glasses. *J. Non-Cryst. Solids* 355 (7): 414–419.

**56** Chung, J., Yanaba, Y., Nakatsuka, Y. et al. (2019). Effect of fluorine on the optical properties of $BaF_2$–BaO–$La_2O_3$–$B_2O_3$ glasses prepared by containerless processing. *Int. J. Appl. Glass Sci.* 10 (2): 181–189.

**57** Pisarski, W.A., Goryczka, T., Pisarska, J., and Ryba-Romanowski, W. (2007). Er-doped lead borate glasses and transparent glass ceramics for near-infrared luminescence and up-conversion applications. *J. Phys. Chem. B* 111 (10): 2427–2430.

**58** Cheng, J., Deng, W., and Wang, M. (2012). Structure of $Na_2O$·MO·$SiO_2$·$CaF_2$ (M = Mg, Ca) oxyfluoride glasses. *Physica B* 407 (14): 2778–2783.

**59** Biswas, K., Karmakar, P., Balaji, S. et al. (2017). Structural modification associated with $Al_2O_3$ addition in oxyfluoride glasses: thermal and mechanical properties. *J. Am. Ceram. Soc.* 100 (12): 5490–5501.

**60** Lin, C., Russel, C., and van Wüllen, L. (2019). Phase separation and nanocrystallization in KF–$ZnF_2$–$SiO_2$ glasses: lessons from solid state NMR. *J. Phys. Chem. B* 123: 1688–1695.

**61** Hendy, S. and Edgar, A. (2006). Structure of fluorochlorozirconate glasses using molecular dynamics. *J. Non-Cryst. Solids* 352 (5): 415–422.

**62** Hayakawa, S., Ohtsuki, C., Matsumoto, S. et al. (1998). Molecular dynamic simulation of heterogeneity and chemical states of fluorine in amorphous alkaline earth silicate systems. *Comput. Mater. Sci.* 9 (3): 337–342.

**63** Boehm, L., Chung, K.-H., and Moynihan, C.T. (1988). Phase separation in lead containing fluorozirconate glasses. *J. Non-Cryst. Solids* 102 (1–3): 150–156.

**64** Bocker, C., Wiemert, J., and Ruessel, C. (2013). The formation of strontium fluoride nano crystals from a phase separated silicate glass. *J. Eur. Ceram. Soc.* 33: 1737–1745.

**65** Drehman, A.J. (1986). Viscous flow vs phase separation during calorimetric analysis of fluoride glasses. *J. Am. Ceram. Soc.* 69 (12): C-306–C-307.

**66** Boehm, L., Chung, K., Crichton, S., and Moynihan, C. (1987). Crystallization and phase separation in fluoride glasses. In: *Infrared Optical Materials and Fibers V*, 10–20. International Society for Optics and Photonics https://www.spiedigitallibrary.org/conference-proceedings-of-spie/0843/0000/Crystallization-And-Phase-Separation-In-Fluoride-Glasses/10.1117/12.968208.full?SSO=1.

**67** Boehm, L., Chung, K., Crichton, S., and Moynihan, C.T. (1987). Observation of phase separation in lead containing $ZrF_4$-based glasses. *Mater. Sci. Forum*, Trans Tech Publ 533–544.

**68** Simmons, C.J. and Simmons, J.H. (1985). Liquid-liquid immiscibility in mixed $ZrF_4$–$ThF_4$ glass. *J. Am. Ceram. Soc.* 68 (9): C-258–C-259.

**69** Bocker, C. and Russel, C. (2009). Self-organized nano-crystallisation of $BaF_2$ from $Na_2O/K_2O/BaF_2/Al_2O_3/SiO_2$ glasses. *J. Eur. Ceram. Soc.* 29 (7): 1221–1225.

**70** de Pablos-Martin, A., Hemono, N., Mather, G.C. et al. (2011). Crystallization kinetics of $LaF_3$ nanocrystals in an oxyfluoride glass. *J. Am. Ceram. Soc.* 94 (8): 2420–2428.

**71** Fedorov, P.P., Luginina, A.A., and Popov, A.I. (2015). Transparent oxyfluoride glass ceramics. *J. Fluorine Chem.* 172: 22–50.

**72** Dymshits, O., Shepilov, M., and Zhilin, A. (2017). Transparent glass-ceramics for optical applications. *MRS Bull.* 42: 200–205.

**73** Bocker, C., Munoz, F., Duran, A., and Russel, C. (2011). Fluorine sites in glasses and transparent glass-ceramics of the system $Na_2O/K_2O/Al_2O_3/SiO_2/BaF_2$. *J. Solid State Chem.* 184: 405–410.

**74** Lin, C., Li, L., Dai, S. et al. (2016). Oxyfluoride glass-ceramics for transition metal ion based photonics: broadband near-IR luminescence of nickel ion dopant and nanocrystallization mechanism. *J. Phys. Chem. C* 120 (8): 4556–4563.

**75** Hemono, N., Pierre, G., Munoz, F. et al. (2009). Processing of transparent glass-ceramics by nanocrystallisation of $LaF_3$. *J. Eur. Ceram. Soc.* 29: 2915–2920.

**76** de Pablos-Martín, A., Pascual, M.J., and Durán, A. (2014). Transparent nano-glass-ceramic for photonic applications. Distribution of RE-doping elements in the fluoride nano-crystals analysed by XAS and HR-TEM. *Adv. Sci. Technol.* 90: 111–120.

**77** Bhattacharyya, S., Höche, T., Hemono, N. et al. (2009). Nano-crystallization in $LaF_3$–$Na_2O$–$Al_2O_3$–$SiO_2$ glass. *J. Cryst. Growth* 311: 4350–4355.

**78** de Pablos-Martin, A., Patzig, C., Höche, T. et al. (2013). Distribution of thulium in $Tm^{3+}$-doped oxyfluoride glasses and glass-ceramics. *Crystengcomm* 15 (35): 6979–6985.

**79** Beall, G. (2004). Anhedral crystallisation in phase separated glasses. *Glass Technol.* 45 (2): 54–58.

**80** Bocker, C., Ruessel, C., and Avramov, I. (2013). Transparent nano crystalline glass-ceramics by interface controlled crystallization. *Int. J. Appl. Glass Sci.* 4 (3): 174–181.

**81** Bhattacharyya, S., Bocker, C., Heil, T. et al. (2009). Experimental evidence of self-limited growth of nanocrystals in glass. *Nano Lett.* 9 (6): 2493–2496.

**82** Häßler, J. and Rüssel, C. (2017). Self-organized growth of sodium borate-rich droplets in a phase-separated sodium borosilicate glass. *Int. J. Appl. Glass Sci.* 8 (1): 124–131.

**83** Stoica, M., Brehl, M., Bocker, C. et al. (2018). Microstructure and luminescence of erbiumdoped $Na_2O/K_2O/CaO/CaF_2/Al_2O_3/SiO_2$ nano glass-ceramics. *Mater. Chem. Phys.* 207: 36–43.

**84** Raghuwanshi, V.S., Hoell, A., Bocker, C., and Russel, C. (2012). Experimental evidence of a diffusion barrier around $BaF_2$ nanocrystals in a silicate glass system by ASAXS. *Crystengcomm* 14 (16): 5215–5223.

**85** Gruenhut, S., Amini, M., Macfarlane, D.R., and Meakin, P. (1997). Structure of Zr/Ba/Na fluoride glass using molecular dynamics. *J. Non-Cryst. Solids* 213: 398–403.

**86** Yasui, I. and Inoue, H. (1985). Molecular dynamic simulations of changes in structure in $ZrF_4$ based glasses. *J. Non-Cryst. Solids* 71 (1–3): 39–47.

**87** Bendow, B., Banerjee, P.K., Lucas, J. et al. (1985). Polarized Raman scattering in rare-earth fluoride glasses. *J. Am. Ceram. Soc* 68 (4): C-92–C-95.

**88** Kawamoto, Y., Horisaka, T., Hirao, K., and Soga, N. (1985). A molecular dynamics study of barium meta-fluorozirconate glass. *J. Chem. Phys.* 83 (5): 2398–2404.

**89** Parker, J.M. (1989). Molecular dynamics simulations of halide glasses. *J. Comput. Aided Mol. Des.* 3 (4): 327–334.

**90** Inoue, H. and Yasui, I. (1987). The study on the structure of $ZrF_4$–$BaF_2$–$RF_3$ (R = Al, La) glasses. *J. Non-Cryst. Solids* 95: 217–224.

**91** Coupé, R., Louër, D., Lucas, J., and Léonard, A.J. (1983). X-ray scattering studies of glasses in the system $ZrF_4$–$BaF^2$. *J. Am. Ceram. Soc.* 66 (7): 523–529.

**92** Soules, T.F. (1990). Computer simulation of glass structures. *J. Non-Cryst. Solids* 123 (1–3): 48–70.

**93** Hamill, L. and Parker, J. (1985). Molecular dynamics simulations of fluoride glass melts. *Mater. Sci. Forum*, Trans Tech Publ 437–440.

94 Harrison, M. and Denning, R. (1996). Site selective spectroscopy and molecular dynamics simulation of Eu (III) ZBLAN glasses. *J. Lumines.* 69 (5–6): 265–285.

95 Uhlherr, A., MacFarlane, D., Moore, L., and Thomas, P. (1991). Molecular dynamics investigation of the structure and stability of zirconium-barium-rare earth fluoride glasses. *Mater. Sci. Forum*, Trans Tech Publ 431–436.

96 Tamai, Y. and Kawamoto, Y. (2000). Structural change in the process of permanent densification of fluorozirconate glass studied with molecular dynamics simulations. *Phys. Rev. B* 62 (2): 865.

97 Messaddeq, Y., Delben, A., Aegerter, M.A. et al. (1993). New fluoroindate glass compositions. *J. Mater. Res.* 8 (4): 885–889.

98 Simmons, J.H., Faith, R., and O'Rear, G. (1987). Molecular dynamics simulations of fluoride glass structures. *Mater. Sci. Forum*, Trans Tech Publ 121–126.

99 Simmons, J.H., O'Rear, G., Swiler, T.P., and Wright, A.C. (1988). Structural modeling of the $ZrF_4$–$BaF_2$ binary using molecular dynamics. *J. Non-Cryst. Solids* 106 (1–3): 325–329.

100 Uhlherr, A. and MacFarlane, D.R. (1993). A systematic survey of experimental and simulated glass formation for the $ZrF_4$–$BaF_2$–NaF system. *J. Non-Cryst. Solids* 161: 98–104.

101 Hamill, L. and Parker, J. (1985). Hydrogen solubility in heavy metal fluoride glasses. *Phys. Chem. Glasses* 26 (2): 52–54.

102 Soga, K., Inoue, H., and Makishima, A. (2001). Simulation of optical spectra of $Eu^{3+}$ ion in fluorozirconate glasses by molecular dynamics simulation and point charge crystal field. *J. Appl. Phys.* 89 (7): 3730–3735.

103 Phifer, C.C., Angell, C.A., Laval, J., and Lucas, J. (1987). A structural model for prototypical fluorozirconate glass. *J. Non-Cryst. Solids* 94 (3): 315–335.

104 Nanba, T., Inoue, H., Arai, Y. et al. (1988). Diffraction studies of $AlF_3$–$BaF_2$–$CaF_2$ glasses. *Mater. Sci. Forum*, Trans Tech Publ 385–390.

105 Angell, C., Cheeseman, P., and Tamaddon, S. (1982). Computer simulation studies of migration mechanisms in ionic glasses and liquids. *J. Phys. Colloques* 43 (C9): C9-381–C9-385.

106 Kawamoto, Y. (1985). Progress in structural study of $ZrF_4$-based glasses. *Mater. Sci. Forum*, Trans Tech Publ 417–425.

107 Inoue, H. and Makishima, A. (1993). Computer simulation of the vibrational spectra and properties of fluoride glasses based on $ZrF_4$. *J. Non-Cryst. Solids* 161: 118–122.

108 Boulard, B., Kieffer, J., Phifer, C., and Angell, C. (1992). Vibrational spectra in fluoride crystals and glasses at normal and high pressures by computer simulation. *J. Non-Cryst. Solids* 140: 350–358.

109 Miyauchi, K., Qiu, J., Shojiya, M. et al. (1999). Peculiar high-pressure behavior in permanent densification of fluorozirconate glass. *Mater. Res. Bull.* 34 (9): 1383–1389.

110 Miyauchi, K., Shojiya, M., Kawamoto, Y., and Kitamura, N. (2001). Structural study on peculiar high-pressure behavior of fluorozirconate glass in permanent densification. *J. Phys. Chem. Solids* 62 (11): 2039–2046.

111 Tamai, Y. and Kawamoto, Y. (2002). Molecular dynamics simulation of high-pressure densification of fluorozirconate glass. *J. Non-Cryst. Solids* 306 (3): 282–291.

112 Tamai, Y. and Kawamoto, Y. (2000). Role of structural relaxation in peculiar permanent densification of fluorozirconate glass. *J. Chem. Phys.* 112 (8): 3875–3883.

113 Tamai, Y. (2003). Effects of pressure on the fragile nature of fluorozirconates studied by molecular dynamics simulations. *Phys. Rev. E* 67 (3): 031504.

114 Lusvardi, G., Malavasi, G., Cortada, M. et al. (2008). Elucidation of the structural role of fluorine in potentially bioactive glasses by experimental and computational investigation. *J. Phys. Chem. B* 112: 12730–12739.

115 Du, J. (2015). Challenges in molecular dynamics simulations of multicomponent oxide glasses. In: *Molecular Dynamics Simulations of Disordered Materials* (ed. C. Massobrio, J. Du, P.S. Salmon and M. Bernasconi), 157–180. Springer.

116 Zhao, J., Nienhuis, E.T., McCloy, J.S., and Du, J. (2020). Structures of fluoride containing aluminosilicate low activity nuclear waste glasses: a molecular dynamics simulations study. *J. Non-Cryst. Solids* 550: 120379.

117 Zhao, J., Xu, X., Li, P. et al. (2019). Structural origins of $RF_3/NaRF_4$ nanocrystals precipitation from phase separated $SiO_2$–$Al_2O_3$–$RF_3$–NaF glasses: a molecular dynamics simulation study. *J. Phys. Chem. B* 123: 3024.

118 Zhao, J., Xu, X., Ren, K. et al. (2019). Molecular dynamic simulations: structural origins of $BaF_2/Ba_{1-x}R_xF_{2+x}/RF_3$ nanocrystals formation from phase separated fluoroaluminosilicate glass: a molecular dynamic simulation study. *Adv. Theory Simul.* 2 (10): 1970034.

119 Xu, X., Zhao, J., Chen, X. et al. (2020). $Ca^{2+}/Sr^{2+}/Ba^{2+}$ dependent phase separation, nanocrystallization and photoluminescence in fluoroaluminosilicate glass. *J. Am. Ceram. Soc.* 103 (10): 5796–5807.

120 Zhao, J., Xu, X., Chen, X. et al. (2019). A structure model for phase separated fluoroaluminosilicate glass system by molecular dynamic simulations. *J. Eur. Ceram. Soc.* 39 (15): 5018–5029.

121 Montorsi, M., Menziani, M., Leonelli, C. et al. (2000). Molecular dynamics simulations of alumina addition in sodium silicate glasses. *Mol. Simul.* 24 (1–3): 157–165.

122 Bocker, C., Wiemert, J., and Ruessel, C. (2014). The effect of viscosity on nanocrystallization of strontium fluoride from a silicate glass. *Solid State Sci.* 30: 55–60.

123 Xu, X., Zhao, J., Luo, X. et al. (2019). Stabilization of fluorescent $[Ag_m]^{n+}$ quantum clusters in multi-phase inorganic glass-ceramics for white LEDs. *ACS Appl. Nano Mater.*

124 Chen, X., Zhao, J., Xu, X. et al. (2018). Phase separation strategy to facilely form fluorescent $[Ag_2]^{2+}/[Ag_m]^{n+}$ quantum clusters in boro-alumino-silicate multiphase glasses. *Phys. Chem. Chem. Phys.* 20: 23942–23947.

125 Christie, J.K., Pedone, A., Menziani, M.C., and Tilocca, A. (2011). Fluorine environment in bioactive glasses: ab initio molecular dynamics simulations. *J. Phys. Chem. B* 115 (9): 2038–2045.

126 Pedone, A., Charpentier, T., and Menziani, M.C. (2012). The structure of fluoride-containing bioactive glasses: new insights from first-principles calculations and solid state NMR spectroscopy. *J. Mater. Chem.* 22 (25): 12599–12608.

127 Shah, F.A. (2016). Fluoride-containing bioactive glasses: glass design, structure, bioactivity, cellular interactions, and recent developments. *Mater. Sci. Eng., C* 58: 1279–1289.

# 14

# Glass Surface Simulations

*Todd R. Zeitler[1] and Alastair N. Cormack[2]*

[1] *Geotechnology & Engineering Department, Sandia National Laboratories, Albuquerque, NM, USA*
[2] *Inamori School of Engineering, New York State College of Ceramics, Alfred University, Alfred, NY, USA*

## 14.1 Introduction

Although the understanding of atomic structures of bulk glasses is important to the understanding of many intrinsic glass properties, surface properties more completely describe the interactions between bulk glasses and the environment. For example, crack propagation, surface dissolution, and ion exchange are all surface phenomena. Bulk glass simulations have been performed for over 50 years and while the first surface simulations were published not long after [1–5], surface simulations continue to be much less common than bulk simulations due to their relative complexity and because surface simulations are typically based on a bulk simulation of the same composition.

Computational approaches to glass surface simulations include classical (Newtonian) and electronic-scale calculations. Classical approaches have typically taken the form of molecular dynamics (MD) [6–9] and Monte Carlo (MC) [10–12] simulations. Electronic-scale approaches include small cluster calculations [13–15], as well as larger, periodic density functional theory (DFT) [16–20] calculations. The current review will focus on classical MD simulations and first principles calculations of silicate glass surfaces, in some cases including water species (i.e. OH groups and $H_2O$ molecules).

## 14.2 Classical MD Surface Simulations

MD is by far the most common computational technique used to study glass surfaces. MD surface structures are almost invariably based on bulk glass structures previously generated by MD. Some early surface structures have been generated by straining a bulk structure through systematically elongating the simulation cell in one dimension, effectively simulating a fracture that resulted in two surfaces [21]. Another option for surface structure creation

*Atomistic Simulations of Glasses: Fundamentals and Applications*, First Edition.
Edited by Jincheng Du and Alastair N. Cormack.

is the construction of nonperiodic cluster "particles" [9]. However, the most conventional technique, pioneered by Garofalini and coworkers [1–5], results in a single surface from a bulk structure via a series of steps in which a vacuum separation gap is introduced, the lower region of the cell is immobilized (to maintain a connection with the bulk structure far from the surface), and the upper region is then subject to an annealing process that allows for rearrangement of atom positions near the surface (Figure 14.1). Local atomic rearrangement of oxygen and network modifiers (and network formers to a much lesser extent) [8] occurs due to the changes to net forces acting on individual atoms when the free surface (i.e. vacuum gap) is introduced [3].

As with all MD simulations, surface simulations have atomic coordinates output over time, which can then be used to calculate structural and dynamical properties of the surface. In some cases, these calculated properties are compared to the properties of the equivalent bulk composition due to changes associated with the conversion of bridging oxygen (BO) atoms to nonbridging oxygen (NBO) atoms. Atomic density profiles are used to distinguish the location of the bulk-to-surface transition (~5–10 Å) [7, 22], as well as identify any changes to the surface structure during the surface creation process (e.g. NBO/BO ratio [23] or alkali enrichment [8, 24]). Other properties that are investigated include atomic diffusion [6], radial distribution function [3, 5], ring statistics [5], and defect statistics [3, 7].

As with simulations of bulk glass structures, surface simulations initially focused on the single component silica composition. With the availability of more general potential sets [24–26], surface simulations of multicomponent oxide glasses became possible and have increased in popularity. Surface simulations have also been extended beyond bulk compositions to include water species, in an attempt to describe some of the surface reactions that occur in humid or liquid conditions. The following three subsections will focus on

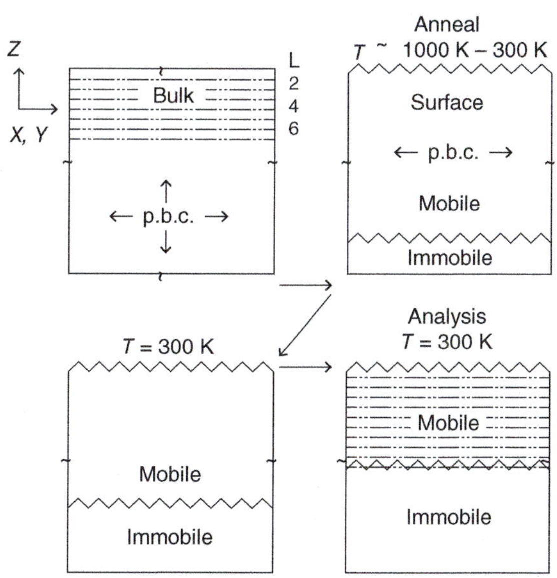


**Figure 14.1** Schematic diagram for creating and analyzing the v-$SiO_2$ surface. Source: Reproduced from Feuston and Garofalini [5].

surface simulations of (i) amorphous silica surfaces; (ii) multicomponent surfaces; and (iii) surfaces in the presence of water. Although a few studies have investigated simulations of glass surfaces in contact with species other than water [27, 28], they will not be covered in the current review. Similarly, the relative lack of surface simulations for nonoxide glasses precludes their review here.

### 14.2.1 Amorphous Silica Surfaces

The first glass surface simulation studies focused on amorphous silica [1–3] and studies of silica surfaces still continue, as newer potentials are developed [29] and MD techniques are improved [30]. Now even a "pure" silica surface created experimentally will not be free of water species on the immediate surface, so simulations of amorphous silica surfaces do not necessarily represent a real surface. Even surfaces created under high vacuum conditions will have some surface hydroxyls present [6]. In addition to their use to investigate the properties of silica surfaces, silica surface simulations provide the important proof of concept on which multicomponent oxide glass surface simulations find their basis.

The ideal bulk structure of amorphous silica consists of fully coordinated silicon atoms, no NBO, and mainly five- and six-membered rings. Simulations of amorphous silica surfaces show the presence of undercoordinated Si, NBO, and small-membered rings at the immediate surface [6, 23]. These species may be described as "defects," are introduced during the surface creation process, and have been seen experimentally [6]. These sites occur at the immediate (top ~3 Å) surface [5] and are typically the most reactive to moisture (as discussed in Section 14.2.3 below).

Shortly following surface creation, undercoordinated defects in the surface undergo rearrangement. Following reconstruction, undercoordinated Si ($^3$Si) become tetrahedrally coordinated [3]. Some "overcoordinated" Si ($^5$Si) may also appear during rearrangement [3]. In general, there is a resultant enrichment of oxygen, rather than silicon, at the surface due to a minimization in the number of "energetically frustrated bonds" at the surface [9, 31]. The surface segregation of oxygen at the immediate surface corresponds to an increased concentration of silicon under the surface [3, 31]. This overall increase in density at the surface is observed in MD simulation and is attributed to the reconsolidation that occurs following fracture, in which the prevalence of small (compact) member rings increases [19].

The breaking of a Si—O bond during surface creation immediately creates an undercoordinated Si and an NBO. Without the presence of water, some localized rearrangement may occur that results in the formation of small-membered rings [5, 7]. The resultant Si–O–Si angles in these rings are smaller than the equilibrium angle expected in bulk silica (~109.5°) and thus underlie an increased amount of localized strain in the surface structure [6]. In addition to the small-membered rings [9, 32], larger rings also exist at the surface and provide a means by which nonglass species (e.g. water, alkali, or other species) might enter the glass structure without breaking of Si—O bonds in the surface rings [33, 34].

Hydrophobicity of the amorphous silica surface has been tested in an MD model that includes a special constraint that results in a surface without NBO, but solely Si–O–Si (siloxane) bridges [35]. This was done in response to some controversy (inconclusive experimental evidence [36, 37]) as to the true hydrophobicity of a pristine silica surface.

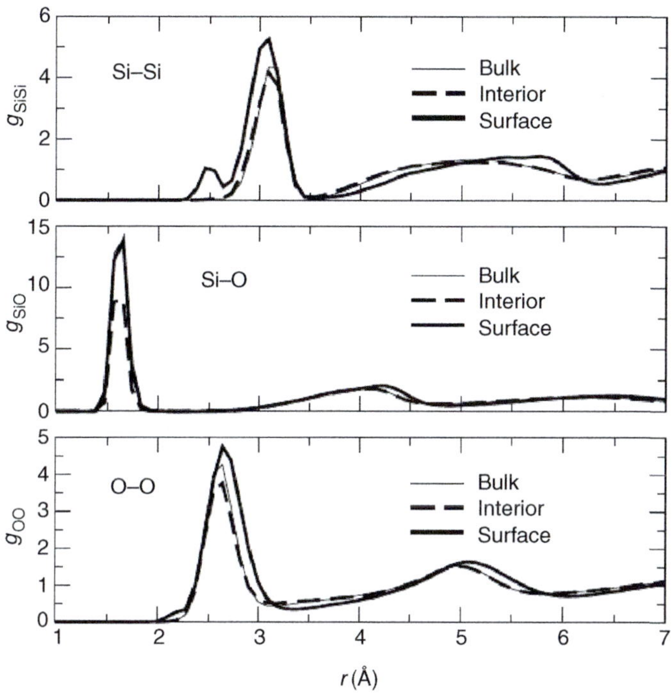


**Figure 14.2** Radial pair distribution functions in a bulk silica sample (thin solid), in the interior region of a silica film (dashed) and in the surface region (solid bold). Source: Reproduced from Rarivomanantsoa et al. [38].

Pair distribution functions (PDFs) comparisons for bulk and surface silica simulations show (following relaxation) some broadening of the first peak, including the presence of a peak at ~1.5 Å (Figure 14.2). The broadening to higher Si–O separations is concurrent with the strained Si—O bond. The peak at 1.5 Å is due to the presence of NBO [3, 38]. A peak at 2.1–2.3 Å exists only at the surface and is attributed to the O–O first neighbor separation (shorter than typical) due to the formation of two member rings [3]. This also corresponds to a shoulder at 2.5 Å in the Si–Si separation [38]. Bond angle distribution differences between bulk and surface also indicate the presence of two member rings [3, 7, 9, 38], but do not always agree quantitatively with ab initio calculations [38, 39] or sometimes even among MD simulations [5, 9].

Some investigation of the diffusion of Si and O at the silica surface has been done [9, 31]. Increased diffusion was found at the surface, as expected due to the lower surface density and presence of smaller $Q_n$ species (i.e. a reduced rigidity of structure). More specifically, NBO motion/migration was found to be more pronounced than for BO, as well as anisotropic with respect to surface versus bulk regions – they move parallel to the surface at the surface, but in the bulk move toward the surface [38].

### 14.2.2 Multicomponent Oxide Glass Surfaces

Building on the successes of amorphous silica glass simulations, multicomponent MD glass surface simulations have become more popular since the review of glass surface simulations provided in 2001 [6]. Two of the initial challenges to the simulation of multicomponent glass surfaces have been (i) the development of potentials for the additional components; and (ii) the necessity of large numbers of atoms in a given simulation in order to be able to adequately account for smaller components. However, with the availability of general force fields and increased capacity of computational power, more complex compositions are available for investigation. From simple alkali-silicate glasses [2, 40] to four-component bioactive glasses [41, 42], studies of multicomponent oxide glass surfaces represent steps forward in increased complexity in glass surface structure and in the analysis of those structures (Figure 14.3).

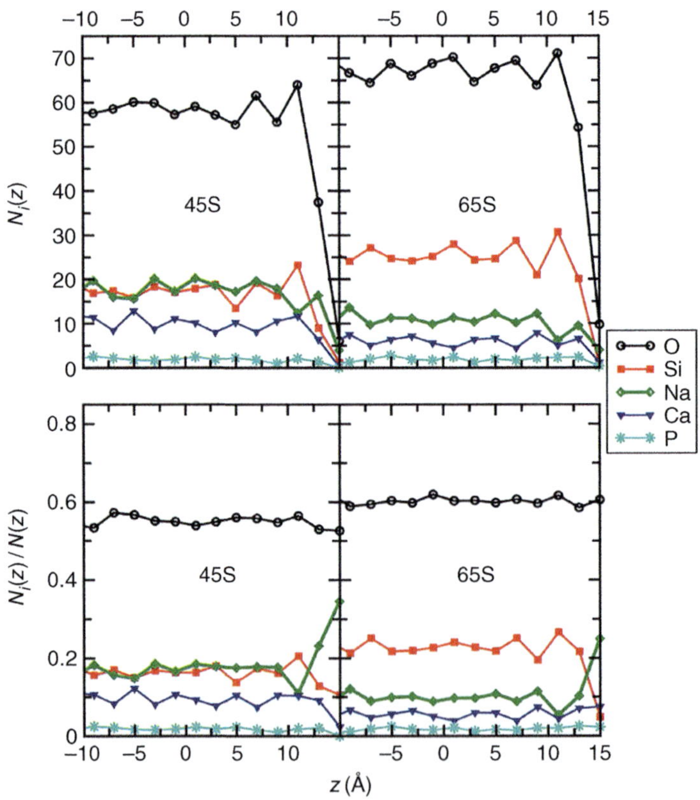


Figure 14.3 Z-profiles of the total number (top panels) and fraction (bottom panels) of O, Si, Na, Ca, and P atoms in the slab models of (left panels) 45S and (right panels) 65S glass. Source: Reproduced from Tilocca and Cormack [41].

The simplest multicomponent silicate glass surfaces investigated are typically those of two-component alkali silicates. The addition of alkali to the structure of amorphous silica serves to disrupt the tetrahedrally composed network framework by introducing NBO. These NBO sites are effectively negatively charged, so the alkali themselves provide local charge balance and are found nearby NBO sites. However, alkali (and thus NBO) clustering has been observed, so the NBO sites are not regularly distributed in the bulk alkali silicate structure [43]. Upon surface creation via fracture, the exposed surfaces exhibit enrichment of sodium and NBO [44]. This has been observed in the simulations of simple alkali silicate glasses [8, 22] and melts [45], as well as glasses of more complex composition [24, 46]. Network depolymerization (i.e. the creation of negatively charged NBO upon fracture) has been proposed as a driving force for Na migration to the surface, as the positively charged alkali ions migrate to the surface in order to achieve surface charge neutrality (Figure 14.4) [44, 45].

While Na ions have been found to be associated with each NBO in bulk simulations, NBO at the surface is coordinated by less than one Na on average [8]. This may be attributed to a lack of thermodynamic equilibrium achieved in the surface simulation, as Na from the bulk have not had sufficient time to fully migrate to the surface upon creation of "excess" NBO there [45]. This is consistent with the observation of alkali-depleted zones immediately below alkali-enriched zones near the surface [22]. The presence of Na at the immediate surface can lead to the interaction with adsorbed water via an ion exchange reaction.

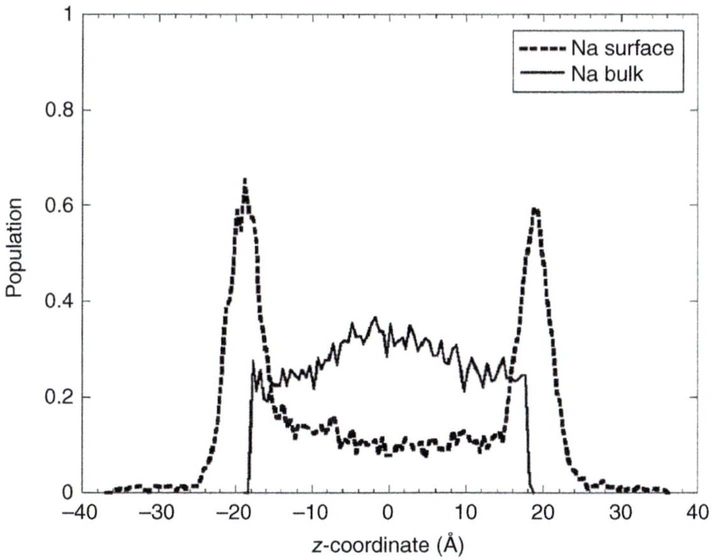


**Figure 14.4** A direct comparison of the sodium ion density profile of the surface and bulk states that reveal a sharp enrichment near the surface. Source: Reproduced from Corrales and Du [45].

Interestingly, surface simulations of amorphous silica glass for which Na has been placed on the surface postsurface-creation show a migration of Na into the surface such that Na are associated with NBO and the local Na environment is similar to that of an as-quenched Na-silicate glass [40]. The local bonding environment for network-forming Si ions was shown to not change, consistent with the modified random network postulation of Greaves [43] in which alkali-rich portions of the glass structure are expected to exist. Additionally, the enrichment of alkali (K and Na, but not Li) at the surface of alkali-silicate glasses has been observed experimentally [47, 48] and via MD simulation solely through the process of creating a surface [2, 8, 24, 49, 50]. However, other ions (e.g. Li, Ca, Mg, Al, B, and Si) do not migrate preferentially to the surface. It is expected that this preferential surface accumulation affects in addition to the surface structure, crack initiation, and leaching behavior [6].

Surface simulations of silicate glasses with up to four to six components have been performed (see also Section 14.2.2.1 below) [22, 24, 41, 46]. The general observations of alkali migration to the surface, depolymerization leading to larger surface rings, and increased surface mobility of ions are consistent with simulations with fewer components [22]. Also, when Al is added as a network former, Na becomes "tied up" in the network as part of a charge-compensating mechanism for Al, and thus less Na is found to migrate to the surface [8]. Although typically Si and Al as network formers have been observed to remain fully four-coordinated at the surfaces in simulations, an increase in three-coordinated B species (and decrease in four-coordinated B species) have been observed at the immediate surface [22]. This is not only consistent with the overall increase in NBO at surfaces but also points to potential differences in the extent of surface rearrangement between silicate and borate glasses in general.

Advanced MD simulations have investigated phenomena as complex as fracture dynamics [51]. Multicomponent glasses have been shown to be somewhat resistant to mechanical deformation via structural rearrangement surrounding modifier cations and structural defects [51]. This is in contrast to the behavior observed for simulations of silica nanowires, in which lower mechanical properties (with respect to bulk models) are seen and are attributed to a slightly higher (compared to soda–silicate nanowires) amount of atomic defects [52]. Surface simulations of multicomponent glasses have also been extended beyond the typical MD simulation cell sizes (via replication) in order to study larger scale mechanical properties such as indentation response (Figure 14.5) [53].

#### 14.2.2.1 Bioactive Glasses

Bioactive glasses are a special subset of multicomponent silicate glasses that have been widely studied for their interesting properties *in vivo* [54, 55]. They have been the subject of a number of bulk simulation studies [56–63] and are typically taken to be phosphosilicate glasses with alkali and alkaline earth modifiers. A well-known series of reactions take place at the surfaces of bioactive glasses in contact with water (or bodily fluid) and that has made them the subject of surface simulation studies as well [24, 46, 64]. The initial reactions, ion exchange of H for Na in the glass and the depolymerization of the network via hydrolysis, are relatively simple and can be studied by both MD and first principles (see Section 14.3.2.1 below) calculations.

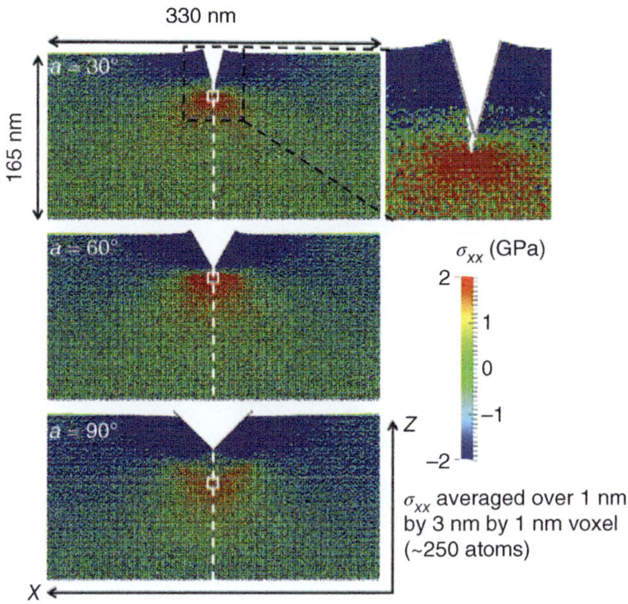


**Figure 14.5** The deformation morphology and $\sigma_{xx}$ stress field under an indentation depth of 36 nm with labeled indenter angle. Source: Reproduced from Luo et al. [53], Fig 02 (p. 03)/Springer Nature.

Atomic density plots of as-created surfaces show enhanced Na presence, as well as NBO/BO ratio, at the immediate surface (Figure 14.6) [24]. The dissolution of network-forming Si—O bonds at the surface was not found to be energetically favorable prior to the ion exchange reaction [24], but it was noted that for some compositions (those shown to be most reactive experimentally), hydrolysis is favorable following ion exchange [65]. The local environment following ion exchange is sufficiently different than that of the as-created glass to allow for a noticeable difference in calculated reaction energies. This was attributed to local strain imposed by the replacement of Na with H on silanols during the ion exchange reaction [65].

Another study investigated the structural effects of the Na/H exchange by incorporating the OH groups from the beginning of the bulk simulation [66]. In fact, surface simulations were not performed, but the structural impacts of the complete exchange on near-surface structure are implied. This technique has the advantage of not assuming details about the exchange process, while resulting in a stable, representative near-surface structure.

### 14.2.3 "Wet" Glass Surfaces

The interaction of water with silicate glass surfaces is of interest to researchers in a number of disciplines, including geochemistry [6], stress corrosion [67, 68], nanoconfined "pore" water [69], and crack propagation [70–72]. Although the processes by which water interacts

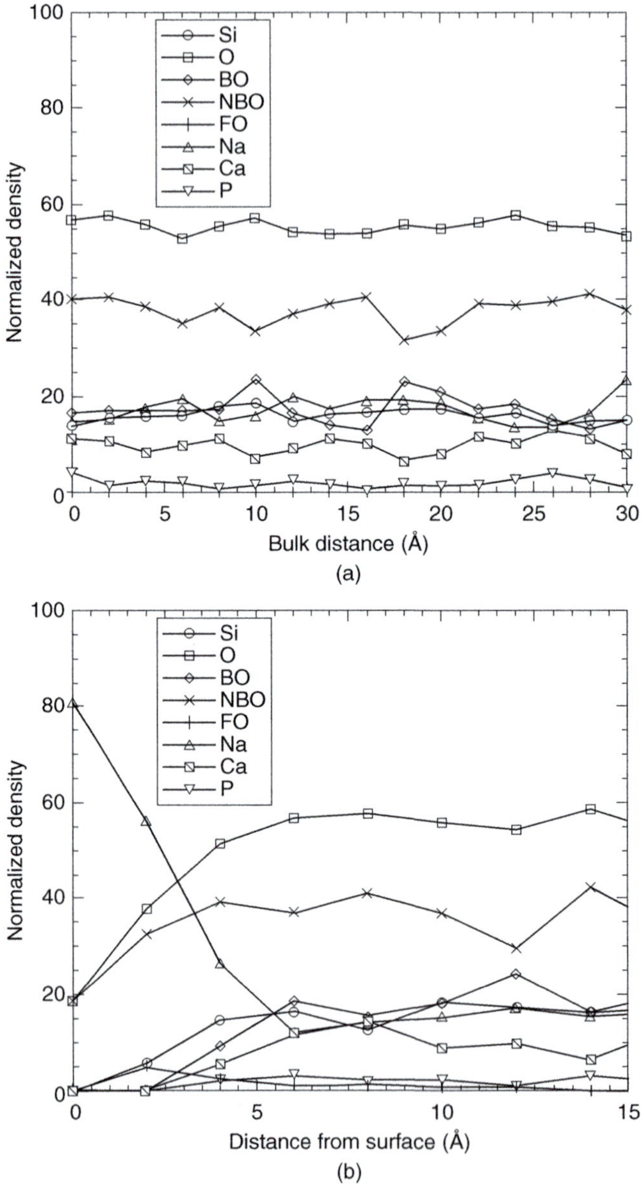


**Figure 14.6** Normalized atomic density plot for (a) bulk and (b) surface of 45S5 glass. Source: Reproduced from Zeitler and Cormack [24].

with glass surfaces is typically thought of occurring on the electronic scale (e.g. proton transfer) [13, 18, 73], water species were incorporated into early classical MD glass surface simulations [74, 75].

These earliest silica–water interactions were captured during the formation of the glass surface and the simulations were considered to represent the formation of surfaces in the presence of water vapor by introducing a specified surface concentration of water molecules at the time of introduction of the vacuum gap [75]. It has been shown experimentally that the structural relaxation associated with fracture is much faster than the process by which the surface absorbs water [76], so simulating a surface that relaxes in the presence of water is not necessary – the relaxation and absorption steps can be considered consecutively (Figure 14.7) [7].

As mentioned in Section 14.2.1, the process of creating a surface via fracture may result in the presence of surface "defects" including undercoordinated Si ($^3$Si or three-coordinated Si), NBO, and small-membered rings [5]. These surface species are not typically found in bulk silica and have been found to be highly reactive to moisture, forming silanols in order to reduce surface energy. For example, a two-membered ring has highly strained Si—O—Si bond angles and thus a single Si–O–Si specie will react with a water molecule to form two silanols (Si–OH). A single $^3$Si–NBO pair may also react with water to form two silanols. Recent simulations have been undertaken in an effort to minimize coordination defects through analysis of the thermal history of the simulated glass to study the presence of defects during the quench cooling process [77].

Following the reaction of $^3$Si, NBO, and small-membered rings, physisorption of water molecules to the hydroxylated surface occurs via hydrogen bonding [6]. The hydrolysis of

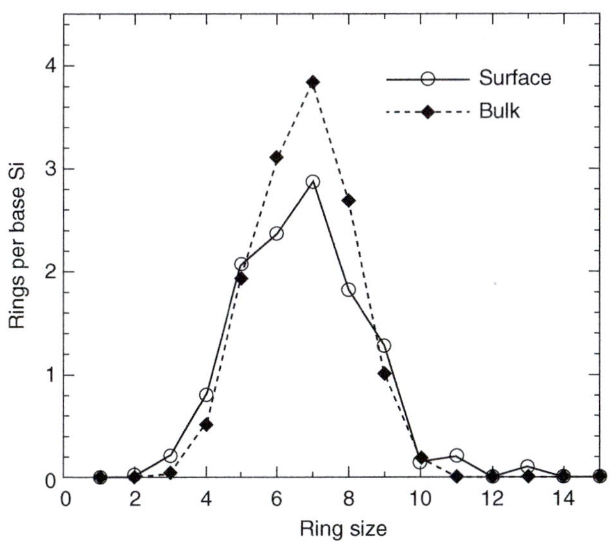


**Figure 14.7** Comparison of ring size distribution of the surface and the bulk of silica glass, showing effect of structural relaxation. Source: Reproduced from Du and Cormack [7].

network-forming Si—O bonds has been shown to occur as a water molecule approaches a surface Si atom, making it pentacoordinated and eventually leading to the formation of two silanols [74]. Amorphous silica surfaces created in the presence of water molecules were found to have fewer surface defects due to silanol formation with adsorbed water [75]. It was concluded therefore that the presence of water reduced the necessary restructuring to create a stable surface. No two member rings were found for surfaces with water present [75].

For multicomponent glasses in the presence of water, in addition to the preferential migration of Na to the surface of a high soda glass, Na migrates even further away into the water phase [6]. The addition of alkali creates a more disruptive structure (more NBO) and more larger rings are formed as a result [22]. A study of gas adsorption showed that Ne penetration is not increased compared to amorphous silica [34]. The presence of alkali creates a more "open" structure via NBO, but there is still enough blocking via the alkali ions themselves to prevent an inert gas Ne from entering further into the surface. Water molecules seem to be able to travel ~10 Å into the surface of amorphous silica via the normal ring structure [78].

The dissolution of a soda-borosilicate glass has been simulated by creating an MD surface and assuming that soluble species (effectively, NBO and neighboring B or Na) directly in contact with the surface will be part of the dissolution process while the silicate network does not [79]. In this way, an "alteration layer" is defined and a "rough" surface (due to inhomogeneities) is formed and analyzed.

For bioactive glasses, repolymerization of the phosphosilicate network during leaching is typically attributed to condensation of silanols (which releases water molecules) [54, 55]. However, MD simulation of the Na/H ion exchange reaction, in which an NBO becomes part of an OH group, shows that repolymerization may occur if those OH groups are no longer associated with the network, but instead solely with other modifier ions (Figure 14.8) [66]. In the presence of moisture, Na at the immediate surface can interact with adsorbed water via an ion exchange mechanism, for example: ≡Si—O—Na + H—O—H → ≡Si—O—H + Na—O—H or ≡Si—O—Na + $H^+$ → ≡Si—O—H + $Na^+$.

#### 14.2.3.1 Reactive Potentials

The Si–O–H interactions of early silica–water simulations were modeled using three-body terms that allowed for a primitive form of dissociation of water molecules but were able to dutifully replicate known structural surface features while maintaining a low concentration of surface defects [5, 75, 80]. A later study used a more accurate dissociable water molecule model but found no dissociation at the silica surface at room temperature [81]. An even more modern dissociative potential has been used to investigate proton dynamics, typically the realm of ab initio calculations, for a water-silica interface [30]. These more modern potentials are able to more accurately represent the liquid water, which provides a greater confidence in the resulting interfacial structures and simulated reactions. Even the presence of $H_3O^+$ ions is found in these surface simulations (Figure 14.9) [30].

These so-called "reactive potentials" can also be used to more accurately calculate activation barriers to the dissolution of the amorphous silica surface, effectively showing the mechanisms by which $Q_n$ Si become $Q_{n-1}$ Si (where $n$ is the number of BO on Si) [82]. Based on MD simulations using a reactive potential, the rate-limiting step in silica dissolution was predicted to be the $Q_{32}$ and $Q_{21}$ reactions [82]. The magnitude of strain (i.e.

**Figure 14.8** OH distribution across the $Na^{+}/H^{+}$-exchanged bioactive glass structures. Free, Si-bonded, and P-bonded OH groups are highlighted in yellow, blue, and purple, respectively. Protons are colored white. For clarity, Si, Na, Ca, P, and nonhydroxyl O atoms are not shown. Source: Reproduced from Tilocca [66].

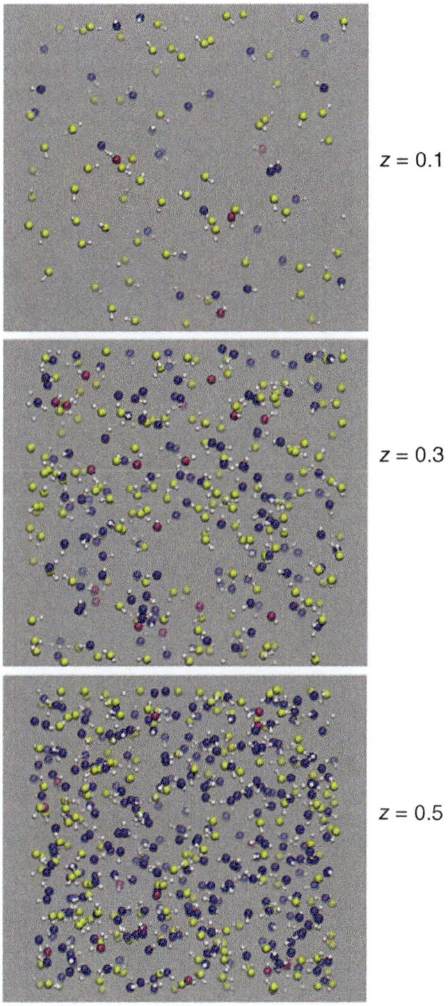


distorted tetrahedra) and accessibility of reaction sites to water molecules were shown to be the principal driving forces for differences in activation energy for bond-breaking [82].

In addition to the reactive potentials put forth by Garofalini and coworkers [30, 82], the ReaxFF model [29] has become very popular for high-quality "reactive MD" simulations that have ab initio calculations as their basis (Figure 14.10) [83, 84]. ReaxFF is based on the principal of a dynamical bond order that changes during the simulation based on interatomic bond distances. The parameterization of ReaxFF potentials is done based on experimental results, as well as a suite of quantum mechanical calculations [29]. The advantage of a ReaxFF approach over that of first principles calculations (see also Section 14.3) is that much larger simulation cell sizes can be accommodated. Also, surface

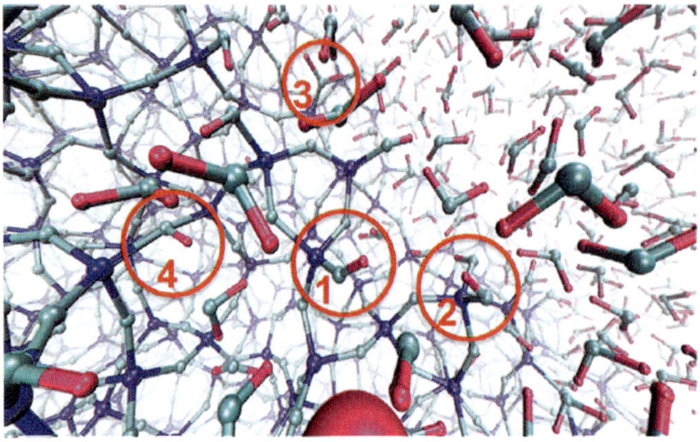


Figure 14.9 Different types of hydroxyls present on the silica surface exposed to water in silica–water simulations using dissociative MD. Source: Reproduced from Lockwood and Garofalini [30].

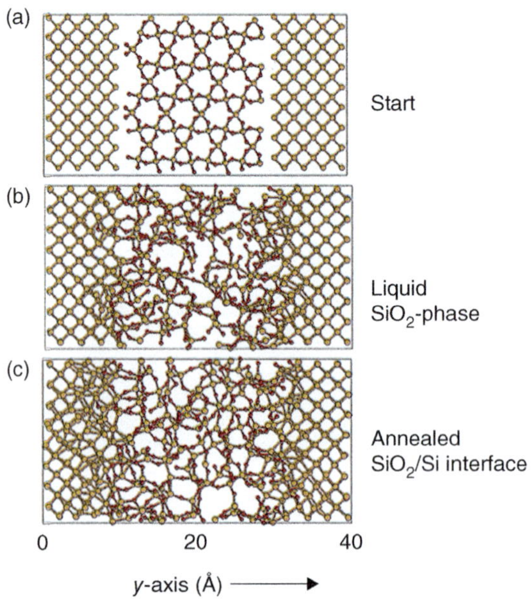


Figure 14.10 (a) Initial structure (top), (b) structure after melting the alpha-quartz phase (middle), and (c) final interface structure (bottom) of silica glass using ReaxFF calculations. Source: Reproduced from van Duin et al. [29].

reactions may take place more "naturally" (compared to MD) during the simulation, which may allow for a deeper understanding of the underlying processes. Dynamical charge assignments can be used with ReaxFF, which focuses on one of the principle features of ab initio calculations [83].

The results of ReaxFF calculations for an amorphous silica surface in contact with water show agreement with ab initio calculations [84]. Starting with pure water in contact with an amorphous silica surface, the result of the calculations is a hydroxylated surface covered in silanol groups (created by water molecules dissociating at the surface). Similar to observations made for MD surface simulations, ReaxFF simulations have undercoordinated Si and O (due to the creation process) that relatively quickly become fully coordinated due to rearrangement at the immediate surface [84]. With time, more water dissociates and silanols are formed while structural surface defects disappear and the surface approaches equilibrium. At the same time, the properties of bulk water and bulk silica can also be represented well by the same force field. In this way, ReaxFF can be used similarly to ab initio calculations, with the advantage of larger length scales and longer time scales, which aids in determining statistical significance of observations.

One criticism of ReaxFF with respect to modeling amorphous structures is that it is parameterized based on crystalline structures [83] – however, that is the case for most general force fields used in MD calculations of glass structure [25, 26]. Another criticism is that ReaxFF is not parameterized for melts, which are a principal part of the melt-quench process used to create bulk glass structures. Despite these criticisms, ReaxFF has been shown to offer description of glass structures not unlike a typical MD-generated structure [83]. A general set of ReaxFF potentials (for all types of alkali, for example) is not available, but the future of ReaxFF as a tool for large-scale simulation of surface reactions on glasses appears promising [83].

## 14.3 First Principles Surface Simulations

High-level quantum mechanical calculations are typically required when details of electronic behavior are desired. Classical interatomic potentials, while they may serve well in their capacity to predict structures, typically do not model details of electronic behavior. Reaction pathways and atomic structures are more accurately described by quantum mechanical methods such as ab initio and DFT levels of theory. For example, properly capturing the strain in an Si—O (which is important to surface hydroxylation and involves some degree of covalency) is not possible with most classical methods. Also, glass–water interfaces are dominated by electronic scale phenomena that may not be described classically.

Because the chemisorption of water at glass surfaces is an important process in determining glass surface structure, a number of studies have attempted to describe the amorphous silica surface, dry and wet. Prior to popular DFT method calculations, cluster and hybrid methods were used. Cluster methods typically involve extracting a portion of a structure from a larger (classical) model, fixing some end points, and then performing higher-level calculations on the smaller system with the constraints on the edges designed to represent the bulk structure. In some cases, hybrid methods have been used in which MD and QM structures are connected as parts of the same overall surface structure – the QM region is imbedded in the larger structure and is used to focus the necessary computational power to extract reaction energy information. With increasing computational availability, full DFT calculations of hundreds of atoms have become possible.

### 14.3.1 Silica Surfaces

One of the principal advantages and purposes of using first principles simulations for glass surfaces is that electronic scale phenomena (e.g. proton transfer from water to the surface) can be investigated. So while the "dry" glass surface may be of some interest, the interaction of water with glass surfaces (see Section 14.3.3) is typical of first principles glass surface simulations. Thus, few "dry" glass surface first principles simulations are reported.

Simulations of dry nanoporous silica surfaces showed the presence of surface defects, including two member rings [16]. In this case, NBO and $^{3}$Si and $^{5}$Si were shown to be removed upon relaxation of the surface, but two member rings showed a higher energy barrier that was only surpassed for some instances of rings at room temperature [16]. For cases in which the two-member ring was "removed," the formation of a $^{5}$Si specie was shown, and this defect was associated with charge transfer from an absorbed water molecule to the silica surface. Surface rings have also been studied via ab initio cluster calculations [39, 85–87]. Surface sites for silica have been investigated via ab initio molecular dynamics (AIMD) simulations of quartz surfaces, which also show smaller rings (and increased densification) at the surface, similarly to amorphous silica surfaces [88, 89] (Figure 14.11).

For a time, hybrid classical and quantum mechanical calculations were the state-of-the-art in glass surface simulations. Du et al. divided the simulations into two regions: one that used classical potentials and one that used DFT pseudopotentials [19]. The structure of the immediate surface was calculated using DFT calculations and was imbedded in a larger region for which classical potentials were used. A pair of studies first developed a complete classical surface model for a dry amorphous silica surface [90], then extracted clusters and investigated hydrolysis reactions on the clusters [91]. The dry surfaces were found to have fewer small rings and more undercoordinated Si and O than typical MD simulations.

Two general problems have been associated with using cluster models to represent larger surfaces: (i) inability to reproduce accurate electronic states; and (ii) inability to describe the mechanical behavior at the surface, particularly as it relates to lack of rigidity in reconstruction of the surface. Calculations for these smaller clusters depend on choices of terminating atoms and may generally overestimate reaction energies and activation barriers [92].

### 14.3.2 Multicomponent Oxide Glass Surfaces

One of the challenges of modeling multicomponent oxide glass surfaces with first principles surface simulations is that of simulation size. These methods typically scale poorly such that only $\sim$100-atom simulations are feasible. The challenge thus becomes especially difficult for multicomponent compositions with relatively small components – perhaps only a few atoms of a particular atom type may be included. For a surface simulation in which the surface is taken from a preexisting bulk model, the choice of where to initiate the surface (will the minor component atom be at or near the surface?) can be unclear. Getting a representative initial surface structure is therefore not always a straightforward process. However, a number of multicomponent oxide glass surface simulation studies exist.

The dissolution process for borosilicate glasses has special importance for considering nuclear waste forms. Because an estimate for the long-term properties of these waste

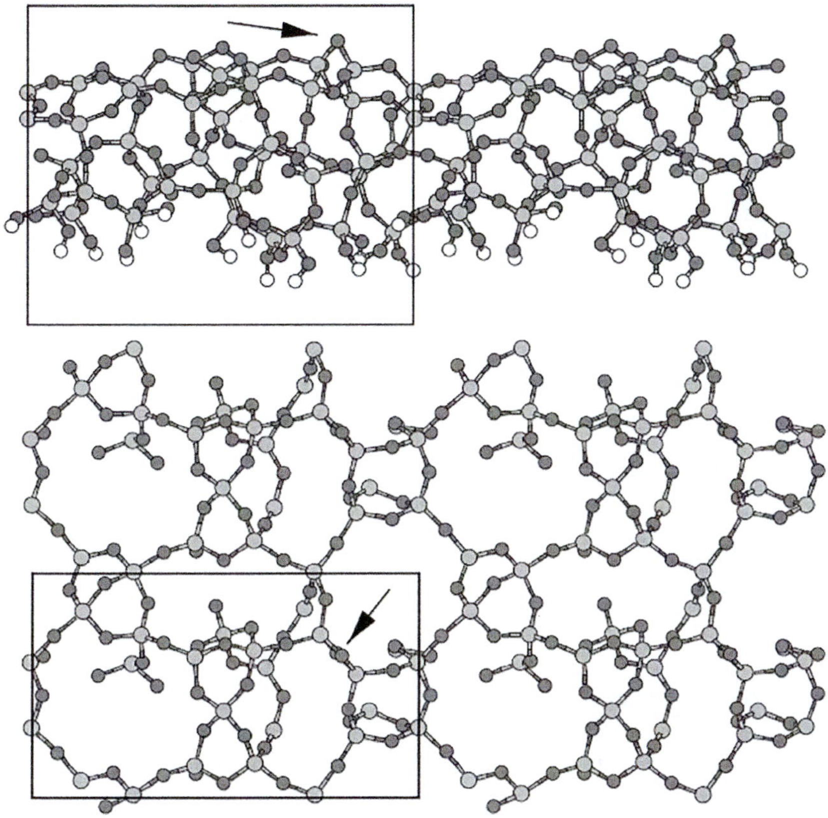

**Figure 14.11** Model of dehydroxylated amorphous surface (top panel: side view of the slab; bottom panel: top view of the slab). Source: Reproduced from Ceresoli et al. [89].

forms is needed for larger-scale repository performance assessment calculations, estimated degradation rates at the atomic scale are important [93]. DFT calculations of sodium borosilicate glass surfaces have been performed, including an implicit ion exchange step achieved by substituting H for Na, creating silanols on nearby NBO [93]. As found for MD simulations of borosilicate glass surfaces (see Section 14.2.2), four-coordinated boron ions at the surface become three-coordinated, even following surface relaxation and rearrangement. A decrease in higher $n$ $Q_n$ species is found for Si, but Si remains fully coordinated [93]. Reaction barriers for hydrolysis reactions at the surface were calculated via DFT calculations, indicating that boron will more readily release from the surface of a borosilicate glass in contact with moisture, as compared to silicon (Figure 14.12) [93].

A study of a calcium aluminosilicate glass surface showed that MD and AIMD structures were very similar [94]. In both cases, Al was found to preferentially exist at the immediate surface, while Ca lay just below the surface. A proposed reason for this surface segregation is that there is an overall lower energetic cost to having Al at the surface compared to

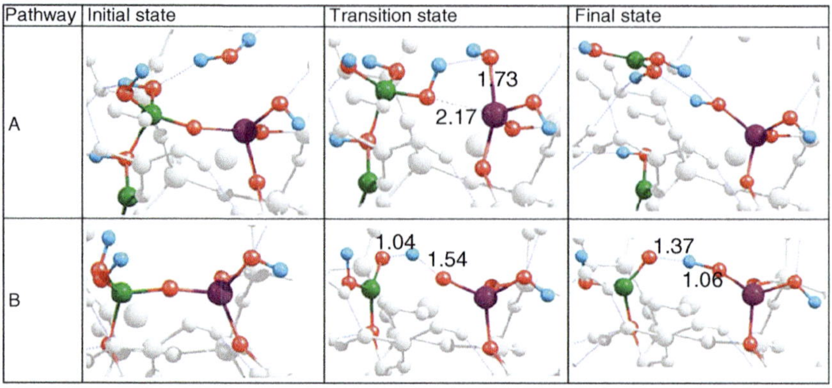


**Figure 14.12** Calculated geometries of initial state, transition state, and final state in reaction pathways for the hydrolysis reactions at the $^3B(Q^2)$-O-Si($Q^3$) bridge under deprotonated conditions. Source: Reproduced from Zapol et al. [93].

Si – that is, NBO may preferentially exist on Al at the surface (compared to Si). Additional first principles studies of the surfaces of multicomponent glasses (those that involve bioactive glasses) are reviewed in Sections 14.3.2.1 and 14.3.3.

#### 14.3.2.1 Bioactive Glasses

A well-studied subset of multicomponent oxide glass surfaces is that of bioactive glasses, which are typically modeled as phosphosilicate glasses containing alkali and alkaline earth modifiers. The surface structures of soda-lime phosphosilicate glasses have been simulated using periodic DFT calculations in order to investigate the key between surface features and bioactivity [20]. While this study did not focus on the hydrolysis of surface network bonds, it did focus on the impact of the ion exchange process on surface structure by considering surfaces for which sodium has been explicitly removed following surface creation. As is the case for many DFT studies, the number of atoms is less than 100, which provides a challenge when dealing with the statistics of contributions from smaller components in multicomponent glasses. Periodic DFT structures of bulk glass have shown that systems this small can still describe local properties of bioactive glasses (Figure 14.13) [95]. A challenge of creating surfaces from small simulation cells is the high dependence of a dipole moment on the choice of position for inserting a vacuum gap [20].

One way of dealing with the heterogeneity of small cells is to make multiple perpendicular cuts to the bulk structure in order to generate multiple representative surface structures [20]. As with other first principles glass simulations, the initial structure was borrowed from a previously melt-quenched MD simulation cell. Upon surface creation, surface ion rearrangement was observed as the network formers were driven to bulk-like local coordination [20]. Alkali enrichment at the immediate surface was not observed, but that is likely due to the static nature of the DFT calculations (compared to the dynamical nature of MD calculations) (Figure 14.14) [20].

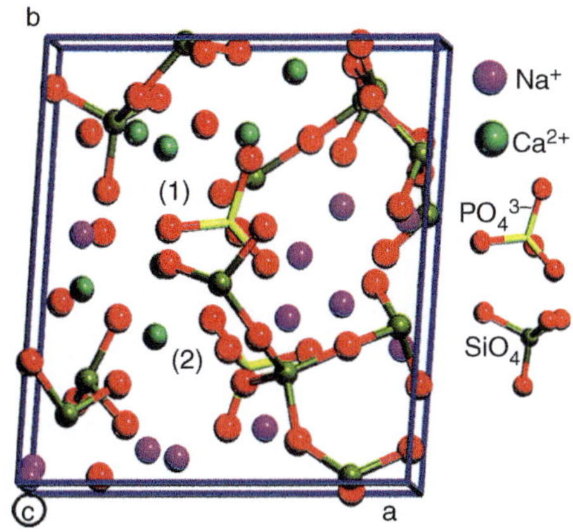


**Figure 14.13** B3LYP Hench 45S5 Bioglass optimized unit cell. Source: Reproduced from Corno et al. [95].

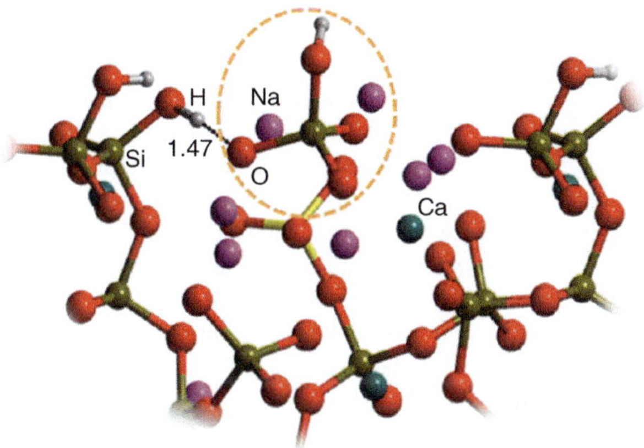


**Figure 14.14** Isolated orthosilicate group (encircled in yellow) exposed on the surface of 45S5. Source: Reproduced from Berardo et al. [20].

Also, highly strained surface rings are found which are not found in the bulk, which is consistent with MD simulations [20]. Lower $n$ $Q_n$ species are also found at the surface, as expected due to the surface creation process [20]. Even with relatively small simulation cells, some relatively minor difference in glass composition resulted in obvious differences in surface OH bond length and can be attributed to the presence of NBO at the surface.

Surface $^3$Si defects (undercoordinated Si) quickly heal during the surface relaxation phase of surface creation using CPMD [46]. This takes place over a matter of picoseconds, more quickly than water can react with the surface, so the assumption that water interacts with a fully relaxed surface (rather than participating in the relaxation process) is valid [46]. This is in agreement with some experimental [76] and MD evidence [96]. Additional first principles studies of bioactive glass surfaces (those that involve water species) are reviewed in Section 14.3.3.

### 14.3.3 "Wet" Glass Surfaces

First principles calculations are well suited for water interactions with glass surfaces, since these interactions typically are described by electronic scale phenomena. Ion exchange, surface dissolution, and repolymerization reactions are able to be investigated by first principles methods such as DFT, Car–Parrinello molecular dynamics (CPMD), and AIMD. Silica water interfaces have been studied by a number of researchers, indicating the great importance of the subject (Figure 14.15) [91, 97–99].

Many mechanisms for the dissolution of silica have been proposed. Hydrolysis has been carried out for a number of different types of surface defects using constrained clusters. Some early dissolution investigations for amorphous silica surfaces were undertaken via the study of small clusters that were meant to represent the full surface. Ab initio simulations have looked at various reaction pathways for water interaction with silica surfaces [13–15, 100]. The relative stability of surface silanols has also been investigated by ab initio calculations of small clusters [101, 102].

One recent study of water interacting with alpha-quartz hypothesizes that hydrogen bonds internal to the surface (rather than surface-water hydrogen bonds) participate in proton transfer [18]. The presence of Na and Mg on the surface serves to disrupt the typical water-quartz hydrogen bond network, reducing the interactions between surface silanols and water. This allows more "intrasurface" hydrogen bonding and leads to a greater dissolution rate. This was investigated via DFT calculations on periodic crystalline systems but may be applicable to amorphous silica.

CPMD simulations of bioactive glass compositions in contact with water have been able to be used to investigate the ion exchange reaction at the surface (i.e. creation of silanol group and concomitant loss of Na) [103]. Because such a large (~10 Å) layer of surface water was included, a distinction between "bulk" water and "surface" water becomes obvious, in terms of water molecule orientation with respect to the glass surface. This was attributed to the hydrogen-bonding network (Figure 14.16) [103].

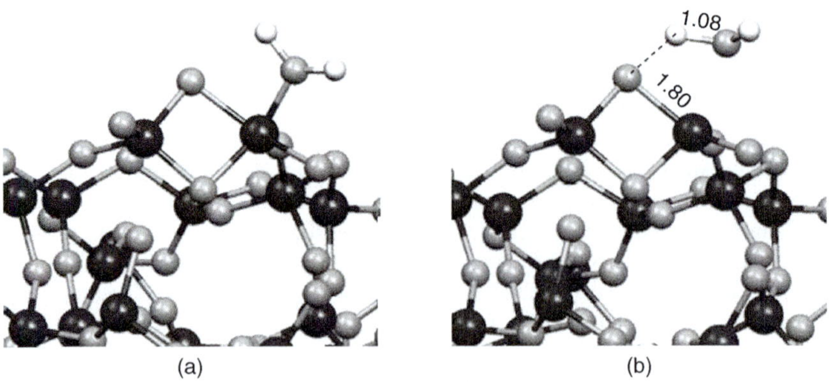


Figure 14.15 Atomic configuration of the transition states for the hydroxylation reaction of the 2M silicon ring along two paths, (a) and (b). Source: Reproduced from Masini and Bernasconi [97].

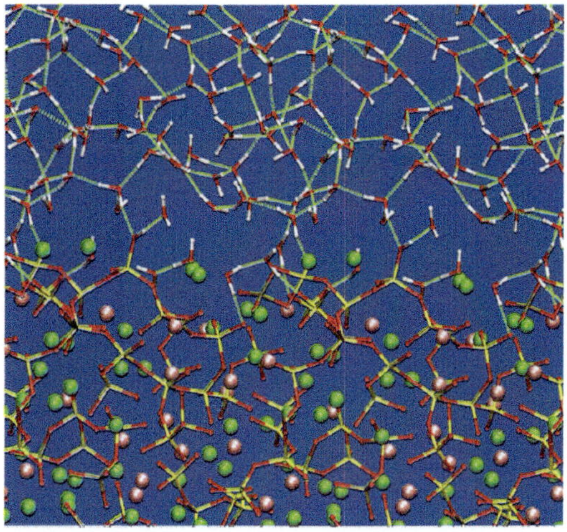

Figure 14.16 Snapshot of the bioglass–water interface, extracted from CPMD trajectory. Phosphosilicate network (bottom) is represented as yellow/red ball-and-stick, with Na and Ca cations as green and brown spheres, respectively. Dashed lines highlight hydrogen bonds. Source: Reproduced from Tilocca and Cormack [103].

In contrast to some of the reactions that occur with water at the surface of bioactive glasses, interstitial water molecules do not behave in the same manner [104]. For example, small rings in the bulk glass do not appear to be favorable sites for network dissolution; this is partly attributed to the reduced strain existing in the "dry" bulk case due to the heavily modified network structure (i.e. increased modifier cation and NBO content).

Another CPMD study has investigated the heterogeneous surfaces of multicomponent bioactive glasses by "mapping" the complex energy surface for reactive sites using a water molecule "probe" [46]. It identified $^{3}Si$ defects (particularly with nearby NBO) and two and three member rings as "very reactive" thermodynamically, although kinetic barriers to water dissociation may be responsible for the stability of those rings [46]. Surface modifiers (Na and Ca) also are strong adsorption sites for water molecules.

## 14.4 Summary

Glass surface simulations have been undertaken for more than three decades with much insight gained into the formation of glass surfaces, as well as their basic atomic structures. Early MD models were based on bulk simulation models and have led to the "proof of concept" that surface properties of amorphous silica could be described well by classical means. These types of simulations were later expanded to multicomponent oxide glass surfaces with much success. Interactions of silica and multicomponent glass surfaces with water have also been the subject of many studies, principally for the investigation of surface reactions and the resultant surface structure.

With the availability of increased computational power, first principles surface simulations have become more feasible and thus the study of electronic-scale surface phenomena, most importantly, reactions with water have become commonplace. However, simulation sizes for these simulation types are still relatively limited, so the use of ReaxFF represents a useful compromise for looking at larger scales of surface phenomena with something approaching the necessary accuracy of first principles models.

## Acknowledgments

Sandia National Laboratories is a multimission laboratory managed and operated by National Technology and Engineering Solutions of Sandia, LLC, a wholly owned subsidiary of Honeywell International Inc., for the U.S. Department of Energy's National Nuclear Security Administration under contract DE-NA0003525.

## References


1 Garofalini, S.H. (1983). A molecular-dynamics simulation of the vitreous silica surface. *J. Chem. Phys.* 78: 2069–2072.

2 Garofalini, S.H. and Levine, S.M. (1985). Differences in surface behavior of alkali ions in $Li_2O{\cdot}3SiO_2$ and $Na_2O{\cdot}3SiO_2$ glasses. *J. Am. Ceram. Soc.* 68: 376–379.

3 Levine, S.M. and Garofalini, S.H. (1987). A structural-analysis of the vitreous silica surface via a molecular-dynamics computer-simulation. *J. Chem. Phys.* 86: 2997–3002.

4 Garofalini, S.H. and Zirl, D.M. (1988). Onset of alkali adsorption on the vitreous silica surface. *J. Vac. Sci. Technol., A* 6: 975–981.

5 Feuston, B.P. and Garofalini, S.H. (1989). Topological and bonding defects in vitreous silica surfaces. *J. Chem. Phys.* 91: 564–570.

6 Garofalini, S.H. (2001). Molecular dynamics simulations of silicate glasses and glass surfaces. In: *Molecular Modelling Theory: Applications in the Geosciences* (ed. R.T. Cygan and J.D. Kubicki), 131–168. Washington, DC: Geochemical Society and Mineralogical Society of America.

7 Du, J. and Cormack, A.N. (2005). Molecular dynamics simulation of the structure and hydroxylation of silica glass surfaces. *J. Am. Ceram. Soc.* 88: 2532–2539.

8 Zirl, D.M. and Garofalini, S.H. (1992). Structure of sodium aluminosilicate glass surfaces. *J. Am. Ceram. Soc.* 75: 2353–2362.

9 Roder, A., Kob, W., and Binder, K. (2001). Structure and dynamics of amorphous silica surfaces. *J. Chem. Phys.* 114: 7602–7614.

10 Kerisit, S. and Pierce, E.M. (2011). Monte Carlo simulations of the dissolution of borosilicate and aluminoborosilicate glasses in dilute aqueous solutions. *Geochim. Cosmochim. Acta* 75: 5296–5309.

11 Kerisit, S., Pierce, E.M., and Ryan, J.V. (2015). Monte Carlo simulations of coupled diffusion and surface reactions during the aqueous corrosion of borosilicate glasses. *J. Non-Cryst. Solids* 408: 142–149.

12 Bogillo, V.I., Pirnach, L.S., and Dabrowski, A. (1997). Monte Carlo simulation of silica surface dehydroxylation under nonisothermal conditions. *Langmuir* 13: 928–935.

13 Criscenti, L.J., Kubicki, J.D., and Brantley, S.L. (2006). Silicate glass and mineral dissolution: calculated reaction paths and activation energies for hydrolysis of a Q3 Si by $H_3O^+$ using ab initio methods. *J. Phys. Chem. A* 110: 198–206.

**14** Xiao, Y.T. and Lasaga, A.C. (1996). Ab initio quantum mechanical studies of the kinetics and mechanisms of quartz dissolution: $OH^-$ catalysis. *Geochim. Cosmochim. Acta* 60: 2283–2295.

**15** Xiao, Y.T. and Lasaga, A.C. (1994). Ab-initio quantum-mechanical studies of the kinetics and mechanisms of silicate dissolution – $H^+(H_3O^+)$ catalysis. *Geochim. Cosmochim. Acta* 58: 5379–5400.

**16** Rimsza, J.M. and Du, J.C. (2015). Ab initio molecular dynamics simulations of the hydroxylation of nanoporous silica. *J. Am. Ceram. Soc.* 98: 3748–3757.

**17** Morrow, C.P., Olsen, A.A., and Kubicki, J.D. (2014). Quantum mechanical modeling of hydrolysis and $H_2O$-exchange in Mg-, Ca-, and Ni- silicate clusters: implications for dissolution mechanisms of olivine minerals. *Am. Miner.* 99: 2303–2312.

**18** Kubicki, J.D., Sofo, J.O., Skelton, A.A., and Bandura, A.V. (2012). A new hypothesis for the dissolution mechanism of silicates. *J. Phys. Chem. C* 116: 17479–17491.

**19** Du, M.H., Kolchin, A., and Cheng, H.P. (2004). Hydrolysis of a two-membered silica ring on the amorphous silica surface. *J. Chem. Phys.* 120: 1044–1054.

**20** Berardo, E., Pedone, A., Ugliengo, P., and Corno, M. (2013). DFT modeling of 45S5 and 77S soda-lime phospho-silicate glass surfaces: clues on different bioactivity mechanism. *Langmuir* 29: 5749–5759.

**21** Yuan, X. (1999). *Molecular Dynamics Simulation of Silicate Glasses and Their Surfaces.* Alfred, NY: Alfred University.

**22** Abbas, A., Delaye, J.-M., Ghaleb, D., and Calas, G. (2003). Molecular dynamics study of the structure and dynamic behavior at the surface of a silicate glass. *J. Non-Cryst. Solids* 315: 187–196.

**23** Beckers, J.V.L. and de Leeuw, S.W. (2000). Molecular dynamics simulation of nanoporous silica. *J. Non-Cryst. Solids* 261: 87–100.

**24** Zeitler, T.R. and Cormack, A.N. (2006). Interaction of water with bioactive glass surfaces. *J. Cryst. Growth* 294: 96–102.

**25** Tilocca, A., de Leeuw, N.H., and Cormack, A.N. (2006). Shell-model molecular dynamics calculations of modified silicate glasses. *Phys. Rev. B* 73: 104209.

**26** Pedone, A., Malavasi, G., Menziani, M.C. et al. (2006). A new self-consistent empirical interatomic potential model for oxides, silicates, and silica-based glasses. *J. Phys. Chem. B* 110: 11780–11795.

**27** Athanasopoulos, D.C. and Garofalini, S.H. (1992). Molecular-dynamics simulations of the effect of adsorption on $SiO_2$ surfaces. *J. Chem. Phys.* 97: 3775–3780.

**28** Min, K., Kim, Y., Goyal, S. et al. (2016). Interfacial adhesion behavior of polyimides on silica glass: a molecular dynamics study. *Polymer* 98: 1–10.

**29** van Duin, A.C.T., Strachan, A., Stewman, S. et al. (2003). ReaxFFSiO reactive force field for silicon and silicon oxide systems. *J. Phys. Chem. A* 107: 3803–3811.

**30** Lockwood, G.K. and Garofalini, S.H. (2014). Proton dynamics at the water–silica interface via dissociative molecular dynamics. *J. Phys. Chem. C* 118: 29750–29759.

**31** Litton, D.A. and Garofalini, S.H. (1997). Vitreous silica bulk and surface self-diffusion analysis by molecular dynamics. *J. Non-Cryst. Solids* 217: 250–263.

**32** Bunker, B.C., Haaland, D.M., Michalske, T.A., and Smith, W.L. (1989). Kinetics of dissociative chemisorption on strained edge-shared surface-defects on dehydroxylated silica. *Surf. Sci.* 222: 95–118.

**33** Levine, S.M. and Garofalini, S.H. (1986). Interactions of Pt single atoms with the vitreous silica surface – adsorption and thermal accommodation. *Surf. Sci.* 177: 157–168.

**34** Kohler, A.E. and Garofalini, S.H. (1994). Effect of composition on the penetration of inert-gases adsorbed onto silicate glass surfaces. *Langmuir* 10: 4664–4669.

**35** Bakaev, V.A. and Steele, W.A. (1999). On the computer simulation of a hydrophobic vitreous silica surface. *J. Chem. Phys.* 111: 9803–9812.

**36** Hochstrasser, G. and Antonini, J.F. (1972). Surface states of pristine silica surfaces 1. ESR studies of Es' dangling bonds and of $CO_2$ adsorbed radicals. *Surf. Sci.* 32: 644–664.

**37** Sprenger, D., Bach, H., Meisel, W., and Gutlich, P. (1990). XPS study of leached glass surfaces. *J. Non-Cryst. Solids* 126: 111–129.

**38** Rarivomanantsoa, M., Jund, P., and Jullien, R. (2001). Classical molecular dynamics simulations of amorphous silica surfaces. *J. Phys. Condes. Matter* 13: 6707–6718.

**39** Lopez, N., Vitiello, M., Illas, F., and Pacchioni, G. (2000). Interaction of $H_2$ with strained rings at the silica surface from ab initio calculations. *J. Non-Cryst. Solids* 271: 56–63.

**40** Rarivomanantsoa, M., Jund, P., and Jullien, R. (2004). Sodium diffusion through amorphous silica surfaces: a molecular dynamics study. *J. Chem. Phys.* 120: 4915–4920.

**41** Tilocca, A. and Cormack, A.N. (2010). Surface signatures of bioactivity: MD simulations of 45S and 65S silicate glasses. *Langmuir* 26: 545–551.

**42** Tilocca, A. (2010). Models of structure, dynamics and reactivity of bioglasses: a review. *J. Mater. Chem.* 20: 6848–6858.

**43** Greaves, G.N. (1985). EXAFS and the structure of glass. *J. Non-Cryst. Solids* 71: 203–217.

**44** Pantano, C.G. (1981). Surface and in-depth analysis of glass and ceramics. *Am. Ceram. Soc. Bull.* 60: 1154–1163.

**45** Corrales, L.R. and Du, J. (2006). Characterization of ion distributions near the surface of sodium-containing and sodium-depleted calcium aluminosilicate melts. *J. Am. Ceram. Soc.* 89: 36–41.

**46** Tilocca, A. and Cormack, A.N. (2008). Exploring the surface of bioactive glasses: water adsorption and reactivity. *J. Phys. Chem. C* 112: 11936–11945.

**47** Kelso, J.F., Pantano, C.G., and Garofalini, S.H. (1983). A comparison of ion-scattering spectra and molecular-dynamics simulations at the surface of silicate-glasses. *Surf. Sci.* 134: L543–L549.

**48** Dickinson, J.T., Langford, S.C., Jensen, L.C. et al. (1988). Fractoemission from fused-silica and sodium-silicate glasses. *J. Vac. Sci. Technol., A* 6: 1084–1089.

**49** Garofalini, S.H. (1984). Behavior of atoms at the surface of a $K_2O{\cdot}3SiO_2$ glass – a molecular-dynamics simulation. *J. Am. Ceram. Soc.* 67: 133–136.

**50** Garofalini, S.H. (1984). Potassium diffusion at the surface of a $K_2O{\cdot}3SiO_2$ glass. *J. Vac. Sci. Technol., A* 2: 79–81.

**51** Pedone, A., Menziani, M.C., and Cormack, A.N. (2015). Dynamics of fracture in silica and soda-silicate glasses: from bulk materials to nanowires. *J. Phys. Chem. C* 119: 25499–25507.

**52** Yuan, F.L. and Huang, L.P. (2012). Molecular dynamics simulation of amorphous silica under uniaxial tension: from bulk to nanowire. *J. Non-Cryst. Solids* 358: 3481–3487.

**53** Luo, J., Vargheese, K.D., Tandia, A. et al. (2016). Crack nucleation criterion and its application to impact indentation in glasses. *Sci. Rep.* 6: 1–10.

**54** Hench, L.L. (2006). The story of bioglass(R). *J. Mater. Sci. – Mater. Med.* 17: 967–978.

**55** Jones, J.R. (2013). Review of bioactive glass: from Hench to hybrids. *Acta Biomater.* 9: 4457–4486.

**56** Malavasi, G., Pedone, A., and Menziani, M.C. (2006). Towards a quantitative rationalization of multicomponent glass properties by means of molecular dynamics simulations. *Mol. Simulat.* 32: 1045–1055.

**57** Tilocca, A. and Cormack, A.N. (2007). Structural effects of phosphorus inclusion in bioactive silicate glasses. *J. Phys. Chem. B* 111: 14256–14264.

**58** Tilocca, A., Cormack, A.N., and de Leeuw, N.H. (2007). The structure of bioactive silicate glasses: new insight from molecular dynamics simulations. *Chem. Mater.* 19: 95–103.

**59** Tilocca, A. (2009). Structural models of bioactive glasses from molecular dynamics simulations. *Proc. R. Soc. A Math. Phys. Eng. Sci.* 465: 1003–1027.

**60** Xiang, Y. and Du, J. (2011). Effect of strontium substitution on the structure of 45S5 bioglasses. *Chem. Mater.* 23: 2703–2717.

**61** Malavasi, G., Menabue, L., Menziani, M.C. et al. (2013). New insights into the bioactivity of $SiO_2$–CaO and $SiO_2$–CaO–$P_2$O5 sol-gel glasses by molecular dynamics simulations. *J. Sol-Gel Sci. Technol.* 67: 208–219.

**62** Lopes, J.H., Mazali, I.O., Landers, R., and Bertran, C.A. (2013). Structural investigation of the surface of bioglass 45S5 enriched with calcium ions. *J. Am. Ceram. Soc.* 96: 1464–1469.

**63** Du, J. and Xiang, Y. (2016). Investigating the structure–diffusion–bioactivity relationship of strontium containing bioactive glasses using molecular dynamics based computer simulations. *J. Non-Cryst. Solids* 432 (Part A): 35–40.

**64** Tilocca, A. (2011). Molecular dynamics simulations of a bioactive glass nanoparticle. *J. Mater. Chem.* 21: 12660–12667.

**65** Zeitler, T.R. (2006). *Structure, Dynamics, and Surface Reactions of Bioactive Glasses.* Alfred, NY: Alfred University.

**66** Tilocca, A. (2015). Atomic-scale models of early-stage alkali depletion and $SiO_2$-rich gel formation in bioactive glasses. *Phys. Chem. Chem. Phys.* 17: 2696–2702.

**67** Wiederhorn, S.M. and Bolz, L.H. (1970). Stress corrosion of glass. *Am. Ceram. Soc. Bull.* 53: 543–548.

**68** Bunker, B.C. (1994). Molecular mechanisms for corrosion of silica and silicate-glasses. *J. Non-Cryst. Solids* 179: 300–308.

**69** Choudhury, N. (2010). Dynamics of water at the nanoscale hydrophobic confinement. *J. Chem. Phys.* 132: 064505.

**70** Wiederhorn, S.M., Dretzke, A., and Rodel, J. (2002). Crack growth in soda-lime-silicate glass near the static fatigue limit. *J. Am. Ceram. Soc.* 85: 2287–2292.

**71** Guin, J.P. and Wiederhorn, S.M. (2003). Crack growth threshold in soda lime silicate glass: role of hold-time. *J. Non-Cryst. Solids* 316: 12–20.

**72** Guin, J.P., Wiederhorn, S.M., and Fett, T. (2005). Crack-tip structure in soda-lime-silicate glass. *J. Am. Ceram. Soc.* 88: 652–659.

**73** Lockwood, G.K. and Garofalini, S.H. (2009). Bridging oxygen as a site for proton adsorption on the vitreous silica surface. *J. Chem. Phys.* 131: 074703-1–074703-8.

**74** Garofalini, S.H. (1990). Molecular dynamics computer simulations of silica surface structure and adsorption of water molecules. *J. Non-Cryst. Solids* 120: 1–12.

**75** Feuston, B.P. and Garofalini, S.H. (1990). Water-induced relaxation of the vitreous silica surface. *J. Appl. Phys.* 68: 4830–4836.

**76** D'Souza, A.S. and Pantano, C.G. (1999). Mechanisms for silanol formation on amorphous silica fracture surfaces. *J. Am. Ceram. Soc.* 82: 1289–1293.

**77** Agnello, G., Hamilton, J., Manley, R. et al. (2015). Investigation of contact-induced charging kinetics on variably modified glass surfaces. *Appl. Surf. Sci.* 356: 1189–1199.

**78** Litton, D.A. and Garofalini, S.H. (2001). Modeling of hydrophilic wafer bonding by molecular dynamics simulations. *J. Appl. Phys.* 89: 6013–6023.

**79** Delaye, J.M., Kerrache, A., and Gin, S. (2013). Topography of borosilicate glass reacting interface under aqueous corrosion. *Chem. Phys. Lett.* 588: 180–183.

**80** Webb, E.B. and Garofalini, S.H. (1998). Relaxation of silica glass surfaces before and after stress modification in a wet and dry atmosphere: molecular dynamics simulations. *J. Non-Cryst. Solids* 226: 47–57.

**81** Mahadevan, T.S. and Garofalini, S.H. (2007). Dissociative water potential for molecular dynamics simulations. *J. Phys. Chem. B* 111: 8919–8927.

**82** Kagan, M., Lockwood, G.K., and Garofalini, S.H. (2014). Reactive simulations of the activation barrier to dissolution of amorphous silica in water. *Phys. Chem. Chem. Phys.* 16: 9294–9301.

**83** Yu, Y., Wang, B., Wang, M. et al. (2016). Revisiting silica with ReaxFF: towards improved predictions of glass structure and properties via reactive molecular dynamics. *J. Non-Cryst. Solids* 443: 148–154.

**84** Fogarty, J.C., Aktulga, H.M., Grama, A.Y. et al. (2010). A reactive molecular dynamics simulation of the silica-water interface. *J. Chem. Phys.* 132: 174704-1–174704-10.

**85** Pacchioni, G. and Vitiello, M. (1998). EPR and IR spectral properties of hydrogen-associated bulk and surface defects in $SiO_2$: ab initio calculations. *Phys. Rev. B* 58: 7745–7752.

**86** Pelmenschikov, A. and Leszczynski, J. (1999). Adsorption of 1,3,5-trinitrobenzene on the siloxane sites of clay minerals: ab initio calculations of molecular models. *J. Phys. Chem. B* 103: 6886–6890.

**87** Wang, L.S., Nicholas, J.B., Dupuis, M. et al. (1997). $Si_3O_y$ ($y = 1$–6) clusters: models for oxidation of silicon surfaces and defect sites in bulk oxide materials. *Phys. Rev. Lett.* 78: 4450–4453.

**88** Rignanese, G.M., De Vita, A., Charlier, J.C. et al. (2000). First-principles molecular-dynamics study of the (0001) alpha-quartz surface. *Phys. Rev. B* 61: 13250–13255.

**89** Ceresoli, D., Bernasconi, M., Iarlori, S. et al. (2000). Two-membered silicon rings on the dehydroxylated surface of silica. *Phys. Rev. Lett.* 84: 3887–3890.

**90** Wilson, M. and Walsh, T.R. (2000). Hydrolysis of the amorphous silica surface. I. Structure and dynamics of the dry surface. *J. Chem. Phys.* 113: 9180–9190.

**91** Walsh, T.R., Wilson, M., and Sutton, A.P. (2000). Hydrolysis of the amorphous silica surface. II. Calculation of activation barriers and mechanisms. *J. Chem. Phys.* 113: 9191–9201.

**92** Penev, E., Kratzer, P., and Scheffler, M. (1999). Effect of the cluster size in modeling the $H_2$ desorption and dissociative adsorption on Si(001). *J. Chem. Phys.* 110: 3986–3994.

**93** Zapol, P., He, H.Y., Kwon, K.D., and Criscenti, L.J. (2013). First-principles study of hydrolysis reaction barriers in a sodium borosilicate glass. *Int. J. Appl. Glass Sci.* 4: 395–407.

**94** Ganster, P., Benoit, M., Delaye, J.M., and Kob, W. (2008). Surface of a calcium aluminosilicate glass by classical and ab initio molecular dynamics simulations. *Surf. Sci.* 602: 114–125.

**95** Corno, M., Pedone, A., Dovesi, R., and Ugliengo, P. (2008). B3LYP simulation of the full vibrational spectrum of 45S5 bioactive silicate glass compared to v-silica. *Chem. Mater.* 20: 5610–5621.

**96** Leed, E.A. and Pantano, C.G. (2003). Computer modeling of water adsorption on silica and silicate glass fracture surfaces. *J. Non-Cryst. Solids* 325: 48–60.

**97** Masini, P. and Bernasconi, M. (2002). Ab initio simulations of hydroxylation and dehydroxylation reactions at surfaces: amorphous silica and brucite. *J. Phys. Condens. Matter* 14: 4133.

**98** Mischler, C., Kob, W., and Binder, K. (2002). Classical and ab-initio molecular dynamic simulation of an amorphous silica surface. *Comput. Phys. Commun.* 147: 222–225.

**99** Mischler, C., Horbach, J., Kob, W., and Binder, K. (2005). Water adsorption on amorphous silica surfaces: a Car-Parrinello simulation study. *J. Phys. Condes. Matter* 17: 4005–4013.

**100** Kubicki, J.D., Xiao, Y., and Lasaga, A.C. (1994). Theoretical reaction pathways for the formation of $Si(OH)_5$ – and the deprotonation of orthosilicic acid in basic solution. *Geochim. Cosmochim. Acta* 58: 2755–2756.

**101** Civalleri, B., Garrone, E., and Ugliengo, P. (1999). Cage-like clusters as models for the hydroxyls of silica: ab initio calculation of $^1$H and $^{29}$Si NMR chemical shifts. *Chem. Phys. Lett.* 299: 443–450.

**102** Giordano, L., Sushko, P.V., Pacchioni, G., and Shluger, A.L. (2007). Optical and EPR properties of point defects at a crystalline silica surface: ab initio embedded-cluster calculations. *Phys. Rev. B* 75.

**103** Tilocca, A. and Cormack, A.N. (2011). The initial stages of bioglass dissolution: a Car–Parrinello molecular-dynamics study of the glass–water interface. *Proc. R. Soc. A* 467: 2102–2111.

**104** Berardo, E., Corno, M., Cormack, A.N. et al. (2014). Probing the fate of interstitial water in bulk bioactive glass by ab initio simulations. *RSC Adv.* 4: 36425–36436.

# 15

# Simulations of Glass–Water Interactions

*Jessica M. Rimsza[1], Thiruvilla S. Mahadevan[2], Lu Deng[2], and Jincheng Du[2]*

[1] *Geochemistry Department, Sandia National Laboratories, Albuquerque, NM, USA*
[2] *Department of Materials Science and Engineering, University of North Texas, Denton, TX, USA*

## 15.1 Introduction

Despite the apparent stability of many common inorganic glasses, their eventual degradation and dissolution in aqueous conditions do occur, which can cause significant issues in packaging, display, architecture, and optical window glasses and their applications in other technological fields that rely on material stability. For example, in nuclear waste disposal where glass is one of the proven waste forms, it relies on the long-term chemical durability of glasses to prevent the release of radioactive elements into the environment [1]. On the other hand, biomedical applications take advantage of easy solubility of some glass compositions to facilitate bonding with the host tissues and the growth of hard and soft tissues [2, 3]. There is thus a wide spectrum of glass durability and solubility in aqueous environments depending on the glass system and composition. It has been found that glasses with high chemical stability, such as those used for nuclear waste disposal, have a common feature of formation of a passivating gel layer when in contact with aqueous solution, as compared to corrosion-resistant alloys and ceramics. In silicate glass dissolution, the hydrolysis reaction that leads to the breaking of the stable Si–O–Si linkages is the rate-limiting step. However, glass corrosion is a very complex process that involves multisequential or concurrent steps including ion exchange, diffusion, and reaction, and there are a lot of unknowns. Therefore, extensive research has been done to advance our understanding of the fundamental mechanism of glass corrosion and to predict the stability of silicate and other glasses. In this chapter, the use of molecular simulations to investigate glass–water interactions is discussed. After discussion the current understanding of the glass dissolution process, first principles-based simulation methods are introduced. This is followed by the reactive potential-based classical simulation methods. Lastly, challenges and future opportunities for atomistic investigations of the glass–water interface are discussed.

*Atomistic Simulations of Glasses: Fundamentals and Applications*, First Edition.
Edited by Jincheng Du and Alastair N. Cormack.

### 15.1.1 Glass Dissolution Process and Experimental Characterizations

Glass dissolution is a complex multistep process, with the amorphous nature of glass structures, multiconcurrent reaction processes, and transient aqueous conditions adding difficulty in predicting water-interactions and dissolution rates. Several mechanisms have been proposed to describe the overall dissolution process, and despite extensive experimental investigations with a wide variety of advanced characterization techniques, some details and mechanisms remain unclear. What is well established though is that glass dissolves in a series of stages, which lead to the formation of a porous hydrated alteration layer, and for some compositions, this can provide passivation to limit further chemical reactions of the bulk glass [4]. The initial stage of glass dissolution is interdiffusion of either protons ($H^+$) or hydronium ions ($H_3O^+$) from the aqueous solution, with alkali ions ($Na^+$, $K^+$, $Li^+$) in the glass [5, 6]. The second stage of glass dissolution is hydrophilic attack of water on its structural linkages (i.e. ≡Si—O—Si≡, ≡Si—O—Al≡, or ≡Si—O—B≡), beginning with the weakest linkages. Successive breakage of structural bonds breaks down the glass network and subsequent reorganization lead to the formation of a hydrated porous silica gel layer [7]. Under certain conditions, the gel structure can be protective, thereby decreasing the overall dissolution rate [8, 9]. The morphology of the gel layer, including features such as thickness, pore structure, and the chemical composition, depends on the nature of the glass and the environmental conditions when the glass dissolves, such as pH, temperature, and solution composition as well as the glass composition [10, 11]. The final result is an alteration layer composed of multiple characteristic regions, including an interdiffusion layer, an amorphous hydrated layer, a variety of gel layers and, in some cases, a crystalline reaction product layer [12, 13]. Similar processes are observed for phosphate glasses, with the final gel structure containing excess ≡P—O—P≡, linkages that are eventually broken, leading to complete glass dissolution [5]. For more information on the dissolution process, please see several review papers on the topic: [3, 4, 11, 14, 15].

The wide interest in the durability of glasses in aqueous conditions has resulted in significant experimental characterizations of the glass–water interactions, including fast reactions between water and unhydroxylated silica surfaces [16–18] and continuing all the way to evaluation of the silica–water solubility limit [19]. Initial hydroxylation of silica surfaces is a nearly instantaneous process, with secondary ion mass spectroscopy by D'Souza and Pantano [16–18] indicating that formation of equilibrium Si–OH concentration occurs within one minute at $1 \times 10^{-9}$ Torr on freshly fractured surfaces. The fast hydroxylation rate of freshly fractured silica surfaces is challenging to measure, requiring high vacuum and continuous monitoring of the surface with a characterization technique sensitive to H atoms. In addition to this kinetic information, Zhuralev has performed comprehensive investigations of the concentration and distribution of Si–OH on silica surfaces [20, 21], providing important benchmarking data for evaluating the results of computational models on surface hydroxylation.

Once a glass surface is fully hydroxylated, the continued presence of water breaks the bonds that form the glass structure. Experimental evaluation for this process captures an average reaction rate or energy barrier, rather than identifying unique features of silica that may be more reactive than others. A field that has been interested in this topic is geochemistry, where dissolution of minerals under varying environmental conditions has been used to derive the Si—O bond strength. Overall, activation energy for Si—O bond breakage is between 50 and 80 kJ/mol depending on the conditions for dissolution and the silicate polymorph studied [22, 23]. Local variation in silica structure, such as silicon coordination number, is known to alter the energy barrier of the Si—O bond [24]. Additional studies of the dissolution rate of silicates in contact with NaCl and NaOH solutions and at varying pH indicate that the activation energy remains relatively unchanged with values between 55 and 94 kJ/mol [25–29]. The lowest activation energies for Si—O bond breakage came from dissolution of silica nanoparticles, at 13.5 kJ/mol, due to the extra strain on the Si—O bonds on the surface [30]. At elevated pH (above 8) and elevated concentrations of dissolved $Na^+$, the silica dissolution rate increases [31]. Lower energy barriers have been attributed to a hydroxide ($OH^-$) or $Na^+$ mediated reaction mechanism but is difficult to characterize with experimental methods. As the solution composition becomes increasingly complex, with mixtures of NaCl, $Na_2SO_4$, and $CaCl_2$ [32], theories for changes in the dissolution rate evolve from local mechanisms, to changes in the electric double layer at the silica–solution interface due to the involvement of charged dissolved species.

Glass surface stability is controlled not only by the composition of the solution and the Si—O bond strength but also the solubility limit of silica in water. During silica–water interactions, the dissolved silica monomers establish an equilibrium with the interface, halting any further dissolution [33, 34]. Amorphous silica has a low solubility in water, with Alexander et al. reporting solubilities between 0.01 and 0.012 mol% that are insensitive to pH values up to 9 [19]. More recent studies at elevated temperatures and pressures have corroborated these values [35, 36] and provided thermodynamic data on the stability of $H_4SiO_4$, a bioproduct of silica dissolution [37]. Changing silica solubilities, especially at high pH [38], perturbs the equilibrium between the silica surface and the solution, which can facilitate further dissolution. Additionally, under elevated dissolved silica concentrations, silica polymerization can occur [39–42], which then deposits on the surface or blocks pore channels, adding another aspect to predicting glass dissolution.

Overall, our knowledge of glass–water interactions has progressed significantly in the last 30 years, with improvements in experimental methods allowing for a deeper understanding of the underlying processes and mechanisms. These results provide insights into overarching trends of the glass dissolution process, such as pH, temperature, or composition dependence, that can then be linked to structural differences or reaction mechanisms. It is difficult to separate unique features of the glass–water interaction process using bulk characterization data. Therefore, there exists an ongoing need for atomistic and molecular scale modeling of glass–water interactions to provide insight into the glass dissolution process. In this chapter, we address past and ongoing efforts to use atomistic and molecular modeling methods to uncover the structural and mechanistic dependence of glass–water interactions.

### 15.1.2 Types of Atomistic Simulation Methods for Studying Glass–Water Interactions

The atomistic simulation methods used to study glass–water interactions include first principles (or quantum mechanical [QM]) methods, molecular dynamics (MD) simulations based on reactive potentials, and kinetic Monte Carlo (KMC). These multiscale approaches can be found in the Figure 15.1. Each method has its advantages and associated limitations. First principles (or ab initio) methods provide the highest accuracy for energetic data,

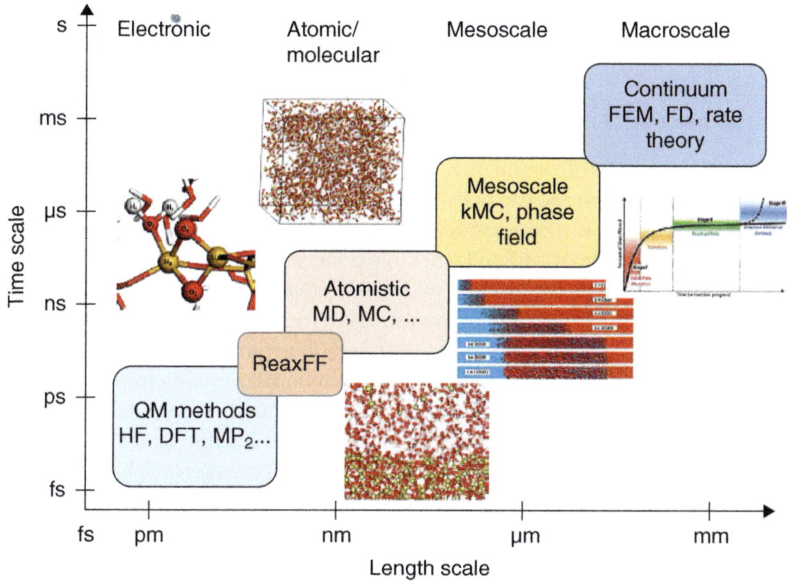


**Figure 15.1** Multiscale simulation methods and applications to glass–water interactions and glass dissolution. These include Quantum Mechanical (QM) methods such as Hartree Fock (HF), Density Functional Theory (DFT), Møller–Plesset perturbation theory (MP2), atomistic simulation methods such as molecular dynamics (MD), Monte Carlo (MC) and Reactive Force Field (ReaxFF) based simulations, mesoscale simulation methods such as kinetic Monte Carlo (kMC) and phase field, and continuum level simulation methods such as finite element analysis (FEM), fluid dynamics (FD) and rate theory based calculations. Source: Reference Du and Rimsza [43].

can naturally handle chemical reactions, and is the default method to study glass–water reactions. The first principles approach can be used to study reaction mechanisms of clusters or small periodic simulation cells that represent the glass systems, but it is generally computationally too expensive for large systems needed to represent the amorphous nature of glass materials. We will introduce both cluster- and periodic-based first principles calculations, including static calculation and ab initio molecular dynamics (AIMD), as well as hybrid methods in Section 15.2.1. Reactive force field-based MD simulations are an intermediate between classical MD and first principles methods, both in terms of accuracy and computational cost, and are a good compromise to investigate glass–water interactions. Systems with tens of thousands of atoms and simulation times up to several nanoseconds can be achieved routinely with current computational efficiencies. With the development and refining of reactive potentials, reactive MD simulations have become a powerful tool in studying glass–water interactions, including silicate glass–water interfaces, structure and diffusion in silica rich gel layer structures, and dissolution of silicate glass. The recent development of reactive empirical potentials will be discussed and examples of applications to various systems are given. Challenges to address longer-term corrosion in atomistic simulations through methods such as KMC to extend the time and length scales are discussed in Section 15.4.

## 15.2 First-Principles Simulations of Glass–Water Interactions

### 15.2.1 Brief Introduction to Methods

Density functional theory (DFT) simulation methods are based on the atoms' fundamental electronic structure [44, 45] and can provide more accurate mechanisms and interaction energies than other atomistic methods, particularly when factors like atomic charges need to be considered. More details about DFT and other first principles methods including ab initio MD simulations can be found in Chapter 2. Given the relatively simple electronic structure of Si, O, B, Na, and K atoms (which are commonly represented in glass compositions), DFT is well suited to study the structure and properties of these materials. The primary drawback of DFT methods is the high computational cost so that only small simulations (hundreds of atoms) and short time scales (10+ ps) can be performed with current computational capabilities. In the case of glass dissolution, which often occurs over months, years, or decades, simulations performed over ps can have challenges in providing relevant insight. Therefore, individual factors of glass dissolution are studied, such as bond energies or reaction mechanisms, which can then be fed into large-scale models. Initially, first principles calculations focused on cluster calculations, using a $SiO_4H_4$ molecule [46–48] or other clusters [49–55] due to the relatively low computational cost. Additionally, due to localized charge transfer, cluster calculations have proven to be a reasonable analog to periodic systems [56], though questions remain about the role of the surface or bulk effects on reaction mechanism and energies. Following cluster models, computational efforts were expanded to include simulation of glass–water interfaces, followed by simulation of more complex structures such as multicomponent glasses, varying solution compositions, or porous glasses.

### 15.2.2 Energy Barriers for Si–O–Si Bond Breakage

Energy barriers for Si–O–Si linkage breakage by water have been extensively studied, since $SiO_2$ is the most common component in oxide glasses and can be represented by simple structural models. The frequent usage of DFT calculations has been to obtain energy barriers for (i) hydroxylation of silica surfaces or (ii) hydrolysis of Si–O–Si linkage. The fastest water–silica reaction is hydroxylation, where a surface nonbridging oxygen (NBO) atom or under coordinated silicon atom reacts with a water molecule to form a silanol (Si–OH) group. The hydroxylation process is so favorable for $H_3O^+$ and $OH^-$ reacting with a silica surface that it has been termed "activationless" by Lowe et al. [57], who found no energy barrier for $H^+$ transfer between the solution and the interface using DFT methods. For Si–O–Si linkage breaking, rather than hydroxylation, there is an energy barrier with values in the range of 70–205 kJ/mol for bonds on unconstrained silica and cristobalite surfaces [43, 58–60]. Variation in the energy barrier is attributed to the number of water molecules that participate in the reaction as well as the $Q_n$ (meaning $n$ bridging oxygen per [$SiO_4$] tetrahedra unit) speciation of Si. Overall, the highest energy barrier for Si—O—Si bond breakage occurs when a single water molecule is involved in the reaction, and when a second water is present the energy barrier drops by 20–40 kJ/mol [51]. Even lower energy barriers are reported for systems with higher numbers of water molecules [61]. The decrease in the energy barrier is due to the extra water molecules providing a mechanism for $H^+$ to transfer to hydroxylate the NBO formed during Si—O—Si bond breakage [61].

Additionally, energy barriers are dependent on the stability of the Si atoms in the ≡Si—O—Si≡ linkage [60–62]. Generally less connected Si atoms, such as Si bound with only one ($Q_1$), two ($Q_2$), or three ($Q_3$) bridging oxygens, have lower energy barriers [60]. While the same general trend is reported by Criscenti et al. [58], there is a variation in the Si—O—Si bond strength in a $Q_2$ configuration, with a $Q_2$ structure having a higher activation energy than other Si coordination structures and therefore being the rate-limiting step in dissolution. Further atomistic studies on the reactivity of different coordination structures will be necessary to provide more insight into how variation in local structures impacts the stability of the glass–water interface.

### 15.2.3 Reaction Mechanism for Si–O–Si Bond Breakage

Two different reaction mechanisms are reported for Si—O—Si bond breakage by water. The first mechanism starts with an O atom in a water molecule binding with a Si atom, creating a short-lived pentacoordinated silica defect ($Si^5$) as an intermediate state [63, 64]. After the $Si^5$ defect is formed the Si—O bond breaks in the ≡Si—O—Si≡ linkage and a H atom from the water transfers to the resulting NBO. The result is the formation of two Si-OH (see Figure 15.2). The energy barrier occurs when one of the H atoms on the adsorbed water molecules breaks off to terminate the new NBO that is formed, with values as high as 44 kJ/mol [66]. This is the most commonly reported mechanism across different structure models and reaction conditions.

In the second mechanism, one of the two H atoms in the water molecule bonds with the bridging oxygen that binds the two Si atoms. This process is followed by breaking a ≡Si—O—Si≡ linkage and formation of two Si-OH [48]. This mechanism is more commonly reported for breaking ≡Si—O—Si≡ linkages in small clusters and when a single water molecule is available to participate in the bond breakage mechanism [51]. Bandura et al. reported no stable structures with an H bonded to a bridging oxygen in simulations of water interactions with a hydroxylated cristobalite surface, which is attributed to O atoms acting as weaker $H^+$ acceptors than O atoms in water or terminal groups [64]. Interestingly, this appears to be a surface effect, with favorable bonding of $H^+$ atoms on O atoms below the surface, though the site was not as stable as surface hydroxyls and can spontaneously transfer to more stable structures [64]. The initial formation of either the $Si^5$ of the protonated bridging oxygen defect generally determines the preference for one-reaction mechanisms over the other [48].

### 15.2.4 Strained Si–O–Si Linkages

To facilitate simulation of bond breaking, highly reactive strained ≡Si—O—Si≡ linkages are often selected for structural models [48, 54, 65, 67, 68]. The most common is a set of edge-sharing tetrahedron, also referred to as a two-membered ring, since they form a ring structure containing two Si atoms and two O atoms (see Figure 15.2). This structure is selected because it is under more strain than in larger rings [69] and have higher relativity with water [63]. It is worth noting that there is limited evidence in literature for the water

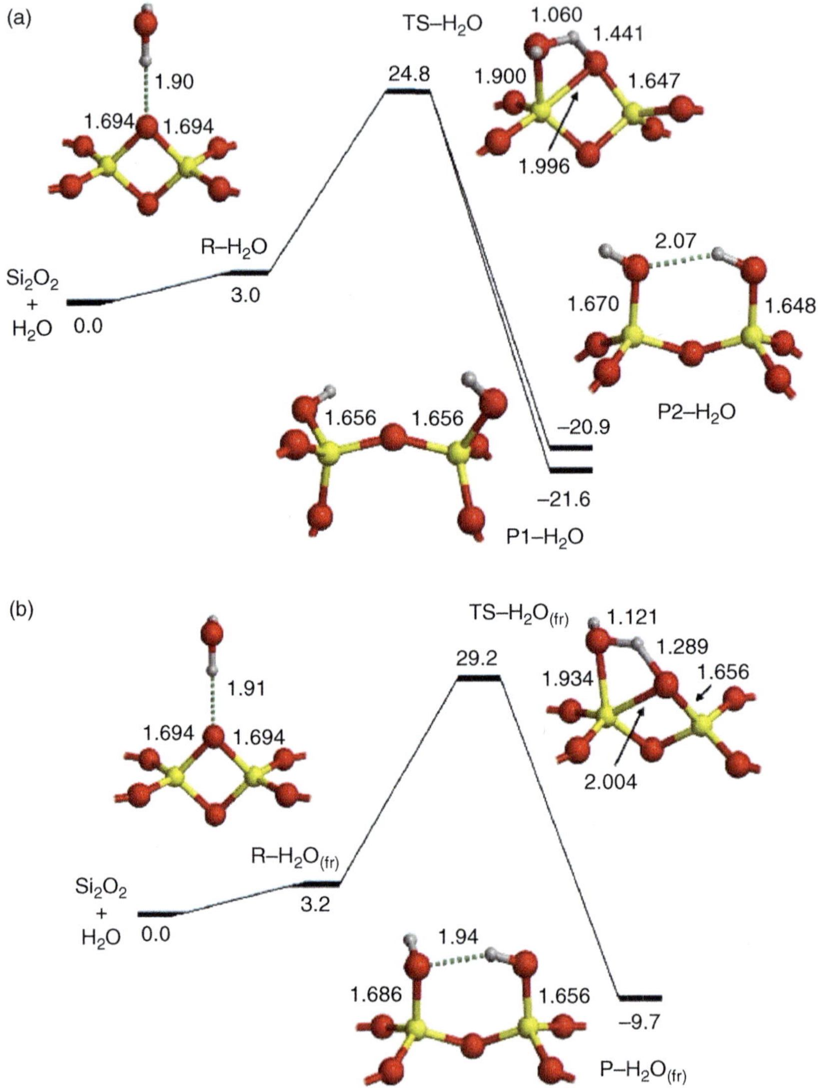


**Figure 15.2** Relative free energy profiles for the reaction of $SiO_2$ with water. (a) Full optimization and (b) with H atoms frozen in the original locations. Energies are relative to $SiO_2$ and water as the reference state. Energies are in kcal/mol and bond distances are in angstrom. Hydrogen bonds are noted in italics. Source: Reference Rimola and Ugliengo [65].

reactions involving two-membered rings in experimental systems. The exceptions are reports by D'Souza and Pantano who provide evidence for a very fast reacting species on the surface of fractured silica surfaces [16–18], which they hypothesize is the result of two-membered rings on the surface. For short molecular dynamics simulations, these highly reactive structures are more likely to break naturally, rather than requiring restrictions in the simulation to facilitate the reaction. Overall, the reaction mechanisms of two-membered rings do not significantly differ from reaction mechanisms of less strained Si—O bonds in ≡Si—O—Si≡ linkages, though the activation energies are lower, at 100–120 kJ/mol for a reaction mechanism with one water molecule and 60 kJ/mol when two or three water molecules are present [63, 65]. The additional water molecules allow for $H^+$ transfer to facilitate bond breakage in ≡Si—O—Si≡ linkages. The reactivity of molecular silica chains composed of linked two-membered rings were also studied, and hydrolysis of the chain began in the center, with each ring exhibiting a slightly different energy barrier for Si—O bond breakage in ≡Si—O—Si≡ linkages [55]. This process further highlights that even for these highly reactive sites there is a structural dependence on the stability of each Si—O bond in a ≡Si—O—Si≡ linkage. Finally, other ring structures, such as three- and four-membered rings also exhibit higher strain than typical six-membered rings, and therefore, are also more vulnerable to hydrolysis [70]. The evaluation of Si—O bond strengths in ≡Si—O—Si≡ linkages based on their ring structure would provide evidence for the role of the intermediate glass structure on glass–water interactions, rather than focusing exclusively on the short-range structure.

### 15.2.5 Reaction Energies for Multicomponent Linkages

For multicomponent glasses, increasingly complex reaction mechanisms and higher variation in energy barriers occur. Historically, investigations of glass–water interactions using DFT and AIMD methods have focused on pure silica compositions. As knowledge of glass–water interactions has progressed, simulations of water interacting with multicomponent glasses, such as aluminosilicates [71–73] and borosilicates [74, 75], have become more common. By using minerals as analogs for multicomponent glasses, it is seen that overall energy barriers for dissolution decrease for complex silicate materials (albite feldspar, anorthite, kaolinite, anthophyllite, enstatite, and forsterite) to 50–126 kJ/mol [58]. Cluster calculations that looked specifically at ≡Si—O—Al≡ linkages in an $Al(SiO_4H_4)_4$ cluster calculated even lower energy barriers of 4–32 kJ/mol [72]. The energy barrier is higher, 171 kJ/mol, when the ≡Si—O—Al≡ linkages form a three-membered ring [71], with the same $Si^5$ reaction mechanism reported for silica systems [48]. The reaction mechanisms in aluminosilicates are more complex than those for ≡Si—O—Si≡ bond breakage, with ions such as $Ca^{2+}$ changing from network modifiers to charge compensation sites and $H^+$ binding with bridging oxygen in ≡Si—O—Al≡ [73]. This mechanism is seen schematically in Figure 15.3. In addition to investigations of ≡Si—O—Al≡ linkages, ≡Si—O—B=, and =B—O—B= linkages have also been studied for their reactivity with water based on the wide application of borosilicate glasses. Overall, the =B—O—B= linkages exhibit energy barriers between 98 and 129 kJ/mol [75], which is comparable with 70–205 kJ/mol reported for ≡Si—O—Si≡ linkages [43, 58–60]. When the linkage includes both B and Si atoms (i.e. ≡Si—O—B=) the energy barriers are slightly higher, between 102 and 151 kJ/mol [75]. There is some inherent variation in the energy barrier due to the coordination structure

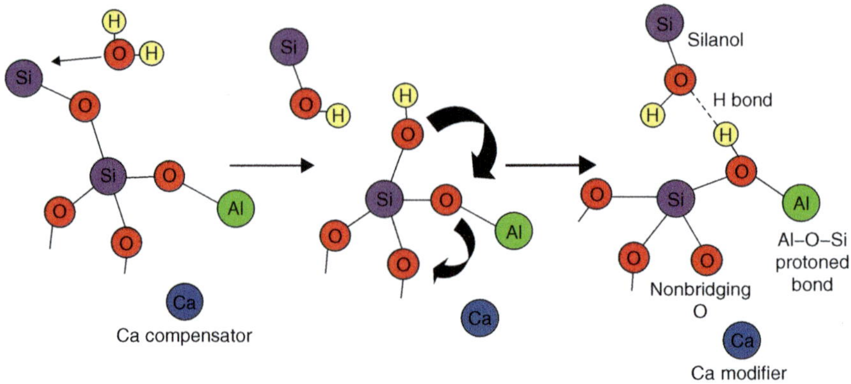


**Figure 15.3** Si–O bond breakage mechanism in a ≡Si–O–Si≡ linkage involving Si-OH, Al-OH, Si-OH-Al groups within aluminosilicate glasses. Source: Reference Bouyer et al. [73].

of the Si and B atoms, with B atoms bound to two bridging oxygen having a lower energy barrier than other configurations [75]. Overall, the ≡Si—O—Si≡ linkage is the most stable, with the addition of Al or B lowering the energy barrier for bond breakage [59].

In addition to the change in energy barrier with different network structures, in glass compositions with significant concentrations of $Na_2O$ the $Na^+/H^+$ ion exchange process can occur on the surface [76]. This is seen in the hydrolysis of glass–water interactions and results in the breakage of ≡Si—O—B≡ linkages [74]. The removal of $Na^+$ allows for free space in the structure for water to diffuse into the bulk, which indicates the beginning of dissolution [77]. The surfaces of bioglasses are also uniquely reactive since they form highly defective surfaces including high concentration of NBOs, exposed network modifiers ($Na^+$, $Ca^{2+}$), and small rings [78]. During AIMD simulations, these reactive surface sites are quickly removed after interaction with water [79]. Interestingly, the two-membered rings in bioglasses are more stable than those in pure silica [79] suggesting that the range of stabilities of different ≡Si—O—Si≡ linkages is a result of competitive reactivity of various sites on the glass surface.

### 15.2.6 Effect of pH on Si–O–Si Hydrolysis Reactions

Given the dependence of silica dissolution on pH, bond breakage mechanisms in ≡Si—O—Si≡ structures that include $H^+/OH^-$ shed light on why elevated pH causes faster dissolution. pH is a difficult feature to capture in atomistic simulations, since the small system sizes mean that the introduction of one or two extra $H^+$ or $OH^-$ results in extremely low or high pH values respectively. There are two methods used to investigate the role of pH on glass–water interactions using DFT methods. The first is altering the structural model to mimic the high or low pH conditions. For example, when an excess of $H^+$ is available in the system, the $H^+$ can bind with a bridging oxygen approximating a low pH condition, or when an excess of $OH^-$ is present the metal site can become undercoordinated, approximating a high pH condition. Nangia and Garrison used this approach and found lower energy barriers for both protonated (65 kJ/mol) and deprotonated (90 kJ/mol) sites compared to the neutral case (147 kJ/mol) for ≡Si—O—Si≡ linkages [80]. Similar results were reported by Zapol et al. for =B—O—Si≡ linkages, with an adsorbed $H^+$ on the bridging oxygen lowering the energy barrier by 50–100 kJ/mol for =B—O—B= linkages and 45–70 kJ/mol for =B—O—Si≡ linkages [75].

The second method of capturing pH effects is by simulation of the Si—O bond breakage mechanisms in a ≡Si—O—Si≡ linkage that includes $H^+/H_3O^+/OH^-$. For ≡Si—O—Si≡ linkages the reaction between $OH^-$ and a neutral Si-OH surface model and a neutral water interaction with a NBO are reported to be very similar, with an energy barrier of ~80 kJ/mol for formation of the $Si^5$ defect and ~20 kJ/mol for breakage of the ≡Si—O—Si≡ linkage [59]. Similar work looking at $H_3O^+$ interaction with ≡Si—O—Si≡ linkages found a decrease in the energy barrier by 20–40 kJ/mol, indicating that the presence of excess $H^+$ significantly weakens the bonds [59]. A compilation of the impact of deprotonated and deprotonated structural models on the energy barrier for ≡Si—O—Si≡, ≡Si—O—B=, =B—O—B=, and ≡Si—O—Al≡ linkages is included in Figure 15.4. Overall, incorporation of pH effects on molecular studies of glass-water interactions is relatively rare in literature and represents an opportunity for future investigations.

## 15.2.7 Nanoconfinement of Water in Porous Materials

The time frames of water-silica interaction have encouraged DFT simulations to focus on initial glass-water interactions, such as mechanisms and activation energies. As

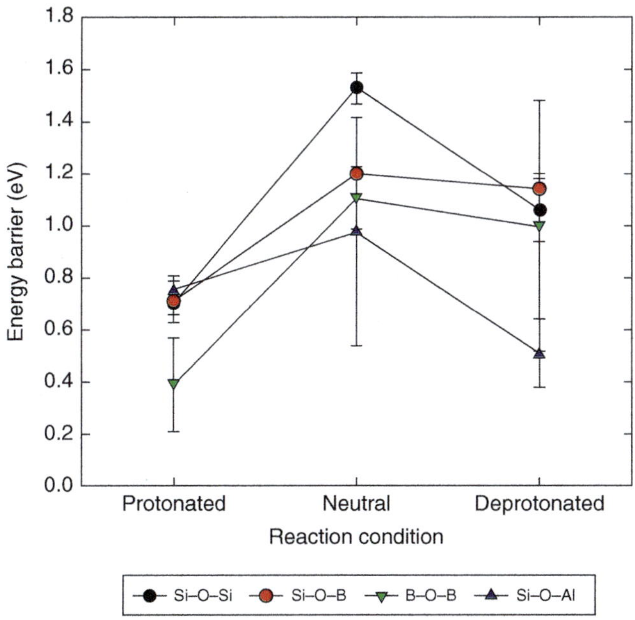


**Figure 15.4** Summary of energy barrier for breakage of network formers under protonated and deprotonated conditions. Source: References Xiao and Lasaga [59], Zapol et al. [75], Nangia and Garrison [80], Rimsza et al. [81].

the knowledge of glass–water interactions has developed, additional work on complex structural models began, which has focused on porous silica. Not only are porous silicates such as aero-gels [82] or sol–gels [83, 84] generally of interest, but porous silica frameworks also form during glass–water interactions due to removal of weakly bound cations or ≡M—O—M≡ linkages [85]. For porous materials, the focus tends to shift from the strength of ≡Si—O—Si≡ linkages, which is more straightforward to investigate with simplified models, to features of the system that are controlled by the extended porous framework, such as diffusion. Due to the increased complexity and size of the models as well as the simulation times necessary for collection of diffusion-based data, much of this work has used classical MD methods. In this section, we will focus on the use of AIMD to address the role of nanoconfinement of water in porous silica, while a discussion of the use of classical MD to address these topics left to other sections in this chapter.

Interactions between water and porous silica can be separated into two sections, one that focuses on ordered pore structures, such as zeolites [86] and molecular sieves [87], and those that focus on disordered network pores. First, in the ordered pore networks, AIMD simulations have been used to identify the role of water interaction with silica nanopores with varying OH concentration and simulated IR spectra were used to validate the porous structure [88]. These simulations were then used as the basis for more complex studies focusing on specific applications such as adsorption of biomolecules [89].

The AIMD simulations of disordered network pores have focused more heavily on ≡Si—O—Si≡ breakage and the mechanisms of water–silica reactions [85, 90]. These AIMD simulations have identified reaction mechanism for two-membered ring bond breakage and hydroxylation through water diffusion and $H^+$ transfer, noting that several different water molecules are involved in the ≡Si—O—Si≡ bond breakage mechanism [91]. Additionally, analysis of a 30 ps AIMD simulation found that $Si^5$ defects are continuously formed by a water molecule interacting with a fully coordinated Si atom. Only a portion of the $Si^5$ defects end up breaking a ≡Si—O—Si≡ linkage, indicating that there is a thermodynamic drive for two-membered ring removal that is kinetically limited [91]. Finally, the interaction between the water and the porous silica framework relaxed the porous structure, including not only removing defects but also narrowing the pair distribution function and bond angle distribution overall [91].

In addition to specific studies of glass–water interactions that provide their own unique chemical insight into the interaction, AIMD simulations can also provide benchmarking data to describe the accuracy of classical MD simulations. For example, AIMD simulations of water in porous silica was used to parametrize ReaxFF MD simulations, specifically the rate of ≡Si—O—Si≡ breakage in two-membered rings and hydroxylation rates [63]. This study was used to further improve the ReaxFF parameterization for water–silica reactions. While the efficiency of computational methods continues to increase, most studies continue to focus on silica–water reactions. Extending the current AIMD studies to capture diffusion and kinetic effects on glass–water interactions would advance our understanding of this ubiquitous system.

### 15.2.8 Oniom or QM/MM Simulations

Another relatively unexplored method of simulating glass–water interactions is through the application of hybrid QM and classical mechanical methods [92]. This method applies QM simulation methods in specific areas of a structure (such as a glass–water interface) while using classical MD methods to model the extended structure (bulk silica or water). For the investigation of structures where the reaction mechanisms are less well characterized, a higher fidelity model such as DFT can be used while still having efficient calculation of relatively simple and nonreactive structures, such as bulk water. This method was successfully used by Du et al. [68] for investigating the reaction of strained two-membered rings on silica surfaces in contact with bulk water, including identification of H atom-controlled mechanisms for ≡Si—O—Si≡ breakage. Additionally, semiempirical methods have been used, which have exhibited comparable hydrolysis reactions and transition states between their molecular orbital methods and those obtained via DFT [51]. Using these semiempirical methods for the study of nanorods they had identified several different initiation steps for hydrolysis reactions and particularly focused on the role of stress in controlling brittle fracture of silica in humid or aqueous conditions [93]. As the size and complexity of structural models continues to limit the use of DFT on complex systems, ONIOM methods may provide additional insight into glass–water interactions through extended structural models.

### 15.2.9 Areas for Improvement/Additional Research

Overall, the use of DFT-based ab initio simulations has resulted in major advancements in our understanding of glass–water interactions, particularly detailed mechanisms and activation energies for breaking ≡Si—O—Si≡ linkages. Additionally, the use of a highly reactive site in the glass structure models, such as a two-membered ring, has enabled AIMD to access kinetics of hydroxylation and ≡Si—O—Si≡ formation. More recently, studies have started to move from investigating water reactions of multicomponent glasses, such as borosilicate and aluminosilicates primarily for highly soluble bioactive glasses. However, there exist large variations of the reported energetics of reactions of multiglass former linkages, and this is clearly shown in Figure 15.4 where large error bars of energy barriers of linkages under different pH conditions. Given the relatively maturity of studies of silica–water interactions, additional work with other glass compositions are necessary to advance our understanding of complex glass systems. Ultimately, the balance between efficiency of the DFT method and the system size necessary to investigate different aspects of the glass–water interfaces will need to be balanced. Therefore, methods that include multiple types of simulations would provide an avenue to capture the accuracy of DFT methods with the efficiency of classical methods in one simulation.

## 15.3 Classical Molecular Dynamics Simulations of Water–Glass Interactions

### 15.3.1 Brief Introduction and History

The QM-based methods outlined in the earlier sections are useful in modeling electronic and atomistic phenomena. However, due to the high computational cost involved, these methods are restricted to a few hundred atoms and smaller length scales as shown in Figure 15.1. MD methods approximate the electronic and nuclear interactions to simpler, many-body interactions, and the resulting improvement of computations speed allows for simulations of millions or even billions [94] of atoms. The common theme in all MD simulations is obtaining the spatiotemporal trajectory of a set of particles by numerical integration of the classical Newtonian equations of motion, based on interaction potential between the particles. With reference to Figure 15.1, the techniques ranging from ReaxFF, atomistic MD, and Mesoscale Particle dynamics including coarse grained simulations can all be classified in this category. In atomistic MD, the particles are the atoms, and the trajectory over time are obtained from the simulations as discrete set of records of positions and velocities. Extrinsic and intrinsic properties of the bulk material are then derived based on the time-averaged properties over the trajectories [95, 96]. For instance, the temperature of an atomic system is calculated based on the time mean kinetic energy of all the atoms, the isochoric heat capacity can be calculated based on the fluctuations around mean total potential energy and the internal energy is calculated as the sum of translational kinetic energy and mean potential energy [97]. Properties that can be easily calculated with MD are the atomistic structure and reactions at the local scale, which are seldom accessible by experimental methods. The atomistic structure is frequently measured by the PDF that can be corroborated by experimental measurements of the structure factor using diffraction techniques [98, 99]. More details of classical MD simulations can be found in Chapter 1 while analysis methods and property calculations in Chapter 4 of this book.

The interaction potential, the gradient of which gives the force on the particle, approximates the electronic and nuclear interactions between the particles. These sets of multibody interactions, also termed as Force-Fields (FFs), are the crucial aspect of MD for obtaining accurate and realistic trajectories of the particles. In the simplest of cases, the potential consists of a pair term like in the case of Lennard–Jones (L–J) liquids. With advancement in computational resources, the potentials have become more sophisticated with the addition of angular, dihedral, and other multibody interactions, charge or electronegativity optimizations, and bond-order/atomic environment-based potentials. Typical MD simulations at the time of this publication involve thousands to tens of thousands of atomic particles of complex systems running over nanosecond time scales. One of the first applications of MD simulations in the 1970s was the simulation of bulk water [100–102] using a combination of partial charges and the L–J potential. This was quickly followed by simulation of silica systems by Woodcock et al. in 1976 [103]. A natural progression from this work was simulation of water–silica systems, which was first undertaken by Feuston and Garofalini in 1990 using a combination of pair and three-body potentials [104–106]. Since then at least 75 subsequent publications have used MD to study water–silica interactions in the following 30 years. While this may not be exhaustive, it nevertheless serves as an indicator of the progress and wide interest in the use of MD methods to study water–silica interactions.

### 15.3.2 Nonreactive Potentials

The most common potentials used for MD simulation of silica–water systems follow a common theme of a partial charge-based coulombic term combined with a repulsion–dispersion term for short-range interactions. The second most common potentials consist of repulsion–dispersion terms that are represented by an exponential repulsion or a combination of several exponential terms [107] with an $r^6$ dispersion variously parameterized as Born–Mayer–Huggins [105] (BMH), Beest–Kramer–Santen (BKS) [108] or Buckingham potential [109]. Additionally, the initial water–silica potentials used by Garofalini [105, 106] also used a Rahman–Stillinger–Lemberg (RSL) term for the water interactions that allowed for dissociation of water molecules and could be considered reactive potentials. Most of the work in the subsequent years have used nondissociable water potentials such as TIP [110] or SPC [111] that are not reactive. These potentials are useful in studying the structure and dynamics of molecular water and phenomena-like wettability on surfaces [112, 113]. Despite the lack of dissociable water potentials, reactive phenomena and transport of alkali and other modifier ions in silicates and clay-like minerals has been well established using nonreactive such as ClayFF [114, 115] and CSH [116, 117] potentials that use a harmonic bond and angle terms besides the coulomb and L–J terms.

### 15.3.3 Reactive Potentials

Recent efforts have focused on the development of reactive potentials that allow for the water molecule to disassociate, breaking into $H^+$ and $OH^-$, and reacting with the surface to form Si–OH groups or participate in ≡Si—O—Si≡ breakage mechanisms. Initial work focused on differentiating between O atoms in water and silica to allow a more accurate representation of the water–silica reactions was reported by Du and Cormack [118] in the FFSiOH potentials [119, 120]. The next interaction has been the development of true reactive potentials [121, 122] that allow for flexible dissociable water molecules that can form silanols on interaction with silicate glasses. The most well known is the ReaxFF potential that was initially parameterized for organic molecules [123, 124]. The same ReaxFF framework was used to develop a Si/O/H forcefield in 2010 by Fogarty et al. [121] followed by several subsequent reparameterizations to add functionality or improve bond energies and resulting structure [125, 126]. The ReaxFF development tree can be seen in Figure 15.5 with the dissociable water branch on the right. Another reactive force field is the Diffuse Charge Reactive (DCR) potentials initially formulated by Mahadevan and Garofalini for water–silica interactions [122] and later parametrized to include sodium [128], calcium, and aluminum silicates [129]. Similar to the ReaxFF potential, the charge optimized many-body (COMB) potential [130] incorporates charge equilibration and bond-order approaches in a many-body formalism for the simulation of silica [130, 131] and more recently for water [132]. Besides these, by adopting fixed ionic charges, Tersoff potentials have also been parameterized [133] and used for silica water systems as also has been the case with discrete [134] or patchy particle interactions [135, 136].

### 15.3.4 Silica Glass–Water Interactions

The relationship between the micro/atomic structure features and the water transport in the gel layer, which factors into the long-term dissolution behavior of glasses, has consistently been of interest but has been difficult to investigate via experiments due to the complexity of the glass compositions. Alternatively, MD simulations can provide atomistic information of such complex systems with reasonable system size and simulation times. Earlier studies of silica–water interfacial reactions using MD simulations start with indissociable

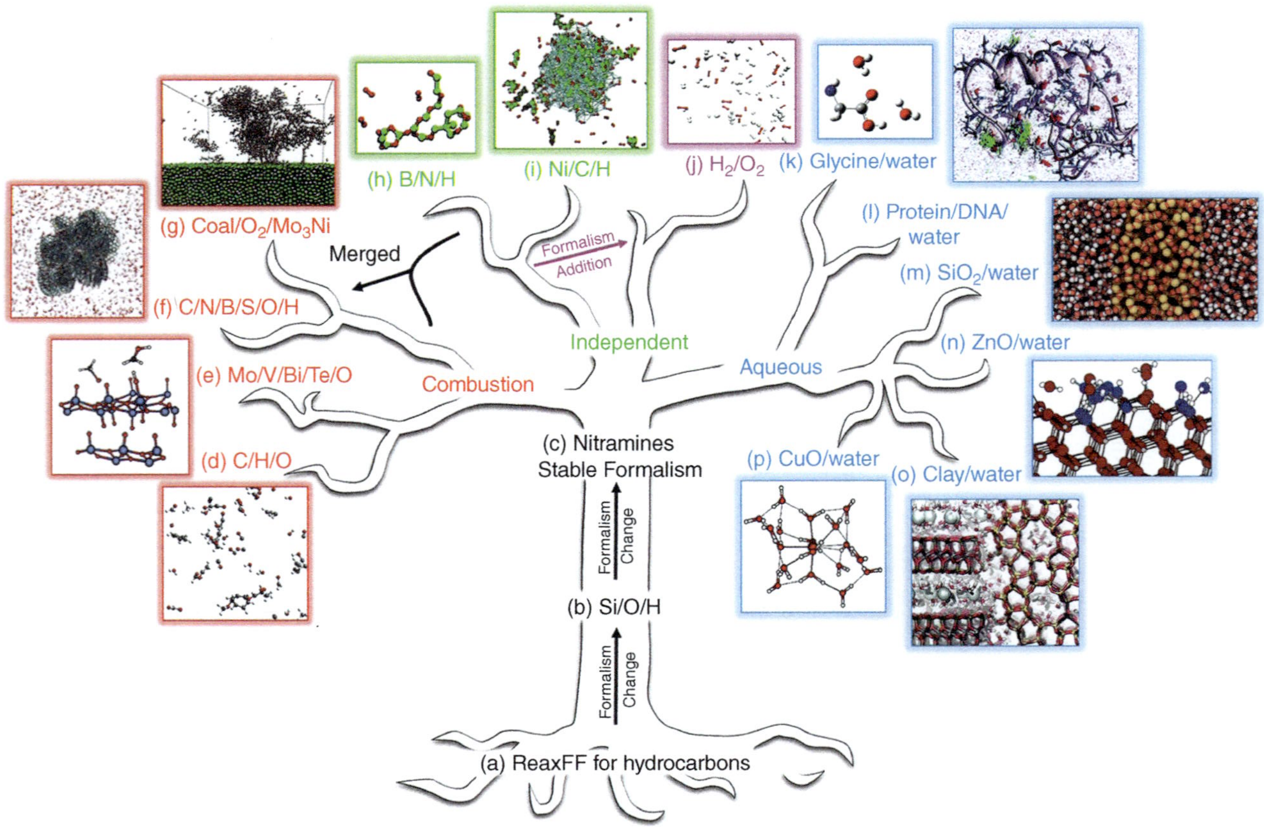


**Figure 15.5** ReaxFF development tree, where parameter sets on a common 'branch' are fully transferable to one another. Progression of ReaxFF began with (a) ReaxFF for hydrocarbons, followed by (b) the development of Si/O/H ReaxFF, and then (c) the nitramine iteration prior to branching based on application. Source: Reference Senftle et al. [127].

water potentials, in which the hydrogen atoms cannot dissociate from the original water molecule. Therefore, studies of glass–water reactions were limited to interactions between the glass surface and water-related species. Specifically, the hydroxylation and hydration of the silica glass surfaces were studied with two different rigid ion potentials [118, 137]. The concentration of hydroxyl group at the surface was estimated to be 4.5 $nm^{-2}$, and the chemisorption energies of water decreases in the following order: three-coordinated silicon and NBO on separate sites, three-coordinated silicon, and NBO on the same sites, and then two- and three-membered rings. Structural relaxation after the hydroxylation was observed [118]. The heat of immersion was calculated at 0.3 $J/m^2$, which is within the experimental range (0.2–0.8 $J/m^2$) [137]. In addition, the strong correlation between the surface density of silanol groups and the character of the surface (either hydrophilic or hydrophobic) are observed through the property perturbation of the interfacial water [138, 139].

After years of exploration and focus on potential development, potential frameworks that can realistically monitor chemical reactions have been established, and dissociable water parameters are available for investigating the reactions between silica glass and bulk water [121, 140]. The structure and properties of pure silica and bulk water, including mass density, bond lengths, PDF, diffusion coefficient, and vibrational frequency spectrum, were calculated and validated with experimental observations [121, 140]. The interfacial reactions between water and glass, including silica surface, strained silica, nanoporous silica, as well as silica gel, were then simulated using these validated potentials [63, 85, 90, 126, 141, 142]. The major products of the reactions between water and siloxane bonds, silanol groups, are observed primarily at the surface of the glass [141]. The total number of the silanol groups suggests that a chemical equilibrium is reached after ~250 ps of simulation time [121]. Silanol formation is also found at the high-strain region of a silica nanorod, which supports the concept that hydroxyl formation prefers the structures with high strain energy [126]. The hydroxylation reaction barriers for strained and nonstrained ≡Si—O—Si≡ structures are reported at ~84 and ~125 kJ/mol, respectively [126].

Various mechanisms in the silica glass–water interfacial reactions are also observed through MD simulations. For example, rather than molecular water diffusion, the major contribution to water molecule penetration in silica is found to be the so-called "proton-transfer process," in which hydrogen atoms passing back and forth between oxygens in bulk silica [121]. The process of surface creation and subsequent defect concentration also plays an important role in the reactivity of an amorphous silica surface. Surface defects such as the dangling bonds and higher energy two-membered rings, formed through initial surface cleavage or nanoporous structure generation, are observed on the silica surfaces [63, 121]. Surface curvature is another important feature of glass that can significantly affect water reactivity (see Figure 15.6). The defect concentration at the surface has a strong correlation to the final silanol concentration, and the undercoordinated Si units are found to be favored to react with the water comparing with the hydrolysis reaction [141]. Another important character of glass–water interfacial reaction is the status of water molecules. The polarization of the water molecules in the surface region and subsequent alignment of the dipoles have been investigated, which leads to a difference of ~10.5 V in electric potential between surface and the water [121]. In addition, the lifetime and rotation of hydrogen bond (H-bond) and proton transfer process at the glass surface are affected

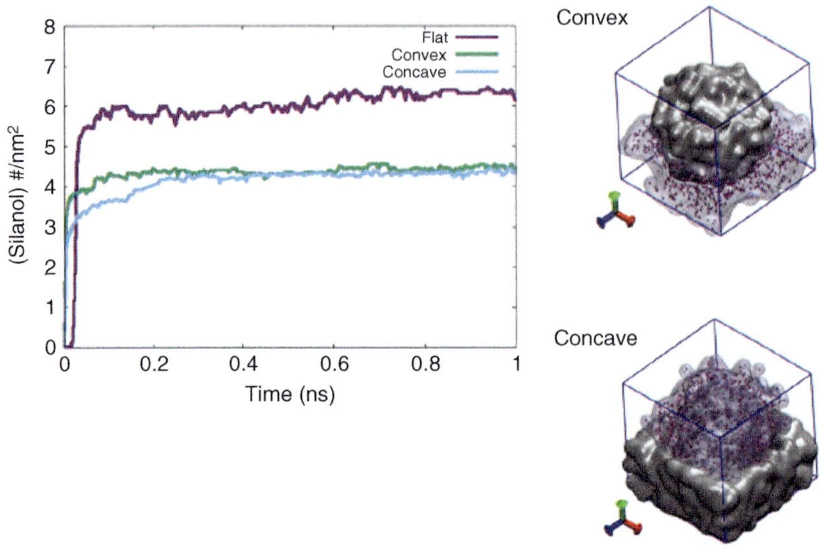


**Figure 15.6** Silanol formation on curved silica surfaces. Snapshots indicate the convex versus concave surfaces that are studied. The silica block has been rendered to show the surface features. Source: Reference Mahadevan and Du [141].

by the change of the H-bond length to the second and third nearest O atoms. Continuous lifetime of the H-bond is much shorter at the glass surface region than in bulk water, while the intermittent lifetime of H-bond shows the opposite trend. The result is a proton transfer process that it is an order of magnitude higher at the surface region than bulk water or deeper subsurface regions. Near-interface hydronium ions ($H_3O^+$) have much lower proton transfer barriers to move back to the surface comparing with the $H_3O^+$ from bulk water. However, the rotation of the $H_3O^+$ dissociated from the initial site contributes to the removal of the excess proton at the surface, and external force can enhance this proton conduction process [142]. Furthermore, the limited internal volume would reduce the water diffusion due to the development of nanoconfined water in nanoporous silica, and hydrogen-hopping mechanisms are found to accelerate the nanoconfined processes as hydrogen atoms diffuse much faster than oxygen in the water [63]. Different morphologies of the pore structures inside the glass can also have impact with the reactivities [85]. More isolated and closed morphology of the porous structure leads to slightly higher network connectivity. It also has slower water diffusivity that can be explained by the nanoconfinement effect of water molecules, as the more isolated pore structures and frequent interactions of the water with nearby silica walls [90].

### 15.3.5 Silicate Glass–Water Interactions

Sodium silicate glass is an architype of multicomponent oxide glasses with wide interests and applications. Understanding how sodium silicate glasses' interaction with water will help explain the corrosion mechanisms of modified silicate glass compositions. Recently developed reactive potentials have enabled the simulation of silicate glass–water interfacial reactions [143–145]. Local glass structures using these potentials are in good agreement with experimental structural information. Furthermore, the hydration of $Na^+$ ions as well as dissociation reactions of NaOH in water are consistent with DFT simulations results [144, 145]. Two processes involving silanol groups are observed during the silicate glass–water interactions: (i) reaction between water and surface NBO (silanol formation process) and (ii) proton transfer between two adjacent NBO sites (silanol re-formation process). [145–147] A schematic indicating where the different reaction mechanisms occur in the sodium-silicate glass–water interface is included in Figure 15.7. The number of silanol groups increases instead of reaching equilibrium when the simulation continues, but the slope of the increase decreases after a certain length of simulation is due to the reversible process of the silanol re-formation process [146, 147]. Sodium ions are bound close to the surface after dissolution during the early stage of reaction, and there is a residence time at the surface prior to diffusion into the bulk solution [147]. The interfacial reactions between silicate glass consist of three reaction stages [146]: (i) the surface region of the alteration layer, dominated by water diffusion and subsequent silanol formation between water and surface NBOs; (ii) the near-bulk region of the alteration layer, driven by the proton transfer process between two nearby NBOs (silanol re-formation); (iii) the intermediate region of the alteration layer (between above two regions) reactions in above two regions all take place. Water transport in the near-surface region is dominated by molecular water diffusion, and in the other two regions proton transfer dominates. Higher reaction temperature accelerates the reactions and leads to deeper water penetration [146]. In addition, simulation of friction between sodium silicate glass and amorphous silica reveals the role of interfacial water molecules in shear-induced chemical reaction. Wear depends on the number of interfacial

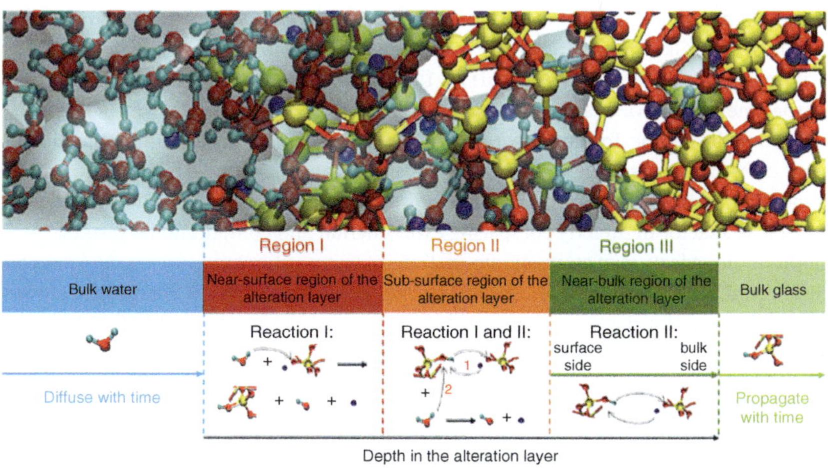


Figure 15.7 Scheme of mechanisms in sodium silicate glass and water interfacial reactions using the ReaxFF potential. Source: Reference Deng et al. [146].

water molecules, and the addition of water reduces the formation of linkages between the substrate and the counter surface to minimize the wear [147].

### 15.3.6 Other Glasses–Water Interactions

For more complicated systems, limited work has been done due to the lack of MD potentials, but a few investigations of multicomponent oxide glass and its reactions with water have been performed. Many of the results are consistent with previous simulations of simpler systems. For example, the concentrations of NBOs and $Na^+$ are higher at the surface than the bulk glass region in the interface between sodium aluminosilicate glass and water [148]. Both hydrolyzed glass networks and molecular water diffusion are observed along with a gradual leaching process of $Na^+$ ions from the glass surface [149]. In more complicated compositions, such as sodium calcium aluminoborosilicate glasses, coordination environment of network formers plays an important role in the reaction. Three-coordinated boron at surface region is favored with a decreasing number of B—O bond from original glass structure. In addition, reaction with water leads to the conversion of $Al^3$ to $Al^4$, while on dry surfaces, the result is reversed. Most of the hydroxyl are SiOH and BOH groups, with concentrations between 2.6 and 4.1 $nm^{-2}$. Mechanisms of B atom coordination changes are included in Figure 15.8. The boron network preferentially breaks following exposure to water and $Na^+$ ions dissolve from the original NBO sites, and this suggests the initial dissolution process of B and Na in the glass corrosion [148].

### 15.3.7 Areas for Improvement

MD simulations of water–silica interactions, even with reactive or dissociable force fields, require different considerations than simulation of bulk glasses. For instance, water molecules do not generally dissociate in the absence of interfaces [140, 150], and if excess $H^+$ is added to bulk water, $H_3O^+$ ions are formed. A few excess protons in a 27 $nm^3$ simulation of bulk water results in a pH of 1.0, which restricts meaningful comparisons with experimental pH values. Despite significant changes in pH, the presence of a silica interfaces results in the formation of several dissociated ionic species, and the silanol concentration compares well with experiments [141, 145]. Therefore, there can be a disconnect between the apparent pH of the solution and the surface condition, which is recognized as a factor in directly comparing experimental and computational results.

Another unique feature of atomistic simulations of water–silica structures is the forces on hydrogen. In the molecular models, the hydrogens and associated point charges are always considered intra molecular – i.e. they only interact within the molecule. In atomistic simulations, on the other hand, forces act between all atomic particles. H atoms with a mass of 1 amu, is many times lighter than other atoms commonly included in the amorphous silicates. Therefore, in MD simulations, H atoms will move faster than other heavier elements for a given timestep. This results in very long displacements for H atoms in each MD step, which impacts the accuracy of the simulation. Therefore, systems that include H atoms need to have significantly smaller timesteps than what is used for other atomistic simulations. In general, simulations containing H atoms use a timestep 10× smaller than the

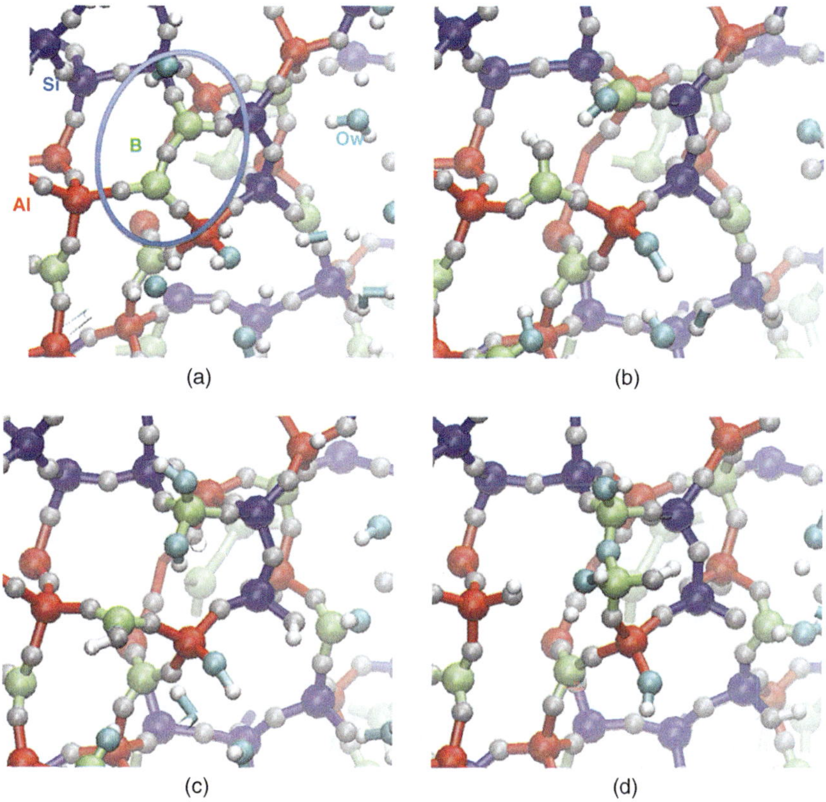


**Figure 15.8** Reactions of B atoms in the surface with water to change the two three-coordinated B atoms in (a) to four-coordinated B atoms by the end of the reaction in (d) through two intermediate states (b) and (c). Colors: Si (blue), Al (red), B (green), O in the glass (gray), O in water (cyan). Source: Reference Garofalini et al. [148].

regular simulations for silica [106, 140, 141]. Because of the reduced timestep, simulations involving dissociable water require more steps to achieve the same time length of trajectory. This can provide additional challenges in accessing reasonable simulation times that can provide physical insight into water–silica systems using existing computational resources. Some types of the reactive potentials are faster than the others. For example, the DCR potentials are about 10 times faster than the ReaxFF based simulations, which would help to reach longer simulation times. The field would benefit from development of computationally more efficient reactive potentials, which have the same level of accuracy of describing the reactions with water.

A key aspect in the development of reactive potentials is the fidelity of the simulated systems to experimentally measured or first principles derived structures and dynamics. The effectiveness of the FF can thus be measured based on how well it reproduces the structure as measured by radial distribution [140] functions and dynamics as measured by the vibrational density of states (VDOS) [151, 152]. Most of the potentials researched reproduce the structure reasonably well and few have explicitly measured the VDOS for glass and water. Applying charge transfer and electronegativity equalization as is done with ReaxFF and COMB could also result in more accurate reactive potentials that can incorporate metals and uncharged components in the matrix [132].

## 15.4 Challenges and Outlook

### 15.4.1 Extending the Length and Time Scales of Atomistic Simulation

A common method of extending the length and time scales of atomistic simulations is the use of Monte Carlo (MC) methods. MC is a stochastic simulation method that calculates numerical solutions to complex processes and has been applied to glass dissolution using coupled hydrolysis/dissolution and condensation reactions [153, 154]. When a time component is included, the method is termed kinetic MC (KMC). In a KMC calculation, the system evolves from state to state based on reaction rates so that infrequent events and extended time scales can be accessed compared to MD simulations [155, 156]. KMC methods have been primarily used for simulation of multicomponent glasses, including alkali silicate [154, 157, 158] and sodium borosilicate glasses [159–162]. Typically, a structure model is used with a variety of lattice sites with different reactivities that then react with water to model the dissolution process [159, 163]. Deverux et al. used a KMC methodology with three different rate constants based on the coordination of the Si atom, and both dissolution and condensation simulations were run [163, 164]. The model reproduced the experimental alteration layer thickness and predicted that the gel layer is free of boron and sodium and is inhomogeneous across the gel thickness [165]. Further KMC studies have used multiple diffusion models and evaluated glass connectivity, diffusion, condensation, liquid connectivity, and coordination to identify explicit treatment of the diffusion of dissolved species in the altered layer that is required for KMC studies to be consistent with experiment [161]. These examples just scratch the surface of the capabilities of KMC to tackle the complex dissolution process and can be an important tool for bridging atomistic scale models to experimental length and time scales.

### 15.4.2 Reactive Potential Development

While significant strides have been taken in the development of reactive potentials for MD simulations, these have primarily been focused on pure silica systems. Only in the past couple of years have Na, Ca, and Al been parameterized into water-reactive silicate systems [144, 145]. Here, several challenges are outlined that need to be overcome to develop a universally applicable and robust FF that can be used to study silicate systems in aqueous environments.

First, an important challenge for the development of reactive potentials is that the parameterization of every additional *N*th element with M parameters would require the optimization of $M \times N^2$ parameters, which creates an optimization challenge.

Additionally, charges on atomic species are considered nonvariant in the simpler potentials. However, atomic charges are themselves a function of the local atomic environment and vary during simulation. The use of separate Lagrangians for charge transfer between atoms is one of the ways to update charges during simulations [166, 167]. Both COMB potentials and ReaxFF incorporate charge equilibration in the potentials. This entails solving additional equations on the fly during MD, leading to slower simulations.

Finally, incorporating multivalent elements such as B, Fe, and P also remain a main challenge in the development of reactive potentials. Ideally, while including the charge equilibration mentioned in Section 15.3.3 above will allow for consideration of multiple valences, complete reproduction of all the electronic interactions in analytical multibody potentials is unlikely. Recent advances in the Neural Network (NN)-based potentials [168] for silica [169, 170] and water [171] show a lot of promise in overcoming many of the challenges listed above. These methods are in the development stage as of the time of this publication but provide promise for future simulation of dissolution of multicomponent glasses.

## 15.5 Concluding Remarks

Atomistic computer simulations including first principles calculations, AIMD simulations, and classical MD simulations and their applications to the study of glass–water interfaces and interfacial reactions have been discussed. Generally, the use of these molecular scale simulation methods provides unique insights into glass–water interactions and, with careful selection of structure models and systems, chemical insight into the glass dissolution process can be obtained. AIMD and first principles calculations have provided a mechanistic understanding of the reaction mechanisms and accurate activation energies for ≡Si—O—Si≡ breakage, as well as hydrolysis, hydration, and transport behaviors during glass dissolution. Classical MD methods that can reach longer time scales and larger system sizes have been used to identify the dissolution properties of silica and sodium silicate glasses, including water–glass reactions, porous gel layer formation, water diffusion, and processes which control the residual rate of glass dissolution.

Additionally, challenges facing the continued use of DFT and MD methods in the study of glass–water interfaces have been identified, including the lack of information on multicomponent glasses based on the lack of available force fields and challenges with accessing longer simulation times and larger length scales to more accurately represent dissolution. Developments in these areas, including new reactive force fields for multicomponent glass compositions [144, 145] and additional molecular scale methods such as ONIOM and KMC, have been presented and discussed. With continued methodological development and increasing computing power, atomistic simulations will continue to be employed for investigations and contribute to the deeper understanding of glass–water interactions.

## Acknowledgments

This research has been made possible by funding from the Center for Performance and Design of Nuclear Waste Forms and Containers, an Energy Frontier Research Center funded by the US DOE, Office of Science, Basic Energy Sciences under Award No. DE-SC0016584 and US National Science Foundation DMR Ceramics Program (#1508001). Sandia National Laboratories is a multimission laboratory managed and operated by National Technology and Engineering Solutions of Sandia, LLC, a wholly owned subsidiary of Honeywell International, Inc., for the US Department of Energy's National Nuclear Security Administration under contract DE-NA0003525. This paper describes objective technical results and analysis. Any subjective views or opinions that might be expressed in the paper do not necessarily represent the views of the US Department of Energy or the US Government. SAND2020-9617 B

## References

1 Gin, S., Abdelouas, A., Criscenti, L.J. et al. (2013). An international initiative on long-term behavior of high-level nuclear waste glass. *Mater. Today* 16: 243–248.
2 Zhang, D., Leppäranta, O., Munukka, E. et al. (2010). Antibacterial effects and dissolution behavior of six bioactive glasses. *J. Biomed. Mater. Res. Part A* 93: 475–483.
3 Hoppe, A., Güldal, N.S., and Boccaccini, A.R. (2011). A review of the biological response to ionic dissolution products from bioactive glasses and glass-ceramics. *Biomaterials* 32: 2757–2774.
4 Frankel, G.S., Vienna, J.D., Lian, J. et al. (2018). A comparative review of the aqueous corrosion of glasses, crystalline ceramics, and metals. *npj Mater. Degrad.* 2: 1–17.
5 Gao, H., Tan, T., and Wang, D. (2004). Dissolution mechanism and release kinetics of phosphate controlled release glasses in aqueous medium. *J. Controlled Release* 96: 29–36.
6 Abraitis, P., Livens, F., Monteith, J. et al. (2000). The kinetics and mechanisms of simulated British Magnox waste glass dissolution as a function of pH, silicic acid activity and time in low temperature aqueous systems. *Appl. Geochem.* 15: 1399–1416.
7 Devreux, F. and Barboux, P. (2001). Numerical modelling of glass dissolution: gel layer morphology. *J. Nucl. Mater.* 298: 145–149.
8 Jégou, C., Gin, S., and Larché, F. (2000). Alteration kinetics of a simplified nuclear glass in an aqueous medium: effects of solution chemistry and of protective gel properties on diminishing the alteration rate. *J. Nucl. Mater.* 280: 216–229.
9 Rebiscoul, D., Frugier, P., Gin, S., and Ayral, A. (2005). Protective properties and dissolution ability of the gel formed during nuclear glass alteration. *J. Nucl. Mater.* 342: 26–34.

**10** Gin, S., Ribet, I., and Couillard, M. (2001). Role and properties of the gel formed during nuclear glass alteration: importance of gel formation conditions. *J. Nucl. Mater.* 298: 1–10.

**11** Gin, S. and Mestre, J. (2001). SON 68 nuclear glass alteration kinetics between pH 7 and pH 11.5. *J. Nucl. Mater.* 295: 83–96.

**12** Rebiscoul, D., Van Der Lee, A., Frugier, P. et al. (2003). X-ray reflectometry characterization of SON 68 glass alteration films. *J. Non-Cryst. Solids* 325: 113–123.

**13** Gin, S., Ryan, J.V., Schreiber, D.K. et al. (2013). Contribution of atom-probe tomography to a better understanding of glass alteration mechanisms: application to a nuclear glass specimen altered 25 years in a granitic environment. *Chem. Geol.* 349: 99–109.

**14** Rimstidt, J.D., Brantley, S.L., and Olsen, A.A. (2012). Systematic review of forsterite dissolution rate data. *Geochim. Cosmochim. Acta* 99: 159–178.

**15** Fournier, M., Gin, S., and Frugier, P. (2014). Resumption of nuclear glass alteration: state of the art. *J. Nucl. Mater.* 448: 348–363.

**16** D'Souza, A.S., Pantano, C.G., and Kallury, K.M. (1997). Determination of the surface silanol concentration of amorphous silica surfaces using static secondary ion mass spectroscopy. *J. Vac. Sci. Technol., A* 15: 526–531.

**17** D'Souza, A.S. and Pantano, C.G. (1999). Mechanisms for silanol formation on amorphous silica fracture surfaces. *J. Am. Ceram. Soc.* 82: 1289–1293.

**18** D'Souza, A.S. and Pantano, C.G. (2002). Hydroxylation and dehydroxylation behavior of silica glass fracture surfaces. *J. Am. Ceram. Soc.* 85: 1499–1504.

**19** Alexander, G.B., Heston, W., and Iler, R.K. (1954). The solubility of amorphous silica in water. *J. Phys. Chem.* 58: 453–455.

**20** Zhuravlev, L. (2000). The surface chemistry of amorphous silica. Zhuravlev model. *Colloids Surf., A* 173: 1–38.

**21** Zhuravlev, L. (1987). Concentration of hydroxyl groups on the surface of amorphous silicas. *Langmuir* 3: 316–318.

**22** Rimstidt, J.D. and Barnes, H. (1980). The kinetics of silica–water reactions. *Geochim. Cosmochim. Acta* 44: 1683–1699.

**23** House, W.A. and Hickinbotham, L.A. (1992). Dissolution kinetics of silica between 5 and 35 °C. Application of a titrimetric method. *J. Chem. Soc., Faraday Trans.* 88: 2021–2026.

**24** Dove, P.M., Han, N., Wallace, A.F., and De Yoreo, J.J. (2008). Kinetics of amorphous silica dissolution and the paradox of the silica polymorphs. *Proc. Natl. Acad. Sci. U.S.A.* 105: 9903–9908.

**25** Fleming, B.A. (1986). Kinetics of reaction between silicic acid and amorphous silica surfaces in NaCl solutions. *J. Colloid Interface Sci.* 110: 40–64.

**26** Icenhower, J.P. and Dove, P.M. (2000). The dissolution kinetics of amorphous silica into sodium chloride solutions: effects of temperature and ionic strength. *Geochim. Cosmochim. Acta* 64: 4193–4203.

**27** Niibori, Y., Kunita, M., Tochiyama, O., and Chida, T. (2000). Dissolution rates of amorphous silica in highly alkaline solution. *J. Nucl. Sci. Technol.* 37: 349–357.

**28** Mazer, J.J. and Walther, J.V. (1994). Dissolution kinetics of silica glass as a function of pH between 40 and 85 °C. *J. Non-Cryst. Solids* 170: 32–45.

**29** Casey, W.H., Lasaga, A.C., and Gibbs, G. (1990). Mechanisms of silica dissolution as inferred from the kinetic isotope effect. *Geochim. Cosmochim. Acta* 54: 3369–3378.

**30** Rimer, J.D., Trofymluk, O., Navrotsky, A. et al. (2007). Kinetic and thermodynamic studies of silica nanoparticle dissolution. *Chem. Mater.* 19: 4189–4197.

**31** Crundwell, F.K. (2017). On the mechanism of the dissolution of quartz and silica in aqueous solutions. *ACS Omega* 2: 1116–1127.

**32** House, W.A. (1994). The role of surface complexation in the dissolution kinetics of silica: effects of monovalent and divalent ions at 25 °C. *J. Colloid Interface Sci.* 163: 379–390.

**33** Techer, I., Advocat, T., Lancelot, J., and Liotard, J.-M. (2001). Dissolution kinetics of basaltic glasses: control by solution chemistry and protective effect of the alteration film. *Chem. Geol.* 176: 235–263.

**34** Neeway, J., Abdelouas, A., Grambow, B., and Schumacher, S. (2011). Dissolution mechanism of the SON68 reference nuclear waste glass: new data in dynamic system in silica saturation conditions. *J. Nucl. Mater.* 415: 31–37.

**35** Gunnarsson, I. and Arnórsson, S. (2000). Amorphous silica solubility and the thermodynamic properties of $H_4SiO_4$ in the range of 0 to 350 °C at $P_{sat}$. *Geochim. Cosmochim. Acta* 64: 2295–2307.

**36** Fournier, R.O. and Rowe, J.J. (1977). The solubility of amorphous silica in water at high temperatures and high pressures. *Am. Mineral.* 62: 1052–1056.

**37** Rimstidt, J.D. (1997). Quartz solubility at low temperatures. *Geochim. Cosmochim. Acta* 61: 2553–2558.

**38** Greenberg, S. and Price, E. (1957). The solubility of silica in solutions of electrolytes. *J. Phys. Chem.* 61: 1539–1541.

**39** Krauskopf, K.B. (1956). Dissolution and precipitation of silica at low temperatures. *Geochim. Cosmochim. Acta* 10: 1–26.

**40** Chan, S. (1989). A review on solubility and polymerization of silica. *Geothermics* 18: 49–56.

**41** Weres, O., Yee, A., and Tsao, L. (1981). Kinetics of silica polymerization. *J. Colloid Interface Sci.* 84: 379–402.

**42** Rothbaum, H. and Rohde, A. (1979). Kinetics of silica polymerization and deposition from dilute solutions between 5 and 180° C. *J. Colloid Interface Sci.* 71: 533–559.

**43** Du, J. and Rimsza, J.M. (2017). Atomistic computer simulations of water interactions and dissolution of inorganic glasses. *npj Mater. Deg.* 1: 1–12.

**44** Kohn, W., Becke, A.D., and Parr, R.G. (1996). Density functional theory of electronic structure. *J. Phys. Chem.* 100: 12974–12980.

**45** Kohn, W. and Sham, L.J. (1965). Self-consistent equations including exchange and correlation effects. *Physiol. Rev.* 140: A1133.

**46** Cheng, H.-P., Barnett, R.N., and Landman, U. (2002). Structure, collective hydrogen transfer, and formation of $Si(OH)_4$ in $SiO_2$–$(H_2O)_n$ clusters. *J. Chem. Phys.* 116: 9300–9304.

**47** Benoit, M., Pöhlmann, M., and Kob, W. (2008). On the nature of native defects in high OH-content silica glasses: a first-principles study. *Europhys. Lett.* 82: 57004.

**48** Masini, P. and Bernasconi, M. (2002). Ab initio simulations of hydroxylation and dehydroxylation reactions at surfaces: amorphous silica and brucite. *J. Phys.: Condens. Matter* 14: 4133.

**49** Schaffer, C.L. and Thomson, K.T. (2008). Density functional theory investigation into structure and reactivity of prenucleation silica species. *J. Phys. Chem. C* 112: 12653–12662.

**50** Leung, K., Nielsen, I.M., and Criscenti, L.J. (2009). Elucidating the bimodal acid–base behavior of the water–silica interface from first principles. *J. Am. Chem. Soc.* 131: 18358–18365.

**51** Laurence, P.R. and Hillier, I.H. (2003). Towards modelling bioactive glasses: quantum chemistry studies of the hydrolysis of some silicate structures. *Comput. Mater. Sci.* 28: 63–75.

**52** Civalleri, B., Garrone, E., and Ugliengo, P. (1999). Cage-like clusters as models for the hydroxyls of silica: ab initio calculation of 1H and 29Si NMR chemical shifts. *Chem. Phys. Lett.* 299: 443–450.

**53** Murashov, V. (2003). Ab initio cluster calculations of silica surface sites. *J. Mol. Struct.* 650: 141–157.

**54** Zhi, L.-L., Zhao, G.-f., Guo, L.-j., and Jing, Q. (2008). Structural, electronic, and vibrational properties of water molecules adsorbed on silica clusters. *Phys. Rev. B: Condens. Matter* 77: 235435.

**55** Wei, X., Zhang, D., and Liu, C. (2008). A DFT study on the hydrolysis stabilities of silica molecular chains and rings based on the two-membered rings. *J. Mol. Struct. Theochem.* 859: 1–6.

**56** Sauer, J., Ugliengo, P., Garrone, E., and Saunders, V. (1994). Theoretical study of van der Waals complexes at surface sites in comparison with the experiment. *Chem. Rev.* 94: 2095–2160.

**57** Lowe, B.M., Skylaris, C.-K., and Green, N.G. (2015). Acid–base dissociation mechanisms and energetics at the silica–water interface: an activationless process. *J. Colloid Interface Sci.* 451: 231–244.

**58** Criscenti, L.J., Kubicki, J.D., and Brantley, S.L. (2006). Silicate glass and mineral dissolution: calculated reaction paths and activation energies for hydrolysis of a Q3 Si by $H_3O^+$ using ab initio methods. *J. Phys. Chem. A* 110: 198–206.

**59** Xiao, Y. and Lasaga, A.C. (1994). Ab initio quantum mechanical studies of the kinetics and mechanisms of silicate dissolution: $H^+$ ($H_3O^+$) catalysis. *Geochim. Cosmochim. Acta* 58: 5379–5400.

**60** Pelmenschikov, A., Strandh, H., Pettersson, L.G., and Leszczynski, J. (2000). Lattice resistance to hydrolysis of Si—O—Si bonds of silicate minerals: ab initio calculations of a single water attack onto the (001) and (111) β-cristobalite surfaces. *J. Phys. Chem. B* 104: 5779–5783.

**61** Ma, Y., Foster, A.S., and Nieminen, R.M. (2005). Reactions and clustering of water with silica surface. *J. Chem. Phys.* 122: 144709.

**62** Walsh, T.R., Wilson, M., and Sutton, A.P. (2000). Hydrolysis of the amorphous silica surface. II. Calculation of activation barriers and mechanisms. *J. Chem. Phys.* 113: 9191–9201.

**63** Rimsza, J., Yeon, J., Van Duin, A., and Du, J. (2016). Water interactions with nanoporous silica: comparison of ReaxFF and ab initio based molecular dynamics simulations. *J. Phys. Chem. C* 120: 24803–24816.

**64** Bandura, A.V., Kubicki, J.D., and Sofo, J.O. (2011). Periodic density functional theory study of water adsorption on the α-quartz (101) surface. *J. Phys. Chem. C* 115: 5756–5766.

**65** Rimola, A. and Ugliengo, P. (2008). A quantum mechanical study of the reactivity of $(SiO)_2$-defective silica surfaces. *J. Chem. Phys.* 128: 204702.

**66** Trinh, T.T., Jansen, A.P., and van Santen, R.A. (2006). Mechanism of oligomerization reactions of silica. *J. Phys. Chem. B* 110: 23099–23106.

**67** Mischler, C., Horbach, J., Kob, W., and Binder, K. (2005). Water adsorption on amorphous silica surfaces: a Car–Parrinello simulation study. *J. Phys.: Condens. Matter* 17: 4005.

**68** Du, M.-H., Kolchin, A., and Cheng, H.-P. (2004). Hydrolysis of a two-membered silica ring on the amorphous silica surface. *J. Chem. Phys.* 120: 1044–1054.

**69** Uchino, T., Kitagawa, Y., and Yoko, T. (2000). Structure, energies, and vibrational properties of silica rings in $SiO_2$ glass. *Phys. Rev. B: Condens. Matter* 61: 234.

**70** Batyrev, I., Tuttle, B., Fleetwood, D. et al. (2008). Reactions of water molecules in silica-based network glasses. *Phys. Rev. Lett.* 100: 105503.

**71** Kubicki, J. and Sykes, D. (1995). Molecular orbital calculations on the vibrational spectra of Q3 T-(OH) species and the hydrolysis of a three-membered aluminosilicate ring. *Geochim. Cosmochim. Acta* 59: 4791–4797.

**72** Criscenti, L.J., Brantley, S.L., Mueller, K.T. et al. (2005). Theoretical and 27Al CPMAS NMR investigation of aluminum coordination changes during aluminosilicate dissolution. *Geochim. Cosmochim. Acta* 69: 2205–2220.

**73** Bouyer, F., Geneste, G., Ispas, S. et al. (2010). Water solubility in calcium aluminosilicate glasses investigated by first principles techniques. *J. Solid State Chem.* 183: 2786–2796.

**74** Geneste, G., Bouyer, F., and Gin, S. (2006). Hydrogen–sodium interdiffusion in borosilicate glasses investigated from first principles. *J. Non-Cryst. Solids* 352: 3147–3152.

**75** Zapol, P., He, H., Kwon, K.D., and Criscenti, L.J. (2013). First-principles study of hydrolysis reaction barriers in a sodium borosilicate glass. *Int. J. Appl. Glass Sci.* 4: 395–407.

**76** Tilocca, A. and Cormack, A.N. (2011). The initial stages of bioglass dissolution: a Car–Parrinello molecular-dynamics study of the glass–water interface. *Proc. R. Soc. A* 467: 2102–2111.

**77** Tilocca, A. and Cormack, A.N. (2009). Modeling the water–bioglass interface by ab initio molecular dynamics simulations. *ACS Appl. Mater. Inter.* 1: 1324–1333.

**78** Berardo, E., Pedone, A., Ugliengo, P., and Corno, M. (2013). DFT modeling of 45S5 and 77S soda-lime phospho-silicate glass surfaces: clues on different bioactivity mechanism. *Langmuir* 29: 5749–5759.

**79** Tilocca, A. and Cormack, A.N. (2008). Exploring the surface of bioactive glasses: water adsorption and reactivity. *J. Phys. Chem. C* 112: 11936–11945.

**80** Nangia, S. and Garrison, B.J. (2008). Reaction rates and dissolution mechanisms of quartz as a function of pH. *J. Phys. Chem. A* 112: 2027–2033.

**81** Rimsza, J., Deng, L., and Du, J. (2016). Molecular dynamics simulations of nanoporous organosilicate glasses using Reactive Force Field (ReaxFF). *J. Non-Cryst. Solids* 431: 103–111.

**82** Dorcheh, A.S. and Abbasi, M. (2008). Silica aerogel; synthesis, properties and characterization. *J. Mater. Process. Technol.* 199: 10–26.

**83** Sacks, M.D. and Sheu, R.-S. (1987). Rheological properties of silica sol-gel materials. *J. Non-Cryst. Solids* 92: 383–396.

**84** Milea, C., Bogatu, C., and Duta, A. (2011). The influence of parameters in silica sol-gel process. *Bull. Transilv. Univ. Brasov Ser. I* 4: 59.

**85** Rimsza, J. and Du, J. (2018). Nanoporous silica gel structures and evolution from reactive force field-based molecular dynamics simulations. *npj Mater. Deg.* 2: 1–10.

**86** Rimsza, J.M., Chapman, K.W., and Nenoff, T.M. (2020). Energetics and structure of ag–water clusters formed in mordenite. *J. Phys. Chem. C* 124: 4517–4524.

**87** Ugliengo, P., Sodupe, M., Musso, F. et al. (2008). Realistic models of hydroxylated amorphous silica surfaces and MCM-41 mesoporous material simulated by large-scale periodic B3LYP calculations. *Adv. Mater.* 20: 4579–4583.

**88** Gierada, M., Petit, I., Handzlik, J., and Tielens, F. (2016). Hydration in silica based mesoporous materials: a DFT model. *Phys. Chem. Chem. Phys.* 18: 32962–32972.

**89** Rimola, A., Costa, D., Sodupe, M. et al. (2013). Silica surface features and their role in the adsorption of biomolecules: computational modeling and experiments. *Chem. Rev.* 113: 4216–4313.

**90** Rimsza, J. and Du, J. (2017). Interfacial structure and evolution of the water–silica gel system by reactive force-field-based molecular dynamics simulations. *J. Phys. Chem. C* 121: 11534–11543.

**91** Rimsza, J. and Du, J. (2015). Ab initio molecular dynamics simulations of the hydroxylation of nanoporous silica. *J. Am. Ceram. Soc.* 98: 3748–3757.

**92** Du, M.-H., Kolchin, A., and Cheng, H.-P. (2003). Water–silica surface interactions: a combined quantum-classical molecular dynamic study of energetics and reaction pathways. *J. Chem. Phys.* 119: 6418–6422.

**93** Zhu, T., Li, J., Lin, X., and Yip, S. (2005). Stress-dependent molecular pathways of silica–water reaction. *J. Mech. Phys. Solids* 53: 1597–1623.

**94** Sharma, A., Nakano, A., Kalia, R.K. et al. (2003). Immersive and interactive exploration of billion-atom systems. *Presence: Teleoperators Virtual Environ.* 12: 85–95.

**95** Frenkel, D. and Smit, B. (2002). *Understanding Molecular Simulation: From Algorithms to Applications*, 1–638. Amsterdam, The Netherlands, in, Elsevier (formerly published by Academic Press): Elsevier.

**96** Allen, M.P. and Tildesley, D.J. (2017). *Computer Simulation of Liquids*. Oxford University Press.

**97** Woodcock, L. (1975). Molecular dynamics calculations on molten ionic salts. In: *Advances in Molten Salt Chemistry* (ed. J. Braunstein, G. Mamantov and G. .P. Smith), 1–74. Springer.

**98** Wright, A.C. (1993). The comparison of molecular dynamics simulations with diffraction experiments. *J. Non-Cryst. Solids* 159: 264–268.

**99** Wright, A.C. (1988). Neutron and X-ray amorphography. *J. Non-Cryst. Solids* 106: 1–16.

**100** Rahman, A., Mandell, M., and McTague, J. (1976). Molecular dynamics study of an amorphous Lennard-Jones system at low temperature. *J. Chem. Phys.* 64: 1564–1568.

**101** Rahman, A., Stillinger, F., and Lemberg, H. (1975). Study of a central force model for liquid water by molecular dynamics. *J. Chem. Phys.* 63: 5223–5230.

**102** Rahman, A. and Stillinger, F.H. (1971). Molecular dynamics study of liquid water. *J. Chem. Phys.* 55: 3336–3359.

**103** Woodcock, L., Angell, C., and Cheeseman, P. (1976). Molecular dynamics studies of the vitreous state: simple ionic systems and silica. *J. Chem. Phys.* 65: 1565–1577.

**104** Garofalini, S.H. (1990). Molecular dynamics computer simulations of silica surface structure and adsorption of water molecules. *J. Non-Cryst. Solids* 120: 1–12.

**105** Feuston, B. and Garofalini, S. (1990). Oligomerization in silica sols. *J. Phys. Chem.* 94: 5351–5356.

**106** Feuston, B. and Garofalini, S.H. (1990). Water-induced relaxation of the vitreous silica surface. *J. Appl. Phys.* 68: 4830–4836.

**107** Taniguchi, T. and Ito, S. (2016). Structure and mechanical property of water-containing soda–alumina–silica glass by molecular dynamics simulation. *J. Non-Cryst. Solids* 432: 177–182.

**108** Van Beest, B., Kramer, G.J., and Van Santen, R. (1990). Force fields for silicas and aluminophosphates based on ab initio calculations. *Phys. Rev. Lett.* 64: 1955.

**109** Sassani, D., Criscenti, L., Zapol, P. et al. (2014). *Waste Form Degradation Model Integration for Engineered Materials Performance*. Albuquerque, NM: Sandia National Laboratories.

**110** Price, D.J. and Brooks, C.L. III, (2004). A modified TIP3P water potential for simulation with Ewald summation. *J. Chem. Phys.* 121: 10096–10103.

**111** Berweger, C.D., van Gunsteren, W.F., and Müller-Plathe, F. (1995). Force field parametrization by weak coupling. Re-engineering SPC water. *Chem. Phys. Lett.* 232: 429–436.

**112** Raj, R., Maroo, S.C., and Wang, E.N. (2013). Wettability of graphene. *Nano Lett.* 13: 1509–1515.

**113** Pafong, E., Geske, J., and Drossel, B. (2016). On the influence of the intermolecular potential on the wetting properties of water on silica surfaces. *J. Chem. Phys.* 145: 114901.

**114** Cygan, R.T., Liang, J.-J., and Kalinichev, A.G. (2004). Molecular models of hydroxide, oxyhydroxide, and clay phases and the development of a general force field. *J. Phys. Chem. B* 108: 1255–1266.

**115** Bourg, I.C. and Steefel, C.I. (2012). Molecular dynamics simulations of water structure and diffusion in silica nanopores. *J. Phys. Chem. C* 116: 11556–11564.

**116** Shahsavari, R., Pellenq, R.J.-M., and Ulm, F.-J. (2011). Empirical force fields for complex hydrated calcio-silicate layered materials. *Phys. Chem. Chem. Phys.* 13: 1002–1011.

**117** Hou, D., Lu, C., Zhao, T. et al. (2015). Structural, dynamic and mechanical evolution of water confined in the nanopores of disordered calcium silicate sheets. *Microfluid. Nanofluid.* 19: 1309–1323.

**118** Du, J. and Cormack, A.N. (2005). Molecular dynamics simulation of the structure and hydroxylation of silica glass surfaces. *J. Am. Ceram. Soc.* 88: 2532–2539.

**119** Pedone, A., Malavasi, G., Menziani, M.C. et al. (2008). FFSiOH: a new force field for silica polymorphs and their hydroxylated surfaces based on periodic B3LYP calculations. *Chem. Mater.* 20: 2522–2531.

**120** Macià Escatllar, A., Ugliengo, P., and Bromley, S.T. (2017). Modeling hydroxylated nanosilica: testing the performance of ReaxFF and FFSiOH force fields. *J. Chem. Phys.* 146: 224704.

**121** Fogarty, J.C., Aktulga, H.M., Grama, A.Y. et al. (2010). A reactive molecular dynamics simulation of the silica-water interface. *J. Chem. Phys.* 132: 174704.

**122** Mahadevan, T. and Garofalini, S. (2008). Dissociative chemisorption of water onto silica surfaces and formation of hydronium ions. *J. Phys. Chem. C* 112: 1507–1515.

**123** Chenoweth, K., Van Duin, A.C., and Goddard, W.A. (2008). ReaxFF reactive force field for molecular dynamics simulations of hydrocarbon oxidation. *J. Phys. Chem. A* 112: 1040–1053.

**124** Van Duin, A.C., Dasgupta, S., Lorant, F., and Goddard, W.A. (2001). ReaxFF: a reactive force field for hydrocarbons. *J. Phys. Chem. A* 105: 9396–9409.

**125** Larsson, H.R., van Duin, A.C., and Hartke, B. (2013). Global optimization of parameters in the reactive force field ReaxFF for SiOH. *J. Comput. Chem.* 34: 2178–2189.

**126** Yeon, J. and van Duin, A.C. (2016). ReaxFF molecular dynamics simulations of hydroxylation kinetics for amorphous and nano-silica structure, and its relations with atomic strain energy. *J. Phys. Chem. C* 120: 305–317.

**127** Senftle, T.P., Hong, S., Islam, M.M. et al. (2016). The ReaxFF reactive force-field: development, applications and future directions. *npj Comput. Mater.* 2: 1–14.

**128** Mahadevan, T.S. and Du, J. (2020). Hydration and reaction mechanisms on sodium silicate glass surfaces from molecular dynamics simulations with reactive force fields. *J. Am. Ceram. Soc.* 103: 3676–3690.

**129** Mahadevan, T.S. and Du, J. (2020). Atomic and micro-structure features of nanoporous aluminosilicate glasses from reactive potential molecular dynamics simulations. *J. Am. Ceram. Soc.* 104: 229–242.

**130** Yu, J., Sinnott, S.B., and Phillpot, S.R. (2007). Charge optimized many-body potential for the Si/$SiO_2$ system. *Phys. Rev. B: Condens. Matter* 75: 085311.

**131** Shan, T.-R., Devine, B.D., Hawkins, J.M. et al. (2010). Second-generation charge-optimized many-body potential for Si/$SiO_2$ and amorphous silica. *Phys. Rev. B: Condens. Matter* 82: 235302.

**132** Phillpot, S.R., Antony, A.C., Shi, L. et al. (2018). Charge optimized many body (COMB) potentials for simulation of nuclear fuel and clad. *Comput. Mater. Sci.* 148: 231–241.

**133** Martinez-Gonzalez, J.A., English, N.J., and Gowen, A.A. (2017). Understanding the interface between silicon-based materials and water: molecular-dynamics exploration of infrared spectra. *AIP Adv.* 7: 115105.

**134** Xu, L., Buldyrev, S.V., Angell, C.A., and Stanley, H.E. (2006). Thermodynamics and dynamics of the two-scale spherically symmetric Jagla ramp model of anomalous liquids. *Phys. Rev. E: Stat. Nonlinear Soft Matter Phys.* 74: 031108.

**135** De Michele, C., Tartaglia, P., and Sciortino, F. (2006). Slow dynamics in a primitive tetrahedral network model. *J. Chem. Phys.* 125: 204710.

**136** Saika-Voivod, I., Smallenburg, F., and Sciortino, F. (2013). Understanding tetrahedral liquids through patchy colloids. *J. Chem. Phys.* 139: 234901.

**137** Hassanali, A.A. and Singer, S.J. (2007). Model for the water–amorphous silica interface: the undissociated surface. *J. Phys. Chem. B* 111: 11181–11193.

**138** Rimsza, J., Jones, R., and Criscenti, L. (2017). Surface structure and stability of partially hydroxylated silica surfaces. *Langmuir* 33: 3882–3891.

**139** Yu, Y., Krishnan, N.A., Smedskjaer, M.M. et al. (2018). The hydrophilic-to-hydrophobic transition in glassy silica is driven by the atomic topology of its surface. *J. Chem. Phys.* 148: 074503.

**140** Mahadevan, T. and Garofalini, S. (2007). Dissociative water potential for molecular dynamics simulations. *J. Phys. Chem. B* 111: 8919–8927.

**141** Mahadevan, T.S. and Du, J. (2018). Evaluating water reactivity at silica surfaces using reactive potentials. *J. Phys. Chem. C* 122: 9875–9885.

**142** Lentz, J. and Garofalini, S.H. (2019). Role of the hydrogen bond lifetimes and rotations at the water/amorphous silica interface on proton transport. *Phys. Chem. Chem. Phys.* 21: 12265–12278.

**143** Yu, Y., Wang, B., Wang, M. et al. (2017). Reactive molecular dynamics simulations of sodium silicate glasses – toward an improved understanding of the structure. *Int. J. Appl. Glass Sci.* 8: 276–284.

**144** Hahn, S.H., Rimsza, J., Criscenti, L. et al. (2018). Development of a ReaxFF reactive force field for $NaSiO_x$/water systems and its application to sodium and proton self-diffusion. *J. Phys. Chem. C* 122: 19613–19624.

**145** Mahadevan, T.S., Sun, W., and Du, J. (2019). Development of water reactive potentials for sodium silicate glasses. *J. Phys. Chem. B* 123: 4452–4461.

**146** Deng, L., Miyatani, K., Amma, S.-i. et al. (2019). Reaction mechanisms and interfacial behaviors of sodium silicate glass in an aqueous environment from reactive force field-based molecular dynamics simulations. *J. Phys. Chem. C* 123: 21538–21547.

**147** Hahn, S.H. and van Duin, A.C. (2019). Surface reactivity and leaching of a sodium silicate glass under an aqueous environment: a ReaxFF molecular dynamics study. *J. Phys. Chem. C* 123: 15606–15617.

**148** Garofalini, S.H., Ha, M.T., and Urraca, J. (2018). Simulations of the surfaces of soda lime aluminoborosilicate glasses exposed to water. *J. Am. Ceram. Soc.* 101: 1135–1148.

**149** Dongol, R., Wang, L., Cormack, A., and Sundaram, S. (2018). Molecular dynamics simulation of sodium aluminosilicate glass structures and glass surface-water reactions using the reactive force field (ReaxFF). *Appl. Surf. Sci.* 439: 1103–1110.

**150** Zhang, W. and Van Duin, A.C. (2017). Second-generation ReaxFF water force field: improvements in the description of water density and OH-anion diffusion. *J. Phys. Chem. B* 121: 6021–6032.

**151** Garofalini, S.H. (1982). Molecular dynamics simulation of the frequency spectrum of amorphous silica. *J. Chem. Phys.* 76: 3189–3192.

**152** Crupi, V., Majolino, D., Migliardo, P. et al. (2003). Structure and dynamics of water confined in a nanoporous sol-gel silica glass: a neutron scattering study. *Mol. Phys.* 101: 3323–3333.

**153** Pierce, E.M., Frugier, P., Criscenti, L.J. et al. (2014). Modeling interfacial glass–water reactions: recent advances and current limitations. *Int. J. Appl. Glass Sci.* 5: 421–435.

**154** Aertsens, M. and Ghaleb, D. (2001). New techniques for modelling glass dissolution. *J. Nucl. Mater.* 298: 37–46.

**155** Mauro, J.C. and Du, J. (2012). Achieving long time scale simulations of glass-forming systems. *Comput. Theor. Chem.* 987: 122–133.

**156** Mei, D., Du, J., and Neurock, M. (2010). First-principles-based kinetic Monte Carlo simulation of nitric oxide reduction over platinum nanoparticles under lean-burn conditions. *Ind. Eng. Chem. Res.* 49: 10364–10373.

**157** Aertsens, M. (2006). The BRAG and GM2003 models for glass dissolution. *Mater. Res. Soc. Symp. Proc.* 985.

**158** Aertsens, M. and Van Iseghem, P. (1995). Modelling glass dissolution with a Monte Carlo technique. *Mater. Res. Soc. Symp. Proc.* 412.

**159** Devreux, F., Barboux, P., Filoche, M., and Sapoval, B. (2001). A simplified model for glass dissolution in water. *J. Mater. Sci.* 36: 1331–1341.

**160** Cailleteau, C., Weigel, C., Ledieu, A. et al. (2008). On the effect of glass composition in the dissolution of glasses by water. *J. Non-Cryst. Solids* 354: 117–123.

**161** Kerisit, S., Pierce, E.M., and Ryan, J.V. (2015). Monte Carlo simulations of coupled diffusion and surface reactions during the aqueous corrosion of borosilicate glasses. *J. Non-Cryst. Solids* 408: 142–149.

**162** Ledieu, A., Devreux, F., Barboux, P., and Minet, Y. (2006). Contribution of Monte Carlo modeling to understanding the alteration of nuclear glasses by water. *Nucl. Sci. Eng.* 153: 285–300.

**163** Devreux, F., Ledieu, A., Barboux, P., and Minet, Y. (2004). Leaching of borosilicate glasses. II. Model and Monte-Carlo simulations. *J. Non-Cryst. Solids* 343: 13–25.

**164** Lobanova, M., Maurer, L., Barboux, P. et al. (2000). Monte Carlo modelling of glass dissolution: comparison with experiments. *Mater. Res. Soc. Symp. Proc.* 663.

**165** Wallace, A.F., Gibbs, G., and Dove, P.M. (2010). Influence of ion-associated water on the hydrolysis of Si—O bonded interactions. *J. Phys. Chem. A* 114: 2534–2542.

**166** Ma, Y. and Garofalini, S. (2005). Modified wolf electrostatic summation: incorporating an empirical charge overlap. *Mol. Simulat.* 31: 739–748.

**167** Ma, Y. and Garofalini, S.H. (2006). Iterative fluctuation charge model: a new variable charge molecular dynamics method. *J. Chem. Phys.* 124: 234102.

**168** Botu, V., Batra, R., Chapman, J., and Ramprasad, R. (2017). Machine learning force fields: construction, validation, and outlook. *J. Phys. Chem. C* 121: 511–522.

**169** Agrawal, P.M., Raff, L.M., Hagan, M.T., and Komanduri, R. (2006). Molecular dynamics investigations of the dissociation of $SiO_2$ on an ab initio potential energy surface obtained using neural network methods. *J. Chem. Phys.* 124: 134306.

**170** Singraber, A., Behler, J.R., and Dellago, C. (2019). Library-based LAMMPS implementation of high-dimensional neural network potentials. *J. Chem. Theory Comput.* 15: 1827–1840.

**171** Morawietz, T., Singraber, A., Dellago, C., and Behler, J. (2016). How van der Waals interactions determine the unique properties of water. *Proc. Natl. Acad. Sci. U.S.A.* 113: 8368–8373.

# Index



*Atomistic Simulations of Glasses: Fundamentals and Applications*, First Edition.
Edited by Jincheng Du and Alastair N. Cormack.

*m*



*n*



*o*

## p



## q



## r



## s

## t



## u



## v

## x



## y



## z

Printed and bound by CPI Group (UK) Ltd, Croydon, CR0 4YY
17/09/2026
14964243-0002